Dr. Peer M. Portner
Vice President
Director, R & D
Andros, Inc.
2332 Fourth Street
Berkeley, CA 94710

ANIMALS FOR MEDICAL RESEARCH
MODELS FOR THE STUDY OF HUMAN DISEASE

ANIMALS FOR MEDICAL RESEARCH
MODELS FOR THE
STUDY OF HUMAN DISEASE

BRIJ M. MITRUKA

Associate Professor, Department of Research Medicine
University of Pennsylvania, School of Medicine
Philadelphia, Pennsylvania

HOWARD M. RAWNSLEY

Professor and Vice-Chairman, Department of Pathology
Dartmouth Medical School
Hanover, New Hampshire

DHARAM V. VADEHRA

Professor and Head, Department of Microbiology
Punjab University
Chandigarh

A Wiley Medical Publication
JOHN WILEY & SONS
New York / London / Sydney / Toronto

Copyright © 1976 by John Wiley & Sons, Inc.

All rights reserved. Published simultaneously in Canada.

Library of Congress Cataloging in Publication Data:

Mitruka, Brij M

    Animals for medical research.

    (A Wiley Medical publication)
    Includes bibliographies.
    1. Diseases—Animal models. 2. Laboratory
animals. 3. Pathology, Experimental. I. Rawnsley,
Howard M., joint author. II. Vadehra, Dharam V.,
1938– joint author. III. Title.

RB125.M57   619   76-7522
ISBN 0-471-61179-4

Printed in the United States of America

10 9 8 7 6 5 4 3 2 1

PREFACE

With the recent concern over the problems and ethics of human experimentation, increasing importance has been placed on the use of animal models to gain an understanding of the pathophysiological mechanisms, etiology, prevention, and control of disease processes in humans. Millions of dollars are spent each year by governmental agencies and private institutions to support research using experimental animals. This book provides valuable information in selecting and using suitable animal species for research so the experimental data may be most meaningful and relevant to human disease. The material has been collected from many different sources of experimental medicine and biology. Since there is such an enormous amount of information in the literature, choice of topics to be covered in this book proved to be a difficult task. Even within the selected topics, it was not possible to include the results reported by many investigators.

A chapter describing, in detail, the nutrition, breeding, and management of fourteen species of animals is included to provide basic information about the techniques in animal husbandry. Common infectious as well as noninfectious diseases occurring in animals under laboratory conditions are briefly reviewed so the experimental scientist may be able to recognize the diseases easily. The methods of selection of laboratory animals and the specific uses of animal models for the study of infectious disease, metabolic disorders, genetic diseases, and other pathological conditions have been discussed in subsequent chapters. Wherever appropriate, spontaneous and induced diseases in experimental animals have been compared with human diseases. The last chapter describes a number of unique animal species that may have potential value in the understanding of specific human diseases.

We hope the book will be useful for everyone involved in experimental biology and medical research.

Brij M. Mitruka
Howard M. Rawnsley

Philadelphia, Pennsylvania
January 1976

ACKNOWLEDGMENTS

For excellent typing and proofreading, we thank Hazel Harber, Irene Soroka, Madeline Griessinger, Lily Chen, Jo Ann Doven, and Dorothea Adams. Special thanks are due Dr. Mary J. Bonner, Surendra Mitruka, Ravendra Mitruka, Polly McGinnis, Eileen Rawnsley, Jo Ann Doven, and L. G. Vasanthi for their help in reviewing portions of the manuscript and checking references cited in this book. We are grateful to Jerry Knee, Dr. D. V. Vadehra, and Dr. M. J. Bonner for preparing illustrations included in this book. We also thank Dr. Mary C. Berwick, Mrs. Valerie A. Pena, and University of Pennsylvania Medical School library, for literature survey with Medline facilities for this book.

It is a pleasure to acknowledge the courtesy of many authors and publishers who promptly granted their permission to utilize previously published material. Sources are acknowledged in the legends of the figures and tables.

The cooperation of the staff of John Wiley and Sons, Inc., and especially of Alan Frankenfield, editor, Biomedical Division, and Ruth Wreschner, assistant editor, Biomedical Division, is also acknowledged gratefully.

The work included in this book was supported in part by a Research Grant GM 21443 to Dr. B. M. Mitruka from the National Institute of General Medical Sciences, U.S. Public Health Service.

B. M. M.
H. M. R.

We dedicate this book to
RAVENDRA, SURENDRA and KIREN MITRUKA;
EILEEN, JILL and SUE RAWNSLEY; and
SURINDER VADEHRA
with love.

AUTHORS

Mary J. Bonner, M.S., Ph.D. Department of Research Medicine, University of Pennsylvania, School of Medicine, Philadelphia, Pennsylvania

William Medway, D.V.M., Ph.D. Clinical Laboratory Medicine, School of Veterinary Medicine, University of Pennsylvania, Philadelphia, Pennsylvania

Brij M. Mitruka, B.V.Sc. and A.H., M.S., Ph.D. Department of Research Medicine, University of Pennsylvania, School of Medicine, Philadelphia, Pennsylvania

Howard M. Rawnsley, M.D. Department of Pathology, Dartmouth Medical School, Hanover, New Hampshire

Dharam V. Vadehra, M.S., Ph.D. Department of Microbiology, Punjab University, Chandigarh

CONTENTS

ANIMALS FOR MEDICAL RESEARCH
MODELS FOR THE STUDY OF HUMAN DISEASE

CHAPTER 1

INTRODUCTION

Brij M. Mitruka

Laboratory animals and farm animals have been used for studies in toxicology, nutrition, microbiology and immunology, psychology, and medicine and surgery for many years. Numerous articles and books have been published describing the uses of animals in these and other medical fields (9, 31, 38, 39, 43, 50, 58, 62, 104). In experimental medicine, certain physiological and psychological processes in humans can be studied using human model systems. However, hypotheses about disease processes in humans often cannot be tested directly. Therefore, various species of animals are required for the studies of pathophysiological processes and the evaluation of drug toxicity. Although animals are used extensively for experimental purposes to understand the mechanism of disease processes, frequently the information obtained is not utilized properly. One of the reasons is that the physiological condition of experimental animals before studies are begun on them is often not defined properly. It is essential that the experimental animals be free of disease in order to derive meaningful interpretation of data obtained by the planned study.

The main objective of this book is to provide some essential information on the care and management of experimental animals for the study of human disease processes. A major portion of this book is devoted to discussion of the appropriate selection of an animal model of human disease based on anatomic, physiological, and biochemical similarities and differences between humans and the experimental animal. Wherever possible, spontaneous diseases of laboratory animals are compared with similar conditions in humans with a view toward their clinical signs and laboratory findings. The potential usefulness of experimental animals in the study of nutritional deficiency diseases, hereditary diseases, infectious diseases, and diseases of the endocrine and reproductive systems have been described in detail. Much of the information on these diseases and on the usefulness of experimental animals in toxicological, behavioral, and cancer research has been derived from the vast literature available in these areas. The last chapter describes some uniquely useful animal species for the study of specific human diseases.

In the following sections, consideration is given to the care, management, and choice of laboratory animals for medical research. More detailed and specific information concerning several animal species is presented in subsequent chapters.

CARE AND MANAGEMENT OF EXPERIMENTAL ANIMALS

The care and management of laboratory animals throughout an experiment are just as important as the design of the study. A researcher should have essential information on how to procure an adequate supply of healthy, uniform, disease-free animals and how to maintain these animals before as well as during the research procedures. Laboratory animals may be procured either by purchasing from a reliable breeding farm as needed or by establishing breeding colonies of the desired species. Depending on the size of program, it is often desirable to have well-trained personnel to handle experimental animals. Not only does the laboratory animal specialist provide basic health care and management programs but he also must be knowledgeable about disease processes, reproductive biology, nutrition, genetics, surgical procedures, and postsurgical care of a wide variety of animal species used in the research laboratory. The knowledge of animal specialists in the areas of techniques for production and care of defined flora or "gnotobiotic" and axenic animals is invaluable when such animals are employed. The investigator should be familiar with methods of producing experimental disease models and utilization of spontaneously occurring animal diseases among various species for studying certain disease states that may occur in other species, including humans.

1. Ordering Laboratory Animals

The animals should be ordered from the appropriate sources with detailed specifications, such as the desired weight ranges, age, sex, strain or breed, color, and freedom from symptoms of infectious, contagious, and parasitic diseases. A specific quarantine period should be established during which the animals must conform to the specifications or be replaced at no expense to the purchaser.

2. Inspection of Animals

The newly arrived animals should be properly inspected as soon as possible by a qualified and competent individual to assure conformance with the specifications. The inspector should be able to recognize a healthy animal as well as the symptoms of various infectious diseases. Animals should also be examined to determine if they were overheated, chilled, or injured during the shipment. Tests for tuberculosis, filariasis, and brucellosis should be made at this time. Animals must be received, examined, and quarantined in an isolated area. This is necessary to protect the breeding colonies as well as those animals on which experiments are performed from contracting diseases that could be carried by the newly acquired animals. Quarantine areas should be isolated and free from insects, rodents, and pathogenic organisms. Proper care and management of purchased animals brought to quarantine areas are essential parts of the program designed to supply healthy, uniform animals for research purposes. These animals should receive the best available diets, under the most favorable environmental conditions because they are likely to be disturbed (or even sick) due to the strains of their transition. Proper medicinal treatment should be given to sick animals as soon as possible. In a newly arrived animal shipment, adequate isolation facilities, properly controlled environmental conditions, proper nutrition, adequate immunization, and a rigid program of sanitation and hygiene are all mandatory for the successful research program. Applicable local, state, or federal regulations pertaining to the health of animals during the quarantine and isolation period must be followed. The duration of the quarantine period varies from species to species of laboratory animals. For example, rats, mice, and hamsters, when obtained from reliable sources, may not have to be quarantined for more than the time required for their inspection upon arrival. On the other hand, primates should be quarantined for at least two weeks while testing the animals for bacterial, viral, and parasitic infestations.

The quarantine period may be used to "condition" the animals for the desired study. During this period, some or all of the following may be performed:

a. Determination as to whether the animals are appropriate for the intended use.

b. Physical examination of the animals, including performance of appropriate clinical and laboratory diagnostic tests and appropriate treatment.

c. Diagnosis and control of diseases transmissible between animals and humans.

d. Stabilization of the nutritional state of the animals.

e. Grooming, including bathing, dipping, drying, and clipping as may be required.

f. Separation of animals by species to protect interspecies disease transmission and to meet environmental requirements.

g. Separation of animals by sex, age groups, and the like as required by the experimental protocol.

h. Recording physical data; numbering or grouping the animals as needed.

3. Housing

The specific requirements for housing a particular animal species are described in detail in Chapter 2. General consideration regarding the facilities and characteristics of housing for animals are given below.

Ideal laboratory animal quarters should be of proper size and should be provided with proper environmental control of temperature, humidity, and nervous stress caused by light, noise, or movement. The proper size of a colony of animal rooms should be determined by the space required to accommodate the individual species of animals in groups or in individual cages. There is little or no definite evidence about ideal or minimum cage sizes for most species, and the allowance of space, whether the animals are caged individually or in groups, is a matter for intelligent conjecture. The length of time during which an animal is to be confined, space, and other facilities available also have a bearing on this. Lane-Petter (49) recommended that the minimum dimensions of a cage in which an animal is to be confined for any length of time may be estimated by the formula

$$A = n\,(3w + 5w) \text{ or } A' = n\,(0.7w' + 6\,w)$$

where A is the floor area in square inches (A' in square centimeters), w the weight of the animal in ounces (w' in grams), and n the number of animals in the cage. This formula gives a good approximation for calculating cage size for rodents, rabbits, and other small laboratory animals. The height of the cage and its width should not be less than the length of the animal, excluding the tail. Since small animals such as mice, rats, and chickens have a tendency to crowd, not more than twenty animals should be caged together. The cages should be capable of restraining animals without causing any harmful effects. They should be easily cleaned and able to withstand the high heats of autoclaving. Preferably, the cages should be made of transparent material and should be functionally versatile. Also, the cages should be adequately insulated to protect the animals against adverse environmental conditions. Cages made of plastic materials and wood are most desirable from this standpoint. However, the need to avoid the difficulties encountered in sterilizing cages made of plastic or wood can be more important to the animal colony than the need for cages with good insulating properties. If good facilities are available to control the room temperatures, the temperature of glass or stainless steel cages can be uniformly maintained without a problem. The prime consideration in housing the animals is that a cage must satisfy the experimental requirements. Sometimes it is necessary to use certain cages for a particular study that may not meet the animals' requirements for health and comfort. Usually, the duration of such experiments is short. Many types of special cages (e.g., metabolic cages, forced exercise cages, etc.) are described in detail elsewhere in this book.

4. Feeding and Watering Experimental Animals

It is advantageous to build into the cage provisions for a water bottle, a food trough or basket, and sometimes a hay rack. These built-in devices make feed and water readily available to animals. These devices should be constructed to prevent the contamination of feed and water and to minimize food waste and water spillage. They should be readily cleaned and sterilized and be easily adapted to a variety of experimental procedures. There are many types and sizes of feeding and watering devices available. The reader is referred to the *Handbook on the Care and Management of Laboratory Animals* (96) for detailed information on this subject. It is sufficient to say that the feeding and watering devices should be designed to make an adequate, uniform, and pathogen-free diet readily available to the animals while at the same time preventing contamination and excessive wastage.

The feeding material for animals varies from species to species, depending on their anatomic, physiological and behavioral differences. The nutritional requirements of each animal species used for experimental purposes are described in Chapter 2. A general discussion regarding the evaluation of feeds, nutritional adequacy and uniformity of feeds, palatability, nontoxicity, and so forth is given in this section. An adequate nutritious diet is essential for the maintenance of experimental animals used in a research program. Nutritional deficiencies can cause numerous physiological and pathological changes in the animals (Chapter 5). They may cause, for example, variations in growth rate, reproduction, lactation, body composition, as well as blood chemistry, sterility, bone formation, nervous disorders, blindness, and cataracts. These changes may occur in varying degrees depending on the severity of the deficiency and may certainly influence the experimental results. The feeding materials for most animal species are available from commercial sources, and they are not only well balanced in the required amounts of nutrients but also are formed in convenient pellets or dry mixtures. These feeds are prepared with care and packaged neatly for reasonable long-time storage. Nevertheless, the chemical analysis of commercially available animal feed is not always a guarantee of a proper diet. Dietary deficiencies may result when certain species are unable to assimilate certain forms of one or more essential nutrients or when feed has been stored improperly. For example, the vitamin A in carotene is readily available to herbivores but is less available to carnivores. Another important consideration is the adequate amount of intake of food by the animal. Besides the easy accessibility factor, the palatability of the food is important to ensure adequate intake. This problem is found particularly in dog and cat colonies, which may have been subjected to nervous stress, exposure to infection, and a change in diet when acquired from an outside source. The palatability of a feed may be altered by such processing procedures as cooking, grinding, chopping, pelleting, baking, and adding certain seasoning materials, antibiotics, and other concentrates. Also, the physical forms of the food may not be desirable to certain animal species. For example, pellets may be too hard for easy consumption, meal feed may be dry and dusty, rolled feed may be excessively bulky, and rations compounded from solvent-extracted vegetable oil meals may be dusty and unattractive to animals. These feeds may be best evaluated by actual feeding trials. Feed should be nontoxic to the animal. The toxicity in feed may result from mold or bacterial spoilage caused by prolonged storage. Also, it may be due to toxic chemicals. These problems are encountered minimally in the feed obtained from a reputable animal feed supplier. However, the investigator should be aware of these problems when special diets are prepared with the ingredients mixed in accordance with a proven formula. Another possibility of introducing toxic material stems from contamination of feed with pathogens. Humans, domestic animals, wild rodents, insects and birds are common carriers of organisms pathogenic to laboratory animals. Commercially prepared meat scraps and bone meals as well as complete diets may be contaminated by any of these carriers. It should be mandatory for feed to be processed so as to destroy the bacteria and properly packaged after processing to prevent this feed from being contaminated. Proper sanitation measures should be taken to control transmission of diseases to experimental animals. The measures should include proper control over animal contact, wild rodents and insects, personnel visitors, feed and water, equipment, airborne infection, and bedding material.

5. Bedding Material

A bedding material used for the experimental animals should be harmless to animals, that is, it should be nontoxic and nonstaining, inedible, and free of pathogenic organisms. Suitable bedding material, usually an industrial or agricultural by-product (e.g., sawdust, wood shavings, or straw), for a particular species of laboratory animal is generally available from commercial suppliers. Bedding material should be autoclaved before using. The material must also be absorbent if laboratory animals are to be maintained in a dry, clean environment, a necessity if such conditions as sore horns, ring worm, foot rot, and various respiratory infections are to be controlled. The growth of bacteria may be prevented by eliminating moisture and autoclaving the materials. The use of absorbent bedding will also help control undesirable odors in the colony room. Bedding material should be readily available, easily stored, and readily disposable. Proper incineration will dispose of all kinds of soiled bedding material with little difficulty. It may also be disposed of by burying or distribution as fertilizer, since the humus content of certain bedding material makes it agriculturally and horticulturally valuable. For example, peat moss, after it has been used as rabbit bedding, is highly valued for nurseries, greenhouses, and golf courses. However, to avoid the possibility of spreading disease, soiled bedding from the cages of infected animals should not be used as fertilizer. Such bedding should be disposed of by incineration as quickly as possible. Another consideration for selecting a particular bedding material for the experimental animal is the cost. However, the final decision as to what material should be used depends on a careful evaluation of all the previously described pertinent factors as they apply to each research program.

6. Identification and Records

Laboratory animals should be identified by placing identification cards in the animal rooms or cages, or on racks; by the use of comfortable collars, bands, or identifying plates or tags on the animals; by stains of various colors; by ear punching or ear tagging; by tatooing; by freeze branding; or by other appropriate means. Identification cards should include such information as the source of the animal, name and location of the responsible investigator, and other pertinent data.

Records on experimental animals are essential not only to good animal care but also to the successful research program. These records should include notations on the source of origin and eventual disposition of each animal in addition to other relevant information.

7. Animal Health Care

The animals used for experimental purpose should have adequate veterinary care to prevent disease and appropriate disease control for each species. It is a usual practice to destroy a colony of small laboratory animals where certain diseases may occur at the endemic proportions. However, often there are many species of laboratory animals which are valuable for a particular research program. In any case, concern for the health of laboratory animals is both a humane and a scientific requirement. A good animal health care program should include the provision of husbandry appropriate for each species, the frequent observation of all animals by a person qualified to verify the health of each animal, the availability of veterinary medical service for animals found to be ill or injured, and the application of currently accepted measures of prophylaxis and therapy. Veterinary medicine should also include humane aspects of animal experimentation, such as the proper use of anesthetics, analgesics, and tranquilizers. All animals should be observed daily for clinical signs of illness, injury, or abnormal behavior. All deviations from normal and deaths from unknown causes should be reported promptly to the person responsible for animal disease control. Animals that develop spontaneous diseases or other abnormalities which cannot be used as an animal model for human disease should be treated or painlessly killed. Diagnostic laboratory service may be required to supplement the physical examination of animals and facilitate the proper diagnosis of abnormal conditions. A well-planned research program should include screening of laboratory animals to establish normal blood chemistries and hematologic values of each species of laboratory animals. Also, it should include necropsy, histological examination of animal tissues, isolation, and identification of specific pathogens and routine and specialized laboratory procedures.

Provision must be made for the emergency health care of animals. Institutional security personnel and fire or police officials should know how to reach the responsible individual, whose name should be prominently posted in the animal facilities or listed with the institution's central telephone center or security department so as to assure that animals will be cared for should an emergency arise.

It must be remembered that the quality of results of a research program is directly related to the quality of the animals used.

THE CHOICE OF THE EXPERIMENTAL ANIMAL FOR MEDICAL RESEARCH

It is important that the animal species be carefully chosen for any experiment in order to prevent the waste of many animal lives and the time, effort, and money of the researcher. Most animals are used in experimental medicine for the purpose of acquiring better understanding of pathophysiological processes in humans. Although an investigation may be well designed and employ the best techniques available, the choice of an inappropriate test animal will lead to obtaining useless information or even more serious misleading information. However, determining the appropriateness of an animal model in itself can be most difficult and time-consuming. One cannot expect a single experimental system to give all the answers one wishes at every level of explanation for even a single experimental type of human disease. Certainly, however, the more completely one knows the experimental system, the greater the possibilities of utilizing it in an intelligent fashion. A researcher should give considerations to the following points before ordering a particular species of animals for his experiments:

1. Is the problem worth solving?

2. Can the alternatives, for example, microbial system, tissue culture, mathematical model system, and so forth be used instead of live animals to solve the problem?

3. Is the animal the best experimental system for the problem?

Assuming that the problem is important enough to spend time, effort, and money on, and it is essential to employ live animal species, the following guidelines may be useful in selecting an experimental animal.

1. Feeding

The animal is easy to feed and thrives well on a diet of well-known composition, the ingredients of which are cheap and easy to obtain. The animals should not require feeding too frequently or at unusual times and not be subject to deficiency diseases. In studying deficiency diseases, it is obvious that susceptibility to them is an advantage.

2. Housing and Management

Temperature, humidity, light, and space for the laboratory animals should be most readily satisfied (65, 96, 97). Such factors are often vital to the success of experimental work. The animals should be able to breed freely in captivity without undue demands for special conditions and with a rapid succession of generation, if desired.

3. Resistance to Infection

The animals should be selected after careful screening for infectious diseases. Epizootics affect some animal species more than others, although experience in this connection varies greatly (51, 66, 92, 94). A number of commonly occurring diseases in laboratory animals are listed in Chapter 3. Effective measures of disease detection and control are discussed elsewhere in this book.

4. Ease of Handling

Animal species that are relatively easier to tame and handle under the laboratory conditions should be selected. Small animals, such as rats, mice, guinea pigs, birds, hamsters, and the like, usually do not pose any particular problem of handling by trained personnel. However, cats, dogs, monkeys, pigs, sheep, cattle, horses, and so forth may sometimes be difficult to handle. It will be a lot simpler if the psychological makeup of the animal renders it tolerant of captivity. Large, shy, or fierce animals may be no more difficult than small ones, for example, the small cats, many reptiles, and some small ungulates. Another point to consider is that if the animals are very difficult to handle, they may easily injure themselves in struggling or striking against the walls of their cages, and keeping them may be very expensive. However, with every species, gentle care and understanding by the animal care attendant and determined efforts to make pets of them is an essential ingredient to success. An unsympathetic attendant will drive the best of animals into a vicious circle of suspicion and moroseness.

5. Suitability for the Experiment in Hand

The prime concern in making the choice of an experimental animal is its suitability to the experiment. Many experimental animals have been studied in great detail, and much is known about their anatomy, physiology, parasitology, deficiency diseases, spontaneous epizootics, hematology, blood chemistries, and many other factors that may be relevant to the experimental studies (15, 16, 17, 18, 19, 29, 48, 52, 79, 83). Important factors the researcher should bear in mind are the anatomic, physiological, and biochemical similarities and differences of the laboratory animal species to that of humans (Table 1-1). Although it is not complete, Table 1-1 shows that many small animals have certain characteristics similar to humans, and other species also have similarities and differences in anatomy, physiology, or metabolic processes. Because of their closeness to humans on the phylogenetic tree, chimpanzees and other primates have been used for certain projects in anatomy, psychology, physiology, and infectious diseases. Other animals often have characteristics that make them more suitable for a particular study. For example, it has been noted that the pig is similar to humans in its nervous system, circulatory system, and gastrointestinal system (25); it has also been reported that the subgross pulmonary anatomy of the horse is closer to that of humans than any other animal studied (63).

Although the use of small laboratory animals is fairly common for economic reasons, ease of handling, and their lesser requirements for laboratory space, the large animals or farm animals have been used frequently for experimental purposes. There are many advantages in using large animals for the selected studies.

a. Size. If the study requires large amounts of fluids or tissues, either normal or pathological, a large animal is a much more efficient source than many small animals.

b. Similarity between diseases to which they are subject and human diseases (see Chapter 3).

c. Similarity of certain parts of their anatomy, physiology, biochemistry, pathology, and pharmacology to that of humans (Table 1-1). A proper selection of species, of course, requires thorough knowledge of comparative medicine. For example, swine develop stomach disorders similar to those of humans, including stomach ulcers following psychogenic stress. Swine also develop atherosclerosis and arteriosclerosis spontaneously without any need for meddling with their thyroids (87).

d. The numbers of farm animals available also constitute a major advantage.

e. Cost factors in the use of large animals. In most studies, initial purchase price of the experimental animals is a minor fraction of the total cost of the experiment. The cost of studies on the large animals is much less than the cost of comparable studies on hospitalized patients. In most cases, studies on pigs, sheep, and even cattle and horses could be conducted at a college of veterinary medicine or agriculture at a total cost that would be less than the cost of doing similar studies on an equal number of dogs in an institution in a large city.

However, there are some disadvantages in using large animals for experimental purposes. The major disadvantages are inconvenience in handling, cost of purchase and long-term maintenance, and accommodation space requirements due to their large size. Another major disadvantage that has prevented widespread use of large animals for experimental medicine is that there is a considerable lack of knowledge on the part of medical investigators about large animal species. Perhaps another factor that may be in-

Table 1-1. Anatomic, Physiological, and Metabolic Similarities and Dissimilarities Between Humans and Laboratory Animals

Animal	Condition, Systems or Structures	
	Similarities	Dissimilarities
Mouse	Senile hepatic change (4)	Spleen (89)
		Liver (54)
Rat	Spleen (89)	Cardiac circulation(34)
	Senile pancreatic changes (5)	Omental circulation(12)
	Senile splenic changes (6)	Gall bladder (lack of) (23)
		Liver (54)
		Sweat glands (59)
Rabbit	Splenic vasculature (53)	Liver (27)
	Spleen (89)	Sweat glands (59)
	Immunity (81)	Pulmonary compliance(20)
	Innervation (103)	Respiratory bronchiolis (55)
	Tensor tympani (8)	
Guinea pig	Spleen (89)	Sweat glands (59)
	Immunity (81)	
Chicken	Sleep (90)	Retinal vessels (67)
	Plica semilunaris (91)	Lymphoid tissue in liver (42)
		Pituitary gland (73)
		Respiratory system (64)
		Oviduct (11)
		Reproductive system(36)
		Acetate metabolism (3)
Cat	Splenic vasculature (53)	Spleen (89)
	Sphenoidal sinus (99)	Reaction to foreign protein (2)
	Liver (27)	Sweat glands (59)
	Epidermis (100)	Laryngeal area (47)
	Clavicle (57)	Mediastinum (78)
	Distribution of epidural fat (77)	Sex cord development(32)
	Tensor tympani (8)	Sensory end organ in subpapillary capillaries (105)
		Sleep (90)
		Heat regulation (23)
Dog	Pituitary vasculature (10)	Cardiac plexuses (71)
	Renal arteries (multiple)(80)	Eosinophil (Greyhound) (41)
	Splenic vasculature (53)	Intestinal circulation (72)
	Sphenoidal sinus (99)	Omental circulation(12)
	Spleen (89)	Renal arteries (80)
	Superficial kidney vasculature (44)	Anal sacs (70)
	Liver (54)	Pancreatic ducts (26)
	Epidermis (100)	Heat regulation (23)
	Nucleic acid metabolism (Dalmation) (24)	Sweat glands (40)
	Adrenal gland innervation(16)	Mediastinum (78)
	Psychotic changes (60)	Laryngeal nerves (101)
		Sleep (90)
Pig	Cardiovascular tree (56)	Lymphocyte predominant (107)
	Erythrocyte maturation (106)	

Table 1-1. (Continued)

| Animal | Condition, Systems or Structures | |
	Similarities	Dissimilarities
Pig	Retinal vessels (25) Gastrointestinal tract (25) Liver (25) Teeth (86) Adrenal gland (108) Skin (74) Penile urethra (30)	Spleen (89) Liver (27) Gamma globulins (new- born) (98) Sweat glands (40)
Sheep	Splenic vasculature (53) Sweat glands (40)	Arteriovenous anastomose (35) Digestion (61) Stomach (23) Vomiting (23) Heat regulation (45) Breeding (33) Sleep (7)
Goat	Ductus venosus (22)	Lymphocyte predominant (107) Digestion (61) Stomach (23) Vomiting (23) Heat regulation (45) Sweat glands (40) Sleep (7)
Primates	Brain vasculature (Orangutan, Gorilla) (102,84) Intestinal circulation (Chimpanzee) (72) Placental circulation (Rhesus monkey) (76) Pancreatic duct (Rhesus monkey) (85) Teeth (primate) (82) Adrenal gland (Rhesus monkey) (37) Innervation (Rhesus monkey) (103) Nucleic acid metabolism (Rhesus monkey) (23) Ischial region (New World monkey) (68) Mandible (primates) (82) Brain (Gorilla) (13) Larynx (Gibbon) (95) Kidney (Spider monkey) (82)	Hemostasis (primates) (75) Inguinal canal (primates) (69) Ischial region (Old World Monkey) (68)

Table 1-1.   (Continued)

Animal	Condition, Systems or Structures	
	Similarities	Dissimilarities
Primates	Placenta (primates)(81) Reproductive performance (Rhesus monkey) (1) Spermatozoon (primates) (14)	
Cattle	Ascending colon (88) Electrolyte excretion (46)	Lymphocyte predominant Digestion (61) (107) Stomach (23) Vomiting (23) Gamma globulins (newborn calf) (28) Apocrine glands (59) Heat regulation (45) Sweat glands (40) Sleep (7)
Horse	Pulmonary vasculature (63) Bile duct (109) Pacreatic duct (109) Lung (63)	Carotid body (110) Spleen (89) Cecum and colon (23) Gall bladder (lack of) (23) Liver (27) Vomiting (23) Sweat glands (59) Mediastinum (78) Sleep (90)

cluded as a disadvantage is the differences in anatomic, physiological, pathological, and pharmacologic conditions of large animals as compared to those of humans.

Modern experimental medicine employs a wide variety of animals (small and large) and continuously searches for "new" experimental animals that may be the best choice in the study of a particular human disease. Table 1-2 lists several more frequently used animal species in various fields of medical research. The usefulness of some of these animal species may be attributed to their unique anatomic, physiological, and biochemical features (Table 1-3). A number of animal species that are not as frequently used at present have been reported to contain certain characteristics suitable for specific experimental work (see Chapter 15). Until the new animal species are found, the choice of animals that are commonly employed for experiments fall into a limited number of the following groups.

1. Rodents and Rabbits

In most institutions, the main stock of the animal houses consists of small rodents and rabbits (rats, mice, etc.) and for economic reasons alone, experiments are tried out on them first. It is important to remember that the rabbit's almost unique gut, with its huge cecum and specialized flora, is quite unlike anything in other animals usually em-

Table 1-2. Uses of Experimental Animals in Medical Research

Species	Areas of Research
The Mouse (Mus musculus)	Genetic research; Cancer research; Infectious Diseases; diagnostic test animal.
The Rat (Rattus norvegicus)	Nutrition research--vitamins, aminoacids (Phenylalanine, histidine, isoleucine, leucine, tryptophan, methionine, lysine and arginine) calcium and phosphorous metabolism; Infectious disease research-CRD, broncho-pneumonia, paratyphoid; Cancer studies--hepatoma, other neoplasms; Purulent lymph adenitis; Poly arthritis; Middle ear diseases and labrinthitis.
The Guinea pig (Cavia porcellus L.)	Nutrition studies--biological assay of vitamin C; Bacteriological research (guinea pigs are susceptible to a wide range of infections e.g. tuberculosis, anthrax, leptospirosis) Diagnosis--test animal for many infectious diseases.
The Hamster (Mesocricitus auratus)	Reproductive physiology: (i) short gestation period-15 1/2 days; (ii) females may be bred 28 days after birth; (iii) from a single cell to parenthood in 60 days, is a unique rate of growth; Nutrition studies--Riboflavin deficiency, Vitamin E deficiency; Inbred strains useful in genetic research; Diabetes mellitus research; Oncology research and drug screening; Toxicological research Hereditary abnormalities

Table 1-2. (Continued)

Species	Areas of Research
The Rabbit (Oryctolagus cuniculus)	Reproductive physiology research: i) As the female ovulates only after mating, the time of ovulation can be determined with great accuracy, ii) dated embryonal material can easily be obtained; Serology--large ear veins which facilitate the withdrawal of blood for serological work; Well suited to teaching purposes; Use for the diagnosis of pregnancy (Friedman test); Useful in experimental physiology; Widely used for studies designated to detect a possible teratogenic effect of drugs or some other evidence of interference with the normal reproductive process; Study of metabolic disorders; Infectious disease.
The Cat (Felis catus)	Mainly employed for acute experiments i.e. those in which the animal is either rendered decerebrate or maintained under anaesthetic throughout the experiment and not allowed to recover; Studies on reflex action, synaptic transmission, the perception of light and sound; secretion of digestive glands and the behavior of cardiovascular, respiratory, excretory and synthetic drugs; Behavioral research, effects of circulatory, digestive and neuromuscular system.
The Dog (Canis familiaris)	Mainly employed for nutritional, pharmacological and physiological studies; Experimental surgery; Cancer research; Toxicology and drug metabolism; Behavior research,e.g. learning processes.
The Sheep (Ovis aries)	Physiological work; Experimental surgery.
The Goat (Capra Hircus)	Nutritional studies; Microbiological studies; Physiology of lactation; Experimental surgery; Radiobiological investigations.

Table 1-2.   (Continued)

Species	Areas of Research
The Pig (Sus scrofa)	Many similarities to man in anatomical, physiological and biochemical aspects; Widely used for nutrition research; Experimental surgery; Atherosclerosis.
The Monkey (Macaca mulatta)	Many similarities to man; Used in the study of bacterial, viral and parasitic diseases; Reproductive biology and endocrine research; Behavioral studies.
Cattle (Bos taurus)	Tissues and fluids are used from normal animals for experimental purposes or media preparation; Study of infectious diseases (Tuberculosis, Brucellosis, Johne's disease); Metabolic disorders; Environmental research--adaptability to increased temperature, and humidity; Also used for genetic research.
The Horse (Equus caballus)	Used for the production of specific antisera or antitoxin such as antidiphtheria serum, tetanus antitoxin etc.; Infectious disease studies.
The Chicken (Gallus domesticus)	Chick embryo is widely used in virology research and in vaccine production; Genetic research; Also used for the study of infectious diseases (Mycoplasma, virus diseases etc.); Arthritidis; Cancer research--Marek's disease, Leukemia.

ployed in the laboratory. Among other rodents, morphological differences occur mostly in various adaptations of the jaw muscles and feet to the environment and mode of life. But other anatomic features are more or less common to all the rodent species; for example, all have a portion of the stomach lined with squamous epithelium. Also, all the rodents and rabbits have a relatively short life and breed freely under proper conditions.

Among other noted differences in rodents and rabbits useful for experimental purposes are their susceptibilities to various infections; for example, susceptibility of the mouse to pneumococcus and the ready response of the guinea pig to diphtheria and tuberculosis. The diphtheria response in the guinea pig is similar to that in the human child, including skin response and antibody formation. Sometimes different species are susceptible only to distinct strains of organisms, for example, pathogenicity of various strains of Pasteurella pseudotuberculosis for different species of host. There may also be great variations in pathogenicity for similar species of laboratory animals. It is well known that geneticists have produced pure lines of mice with marked differences in the susceptibility to infectious diseases, metabolic disorders, toxicological agents, and

Table 1-3. Unique Anatomical, Physiological, and Biochemical Features of Experimental Animals

Species	Anatomy and Physiology	Metabolism
Rats & Mice	Rat liver—Regeneration after 60 to 70% hepatectomy; 90% of the phagocytic activity is due to the Kupffer cell of the liver.	Susceptible to nutritional deficiency (vitamins, aminoacid) and metabolism.
Hamsters	Cheek pouch useful for tissue culture site and heterografts of human cancers; Males have extended serota.	Extremely rapid rate of metabolism and development.
Guinea pigs	Sensitive cochlea useful in hearing experiment; oxygen consumption; Resistant to hypoxia.	Sensitive to penicillin.
Cats	Nerve centres in brain; Stable blood pressure; Stronger vein walls; Highly developed nictitating membrane so that membrane contraction may be used to record intracranial stimulation.	Ability to produce methemoglobin; Suitable for toxicity testing of compounds such as acetanilide; sensitive to all phenols.
Rabbit	Unique gut—huge cecum and specialized flora; Large ear veins.	Bioassay of progestin.

15

Table 1-3. (Continued)

Species	Anatomy and Physiology	Metabolism
Dogs	Discrete pancreas; Good size and easy accessibility of vascular system; Different breeds are of different size.	Sulfonamides; Metabolic processes vary greatly in different breeds.
Cattle & Sheep	Multiple stomach; Slow digestion	Metabolism of cellulose, polysaccharides and other macromolecules.
Frogs	Retina cells—useful for toxicology & drug testing studies.	Metabolism of inorganic ions and salts; Hormones.
Marine animals	Lower nerve axons; electric organs in certain fishes.	Hormones.
Chicken	Nucleated erythrocytes; Alimentary tract.	Demyelinating effects of organophosphorus compounds; High metabolic rate—suitable for the assay of the vitamin B complex expecially vitamin B12; Susceptible to vitamin D deficiency; Purine metabolism.
Duck	Eye lenses; Alimentary tract; Vascular system.	Measurement of lens opacities in dinitrophenol cataract; Detection of aflatoxin in ground nuts; Purine metabolism.

cancers. In reproductive physiology research, rodents are not frequently used except for the use of the mouse in pregnancy tests and the rat in tests for ovulation (Chapter 10). Rodents are often used for the study of parasitic infections; for example, the golden hamster is infested with parasites comparable with those of humans, and many studies have been made on the parasitic diseases of hamsters and other rodents (Chapter 4).

2. Carnivores

Dogs and cats are common laboratory animals frequently used for experimental purposes. These animals are relatively easy to obtain and maintain, and their convenient size and anatomy make them especially attractive for the studies involving physiology and nervous systems. The dog is also used widely for certain types of experimental surgery. Certain viral infections of the carnivores are regarded comparable to those of humans, particularly viruses infecting the nervous system and the brain. Frequently, studies using dogs and cats are made on various types of blood cancers and solid tumors, such as lymphomas, leukemia, adenocarcinomas, endocrine tumors, and so forth (Chapter 11).

The dog is a very valuable animal for the study of drug toxicity, drug metabolism, and chemotherapeutic evaluations of new drugs (Chapter 10). It has become, in recent years, a standard animal to employ in pharmacologic research, especially in toxicology. Other experimental uses of dogs and cats include nutritional research (deficiency diseases), behavioral research (including neurological disorders), bone and joint studies, and pulmonary hemodynamics. The specific uses of these animals in the studies of human diseases are discussed in subsequent chapters.

3. Primates

The primates are expensive animals, but because of their anatomic and physiological closeness to humans, chimpanzees and rhesus monkeys are widely used for many research projects. The primates are not difficult to feed and house, but breeding of these animals in captivity is difficult. The most valuable information obtained by the use of the primate for medical research is in the study of the female reproductive cycle and embryology because of their close similarity to humans. In fact, the most modern ideas on the early development of the human embryo and on the mechanisms of menstruation are derived from early descriptions in the rhesus monkey. The endocrine basis of the change from proliferative to secretory phase in the endometrium at the time of ovulation is also based on observation of these animals. However, the accompanying changes in the sexual organs of the monkey are not similar to those in humans.

Primates, particularly the chimpanzee and the rhesus monkey, are also widely used in behavioral research (Chapter 7). Most work is done on the immature animal to study the behavior patterns of human beings whose development has stopped short in infancy.

Many studies on human infectious diseases have been made using primates because these animals appear to be susceptible to some organisms pathogenic to humans for which no other experimental animal can be found (e.g., the common cold virus). Also, their response to certain microorganisms is closer to the human than that of other susceptible animals; for example, pneumococci will cause pneumonia, meningitis, and septicemia in the chimpanzee and baboon similar to the conditions observed in humans. The Shigella group of bacteria causes enteritis, and mycobacteria cause tuberculosis in the rhesus monkey. Amebic dysentery is also often seen in primates. These protozoa are indistinguishable from those infesting humans.

Primates are occasionally used for experimental work in the fields of toxicology and drug metabolism. However, their biochemical and metabolic changes differ significantly from those of humans (Table 1-1). Certain types of cancer studies are also made using primates as the research animal but not as frequently as carnivores and rodents.

4. Ungulates

This class of experimental animals includes horses, cattle, sheep, goats, and other hoofed animals. Many researchers in the fields of physiology and pathology use these animals for comparative study of diseases in humans. Certain diseases of the domesticated animals are important to humans as a public health concern, for example, foot-and-mouth and rinderprest diseases of cattle, sway-back in horses, scabies of sheep, swine fever, and various parasitic diseases of cattle, swine, and other domestic animals. Also, some ungulates have a response to certain conditions which resembles that of man to some degree, for example toxic conditions of the liver, parasitic diseases of the blood and the gastrointestinal tract, viral and bacterial infections, and so forth. Many ungulate species are used to produce antiserums and antitoxins, for example, the use of the horse for the production of diphtheria and tetanus antitoxins. Cattle are used for public health research in food and dairy microbiology. Many surgical researches, especially chest surgery, use sheep and pigs for experimental purposes. Similarities in anatomy, physiology, and biochemistry of the pig make this animal a good choice for the study of many systemic diseases of humans.

5. Birds

Although many research studies have reported the use of domestic fowl, pigeons, canaries, budgerigars and other birds, chickens are the most frequently used birds for the experimental purposes. Eggs have been used extensively for studies in virology. Bird eggs obviously offer greater ease of manipulation of the embryo than do the embryos of mammals. This has been used in experimental embryology in such areas as grafting the neural crest or other organizers and in cultivation of viruses. The anatomic features of most birds are remarkably similar within the species. The main species differences lie in susceptibility to parasites and in response to hormones. Endocrinologic research employs domestic fowl to study the effects of castration on hormonal behavior changes.

Birds are also used extensively in experimental medicine for the study of parasitic, viral, and bacterial infections. Some of the parasites of birds are of economic importance, for example *Trichomonas meleagridis*; they are of little importance in investigating similar diseases in other types of animals. Bacterial diseases such as streptococcal, mycoplasma, and the like have been studied recently using birds. Psittacosis is one of the rare avian infections that may spread to humans. However, the use of birds as experimental animals may be of value in indirect ways, for example, in the study of the ornithosis group of viruses. Important advances in the knowledge of transmissible tumors (e.g., lymphogranuloma, Marek's disease) have been made by the use of birds as experimental animals.

Other uses of the groups of animal species briefly described above and many other groups of animals, such as reptiles, amphibians, fish, invertebrates, marsupials, and so forth, are described in detail in the following chapters. However, the main objective of the book is to provide information on the comparative medicine of most widely used animal models for the study of disease processes in humans. Within this scope, animal anatomy, physiology, pharmacology, and metabolism have been discussed with a view toward the comparative changes that may occur during the development or treatment of a pathological condition. Particular emphasis on the interpretation of animal models of human disease is given in an effort to point out the best possible utilization of experimental animals as invaluable tools of medical research.

REFERENCES

1. Agate, F. J., Jr., *Am. J. Anat.*, **90:**257 (1952).
2. Akcasu, A., *Int. Arch. Allerg.*, **22:**85 (1963).
3. Allred, J. B., and F. H. Kratzer, *Fed. Proc.*, **21:**86 (1962).
4. Andrew, W., H. M. Brown, and J. B. Johnson, *Am. J. Anat.*, **72:**199 (1943).
5. Andrew, W., *Am. J. Anat.*, **74:**97 (1944).
6. Andrew, W., *Am. J. Anat.*, **79:**1 (1946).
7. Balch, C. C., *Nature* (London), **173:**940 (1955).
8. Blevins, C. E., *Am. J. Anat.*, **113:**287 (1963).
9. Books, S. A., and L. K. Bustad, *J. Anim. Sci.*, **38:**997 (1974).
10. Bosir, M. A., *J. Anat.*, **66:**387 (1932).
11. Bradley, O. C., *J. Anat.*, **62:**339 (1927).
12. Chambers, R., and B. W. Zweifach, *Am. J. Anat.*, **75:**173 (1944).
13. Clark, W. E., *J. Anat.*, **61:**467 (1926).
14. Clermont, Y., and C. P. Lebland, *Am. J. Anat.*, **96:**229 (1955).
15. Cornelius, C. E., *Gastroenterology*, 53:107, 1967.
16. Cornelius, C. E., and I. M. Arias, *Am. J. Med.*, **40:**165 (1966).
17. Cornelius, C. E., *N. Engl. J. Med.*, **281:**934 (1969).
18. Coulouma, P., F. Bastien, and M. Danchy, *Rev. Path. Comp.*, **35:**450; *Abst. Vet. Bull.*, **5:**665 (1935).
19. Creep, R. O., *J. Anim. Sci.*, **31:**1235 (1970).
20. Crosfill, M. L., and J. G. Widdicombe, *J. Physiol,* **158:**1 (1961).
21. Cypress, R. H., and A. I. Hurvitz, in E. C. Melby Jr. and N. H. Altman (eds.), *Handbook of Laboratory Animal Sciences*, Vol. II, CRC Press, Inc., Cleveland, 1974, p. 207.
22. Dickson, A. D., *J. Anat.*, **91:**358 (1957).
23. Dukes, H. H., *The Physiology of Domestic Animals*, Comstock Publishing Association, Ithaca, N.Y., 1955.
24. Duncan, H., K. G. Wakin, and L. W. Ward, *J. Lab. Clin. Med.*, **58:**875 (1961).
25. Earl, F. L., A. S. Tigeris, G. E. Whitmore, R. Morison, and O. G. Fitzhugh, *Ann. N.Y. Acad. Sci.*, **111:**671 (1964).
26. Eichorn, W. P., Jr., and E. A. Boyden, *Am. J. Anat.*, **97:**431 (1955).
27. Elias, H., *Am. J. Anat.*, **85:**379 (1949).
28. Fey, H., and A. Margadant, *Pathol. Microbiol.* (Basel), **24:**970 (1961), *Abstr. Vet. Bull.*, **32:**2530 (1962).
29. Frenkel, J. K., *Fed. Proc.*, **28:**160 (1969).
30. Glenister, T. W., *J. Anat.*, **90:**461 (1956).
31. Green, E. J., and B. Roscoe, in E. L. Green et al. (eds.), *Biology of the Laboratory Mouse*, 2nd ed., Jackson Memorial Laboratory, Bar Harbor, Maine, McGraw-Hill Book Company, Inc., New York, 1966.
32. Gruenwald, P., *Am. J. Anat.*, **70:**359 (1942).
33. Hafez, E. F. E., C. C. O'Mary, and M. E. Ensminger, *J. Hered.*, **40:**111 (1958).
34. Halpern, M. H., *Am. J. Anat.*, **101:**1 (1957).
35. Hamlin, R. L., W. P. Marsland, and C. R. Smith, *Am. J. Physiol.*, **202:**961 (1962).
36. Hewitt, E. A., *J. Am. Vet. Med. Assoc.*, **95:**201 (1939).
37. Hill, B. F., *Charles River Dig.* **13:**1 (1974).
38. Hill, W. C. C. Osman, *Primates*, Edinburgh University Press, Edinburgh, 1953–1974.
39. Homburger, F., and E. Bajusz, *JAMA*, **212:**604 (1970).

40. Hyman, L. H., *Comparative Vertebrate Anatomy,* University of Chicago, Chicago, 1924, pp. 81–82.
41. Jones, R. F., and R. Paris, *J. Small Anim. Pract.,* **4:**29 (1963).
42. Jordan, H. E., *Am. J. Anat.,* **59:**249 (1936).
43. Kalter, S. S., *Lab. Anim. Sci.,* **21:**997 (1971).
44. Kazzaz, D., and W. Shaklin, *J. Anat.,* **85:**163 (1951).
45. Kestner, O., *Vet. Rec.,* **52:**74 (1940).
46. Ketz, H. A., *Abl. Vet. Med.,* **7:**327 (1960), *Abst. Vet. Bull.,* **30k:**3055 (1960).
47. Kingsbury, B. F., *Am. J. Anat.,* **72:**171 (1943).
48. Kitchen, H., *Pediat. Res.,* **2:**215 (1968).
49. Lane-Petter, W., *Proc. R. Soc. Med.,* **65:**343 (1972).
50. Lane-Petter, W., in Worden A. N. and W. Lane-Petter, (eds.), *The UFAW Handbook on the Care and Management of Laboratory Animals,* The Universities Federation for Animal Welfare, London, 1957.
51. Leader, R. W., *Arch. Pathol.,* **73:**390 (1964).
52. Leader, R. W., *Fed. Proc.,* **28:**1804 (1969).
53. Lee, Y. B., M. Elias, and I. Davidsohn, *Proc. Anim. Care Panel,* **10:**25 (1960).
54. Lewis, O. J., *J. Anat.,* **91:**445 (1957).
55. Loosli, C. G., *Am. J. Anat.,* **62:**375 (1938).
56. Lumb, G., and H. Singletary, *Am. J. Pathol.,* **41:**65 (1962).
57. Maksic, D., and E. Small, in E. J. Catcott (ed.), *Feline Medicine and Surgery,* American Veterinary Publications, Inc., Wheaton, Ill., and Santa Barbara, Calif., 1964.
58. Marr, J., *Mich. Med.,* **73:**92 (1974).
59. Marzulli, F. N., and J. F. Callahan, *J. Am. Vet. Med. Assoc.,* **131:**80 (1957).
60. Mason, M. M., and A. M. Scheflen, *Cornell Vet.,* **43:**10 (1953).
61. McAnally, R. A., and A. T. Phillipson, *Biol. Rev.,* **19:**41 (1944).
62. McGrath, J. T., *Neurological Examination of the Dog,* 2nd ed., Lea & Febiger, Philadelphia (1960).
63. McLaughlin, R. E., W. S. Tyler and R. P. Canada, *Am. J. Anat.,* **108:**149 (1960).
64. McLeod, W. M., and R. P. Wagers, *J. Am. Vet. Med. Assoc.,* **95:**59 (1939).
65. Melby, E. C., and N. H. Altman, *Handbook of Laboratory Animal Science,* Vol. I, CRC Press, Inc., Cleveland, 1974.
66. Melby, E. C., and N. H. Altman, Jr., *Handbook of Laboratory Animal Science,* Vol. II, CRC Press, Inc., Cleveland, 1974.
67. Michaelson, I. C., *Retinal Circulation in Man and Animals,* Charles C Thomas, Springfield, Ill., 1954, p. 21.
68. Miller, R. A., *Am. J. Anat.,* **76:**67 (1945).
69. Miller, R. A., *Am. J. Anat.,* **80:**117 (1947).
70. Miller, M. E., G. C. Christensen, and H. E. Evans, *Anatomy of the Dog,* W. B. Saunders Co., Philadelphia and London, 1964, pp. 694–695, 885.
71. Mizeres, N. J., *Am. J. Anat.,* **96:**285 (1955).
72. Noer, R. J., *Am. J. Anat.,* **73:**293 (1943).
73. Oldham, F. K., *Am. J. Anat.,* **68:**293 (1941).
74. Parish, W. G., and J. T. Doen, *J. Comp. Path. Thes.,* **72:**286 (1962).
75. Prydz, H., and R. A. Fosser, *Scand. J. Clin. Lab. Invest.,* **17:**218 (1965).
76. Ramsey, E. M., *Am. J. Anat.,* **98:**159 (1956).
77. Ramsey, H. J., *Am. J. Anat.,* **104:**345 (1959).
78. Rapic, S., *Vet. Arch.,* **7:**581 (1937), *Abst. Vet. Bull.,* **8:**812 (1938).
79. Reinecke, R. D., *Arch Ophthalmol.* (Chicago), **84:**129 (1970).
80. Reis, R. H., and P. Teppe, *Am. J. Anat.,* **99:**1 (1956).
81. Roberts, S. J., *Veterinary Obstetrics and Genital Diseases,* Cornell University Press, Ithaca, N.Y., 1956.

82. Ruch, T. C., *Diseases of Laboratory Primates,* W. B. Saunders Co., Philadelphia and London, 1959.
83. Sacks, A. H., *Angiology,* **25:**120 (1974).
84. Shellshear, J. L., *J. Anat.,* **61:**167 (1926).
85. Singh, I., *J. Anat.,* **97:**107 (1963).
86. Sisson, S., and J. D. Grossman, *The Anatomy of the Domestic Animals,* 4th ed., W. B. Saunders Co., Philadelphia and London, 1953, p. 487.
87. Skold, B. H., and R. Getty, *J. Am. Vet. Med. Assoc.,* **139:**655 (1961).
88. Smith, R. N., and G. W. Meadows, *J. Anat.,* **90:**523 (1956).
89. Snook, T., *Am. J. Anat.,* **87:**31 (1950).
90. Steinhart, P., *Z. Veterinark,* **49:**145, 193 (1937); *Abst. Vet. Bull.,* **8:**254 (1938).
91. Stiebbe, E., *J. Anat.,* **62:**159 (1927).
92. Stout, C., *N. Engl. J. Med.,* **282:**754 (1970).
93. Straus, W., Jr., *J. Anat.,* **69:**93 (1934).
94. Tauraso, N. M., *Lab. Anim. Sci.,* **23:**201 (1973).
95. Thayer, C. B., Personal communication, Animal House, State University of Iowa, Iowa City, Iowa, 1964.
96. *University Federation of Animal Welfare (UFAW) Handbook on the Care and Management of Laboratory Animals,* Williams & Wilkins, Co., Baltimore, 1972.
97. Valerio, D. A., *Lab. Animal Sci.,* **21:**1011 (1971).
98. Van Arsdell, W. C., III, and R. Bogart, *Zbl. Vet. Med.,* **11A:**57 (1964).
99. Van Gilse, P. H. G., *J. Anat.,* **61:**153 (1926).
100. Varicok, T., *Vet. Arch.,* **11:**189 (1941); *Abst. Vet. Bull.,* **18:**581 (1948).
101. Vogel, P. H., *Am. J. Anat.,* **90:**427 (1952).
102. Watts, J. W., *J. Anat.,* **68:**534 (1934).
103. Weddel, G., *Br. Med. Bull.,* **3:**167 (1945).
104. Weisbroth, S. H., R. E. Flatt, and A. L. Kraus, *The Biology of the Laboratory Rabbit,* Academic Press, New York, 1974.
105. Winkelman, R. K., *Am. J. Anat.,* **107:**281 (1960).
106. Wintrobe, M. M., and H. B. Shumacker, Jr., *Am. J. Anat.,* **58:**313 (1936).
107. Wirth, D., and A. I. Littlejohn, *Veterinary Clinical Diagnosis,* 1st English ed., Bailiere, Tindall & Cox, London, 1956, p. 177.
108. Young, F. G., *Br. Med. J.,* **2:**1449 (1961).
109. Zimmerman, A., *Arch. Wiss. Prakt. Tierheilk.,* **68:**112 (1934); *Abst. Vet. Bull.,* **5:**300 (1935).
110. Zimmerman, A., *Math. Term. Ert.,* **60:**900 (1942); *Abst. Vet Bull.,* **15:**2037 (1945).

CHAPTER 2

NUTRITION, BREEDING, AND MANAGEMENT OF EXPERIMENTAL ANIMALS

Experimental animals, large or small, have characteristic nutritional and housing requirements, and they are, to some extent, sensitive to change in their environmental conditions. During the last two decades, extensive research has been carried out to standardize housing and nutritional requirements, breeding techniques, and management of various animals in laboratory conditions. The concern for proper management of experimental animals has resulted in the establishment of the Institute of Laboratory Animal Resources (ILAR) of the National Academy of Sciences in the United States and the Universities Federation of Animal Welfare (UFAW) in Great Britain. The objective of these organizations is to provide information on nutrition, breeding, and management of laboratory animals and on government regulations of animal facilities. The recommendations of these societies include not only the humanitarian aspects in the use of experimental animals but also the necessity of providing healthy animals as valid and dependable subjects for scientific study. Information on the care and management of research animals is also available from research publications, textbooks, and reviews (10, 56, 65, 93, 118, 156, 157). This chapter briefly describes handling techniques, nutritional requirements, and breeding methods of selected animal species that are more frequently used for experimental purposes.

THE MOUSE (*Mus Musculus*)

Among the various animal species, mice are most widely used for medical research (60 to 80 percent) because they are inexpensive and breed easily. Extensive mouse-breeding experience has resulted in the development of special strains for particular experimental purposes. The laboratory mouse is timid, gentle, easy to handle, photophobic, gregarious, and more active at night than during the day. At 4 weeks of age, mice weigh about 18–20 g and older mice weigh as much as 50 g. Male mice tend to be pugnacious after puberty and some strains are vicious fighters. Females are generally docile. The presence of humans tends to inhibit activities in the mouse. Mice do not like being caged singly, and they eat less and weigh less than mice caged in groups.

Mice are used for bioassay and toxicity tests in microbiological research and in radiology. A large number of mice are used for screening new compounds. With the introduction of inbreeding about 40 years ago, 100 to 200 strains of inbred mice are available that are widely used for transplantation and cancer research (145). Several publications (37, 38, 62, 63, 133, 143) are available describing the husbandry, breeding, and uses of the mouse. A mouse newsletter published by MRC Laboratories is another good source of general information on this species.

Housing

Mice are best housed in cages on racks. A wide variety of these, including metal (stainless steel solid bottom), plastic (shoebox type), and disposable cages, are available. The exact space per mouse is difficult to recommend but generally 40 sq in./animal is adequate. Stock mouse cages of 22 × 18 × 6 in. are generally used in four sections, each section containing fifteen to twenty mice, to prevent heaping and suffocation of the lowest layers.

Sawdust and wood shaving are the most common bedding materials, but peat moss, excelsior-treated cellulose fibers, or clean, shredded paper can also be used. The final selection will depend on the type of cage, type of mice, specific property of bedding, and the method of dispensing into cages. In general, bedding should be able to absorb large amounts of urine and water drippings from the bottle.

The animals should preferably be housed in air-conditioned rooms with temperatures ranging from 18–24°C and a relative humidity of 50 to 60 percent. Artificial lighting for 12 hours is necessary, and mechanical ventilation at the rate of six to twelve changes per hour is highly recommended, particularly in large units.

Mice are small animals, and generally there is a tendency to keep many animals in a small area; this can result in outbreaks of epidemics. The colonies should be broken into smaller units and housed in separate, self-contained rooms for better control measures in case of outbreaks. Different strains should be kept separately.

Handling

In general, the mice become tamer with handling. Docile animals grow and breed better. A quieter and more understanding technician is better able to handle the mice. The mice may develop the tendency to bite with rough handling by technicians. These animals should never be provoked. Mice should be handled with bare hands, and the use of gloves and forceps should be avoided unless a particular health hazard is involved.

For experimental purposes, the animals should be held between the thumb and forefinger of the left hand at the back of the neck (Fig. 2-1). The pressure applied should be firm enough that the animal cannot twist and bite, but it should not strangle the mouse. Intravenous inoculations are best done in the tail veins by placing the animals in tubes. The veins can be dilated by placing the tail in warm water before the operation.

Nutrition

Several standard diets are available. A cubed diet is best and should be offered in a hopper. Generally, diet should be kept constant as changes may produce loss of weight and vigor. The diet should be balanced. The gross chemical composition of the diet for mice may vary, for example: protein, 20 to 25 percent; fat, 10 to 12 percent; carbohydrate, 45 to 55 percent; crude fiber, 4 percent or less; ash, 5 to 6 percent. In addition, the diet must contain vitamin A (500–1000 I.U.); vitamin D (1000 I.U.); α-tocopherol, 50 mg; linoleic acid, 5–10 g; thiamine, 3.0 mg; riboflavin, 4.0 mg; pantothenate, 8.0 mg; vitamin B_{12}, 20 μg; pteroylglutamic acid, 1 mg; biotin, 80–200 μg; pyridoxine, 5.0 mg; inositol, 10–1000 mg; and choline, 15 g. All these quantities are per kilogram of diet.

The pellets and water should be provided *ad libitum*; however, water should not be provided in open pots. Plastic, autoclavable bottles are the method of choice. The daily food consumption varies between 3–5 g and water 4.2–7 cc per day (119, 148).

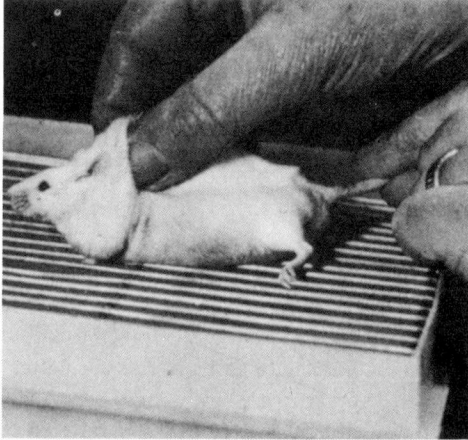

Fig. 2-1. A method of holding the mouse. From (155) with permission.

Breeding

Most strains of mice reach sexual maturity in 6 to 8 weeks; in mature, unmated females, the estrous cycle occurs every 4 to 5 days. Postpartum estrus occurs within 20 to 24 hours after parturation. Weaning induces estrus within 2 to 4 days. The mature male generally weighs 28 g and the female 25 g. Weights in mice are influenced by strain and the number of litters. The gestation period is 20 to 21 days, and the number in a litter varies from three to nine, weighing 1.0–1.5 g. Weight at the time of weaning (18 to 21 days) is 10–12 g.

Different breeding methods including monogamous pair, breeding trios (one male and two females), and harems (one male and four or more females) have been tried, but single pair is probably the best. The males are generally left in the cages with the female throughout her entire breeding cycle; such a practice results in forced breeding and this results in an increase in total number of offspring. The choice of inbreeding, random breeding, or hybrid breeding depends on experimental design. Inbred mice (brother-sister or parent-offspring matings) are of special value in tissue transplantation, radiation protection, cellular grafts, and genetic studies. A minimum of twenty consecutive generations of inbreeding is necessary before the strains can be used for experimental purposes (14, 145, 162).

THE NORWAY RAT (*Ratus norvegicus*)

The original Norway rat was brown and probably originated in eastern Asia. At the present time, three well-known strains are used: (a) hooded with a colored head and shoulder; (b) Wistar, the albino rat originally developed at the Wistar Institute; and (c) Sprague-Dawley, an albino produced by Sprague-Dawley farms (79, 127).

The male rats weigh about 300 g and the females about 200 g at 2 months of age. The laboratory rats are quiet and easy to handle. Male rats are much less prone to fighting than male mice. Rats are less photophobic, less gregarious (as compared to mice), and thrive well singly in cages. A female rat with a litter will tolerate her mate but will not allow other females in the cage (55). Proper handling of all animals in a colony is important; when mishandled, the animal may squeal, warning other rats that may show savage habits.

The rat is a commonly used laboratory animal and constitutes 10 to 15 percent of all experimental animals used in the United States. They are used for nutritional studies, toxicity studies, including long-range experiments, cancer research, infectious diseases, and behavior research (15, 25, 45, 64, 69, 154, 160).

Housing

The animal house should be air-conditioned to maintain a temperature of 22°C; however, a variation of up to 3°C is considered helpful in maintaining vigor. Overheating is more harmful than cooling. Weight gains in rats are better at higher temperatures as compared to that at cold temperatures. Wide ranges (40 to 70 percent) of relative humidity have been used without ill effects. Ringtail generally does not occur at a relative humidity of 50 percent and a higher temperature (50, 111, 152). Twelve-hour lighting arrangement has been found satisfactory in the rooms where rats are housed. Strong light is detrimental to breeding females; however, the females mature faster by 1 to 2 weeks in continuous darkness. The breeding season and extent of breeding are dependent on the variation in exposure to light and temperature.

The rat cages should be provided with bedding unless the experimental procedure requires otherwise. Softwood, sawdust, and peat moss are particularly suitable materials since they delay decomposition and prevent the release of ammonia. Recently, commercial products from cellulose have also been used satisfactorily. Pregnant and lactating rats should be provided nesting material such as paper shreds.

Handling

Rats, like mice, can be handled by the tail; however, the larger rats should be grasped near the base of the tail. This procedure should not be used for pregnant females. Other handling procedure is illustrated in Figure 2-2.

Breeding

The breeding systems used for the rat vary considerably; however, in general, strict hygienic conditions and maintenance of good vigor are important. Males and females for breeding are generally of the same age (90 to 100 days) at the time of mating. Late breeding reduces reproductive ability by increasing the interval between litters. The animals selected for breeding should not be fat or plump but should have a streamlined appearance with a smooth, glossy coat. The selected animals should have high productivity and good mothering ability. High productivity in female rats is desirable because of the usefulness of sex littermates in clinical research. The hooded rats on the average produce ten pups per litter weighing 5.3–6.4 g. The average weaning weight of 51 g for males and 48 g for females is recommended. Sexes are separated at weaning, and up to eight animals are put in each nursing cage.

The mating systems also vary from one male to one female to two males to three to five does. The replacement of bucks is not recommended because it may upset females and there may be a certain amount of fighting. Inbreeding systems are used for specific research purposes. Breeding females should be inspected regularly, at least twice a week, for signs of pregnancy and should be removed to individual cages at least 5 to 7 days before parturation. They should be returned to mating cages about 7 days after the weaning period. The time required for parturation depends on age, litter size, and

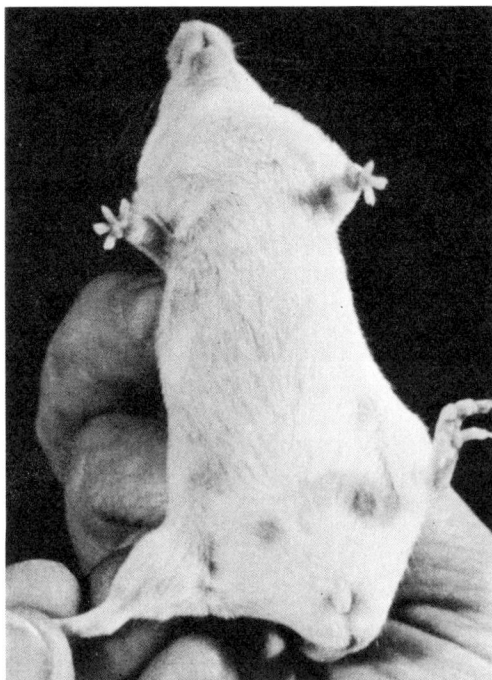

Fig. 2-2. Handling procedure for the mouse (or the rat). From (155) with permission.

physical condition of the female, but usually the litter is born within 1 to 2 hours. Although the females can produce as many as fourteen litters, they should be culled after seven or eight litters unless they are needed for a particular experiment (92, 125).

Nutrition

Rats have long been used as models for nutritional studies for humans; a great deal of information is available on the nutritional requirements of this laboratory animal (1) (Table 2-1). Several commercial companies produce feeds for laboratory rats that are quite adequate. In general, the diet should contain 15.91 percent crude protein, 3.2 percent lipids, 0.67 percent phosphorus, 0.23 percent sodium, 0.84 percent calcium, 0.34 percent chloride, 0.5 percent potassium, and 5.1 percent crude fibers. A practical diet can be prepared with the following compositions.

Diet I		*Diet II*	
Wheat bran	17.7%	Whole wheat ground	50%
Wheat ground	17.7%	Whole barley ground	25%
Oats	17.7%	Fish meal	7%
Barley	8.8%	Meat & bone meal	6%
Maize ground	8.8%	Dried brewer's yeast	5%
Meat & bone meal	8.8%	Dried grass meal	5%
White fish meal	4.5%	Cod liver oil	1%
Dried skim milk	16.8%	Salt	1%
Dried yeast	1.2%		
Cod liver oil	0.4%		

Table 2-1. Nutrient Requirements of the Rat (per 100 g Air Dry Diet)

	Growth	Gestation	Lactation	Maintenance
Feed consumption, g	10	20	30	5-10
Gross energy, Kcal/Kg	4000	4000	4000	4000
Fat, %	5	5	5	5
Crude protein, %	20	20	20	7
Amino acids, %				
L-tryptophan	.15	.20	.20	.07
L-histidine	.30	.54	.54	.07
L-lysine	.90	1.24	1.24	.14
L-leucine	.80	.80	.80	.25
L-isoleucine	.50	.50	.50	.43
L-pheylalanine	.90	.90	.90	.19
L-methionine	.60	.60	1.0	.23
L-threonine	.50	.50	.50	.17
L-valine	.70	.70	.70	.31
L-arginine	.20	.75	.75	
Nonessential amino acids	6.45	5.27	4.87	2.14
Calcium, mg	600	600	600	
Phosphorus, mg	500	500	500	
Sodium, mg	50	500	300	
Potassium, mg	180	140	500	
Chlorine, mg	50	25	18	
Magnesium, mg	40	50	50	
Manganese, mg	5	5	3.3	
Iron, mg	2.5			
Copper, mg	.5			
Zinc, mg	1.2			
Iodine, mcg	15	15	15	
Selenium, mcg	4			
Vitamin A, I	200	1200	1200	
Vitamin E, (alpha-tocopherol), mg	6	3	2	
Vitamin K (menadione), mg	.01			
Thiamine, mg	.125	.4	.4	
Riboflavin, mg	.25	1.0	1.0	
Pyridoxine, mg	.12	.06	.04	
Niacin, mg	1.5			
Ca pentothenate, mg	.8	.8	1.0	
Choline chloride, mg	75	100	100	
Vitamin B12, mcg	.5	.5	.5	

Reproduced from Introduction to Laboratory Animal Science, (10) with permission.

28

THE GUINEA PIG (CAVY)
(Cavia porcellus)

Guinea pigs have long been used for animal experimentation because they are small, tame, and easy to control. Of the three varieties (English, Abyssinian, and Peruvian) of guinea pig, the short haired self, bi- or tricolored English variety is the commonest and most often used in the laboratory.

The English guinea pigs are alert, well fleshed, with smooth, shiny skin. The coat is dense, with hair unsoiled by feces or urine, and there is no discharge from nose or ear. These animals are very consistent in their weight within a strain and age group. The weight at one year of age is about 1000 g for male and 800 g for females.

Guinea pigs are remarkable in their requirements for vitamin C. The animal maintained on vitamin C deficient diet develops scurvylike symptoms. Before the chemical methods were developed, guinea pigs were used for vitamin C assay in nutritional studies (163). The guinea pig is highly susceptible to human and bovine tuberculosis, and it may be used for the diagnosis of human mycobacteriosis. The guinea pig is an excellent model for anaphylaxis and other immunologic procedures (complement).

Housing and Caging

Guinea pigs may be housed indoors in pens on the floor, in fixed or portable tiered compartments, or in cages. Floor pens, despite their disadvantage (such as waste of space and spread of infection) are commonly used because of the simplicity and ease of cleaning and inspection. The general size of a pen is 107 × 244 cm and it can be placed on each side of the room, leaving a passage of about 63 cm wide. Another arrangement uses only one-half of the available space each day.

Tiered Compartments

These may be permanent compartments made of concrete or wooden shelves or portable metal units. The permanent type are not much more economical in space than the floor pens and are generally more expensive and difficult to clean and disinfect. The portable type, up to four tiers, are space saving and easy to clean.

Cages

The initial cost of this system is greater, but it is space saving. Generally, 75 × 75 × 35 cm cages made out of a single aluminum alloy sheet are ideal. These cages can be divided by panels if needed. Each cage can hold a harem of four sows and one boar. The cages will hold twelve youngsters up to 350 g in weight, and the number may be reduced to ten, eight, and six as the weight increases. The cages for experimental animals are 25 × 25 × 25 cm and made up of 1.9 cm wire mesh to hold individual animals, and a 60 × 45 cm floor area may be enough for ten guinea pigs. The most common bedding material is seasoned softwood shavings. Peat moss, though more expensive, is highly recommended. Treated flax and corncobs are also good as they have highly absorbent and deodorant properties. Cereal straw should not be used because it may be irritating to the animals and may act as a source of contamination.

Handling

Guinea pigs are very easy to handle because they are docile animals. The animal should be lifted by grasping the trunk gently but firmly with one hand and supporting the hind limbs with the other (Fig. 2-3). To immobilize it, the animal may be placed on the table and the thorax held between the thumb and fingers of one hand while the hind legs are extended by the other hand.

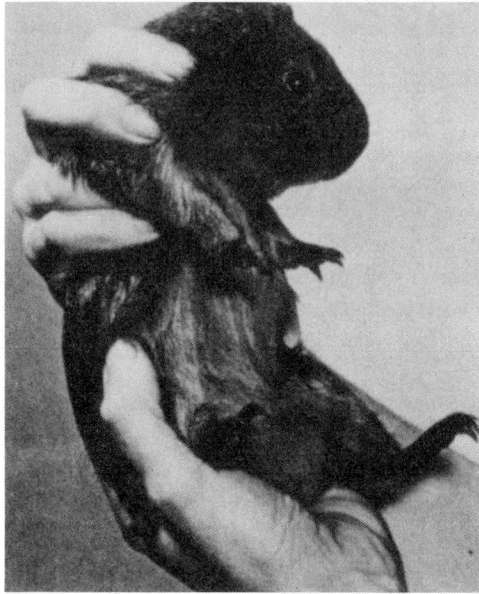

Fig. 2-3. A method of handling the guinea pig. From (155) with permission.

Nutrition

Guinea pigs are vegetarians and thrive well on a variety of diets. However, care should be taken to provide an adequate supply of vitamins A, C, K, and E. Hay provides certain unknown but essential factors; if not provided, the animals will barber each other or consume quantities of bedding. A satisfactory diet consists of a concentrate in either mash or pelleted form. Greens are considered highly essential for this animal species.

Mashes

Probably the simplest mash is two parts of crushed oats mixed with one part of broad bran, which is fed dry or slightly dampened. Bran and sugar beet pulp (2 : 1) may also be used. Only the required amount of mash should be prepared. However, pelleted diets, though more expensive, are preferred because of their uniform quality and convenience. Several diets are available commercially.

Breeding

The breeding stock should be obtained from reliable sources. Soon after weaning, the animals should be quarantined and placed in the breeding quarters early in order to avoid any breeding setback due to environmental changes.

Puberty in the females may occur in 4 to 5 weeks; they weigh about 250 to 400 g at puberty. The males mature at about 8 to 10 weeks, weighing about 400 to 600 g. The first mating, however, should be done around 12 weeks of age, when the female and male weigh about 450 g and 500 g, respectively. The guinea pig experiences postpartum estrus, and mating at this time will considerably reduce the interval between litters. The gestation period varies between 59 to 72 days, with an average of 63 days. The weight of the guinea pig at birth depends on the nutritional status of the sow and the number of litters. For single births, the weight may be 150 g; however, for three to four youngsters, the weight is generally between 85–90 g. If the live birth weight is below average, the chances of survival are generally very poor. There are usually three to four litters per gestation. Development of young guinea pigs is rapid; they gain 3–5 g of weight per day for the first two months. Mature adults may weigh 700–750 g at 5 months of age.

Several breeding methods have been advocated but the nonintensive and intensive methods are most commonly used.

Nonintensive Method

In this procedure, the individual sows (five to ten per boar) may be isolated throughout their breeding span (2 years). The method is wasteful of space, and generally the number of output is lower (twelve per sow per year). However, it has the advantage of keeping correct breeding records, and it is excellent for inbreeding and disease control.

Communal Farrowing

In this system, the heavily pregnant sows are removed from mating pens and allowed to litter and to rear the young communally. When the youngsters are 180 g in weight, the empty sows are removed and returned to mating pens. The preferred system is to have one boar and up to twenty sows; the boar can be rotated every week if desired. Such a system yields about an average of seven litters per two-year life span.

Intensive System

Monogamous or polygamous systems can be followed. Monogamous systems are expensive since large number of boars have to be maintained, and the yields are lower. In the polygamous system, the yields may be as high as fourteen to sixteen young per sow per year.

THE GOLDEN HAMSTER
(Mesocricetus auratus)

The golden hamster, so referred to because of its color, has new varieties that do not conform to this color description. A typical golden hamster is about 5–6 in. long, and 110–140 g in weight. The fur is soft, silky, and reddish gold on the dorsal surface and greyish white below. The eyes and ears are black, and generally there are black and white flash marks on the side.

Hamsters are active during the dark, irrespective of the day/night cycle. During the active period (7 to 13 hours), a hamster may travel the equivalent of 11.5–20 km. In cages, hamsters, particularly females in heat, have a great tendency to escape and can soon learn to open the latches. Basically hamsters spend a considerable amount of time grooming. This habit should not be considered as a sign of ectoparasite infestation. Husbandry, breeding, and behavior of hamsters have been discussed in detail by Richards (124), Robinson (126), Nixon et al. (110), Yager and Sundborg (166), and Hoffman et al. (75). In addition to the golden hamster, another species, the Chinese hamster (Cricetulus griseus), has also become an important laboratory animal. Its care, breeding, and housing have been previously reviewed (104, 168).

Both golden and Chinese hamsters are extensively used in biomedical research. The golden hamster has been used in virology, dental research, and cancer research (76).

The Chinese hamster is used for pathological studies, particularly diabetes mellitus (29, 167). It was among the first species used for parasitological research and has been a favorite model for microbiological studies. Hamsters have been frequently used for transplantation studies (20, 53, 115). Their low chromosome number makes Chinese hamsters a useful model for cytological, genetic, tissue culture, and radiation studies. The short gestation period (15–18 days) makes it especially useful in teratology work (48). Hamsters have also been used for the study of hypothalamus, microcirculation, endocrine, and immunologic responses (53).

Housing

Several types of commercial cages for individual animal housing or stock and breeding cages and metabolic cages are available. Individual cages are 7 × 7 × 9 in. and stock cages of 12 × 24 × 12 in. may hold up to twelve animals. Autoclavable plastics like polypropylene and polycarbonate are good fabrication materials. The floor meshing should be less than ⅓ in. apart. Sawdust, wood chips, peanut hulls, newspaper shreds, straw, or peat moss in treated or untreated form may be used as bedding. Bedding should be replaced at least once a week. Animal room temperature should be maintained at 21–24°C, but the breeding stock should be kept at slightly lower temperature (21–22°C). Air-conditioning is highly desirable. In its absence, cooling of the breeding stock may be done in the refrigerator, and this generally increases the litter size. The relative humidity of 40 to 60 percent and 12-hour light arrangements are adequate for the normal health of the animal, irrespective of the day/night schedule.

Fig. 2-4. The Syrian hamster held in cupped hands. From (155) with permission.

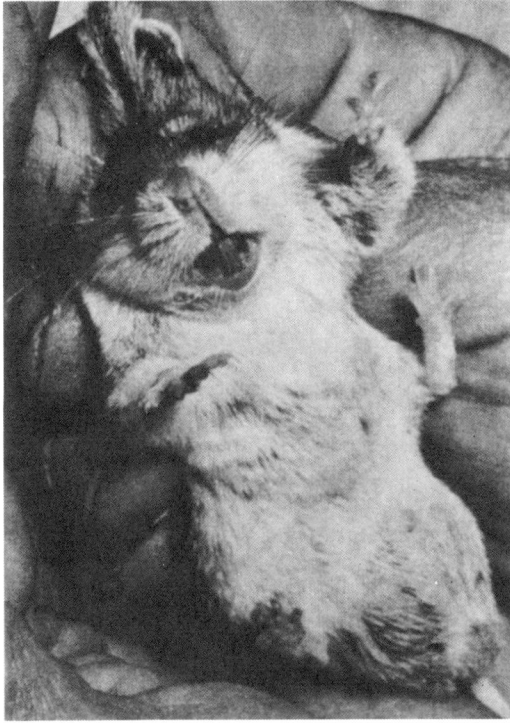

Fig. 2-5. The Syrian hamster held by loose dorsal skin. From (155) with permission.

Handling

Hamsters are aggressive by nature. The Chinese hamster, after frequent handling, becomes more cooperative and rarely bites. The best way to hold these animals is in cupped hands or by the loose dorsal skin (Figs. 2-4 and 2-5). In contrast, golden hamsters resent handling and can be troublesome. A combination of holding in cupped hands and dorsal skin with the little finger placed over the hind leg so that the legs are held between the fourth finger and thumb may be used.

Nutrition

Any of the commercial diets is satisfactory; however, particular attention should be given to the fat level, riboflavin, and biotin concentrations in the diet. Decrease in fat level may result in wet tail, whereas riboflavin deficiency may result in loss of hair, dry flaky skin, and inflamed ocular mucosa. Commercial diets may be supplemented with greens daily or two to three times a week. The average feed consumption is 8–10 g/day. Fresh, clean water should be provided *ad libitum*. The exact amount of water intake by the animals varies with the amount of greens provided, atmospheric temperatures, and urine output. Lack of water supply may result in weight losses (10 g/day).

Breeding

Female hamsters reach puberty early (28 days). The gestation period of hamsters is the shortest (15 to 16 days) among placental laboratory mammals. Ideally, females should be bred at 6 to 8 weeks of age, since older females show a decrease in breeding activity. The breeding can be accomplished either by following the characteristics of the estrous cycle or allowing equal numbers of males and females together for nine days. Then the females can be separated into maternity cages and males are used for the next breeding. Some authors recommend individual mating buckets in which the females are brought; if they are in heat, they should show characteristic lordosis within 10 minutes and mating should result within 15 to 60 minutes. The weight at birth is about 2–3 g; it increases to 35–40 g at weaning (12 to 23 days).

THE RABBIT (*Oryctolagus cuniculus*)

A large number of breeds and varieties of rabbits are available to the researcher. The American Rabbit Breeders Association lists twenty-eight breeds and nearly eighty varieties. The animals vary in size (2–15 lb), ear length (2–13 in.), in color, and conformation (extreme raciness to extreme cobbiness). Very few specialized strains have been developed for laboratory purposes. General aspects of rabbit husbandry have been discussed by several workers (31, 146, 151, 161).

The rabbit has become a favorite model for biomedical research in cardiac surgery, hypertension, arterio- and atherosclerosis, infectious disease, virology, embryology, screening of embryotoxic agents, and teratogens (7, 100, 107). It is used routinely for serological and physiological work and for biologic by-products like blood, plasma, cells, liver, and brain tissues. The rabbit is specially useful for reproductive biology because of nonspontaneous ovulation and the development of oral contraceptives (5, 6, 34, 42, 68). An interesting application is in the breeding of the tsetse fly. Rabbits have also been used as a model for aging research (4, 95).

Housing

In general, rabbits should be housed in metal cages to facilitate proper sanitation. The housing conditions should avoid extreme changes of environment. Water systems should be of the dewdrop type; the floors and walls of the house should be impervious to liquid and moisture and should be verminproof.

Cages are made of smooth, corrosion-resistant, impervious, and easily sterilizable material such as galvanized metal, stainless steel, or other metal alloys. The cages should be movable and have a smooth wire mesh bottom of ⅜ in. square. The space requirement is given below.

Weight of Animal	Square Inches per Animal	Maximum Population/Cage
3 to 5 lb	180	6
Over 5 lb	360	3

Males over 5 lb and females over 6 lb should be caged separately. Nest boxes can be made of wood with inside dimensions of 9 × 17 in. Generally, breeding hutches of 1 sq ft/lb of body weight should be used. Bedding is not required for rabbits. However, for solid-floored hutches, sawdust, hay, or straw is satisfactory, but it is often a source of contamination. Proper artificial lighting is suitable, and windows in rabbit houses are not necessary. The arrival house should preferably be air-conditioned to maintain a temperature between 60–70°F and a relative humidity of 40 to 60 percent. Ventilation and filteration should be provided and recirculation of air should be avoided. However, a minimum of ten air changes per hour is necessary.

Handling

Small rabbits may be lifted and carried by grasping the loin region firmly but gently. The heel of the hand may be placed on the tail region (Fig. 2-6). A medium-sized animal is best handled by grasping the fold of the skin over the shoulder and supporting the rabbit by placing the other hand under the rump. Other convenient methods can be developed, but never lift the rabbit by ears or legs. For injection or other treatment, the rabbit should be "tucked" in a nonslipping towel. If the animal cannot be restrained, then special restrainer boxes may be used.

Nutrition

Rabbits are herbivorous animals, and any of the green foods, roots, or hay and concentrate given to large animals can be given to rabbits. Commercially available feed is satisfactory but should be supplemented with greens. Daily rations of rabbits should contain the following.

	Adult Animals (%)	Pregnant/Lactating Does (%)
Crude protein	12–15	16–20
Fat	2– 3.5	3– 5.5
Fiber	20–27	15–20
Nitrogen-free extract	43–47	44–50
Ash	5– 6.5	4.5–6.5

The feed requirement varies from 3.7 to 4.0 percent of body weight, but lactating does require 4.8 to 5.6 percent. A given amount (125–100 g) of feed may be provided daily or

Fig. 2-6. The rabbit held by grasping the loin region. From (155) with permission.

automated systems of hoppers may be used to allow *ad libitum* feed to the animals. It is important to provide necessary amount of clean, fresh water; automated watering systems, which do not allow backflow, are recommended. Glass or autoclavable plastic bottles can be used. A doe and her litter may consume a gallon of water in 24 hours depending on the room temperature of the animal house. Resting animals may consume 4–6 oz of water daily, which may increase with increase in temperature.

Breeding

Different breeds and strains reach puberty at different ages (4 months for Polish rabbits to 12 months for Flemish). Some variation within breed is also seen. Does usually mature earlier than bucks. The adult sperm levels in bucks are generally achieved at 7 to 8 months, and their breeding life may continue to a maximum of 5 to 6 years, depending on management and frequency of copulation. The best productive life of the doe is up to 3 years; beyond this the litter size decreases. One buck is often kept for eight to twelve does, and mature, vigorous bucks can be used for 2 to 3 successive days. However, in normal practice, bucks should be used for breeding two to three times a week. Among domestic animals, the duration of anestrus varies with the colony and individual rabbit and is generally of 1 to 2 month's duration. The seasonal breeding pattern is also noticed, depending on breed and strain of the rabbit and the locality; for example, New Zealand rabbits in southern California have a maximum conception rate in spring and minimum in fall. Evidently, temperature is an important parameter in conception.

Rabbits do not ovulate spontaneously and a stimulus, either in the form of coitus, luteinizing hormones, or electrical or mechanical stimulation, is needed for ovulation. After coitus, ovulation occurs within 10 to 13 hours. The doe is generally brought to the buck's hutch, and copulation should take place immediately; the doe is then returned to its own cage.

The gestation period is usually 31 to 32 days, but it varies with season and size of litter. The litter size varies with stock, age of the female, and season of the year. The average litter size is generally 8.8 though the number may vary from one to eighteen. Litter analyses indicate that the numbers born alive are greatest in second and third litters and fewer in fourth and fifth litters. The birth weight is 100 g and weaning weight at 6 to 8 weeks varies between 800 to 1500 g.

Rabbits are generally bred 6 to 8 weeks after kindling when rabbits are weaned. A doe can be bred immediately after delivery and the persistence of postpartum pregnancy is inversely related to litter size. After 12 days, most does refuse mating until the mammary glands are active.

Inbreeding (brother-sister, father-daughter, or mother-son) or live breeding are generally used depending on the actual experimental designs.

THE FOWL (*Gallus domesticus*)

The fowl has become an important animal model because of the ease with which pedigree to both dam and sire can be maintained; in addition, the embryo can be studied away from the mother. The fowl can be cheaply housed and can be rapidly multiplied. Chick embryos play an important part in virology research and vaccine production.

Housing

The housing used for commercial poultry can be adapted for experimental work by providing solid partitions to make separate rooms. The room should have a 2.5 m ceiling so that several tiers of brooder and battery cages can be installed. The cages should be made of material that can withstand power hosing.

The raising of day-old chicks can be accomplished in tier brooders for floor brooding. Heating can be provided either centrally (gas or electric) or by infrared lamps. Generally, an infrared lamp (250 w) is enough for twenty-five chicks. The space requirement at 2 weeks is 62.5 sq cm/chick, 155 sq cm/chick between 2 to 4 weeks and 323 sq cm/chick after 6 weeks. The moving and housing of birds after 6 weeks of age presents some problems, and an allowance of 0.27 sq m/bird is absolutely necessary. Single cage laying batteries, if commercially available, are preferred; however, these can be fabricated to fit individual needs. These cages can be fitted on racks. The feed is generally dispensed in trough or tubular feeders. For 1- to 3-day chicks trays in which the depression is filled with food, can be used. Automatic drinking water devices are generally installed. Periodic cleaning is necessary to prevent clogging. No bedding material is needed for the birds in cages; however, for chicks and floor raising, dried wood shavings, chopped straw, peat moss, or sand may be used.

The ventilation system is important for controlling respiratory diseases. A well-designed system should deliver 7.5 cu m/hr/kg body weight; the velocity should be low so that the birds do not feel the draft (40 m/min at 21°C is satisfactory). It is advisable to install roughing filters (5μ) to prevent the ingress of pathogens.

The temperature for brooding should be 32°C for day-old chicks and lowered by 3°C per week to 18°C at 5 weeks of age. The optimum temperature for laying birds is between 13–18°C; however, optimum egg production is at 15.5°C. Lighting has an important effect on the efficiency of fowl. There should be either a constant daylight of 14 hours and darkness of 10 hours or gradual increase from 6 to 8 hours of light for day-old chicks to 18 weeks age and then an increase of 20 minutes per week to a maximum of 16 hours. The lighting system should include a dimmer which decreases light by 50 percent for about one-half hour before darkness.

Handling

The chicks should be grasped loosely by placing the hand over the back and around the body. To catch the adult bird, place hands around the sides of the body so that the thumbs are over the wings, then lift the bird firmly without any struggle or flapping (Fig. 2-7). If the birds are in a pen, they should be first made to huddle in a corner by placing a partition. With the bird facing the handler, slide the palm of the right hand under the breast and place the index finger between the legs and hold them firmly with the hand; place the left hand across the back to prevent the wings from flapping.

Nutrition

The nutritional requirements of poultry have been well studied. Commercial feeds used in poultry farms should be used for experimental animals. Care should be taken to inquire about any additive that might have been placed in the feed. Also, the nutritional requirements for chicks, growers, layers, breeding stock, and broilers are different. Therefore, proper attention should be given to the type of stock. The reviews by Scott et al. (136) and Bolton (23) should be consulted for further details. These reviews can be helpful in compounding feeds on the premises if desired.

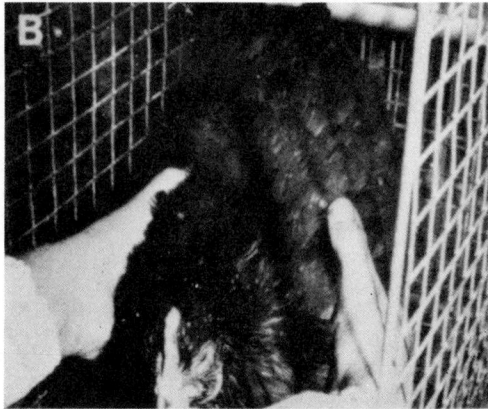

Fig. 2-7. Method of lifting adult bird out of cage. From (155) with permission.

Breeding

Few experimental units maintain breeding stocks of poultry as it is much cheaper and simpler to purchase fertile eggs from reputable breeders. However, if breeding stock is necessary for experimental purposes, it is best to keep one male for nine to ten females (heavy breed) or fourteen to fifteen females (light breed). The animal can be mated in wire-floored or deep litter pens. Natural mating or artificial insemination can be performed without much difference. The fertile eggs are generally hatched in forced air incubators that have automatic temperature, humidity, and turning devices. The temperature during the first, second, and third weeks of incubation is 39.5, 39, and 38°C and a relative humidity of 60 percent. The eggs are transferred to hatching trays of incubators after 18 days of incubation. Candling should be done at this stage to remove the unfertile "clears." The chicks hatch on day 21 and should be removed from the incubator the following day.

Chick Embryo

The chick embryo has been used for scientific investigations in embryology, virology, and genetic studies. Excellent monographs are available on the subject (130, 131). The chick embryo is susceptible to many different viruses. It is easy to handle and provides convenient research material (18). The embryo is also used for investigations on fungi, bacteria, richettsia, protozoa, and on trematode larvae (39, 52, 97, 134). It is a sensitive indicator of toxic and teratogenic properties of drugs and chemicals (82, 90, 106, 158). Chick embryos are also used in cancer research, endocrinology, nutrition, pharmacology, and transplantation research (82, 85). Chorioallantoic membrane in particular has been used for grafting embryonic and neoplastic tissues (36, 41, 83, 122). Different inoculation routes are used for different types of tests (Table 2-2). It is important that eggs from the same genetic stock should be used for a set of experiments since many genetic factors are strain-dependent (116).

Incubation

Fertile eggs may be stored between 13 and 15°C for no more than 10 days at a relative humidity of 80 to 85 percent. The incubation is generally done at a wet bulb temperature of 32°C (RH 60 percent). Candling is necessary during incubation to remove infertile eggs. Normally, commercial hatching eggs have a 90 percent fertility and a hatchability of 80 percent. The eggs are most sensitive at 3 to 5 days and 18 to 20 days of incubation (67). Another period of sensitivity is on days 13 to 14, and mortality at this time is associated with nutritional deficiency in the laying hen. Gross developmental abnormalities of embryos may result from genetic or nutritional disorders (91, 108).

Inoculation Techniques

Several routes of inoculation of fertile embryonated eggs are used according to the desired application (Table 2-2). It is important that the embryology of the egg at various stages of incubation is understood before setting up an experiment (Figs. 2-8 and 2-9.)

Chorioallantoic Inoculation. For virology work, usually 11- to 12-day incubated eggs are used. Chorioallantoic membrane (CAM) is well developed at the equitorial portion, about 15–20 mm below the edge of air sac. A small hole is made in the shell, and the membrane is pierced with a blunt needle. A drop of sterile normal saline facilitates the separation of the CAM from the shell membrane when suction is applied. The inoculum (0.1 ml) is placed on the dropped CAM and the hole sealed. Eggs are placed with the blunt end on top and reincubated. To harvest the CAM, the embryo is killed by chilling (4°C). The egg is cut with scissors and the embryo and yolk discarded. The CAM is removed by forceps.

Technique for Tissue Implantation. This technique is similar to that described above except the hole is larger and generally triangular in shape. The tissue to be implanted is carefully placed on the dropped CAM, and the triangular window is sealed.

Table 2-2. Inoculation Routes Commonly Used in Chick Embryo Studies

Inoculation route	Inoculation material		
	Microorganisms	Chemicals	Tissues
Chorioallantoic membrane	c.u.	c.u.	c.u.
Allantoic cavity	c.u.	c.u.	n.c.u.
Amniotic cavity	c.u.	n.c.u.	n.c.u.
Yolk sac	c.u.	c.u.	c.u.
Albumin	n.c.u.	c.u.	n.c.u.
Air sac	n.c.u.	c.u.	n.c.u.
Embryo body	c.u.	c.u.	c.u.
Blood stream	c.u.	c.u.	c.u.
Brain	c.u.	n.c.u.	n.c.u.
Coelom	n.c.u.	n.c.u.	c.u.
Eye	c.u.	n.c.u.	c.u.
Flank	n.c.u.	n.c.u.	c.u.
Diffusion through shell	c.u.	c.u.	n.c.u.

c.u. = commonly used
n.c.u. = not commonly used

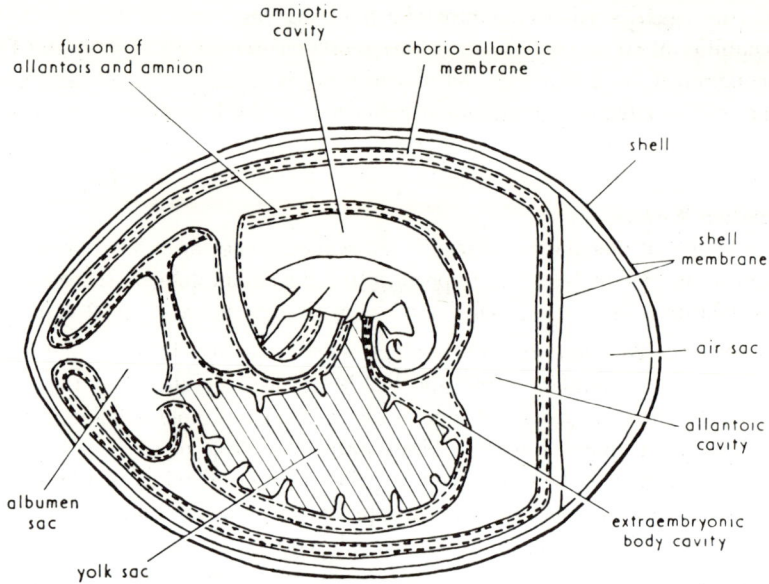

Fig. 2-8. The embryonic membrane on the twelth day of incubation. From (155) with permission.

Several other techniques have been suggested for transplantation studies (11, 72, 114, 169). CAM has also been used to study anti-inflammatory activity (40).

The interpretation of the lesions on CAM is rather difficult because of several non-specific lesions that develop due to physical or chemical trauma (30, 54, 157, 165).

Allantoic Inoculation. The inoculation through the allantoic route is generally done in eggs incubated for 9 to 11 days, and the procedure is illustrated in Figure 2-9 (153). Repeated samples of allantoic fluid can also be made (43, 80).

Amniotic Inoculation. This route has been particularly useful for influenza virus studies because it allows an access to respiratory system. Eggs incubated for 10 to 14 days are generally used. The technique is similar to CAM inoculation except that the dropped CAM is lifted with a fine forceps and a small cut (2–5 mm) is made. The amnion is grasped through this cut and pulled through the CAM opening and 0.05–0.25 ml of the inoculum is injected into the amniotic cavity.

Yolk Sac Inoculation. This route is used for toxicological studies when the agent can be injected at 0, 4, 6, or 12 days during incubation; for virology work 5 to 6 days incubation prior to inoculation is preferred. The yolk sac route may also be used for other experiments like yolk sac replacement studies (60, 87, 134), and growing tumor tissues (149). Other inoculation routes include intravenous injection (13, 58, 59), injection into albumin (21), inoculation into air sac (22, 44, 132), and deembryonated eggs (16, 46). Other less commonly used inoculation routes include intracerebral inoculation, injection into extraembryonic body cavity, introcoelomic implantation, and intraocular implantation. Diffusion of the test chemical through the egg has also been used (57, 66, 141).

Several authors have reported the use of CAM for feeding insects and ticks as well as for testing the effect of chemical agents on bloodsucking insects and ticks (26, 49, 159).

The embryonated eggs have been also used for the production of vaccines. There has been much concern about the normal microflora that may interfere with the production of vaccines or may accidently be transmitted during the use of vaccine. Several reviews are available on this subject (19, 99, 116, 147). Causative agents of certain

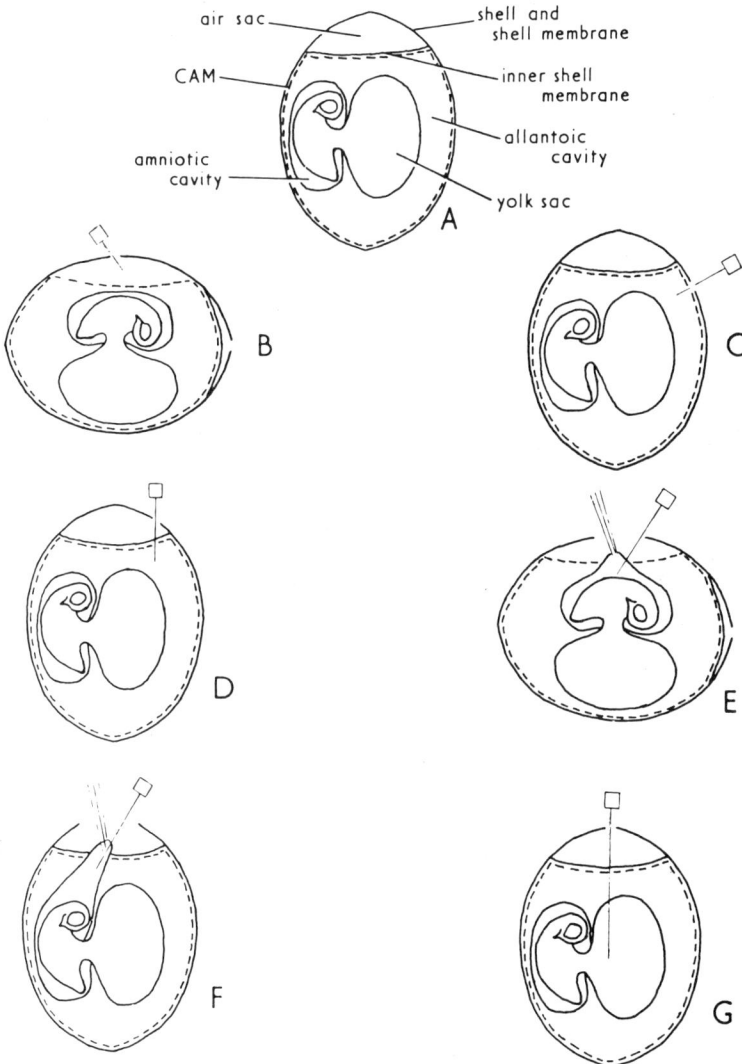

Fig. 2-9. The more commonly used inoculation sites. (A) Principal parts of the embryonated egg. (B) Chorioallantoic inoculation. (C) Allantoic inoculation (method 1). (D) Allantoic inoculation (method 2). (E) Amniotic inoculation (method 1). (F) Amniotic inoculation (method 2). (G) Yolk sac inoculation. From (155) with permission.

diseases of fowl that can be transmitted to eggs are viruses, mycoplasma, and *Salmonella* species (32, 74, 144). Two other agents, CELO (chicken embryo-lethal-orphan) and GAL (Galus adeno-like) have been shown to cause cytopathic effects in chick embryo tissues (27, 84, 89, 150).

THE CAT (*Felis catus*)

Several varieties are available, but *F. catus* appears in two distinct forms (torquata and catus) (112, 117). Domestic cats, as compared to rabbits and dogs, show little inherent variability. The adult male weighs 3.5–5.9 kg and the female weighs 2.25–3.0 kg (94). The genetics of the domestic cat have been well studied, and standardized procedures have been developed (78, 128, 140).

A healthy cat is alert, with straight head, and moves about showing interest in the surroundings. Its skin is clean, free from blemishes, with a thick, bushy undercoat and short or long silky hair. Eyes are shiny, teeth white and free from tartar, gums are clean and pinkish. Healthy cats groom frequently. Over fifty varieties have been reported. Mivart (103) was the first to use the cat as an experimental animal for the study of the mammalian skeleton. The cat is especially suited for acute experiments, and important discoveries concerning reflex action, synaptic transmission, and secretion of digestive glands have been made using the cat as an experimental animal. The cardiovascular, digestive, and neuromuscular systems of the cat closely resemble that of humans. Special use of cats has been made in the study of the effect of drugs on the nervous system and in behavioral studies (88).

Housing

Experimental cats may be confined to cages or allowed to move freely in a room or in a pen formed within a room. The room design should optimize the space per animal and save labor for cleaning of the facilities. Room equipment and supplies (cages, racks, boxes, dirt trays) should be made of sterilizable material. Cats must be provided with softwood posts for scratching; a clean, enamel-topped, short-legged table may serve this purpose. The room temperature should be kept between 18–22°C. An efficient ventilation system with eight to ten changes per hour is important because stagnant air increases the incidence of respiratory diseases. Some workers advocate outdoor housing to overcome serious respiratory disease problems. The housing and care of the cat has been reviewed by several authors (71, 77, 142).

Handling

The cat should be handled gently; stroke the head behind the ears and along the length toward the back. Care should be taken when passing the hand on the lumbar regions because females often have the tendency to bite unless in estrus. The animal should be picked up by placing an arm across its back and passing the hand forward under the chest to support its weight and controlling the uppermost parts of the foreleg with the finger. The body and hind legs should be held firmly by the arm. For holding the cat on the table for examination, the scruff of the neck should be held firmly while holding the arm parallel to the spine (Fig. 2-10). For immobilizing the animal, one assistant should hold the scruff of the neck in one hand and crossed forepaw in the other. The second assistant can hold the upper part of the hind leg in each hand.

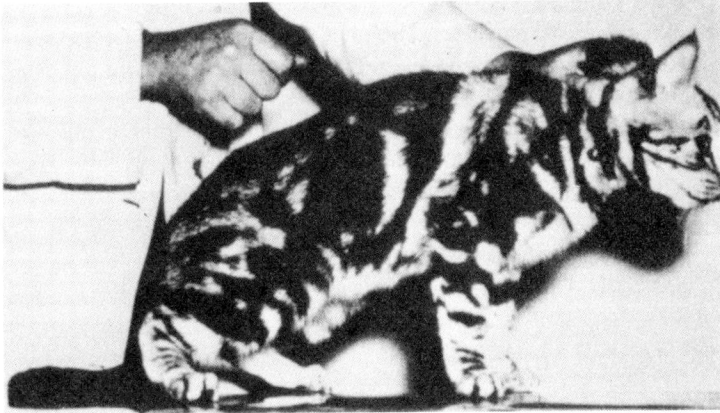

Fig. 2-10. Method of holding the cat on the table for examination. From (155) with permission.

Nutrition

The cat must have a greater proportion of its calorie requirement provided in the form of protein. The exact caloric requirement varies from 250 kcal/kg at weaning to 60–80 kcal/kg for the adult. The requirement in pregnancy is close to 100 kcal/kg, and it further increases during lactation. The commercial diets generally provide 12 to 14 percent protein, 15 to 40 percent fat, and the remainder are starchy materials (61, 96, 102, 137, 139).

The cat is a finicky eater and will die rather than eat a diet it dislikes. Generally, meat, fish, or sweet-tasting materials are preferred. A practical diet can be prepared containing two parts boiled, mashed potatoes, one part protein (steamed fish with bones, liver, lungs) and $\frac{1}{200}$ parts dried yeast. This can constitute the principal meal given in the afternoon at a rate of 100 g per adult. A secondary meal (50–75 g) consisting of one part precooked cereals, one part dried milk powder, and $\frac{1}{200}$ parts dried yeast should be given in the morning.

Fluid intake depends on the type of diet. With normal diets (moisture content 75 percent), the amount of water needed by the adult cat is about 30 ml/day. However, this amount may increase to about 200 ml if the animal is on a dry diet.

Breeding

Sexual maturity is normally attained between 7 and 12 months. At the time of first breeding, the female should weigh about 2.5 kg and the male, 3.4 kg. Cats breed two or three times in a year, typically in spring and autumn. However, under laboratory conditions, first mating occurs in January-March and the second in May-June. Cats are anestrus during October, November, and December. A third mating, usually in September, is possible under controlled conditions. The cat is polyestrus, showing recurrent estrus every 14 days and lasting for 3 to 6 days. The flattening of back and elevation of hind quarters is characteristic of estrus. In the presence of a tomcat, mating may occur several times a day for 3 to 4 days. Estrous cycle can also be identified by the vaginal smears. Practically, it is best to have a single tom for about twenty females. This prevents fighting and provides some genetic relatedness (35, 129, 138). For proper breeding programs, the diet should be rich in animal proteins, vitamin A, calcium, and iodine. The animals should be provided with well-lit, warm, and comfortable quarters. Size of space is not critical, however. The tomcat should not be moved to new quarters because this has inhibitory effects on mating. Females generally do not react to change in environment.

The gestation period is between 56 to 65 days. The variation is due to several matings during estrus over a 3 to 4 day period. The mean litter size is 4.0; larger cats tend to have more kittens. The average lactation period is seven weeks and weaning starts 4 to 5 weeks after birth. After 4 weeks, supplemental feeding is required. The newborn kitten (105 g a/v) should gain at the rate of about 10 g/day.

THE DOG (*Canis familiaris*)

Dogs were used for physiological studies as early as the seventeenth century. Now they are used for a variety of biomedical research projects, including nutritional, behavioral, toxocological, pharmaceutical, and surgical studies (9, 33, 47).

This animal model provides a range of breeds with weights ranging from a few pounds (terrier) to about 45 kg (St. Bernard). Selection of the right type of breed for a particular study is generally based on experience, experimentation, availability, and cost. It is often cheaper to purchase a dog rather than to breed it in the laboratory. Kennel-bred dogs are preferred because they are well suited for laboratory conditions. It is generally possible to select the age and weight of the animal. Among the various breeds available, beagles and spaniels are most frequently used.

A healthy dog is bright and in good, sound bodily condition. There should be no discharge from natural orifices; the skin should be free from disease, and coats should not harbor any parasites. The teeth should be firm and not unduly covered with tartar (8, 24, 51).

Housing

The exact housing practice depends on the breed; however, in general, the dog should be provided with a dry, comfortable bed with adequate space for exercise.

The breeding dogs should have indoor kennels (6 × 3 ft) and an outdoor run twice its size separated by easily controllable popholes. The building should run from north to south and the partitions between the kennels inside should be 3 ft high surmounted by galvanized netting (2 × 2 or 3 × 3 inches). However, the lower 18 in. should have 1 × 1 in. mesh in order to prevent any injury to puppies. The floors and runways of the kennel should be insulated with 1 in. thick polystyrene sheeting, while the passageways and service rooms should have granolithic cement. The building should have good natural and artificial lighting and adequate natural ventilation (four changes per hour). Each pen should also have a wicker basket 4–6 in. and at least 2 in. from the floor for the bitch to whelp and sleep in.

Housing for Experimental Dogs

The type of housing depends on the size of the dog and the type and length of experiment. In general, they should have plenty of room but if the experiment requires confinement, some other opportunity for regular exercise is needed. The dogs used for infectious disease studies need special accommodations for isolation. Special metal autoclavable cages and forced/filtered air ventilation are necessary. Animals for such studies may be reared under special conditions and may be obtained by hysterectomy.

Handling

Dogs respond best to kind but firm handling. Generally, the degree of cooperation is greatly increased if the dog is suitably rewarded after an experiment. Regular grooming with brush and comb contributes to the dog's comfort. A vicious and intractable dog should not be used except for acute nonrecovery experiments. If restraining is necessary, the attendant should stand at some distance, lean over and talk to the animal, then restrain the animal with a dogcatcher. A clove hitch bandage should be used as tape muzzle (Fig. 2-11). A slip noose of soft rope around the neck should be used to lead the dog. The well-established techniques should be used to achieve good restraint and to prevent injury to the animal handler (9, 101).

Nutrition

Although dogs by nature are carnivorous, they can be adapted to various laboratory diets. The simplest adequate ration is raw meat (with 5 percent fat) and an equal quantity of cereal, mostly in the form of dog biscuits. Canned dog foods with the occasional addition of raw meat has also been found adequate. Completely dry dog foods have been marketed but sometimes for initial acceptance these foods have to be moistened. In general, a nutritionally adequate dog food should contain the following.

	Dry Foods (%)	Canned Foods (%)
Protein	22–25	13
Fat	3– 5	5
Fibers	4– 6	2–4
Carbohydrates	16	16
Ash	2.0	2.5

The exact amount of food intake varies, and it can be best judged by the animal at-

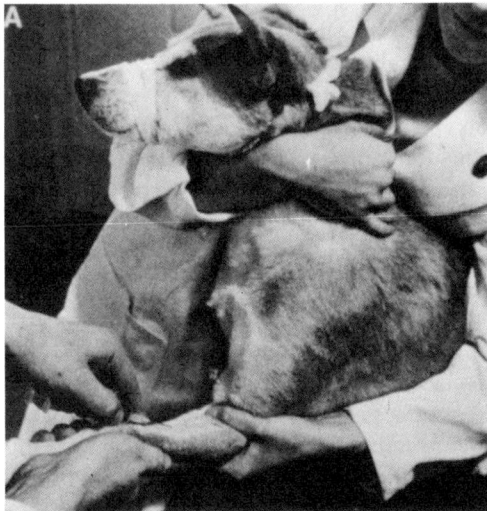

Fig. 2-11. Method of handling the dog. A clove hitch bandage is used as tape muzzle. From (155) with permission.

Table 2-3. Recommended Feed Intake for Dogs

Size	Approximate wt. of dog (lbs)	Approximate feed intake (lbs)
Small	10	0.5
Medium	25	1.25
Large	40	1.75
Very large	100	2.50

tendant. A guideline of daily intake is given in Table 2-3. The diet may be decreased by as much as 50 percent if the dog is not given any exercise. and it may be increased slightly in pregnant and lactating bitches. Puppies should be given small amounts of cow's milk or reconstituted full cream milk when 4 weeks old. This should be mixed with increasing amounts of cereals. The pups can be weaned at 7 to 8 weeks. At first, feed should be given three times a day followed by twice a day; for some dogs, one feed a day is adequate. Water should be always available and automatic devices should be installed. All utensils should be kept clean and any uneaten food should be removed before the end of the day. For details on feeding and nutrition requirements, the reader is referred to Abrams (2, 3), NRC Bulletin (109), and Worden (164).

Breeding

The reproductive physiology of different breeds of dogs is not well understood at present. Most of the information is based on the study of beagles. The first estrous cycle appears at about 7 to 9 months of age. The male dog usually cannot be used for breeding until it is about 8 months old. Male dogs show little variation in sexual interest, but their evaluation as breeders depends on high percentage of conceptivity, physical examination, and semen evaluation. Bitches may be bred when they are about one year old for best results, and the frequency of breeding depends on their estrous cycle (70, 86, 135). Females have a long estrous cycle and ovulate spontaneously. The cycle occurs at intervals of 4 to 18 months, depending on the breed. The estrous cycle consists of four stages (proestrus, estrus, metestrus, and anestrus), each of which can be identified by vaginal smears. The female accepts copulation during estrus only. Breeding in dogs is simple. The bitches are brought to the males pen only during the time of teasing; if the copulation does not occur within one hour, the female is removed and should be brought again after 6 to 8 hours. It is best to have at least one of the pair an experienced breeder. Females should be bred 24 hours after they show signs of accepting service. It is desirable to breed three times at 36 to 48 hour intervals. Pregnant females should receive abundant exercise for muscle tone and the normal maintenance of diet according to weight. The gestation period is about 63 days. Generally, 2 weeks before parturition, the bitch should be removed to whelping quarters containing padding material. This is generally torn to make a nest 24 hours before delivery. At this stage, the animal refuses food and body temperature falls by as much as 1°C. The first pup is generally delivered within 4 to 5 hours after the start of straining; if not, professional help should be sought. Pups are delivered 30 to 90 minutes apart, and the bitch generally bites and eats the umbilical cord during the licking process. The pups should be allowed to nurse as soon as possible. The delivery area should be draft-free with a temperature of 24–27°C.

THE PIG (*Sus scrofa domestica*)

During the past 2 decades, the pig has become an important experimental animal due in part to its many systematic similarities with humans and in part to the development of miniature pigs and germfree gnotobiotic animals.

Pigs are commonly used in nutrition work, radiation effects, experimental surgery, atherosclerosis, and diseases of the species.

There are a large number of breeds available, although the medium-sized, white-skinned breeds are generally selected for experimental purposes. The husbandry and biomedical use of swine have been reviewed (28, 105).

Housing

The type of accommodation depends on the nature of experiment. In some cases, individual pens are required but, in general, up to six pigs may be held in each pen. The floor area varies from 3.5–7.0 sq ft per pig depending on size and 80–100 sq ft are needed for farrowing and rearing pens.

Handling

Pigs are usually restrained in specially designed metabolic types of crates. These crates are suitable for the collection of urine, feces, and blood samples. For larger animals, the crates are used to secure the animals and later turning their backs in the supine position (121). A special harness for long-term experiments and for handling miniature pigs is available (81). The use of the harness has the advantage of allowing the animal to remain standing during most of the procedures. For minor procedures, a slip noose round the upper jaw is useful. A pig may also be held against a wall or in a corner.

Nutrition

The feed for the pig should be rich in digestible nutrients and low in fiber content. Commercially available feeds use cereal grains to provide readily available energy and soybean meal for proteins; however, animal proteins like fish meal or sterilized meat meal may be given. For the young pig, milk powder is generally included in the feed. Copper, up to 250 ppm, should be included in the diet. The nutritional requirement of the pig varies considerably with age. It is necessary to provide different rations to different age groups but care should be taken to change the diets gradually. Commercially available pellets are preferred for young pigs. However, for adults, if there are a large number of experimental animals, it may be cheaper to prepare rations from basic ingredients. Table 2-4 summarizes several rations which have been found satisfactory (98).

Breeding

The nulliparous female (gilt) reaches puberty at 7 months of age; however, the time of year of birth, inbreeding, and nutritional factors greatly influence maturity. After maturity, the estrous cycle (21 days) is regularly repeated in the absence of pregnancy without anestrus. The duration of estrus is 40 to 60 hours, and this can be recognized by enlargement of the vulva, mounting behavior, and the characteristic lordotic behavior when pressed at the back. The follicles for ovulation develop 2 to 3 days before estrus, and ovulation occurs over a period of 6 to 7 hours. The number of

Table 2-4.   Basic Mixtures for Feeding Pigs

	A	B	C
Ground barley	85-65%	87-75%	90-62%
Milling offal	0-12	0-12	25-30
White fish meal	7 -8	5-8	0-2.5
Extracted soybean meal	6-8	0-8	5-7.5
Salt (NaCl)	0.5	0.5	0.5
Sterilized bone meal or ground limestone	0.5	1.0	1.0
Vitamin supplement*	0.5	0.5	0.5
Minerals**	–	–	–

*Vit. A, 1,000 I.U.; Vit. D, 200 I.U. per lb of meal, Vit. B2.

**Zinc and Copper should be added.

A = 40-120 lb;   B = 120-210;   C = 170-260 lb.

From <u>Vet. Record</u>, (98) with permission.

ova increase with age from eleven to fourteen ova in gilt to fifteen to twenty-five at 15 to 17 months of age. The litter size varies considerably with a general range of from eight to sixteen. Generally, there is a decline in litter size in older animals probably due to intrauterine deaths. Females have been known to reproduce up to 10 years of age but normally few are kept after 4 years.

The males reach puberty at 6 to 7 months, although spermatozoa appear in the testis as early as 4 months. However, the quality and quantity of semen does not reach normal adult values until 11 to 12 months of age. Boar ejaculation is characterized by large volume (200–500 ml), low sperm count, and long ejaculation time (30 minutes). Mating 24 hours after the beginning of estrus should be fertile; however, for best results a second mating in the second day should be practiced.

THE SHEEP (*Ovis aries*)

Several breed of sheep are available but merino, originally a Spanish breed, are now used worldwide for experimental purposes. Sheep are used for physiology experiments, surgical technique, and investigation of the disease of the species.

Housing

Sheep are hardy animals and do not need elaborate housing, but it should be well ventilated with a constant temperature. Sheep are best housed in pens with a floor space of 12, 16, or 18 sq ft, depending on the breed. Concrete flooring with a rough finish is most desirable, and it need not be isolated. Wood shaving, straw, or sawdust should be comfortable as bedding material, or 1 in. sq slats about ¾ in. apart on a 4 in. deep backing are also suitable. Sawdust should be provided to soak the urine. The partitions between pens should be low (3½ ft) as sheep are gregarious and like company. The pens should have no projections or rough surfaces to prevent injury and to avoid catching of any experimental devices attached to sheep.

Handling

The untrained individual may instinctively catch the sheep by holding the wool; this is absolutely incorrect because it can damage the wool, hurt the sheep, and allow entry of water into deeper layers. The best way to catch the sheep in a pen is to approach it gently and to slide one hand under the base of the neck and then move it up to chin to hold the head up. A wild animal will need to be held by the horns, and once the head is held up the animal will back up to move the head down; allow the animal to do so until it reaches the side or end of pen. The animal can be held against the side of the pen by pressing with the knee behind the shoulder or by straddling with the knees behind the shoulder. The important point is to keep the head of the animal up. Sheep, by gentle handling, become used to the routines. The animals can be carried if necessary by placing one arm around the hind quarters and the other around the front of the thorax; handling heavier animals may require two men to lift and carry.

Nutrition

The feeding and water trough should be kept just outside the door of the pen and a small hole provided for the animal to put its head out for feeding. The removable feeding box with a hood (20 × 10 in.) has the advantage that the feed intake can be closely monitored. The normal ration for sheep includes roughage like hay, oat, or wheat straw. The straws should be chaffed and mixed with concentrates. Good hay can serve as complete rations by itself. An adult sheep (68.0 kg) will eat up to 1200 g of hay daily, and if concentrates are provided the quantity of hay may be decreased to 200–400 g. Several concentrate mixtures are available and one is given below.

Component	Parts
Ground corn	4
Crushed oats	1
Wheat bran	1
Linseed cake	0.5
White fish meal	0.5

High proteins in starch diets should be introduced gradually as it may otherwise cause gastrointestinal problems. Common salt (1 percent) may be provided in the concentrate or salt licks may be provided in the pen.

The sheep may be fed twice daily; however, a single feeding in the morning may be sufficient. The concentrates are usually fed in the morning and the hay in the late afternoon. The concentrate should generally be finished within 20 to 40 minutes and the hay by evening. The animal should be fed at a regular feeding time every day.

Lambs, when needed for experimental purposes, should be weaned within 48 hours and brought up by hand; however, it is important that the lambs receive ewes' clostrum. Lambs should be fed 4 hours apart, five times a day (6, 10, 14, 18, 22 hours) with cows' milk which is supplemented with 2.5 percent casein (for the first 3 months). The number of feeds may be reduced as the animal starts eating solid food and should be weaned at 4 to 5 months from bottle feeding.

Breeding

The first estrus in females occurs between 5 and 10 months of age with live weight of 40 to 60 percent of mature adult weight. Many ewe lambs do not have estrus until the second year. Ewes showing no estrus during the first breeding season (fall) will not do so until the second season. At that time, sheep may be almost 2 years of age. Low plane of nutrition may also delay estrus. Under good managerial conditions, about 50 to 60 percent of ewe lambs exposed will become pregnant during the first breeding season.

The male lamb shows puberty at an age of 100 to 150 days. Spermatogenesis commences about 80 to 90 days and live sperms are ejaculated by the 150th day. The breakdown of preputial adhesions of the penis is a good indication of the maturity of the male. At the age of 6 to 8 months, 10 to 50 percent of rams are highly fertile, but first matings are delayed until 18 months of age. The male does not show a restricted breeding season as seen in the female; however, seasonal variation in semen production is normal. In the United States, there is a decline in the normal sperm in spring and summer and an increase in the fall and winter. This is related to environmental temperature and extent of daylight.

Estrus in ewes is difficult to detect. Their willingness to stand and allow the ram to mate are about the only signs. The duration of estrus varies from a few hours to 3 to 4 days (24 to 48 hour average). Ovulation may be separated by 2 to 7 hours. The estrous cycle varies from 14 to 19 days. Ewes may show postpartum estrus within a few days while the lactation anestrus lasts for 4 to 10 weeks. The mean length varies with breed but the normal range is 144 to 155 days. The birth weight of the lamb varies with the breed, sex, and numbers; however, in general, the male weighs approximately 4.5–5.0 kg and the female weighs 3.2–4.1 kg. For twins, the weights are slightly lower. A 50:50 sex ratio is generally produced.

THE GOAT (Capra hircus)

Goats as experimental subjects are easy to feed, handle, and house, and hence are increasingly being used as miniature ruminant models. Goats are used to study nutrition, rumen metabolism, physiology of lactation, and mastitis diseases. More recently, their use in experimental surgery and radiobiological investigations has increased considerably.

A healthy goat is active and inquisitive and hence disease conditions can be easily detected. The gait should be normal; skin should be clean with a good bloom and free from any evidence of scratching, which indicates the presence of ectoparasites. Of the three main types of goat—meat, milk, and wool—the milk type is most commonly used for experimental purposes.

Housing

Goats can be housed in goat houses with pens, stalls, or in large loose boxes. The choice will depend on the number of goats, the type of experimental work, and economics. The house should have a minimum height of 6 ft, and pens 4½ × 4½ with 4 ft high wall are considered adequate. Stalls $2\frac{1}{2} \times 3\frac{1}{2}$ ft are satisfactory. The goats are generally tied in stalls with sliding chains attached to a vertical bar. The pen/stall should preferably be constructed on washable concrete and should be on either side of a control gangway. The concrete flooring should have insulation and drainage. The bedding material generally consists of straw or peat moss. Each pen should have hay rack and rings (9 in.) to contain buckets for water and feed. The loose box system of housing several goats in one room is not very satisfactory for experimental purposes. Good ventilation should be provided in the house with provisions and regular change of filters in the inlet and outlet ducts.

Handling

Goats are docile, sociable, and appear to enjoy handling. The experimental procedures are facilitated if constant handling of the animal is practiced before the start of the experiment. However, males can, at times, be difficult. The kids are best held by holding in arms. The adult females can be restrained in the standing position by straddling over the back and holding the head under the jaw with one hand and the base of the ear with the other hand. Cornering the animal for most procedures is satisfactory. A special cradle may be useful for several procedures.

Nutrition

The feeding requirements of goats are generally considered similar to sheep. Reid (123) recommends a starch equivalent of 640 g and digestible protein, 60 g/56.6 kg of body weight. The ratio should contain succulents and concentrated foods. As a general guide, 23.1 kg hay, 23.1 kg succulents, and 0.45 kg concentrate should be given to a goat weighing 59.1 kg. The concentrate mixture can be formulated with the following composition.

Mixture A		Mixture B	
Component	Parts of Weight	Component	Parts of Weight
Linseed cake	4	Meat meal	1
Flaked maize	3	Decorticated ground nut cake	1
Bran	2	Bran	1
Crushed oats	1	Flaked corn	4
		Oats	4
		Barley	6
		Mineral mixture	0.1

The concentrate and the hay are generally given in the morning and afternoon. Succulents are given between the milkings. Fresh water should be provided in buckets *ad libitum*.

Breeding

Both male and female reach puberty at about 6 months. The animals are seasonally polyestrous and reach peak with reproductive activity in September and the cycles continue through December. The estrous cycle lasts about 21 days with wide variations ranging from 12 to 24 days, and the duration of estrus is 2 to 3 days. The main signs of heat are restlessness, tail switching, and bellowing. The vulva is swollen and red. Ovulation occurs toward the end of estrus, and the gestation period varies between 140 to 160 days, with an average of 150 days. The number of kids vary but the usual number is two.

NONHUMAN PRIMATES

Various species of nonhuman primates are frequently used in biomedical research because of their anatomic similarities to humans. However, primates are expensive to purchase, breed, and maintain. Imported animals are of uncertain quality and our natural resources are diminishing. Even laboratory-bred animals have their limitations because most of the species produce one offspring per year. Research with primates should be restricted to those experiments for which other animals are not suitable.

Primates vary enormously in size, from the smallest mouse lemurs weighing a few grams to the gorilla weighing up to 453 kg. Table 2-5 summarizes some of the commonly used primates in biomedical research.

Primates are generally used in virus research for the production of cell lines, particularly those derived from kidney. Simian cell cultures have also been used for the production of vaccines (polio); however, their use in vaccine production has decreased considerably since the cell cultures were found to contain latent viruses such as SV40. Fetal and newborn primates of different ages have been used to isolate oncogenic viruses. Primates are also used in immunologic and immunosuppression research; this has become important, particularly with the availability of surgical techniques for transplantation. Primates are used for nutrition, general pathology, drug testing and toxicology, reproduction, aging, irradiation, and behavior studies and in the development of surgical techniques. Several reviews and books are available describing the nutrition, management, breeding, and uses of nonhuman primates for biomedical research (17, 73, 120).

Gibbons

Gibbons are light in weight—the adult male weighs 6 kg, the female 5 kg—and they have been used as experimental animals. They are very active, intractable, aggressive, and difficult to control. They breed well in captivity, producing one offspring annually. Gibbons are monogamous, and a strange, new animal introduced in the cage is attacked and killed. These animals live mostly on a vegetarian diet but at times may eat insects, eggs, and birds.

Orangutans and Gorillas

These are useful experimental animals, but they are scarce, difficult to obtain, and expensive. They do not breed well in captivity; however, some births have been recorded in zoos and laboratories.

Baboons

These animals are considered to be vermin in Africa; hence, import to the United States is assured for many years. Male baboons weigh 25 kg and are about 4 ft tall; the

Table 2-5.   Frequently Used Species of Primates in Biochemical Research

Genus	Common Name	No. of Species	Place of Origin
Lemur	Lemurs	5-8	Madagascar
Microcrebus	Mouse Lemur	2-4	Madagascar
Aotus	Night monkey	1	S. America
Saimiri	Squirrel monkey	2	C. and S. America
Pithecia	Saki	4	S. America
Cebus	Capuchin monkey	4	S. America
Alouatta	Howler monkey	5	C. & S. America
Lagothrix	Wooly monkey	2	S. America
Ateles	Spider monkey	4	C. & S. America
Colobus	Colobus monkey	4-8	Africa
Presbytis	Langur	13	S.E. Asia
Nasalis	Proboscia monkey	2	Borneo Islands
Papio	Baboons	3-4	Africa
Ceropithecus	Guenons	12-20	Africa
Macaca	Macaques	10-16	S.E. Asia
Hylobates	Gibbons	6	S.E. Asia
Pongo	Orangutan	1	Borneo
Pan	Chimpanzee	1	Africa
Gorilla	Gorilla	2	Africa

females are about one-half the male weight. Baboons have a mild temperament and are easy to handle. They have been commonly used for surgical experiments (transplantation and cross circulation), immunologic research, and trachoma investigations. Young animals reach puberty at 4 years of age but are not full grown until at least 7 years. They may live over 30 years. The reproductive life of the female goes beyond 20 years and she produces one offspring annually. They breed well in captivity. Once they reach adulthood, their appearance is fierce; the female has a large swelling in the female sexual skin and the males' rump and genitalia have bright coloration.

Macaques

These animals are used most frequently for the study of human pathogens and their immunologic response is similar to humans. The rhesus monkey is the most important experimental model in this group. The male reaches puberty at 3 to 4 years, while the female matures at 1.5 to 2.5 years. The adult rhesus weighs about 12 kg and the female weighs 10 kg. The birth weight is 500–700 g and the weight reaches approximately 1 kg at weaning (3 to 6 months). The animals thrive well in captivity and breed without any problem. This species had been widely used for vaccine production and providing the cell lines for tissue culture.

Chimpanzee

These are commonly used primates. At birth, their average weight is 1.85 kg, and they usually (but not always) are covered with a coat of black hair. Weight gain with age is progressive and at times has been used to determine age; however, estimation of age is best made with the examination of the teeth. The young develop more quickly than human babies. They come to sexual maturity around 8 years of age and the average life span is around 30 to 40 years. The average adult male weighs 50 kg and female weighs about 40 kg. The young chimpanzees are docile, malleable, and affectionate but these characteristics may change after puberty. These animals like to live in troops; generally, males do not fight and the females are promiscuous. The animals breed well in captivity especially in groups. Females sometimes fight particularly over babies, and the female with a dead baby will try to steal one from other females. Their diet is mainly vegetarian.

Housing

The housing for simians should be based on units, needs for quarantine, and experiment of designs. It should also be kept in mind that simians may carry pathogens which can be transmitted to humans (Chapter 3).

The animal rooms should be adequately ventilated (ten to fifteen changes per hour) and provide a minimum of 1.3 cu m per animal weighing about 3 to 4 kg. The rooms should have a negative pressure and temperature of 21–24°C at 50 percent RH. The temperature and relative humidity will vary slightly with the experimental species. The room height should be less than 3 m, and the door should be at least 1.3 m in. and corridors about 1.8 m. The doors should always be locked and security traps provided.

Cages

The size of the cage depends on the animal species and experimental design. Macaque species weighing 4 kg can be held in pairs in cages measuring 78 × 78 × 91 cm. Animals weighing up to 8 kg can be held singly in these cages. For baboons and chimpanzees, cages measuring 122 × 91 cm and 183 cm high can be used. Several commercial designs are available, but cages with movable partition to "squeeze" the animal are preferable for handling.

Pelleted food may be offered in hoppers and greens or fruits can be given through the mesh.

Nutrition

Most primates anatomically and physiologically have omnivorous capabilities. However, under natural conditions, they eat vegetarian diets (grasses, seeds, roots, fruits, bulbs, and crops) and occasionally they may eat eggs, insects, birds, or lizards. However, in captivity, it is very difficult to provide the variety of foods to fulfill the nutritional requirements of the monkey. It is suggested that these animals should be given compound diets with known composition and some some roots, fruits, and vegetables to keep them busy and occupied. Several commercial diets are available. The gross composition and formulation of monkey diets are presented in Table 2-6.

Table 2-6. Composition of Monkey Diet

Ingredient	Percent	
Ground corn	16	Gross composition
Ground wheat	12	Protein 24
Ground barley	10	Crude fat
Rolled oats	10	Crude fibre 2.5
Ground nut meal	10	Calcium 0.9
Soybean oil meal	8	Phosphorus 0.8
Wheat germ	5	
Dried Yeast	10	
Skimmed milk powder	10	
Fat	5	
Bone meal	2	
Salt	0.9	
*Trace elements	0.1	
**Vitamin mixture	1.0	

*mg/kg of feed: Fe, 20; Cu, 2; Mn, 50; Iodine, 2; Zn, 10; Co, 1.

**Vit. A. 40,000 I.U.; D3, 6,000 I.U.; B2, 40 mg per kg of feed.

NOTE: Other diets contain additional amounts of Vitamin B1, B12, Panthothenic acid, Nicotinic acid, Vitamin E and K.

From Symposium Zoological Society London, (17, 113) with permission.

Breeding

The breeding stock should be checked for good health, and an unsedated monkey should be alert, active, and have bright eyes and lustrous coat. Sexual maturity is generally not reached until they have twenty-eight teeth.

The reproductive status of the female should be judged nulliparous; primiparous females have small, pink nipples and multiparous females have large, elongated, and pigmented nipples. The nonpregnant uterus is about 2.5–4 cm in length, dependent on the number of pregnancies. In the rhesus monkey, the average age at the time of first menstruation is 3 years and 7 months and sexual maturity is reached at 4½ years when a regular menstrual cycle of 28-day duration is established. Ovulation can occur between 8 and 21 days of the cycle but most frequently occurs on the 12 and 13 days. The detection of exact ovulation time is difficult. However, rectal palpation of the ovaries and uterus indicates that at the time of ovulation, the uterus becomes extremely resilient and the ovaries undergo a sudden reduction in size.

Table 2-7. Summary of Life Cycle of Commonly Used Nonhuman Primates.

Parameter	Baboon	Chimpanzee	Rhesus	Squirrel Monkey
Average wt (kg)				
Male	25	50	12	1
Female	13	40	10	0.8
Female breeding				
Age (yrs)	5	10	5	3
Weight (kg)	11-12.5	9-10.5	8-9.5	0.6-0.7
Estrus cycle				
(days)	34	36	28	6-19
Gestation	164-186	216-260	150-180	170

The males reach sexual maturity at about the same time as the female. The testis should be permanently in the scrotal region and a full examination of semen should be done. The normal rhesus male should ejaculate approximately 1.5 ml of semen with a sperm count of 200–500 million/cu cm.

Both monogamous and polygamous systems of breeding are used, depending on the species and the facilities available. In the polygamous system, generally one male is provided for three to five females and the system provides 73 to 83 percent conception in a semifree colony. The fertility figures are about two to three times higher as compared to indoor monogamous pairs (12). However, fighting may erupt between males and their less favorite females. It is also difficult to estimate the date of conception.

In the monogamous system, a female is caged with the male for a restricted or prolonged period of time. The system calls for a high ratio of males to females. It is more expensive and less productive. Correct menstrual dates can be kept, and best results are obtained if females are kept with males from the morning of day 2 of the menstrual cycle. Almost 70 percent conception rate can be achieved but it requires two to three consecutive matings. The exact diagnosis of pregnancy is important. Physical examination, implantation bleeding 15-21 days after conception, and biological and immunoassay have been used with some success. Table 2-7 summarizes comparative biological data on the most commonly used nonhuman primates.

CATTLE (*Bos taurus*)

A large number of varieties (dairy, beef, dual purpose) and breeds of cattle are available. However, their use as experimental animals is restricted except for studies (nutrition, physiology, disease, and metabolism) on their own species. Because of the initial cost and the cost of housing and maintenance, these animals are seldom maintained for the collection of fluids or tissue, as these can be easily obtained from the slaughterhouses.

Cattle as experimental models are used for the study of infectious diseases; effects of environmental conditions, particularly temperature; and, more recently, the intake of radioactivity and pesticides from the environment and their subsequent excretion in milk.

The experimental animals should be purchased from a reliable breeder with known records of good health and performance (genetic). Unless specifically needed, cattle should not be purchased under 6 months of age.

Housing

The housing of cattle could be a major problem because of the space requirement and isolation needs. If there are a few animals, they can be kept in loose boxes or single stalls, but if large numbers are to be housed, it is preferable to have specially constructed buildings. Several different commercial designs are available; however, each building should preferably hold twenty-four cattle in two groups of twelve each with facilities for isolation of the two groups. Exercise yards should be provided, and the animals should be tied up in lines like ordinary dairy sheds for feeding and milking.

Within the area, there should be single isolation units. Other facilities like cleaning rooms, sterilization rooms, food stores, milking machines, milk storage, and disposal of manure should be provided. A postmortem room, slaughterhouse, cold store, and incinerator may be needed, depending on the size of the experimental facility.

The bedding provided for the animals is generally wheat straw. Barley straw should be avoided as it can cause irritation. Peat moss is the best choice in feeding experiments because of the possibility of cattle eating the straw bedding. However, peat moss is expensive and it may be dusty at times. The soiled bedding should be removed regularly.

Handling

The animals should be handled quietly, efficiently, and correctly. Most cattle are quiet and cooperative; animals that are difficult to handle should be culled at the earliest opportunity.

The simplest form of restraint is the use of a halter or even holding the head, and if this does not work the halter should be tied to a fixed object and the nose held by fingers. "Bulldogs" may be used in difficult cases but ordinarily they should be avoided; half-hitch of the halter rope around the nose will often steady the animals. Heifers may cause some problems with kicking during the milking procedure; such animals can be steadied by holding the tail from the rear. In bad cases, a looped rope around the legs near the udder may be used. Other elaborate techniques such as casting are given in details in animal husbandry textbooks. It is best that the herd is dehorned to prevent injury to other cattle and to attendants. The bull calves should be castrated if they are to be kept for long periods of time.

Nutrition

Several commercial cattle feeds are available, and it is best to use these for experimental animals. Roughages can be provided with grass, hay, and oat straw. It is not necessary to let the animals out in the field for grass, as animals kept carefully indoors and supplied with grass perform as well in their breeding and milk records. The changes in the ration or feeding regimen should be done gradually; however, abrupt changes, if needed for experimental purposes, may not be as harmful as originally thought.

Water and feed should be supplied *ad libitum*.

Breeding

Cattle are polyestrous and ovulate spontaneously. The age at estrus in heifer varies from 6 to 18 months, depending on breed, feed intake, and the season of birth. Estrus depends more on weight (270 kg for Holsteins) than age. Animals are known to ovulate before any signs of estrus. In males, the differentiation of testicular tissues takes place at the age of 3 to 4 months and sperm production starts by 6 months, but abundance of free sperm is not observed in the tubules until 8 to 10 months of age.

A nonpregnant cow in a temperate climate exhibits estrus every 17 to 24 days throughout her lifetime, with the exception of the postpartum period (30 to 90 days). The postpartum estrus is greatly influenced by the frequency of milking, nursing of calves, nutrition, and the season. The estrous period in the cow is rather short but can be easily detected from the behavior of the females, and the cervical mucous in the dry cow will show the characteristic fern pattern. Normally, one ovum is shed per ovulation and 60 percent of the time, it is from the right ovary. The incidence of twins is only 2 percent in dairy cattle and less (0.5 percent) in beef cattle. The fertilization rate is almost 100 percent in virgin heifers. However, it decreases considerably in "repeat breeders." Improvement in the fertility rate can be achieved by culling the females with obvious structural and physiological genital abnormalities. Similarly, good quality bulls both for artificial insemination and natural breeding should be kept in the herd. Generally, the bull semen is of good quality and its spermatozoa have an exceptionally long life span. Therefore, it can be preserved for long periods of time.

THE HORSE (*Equus caballus*)

Like other large animals, the use of the horse as an experimental animal is generally restricted to certain centers because of the high initial cost and their high feeding and housing costs. However, their primary use for the production of antisera and antitoxin still has not been matched by any other animal.

There is a wide choice of breeds available ranging in weight from 160–180 kg (Shetland ponies) to large breeds of horses weighing up to 900 kg. The horse should be in good health and need not necessarily be sound (from the standpoint of work or racing). Ponies are generally preferred because of their size, initial cost, and cost of housing and feeding. However, for the production of antisera, larger animals may be more desirable. It is advisable to purchase well-broken and easy to handle geldings or mares about 4 years of age in medium condition. The newly purchased animals should be kept in quarantine for 3 weeks or more; during this period vermifuge treatment should be given, if needed.

Housing

Horses can be kept outdoors in grassland for most of the year. However, if space is a problem, then individual loose boxes or stables with partitions can be used. In fact, for most experimental purposes where routine data on temperature, pulse, and respiration are needed, indoor housing is preferred. Regular plans for the construction of indoor stables are available. A proper and suitable stable routine should be established and adhered to as strictly as possible. Horses are creatures of habit; regular feeding, watering, and exercising maintain their health and avoid abnormal dysfunctions. A general feeding routine may be as follows.

- *7:00 A.M.* Water and feed with about half the daily allowance of concentrates.
- *9:00 A.M.* Groom, pick out feet, wash or sponge eyes, nostrils and dock, remove dung and soiled bedding.
- *10:00 A.M.* Exercise with hand or turn out to graze.
- *12:30 P.M.* Water; feed half the daily hay
- *4:30 P.M.* Water, feed remainder of hay and concentrates, make up bedding for night.
- *8:00 P.M.* Every horse checked by attendant.

The horses should be exercised every day unless a particular experimental procedure specifically calls for rest. Similarly, grazing should be a part of the daily activities for the horses; the exact time for grazing depends on the environmental temperature and the population of biting flies.

Handling

Generally, a well-tended horse is not difficult to manage. Special procedures such as twitches, slings, side stocks, neck cradles, and casting may be needed for difficult horses or special circumstances. For minor manipulations, raising the foot and holding it flexed will be enough for most of the routine work (rectal temperature, dressing of wounds); however, for other procedures, the application of a twitch to the upper lip may be desirable.

Nutrition

Horses apparently do well on a wide variety of rations. The choice of ingredients depends on local availability and cost. However, most diets should contain oats, bran, and hay. Other grains often used are corn, millets, sorghum, beans, and barley. Commercially available concentrates are generally prepared for racehorses and less rich diets are preferable for experimental animals. Grass or grass products (hay, silage, green or dried cut grass) constitutes up to 75 percent of the total nutrition intake. The grain foods may be given whole, crushed, rolled, or ground into a meal. A few suggested rations are summarized in Table 2-8.

In most of the diets, the protein, starch, and dry matter equivalents are 1.35, 9.5 and 16.0 units. The maintenance rations may be further modified according to needs.

Water must be provided in the paddock, preferably in a trough with an automatic level maintenance device. The water trough should be recessed or protected to prevent injury to galloping horses. The exact amount of water intake will depend on the percentage of grass in the feed and amount of exercise.

Breeding

Stallions attain sexual maturity at 2 years of age, although sperm production starts at the age of 1 year. Males are not generally used for natural breeding till the age of 3 to 4 years. Sometimes, semen may be collected for artificial insemination. Females reach puberty at 15 to 18 months of age. In areas where there is a definite breeding season, females may not reach puberty until they are 25 to 28 months old.

Sexual activity is greatest during spring and summer when the amount of daylight is maximum. In tropical and subtropical countries, the breeding season may be extended

Table 2-8. Recommended Rations for the Horse

Ration A		Ration B	
Rolled or crushed oats	4 part per 1,000 body weight	Crushed oats	1 part/1,000 part body weight
Wheat bran	1 part per 1,000 body weight	Flaked corn	2 part/1,000 part body weight
Hay (mixed quality)	31 part per 1,000 body weight	Bran	1 part/1,000 part body weight
		Hay	15 part/1,000 part body weight

throughout the year. In males, the breeding season is not well marked and good quality semen can be collected throughout the year. However, in certain areas, depending on the latitude, the amount of total semen and gelatinous material varies during the year. It is evidently related to the length of daylight since provision of artificial light rectifies this problem. In females, there are much greater differences in breeding patterns which may be classified into:

1. Defined breeding season generally during the longest days of the year. Foals are born during restricted time in the year.

2. Transitory breeding season; the females show estrous cycle throughout the year but ovulate only during the breeding season. Foals are born during limited foaling season.

3. Year-round breeding.

The females in estrus can be easily detected from their behavior, the enlargement of labial folds of the vulva, congestion and vascularization of the mucous membrane of the vulva and vagina, and the extensive dilation of the cervix.

For best conception rate, the time of ovulation should be determined before servicing, since the time interval between estrus and ovulation varies considerably. The conception rate also increases with the number of inseminations during estrus.

REFERENCES

1. Abrams, J. T., *Animal Nutrition and Veterinary Dietics,* 4th ed., Green, Edinburgh (1971).
2. Abrams, J. T., in Graham-Jones, O. (ed.), *Canine and Feline Nutritional Requirements,* Pergamon Press, Oxford, 1964.

3. Abrams, J. T., *The Feeding of Dogs,* Green & Co., Edinburgh, 1962.
4. Adams, C. E., *J. Embryol. Exp. Morphol.,* **23:**21 (1970).
5. Adams, C. E., *J. Reprod. Fertil.,* **14:**351 (1967).
6. Adams, C. E., *J. Reprod. Fertil. Suppl.,* **12:**1 (1970).
7. Adams, C. E., M. F. Hay, and C. Lutwah-Mann, *J. Embryol. Exp. Morphol.,* **9:**468 (1961).
8. Anderson, A. C., *The Beagle as an Experimental Dog,* Iowa State University, Ames, Iowa, 1970.
9. Archibal, J., *Canine Surgery,* American Veterinary Publications, Illinois, 1965.
10. Arlington, L. R., *Introduction to Laboratory Animal Science,* The Interstate Printers and Publishers, Illinois, 1972.
11. Ballantyne, D. L., *Transplant. Bull.,* **7:**110 (1959).
12. Baneyee, B. N., and Woodward, *Lab. Anim. Care,* **20:**82, 1970.
13. Barnes, C. M., and L. M. Julian, *Am. J. Vet. Res.,* **19:**759 (1958).
14. Barnett, S. A., *Proc. R. Soc. Series B.,* **151:**87 (1960).
15. Barnett, S. A., *A Study in Behaviour. Principals of Ethology and Behavioural Physiology, Displayed Mainly in the Rat,* Methuen, London, 1963.
16. Beard, C. W., *Avian Dis.,* **13:**309 (1969).
17. Beveridge, W. I. B., *CIBA Found. Symp. Health Mankind,* **63:** (1967).
18. Beveridge, W. I. B., and F. M. Burnet, *Spec. Rep. Ser. Med. Res. Coun. No. 256,* Australia (1946).
19. Biggs, P. M., *Lab. Pract.,* **19:**45 (1970).
20. Billingham, R. E., G. H. Sawchuck, and W. K. Silvers, *PNAS (USA),* **46:**1079 (1960).
21. Blackwood, U. B., *Proc. Soc. Exp. Biol. Med.,* **104:**373 (1960).
22. Bloom, H. H., and F. B. Gordon, *J. Bact.,* **70:**260 (1955).
23. Bolton, W., *Poultry Nutrition MAFF Bull. No. 174,* Her Majesty Stationery Office, London, 1967.
24. Bowden, N. L. R., in Tavernor, W. D. (ed.), *Symposium of Nutrition and Diseases in Experimental Animals,* Bailliere, Tindall & Cassell, London, 1970.
25. Broadhurst, P. L., *Lab. Anim. Cent. Coll. Pap.,* **12:**65 (1963).
26. Burgdorfer, W., and E. G. Pickens, *J. Infect. Dis.,* **94:**84 (1954).
27. Burmester, B. R., G. R. Sharpless, and A. K. Fontes, *J. Nat. Cancer Inst.,* **24:**1443 (1960).
28. Bustad, L. K., R. O. McClellan, and M. Paul Burns, *Swine in Biomedical Research,* Battelle Memorial Institute, Pacific Northwest Laboratory, Richland, Wash., 1965.
29. Butler, L., *Diabetogia,* **3:**124 (1967).
30. Camain, R., P. Bres, and H. Plagnol, *Ann. Inst. Pasteur, Paris,* **98:**846 (1960).
31. Casady, R. B., P. B. Sawin, and J. van Dam, *Commercial Rabbit Raising Handbook,* No. 309, USDA, Washington D.C., 1966.
32. Casorso, D. R. and E. L. Jungherr, *Am. J. Vet. Res.,* **20:**547 (1959).
33. Catcott, E. J., *Canine Medicine: Text and Reference Work,* American Veterinary Publications, 1968.
34. Chang, M. C., *Nature (Land),* **184:**466 (1959).
35. Colby, E. D., *Lab. Anim. Care,* **20:**1075 (1970).
36. Converse, J. M., D. L. Ballantyne, B. O. Rogers and A. P. Raisbeck, *Transplant. Bull.,* **5:**108 (1958).
37. Cook, M. G., *The Anatomy of the Laboratory Mouse,* Academic Press, New York, 1965.
38. Cotchin, E., and F. J. C. Roe, *Pathology of the Laboratory Rat and Mice,* Blackwell, Oxford, 1967.
39. Cox, H. R., *Ann. N.Y. Acad. Sci.,* **55:**221 (1952).
40. D'Arey, P. F. and E. H. Howard, *Br. J. Pharmacol. Chemother.,* **29:**378 (1967).
41. Dagg, C. P., D. A. Karnofsky, and J. Roddy, *Cancer Res.,* **16:**589 (1956).
42. Desjardins, C., K. T. Kirton, and H. D. Hafs, *J. Reprod. Fertil.,* **15:**27 (1968).
43. Dubose, R. T., *Avian Dis.,* **9:**598 (1965).
44. Ehmann, B., *Lancet,* **1:**772 (1963).
45. Farris, E. J., and J. Q. Griffiths, *The Rat in Laboratory Investigations,* Hafner, New York, 1962.
46. Fazekas, S. de St. Groth, and D. O. White, *J. Hyg. Camb.,* **56:**523 (1958).
47. Felson, B., *Roentgen Techniques in Laboratory Animals,* Saunders, Philadelphia, 1968.
48. Ferm, V. H., *Lab. Anim. Care,* **17:**452 (1967).

49. Ferris, D. H., and R. P. Hanson, *Cornell Vet.*, **42**:389 (1952).
50. Flynn, R. J., In M. L. Conalty (ed.), *Husbandry of Laboratory Animals,* Academic Press, New York, 1967.
51. Fox, M. W., *Canine Behaviour,* Charles C. Thomas, Springfield, Ill., 1965.
52. Fried, B., *J. Parasitol.,* **48**:545 (1962).
53. Fulton, G. P., in P. F. Robinson and H. Magalhaes (eds.), *The Golden Hamster,* Iowa State University Press, Ames, Iowa, 1968.
54. Ganote, C. E., D. L. Beaver, and H. L. Moses, *Lab. Invest.,* **13**:1575 (1964).
55. Gartner, K., *Zentbl. Vet. Med.,* **17**:18 (1970).
56. Gay, W. I., *Methods of Animal Experimentation,* Vols. 1–4, Academic Press, New York, 1965–1973.
57. Glick, B., and C. R. Sadler, *Poult. Sci.,* **40**:185 (1961).
58. Goerttler, K., *Klin. Wschr.,* **40**:809 (1962).
59. Goldwasser, R., and M. C. Shelesnyak, *Science* (N.Y.), **118**:47 (1953).
60. Grau, C. R., H. I. Fritz, N. E. Walker, and N. W. Klein, *J. Exp. Zool.,* **150**:185 (1962).
61. Greaves, J. P., and P. P. Scott, *Br. J. Nutr.,* **14**:361 (1960).
62. Green, E. L., *Biology of the Laboratory Mouse,* McGraw-Hill Co., New York, 1966.
63. Grüneberg, H., *The Genetics of the Mouse,* Nijhoff, The Hague, 1952.
64. Hacking, M. R., and Lane-Petter, *Lab. Anim.,* **2**:131 (1968).
65. Hafez, E. S. E., *Reproduction and Breeding Techniques for Laboratory Animals,* Lea & Febiger, Philadelphia, 1970.
66. Hall, C. F., A. I. Flowers, and L. C. Grumbles, *Avian Dis.,* **7**:178 (1963).
67. Hamilton, H. L., *Lillie's Development of the Chick,* 3rd ed., Holt, New York, 1952.
68. Harper, M. J. K., *J. Enocr.,* **22**:147 (1961).
69. Harris, G. W., Private communication, Cambridge.
70. Harrop, A. E., *Reproduction in the Dog,* Bailliere, Tindall and Cox, London, 1960.
71. Hediger, H., *Wild Animals in Captivity,* Butterworth, London, 1950.
72. Hellman, K., and D. F. Tucker, *Cancer Chemother. Rep.,* **47**:73 (1965).
73. Hill, B. F. (ed.), *Some Physiological Parameters of Small Animals,* Charles Rivers Dig., **10** (4) (1971).
74. Hlozanek, I., and P. Vesely, *Inst. Exp. Biol. Genet. Monogr.,* Prague (1969).
75. Hoffman, R. A., R. P. Robinson, and Magalhaes (eds.), *The Golden Hamster: Biology and Uses in Biomedical Research,* Iowa State University Press, Ames, Iowa, 1968.
76. Hornburger, F., *Cancer,* **23**:313 (1969).
77. Jackson, O. F., and P. P. Scott, *Lab Animals,* **4**:135 (1970).
78. James, C. C. H., L. P. Lassman, and B. E. Tomlinson, *J. Pathol.,* **97**:269 (1969).
79. Jay, G. E., in W. J. Burdette (ed.), *Methodology in Mammalian Genetics,* Holden-Day, San Francisco, 1963.
80. Jorden, C. E., B. M. Monk, and R. T. Dubose, *Avian Dis.,* **11**:255 (1967).
81. Karagianes, M. T., *J. Appl. Physiol.,* **25**:641, 1968.
82. Karnofsky, D. A., *Stanford Med. Bull.,* **13**:247 (1955).
83. Karnofsky, D. A., L. P. Ridgway, and P. A. Patterson, *N.Y. Acad. Sci.,* **55**:313 (1952).
84. Kawamura, H., F. Shimizu, and H. Tsubahara, *Nat. Inst. Anim. Hlth. Q.,* Tokyo, **4**:183 (1964).
85. Khera, K. S., and D. A. Lyon, *Toxicol. Appl. Pharmacol.,* **13**:1 (1968).
86. Kirk, R. W., in E. S. E. Hafez (ed.), *Reproduction and Breeding Techniques in Laboratory Animals,* Lea & Febiger, Philadelphia, 1970.
87. Klein, N. W., C. R. Grau, and N. J. Green, *Proc. Soc. Exp. Biol. Med.,* **97**:425 (1958).
88. Kling, A., J. Kovach, and T. Tuker, in E. S. E. Hafez (ed.), *Behavior of Domestic Animals,* Lea & Febiger, Philadelphia, 1969.
89. Kohn, A., *Am. J. Vet. Res.,* **23**:562 (1962).
90. Landauer, W., *J. Cell Comp. Physiol.,* **43**:261 (1954).
91. Landauer, W., "The Hatchability of Chicken Eggs as Influenced by Environment and Heredity," *Stross Agri. Exp. Stn. Monograph 1,* University of Connecticut, Storrs, Conn. 1967.
92. Lane-Petter, W. and M. E. Lane-Petter, in *Defining the Laboratory Animal,* National, Washington, 1971.

93. Lane-Petter, W. and A. E. G. Pearson, *The Laboratory Animal-Principles and Practice,* Academic Press, London, 1971.
94. Latimer, H. B., *Anat. Rec.,* **157:**449 (1967).
95. Little, R. A., *J. Physiol.,* **208:**485 (1970).
96. Loew, F. M., C. L. Martin, R. H. Dunlop, R. J. Mapletoft, and S. I. Smith, *Can. Vet. J.,* **11:**109 (1970).
97. Long, P. L., *Parasitology,* **56:**575 (1966).
98. Lucas, I. A. M., *Vet. Rec.,* **76:**101 (1964).
99. Luginbuhl, R. E., *Nat. Cancer Inst. Monogr.,* **29:**109 (1968).
100. Makepiease, L. I., *Rabbits: A Subject Bibliography,* Colorado University Press, Fort Collins, Colo., 1956.
101. Markowitz, J., J. Archibald, and H. G. Downie, *Experimental Surgery, Including Surgical Physiology,* Williams & Wilkins, Baltimore, 1964.
102. Miller, S. A., and J. B. Allison, *J. Nut.,* **64:**493 (1958).
103. Mivart, ST. G. *The Cat—An Introduction to the Study of Backboned Animals, Especially Mammals,* J. Murray, London, 1881.
104. Moore, W. Jr., *Lab. Animal Care,* **15:**94 (1965).
105. Mount, L. E., and D. L. Ingram, *The Pig as a Laboratory Animal,* Academic Press, New York, 1971.
106. Murphy, M. L., C. P. Dagg, and D. A. Karnofsky, *Pediatrics,* **19:**701 (1957).
107. Napler, R. A. N., in W. Lane-Petter (ed.) *Animals for Research,* Academic Press, New York, 1963.
108. National Academy of Sciences, *Methods for the Examination of Poultry Biologics and for Identifying and Quantifying Avian Pathogens,* Washington D.C. 1971.
109. National Research Council, *Nutritional Requirements of Dogs,* National Academy of Sciences, Washington, D.C., 1962.
110. Nixon, C. W., R. Whitney, J. Beaumont, and M. E. Connelly, *J. Hered.,* **60:**299 (1969).
111. Njaa, L. R., F. Utne, and O. R. Braekkan, *Nature (Land),* **180:**290 (1957).
112. O'Reilly, K. J., and A. M. Whitaker, *J. Hyg. Camb.,* **67:**115 (1969).
113. Ottewell, D., in R. Hare and P. N. O'Donoghue (eds.), *Design & Function of Laboratory Houses,* Lab. Anim. Symp., I, 1968, p. 97.
114. Owen, J. J., and G. A. Harrison, *Transplantation,* **5:**643 (1967).
115. Palm, J., W. K. Silver, and R. E. Billingham, *J. Hered.,* **58:**40 (1967).
116. Payne, L. N., in T. C. Carter (ed.), *Eggs in Virology in Egg Quality, A Study of Hen's Egg* Oliver, & Boyd, Edinburgh, 1968.
117. Pockock, R. I., *Catalogue of the Genus Felis,* British Museum of Natural History, London, 1951.
118. Poiley, S. M., *Adv. Pharmacol. Chemother.,* **12:**125 (1975).
119. Porter, G., W. Lane-Petter, and N. Horne, *Z. Versuchstierk,* **2:**75 (1963).
120. *Primate News Letter,* Brown University, Providence, R.I.
121. Pugh, O. L., and R. H. C. Penny, *Vet. Rec.,* **79:**390 (1966).
122. Rawles, M. E., *Ann. N.Y. Acad. Sci.,* **55:**302 (1952).
123. Reid, T. T., in *The Mammary Glands and Its Excretions,* E. Kon and C. Courie (eds.), Academic Press, New York, 1961.
124. Richards, M. P. M., *Anim. Behav.,* **14:**450 (1966).
125. Ritchie, D. H., and J. K. Humphrey, *J. Inst. Anim. Techn.,* **21:**100 (1970).
126. Robinson, P. F., in R. A. Hoffman, P. F. Robinson, and H. Magalhaes, *The Golden Hamster,* Iowa State University Press, Ames, Iowa, 1968.
127. Robinson, R., *Genetics of the Norway Rat,* Pergamon Press, Oxford, 1965.
128. Robinson, R., *Genetics for Cat Breeders,* Pergamon Press, Oxford, 1971.
129. Robinson, R., and H. W. Cox, *Lab. Animal,* **4:**99 (1970).
130. Romanoff, A. L., *The Avian Embryo,* Macmillan & Co., New York, 1960.
131. Romanoff, A. L., and A. J. Romanoff, *The Avian Egg,* John Wiley & Sons, New York, 1949.
132. Ruggiero, J. S., and D. M. Skauen, *J. Pharm. Sci.,* **51:**233 (1962).
133. Rugh, R., *The Mouse,* Burgess Publishing Co., Minneapolis, 1968.
134. Ryle, M., *J. Exp. Biol.,* **34:**539 (1957).

135. Schumacher, H., and W. Strasser, *J. Small Anim. Pract.*, **9**:597 (1968).
136. Scott, M. L., M. Neishui, and R. J. Young, *The Nutrition of Chicken*, Scott Association, Ithaca, N.Y., 1971.
137. Scott, P. P., in Grahm-Jones (ed.), *Canine and Feline Nutritional Requirements*, Pergamon Press, London, 1965.
138. Scott, P. P., in E. S. E. Hafez (ed.), *Reproduction and Breeding Techniques for Laboratory Animals*, Lea & Febiger, Philadelphia, 1970.
139. Scott, P. P. and M. G. Scott, in M. L. Conalty, *Husbandry of Laboratory Animals*, Academic Press, New York, 1967.
140. Searle, A. G., *Comparative Genetics of Coat Color in Mammals*, Academic Press, London, 1968.
141. Seltzer, W., *USP at Syst. Leafl.*, **2**:734.
142. Short, D. J., and D. P. Woodnott, *The ATA Manual of Laboratory Animal Practice and Technique*, Crosby Lockwood, London, 1965.
143. Simmons, M. L., and J. O. Brick, *The Laboratory Mouse*, Prentice Hall, Inc., Englewood Cliffs, N.J., 1970.
144. Solomon, J. J., R. L. Witter, H. A. Stone, and L. R. Champion, *Avian Dis.*, **14**:752 (1970).
145. Staats, J., *Cancer Res.*, **32**:1609 (1972).
146. Standford, J. C., *The Domestic Rabbit*, Crosby Lockwood, London, 1957.
147. Starke, G., H. Glathe, and P. Hlinak, *Pharmezie*, **23**:669 (1968).
148. Tavernor, W. D., *Nutrition and Disease in Experimental Animals*, Bailliere, Tindall & Cassell, London, 1970.
149. Taylor, A., N. Carmichael, and Norris, *Cancer Res.*, **8**:264 (1948).
150. Taylor, P. J., and B. W. Calneck, *Avian Dis.*, **6**:51 (1962).
151. Templeton, G. S., *Domestic Rabbit Production*, 3rd. ed., Danville Interstate Publication, Danville, Ill., 1962.
152. Totton, M., *J. Hyg. Camb.*, **56**:190 (1958).
153. Tuder, D. C., and H. Woodward, *Avian Dis.*, **12**:379 (1968).
154. Tyrrell, D. A. J., and C. R. Coid, *Vet. Rec.*, **86**:164 (1970).
155. UFAW Handbook, *Care and Management of Laboratory Animals*, T. W. Fiennes, F. A. Harrison, R. Scott, P. Scott, and W. N. Scott (eds.), Williams & Wilkins Co., Baltimore, 1972.
156. UFAW (ed.), UFAW Handbook, *Care and Management of Farm Animals*, Edinburgh, Livingstone, 1971.
157. Voss, H., and G. Henneberg, *Virchows Arch. Pathol. Anat. Physiol.*, **329**:765 (1957).
158. Walker, N. E., *Toxicol. Appl. Pharmacol.*, **10**:290 (1967).
159. Wallis, R. C., *Mosquito News*, **22**:305 (1962).
160. Weishbroth, S. H., *Lab. Anim. Care*, **19**:699 (1970).
161. Weisbroth, S. H., R. E. Flatt, A. L. Kans, *The Biology of the Laboratory Rabbit*, Academic Press, New York, 1974.
162. Whitten, W. K., *Endocrinol.*, **17**:307 (1958).
163. Wickham, N., *Ans. Vet. J.*, **34**:244 (1959).
164. Worden, A. N., in Grahm-Jones (ed.), *Canine and Feline Nutritional Requirements*, Pergamon Press, Oxford, 1964.
165. Wyler, R., and H. A. E. Van Tongeren, *J. Pathol. Bacteriol.*, **74**:275 (1957).
166. Yager, R. H., and M. B. Sunborg, *Animals for Research*, 6th ed., Publ. No. 1678, National Academy of Science, Washington, D.C., 1968.
167. Yerganian, G., in M. P. Cameron and C. M. O'Conner (eds.), *Ciba Foundation Colloquia on Endocrinology*, Churchill, London, 1964.
168. Yerganian, G., in the UFAW Handbook, *Care and Management of Laboratory Animals*, 3rd ed., Edinburg, Livingstone, 1967.
169. Zwilling, E., *Transplant. Bull.*, **6**:115 (1959).

CHAPTER 3

FREQUENTLY OCCURRING DISEASES OF ANIMALS UNDER LABORATORY CONDITIONS

B. M. Mitruka and W. Medway

Many infectious and noninfectious diseases can occur in animals used for medical research. Various diseases may be induced in animals for experimental purposes. However, certain diseases affect a particular species of animal more frequently than others. With modern practices in the care and management of laboratory animals, many of the diseases, particularly infectious diseases, can be prevented. Practical routines have evolved for raising animals free of common pathogens. The so called specific pathogen-free (SPF) animals are increasingly available from various commercial sources. Production colonies have been started by foster suckling cesarean-derived young on germfree mothers, or by hand raising these young on sterile diets while maintaining the animals in specially designed isolation facilities. By rigid application of the principles of medical asepsis, losses from overt disease have virtually been eliminated and the quality of laboratory animals for research has improved.

In spite of the availability of SPF animals for medical research, a researcher should have sufficient knowledge of some frequently occurring diseases of animals under laboratory conditions. The recognition of these diseases is relevant not only to the experimental studies but also for their possible transmission to humans. Early diagnosis of these diseases may make it possible to take preventive measures. Sacrifice of an individual animal or the entire colony frequently may be the safest procedure to follow when unplanned infections occur in the laboratory. Treatment should not be attempted when there is great danger of spreading infection or creating carrier animals. Thus, monkeys that are positive after tuberculin testing should be destroyed. Similarly, chronic respiratory diseases in rat or mouse colonies can be best eliminated by destroying infected animals. Colonies of mice sometimes suffer outbreaks of mouse pox, or ectromelia, or acute generalized virus disease which may result in a high mortality. This disease also occurs in chronic form and is marked by frequent necrosis of the extremities, but many symptomless carriers also exist. The only quick, effective control measure is destruction of the colony and thorough cleaning and sterilization of quarters and equipment before restocking. On the other hand, valuable inbred colonies may be salvaged by routine vaccination for several generations of animals.

This chapter briefly describes some of the frequently occurring diseases of animals, particularly emphasizing their etiology, incidence of occurrence, signs of disease, diagnostic methods, postmortem lesions, control measures, and transmissibility of the disease to humans. It is beyond the scope of this book to describe in detail all the spontaneous diseases and potential susceptibility of animals to various other diseases. Only those diseases that affect the animals frequently will be described with a view toward their effects on the experimental studies or transmission to humans. A few key references are listed with the description of each disease in an animal species for further information on the subject.

DISEASES OF MICE

1. Infectious Ectromelia (Mouse pox)

A. Etiology
Filterable virus.

B. Incidence
Uncommon in the United States.

C. Signs
Acute form: mice die shortly after showing the signs of general malaise. Chronic form: mice are ill for longer periods and skin eruptions and ulcerating lesions with swelling of the legs and tail appear. These lesions may proceed to amputation. Deaths occur sporadically.

D. Diagnosis
Diagnosis on clinical grounds is not sufficient; in particular, infection with *Streptobacillus moniliformis* may mimic the cutaneous manifestation of ectromelia. Diagnosis may be made by histopathologic examination, by virus isolation in mice or embryonated eggs, or by the demonstration of hemagglutinin-inhibiting bodies.

E. Postmortem Lesions
Acute form: few specific lesions including congestion and small hemorrhages of internal organs, excessive body fluids and necrotic lesions of liver and spleen are found. Chronic form: in these mice, dirty white necrotic lesions occur in the liver and spleen. Eosinophilic intracytoplasmic inclusions can be readily demonstrated in the skin lesions and other diseased tissues.

F. Control
Destruction of the infected colony is recommended. However, the slaughter of a mouse colony infected with ectromelia may not always be practical, and in these circumstances mice may be immunized with vaccinia virus vaccine.

G. Transmissibility to Humans
None.

H. References
Mims (83, 84); Seamer (111, 112).

2. Tyzzer's Disease

A. Etiology
The causative organism of Tyzzer's disease is *Bacillus piliformis*.

B. Incidence
The disease is more frequent under overcrowded, unsanitary, or other adverse conditions and is a rare occurrence today.

C. Signs
The disease is first manifested by diarrhea followed by anorexia and weight loss, leading to death in a few days.

D. Diagnosis

By demonstration of the organisms in intestinal epithelial cells, in the gall bladder, in bile or epithelium of the cecum.

E. Postmortem Lesions

Gross lesions are found only in the liver and consist of round nodules. These nodules are either a pale gray color, opalescent grayish, or have a dark brown central point. Fatty change may be evident in nearby hepatic cells, which may contain the bacteria.

F. Control

The infected colonies should be destroyed and new colonies started with disease-free stock in isolated quarters.

G. Transmissibility to Humans

None.

H. References

Tuffery (117, 119); Fujiwara et al. (37, 38, 39); Craigie (23).

3. Pseudotuberculosis

A. Etiology

The causative organisms of this disease are *Corynebacterium pseudotuberculosis* and *C. kutscheri*.

B. Incidence

The incidence of pseudotuberculosis is common in mouse colonies.

C. Signs

The signs are inapparent. Nonspecific signs of weakness, debility, and increased respiration may appear in the infected animals.

D. Diagnosis

By the identification of corynebacteria in characteristic lesions.

E. Postmortem Lesions

Abscesses may be present in kidneys, myocardium, or liver. However, frank abscesses are not always characteristic; the lesions in the liver may consist of sharply circumscribed granulomas from which corynebacteria may be isolated. Frequently, tubercle-like lesions filled with caseous material in lungs and lymph glands may be present.

F. Control

The infected colonies should be destroyed.

G. Transmissibility to Humans

None.

H. References

Shechmeister (113); Pierce-Chase et al. (101); Fauve et al. (31).

4. Salmonellosis (Mouse Typhoid)

A. Etiology

Salmonella typhimurium or *S. enteritidis* are the causative organisms of salmonellosis in mice.

B. Incidence

This disease occurs frequently in mice.

C. Signs

The common signs are diarrhea, roughened hair, weight loss, debility, and widely different mortality rates (100 percent mortality in susceptible strains and 52 percent mortality in resistant strains of mice).

D. Diagnosis

By isolation and identification of organisms from feces, blood, liver, or spleen.

E. Postmortem Lesions

Moderate inflammation occurs in the mucosa of the small intestine. The organisms pass via the portal vein to the liver, where necrosis of islands of hepatocytes and proliferation of macrophages ensue. The organisms may be isolated in large numbers from the mesenteric and, less frequently, the tracheal, bronchial, and cervical lymph nodes. The spleen is enlarged, engorged with blood, and contains areas of focal necrosis. In cases of longer duration of infection, the lesions are more distinctive. The liver is enlarged, deep brown to brownish yellow in color, and often friable. Scattered yellow foci of various sizes may be seen in several lobes. Microscopically, necrotic foci, lymphoid, histocytic-monocytic nodules, swelling of Küpffer cells, and widespread thrombosis of branches of the portal vein are found.

F. Control

Destruction of the infected colonies and sterilization of the feed, bedding, and equipment is recommended. The new colonies should be started with disease-free stock. The newly arrived animals should be held in isolated quarters.

G. Transmissibility to Humans

Salmonella is transmissible to humans if ingested through contaminated food or water.

H. References

Habermann and Williams (49); Böhme et al. (13).

5. Lymphocytic Choriomeningitis (LCM)

A. Etiology

LCM virus.

B. Incidence

The LCM infection is quite common in colonies of breeding mice; the infection can be expected to spread to young mice by intranasal, subcutaneous, or oral routes. In a population containing only adult mice, LCM virus infection might be self-limiting.

C. Signs

Young mice infected with certain strains of the virus show few signs of sickness and may develop into apparently normal adults. But in other mice, varying degrees of pathogenicity and a few deaths might occur. In the acute phase of the disease in young animals, convulsions, paralysis of the hind legs, and conjunctivitis are noticed.

D. Diagnosis

Diagnosis of LCM infection is best made by virus isolation in guinea pigs or by intracerebral inoculation of mice. Both species usually sicken and die in about a week. Once isolated, the virus can be identified by the use of known antisera. The antibodies can be demonstrated serologically in the infected mice.

E. Postmortem Lesions

Spontaneous infection of mice with LCM does not show characteristic lesions. Lymphocytic infiltrations of the brain and meninges are the principal characteristics of this infection. However, infiltration of the choroid plexus is not always prominent. Most infected mice have changes in the liver, and infiltrations of lymphocytes may also be found in various other tissues. During the course of infection, viral antigen can be detected in blood leukocytes.

F. Control

The mice colonies should be started with serologically screened stock and maintained in isolated quarters.

G. Transmissibility to Humans

Human infection with LCM virus can arise from laboratory mice or from wild mice.

H. References

Pfau et al. (100); Lewis et al. (74); Ackermann et al. (1); Bartawidjaja et al. (6); Seamer and Gledhill (111).

6. Infantile Diarrhea

A. Etiology

The causative agent of epidemic diarrhea of infant mice is believed to be a filterable virus.

B. Incidence

It is quite common in mice between 7 and 17 days of age; adult mice show no signs of infection.

C. Signs

The skin and coat of the infected animals become soiled with pasty, mustard-colored feces. The mice feed hungrily and may have abdominal distension, but growth stops. The animals show scaly backs and after 2 to 3 days, obstipation occurs. After about a week, accumulation of feces around the anus may occur, and if the encrustations are not removed manually, death ensues. Mortality is about 35 to 50 percent.

D. Diagnosis

Infantile diarrhea is diagnosed by the signs and history.

E. Postmortem Lesions

Infected mice have gaseous, yellow-colored feces of more or less fluid consistency in their colons. Histological changes in epidemic diarrhea are not marked and are confined to the small intestine, where sloughing of cells from the jejunal and ileal villi is more pronounced than in a normal mouse. Acidophilic intracytoplasmic inclusions may be detected between the nucleus and the lumen of some intestinal epithelial cells.

F. Control

The mouse colony should be started from disease-free stock and the new arrivals should be held in isolated quarters. The feed, bedding, and equipment should be sterilized.

G. Transmissibility to Humans

None.

H. References

Kraft (67, 68); Pappenheimer and Cheever (93); Runner and Palm (107).

7. Mouse Septicemia

A. Etiology

The causative organism of mouse septicemia is *Erysipelothrix muriseptica.*

B. Incidence

The disease affects mouse colonies occasionally.

C. Signs

The infected animals show signs of serous conjunctivitis followed by a purulent discharge with gluing together of the eyelids.

D. Diagnosis

Isolation and identification of organisms in characteristic lesions.

E. Postmortem Lesions

Small, grayish white necrotic lesions on surface of enlarged spleen are found.

F. Control

The mouse colonies should be started with disease-free stock in isolated quarters.

G. Transmissibility to Humans

None.

H. References

Brumby (14); Cotchin (22); Handler (51).

8. Nematode Infestation

A. Etiology

Aspicularis tetraptera, Syphacia obvelata, or *Trichosomoides crassicauda* may be the causative parasite.

B. Incidence

It is commonly found in mouse colonies where the animal management is not adequate. The infective eggs of nematode worms may be ingested by the animals, resulting in the development of adult worms in the cecum and colon in 15 to 20 days.

C. Signs

Enteritis and diarrhea may be seen in the infected animals.

D. Diagnosis

Fecal examination will reveal the presence of eggs.

E. Postmortem Lesions.

Enteritis is usually found in infestations and occasionally rectal prolapse may be seen.

F. Control

Contamination of feed, water, and bedding should be prevented. The colony should be started with clean stock.

G. Transmissibility to Humans

None.

H. References

Habermann (48, 49); National Academy of Sciences (88); U.S. Department of the Army and the Air Force (122).

9. Lice Infestation

A. Etiology

The common lice of laboratory mice are *Polyplax serrata*.

B. Incidence

It is fairly common in the mice colonies which lack adequate sanitation. Ova or nits of lice are attached to mouse hair and develop on the same host in about 2 weeks.

C. Signs

Loss of blood in severe infestation, which may be fatal.

D. Diagnosis

Identification of eggs or adult lice on animal.

E. Postmortem Lesions

Loss of blood and general debility is noticeable. Usually, there are no internal lesions.

F. Control

The bedding should be treated with nontoxic insecticides. Measures should be taken to prevent reinfestation of equipment, feed, and bedding of the animals.

G. Transmissibility to Humans

The personnel handling the infected mice may acquire these parasites.

H. References

Flynn (33); UFAW Handbook (120).

10. Flea Infestation

A. Etiology

There are various species of fleas that may infest mice colonies. Among the common fleas are *Ctenopsyllus segnis* (the mouse flea), *Leptopsylla musculi* (the European mouse flea) and *Nosopsyllus fasciata* (the American rat flea).

B. Incidence

The mouse colonies with inadequate sanitary conditions are commonly infested with these fleas. Ova are deposited in bedding and the larvae molt three times before becoming adult and start feeding on the animals.

C. Signs

General signs of irritation are noticeable.

D. Diagnosis

Diagnosis is made by identification of the fleas on the animals.

E. Postmortem Lesions

None.

F. Control

Bedding should be sanitized by nontoxic insecticides. Reinfestation should be prevented by sterilization of feed, bedding, and equipment.

G. Transmissibility to Humans

The personnel handling the infested animals may acquire the fleas.

H. References

Belding (8); National Academy of Sciences (88).

11. Mites

A. Etiology

Several species of mites including *Bdellonyssus bacoti, Echinolaelaps echidninus, Myobia musculi, Psorergates simplex, Myocoptes musculimes, M. rombousti,* and *Radfordia affinis* may infest mice. Ova are deposited in bedding, which develops into nymphs that require several blood meals. The mites leave the host after each feeding. *Psorergates simplex* may encyst in the skin. Others complete their life cycles on the animals.

B. Incidence

Common in poorly managed mouse colonies.

C. Signs

Usually signs of irritation are presented. The hair coat of infested animals appears scabby.

D. Diagnosis

By identification of mites on animals.

E. Postmortem Lesions

None.

F. Control

The infested animals should be dipped in nontoxic insecticides. The bedding should be sterilized with nontoxic insecticides, and reinfestation of feed, bedding, and equipment should be prevented.

G. Transmissibility to Humans

The personnel handling the infested animals may contract the mites.

H. References

Flynn (33); National Academy of Sciences (88).

Other important diseases of mice are *Streptobacillus moniliformis* infections (chronic respiratory disease), Pasteurellosis, pulmonary phlebitis with myocarditis, viral pneumonia, hereditary muscular dystrophy in inbred strains, amyloidosis, mouse encephalitis, and a variety of tumors. Most of these occur in certain strains of inbred animals.

By the use of specific pathogen-free (SPF) animals, many spontaneous diseases, particularly infectious diseases, of animals are limited. However, Tucker and Baker (116) reported that various tumors were a major cause of death once the more common diseases of laboratory mice have been eliminated, and that there was a sharp upward trend in the incidence after 18 months of age. The types of tumor occurring vary from strain to strain. In some cases, virus etiology may be involved in the genesis of tumors in experimental mice. Among the common tumors found were endocrine tumors (pituitary, mammary tumors), adenocarcinomas of salivary glands and liver, lung adenomas, generalized lymphosarcomas, and other tumors of reticuloendothelial system.

DISEASES OF RATS

The literature on the rat shows that there is a formidable list of diseases which require consideration and which occur commonly in colonies not established as "clean" ones. A few generalizations regarding the diseases that may frequently occur (or do not occur) can be made.

a. The rat is infected by remarkably few spontaneous bacterial diseases and has a resistance to many bacteria which are highly pathogenic for other species.

b. It suffers from very few viral diseases of any organ system and is not highly susceptible to many strains of viruses pathogenic to other hosts.

c. There are also few protozoan infections of importance in the rat, although it is experimentally susceptible to some trypanosomes.

d. The major helminthic infestations of the laboratory rat are largely alimentary and can be controlled.

e. The rat has been the standard animal for the study of experimental nutritional problems (see Chapter 5); however, the occurrence of such natural conditions as rickets and vitamin A or tocopherol deficiency is not known in this animal species.

f. The inherited diseases affecting rats are few as compared with the divergent conditions well recognized in many strains of mice.

g. The rat does not suffer from a great variety of neurological disorders. Also, the rat (unlike the mouse) is not susceptible to the viruses from other hosts which affect the brain and spinal cord.

However, there are certain diseases that occur frequently in rats under laboratory conditions; these diseases are briefly described below.

1. Infectious Catarrh (*Middle Ear Disease*) (*Streptobacillus moniliformis—* Rat-Bite Fever—and *Mycoplasma Infection*)

A. Etiology
Pleuropneumonia-like (PPLO) organisms, other pathogenic organisms such as *Streptobacillus moniliformis* and *Mycoplasma* are the causative organisms of infectious catarrh. In culture, *S. moniliformis* gives rise to variants termed L-forms, which are morphologically similar to PPLO.

B. Incidence
Infectious catarrh is widespread in rat colonies not established by techniques specifically designed to eliminate it. The disease is chronically seen in animals over 3 months old. Mice kept in the same room as rats act as reservoirs of *Streptobacillus*. Mice may be infected directly as a result of inoculation with rat-derived material.

C. Signs
Usually, there are no specific signs other than snuffling. Occasionally, the animal circles, staggers, and holds its head to one side. It is usually associated with pneumonia. In mice, the most characteristic sign of the disease is a chattering or clicking noise, but rats do not chatter; they snuffle, and there may be blood-stained encrustations around the nose. As the disease progresses, the animal becomes less and less active, the coat becomes rough, and the animal becomes very ill. Rhinitis, although usually present, never shows externally.

D. Diagnosis
Diagnosis is established by the isolation and identification of the organisms from the middle ear.

E. Postmortem Lesions

Lesions are confined to the nasal passages, middle ear, and lungs. Rhinitis is manifest by a profuse semifluid exudate in the nasal passages, but sometimes there is little exudate. Inflammation is indicated by the presence of polymorphonuclear leukocytes and desquamated epithelial cells. Infection may be unilateral or bilateral. *Streptobacillus moniliformis,* causing a middle ear infection of the rat, is also capable of producing inflammation of bones and periarticular tissues. Swelling and reddening of joints indicates the involvement of extra-articular tissues, bones, and synovial tissues. Lesions of infectious osteoarthritis and osteomyelitis may be seen.

F. Control

The absolute method of control is to take young animals by cesarean section and raise them in strict isolation by hand or on clean nurse stock.

G. Transmissibility to Humans

Streptobacillus is transmissible to humans by the bites of rats. It is called rat-bite fever (Haverhill fever).

H. References

Kleinberger-Nobel (66); Tuffery and Innes (118); Lerner and Sokoloff (73); Pankevicius (92); Wilson and Miles (126).

2. Chronic Murine Pneumonia (*Enzootic Bronchiectasis*)

A. Etiology

Chronic pulmonary disease is initiated in suckling rats by a virus transmitted by the mother. Confusion of the etiology may be caused by the common isolation of *Mycoplasma* from infected lungs. With the establishment of chronic lesions and bronchiectatic cavitation, secondary invasions by bacteria (*Bacillus muris, B. bronchiseptica, P. multocida muris, S. moniliformis* and *Mycoplasmatales*) may occur.

B. Incidence

It is a fairly common disease and an important hazard in long-term experimental work with rats. In some experimental work on longevity of animals, chronic pulmonary disease may affect nearly 100 percent of 1- to 2-year-old rats. Mortality at a young age is low, but morbidity can be very high in conventional breeding colonies.

C. Signs

The signs are nonspecific, and it is very hard to detect the early stages of disease. Pneumonitis develops slowly. Respiratory involvement may be seen following undue stress or temperature changes.

D. Diagnosis

The diagnosis can be established by intranasal injection of rat lung filtrate into mice (Nelson test).

E. Postmortem Lesions

The macroscopic lesions are gray to red areas of part or all of the pulmonary lobes. The whole lobe may present a shrunken, cobbled surface and rubbery appearance. Disseminated small foci, sharply circumscribed and reddish brown in color, may be seen. There are cystlike apices which are irregularly dilated bronchi filled with mucoid and mucopurulent material. The lesions progress slowly and are of long duration before the lung is affected to the extent of causing respiratory incapacity. In very old lesions, the affected lobe or lobes are distorted by nodular pinkish or pearl gray masses. These are gross bronchiectatic cavities filled with caseous debris or inspissated exudate and superficially resemble abscesses.

F. Control

Only possible control is to obtain rats by cesarean section and raise them in strict isolation.

G. Transmissibility to Humans

None.

H. References

Farris and Griffith (30); Nelson (89); Parish (95); Innes et al. (58); Gray (43).

3. Paratyphoid Disease (*Salmonellosis*)

A. Etiology

The disease is caused by the *Salmonella* group of organisms, mainly by *S. enteritidis* and *S. typhimurium*.

B. Incidence

It occurs quite frequently in rat colonies with inadequate sanitary conditions. Improved sanitary methods in animal colony management have led to a decreased frequency of Salmonellosis in recent years.

C. Signs

In the acute infections, diarrhea and severe anemia with general loss of health conditions is presented. In chronic cases, signs of diarrhea and anemia are milder. Many animals may be carriers and do not show any signs.

D. Diagnosis

Diagnosis is established by the isolation and identification of *Salmonella* in the infected animals.

E. Postmortem Lesions

Usually, necrosis of islands of hepatocytes and proliferation of macrophages are found in the liver. In cases of longer duration, the lesions are more distinctive. The liver is enlarged, deep brown to brownish yellow in color and often friable. Scattered yellow foci of various sizes may be seen in several lobes. Microscopically, necrotic foci, lymphoid, histiocytic-monocytic nodules, swelling of Küpffer cells, and widespread thrombosis of branches of the portal veins are found. These veins are often surrounded by zones of necrosis with hemorrhage in adjacent parts of the liver. The periphery of the necrotic foci is often invaded by polymorphonuclear cells. Clumps of bacteria may be seen in these lesions. The spleen is usually enlarged, engorged with blood, and contains areas of focal necrosis similar to those in the liver.

F. Control

The infected animals should be destroyed. The new colony should be started from disease-free stock and the new arrivals should be held in isolated quarters.

G. Transmissibility to Humans

The infection may possibly be transmitted to humans by contamination through food, material, and equipment.

H. References

Habermann and Williams (49); Böhme et al. (13).

4. Coccidiosis

A. Etiology

Many species of *Eimeria* infest the rat. Common protozoan parasites are *Eimeria migairii, E. separata,* and *E. carinii.*

B. Incidence

The disease is common in rat colonies.

C. Signs

Diarrhea may occur following shipment or unusual stress or in heavy infestation.

D. Diagnosis

Diagnosis is confirmed by the identification of oocytes upon fecal examination.

E. Postmortem Lesions

The endogenous development of *E. miyarii* occurs in the small intestine, cecum, and colon. The parasites settle in the mucosa of the small intestine, most of them superficially, although occasionally they may also occur in epithelial cells at the fundus of the glands and occasionally more than one stage invades a single host cell. *Eimeria separata* develops in luminal cells of the cecum and colon, leaving the cells at the fundus of the glands intact, and three generations of schizonts are produced. *Eimeria carinii* invades mucous epithelial cells of the villi of the small intestines.

F. Control

Contamination of feed, water and bedding by rodent species should be prevented. All feed, bedding, and equipment should be sterilized.

G. Transmissibility to Humans

None.

H. References

Davies et al. (25); Farris and Griffith (30); National Academy of Science (88).

5. Liver Cysts

A. Etiology

Cysts of the liver of rats are caused by the larval stages of the cat tapeworm *Taenia crassicollis.*

B. Incidence

It is a common disease in rat colonies.

C. Signs

Ivory white cysts are seen in liver and mesentery.

D. Diagnosis

Definitive diagnosis is made by finding the cysts at autopsy.

E. Postmortem Lesions

Cysts in liver and mesentery.

F. Control

All feed, bedding, and equipment should be sterilized and contamination of feed and bedding with cat feces should be prevented.

G. Transmissibility to Humans

None.

H. References
Brumley (14); Farris and Griffith (30).

6. Rat Tapeworm

A. Etiology
Hymenolepis diminuta is a common tapeworm infesting rats; *H. nana* rarely infests. Usually, roaches, beetles, or fleas are intermediate hosts. Within the gut of the arthropod host, the egg hatches and the escaping oncosphere penetrates the gut wall and enters the hemocoel where it develops within about 180 hours at 30°C into a cercocystis larva. When an infected arthropod is eaten by a suitable rodent host, the larvae are liberated by the digestion of the arthropod and after eversion of the scolex, they attach themselves to the intestinal mucosa and develop into adult worms.

B. Incidence
About 5 to 15 percent of rat colonies are infested by the tapeworms.

C. Signs
Enteritis and diarrhea may be caused by heavy infestations.

D. Diagnosis
Diagnosis is made by the identification of ova in fecal samples.

E. Postmortem Lesions
Enteritis and diarrhea may be caused by heavy infestations.

F. Control
Roaches, fleas, and beetles should be eliminated from feed, bedding, and equipment.

G. Transmissibility to Humans
Humans are susceptible to *H. nana.*

H. References
Cotchin and Roe (22); Farris and Griffith (30); U.S. Department of Health, Education and Welfare (121).

7. Blood Parasites

A. Etiology
Hepatozoon muris, Babesia muris, and *Bartonella muris (Hemobartonella muris)* infest blood of rats.

B. Incidence
The blood parasites are common in rat colonies where the sanitary conditions are inadequate.

C. Signs
Signs most often follow splenectomy which results in a severe, often fatal, anemia of the macrocytic hypochromic type.

D. Diagnosis
Diagnosis is made by examination of blood smears for parasites in red blood cells.

E. Postmortem Lesions
The parasites are found in blood cells causing anemia. The life cycle involves an

intermediate host of fleas, lice, and ticks. Infected insects are eaten by the rat, and the parasite penetrates into the bloodstream invading the red blood cells. Liver damage is usually found in the infested animals.

F. Control
All insects should be eliminated from the animals and colony room. All feed, bedding, and equipment should be sterilized.

G. Transmissibility to Humans
None.

H. References
Farris and Griffith (30); Oldham (91).

8. Nematodes

A. Etiology
Syphacia obvelata, *Trichosomoides crassicauda*, and *Heterakis spumosa* are the roundworms causing infestations in rats.

B. Incidence
The parasites infest rat colonies frequently.

C. Signs
Heavy infestation causes diarrhea and loss of general health condition.

D. Diagnosis
By identification of eggs upon microscopic examination of feces.

E. Postmortem Lesions
The parasites infest the cecum and colon. No pathological lesions are seen, but gross infestation is often associated with various forms of malfunction of the gut. At autopsy, there may be some evidence of gut inflammation, but, on the other hand, large numbers of pinworms may be found in the cecum and colon with no associated macroscopic lesions. *Trichosomoides crassicauda* lives in the bladder, kidney, pelvis, and uterus of rats. Larvae or immature worms may invade the lungs producing granulomatous lesions.

F. Control
All feed, bedding, and equipment should be sterilized to prevent contamination by infested animals.

G. Transmissibility to Humans
None.

H. References
Innes et al. (60); Farris and Griffith (30); Habermann (48).

9. Lice, Mites, and Flea Infestations

A. Etiology
Polyplax spinulosa (lice); *Radfordia ensifera* (mange mite); *Bdellonyssus bacoti* (the rat mite); *Notoedres minor* var. *cati* (mange mite); *Otodectes cynotis* (ear mite); and a number of species of fleas including *Nosopsyllus segnis*, *Echidnophaga gallinacea*, *Xenopsylla cheopis*, *Ctenopsyllus segnis* and *Hematopinis spinulosis* cause infestations in rats.

B. Incidence

Fairly common in rat colonies with inadequate sanitary conditions.

C. Signs

General loss of health condition and irritation.

D. Diagnosis

Diagnosis is made by identification of nits, adult lice, or fleas on the animals. Microscopic examination of skin scrapings are made by thoroughly wetting the hair on the animal and by examining the small red objects (mites) with a magnifying glass or under a microscope.

E. Postmortem Lesions

None except skin lesions may be found.

F. Control

All feed, bedding, and equipment should be sterilized; recolonize with noninfested breeding stock.

G. Transmissibility to Humans

The personnel handling infested animals may acquire the parasites.

H. References

Farris and Griffith (30); Flynn (33); National Academy of Sciences (88).

In addition to the diseases described above, many spontaneous infectious diseases of rats have been recorded which include Pasteurellosis (*Pasteurella multocida, P. muriseptica, P. muricida*), *Streptobacillus moniliformis* infection, Haverhill fever (rat-bite fever), protozoan infection of the kidney (*Klossiella muris*), Trichomoniasis, Coccidiosis (*Eimeria falciformis, Cryptosporidium muris,* and *C. parvum*), tapeworm infections (*Hymenolepis diminuta, H. nana*), ringworm infections, histoplasmosis, coccidiodomycosis, adiaspiromycosis, sporotrichosis, cryptococcosis, aspergillosis, penicilliosis and phycomycosis. Of these, some of the infections (e.g., ringworm) are important because these animals serve as reservoirs of human infection. In most cases, a few spontaneous diseases have been found in mice and rats under laboratory conditions. Rats and mice may be resistant to many of the infections such as aspergillosis, pencilliosis, and phycomycosis.

By the use of specific pathogen-free (SPF) animals, many spontaneous diseases, particularly infectious diseases, of animals are limited. However, Tucker and Baker (116) reported that various tumors were a major cause of death and that there was a sharp upward trend in the incidence after 18 months of age. The types of tumor vary from strain to strain. In some cases, virus etiology may be involved in the genesis of tumors in experimental mice. Among the common tumors found were endocrine tumors (pituitary, mammary tumors), adenocarcinomas of salivary glands and liver, lung adenomas, generalized lymphosarcomas and other tumors of reticuloendothelial system.

Since rats (and mice) are used quite frequently in biomedical research, the importance of knowing the details of behavior and spontaneous diseases of these experimental animals cannot be overemphasised. One possibility, which is at times difficult to evaluate, is whether treatment has made the spontaneous disease worse. Many strains of viruses infect mice and rats and cause deaths in breeding colonies or among animals on experiments. Usually the virus etiology is obvious by the examination of lesions. Under appropriate conditions, most viruses are able to produce inapparent or latent infections in their hosts. Such infections may remain concealed for long periods. Nevertheless, latent virus infections may lead to the misinterpretation of results and to other untoward effects that are as serious as those produced by overt infection. Because they rarely give rise to marked histopathologic changes, the detection of inapparent infections frequently depends on serological techniques and facilities for virus isolation. Also important are the frequent chemical and hematologic determinations of the animals which may give the researchers a clue to the detection of inapparent infections. Under these circumstances, a knowledge of pathogenesis may prove helpful in keeping an investigator alert to the possible occurrence of inapparent infections. Similarly, knowledge of the ways in which viruses affect the responses of their hosts will assist in the correct interpretation of results.

DISEASES OF RABBITS

1. Pasteurellosis (Hemorrhagic Septicemia)

A. Etiology

Infections are caused by *Pasteurella lepiseptica, P. multocida,* or *P. cuniculicida.*

B. Incidence

This disease frequently causes infections in laboratory rabbits. It is highly contagious and can be transmitted by either direct or indirect contact. Apparently, rabbits develop little immunity following infection. Some animals are healthy carriers and probably perpetuate the disease in rabbit colonies.

C. Signs

In the chronic course of disease, the signs are a thin or purulent exudate from the nose and, to a lesser extent, the eyes. The fur on the inside of the front legs just above the paws will be matted and caked with dried exudate. The infected animals usually sneeze and cough. Snuffles, in general, occur when the resistance of the rabbit is low at kindling time. The disease often causes death; those animals that recover may become carriers. In acute cases of this disease, deaths occur suddenly.

D. Diagnosis

Definitive diagnosis can be made by the isolation of organisms from the lungs of infected animals.

E. Postmortem Lesions

There is an acute, subacute, or chronic inflammation of the mucous membranes of the air passages and lungs. Usually, the lesions are associated with rhinitis, sinusitis, occasional otitis, meningitis, and bronchopneumonia. Abscesses may be found in any part of the body or head. In the acute cases, there is septicemia which usually is fatal within 48 hours. Necropsy reveals bronchial congestion, tracheitis, splenomegaly, and subcutaneous hemorrhage.

F. Control

The infected animals should be destroyed since treatment of this type of infection is unsatisfactory. The animal colony and its equipment should be thoroughly disinfected. In the case of a valuable breeder, antibiotics may be used to treat the infection; however, a poor prognosis should be given.

G. Transmissibility to Humans

It is transmissible to humans, but it is highly contagious within the animal species and other species of animals. The infection can be transmitted either by direct or indirect contact.

H. References

Alexander et al. (3); Griffin (47); Parish (95).

2. Vent Disease (*Rabbit Syphilis*)

A. Etiology

It is a true venereal disease of rabbits caused by *Treponema cuniculi.*

B. Incidence

It occurs fairly commonly in the rabbit colonies where sanitary conditions are inadequate. It is found in both sexes and is transmitted by coitus.

C. Signs

The disease is characterized by appearance of denuded or scab-covered areas about the external genitalia.

D. Diagnosis

Diagnosis is made by the clinical signs.

E. Postmortem lesions

Small vesicles or ulcers are formed which ultimately become covered with a heavy scab. These lesions are usually confined to the genital region, but in some cases ulcers on the mucous membranes of the lips and nose are seen.

F. Control

Infected animals should not be mated. Penicillin in daily doses of 50,000 units appears to be a specific treatment. Lesions usually heal within 10 to 14 days and the recovered animals can be bred without danger of transmitting the infection.

G. Transmissibility to Humans

It is not transmissible to humans or other domestic animals.

H. References

Adams et al. (2); Lane-Petter (70).

3. Oral Papillomatosis

A. Etiology

It is caused by a filterable virus which is different from the Shope papilloma virus.

B. Incidence

It is a chronic disease found frequently in rabbit colonies.

C. Signs

Papillomas may occur in the mouth of rabbits, particularly on the lower surface of the tongue and floor of the mouth.

D. Diagnosis

Diagnosis may be made by the clinical signs.

E. Postmortem Lesions

Usually, no lesions are seen other than papillomas. They consist of small, grayish white, pedunculated nodules found on the under surface of the tongue or on the floor of the mouth. These are benign tumors.

F. Control

The virus-free breeding stock should be completely isolated from the infected ones. Individual animals are not treated but the colony may be vaccinated with an autogenous vaccine (a 10 percent suspension of papilloma tissue in physiological saline solution, 0.4 percent formalin added), inoculated subcutaneously in 0.5 ml quantities at 7-day intervals for 3 weeks.

G. Transmissibility to Humans
None.

H. References
Ribelin and McCoy (104); Cotchin and Roe (22).

4. Coccidiosis

A. Etiology

Liver coccidiosis is caused by *Eimeria stiedae* and intestinal coccidiosis is caused by *E. magna, E. perforans, E. media,* or *E. irresidua.*

B. Incidence

The infections due to *E. magna, E. media,* and *E. irresidua* are common, but the other species of *Eimeria* listed above are also frequent. Those animals recovering from the protozoan infection frequently become carriers and perpetuate the disease.

C. Signs

The severity of signs depends on the number of oocysts ingested. There may be an infection with no apparent signs, or the infection may cause death after a short course. Young rabbits are most susceptible to liver coccidiosis. Affected animals exhibit diarrhea, anorexia, and a rough hair coat. Growing rabbits fail to make normal gains. The animals usually succumb within 30 days after a severe experimental exposure.

In intestinal coccidiosis, the signs are inability to gain weight, anorexia, and a "pot-bellied" condition.

D. Diagnosis

Diagnosis can be made by identification of oocysts in feces or histologic sections of bile duct or intestine.

E. Postmortem lesions

In most affected livers, the lesions are small, grayish white nodules or cysts present throughout the parenchyma of the hepatic tissue. They may be sharply demarcated in the early cases, while in the later stages, they coalesce with other affected areas. Microscopically, the nodules are composed of hypertrophied bile ducts and large numbers of oocysts. The lesions are inconsistent in the intestinal type of coccidiosis. The early cases show little, while in the advanced cases, the intestine may be thickened and pale. The protozoa develop in intestine.

F. Control

Clinical signs may be controlled by proper management and treatment with sulfaquinoxaline in drinking water in a concentration of 0.025 percent for 30 days. Coccidiosis-free breeding stock must be properly isolated and the feed, bedding, and equipment sterilized. For intestinal coccidiosis, sulfaquinoxaline fed continuously for 2 weeks at 0.1 percent of the feed is an effective treatment.

G. Transmissibility to Humans
None.

H. References

Habermann (48); Morgan and Hawkins (86); Parish (95); U.S. Department of Health, Education and Welfare (121).

5. Lice and Mite Infestations

A. Etiology

A common louse of rabbits is *Haemodipsus ventricosus* and the following mite species: *Chorioptes cuniculi, Psoroptis cuniculi* (ear mite), *Sarcoptes scabiei,* and *Notoedris minor* (mange mites).

B. Incidence

Fairly common in rabbit colonies where the sanitary conditions are inadequate.

C. Signs

Louse-infested animals show signs of irritation, hair loss, anemia, and loss of general health condition. Infestation with ear mites causes head shaking, ear flapping, and scratching at their ears with the hind feet. Torticollis and spasms of the eye muscles may be observed in some animals. Affected rabbits lose flesh, fail to reproduce, and succumb to secondary infections. In mange mite infestation, the rabbits scratch at themselves almost continually. There is a loss of hair on the chin, nose, head, base of the ears, and around the eyes. The condition is highly contagious.

D. Diagnosis

Louse infestations may be diagnosed by identification of nits or adult lice on the animals. The diagnosis of mite infestation is made by the identification of mites on microscopic examination of scrapings from the ear canal or skin.

E. Postmortem lesions

No internal lesions are seen in louse infestations. Ear mite infection may frequently damage the inner ear and may reach the central nervous system. The mites irritate the lining of the ear, causing serum and thick crusts to accumulate. In mange mites, skin lesions are usually present.

F. Control

The parasites should be eliminated from animals and animal rooms. The infested animals should be isolated and treated with limesulfur or Canex in mineral or vegetable oil. The feed, bedding, and equipment should be sterilized.

G. Transmissibility to Humans

The personnel handling the infested animals may acquire these parasites.

H. References

Flynn (33); Parish (95).

In addition to the disease discussed above, there are a wide variety of spontaneous diseases that affect the laboratory rabbit infrequently. These diseases have been studied primarily from a point of view of pathogenesis. Among the diseases of rabbits, disorders based on hereditary variations have been noted with the greatest frequency. Many of these disorders bear a close resemblance to well-organized disorders in humans, and, because they can be reproduced conveniently, offer a unique opportunity for investigation. Such disorders are briefly described below:

Oxycephaly

It occurs in rabbits as heredity variation based on a premature fusion of the sagittal and coronal sutures. Compensatory growth occurs at other suture lines and as a result the bregma is pushed up in a peak, the bones become distorted, and the base of the skull deformed. In humans, oxycephaly is a craniosynostotic deformity often associated with blindness and pronounced distortion of features. Injuries in embryonic life, disease of the cranial bones, and meningitis have been considered as etiological factors. Tuberculosis and syphilis have been emphasized in the theories supporting an inflammatory origin. In rabbits, it is definitely of hereditary character and can be produced by inbreeding.

Brachydactylia

This deformity is characterized by a shortening or absence of the component parts of the feet. Abnormalities of the ear may be associated with malformations of the feet. During the latter half of the gestation, fetuses derived from the interbreeding of brachydactylous animals show pathological changes at sites corresponding to the deformities at birth. Although the deformities "breed true" and the presence of some degree of abnormality in one or the other of the feet is regularly transmitted from parents to progeny, the distribution and extent of the malformations are subject to extreme variations in succeeding generations.

Dwarfism

It is similar to a human condition known as nanosomia premordialis. In the homozygous form, the variation is lethal and produces a miniature individual one-third the size of its normal sibs. Heterozygous animals are approximately two-thirds the size of normal sibs at birth and never attain an equal stature. In addition to the small size, the dwarf animals are distinguished by an abnormally bombose configuration of the head, resulting from a disproportionate reduction in the size of the brain. The actual weights of all organs are decreased in heterozygotes but the decrease is not proportionate. The decrease in relative weights is most pronounced in the case of thyroid, pituitary, and adrenal dwarfism, while the increase is greatest in the case of brain dwarfism.

Cretinism

The distinctive features of this disorder usually develop toward the end of the first or second week of life. A faint redness with an edematous thickening of the skin appears first in the nape of the neck, spreads rapidly over the whole surface of the body, and becomes particularly prominent about the genital and anal regions. The edematous thickened skin is thrown into loose, transverse folds, at first having a red color, but later becoming stiff, indurated, and covered with white scales and thick crusts. Growth of the affected animal ceases within a few days of the development of signs, and the disease progresses to a lethal termination in the course of a week or 10 days. However, autopsy and microscopic examination fail to show indicative changes in the thyroid or pituitary, and thorough morphological study of all organs and tissues gives no clue to the etiology or pathogenesis of the disorder.

Hydrocephalus

Internal hydrocephalus occurs in two forms in rabbits. One form is present at birth and is rapidly lethal. The other form arises from a deficiency of vitamin A-containing substances in the dam's diet and occurs without relationship to genetic constitution. The latter type usually occurs in epidemic form in a colony of breeding rabbits following a period of dietary deficiency. The incidence of hydrocephalus in the colony returns to a normal figure (about 0.2 percent) with the institution of an adequate diet.

Rabbit Pox

This is an acute, generalized disease of laboratory rabbits that occurs in the animal colonies. It is characterized by pyrexia, nasal and conjunctival discharge, and a skin rash. The mortality rate varies but is usually high. Clinically, the disease resembles smallpox in humans, and it is produced by a virus of the vaccinia group. The most distinctive sign of this highly contagious disease of rabbits is a pocklike eruption which is often prominent and widespread over the body but is sometimes poorly defined. The incubation period varies between 5 and 7 days. The dominant gross lesion in all organs and tissues is a small nodule or papule consisting of a mononuclear infiltration and necrosis. Diffuse lesions are also found in which the infiltration is widespread and accompanied by edema, hemorrhage, and extensive necrosis of the affected tissues. Death may occur within a few hours after infection or be delayed for days or weeks.

Infectious Myxomatosis

It is a highly fatal disease of rabbits and not transmissible to human or other animal species. The virus causing the disease is transmitted by mosquitoes, biting flies, and by direct contact. In the United States, myxomatosis is restricted largely to the coastal area of California, where epizootics occur every 8 to 10 years during the months of May, June, July, and August, which correspond to the height of the mosquito season. Rabbits of all ages are susceptible, although the young up to the age of 1 month appear to be more resistant than the adult animals. The first characteristic sign to appear is conjunctivitis, which rapidly becomes marked and accompanied by a milky discharge from the inflamed eyes. The animal appears listless, is anorectic, and its temperature frequently reaches 108°F. In extremely acute outbreaks, some of the animals may die within 48 hours after showing signs. In less severe cases, the animals become progressively depressed, develop a rough coat, and the eyelids, nose, lips, and ears become edematous, giving a swollen appearance to the head. The vent becomes inflamed and edematous, and in the male animal, swelling of the scrotum occurs. A very characteristic sign at this stage is the drooping of the ears caused by their edematous condition. A purulent nasal discharge invariably appears, the breathing becomes labored, and the animal becomes comatose just before death, which usually occurs within 1 to 2 weeks after the appearance of clinical signs. Occasionally, the animal will survive for several weeks before death ensues. In these cases, fibrotic nodules appear on the nose, ears, and fore-feet. No characteristic lesions are found at autopsy. The spleen is occasionally enlarged and black in color but this is not a specific lesion of myxomatosis. The seasonal incidence of the disease, the clinical appearance of the infected animals, and the high mortality are all of diagnostic significance. There is no known treatment for this disease. Mosquito control and the prompt slaughter of all clinically affected animals appear to offer the best means of controlling the infection.

The diseases described above are more common in laboratory rabbits; however, there are many other infectious and noninfectious diseases that may occasionally cause problems in rabbit colonies. These include Listeriosis of young rabbits; mastitis (blue breasts) due to *Streptococcus* species; pneumonia due to *Pasteurella* sp., *Klebsiella pneumoniae, Bordetella (Brucella) bronchiseptica* or *pneumococci* (*Diplococcus pneumoniae*); conjunctivitis (weepy eyes) caused by a variety of bacteria; mucoid enteritis (unknown etiology); tapeworm larvae infections (caused by *Taenia pisiformis* or *Multiceps serialis* or *T. taeniaeformis*). Among the noninfectious diseases that sometimes occur in rabbit colonies are wet dewlap (heavy fold of skin on the ventral aspect of the neck); cannibalism; wool-eating; heat exhaustion; ketosis; milkweed poisoning; dystocia; toxemia of pregnancy (in certain strains of rabbits such as Polish and Dutch breeds); and sore hocks.

Tumors occur in laboratory rabbits with rapidly increasing frequency after the third year of life. The most common in order of decreasing incidence are adenocarcinoma of the uterus and mammary gland, epidermoid carcinoma of the vaginal squamous-columnar junction, embryonal nephroma, epidermoid carcinoma of the skin, and leiomyosarcoma of the uterus. Leiomyosarcomas of the stomach and intestine, generalized lymphosarcomas, granulosa cell tumors of the ovary, and thymomas have also been reported, but they seem to occur in rabbits with much less frequency.

DISEASES OF GUINEA PIGS

1. Salmonellosis or Paratyphoid

A. Etiology

The causative organisms are usually either *Salmonella typhimurium* or *S. enteritidis,* but other species of *Salmonella* may infect also.

B. Incidence

Guinea pig paratyphoid is fairly common in animal colonies with inadequate sanitary conditions. It may cause widespread and severe epidemics. The infection can remain latent in colonies indefinitely, but may be provoked into causing a major outbreak of disease by adverse environmental factors, the most important of which is probably hypovitaminosis C.

C. Signs

The signs of salmonellosis are roughening of the fur, emaciation, weakness, wasting, diarrhea, and death. Salmonellosis may be completely inapparent in the animals who are carriers of *S. typhimurium.*

D. Diagnosis

Diagnosis is made by isolation of organisms from the spleen, heart blood, liver or gut, and from the feces. It may also be found in the water bottle.

E. Postmortem Lesions

In acute cases, the primary lesions are inflammation of the intestines with slight enlargement of spleen and intestine. In some cases, patchy parenchymal necrosis of liver cells are noted which may become abscesses at a later stage. The abscesses may progress to nodules of histiocytes. In chronic infections, gallbladder and liver are enlarged and a slight enlargement of lymph nodes is also seen.

F. Control

An infected colony should be completely destroyed if the infection is widespread and long standing. New colonies should be started with disease-free animals in isolated quarters.

G. Transmissibility to Humans
Man is susceptible to *Salmonella* infection if ingested in food or water.

H. References
Brunner and Edwards (15); Meyer and Eddie (81); Parish (95); Lane-Petter and Porter (69); Habermann and Williams (49); Böhme et al. (13).

2. Pseudotuberculosis

A. Etiology
The causative organism is *Pasteurella pseudotuberculosis* var. *rodentium*.

B. Incidence
It is a fairly common disease in animal colonies where the sanitary conditions are inadequate.

C. Signs
Pseudotuberculosis in guinea pigs occurs in three forms. In the septicemic form of the disease, major signs are coughing, rapid breathing, and death in 24 to 48 hours.

In the classical form, emaciation, diarrhea, and palpable nodules are observed, and death ensues in 3 to 4 weeks.

In the granular form of the disease, the signs are indefinite. However, the enlargement of cervical and thoracic lymph nodes are generally found.

D. Diagnosis
Isolation and identification of bacteria from abscesses.

E. Postmortem lesions
In the granular form, spleen and liver are covered with abscesses; occasional peritonitis with adhesions to liver is seen. Also, well-defined abscesses in lungs are present. In septicemic pseudotuberculosis, lungs are usually congested.

F. Control
The colony should be started with disease-free stock. The infected animals should be destroyed and the feed, bedding, and equipment sterilized. The animals should have an adequate supply of vitamin C in their diet and be maintained in good hygienic conditions.

G. Transmissibility to Humans
None.

H. References
Paterson (96, 97).

2. Pneumonia—Bacterial infections

A. Etiology
The causative organisms include pneumococcii type III, IV and XIX, *Klebsiella pneumoniae, Bordetella bronchiseptica,* or streptococcal species.

B. Incidence
The respiratory disease due to these bacteria is fairly common in guinea pigs where the colony management is poor. Although any of these infections may give rise to fatal epidemics, they are not common in well-managed colonies. As with virtually all bacterial diseases of guinea pigs, a shortage of vitamin C in the diet may provoke the ap-

pearance of these diseases. Conditions of poor ventilation and heat control seem especially to contribute to the occurrence of these diseases.

C. Signs

The signs vary from a slight nasal discharge to pneumonia. In pneumococcal pneumonia, nasal discharge and enlarged cervical lymph nodes are classical signs. *Klebsiella pneumoniae* infection causes emaciation, wheezing, and sticky, purulent nasal discharges. Signs of respiratory distress are observed in *B. bronchiseptica* and streptococcal infections.

D. Diagnosis

Isolation and identification of organisms from lungs, bronchi, nasal passages, or blood specimens provides definitive diagnosis.

E. Postmortem Lesions

Pneumococcal infections show lesions of pleurisy, acute lobar pneumonia, or chronic bronchopneumonia, whereas patches of pneumonia accompanied by gelatinous exudate are present due to *Klebsiella* infections. Sometimes pneumonia with adhesions are seen and occasionally pericarditis is present. Consolidated areas and necrotic foci are usually found in the lungs of animals infected with *B. bronchiseptica* infections. In streptococcal infections, the lesions are abscesses of the lymph nodes. Also rarely, consolidation of lungs, pericarditis, and inflammation of subcutaneous tissues are found.

F. Control

Infected and exposed animals should be destroyed and feed, bedding, and equipment sterilized. The new colony should be started with disease-free stock in isolated quarters. Rabbits act as carriers of *B. bronchiseptica;* therefore guinea pigs should not be housed in the same room with rabbits.

G. Transmissibility to Humans

These bacterial species are infective to humans, so proper precautions should be taken when handling infected animals.

H. References

Lane-Petter and Porter (69); Homburger et al. (54); Meyer and Eddie (81); Parish (95); Innes (59).

Meyer and Eddie (81) and Rhodes and Van Rooyen (103) reported acute pneumonia in guinea pigs caused by a filterable virus. The route of infection is by inhalation or ingestion of virus. The animals show signs of emaciation, with ruffled hair. The main pathological lesions are extensive congestion and consolidation of lungs. The fatality rate is very high. The diagnosis is made by mouse inoculation. The infected and exposed animals should be destroyed. Other viruses that have been described by these workers in guinea pigs are *salivary gland virus* which is usually an acute infection. Sometimes progressive posterior paralysis occurs in the infected animals. Postmortem lesions show inclusion bodies in ducts, salivary glands and kidneys. *Pneumopathy,* also normally with few or no signs, has been found, as well as the so called guinea-pig plague which is always associated with, and may be due to, infection with *Salmonella suipstifer.* Blanc et al. (12) reported *lymphocytic chroiomeningitis,* a rarely but naturally occurring virus infection of guinea pigs.

Certain bacterial diseases, other than described before, also infect guinea pigs rarely; these include *Pasteurella muriseptica* causing pasteurellosis; *Streptobacillus moniliformis* or *Actinomyces muris* or *Bacteroides cavaiae,* causing cervical adenitis (enlargement of cervical lymph glands followed by regression or rupture); *Pseudomonas caviae* causing *Pseudomonas* infection which is characterized by enlarged lymph glands and adherent eyelids, listlessness and prostration followed by death in 4 to 6 hours. Several other bacterial diseases of rare occurrence in guinea pigs are described by Dumas (28).

4. Wasting Disease

A. Etiology
The etiology is now known; a virus is suspected.

B. Incidence
Fairly common in guinea pig colonies.

C. Signs
The infected animals show sluggishness, loss of appetite, and marked emaciation. Marked drooling from the mouth is a characteristic sign. Death ensues within 6 days.

D. Diagnosis
Diagnosis may be made by the clinical signs.

E. Postmortem Lesions
None.

F. Control
Destruction of infected and exposed animals.

G. Transmissibility to Humans
None.

H. References
Meyer and Eddie (81).

5. Coccidiosis

A. Etiology
Eimeria cavaiae is the common cause of coccidiosis in guinea pigs; *Balantidium cavaiae* infestation is rather uncommon.

B. Incidence
The disease is normally uncommon, but can cause epidemic losses, especially among young stock.

C. Signs
Heavy infestations produce enteritis, hemorrhages, diarrhea, and loss of general health condition.

D. Diagnosis
Diagnosis can be made by the identification of oocysts upon microscopic examination of feces.

E. Postmortem Lesions
No specific lesions except the inflammation of intestinal mucosa and hemorrhages.

F. Control
All feed and bedding should be sterilized and contamination by feces of infested animals should be prevented.

G. Transmissibility to Humans
None.

H. References
Dumas (28).

6. Louse Infestation

A. Etiology
Guinea pigs are frequently infested with *Gyropus ovalis, Trimenopon jenningsi,* or *Gliricola porcelli* species.

B. Incidence
Gliricola porcelli is the most common louse of guinea pigs. In healthy stocks, their presence is not suspected, and it may be necessary to search carefully in the fur, especially behind the ears. Both the lice and the eggs that are attached to the shafts of the hair near the base are pearly white in color; they are difficult to detect in white-coated animals but show up well against black or dark fur. Guinea pigs in poor health may become heavily infested with lice, which may be found all over the body, especially around the nipple areas on the abdomen.

C. Signs
Common signs of louse infestation are irritation, loss of hair, anemia, and loss of condition.

D. Diagnosis
Diagnosis is made by the identification of nits (eggs) or adult lice on animals.

E. Postmortem lesions
None.

F. Control
Eliminate lice from animals and animal rooms, and sterilize feed, bedding, and equipment.

G. Transmissibility to Humans
The personnel handling the infested animals may acquire the parasites.

H. References
Paterson (96); Lane-Petter (70).

In addition to the diseases discussed above, there are certain other diseases that affect guinea pig occasionally. Among them are *soft tissue calcifications* and *pregnancy toxemia* (both of them are nutritional in origin); *dystocia* (may result from inbreeding); *heat exhaustion* (heavily pregnant females are especially susceptible in hot weather or hot climates). The guinea pig is subject to comparatively few diseases. Those of genetic origin are rare. Diseases of nutritional origin (e.g., vitamin C deficiency) or from the ingestion of toxic substances occur from time to time. Neoplasms are rare and of little practical importance. However, spontaneous development of sarcomas, adenomas, and lipomas has been reported in guinea pigs. Mammary carcinoma, malignant tumors of most organs, and leukemia have also been described. Another frequent occurrence in guinea pig colonies is death due to trauma from the stampeding of young animals or of adults as a result of fighting. The reader is referred to Parish (94), Paterson (96, 97), Lane-Petter and Porter (69), and Ribelin and McCoy (104) for further details on the diseases of laboratory guinea pigs.

DISEASES OF HAMSTERS

The Syrian golden hamster (*Mesocricetus auratus*) is relatively new to medical research. There are other varieties of hamsters that are widely used for research purposes; they are the European or black hamster (*Cricetus cricetus*) and the Chinese grey hamster (*Cricetulus griseus, C. barabensis*). In the following section, frequently occurring diseases, primarily of the Syrian hamster, will be described.

One of the great advantages of the hamster as a laboratory animal is its resistance to most diseases occurring naturally in other species such as rats and mice. However, the animals are susceptible to most bacterial, parasitic, and such viral diseases as polio-myelitis, foot-and-mouth disease, mumps, influenza, West-Nile virus disease, psittacosis, and several so-called oncogenic viruses (e.g., mouse parotid tumors (polyoma virus), Simian (SV-40) virus and human adenovirus). Hamsters are also resistant to the toxic effects of various drugs and toxicologic agents, such as Colchicine.

However, there are some disorders to which hamsters are subject; these are briefly described below.

1. Wet-Tail Disease

A. Etiology

Unknown. Inadequate care, management and diet of animals may be responsible for this disorder. Also, *Escherichia coli* infection may be a possible causative agent.

B. Incidence

It occurs frequently in hamsters and has become a major concern to animal breeders and investigators. Animals young or old are subject to this disorder. Wet-tail in hamsters resembles infantile diarrhea in mice, scouring in piglets and chickens, white scours in calves and enteritis of rabbits.

C. Signs

The diseased animals become emaciated and weak. The area about the anus becomes wet and discolored as though the animals were suffering from diarrhea. The signs after the tail area becomes moistened are lethargy, irritability, anorexia, and finally emaciation. Animals usually die in 48 hours to a week after onset of signs.

D. Diagnosis

Diagnosis can be made by clinical signs.

E. Postmortem Lesions

The ileum, jejunum, colon, and rectum contain yellow semifluid material and some gas. Many of the hamsters have inflammatory lesions of the ileum and in the cecum. In some instances the cecal mucosa is ulcerated. There may be trichomonal infestation in the intestine with hemorrhagic areas starting at the junction of the intestine and cecum which progresses cephalad towards the stomach. In some animals, thickening of the walls of lower portion of the small intestine is found and contains a thick, white, caseous material. There may be ulcerations and hemorrhage present in the ascending and transverse colon. Heavy growths of *E. coli* are found in the gut contents of affected hamsters in contrast to a few *E. coli*, staphylococci and bacilli in healthy animals. *Escherichia coli* may be the cause of fatal diarrhea since neomycin sulfate appears to be effective in the prophylaxis of "wet-tail" and *in vitro* inhibits the growth *E. coli* strains isolated from hamsters.

F. Control

Hamsters should be housed in clean, dry cages with plenty of clean water and food supply. The sick animals should be destroyed.

G. Transmissibility to Humans

None.

H. References

Handler (51); Handler and Chesterman (56); Whitney (127).

2. Muscular Dystrophy

A. Etiology

Continued deficiency of vitamin E in the diet of hamsters from weaning stage is the cause of muscular dystrophy.

B. Incidence

The gross appearance of dystrophy in hamsters does not usually occur before the tenth or twelfth month of life and has never been known to result in the marked type of disability observed in long-term vitamin deficient rats.

C. Signs

Muscular changes in limbs, unsteady gait, and death due to cardiac necrosis are the signs of the acute disease.

D. Diagnosis

By clinical signs.

E. Postmortem Lesions

The lesions in skeletal muscles are pleomorphic and represent several kinds of response to injury in contrast to a segmental process of disease. Involvement of the long segments of the muscle fiber generally occurs. Reactions are characterized by the alignment of muscle nuclei in chains within the fiber. Irreversible changes in focal degeneration of myofibrils and formation of contraction clots and coagulation necrosis occur. In the final stages of irreversible injury and necrosis in the involved muscle segment, there is a conversion of myofibrils to a granular mass composed of surviving muscle nuclei, macrophages, and a few leukocytes. When regeneration occurs, it does so through terminal budding or plasmodic outgrowth from the viable portions of the diseased fibers. Vitamin E therapy prevents further degenerative changes and all evidence of previous necrosis is absolished in five to ten days. Sarcoplasmic basophilia due to ribonucleic acid in some of the dystrophic fibers are sometimes noted.

F. Control

The animals should be provided with sufficient amounts of vitamin E in their diet.

G. Transmissibility to Humans

None.

H. References

West and Mason (124); Homburger et al. (55); Handler and Chesterman (52).

3. Diabetes Mellitus (*in Chinese Hamster*)

A. Etiology

Probably transmitted by a recessive gene in inbred strains.

B. Incidence

The disease occurs frequently in some intensively inbred strains in aged animals.

C. Signs

Blood glucose levels are between 200 to 800 mg/dl as compared to normal levels of 110 ± 6 mg/dl.

D. Diagnosis

By the determination of sugar and protein levels in serum of the animals.

E. Postmortem Lesions

The disease is pancreatogenic with degranulation. Hydropic degeneration and deficiency of β cells. The α_2 fraction of the serum protein in diabetic hamsters is elevated to 12 to 30 percent as compared to 5 to 10 percent in normal hamsters.

F. Control

The inbreeding should not be intensive.

G. Transmissibility to Humans

None.

H. References

Meier (79); Meier and Yerganian (78); Green et al. (44, 45); Green and Yerganian (46).

Other spontaneous diseases of hamsters include *osteoarthritis*. The disease is seen only in aged animals and has never appeared in animals under 2 years old. This disease is similar to that observed in mice and it is characterized by a deformity about the stifle. There is sclerosis and dislocation of bone, with fibrillation of ligaments and fibrosis of the synovial membrane [Sokoloff (115)].

Amyloidosis

There is deposition frequently of amyloid in the adrenals and kidney in normal aging hamsters. In experimental animals, massive amyloidosis have been observed in leishmaniasis or tuberculosis, and also in hamsters treated with diethylstilbesterol and when certain types of tumors (e.g., pituitary tumors) are present. Amyloidosis is characterized by an elevation of α_2- (rather than α_1) serum globulin in hamsters. It seems that the component which has the position of α_1-globulin in the golden hamster behaves physiologically like an α_2-globulin [Dontenwill et al. (26); Chute et al. (19); Russfield and Green (108)].

Colloid Goiter

Hamsters suffer frequently with colloid goiter when they are fed a diet deficient in iodine. Extreme hyperplasia of thyroid glands results. When iodine is supplemented, large amounts of colloid accumulate in the thyroid follicles giving rise to a classical, morphological picture of colloid goiter resembling those occurring under similar conditions in man [Follis (34)].

Gallstones

The hamsters fed a cholesterol-free and nearly fat-free diet may deposit gallstones in their gallbladder. Such gallstones usually contain from 10.5 to 51.0 percent cholesterol. Also, the incidence of gallstones in hamsters can be increased by feeding them a diet deficient in vitamin A. The gallstones are often imbedded in a gelatinous mucoid mass in the lumen of the gallbladder [Dam and Christensen (24); Fortner (35)].

Many other conditions are reported to occur sometimes in hamsters. The reader is referred to the works of Frenkel et al. (36); Herrold and Damham (53); Handler (51); Handler and Chesterman (52).

Parasitic, Bacterial, Fungal, and Viral Diseases in Hamsters

Exernal parasites are known to cause or transmit skin lesions and a variety of inflammatory disorders. Some species of lice, fleas, mites, and bedbugs, which are common to mice and rats, infest hamsters causing alopecia and dermatitis by biting, sucking, or burrowing.

Internal parasites found in other rodents are also involved in the causation and transmission of disease in hamsters. A cestode, *Hymenolepis diminuta,* is a common parasite that infests hamsters, causing marked catarrhal enteritis. The enteritis may be either acute or chronic which produces both gastroenteritis and enterocolitis. *Hymenolepis nana* has also been isolated from hamsters. Occasionally, the animals are heavily infested with this parasite and die from intestinal impaction and obstruction. *Trichomonas muris* infection in the cheek pouches, esophagus, stomach, and intestines of hamsters has been reported. The morphology differs somewhat from region to region. The greatest population of these parasites are found in the cecum. The same strain of parasite occurs in both rats and hamsters and sometimes also in the mouse [Bjotvedt and Tufts (11); Habermann (50); Ratcliffe (102); Habermann and Williams (49); Soave (114); Handler (51); Lane-Petter et al. (71); Wantland (123)].

Hamsters do not suffer from chronic respiratory infections as do rats but sudden chilling can bring about a pneumonia-like disease, which usually kills the animal in about 3 days. Salmonellosis occurs in hamsters occasionally. Innes et al. (58) have described the pathological picture presented in an infection by *Salmonella enteritidis.* Death follows from septic pulmonary phlebothrombosis. *Salmonella typhimurium* has also been isolated from hamsters. The lesions of Salmonellosis in hamsters are not unlike those found in the mouse, rat, guinea-pig or rabbit [Habermann and Williams (49)].

Bartonellosis, a severe disease in rats, is rare in hamsters. It is produced by an organism resembling *Hemobartonella muris* causing a hemolytic anemia. Among the viral diseases, hamsters are not infected with ectromelia, a serious disease of mice. However, certain virus infections produce pneumonia in hamsters (Pearson and Eaton, (99)]. This virus was identified by cross-immunity and neutralization test and was related to *mouse pneumonia virus* (PVM). For further information on the virus infections of hamsters, the reader is referred to Handler and Chesterman (52).

Various forms of tumors have been reported in hamsters. The reader is referred to the publications by Handler (51), Handler and Chesterman (56), and Kirkman and Algard (65) for a detailed description of induced and spontaneously occurring neoplasms in hamsters.

DISEASES OF CHICKENS

There is a voluminous literature on this subject and a number of authoritative surveys are available, for example, Gordon and Horton-Smith (42) and Biester and Schwarte (10). A group of frequently occurring diseases of poultry is listed in Table 3-1.

Table 3-1. Frequently Occurring Diseases of Poultry

Disease	Etiology	Characteristics
1. THE AVIAN LEUKOSIS COMPLEX	PROBABLE VIRAL AGENTS	
A. VISCERAL LYMPHOMA-TOSIS (Big-liver disease)		A contagious neoplastic disease of the avian leukosis complex characterized by enlargement of liver or other internal organs with either a diffuse or focal type of involvement.
B. NEURAL LYMPHOMATOSIS (Range paralysis, Fowl paralysis, Marek's disease)		A form of the avian leukosis complex characterized by paralytic signs usually involving the legs, wings but occasionally other parts and organs.
C. OCULAR LYMPHOMATOSIS (White eye, Pearly eye, Epidemic blindness)		A disease of the leukosis series characterized by blindness and misshapen pupil or irregular depigmentation of the iris of one and occasionally both eyes.
D. OSTEOPETROTIC LYMPHOMATOSIS (Marble bone, Osteopetrosis galli-narum, Hypertrophic osteitis, Thick-legs disease)		Characterized by enlargement of the long bones.
E. ERYTHROBLASTOSIS and GRANULOBLASTOSIS (Myelotic leukosis)		Characterized by the presence of large numbers of immature cells of the erythroid and myeloid series in the circulating blood. Anemia often accompanies these leukemic diseases which are a result of a neoplasia of the respective immature cell forming elements in the bone marrow and other hematopoietic tissues.
F. MYELOCYTOMATOSIS (Myeloma, White tumors)		White tumor formation principally along the sternum and in the liver.

Table 3-1.   *(Continued)*

Disease	Etiology	Characteristics
2. FOWL POX	VIRUS	A slow spreading virus characterized by the formation of wart-like nodules on the skin and diphtheritic necrotic masses (cankers) in the upper digestive and respiratory tracts.
3. LARYNGOTRACHEITIS	VIRUS	An acute, highly contagious respiratory disease of chickens characterized by severe dyspnea, coughing and râles.
4. INFECTIOUS BRONCHITIS	VIRUS	An acute, rapidly spreading infection of chickens characterized by râles, coughing and sneezing without the accompaniment or subsequent development of nervous symptoms.
5. NEWCASTLE DISEASE (Avian Pneumo-encephalitis)	VIRUS	An acute, rapidly spreading disease of poultry and other birds in which coughing, sneezing and râles are often accompanied or followed by nervous manifestations.
6. FOWL PLAGUE (Fowl Pest)	VIRUS	Sudden onset, with early symptoms of weakness and lethargy. Cyanosis and edema of the head, wattles, intermandibular space and larynx are often prominent. Difficult breathing and blood stained mouth and nasal discharges are common manifestations. Extremely high mortality among birds of all ages.
7. AVIAN ENCEPHALOMYELITIS (Epidemic tremor)	VIRUS	A virus disease affecting particularly the central nervous system of young chickens and characterized by ataxia and by rapid tremors of the head, neck and limbs.
8. PSITTACOSIS and ORNITHOSIS	VIRUS (many types)	A chronic disease characterized by respiratory infection and systemic reactions.

Table 3-1. (Continued)

sease	Etiology	Characteristics
INFECTIOUS SYNOVITIS (Infectious arthritis)	PROBABLY A VIRUS	Inflammatory processes in synovial membranes. It occurs most frequently in young chickens 4-10 weeks of age. Liver is enlarged, sometimes discolored green.
SALMONELLOSIS (Paratyphoid infections)	BACTERIA (Salmonella pullorum and S. gallinarum)	Pullorum is highly fatal for young birds while adult birds are more resistant. The common signs are that the infected birds huddle near a source of heat, do not eat, appear sleepy, have whitish diarrhea and many die within a week. S. gallinarum causes fowl typhoid in adult birds. Signs are weakness, drowsiness, diarrhea, and a rapidly developing anemia.
FOWL CHOLERA	BACTERIA (Pasteurella multocida)	An acute or chronic, generalized or local infectious disease of poultry characterized by sudden onset with high morbidity and high mortality, showing enteritis, submucous and subserous hemorrhages, and vascular congestion.
TUBERCULOSIS	BACTERIA (Mycobacterium avium)	It is most prevalent in chickens. Deep ulcers in intestinal tract filled with caseous material containing many organisms which are discharged into the lumen and appear in feces. Also, the typical caseous lesions found in the spleen and liver. Lungs and other tissues are free of lesions.
LISTERIOSIS	BACTERIA (Listeria monocytogenes)	Occurs most frequently as a septicemia, but localized encephalitis similar to that seen in ruminants has been reported in chickens and turkeys. No specific signs.
SPIROCHETOSIS	BACTERIA (Borrelia anserina or Spirochaeta anserina)	A highly fatal disease caused by blood inhabiting spirochetes. It is an enzootic disease in many parts of the world. The signs are listlessness, cyanosis, fever,

Table 3-1. (Continued)

Disease	Etiology	Characteristics
		accompanied by increased thirst and a greenish yellow diarrhea with an abundance of urates. Th[e] temperature reaches 109 to 111.0°F in 4 to 5 days and returns to normal on 7th or 8th day. Infected birds sit on their hocks with their eyes closed. Complete paralysis may occur.
15. AVIAN VIBRIONIC HEPATITIS (Avian Infectious Hepatitis)	BACTERIA (Vibrio species)	A chronic, sometimes subacute disease of chickens character-ized by parenchymal degenera-tion and necrosis of liver.
16. INFECTIOUS CORYZA	BACTERIA (Hemophilus gallinarum)	An acute or chronic disease characterized by nasal dis-charge, sneezing and swelling or edema of the face and less frequently by lower respiratory tract infection.
17. CHRONIC RESPIRATORY DISEASE	BACTERIA (Mycoplasma gallinarum)	Involvement of the upper and lower respiratory system including the air sacs. Affected birds have nasal discharges, swelling of the face, sneezing and rales.
18. OMPHALITIS (Navel Ill or Mushy Chick)	EXCESSIVE HUMIDITY AND HEAVY CONTAM-INATION WITH COLIFORMS, PSEUDOMONAS STAPHYLOCOCCI AND PROTEUS SPECIES	Non-contagious, the animals appear normal until death. Depression, drooping of head and huddling near the source of heat usually are the only signs.
19. BLUE COMB DISEASE (Pullet disease, Avian monocytosis, Summer disease or Housing disease)	NON-BACTERIAL (Unknown)	An acute or subacute condition primarily of young laying chickens characterized by dehy-dration, fish-flesh-like areas of degeneration associated with congestive phenomena in the skeletal musculature, sparse petchiation of serous membranes,

Table 3-1. (Continued)

Disease	Etiology	Characteristics
		delicate evenly spaced foci in liver, mucoid enteritis, chalky appearance of pancreas, soft or broken ovarian follicles and various renal changes from tumefaction to uric nephritis.
20. ASPERGILLOSIS (Mycotic pneumonia or Pneumomycosis)	FUNGUS (Aspergillus fumigatus)	A disease usually with respiratory systems, dyspnea, gasping, accelerated breathing and nervous signs. Somnolence, inappetence, emaciation and increased thirst are also present.
21. COCCIDIOSIS	PROTOZOA (Eimeria tenella, E. necatrix or E. acervulina)	Involvement of the digestive tract. Accumulation of blood in the ceca; spotted appearance of the unopened intestine; numerous gray or whitish transverse patches in the upper half of the small intestine.
22. HISTOMONIASIS (Infectious enterohepatitis, Blackhead)	PROTOZOA (Histomonas meleagridis)	Listlessness, drooping wings, unkempt feathers and sulfur colored droppings are seen. The greatest mortality occurs in birds under 12 weeks old.
23. TRICHOMONIASIS (Canker or Roup)	PROTOZOA (Trichomonas gallinae)	Caseous accumulations in the throat, usually accompanied by loss of weight.
24. LEUCOCYTOZOON DISEASE	PROTOZOA (Leucocytozoon sabrazesi, L. caulleryi or L. andrewsi)	Relatively sudden onset with anemia, leukocytosis, splenomegaly and hypertrophy of the liver commencing about 1 week after infection. Fatalities usually occur 1 week after infection has appeared in blood. Birds with heavy infestations show lack of appetite, dullness, inability to run, loss of equilibrium, lameness and weakness.
25. BUMBLEFOOT (Abscess of the foot pad)	INJURY FOLLOWED BY BACTERIAL INFECTION (e.g. Staphylococci)	A sporadic local infection of the feet of chickens and turkeys characterized by enlargement and lameness in one or both feet.

Table 3-1.   (Continued)

Disease	Etiology	Characteristics
26. BOTULISM (Limberneck or Westernduck sickness)	BACTERIA (Clostridium botulinum type C)	Intoxication due to ingestion of food material containing toxin. No characteristic lesions. Toxin affects nervous system.
27. HEMORRHAGIC SYNDROME (Aplastic anemia or Hemorrhagic anemia)	UNKNOWN (Probably not infectious)	A disorder of growing chickens characterized by fatty bone marrow, delayed blood clotting time, anemia, often hemorrhage and variable mortality.
28. INTESTINAL HELMINTHIASIS	MANY SPECIES OF WORM PARASITES	Alterations in mucosa of the digestive tract. General unthriftiness, retarded growth and lowered production.
29. ECTOPARASITE INFESTATIONS	SPECIES OF: Lice, Ticks, Mites, Bedbugs, Fleas, etc.	Skin irritation, restlessness, low production, etc.

DISEASES OF CATS

1. Infectious Panleukopenia

Feline distemper, feline infectious enteritis, feline agranulocytosis, cat plague, cat fever.

A. Etiology

The disease is caused by a virus that attacks primarily members of cat family.

B. Incidence

It is a frequently occurring disease of cats and spreads rapidly in the form of an epizootic. All secretions and excretions of affected animals contain virus and the infection spreads through direct contact or by means of material contaminated with virus. The disease is acute and the incubation period is 3 to 10 days.

C. Signs

Common signs are elevated temperature, diarrhea with blood, extreme lethargy, conjunctivitis, dehydration, anorexia, and typical yellow-colored vomitus. Shortly before the temperature becomes elevated, there is a decrease in leukocytes. As the course of the illness develops, the leukopenia increases markedly so that sometimes no leukocytes can be found. Mortality parallels leukopenia fairly accurately. The mortality rate is high, especially in young cats, and losses from 60 to 90 percent are reported.

D. Diagnosis

Diagnosis can be based on leukopenia and clinical signs. Also, finding inclusion bodies in lymph nodes, spleen, and red bone marrow is a good indication of infectious panleukopenia. In the event of death, autopsy findings offer confirmatory evidence.

E. Postmortem Lesions

The first changes are found in lymph nodes and consist of hyperplasia, edema, and necrosis. Later, the red marrow of the long bones may become semifluid in consistency and appear fatty. Reddening may be seen in the terminal portion of the ileum and sometimes extends to involve most of the small intestine. The liver, kidneys, and spleen may appear slightly swollen. Microscopically, the epithelium of the villi of the affected portions of the intestine shows degeneration and the intestinal wall may be edematous. Degeneration of liver cells and tubular epithelium of the kidney is seen. Intranuclear inclusion bodies are found in the cells of the intestinal epithelium, lymph nodes, spleen, and in red bone marrow. When recovery occurs, a marked myelogenous cellular response is seen.

F. Control

Homologous antiserum offers good protection for about 2 weeks. A vaccine made from tissues of infected cats and inactivated by chemical means creates a measure of immunity to panleukopenia. Repeated annual or biannual vaccination increases the immunity. The infected cats should be isolated and proper sanitary conditions should be used. Antibiotic therapy is not effective against panleukopenia.

G. Transmissibility to Humans
None.

H. References
Gledhill (40); UFAW Handbook (121); Rhodes and Van Rooyen (103); Baker and York (4).

2. Bacillary Diarrhea

A. Etiology
Bacillary diarrhea in cats is caused by various enteric organisms including *Salmonella* species.

B. Incidence
The disease is common in cats where the sanitary conditions are not adequate. It is an acute disease with the incubation period of 3 to 10 days and acquired by the animal through ingestion of the organisms.

C. Signs
The signs are anorexia accompanied by pyrexia, followed by listlessness, diarrhea in many cases, and vomiting in some. Acute pain on handling due to peritonitis, dehydration and prostration are present. Death ensues in 48 hours or less from the time of onset. About 30 to 80 percent recover and are then apparently immune from further attacks.

D. Diagnosis
Diagnosis can be made by the clinical signs and isolation of the organisms from fecal specimens. White cell counts fall below 2000/cu mm of blood at the height of disease. Neutropenia is especially marked.

E. Postmortem Lesions
Inflammation of the alimentary tract, especially of the small intestine is observed in

many cases with enlargement of the abdominal lymph glands. Eosinophilic inclusion bodies have been found in the intestinal mucosa, abdominal lymph glands, liver, and spleen of cats which have died or been killed at the height of the disease. The presence of inclusion bodies suggests a concurrent viral infection.

F. Control
Good sanitary conditions are important in controlling spread of the disease. The infected animals should be isolated and their diets controlled.

G. Transmissibility to Humans
None.

H. References
Brumley (14); Jennings (61, 62); Scott (110).

3. Feline Pneumonitis (Coryza)

A. Etiology
Causative agent of feline pneumonitis is a virus of the psittacosis lymphogranuloma group.

B. Incidence
The disease is common in cats. The incubation period varies between 5 and 8 days, and the disease is manifest in a mild and a severe form. The animals are infected by ingestion or inhalation of the organism.

C. Signs
The early signs of the disease are a slight rise in temperature, generally accompanied by mild photophobia, and occasional sneezing. Manifestations of illness develop 6 to 10 days after exposure and consists of increased lacrimation followed after a day by conjunctivitis and a mucopurulent discharge from the eyes. A nasal discharge accompanies the eye changes. The respiratory signs also include coughing. Cats rarely die if proper care is taken.

D. Diagnosis
Diagnosis can be made by clinical signs and mouse inoculation.

E. Postmortem Lesions
Progression of the disease may lead to an inflammation of the lower respiratory tract. The larynx and trachea are reddened and the accumulation of small amounts of thick, cloudy, mucus may be noted. Pneumonia is generally characterized by dense consolidations of a pinkish gray color in the anterior pulmonary lobes and occasionally in the diaphragmatic lobes. Small consolidations near the hilus are sometimes visible. The bronchial lymph nodes are not noticeably enlarged. Occasionally there is a slight enlargement of the spleen. The other organs appear normal.

F. Control
Isolation of infected animals and sanitary conditions are effective control measures.

G. Transmissibility to Humans
None.

H. References
Paterson (98); Baker and York (4); Rhodes and Van Rooyen (103).

4. Feline Pneumonia

A. Etiology

A filterable virus causes pneumonia in cats.

B. Incidence

It is an acute disease contracted commonly by the animals through ingestion or inhalation of the organisms.

C. Signs

The signs of viral pneumonia include conjunctivitis, rhinitis, sneezing and coughing. Death ensues within 3 to 5 days.

D. Diagnosis

Diagnosis may be made by clinical signs and serological tests.

E. Postmortem Lesions

Usual postmortem findings are congestion of respiratory tract and areas of consolidation in the lungs.

F. Control

The infected animals should be isolated and the contaminated feed, water, equipment, and room should be disinfected.

G. Transmissibility to Humans

None.

H. References

Baker and York (4); Rhodes and Van Rooyen (103).

5. Feline Infectious Anemia

A. Etiology

It is caused by an organism *Hemobartonella felis*.

B. Incidence

An acute or chronic hemolytic anemia is commonly found in cats of 1 to 3 years of age. The disease can be transmitted either parenterally or orally. A significant portion of the cat population may carry the infection in latent form which becomes exacerbated in the presence of various debilitating diseases or stresses.

C. Signs

In acute cases there is fever of 103° to 106°F. Jaundice, anorexia, depression, weakness, and splenomegaly are common signs. Dyspnea varies with the degree of anemia.

D. Diagnosis

Diagnosis can be confirmed by identification of the parasite in red blood cells. A series of smears over a period of several days may be required for an accurate diagnosis since the erythrocyte bodies show up periodically.

E. Postmortem Lesions

Gross autopsy findings are not consistent. The spleen and mesenteric lymph nodes may be enlarged, and a bone marrow hyperplasia may be present.

F. Control

Care should be taken in selecting blood donor cats for general transfusion. The infected animals may be treated by blood transfusion and broad-spectrum antibiotics.

G. Transmissibility to Humans
None.

H. References
The Merck Manual (80); Lane-Petter (71).

5. Coccidiosis

A. Etiology
The disease is caused by *Isospora bigemina, I. felis,* and/or *I. rivolta.*

B. Incidence
Coccidiosis is a frequently occurring diseases in cats. The localization of infection is mainly in the small intestine, occasionally also involving the cecum. Some species penetrate into the subepithelial tissues. Oocysts are passed in the feces.

C. Signs
The severe infections cause blood diarrhea, anorexia, weakness, and emaciation. The course may vary from a few days to a week or 10 days and seldom is fatal unless complicated by other factors.

D. Diagnosis
Diagnosis can be made by identification of oocysts upon microscopic examination of feces.

E. Postmortem Lesions
Heavy infestations produce hemorrhages and desquamation of intestinal epithelial cells with resulting diarrhea.

F. Control
Contamination of feed, water, bedding, and equipment with feces from infested animals should be prevented.

G. Transmissibility to Humans
None.

H. Reference
Becker (7); Brumley (14); Habermann (48); Morgan and Hawkins (86).

6. Ringworm (*Dermatomycosis*)

A. Etiology
The causative agents are *Epidermophyton floccosum, Microsporum audouini, M. canis, M. gypseum, Trichophyton gypseum, T. rubrum, T. rosaceum, T. faviforme,* and *T. crateriforme.*

B. Incidence
Ringworm infection is common in cats. The fungi are invaders of keratinized tissues (nails, hair, skin, etc.).

C. Signs
Common signs are loss of hair with crusted bald spots, broken hairs, and otitis.

D. Diagnosis
Diagnosis may be made by microscopic examination of skin scrapings. Also, some species fluoresce under ultraviolet light.

E. Postmortem Lesions
None.

F. Control
All bedding material and equipment should be sterilized. The infested animals should be isolated. Treatment of ringworm in cats is difficult.

G. Transmissibility to Humans
Many of the causative organisms can be infective to humans.

H. References
Meenan (77); NAS-NRC Publication 317 (88).

8. Ectoparasites

A. Etiology
Lice: *Felicola subrostrata*. Ear mites and mange mites: *Notoedres minor* var. *cati* (mange mite); *Otodectes cynotis* (ear mite). Fleas: *Pulex irritans* (human flea); *Ctenocephalides canis* (dog flea); *C. felis* (cat flea).

B. Incidence
The ectoparasite infestation is common in cats kept under inadequate sanitary conditions.

C. Signs
Common signs are skin irritation, loss of hair, loss of condition, and infected ear canals (from the ear mite infestation). In heavy infestation, the animal may develop anemia.

D. Diagnosis
Diagnosis is made by identification of the parasite on the animal.

E. Postmortem Lesions
None.

F. Control
The parasites should be eliminated from animals and animal rooms. All bedding and equipment should be sterilized.

G. Transmissibility to Humans
Man is susceptible to infestation by the ectoparasites.

H. References
Monning (85); NAS-NRC Publication 317 (88).

9. Endoparasites
Cats are susceptible to various endoparasites listed below.

A. Hookworm (Ancylostoma caninum)
It is a common disease causing enteritis, ulceration of intestines, diarrhea and anemia from heavy infestations. The parasite eggs can be identified by microscopic examination of feces.

B. Ascariasis
It is caused by *Toxocara canis, T. cati,* or *Toxascaris leonina*. It is a commonly occurring disease in cats resulting in pneumonia due to destruction of lung tissue by migrating larvae. Also, bloating and loss of condition occurs from heavy infestation.

C. Tapeworms
Dipylidium caninum causes colic, chronic enteritis, skin lesions and loss of condi-

tion. *Taenia pisiformis* and *T. taeniaformis* are also common parasites causing similar signs.

D. Roundworms

Toxascaris leonina infestation occurs frequently in cats. The worms may clog bile ducts or the lumen of the small intestine. Enteritis and liver damage may result.

E. References

Belding (8); Matheson (76); NAS-NRC Publication 317 (88); Paterson (98), Michael (82).

10. Neoplasms of Cats

Neoplastic diseases seem to occur less frequently in the cat. Most cases involve the skin, mammary glands, and bones. The range of incidence is about 1 to 6 percent. The types of tumors reported include squamous cell carcinoma, myxosarcoma, adenocarcinoma of the pancreas, liver, and mammary glands, osteogenic sarcoma, spindle cell sarcoma, lymphosarcoma, chrondrosarcoma, fibrosarcoma, basal cell carcinoma of the skin, bronchogenic carcinoma, and angiomatosis of the liver.

References

Ruebner et al. (105); UFAW Handbook (120).

In addition to the diseases described above, there are many upper respiratory diseases of cats caused by viruses. Also, toxoplasmosis occurs in cats frequently.

DISEASES OF DOGS

1. Canine Distemper

A. Etiology

Canine distemper is caused by a virus that produces inclusion bodies usually found in the cytoplasm of epithelial cells of the respiratory and urinary tracts.

B. Incidence

Canine distemper is highly prevalent in dogs. The animals are usually exposed early in life to this rapidly spreading disease unless they are reared in complete isolation. Aged dogs that have led comparatively sheltered lives may be very susceptible to this disease, even though vaccinated when young. The virus is airborne, being spread by droplets of secretions from the eyes and nose. Indirect transmission may also occur. Dogs can develop encephalitis 45 days after exposure, and virus can be isolated from the brain. The disease may be acute or subacute.

C. Signs

Initially there is an elevated temperature 6 to 9 days after the animals have been exposed by contact. Typically, the temperature follows a diphasic course with an initial rise lasting for 1 to 3 days, followed by an apparently normal temperature for a day or two and then a secondary rise lasting for a week or longer. With the diphasic fever, signs of conjunctivitis are common. Secondary signs include gastritis, enteritis, emaciation, cough, purulent ocular nasal discharge, and occasionally nervous involvements. A leukopenia accompanies the fever and may fluctuate with the temperature or remain low throughout the course of illness. A diarrhea usually develops that may persist after acute signs have disappeared. The feces become fluid and contain mucus and often blood. This condition leads to dehydration, and the dog rapidly loses weight. Most cases, unless complicated, may recover. Pneumonia is not an uncommon complication. Respiration at first becomes rapid and with extensive involvement, labored breathing develops. Nervous involvement in young animals are manifested by convulsive attacks. In an attack, the limbs and body become rigid, the jaws open and close rapidly, and the head jerks rhythmically. Saliva made foamy by the jaw movements appears at the corners of the mouth. An attack may last several minutes leaving the animal exhausted.

D. Diagnosis

Diagnosis may be made by the typical diphasic fever and other clinical signs described above. Intracytoplasmic inclusion bodies in the respiratory system and mucus membrane of the bladder are good indications of distemper. In the absence of typical nervous signs, or a serologic finding based on the development of specific distemper antibodies, the positive diagnosis of distemper is difficult.

E. Postmortem Lesions

The postmortem findings depend on the course of the disease. In uncomplicated cases, mucus membranes appear reddened and covered with a mucopurulent exudate. Gastrointestinal mucus membranes are reddened and inflamed. Peyer's patches are swollen and may be ulcerated. Splenic enlargement may be found. When pneumonia develops, it is of the bronchial or the interstitial type. The interstitial pneumonia is regarded as primarily caused by a pure virus infection. Grossly, the lungs show a grayish red mottling and are doughy in consistency. If the lungs are secondarily infected with bacteria, there is a superimposed bronchial pneumonia characterized by plum red consolidations on the anterior and lower portions. In severe cases, the pneumonic areas may contain purulent foci.

F. Control

In the interval before vaccination, protection is usually provided by canine distemper antiserum inoculated subcutaneously. Dogs showing signs of illness should be placed in warm, dry quarters. Good nursing care and symptomatic treatments are important. Antibiotics (broad spectrum) should be given when evidence of secondary bacterial infection is present.

G. Transmissibility to Humans
None.

H. Reference
Worden et al. (128).

2. Canine Hepatitis

A. Etiology

Infectious hepatitis is caused by canine hepatitis virus which mainly affects the liver, although it is present throughout the body during the acute phase of the disease.

B. Incidence

It is fairly common in dogs. The disease may be acute or chronic. Death may occur with 24 to 48 hours. Dogs of all ages are susceptible and infection occurs by ingestion. The virus is not airborne.

C. Signs

A high temperature (104°F) is the first sign, and this occurs a week or 10 days after the infection has been acquired. Sometimes a transient rise of temperature is the only sign of the disease. The disease varies from a slight fever to a moderate or severe reaction, which may result in 10 percent mortality. Serological tests have shown that an average of 50 percent of dogs over 1 year of age have had hepatitis. On the day after initial temperature, leukopenia develops, and the count remains low throughout the febrile period. General signs of illness are apathy with partial or complete loss of appetite and intense thirst. Conjunctivitis accompanied by a serous discharge from the eyes and nose develops. An intense hypermia of the oral mucous membrane is seen. Enlargement of tonsils may be present, but the degree of involvement varies. Signs of abdominal tenderness sometimes can be elicited by palpation of the abdomen. Vomiting may also be observed. Subcutaneous edema of the head, neck, and trunk is seen occasionally and at times causes disfigurement. Petechial and ecchymotic hemorrhage of the skin occur in a few cases primarily in the region of the abdomen. When dogs injure themselves during illness, they usually bleed profusely.

D. Diagnosis

Usually the rapid onset of the disease and prolonged bleeding time (more than 2 minutes) suggest infectious hepatitis. Clinical evidence is not always sufficient to differentiate infectious hepatitis from distemper. In the laboratory, examination of a sample of liver obtained by biopsy will confirm a diagnosis if the characteristic intranuclear inclusion bodies are found.

E. Postmortem Lesions

The primary lesions appear in the endothelium and the hepatic cells. The endothelial damage of blood vessels often leads to intraocular hemorrhages, bleeding in the mouth, paintbrush hemorrhages on the gastric serosa, and hemorrhagic lymph nodes. The liver appears normal in size, but it usually is tan colored, with darker, mottled variegations. Occasionally, it is tannish red and slightly hyperemic. The gallbladder wall becomes edematous and thickened. Edema of thymus is usually found with or without petechial hemorrhages. Other lesions may be found as a result of secondary infections.

F. Control

Infectious hepatitis is transmitted through the urine or saliva of an infected dog. Since the live virus can be excreted in the urine of recovered dogs for many months, such animals, although themselves immune, are a source of infection. The sick animals should be isolated. The susceptible animals should be vaccinated. Vaccines are available and usually contain a live but attenuated virus. It is now a common practice to combine the hepatitis vaccine with a distemper vaccine so that protection against the two diseases can be given simultaneously. It is also useful to give homologous distemper serum to prevent simultaneous infection with this disease when using univalent vaccine.

G. Transmissibility to Humans

None.

H. References

Worden et al. (128); Baker (5).

3. Leptospirosis

Stuttgart's disease, Weil's disease, canine typhus, infectious jaundice.

A. Etiology

Leptospira canicola is much more prevalent than *L. icterohemorrhagiae,* spreading rapidly from dog to dog and frequently producing epizootics of hemorrhagic type of leptospirosis.

B. Incidence

Leptospirosis occurs occasionally to commonly in dogs depending upon the area. During the active disease and following recovery, many dogs act as urinary shedders of *L. canicola*. Infection occurs following contamination of the oral and nasal mucus membranes with infective urine. Rats are a reservoir of *L. icterohemorrhagiae* infection and the disease is a result of accidental infection.

C. Signs

Common signs of leptospirosis are high temperature (103° to 105°F), vomiting, diarrhea, dehydration, usually icterus, dark urine, and muscular soreness. There is often a mild congestion of the conjunctivae in the infected animals. At this stage clinical diagnosis is difficult, however. On the following day there is a sharp decrease in temperature, but depression is more pronounced, breathing is labored, and thirst is marked. Muscular stiffness and soreness, particularly of the hind legs, as indicated by the unwillingness to rise from a sitting position and pain on palpation, usually are present. The mucous membranes of the oral cavity may show at first irregular hemorrhagic patches resembling abrasions or burns which later become dry and necrotic and slough off in sections. A slimy salivary secretion around the gums is at times tinged with blood. In some cases, the tongue also may show necrotic patches. More severe cases show intense depression and muscular tremors with the temperature dropping gradually to subnormal (97°F). Bloody vomitus and bleeding gums are observed, and dehydration is marked. Severe leukocytosis occurs. There is abdominal pain on palpation, and the feces often become liquid and tinged with blood, indicating severe hemorrhagic gastroenteritis. Frequent urination is often observed and the urine shows albumin, pus cells, and casts indicative of nephritis. The eyes become sunken in their orbits and the small vessels of the conjunctivae are injected. In severe cases, uremia develops. Mortality is about 10 percent, and in fatal cases, death occurs usually 5 to 10 days after onset.

D. Diagnosis

Diagnosis can be made on the basis of clinical findings, postmortem examination, histopathological demonstration of leptospirae in the kidneys or liver, demonstration of leptospiral and serological tests.

E. Postmortem Lesions

Hemorrhagic gastroenteritis is the predominant postmortem lesion. The contents of the intestines and sometimes the stomach are bloody and chocolate colored. The tissues may be uniformly bile stained. The liver is engorged, the lymph nodes often are hemorrhagic and an acute interstitial nephritis is present. The myocardium shows diffuse hemorrhages. The lungs may present hemorrhagic and pneumonic areas. There may be a urine-like odor from the organs, due to uremia. The kidneys may be much enlarged in the acute phase and a yellow band is often present at the corticomedulary junction. Chronic cases present a pale, fibriotic kidney. Hemorrhagic lesions may be present throughout the body.

F. Control

Contamination of feed and water by urine of rodents and potential carriers should be prevented. Early and adequate use of antibiotics can reduce mortality.

G. Transmissibility to Humans

None.

H. References

Hoskins et al. (56); Worden et al. (128).

4. Toxoplasmosis

A. Etiology

A protozoan *Toxoplasma gondii* is the causative agent to toxoplasmosis which is an obligate intracellular parasite capable of invading a large variety of cell types, such as neurons, muscle cells, parenchyma cells of liver, alveolar cells of lungs, epithelial cells, and leukocytes.

B. Incidence

Toxoplasmosis occurs frequently in laboratory dogs. It may also occur simultaneously in women or children and dogs in the same household. It is transmitted through the placenta from infected mothers to their unborn offspring. This has been reported in dogs, cattle, swine, sheep, and humans by isolation of *T. gondii* from infected fetuses at the time of birth. Another mode of transmission is by ingestion.

C. Signs

Clinical signs of the disease are anorexia, weakness, depression, emaciation, coughing, vomiting, dyspnea, fever, tremors, incoordination, irritability, paralysis, and other nervous disturbances, premature birth, and abortion. There may be a transitory exanthema on the abdomen and the medial surface of the thighs. Sudden onset of high temperature in infected animals makes the disease sometimes difficult to distinguish from distemper. Also, the lymph glands are enlarged.

D. Diagnosis

Laboratory examination of exudates and internal lesions may result in cultivation of the organism. Toxoplasmosis is suspected in deaths of animals with unexplained nervous and respiratory disease. Serologic tests are valuable in making diagnosis of toxoplasmosis.

E. Postmortem Lesions

Frequently necropsy findings include pneumonitis, encephalomyelitis, focal necrosis, and swelling of the liver. Lymphadenitis, ulceration of the oral and gastrointestinal mucosa, hydrothorax, peritonitis, and pancreatitis are also common findings upon necropsy. The lesions may be changed by intercurrent distemper, hepatitis or other diseases. Toxoplasmic myocarditis and encephalitis may account for an occasional sudden death.

F. Control

Strict sanitary measures should be taken to control toxoplasmosis in dogs. The animals may be treated with sulfa drugs and antibiotics.

G. Transmissibility to Humans

Humans are susceptible to toxoplasmosis. Particularly, women and children may acquire the disease by ingestion and inhalation of contaminated material from dogs.

H. References

Cole (21); Hoskins et al. (56).

5. Rabies

A. Etiology

Rabies is caused by a virus. The disease is important in dogs, cats, and wild animals. It is also important to public health and to agriculture.

B. Incidence

It is enzootic and sometimes epizootic in dogs. The virus may be recovered from the central nervous system and also from the salivary glands, lacrimal glands, pancreas, kidney and adrenal tissues of infected animals. It is transmitted from animal to animal or to humans by means of bite and contamination of the wound with virus-bearing saliva. Animals in very early stage of rabies may transmit the disease. The course of disease is acute; however the incubation period may be 10 to 14 months.

C. Signs

Change in mental attitude is the first sign that may be indistinguishable from a digestive disorder, injury, foreign body in the mouth, poisoning, or early infectious disease. Animals usually stop eating and drinking and may seek solitude. A change in vocalization is noticed. The animal may be extremely aggressive or dumb and staring. Increased salivation is common in rabid dogs. There is frequently irritation or stimulation to the genitourinary system as evidenced by frequent urination, erection in the male, and sexual desire. With rare exceptions, prostration, coma, and death follow in 2 to 3 days after onset of the encephalitic syndrome.

D. Diagnosis

Clinical diagnosis may be difficult because of the lack of classical syndrome in rabies even in advanced furious cases. Rabies progresses rapidly and usually typical signs appear in a day or two. A complete history and holding the animal for observation are important in diagnosis. The positive diagnosis can be made by the demonstration of Negri bodies in the brain after natural death. However, if the brain examination is negative or unsatisfactory, inoculation of mice will give a definite answer.

E. Postmortem Lesions

No specific lesions are found upon necropsy, except that Negri bodies be found in the brain.

F. Control

Effective control measures include vaccination of all dogs and elimination of stray dogs. Chemically killed virus vaccine confers a relatively short immunity. However, this type of vaccine has proved useful in the control of rabies in dogs but the animals must be vaccinated annually. Chick embryo live-virus is an important recent development and will produce satisfactory immunity in dogs for more than 3 years.

G. Transmissibility to Humans

Highly contagious.

H. References

Baker and York (4); Hoskins et al. (56); Worden et al. (128).

6. Histoplasmosis

A. Etiology

A fungus disease caused by *Histoplasma capsulatum*.

B. Incidence

In the central United States, histoplasmosis is a frequent systemic fungus disease encountered in dogs and humans. Cats, cattle, and wild animals also become infected, but it is not a problem in these species. The animals become infected by inhalation of the fungus. Dogs are easily infected by intratracheal inoculation. The disease is not common in laboratory animals kept under good controlled conditions. The disease may be chronic or acute with incubation period 3 months to 3 years.

C. Signs

The infected animals have a chronic cough or chronic diarrhea which fails to respond to usual treatment. The bronchial lymph nodes and spleen are enlarged. Other signs include anorexia, emaciation, vomiting, dermatitis, irregular fever, and enlarged visceral lymph nodes. Acute histoplasmosis is nearly always fatal after a course of 2 to 5 weeks.

D. Diagnosis

The histoplasmin intradermal test and chest x-rays showing enlarged bronchial lymph nodes are good indications of histoplasmosis. The presence of calcified pulmonary lesions on radiographs is also a useful means of differential diagnosis. Further information in diagnosis may be obtained by serological tests.

E. Postmortem Lesions

Necropsy shows ulcerations of gastrointestinal mucosa, enlarged lymph nodes and vessels of mesentery and bronchial regions.

F. Control

Since the *H. capsulatum* is disseminated through the saliva, feces, vomitus, and urine, the environment of infected animals contaminated with fungus should be decontaminated.

G. Transmissibility to Humans

Direct transmission of *H. capsulatum* from dog to humans has not been established. However, humans are susceptible to the infection. Therefore, careful handling of infected dogs is recommended.

H. References
Schwarz (109).

7. Ectoparasites

Dogs are commonly infested with fleas, lice, ticks, and mites. The following are among the common species of ectoparasites infesting dogs. Lice: *Linognathus piliferus* (the sucking louse); *Trichodectes canis* (the biting louse). Mites: *Sarcoptes scabiei* (Sarcoptes mange), *Demodex canis* (red mange), *Otodectes cynotis* (ear mange). Ticks: (American dog tick), *Dermacentor variabilis, Rhipicephalus sanguineus* (brown dog tick). Fleas: *Pulex irritans* (human flea), *Ctenocephalides canis* (dog flea), *C. felis, Echidnophaga gallinacea* (stick tight flea).

Common signs of these ectoparasite infestations are skin irritation, loss of hair, anemia, restlessness, scratching, and development of skin lesions and loss of general health condition.

Both fleas and ticks infest the environment of the dog and merely use the animal for taking meals. In attempting control, therefore, measures must be taken to eradicate the parasite from the environment as well as from the dog. The application of any good insecticide to the coat of the dog, by wash or powder will destroy fleas and ticks present on the dog at that time. The cleaning should be repeated. Thorough cleaning of the kennel area with an insecticide wash is recommended. Lice live permanently on the dog and the problem of control is that the eggs of the louse, which are attached to the hair are not affected by insecticide. For this reason, it is usual to treat the dog again when the eggs have hatched but before the emergent lice have matured and commenced egg laying, that is, at about 14-day intervals.

These ectoparasites are transmissible to humans.

References
Worden et al. (128); Monnig (85).

8. Ringworm (Dermatomycosis)

A. Etiology
The common ringworm fungus infection of dogs are *Epidermophyton floccosum, Microsporum audouini, M. canis, M. gypseum, Trichophyton gypseum, T. rubrum, T. rosaceum, T. faviforme,* and *T. crateriforme.*

B. Incidence
Ringworm infection is common in animals where the sanitary conditions are not adequate.

C. Signs
These fungi are invaders of keratinized tissues (nails, hair, skin). The infected animals have loss of hair with crusted bald spots, broken hair, and otitis. The ringworm often appears on the head around eyes and on the paws, but it can grow anywhere on the body.

D. Diagnosis
By clinical signs and demonstration of fungus on the scraping of infected areas.

E. Postmortem Lesions
None.

F. Control
All bedding and equipment should be sterilized and the infected animals should be isolated. The ringworm is usually responsive to treatment and will sometimes rapidly disappear without any dressing.

G. Transmissibility to Humans
It can be transmitted to humans.

H. References
Matheson (76); Worden et al. (128).

9. Coccidiosis

A. Etiology
The protozoan parasites causing coccidiosis in dogs are *Isospora canis, I. felis,* and *I. rivolta.*

B. Incidence
It is a frequently occurring disease in dogs, particularly young dogs.

C. Signs
The developmental cycles occur in the epithelium of the small intestines and oocysts are passed in the feces. Heavy infestation produces hemorrhages and desquamation of epithelial cells with resulting diarrhea.

D. Diagnosis
Diagnosis can be made by the identification of oocysts on microscopic examination of feces.

E. Postmortem Lesions
Hemorrhagic patches and desquamation of epithelial cells of the small intestine are common lesions.

F. Control
Contamination of feed, water, bedding, and equipment with feces from infested animals should be prevented.

G. Transmissibility to Humans
None.

H. References
Becker (7); Brumley (14); HEW Publication No. 343 (121).

10. Endoparasites

A. Etiology
Dogs may be infested with a wide variety of endoparasites, some of which are listed below.

1. Tapeworms. Taenia taeniaformis (cat tapeworm). Rodents are intermediate host. Dogs become infested by eating infested rodents.
Dipylidium caninum. Eggs are ingested by fleas (larvae). Fleas mature and are ingested by animals. Eggs develop to mature tapeworm in intestine.
Taenia pisiformis. Rodents act as intermediate hosts. The common signs of these tapeworms are colic, enteritis, skin lesions, and loss of general health condition.

2. Heartworms (Dirofilaria immitis). It is common in the southern states. The adult worms are found in the lumen of the heart. Larval stages are found in the circulating blood. Mosquitos act as intermediate hosts and spread infestation after feeding on infested animals. Common signs are chronic endocarditis, dilation of the heart, congestion of the lungs and enlargement of the spleen (results of a failing heart).

3. Lungworms. The causative agent is *Capillaria aerophila.* Heavy infestation causes bronchitis and bronchopneumonia. It occurs rarely.

4. Bladderworms. Capillaria plica sometimes causes bladder infestation in dogs. It may cause irritation of the bladder mucosa.

5. Whipworms. Trichuris vulpis is a common worm of dogs causing thickening and inflammation of the mucosa of the cecum and colon, colic, anemia, and chronic enteritis. The embryonated eggs are ingested, and larvae develop to penetrate the lining of the intestines. They return to the lumen of the cecum and colon and mature in 70 to 90 days.

6. Hookworms. Ancylostoma caninum is the common species. Infestation of the host takes place by ingestion of eggs in feed or water as well as by active penetration of the skin by the larvae. Larvae develop to maturity in the intestines. Prenatal infestation may also occur. The parasites attach themselves to the walls of the intestines and suck blood. Enteritis, ulceration of the intestines, diarrhea and anemia result from heavy infestations. Neonatal puppies may also become infested via the colostrum.

7. Roundworms. Toxocara canis and *T. cati* are common roundworms and infest dogs frequently. Larvae develop from ingested eggs and penetrate into the blood stream. These are carried to the lungs, then travel up the tracheae, are swallowed and develop to maturity in the small intestines. Pneumonia may result from destruction of lung tissue by the migrating larvae. Bloating, enteritis and loss of condition results in young dogs. Pups may be born with infestation or become infested via the colostrum.

The control of endoparasites in dogs is effective by the isolation of infested animals;

prevention of feed, water and equipment from contamination by feces of infested animals.

References

HEW Publication No. 343 (121); Worden et al. (128); Mercks Veterinary Manual (80); Monnig (85); NAS-NRC Publication No. 317 (88); Belding (8).

11. Neoplasms of Dogs

Cancers usually occur in older dogs. Malignant lymphoma and more undifferentiated or embryonal sarcomas have been reported in dogs during the fourth and fifth years of their life. However, a great majority of the cases of malignancy are seen between the seventh and twelfth years. Although the occurrence of neoplasms in dogs are generally rare, females seem to have a higher frequency. In laboratory animals, tumors are not found in significant numbers because of use of younger animals in the experimentation and also because of the generally short duration of the experiments. However, a number of types of tumors can be induced in dogs. In pharmacology and cancer research, various types of cancers are being studied in laboratory dogs. The types reported include squamous cell carcinoma of the skin and tongue, spindle cell sarcoma, myxosarcoma, adenocarcinoma of the pancreas, liver, and mammary gland, lymphosarcoma, osteogenic sarcoma, chondrosarcoma, fibrosarcoma, basal cell carcinoma of the skin, bronchogenic carcinoma, angiomatosis of the liver, and many others. The literature on the neoplasms of dogs is too vast to cover in this section. Some topics on selected tumors in dogs are discussed in Chapter 11.

DISEASES OF MONKEYS

1. Tuberculosis

A. Etiology

Mycobacterium tuberculosis or *M. bovis* are the causative agents of tuberculosis in the monkey.

B. Incidence

Tuberculosis is a common disease of monkeys in captivity but is of low incidence in monkeys in their natural environment. The disease may be chronic or subacute and the incubation period is usually 1 to 3 months. The infection may be acquired from contaminated food or water or through contact with infected monkeys or humans.

C. Signs

The clinical signs of tuberculosis are not evident until the condition is quite far advanced. Animals which appear in good condition may be severely affected. The usual signs are coughing, progressive loss of weight and health condition, evidence of fatigue and prostration following exercise. Tuberculin testing is of some value, but false positive and false negative reactions can occur.

D. Diagnosis

Diagnosis may be made by clinical signs, but the definitive diagnosis is established by isolation of *M. tuberculosis* from typical lesions. Tuberculin test should be performed by the injection of 0.1 ml of a 1:10 dilution of Koch's Old Tuberculin into the upper eyelid of the animal. The test is read after intervals of 48 and 72 hours. An edematous discolored swelling of the right eyelid is a sign of a positive reaction. Chest x-ray is also used as a diagnostic test. It is quite useful for the diagnosis of advanced tuberculosis.

E. Postmortem Lesions

Tuberculosis in monkeys usually has the characteristics of the acute hematagenous or childhood variety, if compared with humans. Typical tuberculosis lesions are seen in the lungs and intestines.

F. Control

All animals showing positive tuberculin reactions should be destroyed and all carcasses incinerated. The tuberculin tests should be repeated at 30-day intervals on all primates until the entire group proves to be negative on two consecutive tests. The test is repeated on this entire group at 90-day intervals. If positive reactors are found, the test should be repeated at 30-day intervals until the entire group proves to be negative on two consecutive tests. All cages and equipment contacted by infected animals should be autoclaved as soon as this condition is detected.

G. Transmissibility to Humans

Tuberculosis is highly transmissible to humans. All personnel handling primates should be required to have chest x-rays at 6-month intervals. This is necessary to eliminate the possibility of human carriers transmitting tuberculosis to the primates and vice versa.

H. References

Benson et al. (9); Ratcliffe et al. (102); Weston (125); Ruch (106).

2. Pneumonia

A. Etiology

Various organisms may cause pneumonia in monkeys.

B. Incidence

The most common infection encountered in monkeys is in the respiratory system. Interstitial pneumonitis, acute, subacute, and chronic lobular or lobar pneumonias occur in monkeys, caused by bacteria or viruses. The animals die within a few weeks of delivery. Aspiration pneumonia also occurs in monkeys and is due to recognizable sawdust shavings or food particles in the exudate. The course of infection is variable and it may be secondary to other respiratory disorders. Route of infection is most commonly by ingestion or inhalation.

C. Signs

Characteristic signs of pneumonia are rapid onset, fever, anorexia, and respiratory distress. It may also occur secondary to debilitating conditions.

D. Diagnosis

Diagnosis is made by clinical signs and by isolation of organisms from the naso-pharynx or trachea.

E. Postmortem Lesions

The lesions are histologically comparable to those seen in man with respiratory infections. Pneumonic areas of the lungs with areas of consolidation are typical lesions found.

F. Control

The probably debilitating conditions should be corrected and the animals may be treated with antibiotics. Temperature fluctuations, humidity, and so forth should be properly controlled in animal quarters.

G. Transmissibility to Humans

Many forms of bacterial and viral pneumonias are transmissible to humans. Proper precautions should be taken by the personnel handling the infected primates.

H. References

Ratcliffe et al (102).

3. Gastroenteritis

A. Etiology

Salmonella enteritidis, S. typhimurium, Endamoeba histolytica, and *Balantidium coli* are the main causative agents of gastroenteritis in monkeys. Also, *Shigella* species may cause gastroenteritis occasionally.

B. Incidence

Gastrointestinal infections are very common in newly arrived monkeys, and they may be associated with pulmonary infections. The condition takes a variable course and the incubation period depends upon diet, stress, and the type of organism involved. Route of infection is primarily through ingestion.

C. Signs

Characteristic signs of gastroenteritis are nausea, vomiting, diarrhea, abdominal pain, and fever.

D. Diagnosis

Diagnosis is made by the clinical signs and identification of the infective agent.

E. Postmortem Lesions

Severe enteritis and ulceration of the colon are seen. Both the liver and kidney also frequently show evidence of preceding bacterial (or viral) infection. This is most commonly characterized by varying amounts of infiltration by predominantly mononuclear type cells. To a lesser degree, this type of infiltration has also been seen in almost every organ but particularly in pancreas, gut, salivary glands and brain. It is frequently restricted to the area adjacent to blood vessels.

F. Control

The infected animals should be isolated and treated with appropriate antibiotics. Strict sanitary practices should be observed, and diet should be controlled. Change of diet and water and food of poor grade usually accompany capture and transportation. Corrective measures must be taken as soon as the shipment is received.

G. Transmissibility to Humans

The organisms can infect humans when ingested in food or water.

H. References

Ratcliffe et al. (102); Weston (125).

4. Virus Infections

B-virus and other viruses.

A. Etiology

Virus infection, B-virus, or reticuloencephalitis virus and lymphocytic choriomeningitis are some of the viruses infecting the central nervous system of monkeys.

B. Incidence

The occurrence in monkeys is not exactly known. Usually these infections are rare and vary from chronic to acute disease.

C. Signs

B-virus infection causes rigidity and incoordination in the animal. The animal becomes weak and debilitated with diarrhea, edema, nasal discharge, conjunctivitis, and a high temperature. Animals infected with lymphocytic choriomeningitis virus also have a high temperature.

D. Diagnosis

Diagnosis is made by serological determinations and also by the identification of inclusion bodies.

E. Postmortem Lesions

Perivascular and diffuse mononuclear cell infiltration, especially in the dorsolateral medulla in proximity to the V, VII, IX, and X cranial nerve components are the most commonly encountered lesions. Such lesions, when confined to this distribution, are characterized as viral radiculoencephalitis. When a wider distribution in the forebrain is observed and depending on the extent of this distribution, the diagnosis of subacute encephalomyelitis (virus type) or meningioencephalitis is applied. All gradations of these lesions to glial scarring have been encountered in various parts of the brain or spinal cord. Perivascular infiltration of a similar type but confined to the meninges is also frequently encountered in some monkeys.

The postmortem lesions of B-virus infections are meningitis, infiltration of brain with fibrinous exudate, and inclusions in meningeal vessel walls. The lesions are similar to the above in many respects but more widespread in distribution and with the admixture of a more acute cellular infiltrate in the few cases seen.

F. Control

The infected animals should be destroyed.

G. Transmissibility to Humans

B-virus is contagious to humans. Proper precautions should be taken in handling infected animals.

H. References

Rhodes and Van Rooyen (103); Baker and York (4); Weston (125); Ruch (106); Lapin and Yakolvelva (72); Coid and Laursen (20); Keeble (63).

5. Protozoan Infection

A. Etiology

Entamoeba histolytica or *Balantidium coli* may be the cause of protozoan infection in monkeys.

B. Incidence

Entamoeba histolytica is fairly common in monkeys, especially in the recently captured "old-world" primates. Flies may act as mechanical carriers and thus may spread the infection in a monkey colony. Infections from *B. coli* occur less frequently.

C. Signs

General signs of these protozoan infections are diarrhea or dysentery. Infected animals usually show few signs unless under stress. The infection may cause a wide range of signs including debility, sluggishness, anorexia, and bloody diarrhea.

D. Diagnosis

Diagnosis can be established by the examination of fecal material and isolation of cystic forms or by the identification of protozoa upon examination of fecal smears.

E. Postmortem Lesions

Not clearly defined in monkeys.

F. Control

The contamination of feed and water by feces of carrier animals should be prevented. The animal cages and room should be decontaminated. The carriers and infected animals should be isolated.

G. Transmissibility to Humans

These protozoa are transmissible to humans, causing dysentery and abscesses of the liver.

H. References

Belding (8); Cass (17); Ratcliffe et al. (102).

6. Lung Mites (Pulmonary Acariasis)

A. Etiology

Pneumonyssus simicola is the causative agent of pulmonary acariasis in monkeys.

B. Incidence

The frequency of lung mite infestation is high in monkeys.

C. Signs

No external signs are produced in many cases. In some cases, ulcers and diarrhea may be caused by this infestation.

D. Diagnosis

Diagnosis is made by the isolation of lung larvae and adult mites in lung washings. Also, the postmortem lesions are indicative of lung mite infestation.

E. Postmortem Lesions

The presence of lung mites and the irritation resulting therefrom cause the development of localized chronic granulomatous lesions, varying in number and extent. They may be misdiagnosed as tuberculosis. Nodular lesions may also be mistaken for pneumonia.

F. Control

The infected animals should be destroyed. An effective way to control lung mite infestation is not known.

G. Transmissibility to Humans

Although no case of pulmonary acariasis has been reported in humans, there is the possibility of transmission from monkey to human. The infection can be transmitted from monkey to monkey.

H. References

Weston (125); Innes et al. (57).

7. Roundworm Infestation (*Black disease*)

A. Etiology

Oesophagostomum brumptii and *O. apiostomum* are the common causes of roundworm infestations in monkeys.

B. Incidence
Roundworm infestation is fairly common in monkey colonies and as much as 30 percent of the animals may be infested.

C. Signs
Common signs are diarrhea, loss of appetite, and loss of general health condition.

D. Diagnosis
Diagnosis can be made by the identification of larvae in fecal specimens.

E. Postmortem Lesions
Lesions due to *Oesophagostomum* may be encountered almost anywhere along the gastrointestinal tract but particularly along the large intestine. Many are grossly easily recognizable as cysts attached to the peritoneal surface. Histologically as with most worm parasites, the surrounding tissues exhibit a varying degree of acute, subacute, or chronic inflammatory reaction frequently with some part of the worm being sectioned and recognizable. Usually, multiple, hemorrhagic nodules, 3–5 mm in diameter in walls of large intestine are present.

F. Control
Contamination of feed and water with infective larvae should be prevented. All bedding and equipment should be sterilized.

G. Transmissibility to Humans
Possibility of transmission from monkey to humans should be kept in mind.

H. References
Weston (125).

In addition to the parasitic infections described, monkeys may be infested with the following parasitic diseases, though much less frequently.

Myiasis, filariasis, gallbladder infections, strongyloidiasis, *Nochtia nochti* infection, cysticercosis, gongylonemiasis, echinococcosis, and so forth. Undoubtedly, the monkey is susceptible to many other parasites, which may be diagnosed with the cooperation of an expert parasitologist.

The incidence of tumors reported in primates is very low when compared with the incidence in humans and other species. However, various forms of spontaneous tumors are found in monkeys and many tumors can be induced in the experimental animals. It is beyond the scope of this chapter to describe the neoplasms of monkeys; some topics on this subject are covered in Chapter 11. For further information, the reader is referred to Ruch (106), New York Academy of Sciences (90), Lapin and Yokovleva (72), Kent and Pickering (64).

DISEASES OF SHEEP AND GOATS

Many diseases affect all ruminants including sheep and goats. Among these diseases are rinderprest, rabies, anthrax, foot-and-mouth disease, mastitis, tuberculosis, and various parasitic, viral and bacterial diseases. A brief description of some of these diseases is given in Table 3-2.

Table 3-2.   Frequently Occurring Diseases of Sheep and Goats

Disease	Etiology	Characteristics
1. INFECTIOUS ABORTION (Enzootic virus abortion)	VIRUS (Psittacosis-lymphogranu-loma group <u>Miyagwanella ovis</u>)	Premature birth of fetus which is usually dead but no specific signs.
2. SCRAPIE	PROBABLY A FILTERABLE VIRUS	A disease of the nervous system characterized by a progressive syndrome of severe pruritus, debility and locomotive incoordination. It is a highly fatal disease of older sheep between ages of 2 to 4 years.
3. SHEEP POX and GOAT POX	VIRUS	A highly infectious and frequently fatal exanthematous disease. The eruptions are prevalent on the cheeks, nostrils, lips and areas devoid of fleece. The vascular form may be hemorrhagic; it is followed by development of pustules and generalized lesions. The lesions are not as widespread in goats.
4. FOOT AND MOUTH DISEASE	VIRUS	Vesicles in the mouth, particularly on the tongue and on the thin skin around the hoofs and between the claws. These lesions cause drooling and lameness.
5. BLUE TONGUE	VIRUS	An infectious disease of sheep. Signs vary; some cases exhibit dyspnea and nasal frothing, others show mainly edema of the ears, lips, head and neck.
6. CONTAGIOUS ECTHYMA (Contagious pustular dermatitis, sore mouth)	VIRUS	An infectious dermatitis of sheep and goats, affecting primarily the lips of young animals. The condition may occur in very young lambs in the spring and rarely in mature sheep.

Table 3-2. (Continued)

Disease	Etiology	Characteristics
7. ULCERATIVE DERMATOSIS OF SHEEP (Lip and Leg Ulceration)	VIRUS	An infectious disease of sheep manifesting in two forms, one characterized by the formation of ulcers around the mouth and nose or on the legs, and the other as a venereally transmitted ulceration of the prepuce and penis or the vulva.
8. ANTHRAX (Splenic fever)	BACTERIA (Bacillus anthracis)	Sudden death without obvious illness. A general picture of septicemia is commonly observed in carcasses of animals dead of anthrax.
9. MALIGNANT EDEMA (Gas phlegmon, Gas edema)	BACTERIA (Clostridium septicum)	An acute, usually fatal toxemia of sheep and goats. General signs include anorexia, intoxication and high fever which develop rather suddenly. Local lesions are characterized by rapidly extending soft swellings which pit on pressure.
10. BLACKLEG (Black quarter, Emphysematous gangrene)	BACTERIA (Clostridium chauvoei)	An enzootic, febrile disease of sheep. Emphysematous, sero-hemorrhagic swellings in the heavy muscles.
11. BLACK DISEASE (Infectious necrotic hepatitis)	BACTERIA (C. novyi)	An acute infectious disease of sheep. The organisms multiply in the areas of liver resulting from the migration of liver flukes and produces a powerful necrotizing exotoxin.
12. THE ENTEROTOXEMIAS	BACTERIA (Clostridium perfringens)	Type A - characterized by sudden onset, hemoglobinuria and icterus. Type B - lamb dysentery Type C - Acute intoxication ("Struck") of mature sheep, causing hemorrhagic enteritis and peritonitis. Type D - An enterotoxemia of sheep and less frequently of goats. An acute afebrile disease of larger, more vigorous and fast growing lambs.

Table 3-2. (*Continued*)

Disease	Etiology	Characteristics
13. LEPTOSPIROSIS	BACTERIA (Leptospira pomona)	Similar to those in cattle.
14. TUBERCULOSIS	BACTERIA (Mycobacterium bovis)	Similar to that in cattle. It is a rare disease in sheep and goats.
15. NONSUPPURATIVE POLY-ARTHRITIS IN LAMBS	BACTERIA (Erysipelothrix insidiosa)	An acute arthritis of one or more of the diarthroidal joints usually of the limbs.
16. SALMONELLOSIS	BACTERIA (Salmonella typhimurium)	Characterized by fever, loss of appetite and diarrhea.
17. SHIPPING FEVER (Pasteurellosis, Hemorrhagic septicemia)	BACTERIA (Pasteurella species)	An acute or subacute febrile disease of sheep characterized by pneumonia or septicemia.
18. TULAREMIA	BACTERIA (Pasteurella tularensis)	Occurrence of small miliary whitish or grayish yellow foci in lymph nodes, liver, spleen, lung and kidney.
19. PARATUBERCULOSIS (Johne's Disease)	BACTERIA (Mycobacterium paratuberculo-sis)	Progressive loss of condition with diarrhea which may be intermittent.
20. VIBRIOSIS OF SHEEP	BACTERIA (Vibrio fetus)	An infectious disease characterized by abortion in late pregnancy.
21. EPIDIDYMITIS OF RAMS	BACTERIA (Brucella ovis)	A specific, infectious disease characterized by epididymitis, orchitis and impaired fertility, in the ewe by placentitis and abortion; and in the lamb by septicemia and perinatal mortality.
22. CASEOUS LYMPHADENITIS	BACTERIA (Corynebacter-ium pseudo-tuberculosis ovis)	It is a chronic disease in which signs may not be observed until several months after infection. The superficial, particularly the precrural and prescapular

Table 3-2.   (Continued)

Disease	Etiology	Characteristics
		lymph nodes usually are the primary site of the lesions. Later, the visceral nodes may be involved.
23. COCCIDIODOMYCOSIS	FUNGUS (Coccidioides immitis)	A highly infectious but not contagious disease affecting sheep. It is characterized by single or multiple pulmonary and thoracic lymph node granulomas and a tendency to disseminate to other tissues.
24. OVINE BABESIASIS (Piroplasmosis)	PROTOZOA (Babesia ovis and B. motasi)	Acute febrile onset with depression and inappetence followed soon by anemia and icterus, and hemoglobinuria in some cases. Hindquarter paresis is frequent.
25. COCCIDIOSIS	PROTOZOA (Eimeria arloingi, E. faurei and E. nina-kohl-yakimovi)	It occurs in sheep and goats 1 to 3 months old. It may be severe and sometimes fatal. The infected animal may develop sudden illness with constipation and slimy feces streaked with blood.
26. TOXOPLASMOSIS OF SHEEP	PROTOZOA (Toxoplasma gondii)	It should be suspected in sheep when there are abortions, stillbirths or premature births associated with central nervous disturbances. Extreme depression and circling movements are the most frequent signs.
27. SARCOSPORIDIOSIS	PROTOZOA (Sarcocystis tenella or S. moulei (goat))	Heavily infected animals may die. No signs until death ensues.

Metabolic Disorders of Sheep and Goats

Milkfever (hypocalcemia) is characterized by paresis passing into coma in lactating females shortly after parturition.

Grass tetany is characterized by restlessness, nervousness, unsteady gaits, muscle twitching, and convulsive fits in animals turned out on spring pasture.

Ketosis is manifested by steadily diminishing milk production with poor appetite and loss of condition a few days or weeks after parturition in high yielding animals. The rare sign is an odor of acetone in the breath, urine, and milk.

Pregnancy toxemia is characterized by weakness, loss of appetite, disturbed vision and tremors leading to prostration, coma, and death in females toward the end of pregnancy.

Nutrition Deficiency Diseases

Rickets: In young animals characterized by lameness and distortion of the long bones.

Osteodystrophia: Softening of the bones of the lower jaw in adult animals.

Iodine deficiency: Characterized by thyroid enlargement (goiter) in young animals.

Vitamin A deficiency: The fetus is born dead; scouring in young animal and night-blindness in adults.

Miscellaneous Conditions

Urinary Calculi

Loss of appetite, abdominal distension, frequent adoption of urination posture with urine dribbling from prepuce, principally in castrated males.

Conjunctivitis

Pain, reddening, and swelling of conjunctivae with watery or purulent discharge. Usually a foreign body in the eye.

Eczema

Scabby condition of skin without irritation or loss of condition.

External Parasites (Lice and Mange Mites)

Skin lesions, loss of condition, loss of hair, dry skin and severe irritation. Mange lesions are confined around the head and ears.

Internal Parasites

Tapeworms and *liver flukes* are common internal parasites of these animals.

Neoplasms

Chondroma and chondrosarcoma also mixed osseous and cartilaginous tumors, benign and malignant, occur in sheep. They arise chiefly in relation to the scapula or ribs. The malignant varieties may metastasize to the lungs. Other tumors that occasionally occur are malignant lymphoma, thymoma, carcinoma and adenoma of the liver, adenocarcinoma of the kidney and benign papilloma of the bladder.

References

UFAW Handbook (120); Merck Veterinary Manual (80); Mackenzie (75).

DISEASES OF PIGS

By using specific pathogen-free animals, most commonly occurring diseases of pigs are excluded at the outset, and experimental results are not influenced by subclinical disease. There is also less likelihood of scientists using animals which, although affected by disorders producing only mild signs, are markedly abnormal in a physiological sense. In pigs, the period of highest morbidity and mortality extends from birth to 4 months of age in an experimental colony. Usually, starvation, congenital abnormalities in the neonatal period and generalized infections and enteritis in pigs up to 4 months old result in the highest rate of mortality. Occasional cases of injury, sepsis, pneumonia, and generalized bacterial infection occur within all age groups; however, certain diseases are more common to a particular age group as described briefly in the following section.

1. Swine Erysipelas

It is an infectious disease, caused by *Erysipelothrix insidosa*. It is manifested in a variety of forms, affecting principally young swine. However, erysipelas may affect pigs of all ages. In an unvaccinated herd, it is most common in spring and autumn. Acute septicemic, subacute cutaneous, and chronic forms of the disease are recognized. The cutaneous form is characterized by raised reddened areas of skin, approximately diamond-shaped, raised body temperature, and loss of appetite. The acute septicemic form is often fatal, but subacute cases after recovery may later exhibit arthritic signs and death may result from lesions on the heart valves following endothelial damage. All pigs that are in contact should receive antiserum, vaccination should be carried out twice each year. It is virtually 100 percent effective in preventing the disease. Humans are susceptible to erysipelas infection.

2. Swine Fever (*Hog Cholera*)

It is an acute, highly infectious virus disease of swine characterized by sudden onset and a high morbidity and mortality. The disease may appear or be noticed first as scouring in a number of animals in a group. In the early stages, the pigs are dejected, body temperature may reach 106° to 107°F, and although they refuse food, the animals may be excessively thirsty. Some cases die within 2 to 3 days, while others develop oculonasal discharge, pinpoint areas of skin hemorrhage, reddish discoloration of the ears and skin of the belly, and sometimes pneumonia. Increased sensitivity, nervous excitability, and other central nervous manifestations seem to be more common. Postmortem examination may show only enlarged edematous lymph nodes and hemorrhages on the mucosa of the bladder, the typical kidney and spleen lesions being few in number, vague, or entirely absent. Chronic cases which linger for several weeks have also been observed. The animals should be immunized against hog cholera with an attenuated live virus vaccine preparation as a prophylactic measure.

3. Diarrhea in Young Pigs

Diarrhea is a sign of several diseases both infectious and nutritional which may affect young pigs. In addition, primary diarrhea of more or less sporadic nature is relatively common in newborn pigs. Alteration in the quantity or quality of sow's milk or poor sanitation in the pen are probably the principal causative factors involved. Coliform infection, while less common than in newborn calves, may cause heavy death loss in some herds. Enteritis associated with *E. coli* commonly occurs 7 to 14 days after weaning, irrespective of the age at which weaning is carried out. In most sporadic outbreaks, partial loss of appetite and various degrees of diarrhea are the only signs. There are few deaths. Infectious diarrhea due to coliforms responds well to the oral administration of antibiotics. Vaccination of the sows before farrowing with an autogenous bacteria is valuable.

4. Salmonellosis

Scouring is the cardinal feature of *Salmonella* enteritis, followed by chronic incontinence due to progressive necrosis of the bowel wall. Bacteriological examination is necessary to detect the affected animals to eliminate carriers and to prevent the spread of infection. Humans are susceptible to *Salmonella* infections.

5. Transmissible Gastroenteritis

A rapidly spreading viral disease that affects pigs of all ages, it is usually fatal in the newborn and less severe in adult pigs. The disease is characterized by signs of inflammation of the stomach and intestine. Many cases occur in which there is diarrhea in the absence of gross inflammatory changes in the gastrointestinal tract. The disease is accompanied by marked scouring in all cases, by vomiting in some cases, and by high mortality in pigs less than 7 to 10 days of age. The great majority of outbreaks occur in the late winter and early spring. The gross pathological changes are found in the gastrointestinal tract. The most constant finding is a distended intestine which is filled with liquid contents. The stomach is markedly congested. The kidneys usually show gross evidence of degenerative changes and frequently contain urates. As a rule, the renal medulla is congested. The infected animals should be isolated. Should an outbreak occur, pregnant sows should be mixed with the infected animals in the hope that by contacting the disease and developing immunity before farrowing, the sow and litter will not be infected.

6. Swine Pox

An acute infectious disease characterized mainly by skin lesions that resemble those of pox. It occurs most frequently in pigs at the late preweaning stage. The lesions, initially raised red areas on the skin of the belly, thighs, and flanks, progress to scabby patches, sometimes fairly extensive, and occasionally death occurs. Isolation of the sow and the infected litter is advisable.

7. Vesicular Exanthema

An acute highly infectious disease of swine characterized by the formation of vesicles on the snout and mucous membranes of the mouth, on the feet between the toes and on the soles, the coronary band, and the dew claws. It is clinically indistinguishable from foot-and-mouth disease. The animal may show a fever of 104° to 107°F and some reluctance to eat. The primary lesions usually heal in about 5 to 7 days but the development of complications even in mild outbreaks may delay resolution by 2 to 3 weeks. The ruptured vesicles may become infected with pyogenic bacteria and secondary infection may become more serious than the primary disease. Badly infected feet, pneumonia, and septicemia are complications sometimes observed. Control of the disease is based on segregation and quarantine.

8. Teschen Disease (*Porcine Encephalomyelitis*)

An infection caused by a filterable virus producing severe pathological changes in the nervous system of swine. The virus may be found in the brain and spinal cord of infected pigs. It may appear transiently in the blood but rarely is identifiable in other organs or excretions. It destroys the nerve cells of the brain and spinal cord. The degeneration of neurons may be seen in various stages terminating in necrosis of these cells and followed by neuronophagia, proliferation of glial cells, and vascular congestion. The accumulation of the lesions through the brain and cord is extensive and quite characteristic. The disease resembles human poliomyelitis in that the motor neurons of the ventral gray columns of the cord and cerebral motor cortex are affected but the lesions are much more widespread in Teschen disease. The disease is manifested as anorexia, depression, lassitude, slight incoordination of limbs and fever (104° to 105°F). Within a few hours a stage of irritability ensues, followed by stiffness of the extremities, stumbling, falling, tremors, nystagmus, violent convulsions, prostration, and coma. The temperature may fall rapidly, terminating in death only a few days after onset. No effective treatment is available for the disease.

9. Virus Pneumonia of Pigs (VPP)

Infectious pneumonia of pigs is a chronic respiratory disease characterized mainly by depressed growth and a persistent dry cough. The etiologic agent has the characteristics of the psittacosis-lymphogranuloma group. The virus may remain in the lungs and be capable of exciting an infection 6 months or more after a pig is first infected. The lung lesions of VPP can not be distinguished grossly from those caused by swine influenza (see next section). Diarrhea usually occurs when the affected pigs first begin to cough. It lasts only a few days whereas the coughing persists. Temperature is only slightly elevated. Control by eradication of infection is possible.

10. Swine Influenza (Hog Flu)

It is an acute, highly contagious, febrile disease primarily of the respiratory organs, occurring usually in the fall and winter. The typical disease is known to be caused by the combined action of a virus and a bacterium called *Haemophilus suis*. The lesions are usually confined to the chest cavity. Some cases show a hyperemia of the gastric mucosa. The spleen may be moderately enlarged. The lung lesions are quite characteristic and consist of an atelectasis which involves all or portions of the anterior lobes and sometimes the anterior portions of the diaphragmatic lobes. Affected portions of the lungs are dark purplish red in color, while the nonaffected portions are pale and usually emphysematous. In fatal cases, the nonpneumonic portions of lungs are bloody, edematous and heavy. Ordinarily, the trachea and bronchi contain a thick, tenacious, mucopurulent exudate in moderate to copious amounts and there is sometimes a serous or serofibrinous pleuritis. The bronchial and medistanal lymph nodes are swollen and edematous. Among the common signs are listlessness, fever (104°–107°F), anorexia, and marked weakness and prostration. Prominent signs at the height of disease are coughing, a jerky type of respiration ("thumps"), and a mucus discharge from the eyes and nose. The course of disease varies from 3 to 7 days, with recovery of the herd occurring almost as suddenly as the onset. The most effective control measure is to provide dry, well-bedded, relatively warm pens that are free from drafts and supplied with clean, fresh drinking water.

11. Leptospirosis

A febrile disease caused by certain species of leptospirae (*L. pomona*). Swine act as reservoirs of infection for other animals and humans because apparently healthy animals can excrete large numbers of organisms in their urine. Leptospiral infection in swine is possible where rats are abundant or where they are able to contaminate food or water. Vermin control is the essential preventive measure and this is important since human infection is often fatal despite treatment. The organisms can gain entry through unnoticed fissures in the skin. Normally more than one member of a litter is affected and a febrile illness results with jaundice appearing 3 to 4 days after the infection which is usually fatal. Treatment involves dosing with the appropriate antibiotic.

12. Listeriosis (*Circling Disease*)

Listeriosis in pigs is important because humans are susceptible to the causative agent, *Listeria monocytogenes,* through the animal contact. Despite the low invasiveness of the organisms, all suspected material should be handled with caution. The lesions are associated with central nervous system in pigs, can cause a fatal meningitis or meningoencephalitis in humans, and have been isolated from premature infants.

13. Parasitic Diseases

Ascarid infection predispose pigs to pneumonia due to larval forms of *A. lumbricoides* migrating through the lungs. Jaundice and enteritis may result from migrations in the digestive tract. The examination of feces for worms and worm eggs assists in the diagnosis. Elimination of worms from the herding stock is the best method of control.

Swine are hosts to at least six species of coccidia, but only *Eimeria scabra* and *E. debliecki* appear to be of pathogenic significance.

Toxoplasma gondii causes clinical illness and death in newborn pigs but in sows it is usually a subclinical infection. Pregnant, apparently healthy sows carry *Toxoplasma* and transmit the infection to their pigs in utero. The occurrence of weakness, tremors, incoordination, coughing, and pyrexia in newborn pigs suggests toxoplasmosis but should be confirmed by serologic tests.

14. Neoplasms

Neoplasms are rare in swine, probably because the great majority are slaughtered at 6 to 8 months of age. The most common tumor in swine is an embryonal growth in the kidney called embryonal nephroma (embryonal adenocarcinoma). It usually remains localized; occasionally metastases are found chiefly in the lungs. Other tumors that occur occasionally are malignant lymphomas, malignant melanomas (in white hogs), and benign papillomas of the tongue, esophagus, and bladder. Hemangiomas of the skin are also fairly common. They occur as multiple, small elevated, reddish tumors in adults of either sex and are not fatal.

References
Dunne (29); Field (32); Mount and Ingram (87); Bustad and McClellan (16).

DISEASES OF CATTLE

Experimental cattle are liable to develop any of the diseases usually associated with those living under ordinary farm conditions. However, because of their isolation from other animals, the occurrence of many diseases are prevented. Infectious or highly contagious diseases (e.g. tuberculosis, foot-and-mouth, contagious bovine pleuropneumonia, cattle plague, anthrax, etc.) may not affect the cattle under laboratory conditions, although the possibility of their occurrence should be borne in mind. Among the more commonly occurring diseases of cattle under laboratory conditions are the problems associated with the mammary gland, the genital tract, the alimentary tract, and the feet and legs of the dairy cow. Mastitis due to staphylococcal infections, milkfever, abortions, tympany and impaction, cetonemia, traumatic injury of legs and feet, and ectoparasites are some of the examples of these diseases that may occur in experimental cattle. It is beyond the scope of this chapter to describe separately all the diseases occurring in cattle; a list of commonly occurring diseases are given in the Table 3-3. For further details on this subject the reader is referred to The Merck Veterinary Manual (80), the UFAW Handbook (120); and Doyle et al. (27).

Table 3-3. Frequently Occurring Diseases of Cattle

Disease	Etiology	Characteristics
1. PSEUDORABIES DISEASE (Aujesky's, Mad Itch)	VIRUS	Localized pruritis.
2. COWPOX (Variola vaccinia)	VIRUS	A mild eruptive disease of lactating cows usually restricted to the skin of the udder and teats.
3. INFECTIOUS PAPILLOMAS (Warts)	PROBABLY VIRUS	Benign tumors consisting of fibrous cores covered to a variable depth with squamous epithelium the outer layers of which become hyperkeratinized. Usually found on the external body surface, esophagus and omasum.
4. VESICULAR STOMATITIS	VIRUS	Vesicular eruption principally on the mucosa of the mouth and sometimes on the skin of the feet.
5. FOOT AND MOUTH DISEASE (Aphthous fever, Epizootic aphthae)	VIRUS	Eruption of vesicles of variable size in the mouth and on the feet.
6. RINDERPREST (Cattle plague)	VIRUS	Inflammation, hemorrhage, erosion and necrosis of the mucous membranes of the digestive tract with diarrhea.
7. THE VIRUS DIARRHEA-MUCOSAL DISEASE COMPLEX	VIRUS	A febrile reaction, nasal discharge and diarrhea, hyperemia and erosions of and alterations of alimentary tract mucosa from the lips to anus.
8. MALIGNANT CATARRHAL FEVER (Gangrenous coryza, Snotsiekti)	VIRUS	Encephalitis, monophasic temperature rise and signs which vary greatly in degree and extent. Signs associated with lesions of the mucous membranes of the head and gastrointestinal tract usually predominate.

Table 3-3. (Continued)

Disease	Etiology	Characteristics
9. INFECTIOUS BOVINE RHINOTRACHEITIS (IBR)	VIRUS	Acute febrile disease characterized by inflammation of the upper respiratory tract leading to coughing and a profuse nasal discharge. Pneumonic signs are followed by involvement of the gastrointestinal tract.
10. INFECTIOUS PUSTULAR VULVOVAGINITIS (IPV)	VIRUS	Acute contagious disease, inflammation, necrosis and pustule formation in the vulva and vagina and in males by similar lesions on the skin of the penis and prepuce.
11. EPHEMERAL FEVER (Three day sickness)	VIRUS	A benign viral disease, characterized by fever, stiffness and lameness. Spontaneous recovery within a few days.
12. LUMPY SKIN DISEASE	VIRUS	An infectious, eruptive pox-like disease of cattle characterized by the appearance of nodules on the skin and other parts of the body. Secondary infections often aggravate the condition.
13. SPORADIC BOVINE ENCEPHALITIS (Buss disease)	VIRUS OF PSITTACOSIS-LYMPHOGRANULOMA VENERUM GROUP	An acute infectious disease of cattle which in severe and fatal cases involves the serous surfaces in addition to the CNS.
14. ANAPLASMOSIS (Gallsickness)	ANAPLASMATA	An acute or chronic infectious disease, characterized by the presence on erythrocytes of a body referred to as Anaplasma marginale.
15. HEARTWATER	RICKETTSIA (Cowdria ruminatum)	A tickborne, septicemic, rickettsial disease, characterized by hyperthermia and nervous signs.
16. RIFT VALLEY FEVER	VIRUS	An acute disease characterized by a short incubation period, a high mortality in calves, abortion and liver lesions.
17. ANTHRAX (Splenic fever, Charbon, Milzbrand)	BACTERIA (Bacillus anthracis)	Acute, infectious, febrile disease. Essentially a septicemia characterized by its rapidly fatal course.

Table 3-3. (Continued)

Disease	Etiology	Characteristics
18. BOTULISM (Lamzietke, Lion Disease)	BACTERIA (Clostridium botulinum)	Food poisoning caused by ingestion of powerful toxin which affects nervous system (motor paralysis).
19. MALIGNANT EDEMA (Gas phlegmon, Gas edema)	BACTERIA (Clostridium septicum)	Acute, usually fatal, toxemia characterized by rapidly extending soft swellings which pit on pressure.
20. BLACKLEG (Black quarter, Emphysematous gangrene, symptomatic anthrax)	BACTERIA (Clostridium chauvoei)	An enzootic, acute, febrile disease characterized by emphysematous, serohemorrhagic swellings in the heavy muscles.
21. BACILLARY HEMOGLOBINURIA	BACTERIA (Clostridium hemolyticum)	An acute, infectious disease; the organisms grow in liver, giving rise to a characteristic infarct and the production of a hemolytic toxin.
22. LEPTOSPIROSIS (Redwater of calves, Asymptomatic abortion, Leptospiral mastitis)	BACTERIA (Leptospira pomona, L. canicola and L. icterohaemorrhagiae)	Acute hemolytic form characterized by anemia and icterus. The urine is a clear red or port wine color.
23. BRUCELLOSIS (Bang's Disease, Contagious abortion)	BACTERIA (Brucella abortus)	Localization of organisms in the uterus and placenta and the resulting death and abortion of fetus.
24. TUBERCULOSIS	BACTERIA (Mycobacterium bovis)	Weakness, anorexia, emaciation, low grade fever, cough.
25. PARATYPHOID INFECTION (Salmonellosis)	BACTERIA (Salmonella typhimurium, S. enteritidis and S. dublin)	Severe enteritis and septicemia in calves.
26. SHIPPING FEVER (Pasteurellosis, Hemorrhagic septicemia, Rinderseuche)	BACTERIA (Pasteurella sp.)	An acute or subacute disease characterized by pneumonia or septicemia.

Table 3-3. (Continued)

Disease	Etiology	Characteristics
27. ACTINOBACILLOSIS	BACTERIA (Actinobacillus lignieresii)	Hard nodular tumor which develops slowly on the lower jaw and neck.
28. PARATUBERCULOSIS (Johne's Disease)	BACTERIA (Mycobacterium paratuberculo- sis)	A chronic infectious disease characterized by thickening of intestinal wall and a recurrent fetid diarrhea that may persist for months causing a gradual loss of flesh.
29. BOVINE WINTER DYSENTERY (Winter Scours)	BACTERIA (Vibrio jejuni; perhaps a virus involved)	An acute infectious, enzootic disease of stabled cattle occurring between the months of November and March. Very high morbidity rate but an exception- ally low mortality rate. It leads to dehydration, loss of weight and condition and sharp fall in milk production in lactating cows.
30. CONTAGIOUS BOVINE PLEUROPNEUMONIA	BACTERIA (Mycoplasma mycoides)	A highly contagious pneumonia generally accompanied by pleurisy. Rare occurrence in the U.S.A.
31. VIBRIOSIS (Epizootic bovine infertility)	BACTERIA (Vibrio fetus)	A venereal disease characterized by infertility and early embryon- ic death. Abortion occurs in a relatively small percentage of infected cows.
32. ACTINOMYCOSIS (Lumpy Jaw)	FUNGUS (Actinomyces bovis)	A local or systemic chronic suppurative and granulomatous disease. It affects mandible, maxilla or other bony tissue of the head.
33. NOCARDIOSIS	FUNGUS (Nocardia asteroides)	Generalized purulogranulomatous nodular lesions; clinical signs include fever, soreness, lame- ness, dyspnea, empyema, enlarged abdomen, lymphadenitis and fluctuating subcutaneous or salivary gland abscesses.

Table 3-3. (Continued)

Disease	Etiology	Characteristics
34. CRYPTOCOCCOSIS	FUNGUS (Cryptococcus neoformans)	A subacute or chronic disease characterized by decreased milk production, anorexia, severe swelling and firmness of the udder and enlarged lymph nodes.
35. BOVINE TRICHOMONIASIS	PROTOZOA (Trichomonas foetus)	A contagious, venereal disease characterized by sterility, pyometra and abortion.
36. CATTLE TICK FEVER (Bovine piroplasmosis, Tick fever, Texas fever, splenetic fever, Redwater, Redwater fever, Southern fever)	PROTOZOA (Babesia bigemina)	Acute onset, characterized by temperature rise to 106 to 107°F, inappetence and marked depression. Anemia and hemoglobinuria are frequent.
37. EAST COAST FEVER (Theileriasis, Coastal fever)	PROTOZOA (Theileria parva)	An acute disease characterized by high fever, swelling of lymph nodes, emaciation and high mortality.
38. COCCIDIOSIS	PROTOZOA (Eimeria zurnii, E. bovis, E. ellipsoidalis)	An acute or chronic intestinal infection caused by the invasion and destruction of the intestinal mucosa. It is characterized by diarrhea, hemorrhage and emaciation.
39. TOXOPLASMOSIS	PROTOZOA (Toxoplasma gondii)	Hyperexcitability or extreme depression, ataxia, muscular tremors. Coughing, nasal discharge, dyspnea, fever and prostration are also characteristic signs.
40. SARCOSPORIDIOSIS	PROTOZOA (Sarcocystis cruzi)	A disease caused by the invasion of the skeletal musculature by the organism.

B. OTHER DISEASES

Disease	Etiology	Characteristics
41. CANCER EYE (Ocular squamous carcinoma)	HEREDITARY -- UNKNOWN ETIOLOGY	Small cutaneous horny growths (papillomatous growths) or "Wickers" on the skin near the lid margins of older cattle.

Table 3-3. (*Continued*)

Disease	Etiology	Characteristics
42. BOVINE MALIGNANT LYMPHOMA (Lymphoblastoma, Lymphosarcoma, Lymphocytoma, Lymphatic leukemia, Pseudoleukemia)	UNKNOWN	A progressively fatal disease characterized by neoplasia of lymphoid elements of the hemo-poietic system with involvement of various internal organs. A transitory elevation of white blood cell count.
43. DISPLACEMENT OF THE ABOMASUM	HEAVY FEEDING OF CONCENTRATES, AN ABDOMEN OF LARGE CAPACITY, LATE PREGNANCY OR PARTURITION AND DAIRY TYPE CONFORMATION	A common disease of mature, high-producing dairy cows in which abomasum become displaced from its normal position on the right anterior abdominal floor.

DISEASES OF HORSES

There are a wide variety of diseases of horses that may affect the animals under natural farm conditions as well as under laboratory conditions. Some of these diseases are important as public health concerns, for example, epizootic lymphangitis, glanders, anthrax, parasitic mange, rabies, and so forth, because they are transmissible to humans. However, it is very rare for a horse to be affected with any of these at present. Literature on the diseases of horses is too vast to cover in this chapter. In general, when a horse under experimental conditions shows the following signs, it is advisable to consult a veterinary surgeon for specific diagnosis and treatment of the animal.

1. Refusing to eat or drink for 24 hours; possibly coughing repeatedly.

2. A rise in temperature from normal (100.5°F) to above 103°F and persisting for 12 hours; exhibiting signs of shivering or muscular tremors.

3. Sweating for no obvious reasons; increase in respiration rate from the normal (12 to 20) up to 30 or more per minute.

4. Standing and looking anxious and worried, refusing to move or doing so only with great difficulty; frequently lying down and rolling with or without grunting or groaning, thrashing about on the floor, rising to its feet, then lying down and rolling again.

5. Severe injury with obvious damage to skin and muscles, is lame on one or more limbs or is unable to place any weight on one leg.

6. Scouring without any apparent reason; diarrhea continues for 12 hours or more.

A few selected infectious diseases which commonly occur in horses are listed in Table 3-4.

Table 3-4. Frequently Occurring Diseases of Horses

Disease	Etiology	Characteristics
1. HORSE POX (Contagious Pustular Stomatitis)	VIRUS	A highly contagious pustular disease. Not reported in the U.S.A.
2. THE EQUINE RESPIRATORY DISEASE COMPLEX (Influenza)	VIRUS - primary BACTERIA - secondary	Respiratory distress
A. EQUINE RHINO-PNEUMONITIS (equine virus abortion)	VIRUS	A highly contagious virus disease producing rhino-pneumonitis in young animals and abortion in mares.
B. EQUINE VIRAL ARTERITIS	VIRUS	An acute contagious disease characterized by fever and other signs of respiratory and digestive tract disorders, and also ventral edema.
C. EQUINE INFLUENZA	INFLUENZA VIRUS (A56)	Respiratory catarrh and fever with a rise in temperature of 1 to 4°F for a period of 2 to 5 days.
3. EQUINE ENCEPHALOMYELITIS (Equine encephalitis, Virus encephalitis, Arthopod borne encephalitis)	VIRUS	An acute viral disease of horses and mules characterized by central nervous disturbances and high mortality. Transmissible to man.
4. EQUINE INFECTIOUS ANEMIA (Swamp fever)	VIRUS (Equine infectious Anemia virus EIA)	An acute or chronic virus disease characterized by intermittent fever, depression, progressive weakness, loss of weight, edema and anemia of a progressive or transitory type.
5. AFRICAN HORSE SICKNESS	VIRUS	Widespread edema and serous effusions. It is transmitted by biting flies.

Table 3-4. (Continued)

Disease	Etiology	Characteristics
6. VESICULAR STOMATITIS	VIRUS	Raised vesicular eruptions principally on the mucosa of the mouth and sometimes on the skin of the feet. Transmissible to man.
7. MALIGNANT EDEMA (Gas phlegmon, Gas edema)	BACTERIA (Clostridium septicum)	Anorexia, intoxication, and sudden onset of high fever. Local lesions are character-ized by rapidly extending, soft swellings which pit on pressure.
8. LEPTOSPIROSIS	BACTERIA (Leptospira pomona)	An elevation of body tempera-ture to 103.5°F to 105°F for 2 or 3 days, depression or dullness, anorexia, icterus and neutrophilia.
9. PARATYPHOID INFECTION (Salmonellosis)	BACTERIA (Salmonella abortivo-equina and S. typhimurium)	Salmonellosis in young foals is highly fatal and in mares it causes infectious abortion.
10. GLANDERS (Farcy)	BACTERIA (Actinobacillus mallei)	Formation of nodules or tuber-cules which tend to break down and form ulcers. The upper respiratory tract, skin and lungs are affected. Man is susceptible.
11. STRANGLES ("Shipping fever")	BACTERIA (Streptococcus equi)	An acute contagious disease characterized by a mucopurulent inflammation of the nasal and pharyngeal mucosa and abscess formation of the regional lymph nodes and occasionally other parts of the body.
12. CORYNEBACTERIUM INFECTION	BACTERIA (C. equi)	Purulent pneumonia in foals, no lesions in the lymph nodes of the head.
13. SALMONELLA ABORTIVO-EQUINA ABORTION	BACTERIA (S. abortivo-equina)	Prior to abortion, the mare may show a febrile reaction possibly from transient septicemia. There may be vaginal discharge following abortion. Abortion occurs during late pregnancies (8th, 9th or 10th month).

Table 3-4. (Continued)

Disease	Etiology	Characteristics
14. VISCOSUM INFECTION (Shigellosis)	BACTERIA (Actinobacillus equuli)	An acute septicemic infection usually of young foals characterized by sudden onset, a rapid course and a high mortality.
15. ULCERATIVE LYMPHANGITIS	BACTERIA (Corynebacterium pseudotuberculosis)	Resembles the cutaneous form of glanders. It is characterized by nodules, ulcers and inflammation of lymph vessels of the fetlock and leg.
16. COCCIDIODOMYCOSIS	FUNGUS (Coccidiodes immitis)	A highly infectious but noncontagious disease characterized primarily by single or multiple pulmonary and thoracic lymph node granulomas and tendency to disseminate to other tissues. Man is susceptible.
17. NORTH AMERICAN BLASTOMYCOSIS	FUNGUS (Blastomyces dermatitidis)	A chronic disease characterized by granulomas, abscesses and ulceration in the lungs, skin and other organs.
18. CRYPTOCOCCOSIS (Epizootic lymphangitis)	FUNGUS (Cryptococcus neoformans)	A chronic nodular and suppurative disease of the skin, superficial lymph vessels, lymph nodes and mucous membranes. Respiratory signs and nasal discharge result from nasal granulomas. Granulomatous foci and liquefaction occur in lungs or may be generalized in viscera.
19. SPOROTRICHOSIS	FUNGUS (Sporotrichum schenckii)	Nodules and ulcers in the skin and sometimes involving internal organs.
20. DOURINE (Mal de Coit)	PROTOZOA (Trypanosoma equiperdum)	A subacute or chronic contagious disease of equine species characterized by edema and cutaneous eruption of the external genitalia, by edematous plaques on the skin of other parts of the body and subsequently by nervous manifestation.
21. TOXOPLASMOSIS	PROTOZOA (Toxoplasma gondii)	Fever, depression, anorexia and dyspnea.

ZOONOSES

The principle characteristic of zoonoses is that they attack both humans and laboratory animals; the transmission can be either way—animal to human or human to animal. The diseases of laboratory animals most hazardous to animal handlers in the United States are those transmitted by direct contact, especially during handling. The closer the association with sick animals, the more likely the spread of disease. This is especially true for external infections such as ringworm. Table 3-5 lists some of the animal diseases that are most hazardous to humans.

There are certain human diseases that may be transmitted to laboratory animals and thus may infect those working in laboratory animals (Table 3-6). Table 3-7 lists certain frequently occurring diseases of laboratory animals that are not reported to be transmissible to humans.

Control of infection is more difficult to achieve when dealing with pathogens which can survive in both humans and animals, for example, tuberculosis, *Pseudomonas* infection, dermatomycoses and bacterial skin disorders, and certain parasites. The incidence of human zoonoses is rather low among the staff handling laboratory animals in this country due to effective laboratory animal disease control programs, such as laboratory animal disease diagnostic laboratories, institutional safety practices, and personnel training programs, and the like. However, there is a potential danger of zoonoses in dealing with unknown stocks of laboratory animals and many new potential laboratory animals (especially wild animals).

Table 3-5.   Laboratory Animal Diseases Most Hazardous to Humans

Disease	Etiology	Animal Species
Lymphocytic choriomeningitis	Virus	Mice
Ringworm	Fungus	Rats, Cats, Dogs
Leptospirosis	Bacterial	Rats
Bite infections	Bacterial	Rats, Cats, Monkeys, Chimpanzees
Anthrax	Bacterial	Cattle, Sheep, Pigs
Pneumonia	Bacterial	Guinea pigs
Listeriosis	Bacterial	Guinea pigs, Rabbits
Pasteurellosis	Bacterial	Rabbits
Cat scratch fever	Bacterial	Cats
Rabies	Virus	Dogs
Shigella	Bacterial	Monkeys
Salmonellosis	Bacterial	Pigs, Other animal species
Tuberculosis	Bacterial	Monkeys, Chimpanzees
Swine erysipelas	Bacterial	Pigs
B virus	Virus	Monkeys
Infectious hepatitis	Virus	Chimpanzees

Table 3-6. Certain Diseases of Humans Transmissible to Laboratory Animals

Disease	Animal Species
Lymphocytic choriomeningitis	Mouse
Streptococcal infections	Mouse
Pasteurella infections	Rabbit
Staphylococcal infections	Cats and Dogs
Tuberculosis	Chimpanzees
Measles	Chimpanzees
Infectious hepatitis	Chimpanzees
Poliomyelitis	Chimpanzees and Orangutan

Table 3-7. Very Commonly Occurring Diseases of Laboratory Animals That Are Not Transmissible to Humans

Animal Species	Diseases	Animal Species	Diseases
Mice	Infantile diarrhea Tyzzer's disease Pseudomonas infections Hepatitis Pin worms Lice and Mite infestations Tumors	Rabbits	Snuffles Rabbit pox Rabbit Tyzzer's disease Tumors
Rats	Otitis media infections Ringtail Bartonellosis (Hemobartonella) Nephritis Tumors	Monkeys	Ascariasis Oesophagostomiasis Rickets Scurvy Pediculosis Diplococcus pneumoniae infec Tumors
Guinea pigs	Pregnancy toxemia Scurvy Lymphadenitis Tumors	Dogs	Canine distemper Canine hepatitis Tumors
Hamsters	Ileitis Salmonellosis Wet-tail Tumors	Pigs	Swine fever (Hog cholera) Swine pox Vesicular exanthema Teschen disease (Porcine encephalomyelitis) Virus Pneumonia of Pigs (VPP) Swine influenza (Hog flu) Tumors
Rabbits	Coccidiosis		

REFERENCES

1. Ackermann, R., H. Bleedhorn, B. Kupper, I. Winkens, and W. Sheid, *Zbl. Bakt.*, **194:**407 (1964).
2. Adams, D. K., D. F. Cappell, and J. A. W. McCluskie, *J. Pathol. Bacteriol.*, **31:**157 (1928).
3. Alexander, M. M., P. B. Sawin, and D. A. Roehm, *J. Infect. Dis.*, **90:**30 (1952).
4. Baker, J. A., and J. C. York, *Adv. Vet. Sci.*, **1:**65 (1953).
5. Baker, R. K., *Cancer Res.*, **13:**137 (1953).
6. Bartawidjaja, R. K., L. P. Morrissey, and N. A. Labzoffsky, *Arch. Virusforsch*, **17:**273 (1965).
7. Becker, E. R., *Coccidia and Coccidiosis of Domesticated Animals and Man*, Iowa State Press, Ames, Iowa, 1934.
8. Belding, D. L., *Textbook of Clinical Parastology*, 2nd ed., Appleton-Century-Crofts, Inc., New York, 1952.
9. Benson, R. E., B. D. Fremming, and R. J. Young, *Am. Rev. Tuberc.*, **72:**204, (1955).
10. Biester, H. E., and L. H. Schwarte, *Diseases of Poultry*, Iowa State College Press, Ames, Iowa, 1959.
11. Bjotvdt, G., and J. Tufts, *Manual for Laboratory Animal Care*, Ralston Purina Co., 1963.
12. Blanc, G., J. Bruneau, B. Delage, and R. Poitrot, *Bull. Acad. Nat. Med. Par.*, **35:**255 (1951).
13. Bohme, D. H., H. A. Schneider, and J. M. Lee, *J. Exp. Med.*, **110:**9 (1959).
14. Brumley, D. V., *A Textbook of the Diseases of the Small Domestic Animals*, 4th ed., Lea & Febiger, Philadelphia, 1943.
15. Bruner, D. W., and P. R. Edwards, *Salmonellosis of Domestic Animals*, Proceedings 50th Annual Meeting U.S. Livestock Sanitary Assoc., December 1946, p. 194.
16. Bustad, L. K., and R. O. McClellan, *Swine in Biomedical Research*, Battle Memorial Institute, 1966.
17. Cass, J. S., *Enteric Infections in the Monkey*, Proceedings of the 3rd Annual Meeting of the Animal Care Panel, December 1952, Chicago, p. 18.
18. Chesterman, F. C., and A. Pomerance, *Med. T. Geneesk*, **108:**49 (1964).
19. Chute, R. N., H. B. Kenton, and S. C. Sommers, *Am. J. Clin. Pathol.*, **24:**223, (1954).
20. Coid, C. R., and A. C. Laursen, in W. Lane-Petter (ed.) *Animals for Research*, Academic Press, London and New York, 1963, p. 437.
21. Cole, C. R., *Toxoplasmosis in Animals and Man*, Proceedings of the Public Health Veterinary Meeting, June 1953, U.S. Public Health Service Communicable Disease Center, Atlanta, Georgia, p. 56.
22. Cotchin, E., and F. J. C. Roe, *Pathology of Laboratory Rats and Mice*, F. A. Davis Co., Philadelphia, 1967.
23. Craigie, J., *Proc. Roy. Soc. B.*, **165:**35 (1966).
24. Dam, H., and F. Christensen, *Acta Pathol. Microbiol. Scand.*, **30:**236, *Vet. Bull.*, **23:**173 (1952).
25. Davies, S. F. M., L. P. Joyner, and S. B. Kendall, *Coccidiosis*, Oliver & Boyd, Edinburgh, 1963.
26. Dontenwill, W., H. Ranz, and V. Mohr, *Beitr. Pathol. Anat.*, **122:**390, (July 1960).
27. Doyle, R. E., S. Garb, L. E. Davis, D. K. Meyer, and F. W. Clayton, *Ann. N.Y. Acad. Sci.*, **147:**129 (1968).
28. Dumas, J., *Les Animaux de Laboratoire*, Editions Medicales Flammarion, Paris, 1953.
29. Dunne, H. W., *Diseases of Swine*, 3rd ed., Iowa State University Press, Ames, Iowa, 1970.
30. Farris, E. J., and J. O. Griffith, *The Rat in Laboratory Investigation*, 2nd ed., J. B. Lippincot, Philadelphia, 1949.
31. Fauve, R. M., C. Peirce, and R. Dubos, *J. Exp. Med.*, **120:**283 (1964).
32. Field, H. I., *Ministry of Agri, Fisheries and Food Bull.*, H.M.S.O., London, (1964).
33. Flynn, R. J., *Mouse Mange*, Proceedings of the 5th Annual Meeting of the Animal Care Panel, December 1954, Chicago.
34. Follis, R. H., Jr., *Nature (London)*, **183:**1817 (June 27, 1959).
35. Fortner, J. G., *Surgery*, **36:**932 (November 1954).

36. Frenkel, J. K., P. Rasmussen, and O. D. Smith, *Abst. Fed. Proc.,* **18 (I):**477 (March 1959).
37. Fujiwara, K., Y. Takagaki, K. Maejima, K. Kato, M. Naiki, and Y. Tajima, *Japan. J. Exp. Med.,* **33:**183 (1963).
38. Fujiwara, K., S. Ukuda, Y. Takagaki, and Y. Tajima, *Japan. J. Exp. Med.,* **33:**203 (1963).
39. Fujiwara, K., K. Maejima, Y. Takagaki, M. Naiki, Y. Tajima, and R. Takahashi, *C.R.S. Biol.,* **158:**407 (1964).
40. Gledhill, A. W., *Vet Rec.,* **64:**723 (1952).
41. Gladhill, A. W., in R. J. C. Harris (ed.), *The Problems of Laboratory Animal Disease,* Academic Press, London and New York, 1962.
42. Gordon, R. F., and Horton-Smith, in the UFAW Handbook, *The Care and Management of Laboratory Animals,* UFAW, London, 1957, p. 691.
43. Gray, J. E., *Am. J. Vet. Res.,* **24:**1044 (1963).
44. Green, M. N., G. Yerganian, and H. J. Gagnon, *Nature* **197:**396 (1963).
45. Green, I. J., S. H. Hsu, and H. M. Yu, *J. Formosan Med. Ass.,* **62:**41 (Jan. 1963).
46. Green, M. N. G. Yerganian and H. Meir, *Experientia,* **16:**503 (1960).
47. Griffin, C. A., *Proc. of the Third Ann. Meeting of the Anim. Care Panel,* Chicago (1952), p. 3.
48. Haberman, R. J., *Proc. of the Second Ann. Meeting of the Anim. Care Panel* (1951), p. 36.
49. Haberman, R. T., and F. P. Williams, *J. Nat. Cancer Inst.,* **20:**933 (1958).
50. Habermann, R. T., *Public Health Rep.,* **74:**165 (Feb. 1959), *Abst. 325 Fed. Proc.* **19(Suppl 6):**73 (1960).
51. Handler, A. H., in W. E. Ribelin and J. R. McCoy (eds.), *Pathology of Laboratory Animals,* Charles C Thomas, Springfield, Ill., 1965, p. 210.
52. Handler, A. H., and F. C. Chesterman, in R. A. Hoffman, P. F. Robinson and H. Magalhaes (eds.), *The Golden Hamster: Its Biology and Use in Medical Research,* Iowa State University Press, Ames, Iowa, 1968.
53. Herrold, K. M., and L. J. Dunham, *Cancer Res.,* **23:**773 (1973).
54. Homburger, F., C. Wilcox, M. W. Barnes, and M. Finland, *Science,* **102:**449 (1945).
55. Homburger, F., C. W. Nixon, and J. Harrop, *Abst. Fed. Proc.,* **20:**195 (1963), *Biol. Abst.,* **43:**14082.
56. Hoskins, H. P., J. V. LaCroix, and Karl Mayer (eds.), *Am. Vet. Pub.,* Evanston, Ill. (1953).
57. Innes, J. R. M., M. W. Colton, P. P. Yerrick, and C. L. Smith, *Am. J. Pathol.,* **30:**813 (1954).
58. Innes, J. R. M., A. J. McAdams, and P. P. Yevich, *Am. J. Pathol.,* **32:**141 (1956).
59. Innes, J. R. M., in, W. E. Ribelin and J. R. McCoy (eds.), *Pathology of Laboratory Animals,* Charles C Thomas, Springfield, Ill., 1965.
60. Innes, J. R., F. M. Garner, and J. L. Stookey, in, E. Cotchin & F. J. C. Roe (eds.), *Pathology of Laboratory Rats and Mice,* F. A. Davis Co., Philadelphia, 1967.
61. Jennings, A. R., *Br. Vet. J.,* **105:**89 (1949).
62. Jennings, A. R., *J. Comp. Pathol. Ther.,* **62:**161 (1952).
63. Keeble, S. A., *Vet. Rec.,* **73:**618 (1961).
64. Kent, S. P., and J. E. Pickering, *Cancer,* **11:**138 (1958).
65. Kirkman, H. and F. T. Algard, *Vivo Cancer Res.,* **25:**141 (1965).
66. Kleinberger-Nobel, E., *Mycoplasmataceae,* Academic Press, London and New York, p. 157 (1962).
67. Kraft, L. M., *J. Exp. Med.,* **106:**743 (1957).
68. Kraft, L. M., *Science,* **137:**282 (1962).
69. Lane-Petter W., and G. Porter, in W. Lane-Petter (ed.), *Animals for Research,* Academic Press, London & New York, 1963.
70. Lane-Petter W., *Animals for Research Principles of Breeding & Management,* Academic Press, London & New York, 1963.
71. Lane-Petter, W., in *UFAW Handbook on the Care & Management of Laboratory Animals,* Williams & Wilkins Co., Baltimore, 1967.
72. Lapin, B. A., and Yakovela, in W. F. Windle (ed.), *Comparative Pathology in Monkeys,* Charles C Thomas, Springfield, Ill., 1963.
73. Lerner, E. M., and L. Sokoloff, *Arch Pathol.* **67:**364 (1959).

74. Lewis, A. M., W. P. Rowe, H. C. Turner, and R. J. Huebner, *Science* **150**:363 (1965).

75. Mackenzie, D., *Goat Husbandry,* Fabir, London, 1957.

76. Matheon, R., *Medical Entomology,* 2nd ed., Comstock, Ithaca, N.Y., 1950.

77. Meenen, F. O. C. *Vet. Record,* **67**:666 (1955).

78. Meier, H., and G. Yerganian, *Diabetes,* **10**:19 (1961).

79. Meier, H., *Nature,* **188**:506 (1969).

80. *Mercks Veterinary Manual,* Merck and Co. Inc., Rahway, N.J., 1974.

81. Meyer, K. F., and B. Eddie, *Proc. of the 3rd Ann. Meeting of the Anim. Care Panel,* Chicago, 1952, p. 23.

82. Michael, J. F. *Nature* (London), **169**:881 (1952).

83. Mims, C. A., *Br. J. Exp. Pathol.,* **40**:543 (1959).

84. Mims, C. A. *Bact. Rev.,* **28**:30 (1964).

85. Monning, H. O., *Veterinary Helminthology and Entomology,* 3rd ed., Bailliere, Tindall and Cox, London, 1949.

86. Morgan, B. B., and P. A. Hawkins, *Veterinary Protozoology,* 2nd ed., Burgess, Minneapolis, 1952.

87. Mount, L. E., and D. L. Ingram, *The Pig as a Laboratory Animal,* Academic Press, London and New York, 1971.

88. National Academy of Sciences, *National Research Council: A Handbook of Laboratory Animals,* Publication 317, Washington D.C., 1954.

89. Nelson, J. B., *J. Exp. Med.,* **94**:377 (1951).

90. New York Academy of Sciences, Conference on the Care and Diseases of the Research Monkey. *Ann. N.Y. Acad. Sci.* **85** (Art. 3) 1960.

91. Oldham, J. N., in E. Cotchin & F. J. C. Roe (eds.), *Pathology of Laboratory Rats & Mice,* F. A. Davis Co., Philadelphia, 1967, p. 641.

92. Pankevicius, J. A., C. E. Wilson, and J. F. Farber, *Cornell Vet.,* **47**:317 (1957).

93. Pappenheimer, A. M., and F. S. Cheever, *J. Exp. Med.,* **83**:317 (1948).

94. Parish, H. J., *Notes on Communicable Diseases of Laboratory Animals,* Edinburgh, Livingstone, 1950, p. 69, Abstr. 27 Fed. Proc. 19 (Suppl 6), 1960).

95. Parish, J. J., *Notes on Commucicable Diseases of Laboratory Animals,* E. & S. Lingstone, Edinburgh, 1953.

96. Paterson, J. S., in *UFAW Handbook on the Care & Management of Laboratory Animals,* Universities Federation for Animal Welfare, London, 1957, p. 203.

97. Paterson, J. S., in R. J. C. Harris (ed.), *The Problems of Laboratory Animal Disease,* Academic Press, London, 1962.

98. Paterson, J. S., in W. Lane-Petter, (ed.), *Animals for Research: Principles of Breeding & Management,* Academic Press, London and New York, 1963.

99. Pearson, H. E., and M. D. Eaton, *Biol. Abstr.,* **15**:4728, *Proc. Soc. Exp. Biol. Med.,* **45**:677, (1940), *Vet. Bull.,* **12**:26.

100. Pfau, C. J., I. R. Pedersen, and M. Volkert, *Acta Pathol. Microbiol. Scand.,* **63**:181 (1965).

101. Pierce-Chase, C. H., R. M. Fauve, and R. Dubos, *J. Exp. Med.,* **120**:267 (1964).

102. Ratchiffe, H. L., et al., *Proc. of the 5th Ann. Meeting of the Anim. Care Panel,* Chicago, December 1954.

103. Rhodes, A. J., and C. E. Van Rooyen, *Textbook of Virology,* Williams & Wilkins, Baltimore, 1953.

104. Ribelin, W. E. and J. R. McCoy, *Pathology of Laboratory Animals,* Charles C. Thomas, Springfield, Ill., 1965.

105. Ruebner, B. H., J. R. Linsey and E. C. Melby Jr, in W. E. Rebelin and J. R. McCoy (eds.), *The Pathology of Laboratory Animals,* Charles C Thomas, Springfield, Ill., 1965.

106. Ruch, T. C., *Diseases of Laboratory Primates,* W. B. Saunders, Philadelphia & London, 1959.

107. Runner, M. N. M., and J. Palm, *Proc. Soc. Exp. Biol. Med.,* **82**:147 (1953).

108. Russfield, A. B., and M. N. Green, *Am. J. Pathol.,* **46**:59 (1965).

109. Schwarz, J., *Proc. of the 5th Ann. Meeting of the Anim. Care Panel,* Chicago, December 1954.

110. Scott, P. P., in *UFAW Handbook on the Care & Management of Laboratory Animals,* 3rd ed., Williams & Wilkins Co., Baltimore, 1967.

111. Seamer, J., and A. W. Gledhill, *Arch. Virusforsch.,* **17:**664 (1965).

112. Seamer, J. H., in E. Cotchin, and F. J. C. Roe (eds.), *Pathology of Laboratory Rats and Mice,* F. A. Davis Co., Philadelphia, 1967, p. 537.

113. Shechmeister, I. L., *Science,* **123:**463 (1956).

114. Soave, O. A., *J. Am. Vet Med. Assoc.,* **142:**285 (1963).

115. Sokoloff, F., *J. Nat. Cancer Inst.,* **20:**965 (1958).

116. Tucker, M. J., and S. B. de C. Baker, in E. Cotchin and F. J. C. Roe (eds.), *Pathology of Laboratory Rats and Mice,* F. A. Davis, Co., Philadelphia, 1967, p. 787.

117. Tuffery, A. A., *Vet. Rec.,* **68:**511 (1956).

118. Tuffery, A. A. and J. R. M. Innes, in W. Lane-Petter (ed.), *Animals in Research,* Academic Press, London & New York, 1963, Chap. 3.

119. Tuffery, A. A., *J. Hyg. Camb.,* **57:**386 (1959).

120. UFAW (ed.) *The UFAW Handbook on the Care and Management of Laboratory Animals,* 4th ed., Churchill Livingstone, London, 1972.

121. U.S. Department of Health, Education & Welfare, *Public Health Service Publication No. 343,* Washington, D.C., 1954.

122. U.S. Department of the Army & the Air Force, *Tech. Bull. No. Med. 255, Air Force Pamphlet No. 160-12-8,* Washington, D.C., 1958.

123. Wantland, W. W., in R. A. Hoffman, P. F. Robinson, and H. Magalhaes (eds.), *Golden Hamster,* Iowa State University Press, Ames, Iowa, 1968.

124. West, W. T., and K. E. Mason, *Anur. J. Anat.,* **102:**323 (1958).

125. Weston, J. K., in W. E. Ribelin and J. R. McCoy (eds.), *Pathology of Laboratory Animals,* Charles C Thomas, Springfield, Ill., 1965, p. 351.

126. Wilson, G. S., and A. A. Miles, in Topley and Wilson (eds.), *Principles of Bacteriology and Immunology,* 4th ed., Arnold, London (1970).

127. Whitney, R., in W. Lane-Petter (ed.), *Animals for Research,* Academic Press, London and New York, 1963.

128. Worden, A. N., P. Noel, and D. Jolly, in W. Lane-Petter (ed.), *Animals for Research: Principles of Breeding & Management,* Academic Press, London & New York, 1963, p. 393.

CHAPTER 4
USE OF EXPERIMENTAL ANIMALS FOR THE STUDY OF INFECTIOUS DISEASES

The use of experimental animals began in the midnineteenth century when the microorganisms as causative agents of infectious diseases were discovered. In the early days, many spontaneous infectious diseases of animals were discovered by the isolation and identification of infectious agents. Human diseases were induced in laboratory or farm animals by inoculation of isolated organisms or of clinical material from infected patients. At the present time, research in human infection is being carried out on spontaneous animal disease models or by producing infectious diseases in suitable animal species (32, 50, 63, 65, 66, 102, 103, 117, 122, 137, 145, 147, 155, 158, 188, 205, 211, 232, 241, 252, 264, 267, 268, 280, 291). A model that is useful for understanding infectious disease processes should have the following properties: (a) it should be simpler than the natural infection and therefore easier to manipulate and to reproduce; (b) it should simulate the important features of the natural host-parasite system; (c) it should demonstrate the features of infection at a cellular and molecular level.

There are a number of fundamental problems of infectious diseases that can be solved by using animal model system. These include: (a) determinants of the virulence of a parasite; (b) mechanism of host resistance; (c) mechanism of pathogenicity; (d) establishment and regulation of chronic infection; and, (e) antimicrobial and chemotherapeutic actions of drugs on infectious agents.

The laboratory investigator working with an animal model of an infectious disease has several advantages over his colleagues who are clinically investigating the disease in patients. The laboratory investigator can manipulate the exact moment and route of infection and can determine the quantity and virulence of the microbe and other variables. By sacrificing animals drawn at random from the infected population, the detailed unfolding of the events in the infectious process can be followed. Laboratory investigators can also apply or withhold experimental treatments according to protocols that would not be permissible in clinical studies. A major drawback of an animal model of infectious disease is the risk that the model may inaccurately reflect events in human disease. Appropriate controlled studies of the pathogenesis, therapy, and prophylaxis of serious infections in humans are usually not feasible because untreated controls cannot and should not be employed. Thus, there are frequent attempts to extrapolate data from various animal models to humans. A simple host-parasite model can demonstrate the potentialities of a parasite for causing disease, but the model can say nothing about whether or not these potentialities are realized in a natural infection. Thus, the model may succeed if the infectious process found in the model also takes place in the natural infection, and the model will fail if it does not. Conversely, failure is assured if an important aspect of the natural infection is absent in the model. Many spontaneous animal disease models and experimentally induced models of infectious disease have been reported in the literature. It is beyond the scope of this chapter to include in the discussion all the information available on infectious diseases. Therefore, selected human infectious diseases that occur in experimental animals spontaneously and also certain infections that can be produced in experimental animals similar to human conditions are described in this chapter.

EXPERIMENTAL INFECTIOUS DISEASE MODELS

BACTERIAL DISEASES

Leprosy

Leprosy is an infectious disease caused by *Mycobacterium leprae* and manifests itself most commonly in the skin, nose, upper respiratory tract, and peripheral nerves. The effect on nerves is responsible for serious disabilities and deformities of the feet, hands, and face. There are two forms of the disease: (a) tuberculoid leprosy, in which there are few bacilli; and (b) lepromatous leprosy, in which the number of bacilli is extremely high. Because it is still not possible to grow *M. leprae* in vitro, all experimental models for studying leprosy have had to be undertaken in vivo. Leprosy could not be transmitted to experimental animals until 1969. Previously, it was possible, by using *M. lepraemurium* in the mouse as a model, to develop indirect methods for determining the viability of the leprosy bacilli. Systematic development of this method for measuring viability completely revolutionized the assessment of progressive infection in humans and animals.

The important advancement came in 1960 when it was discovered that limited growth of *M. leprae* could be sustained in the footpads and the ears of mice and other rodents. The experimental infection models may be used for the determination of generation of time of *M. leprae* and for the assessment of antileprosy drugs. The model is also useful for the determination of the sensitivity of strains of *M. leprae* to DDS, the identification of drug-resistant strains, and the protective effect of BCG vaccination.

Despite the tremendous advances made with this animal model, the limited and localized nature of the infection and the failure to reproduce the human disease pattern (particularly the involvement of nerves) restricts its potential value (153, 234, 235). The behavior of organisms in mice whose immunologic capacity has been reduced by thymectomy plus total body irradiation parallels in every possible way its behavior in humans, including the histological picture of lepromatous leprosy. This model has also demonstrated the predilection of *M. leprae* for muscle and nerve. By replacing immunologically competent lymphoid tissue in T/900r mice at the time of inoculation or by following establishment of the infection, changes result in the infection and histological picture closely resembling the intermediate or borderline types of leprosy seen in humans.

Leprosy qualifies as a complete model for the study of granulomatous disease. Resistance to leprosy appears to be primarily cell mediated immunity—the same immunogenic machinery that is sought to control neoplastic growth and the rejection of foreign tissue transplants and other infectious agents (viruses, fungi, etc.).

Storrs (270, 272), Kirchheimer and Storrs (154), and Couvit (49) reported the development of leprosy in a single nine-banded armadillo (*Dasypus novencincta*) 15 months after inoculation with leprosy bacilli obtained from a human patient. Further studies reported by Storrs, et al. (271) showed that eight of twenty armadillos developed severe lepramatous leprosy 3 to 5 years after inoculation with viable *M. leprae*. A total of 988 g of lepromas containing an estimated 15–20 gms of leprosy bacilli has been harvested from these animals. The large amounts of material now available should make possible definitive studies on the immunology, chemotherapy, and epidemiology of the disease.

Tuberculosis

Tuberculosis is still a disease of major importance in humans and in animals. Current figures suggest that 30,000 people contract tuberculosis in the United States each year. In addition, research animals, especially old-world primates, are exquisitely sensitive to *M. tuberculosis,* and the potential dangers to colonies of such animals are well known to research workers in this field. Pulmonary tuberculosis induced by instillation of small numbers of tubercle bacilli into the main stem bronchi of the rhesus monkey (*Macaca mulatta*) provides a model of proven utility for evaluation of the prophylactic and therapeutic potentialities of new antituberculous drugs, and combination thereof with existing agents (255). The values of this model for appraising the merits of immunizing agents is equally well established but less well known (108). Studies in immunization against tuberculosis have been carried out in three main areas. The first dealt with evaluating the capacity of BCG to protect the simian host against otherwise fatal disease. The variables examined were size of the vaccine dose; need for multiple doses; importance of the time interval between such doses; duration of immunity after dosage; influence of the routes of vaccine after administration; and lastly, the relation between size of the challenge inoculum and degree of protection achieved. The second area compared various vaccines, such as, tubercle bacilli killed by ultraviolet light, particulate and cell wall fraction of H37Ra of BCG, and cell wall fractionation isoniazid-resistant BCG. The third area of activity was concerned with the assessment of isoniazid prophylaxis.

Other animals such as guinea pigs, chickens, cats, dogs, and so forth have been used to induce tuberculosis with bovine or human strains (31, 33, 216, 222, 262). Guinea pigs may provide a good model for the acute phase of the disease in human beings. However, the nonhuman primates are by far the best models for the study of pathogenicity of tuberculosis and to study various aspects of disease mechanisms in humans.

Mycoplasma Infections

Most of the diseases known to be caused by mycoplasma occur in mammals and birds, for example, pneumonia and arthritis (266). The human strains of *M. pneumoniae* have been used extensively in experimental animals to develop information about host-parasite interactions. This is necessitated in part by the exceedingly low mortality of the natural disease in humans so there is little opportunity for adequate assessment of the pathological processes. Even at best, the changes seen in human postmortem specimens represent end-stage disease that may be complicated by other extrinsic or intrinsic factors. Some of the models used in research on pathogenesis of *M. pneumoniae* illustrate the usefulness of systems differing in complexity (47, 48).

From the clinical point of view, pneumonia caused by *M. pneumoniae* is usually a benign, self-limiting disease in human beings. Mortality and morbidity of the illness are extremely low. In comparison, pneumococcal lobar pneumonia with bacteremia is a severe infection with significant morbidity and mortality. Pathogenesis of *M. pneu-moniae* has been extensively studied in the experimental hamster model (164, 257). However, many aspects of the immune response in *M. pneumoniae* disease is relatively unexplored. In comparative studies, Fernald and Clyde (80) found that resistance to the development of lung lesions was induced most effectively by intranasal administration of live *M. penumoniae*. Although it induced higher levels of serum antibody, killed vaccine did not provide effective protection to intranasal challenge with virulent organisms. These findings suggested that mechanisms other than humoral antibody, perhaps secretory IgA, were involved in resistance to *M. pneumoniae*.

Young guinea pigs were experimentally infected with *M. pneumoniae* by Brunner et al. (36). The organisms were isolated from the lungs of infected animals for at least 9 weeks thereafter. Mycoplasmicidal antibody developed 5 weeks after infection while lung lesions were detected within 2 weeks after inoculation. The gross pathological lesions of the affected lungs were similar to those observed in infected hamsters (Figs. 4-1, 4-2). Guinea pigs may prove to be useful models for studies on the pathogenesis of the disease and the cell-mediated immune response during infection with *M. pneumoniae*.

Rheumatoid Arthritis

Arthritis induced in rats by *M. arthritidis* is an acute suppurative disease that readily resolves with complete healing after several weeks. *M. arthritidis* responds quite differently in mice, producing a disease characterized by periods of remission and exacerbation (118). Although the course of the disease in individual mice vary greatly, arthritis usually declines somewhat by 70 days and may exhibit a marked manifestation between 70 and 150 days after initial infection. The arthritis may persist through at least 269 days. The subcutaneous injection of mice with *M. arthritidis* causes a spreading necrotic abscess and joint lesions. The early lesions observed in the joints of mice resemble those of rat arthritis induced by *M. arthritidis*. Polymorphonuclear cells pre-

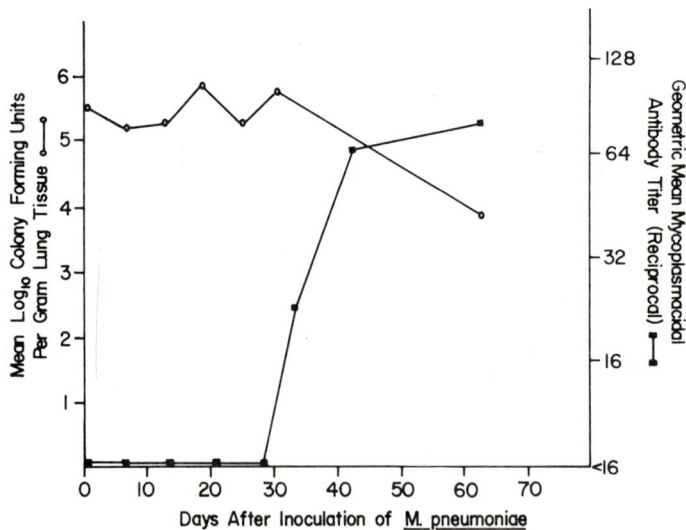

Fig. 4-1. Experimental *Mycoplasma pneumoniae* infections of young guinea pigs. Reproduced from *The Journal of Infectious Diseases* (36) with permission.

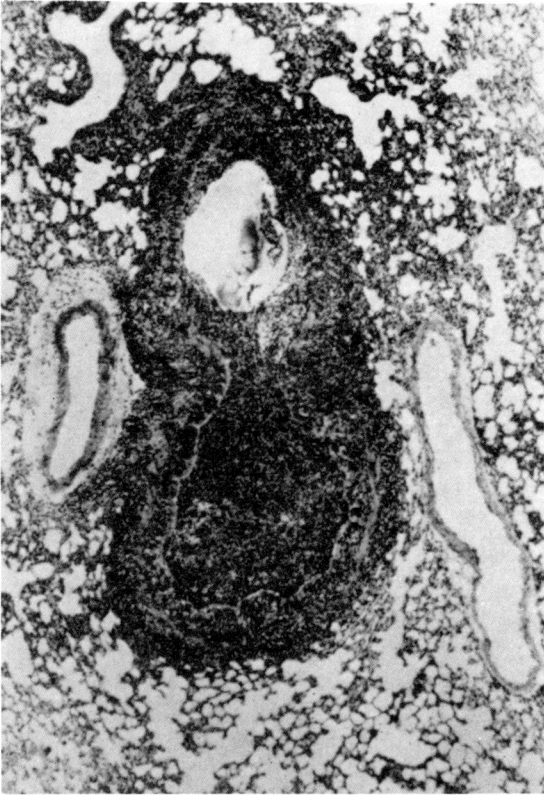

Fig. 4-2. Appearance of guinea pig lungs 14 days after inoculation with *Mycoplasma pneumoniae.* Note the peribronchial cellular infilteration and the polymorphonuclear leukocytic exudate (hemotoxylin and eosin, × 36). Reproduced from *The Journal of Infectious Diseases* (36) with permission.

dominate over mononuclear cells. There is some early proliferation of the synovial membrane, and a proteinaceous exudate can be seen within the joint spaces. The lesions intensify and progress to a chronic inflammatory reaction of the periarticular tissues. Massive villus proliferation and fibroplasia of the synovial membrane with infilteration of mononuclear cells occurs. Bone damage is minor, although some remodeling can be observed. Pannus formation is common in the chronic stage of the disease. While the later lesions are characteristic of a chronic nature, occasional foci of polymorphonuclear cells and proteinaceous exudate can be seen in some animals. The chronic lesions observed bear a close resemblance to those of rheumatoid arthritis of humans; thus, the murine disease would appear to be an excellent model of experimental chronic infection.

Intrarenal inoculation of rats with T-strain of mycoplasma leads to the formation of bladder stones due to a urolithic property of the organism.

Arthritis in rats and swine have been studied by Sokoloff (266). In most cases of rheumatoid arthritis, rheumatoid factors in serum are present. Those factors are 7S or larger globulins characterized by their reactivity with other immunoglobulins. They are not identified in *M. hominis* infections. The principal value of these postinfectious arthritides as models of rheumatoid arthritis is to elucidate the mechanisms by which inflammation may persist in articular tissues even after the primary infection apparently has disappeared. A low-grade infection or isolated pockets of it may continue to reside in tissues. Another possibility is that the antigenic derivative of the dead organisms resists phagocytosis or degradation. The residual antigen in the synovial cavity might then by immunological means perpetuate the inflammatory reaction. Thomas (276) reported that *M. gallispeticum* infections in chickens provide lesions that quite closely resemble those that characterize human disease. The S6 strain of *M. gallisepticum* causes poly-arteritis nodosa with extraordinary reproducibility in virtually all of the arteries of the central nervous system. Also there is a variety of arthritis similar to rheumatoid arthritis, regularly associated with polyarteritis and fibrinoid necrosis in the walls of arteries in the periarticular tissue. Chickens given a large enough dose of the S6 strain will also undergo arteritis in the central nervous system, but by far the most frequently involved arteries are those in periarticular tissues. The arthritis is an outstanding lesion that is characteristically progressive and can become a chronic destructive disease.

Another reason for interest in the lesions produced by *M. gallisepticum* in consideration of their possible relationship to human disease is that all of the lesions, the polyarthritis and arthritis, are prevented or can be cured by treatment of the animals with gold salts. Two drugs, aspirin and gold salts, are significantly effective against arthritis.

Sabin (244) was the first to show that gold salts reversed mycoplasma-induced tissue damage and infection. His initial studies were based on the effects of gold salts in rheumatoid arthritis in humans. Working with *M. arthritidis* in the late 1930s, he undertook empirically to see what effect gold thiomalate would have on this disease and found it prevented arthritis in rats and also would bring about reversal of the joint lesions if given after the onset of the disease. Marmion and Goodburn (179) used gold thiomalate in their initial explorations of possibilities that Eaton agent might be a mycoplasma in the chick embryo and hamster models, and it provides an active and completely effective therapeutic agent in the reversal of arteritis in the turkey and a reversal of the chronic progressive arthritis in the chicken. Moreover, the treatment of mice with gold salts (gold thiomalate) will protect them against the effects of live *M. neurolyticum*. It will not protect or influence in any way the development of the brain lesion produced by the exotoxin of *M. neurolyticum,* but it will prevent the onset of rolling disease and development of brain lesions after intravenous injection of live concentrated suspensions of the mycoplasma.

Gonorrhea

Gonorrhea is currently the most frequently reported infectious disease in the United States. Research on gonorrhea has been handicapped by lack of a suitable model in laboratory animals. The first confirmed animal infection with *Neisseria gonorrhoea* were reported by Miller (190) using the rabbit as an animal model. Subsequent work have utilized mice and chimpanzees (12, 13, 14, 169). Experimental gonococcal urethritis and cervicitis were produced in chimpanzees by Lucas et al. (169). It is an expensive experimental animal to use and the supply is limited.

Miller reported the ocular infection of rabbits by direct inoculation of gonococci into the anterior chamber, but because of the discomfort to the animal and the limited access to the infection site, this technique has not been used to any great extent.

The controlled in vivo testing of gonococcal antigens, which in the past has been limited by the absence of a suitable model in laboratory animals, is important to the development of an effective immunizing agent for gonorrhea. Experimental infections with *N. gonorrhoeae* have been recently reported in four species of laboratory animals: rabbits, guinea pigs, hamsters, and mice (12). When the infection in each of these species were considered and the potential use of the species as a laboratory animal model was evaluated, the guinea pig was found to be most satisfactory for the study of the immune mechanism of gonococcal infections (14). Both active and serum-mediated passive immunity were demonstrated in the guinea pig model.

Bacterial Endocarditis

The diagnosis of bacterial endocarditis in humans poses significant problems. Febrile illness in a patient with valvular heart disease, whose blood cultures fail to yield an organism, not infrequently leads to treatment for unconfirmed bacterial endocarditis in the absence of a definitive diagnosis because of the threat of the valvular infection to life. Bacterial vegetations have been observed at autopsy on patients from whose blood no organism could be grown during life. In other instances, the differentiation of reactivation of rheumatic fever from endocardial infection poses a significant diagnostic dilemma. Differentiation of drug fever from continuing endocardial infection during the treatment at a time when microbial growth may be suppressed by therapy is an additional problem in the management of patients.

A number of animal species including rat, chicken, rabbits, dogs, cattle, and the like have been used to produce bacterial endocarditis disease with clinical symptoms similar to those seen in humans (8, 61, 99, 148, 152, 192). Despite advances in chemotherapy, bacterial endocarditis continues to be a serious disease with significant morbidity and mortality. Hormonal and physiological factors as well as anatomic cardiovascular lesions appear to play a role in the development of bacterial endocarditis (8, 210).

Keys et al. (152) used eighteen medium-sized mongrel dogs to study experimental enterococcal endocarditis. Aortic valve injury was accomplished by needle punctures through a catheter in the left carotid artery. Thirteen dogs survived the procedure and a week later received a suspension of ten enterococci intravenously. In ten dogs, a sustained bacteremia of 10^2 organisms/ml of blood or greater occurred within 1 week and rose to 10^3 cells/ml in all but one animal shortly before death or sacrifice. Necropsy was performed in the dogs from 3 to 103 days after inoculation. Vegetative endocarditis involving both aortic and mitral valves occurred in nine of ten canine subjects. Bacterial counts of affected valves ranged from 10^7 to 10^9 cells/g of tissue. Thermoembolic and hemorrhagic lesions were noted in the kidneys, myocardium, spleen, and lungs. The simplicity of this technique should prove this a useful model for the studies of therapeutic and pathophysiologic processes of endocarditis in humans.

Durack et al. (62) produced nonbacterial endocarditis on either side of the heart of rabbits by introducing a polyethylene catheter through the juglar vein. One day later, this was converted into bacterial endocarditis by single intravenous injection of streptococci, staphylococci, *Proteus,* or *candida* organisms. However, no infection resulted from injection of L-forms or virus. Reductions of inoculum size or withdrawal of the catheter reduced the incidence of bacterial endocarditis, but the presence of a catheter in the heart for only a few minutes predisposed to infection. Left-sided *S. viridans* infection was uniformly fatal with average survival of about 2 weeks. Right-sided infection was not fatal; 25 percent of infected vegetations healed spontaneously. Streptococci could be cultured from valves within 30 minutes of injection, and thereafter their growth curve could be predictable. They may be identified in sections within 6 hours (61). There is reduced incidence of bacterial endocarditis when injection of bacteria are delayed for some weeks permitting endothelialization of the vegetations. The rabbit model has many advantages over the dog model, where thoracotomy is performed to allow perforation of the aortic valve via an incision in the thoracic aorta. The rabbit is less expensive and easier to handle and maintain. The procedure of catheterization through the juglar vein of the rabbit is rather simple and rapid. Streptococcal endocarditis is easier to produce in rabbits.

In staphylococcal endocarditis produced in rabbits, the organisms could not be cultured from valves 18 hours postinfection with a variable pattern of growth; thereafter staphylococci could not be identified in tissue sections until 44 hours postinfection (116).

Staphylococcal Infection

A topical staphylococcal infection model for the study of the effectiveness of antimicrobial was developed by Goldschmidt (107). Animals were laparotomized, sutured with braided silk, and inoculated with a strain of *S. aureus*. Concentrations of bacteria from 5×10^4 to 10^8 cells per incision produced large body wall stitch abscesses with occasional drainage through the skin. This wound infection is readily reproducible and can be used for evaluation of the ability of topical antimicrobials to prevent *S. aureus* stitch abscesses. An animal model for staphylococcal epidermal necrolysis was reported by Dick (56).

Andriole et al. (6) developed an experimental rabbit model of chronic staphylococcal osteomyelitis. The inoculation of *S. aureus* into the tibial marrow cavity of rabbits following tibial fracture and rodding as well as into the rodded tibia without fractures, resulted in chronic staphylococcal osteomyelitis in high percentages of these animals. Staphylococci were easily recovered from the tibial marrow cavity for as long as 18 months after onset of the infection. This model closely resembles the human disease in which chronic osteomyelitis develops as a complication of internal fixation devices and provides a reliable method for evaluating various specific approaches to the treatment of this difficult management problem. Chronic osteomyelitis can be produced in the rabbit tibia in the presence of a metallic implant. This model closely approximates the condition of chronic osteomyelitis in patients treated with intramedullary nails. As in human disease, this model of osteomyelitis in the rabbit results in fever, local signs of early inflammation, abscess formation, spontaneous drainage and healing, but with slowly progressive destruction of the tibial shaft manifested by periosteal reaction cavity formation at the metal bone interphase and frequent sequestra and new bone formation. Although the draining sinuses heal, the tibia continues to show evidence of severe chronic osteomyelitis. In the study reported by Andriole et al. (6), both strains of staphylococci (Giorgio strain-hemolytic, coagulase positive, penicillin resistant, and phage type; and Phage type 80–81) tested were capable of producing chronic osteomyelitis but at different doses of inoculum.

Bacterial Meningitis

Bacterial meningitis in young children continues to be a serious problem. There are approximately 40,000 cases each year in the United States, of which the majority are in children less than 5 years of age. *Haemophilus influenzae B* accounts for more than 50 percent of cases of bacterial meningitis. The mechanisms of the pathogenesis of bacterial meningitis are not known. A suitable animal model would facilitate investigation of the pathophysiology of meningitis and might uncover the mechanisms. An animal model permitting study of bacterial meningitis was developed by Moxon et al. (206) in infant rats by the intranasal inoculation of *H. influenzae B*. Seventy-three percent (twenty-nine out of forty) of 5-day-old animals were bacteremic 48 hours after inoculation with 10^7 *H. influenzae B* cells; of those with bacteremia, 79 percent had documented meningitis. Studies with fluorescein-labeled specific antiserum suggested that bacteria were widespread in the upper respiratory tract, penetrated the nasal mucosa, entered the systemic circulation, and spread to the meninges from the dorsal longitudinal and lateral dural sinuses. The histologic appearance of the meningitis in infant rats was similar to that reported from human specimens taken at autopsy (Fig. 4-3).

Pyelonephritis

Epidemiological studies in humans have clearly demonstrated an increase in the prevalence of urinary infections with increasing age in both sexes. Although there are many possible explanations for this finding, an important factor in the increased susceptibility to urinary infection is the increase in abnormalities of the urinary tract with advancing age. Freedman (86) demonstrated that rats are more susceptible to urinary infections in old age and that there are at least two mechanisms involved: increased susceptibility to pylonephritis following the intravenous inoculation of bacteria and decreased effectiveness of bacterial clearance mechanisms in the bladder lumen. Intravenous inoculation of *Escherichia coli* produced pyelonephritis in a small percentage of 1-year-old rats whereas infection was never noted in young rats. When *E. coli* were injected into the bladder cavity of old rats, there was a decreased ability of these animals to rid the urine of bacteria as compared with young rats.

A basic problem in the studies of acute and chronic experimental pyelonephritis has been the induction of a disease that resembles the natural disease in humans. Both the

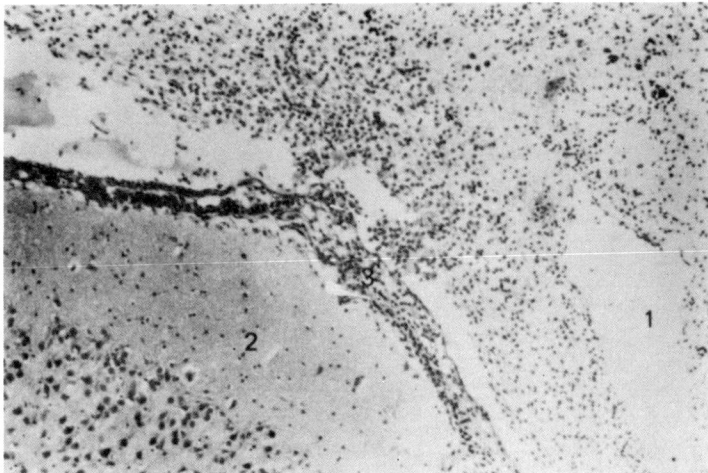

Fig. 4-3. Coronal section of dura mater of a meningitic 5-day-old rat depicting dorsal longitudinal sinus (1), left frontal cortex (2), and the leptomeninges (3). The dura is filled with inflammatory cells and fibrin, while the leptomeninges are mildly inflamed (hematoxylin and eosin × 100). Reproduced from *The Journal of Infectious Diseases* (206) with permission.

upper and lower urinary tracts of most experimental animals are extremely resistant to infection, and it has been necessary to obstruct the urinary tract to produce a focal injury with obstruction (111, 119, 226, 228, 240). During diuresis in mice, it has been possible to induce *E. coli* pyelonephritis in an unobstructed renal tract (149). Miller and Robinson (191) developed a new method which was consistent and reproducible. Infection was induced in the rat by the use of a micropipette that introduced bacteria directly into the kidney. The gross appearance and histopathological changes were similar to these seen during the course of pyelonephritis in man and were confined to the site of inoculation. Bacteria persisted for extended periods in the kidney, but in contrast to other reports, infection was not confined to the medulary region. With 3×10^5 organisms, approximately 30 percent of the inoculum remained in the kidney 5 minutes after challenge; bacteria were also recovered from the liver, blood, and perineal nodes immediately after challenge (Fig. 4-4).

Burrows and Cower (39) described *Proteus* and *Escherichia* pyelonephritis in the rat in which reproducible chronic infections were achieved. The efficacy of clinically effective agents was reliably demonstrable with five rats in 17 days. The suspension was introduced into the cortex of the kidney by means of a 0.25 in. (0.63 cm) 27-gauge needle attached to a 0.5 ml syringe. Entrance into the abdominal cavity and left kidney was gained by a quick jab of the needle. This manner of inoculation allows a differentiation of the pathophysiology of infection and trauma and thus facilitates the evaluation of antibiotic therapy. The model can serve as primary screen in which one technician can evaluate 180 compounds per month with 900 rats.

Enteric Infections

One of the more common causes of diarrhea in humans is enteric infection with *Salmonella typhimurium*. Since this form of gastroenteritis is generally self-limiting and of short duration, little is known of its pathology. Experimentally, a wide variety of clinical states has been produced in the mouse, rat, and guinea pig (3, 7, 112, 138, 227). A reproducible model of infectious diarrhea was produced in the rat by peroral

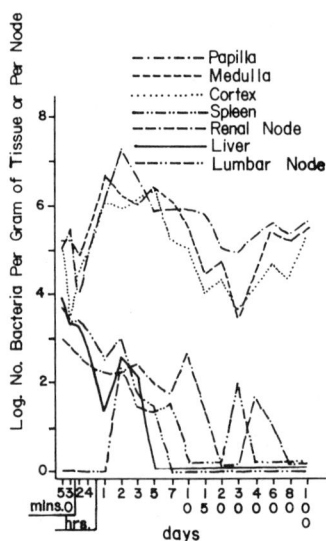

Fig. 4-4. Distribution of bacteria in the kidney, spleen, lymph nodes, and liver in animals with pyelonephritis due to *Escherichia coli*. Animals were studied at intervals from 5 minutes to 100 days after challenge with a total of 3×10^5 organisms inoculated directly into the kidney. Reproduced from *The Journal of Infectious Diseases* (191) with permission.

challenge with *S. typhimurium* by Maenza et al. (171). The evaluation of morphological changes were recorded and correlated with functional alterations. A single intragastric inoculum of 2×10^9 *S. typhimurium* was used in rats weighing 250–300 g. The animals were fasted for 18 hours prior to the challenge. The syndrome carries a low mortality and is characterized by diarrhea of varying severity, starting as early as 2 days after inoculation and lasting from 1 day to 3 weeks. Pathological lesions were established in the ileum at 2 days and in the cecum by 3 days after challenge. In animals with diarrhea, lesions progress in extent and severity, generally being most severe at 6 to 7 days, with involvement of the ileum and cecum predominantly. Histopathologically, there is pronounced cryptal elongation and distortion of the ileal villi, damage to the surface epithelium, and marked cellular infiltration of the lamina propria. Few microulcers are seen in the ileum, but larger ulcers are common in the cecum. A positive correlation exists between severity and extent of ileal lesions and clinical symptoms of diarrhea, with the symptoms subsiding as the lesions regress. This model shares certain histological and clinical features with other lesions of the small bowel such as parasitic infestation, sprue, or the early recovery phase in sublethal radiation. Thus the possibility is raised for the use of this model for the study of general pathophysiologic mechanism in diarrhea.

Cholera

A good experimental animal model of cholera should have the following features.

1. Experimental cholera can be evoked with minimal artificial manipulation.

2. It should simulate clinical symptoms of human cholera.

3. The immunogenic machinery is mature and functional.

4. The disease lasts for several days.

Sack and Carpenter (245, 246) reported a canine model of cholera that seems to satisfy all of these requirements. However, because of the cost and labor involved, the dog is an expensive model for studies requiring a large number of animals.

Blackman (29) developed a chinchilla model in which experimental cholera could be predictably evoked by innoculation with organisms in animals after ligation of the gut. Chinchillas inoculated intraintestinally with 10^6 *Vibrio cholera* (569B) organisms developed a broad spectrum of symptoms ranging from asymptomatic infection to severe disease. Often fatal disease occurred when larger doses were used. Although diarrhea was not evident, large amounts of fluid accumulated throughout the gastrointestinal tract. The severity of symptoms was related to the amount of fluid accumulated and was directly proportional to the number of bacteria found in the small intestine at necropsy. Intestinal fluid was isotonic to serum, rich in potassium, and almost free of proteins. Histologically the intestinal mucosa appeared to be intact and goblet cells were only infrequently found. Acidosis and hemoconcentration were found in moribund animals. Since the chinchilla is small, inexpensive, immunologically competent, and reproducibly susceptible to infection with *V. cholerae,* its use as an animal model for the study of the pathogenic processes in cholera can be valuable.

Anaerobic Bacterial Infections

Increased recognition of the importance of anaerobic bacteria in production of clinical diseases has indicated the need to reexamine pathogenicity of single and mixed cultures anaerobic bacteria. Development of suitable model animal infections are necessary to better understand pathogenicity and synergism between species and to evaluate potential therapeutic agents. Data from clinical specimens indicate that many human infections are caused by mixed anaerobic or anaerobic and faultative species. For example, Hill et al. (124) reported a model anaerobic infection consisting of progressive intrahepatic abscesses which was produced in mice by injection of single and combined species of nonspore-forming anaerobic bacteria (*Bacteroides fragilis, B. melanino-genicus, Fusobacterium necrophorum* and *Peptostreptococcus species*) (126). Models of progressive anaerobic disease involving the viscera which mimic some of the clinical infections commonly seen would be particularly valuable.

Other Bacterial Disease Models

Numerous studies have been reported on experimental production of bacterial diseases in animal species (Table 4-1). Since appropriately controlled studies of pathogenesis, therapy, and prophylaxis of serious infections in humans are usually not feasible, there are frequent attempts at extrapolation of data from various animal models to humans. Saslaw and Carlisle (251) reported that certain serious infections that occur in humans can be studied in depth with the use of the rhesus monkey as a model. The various clinical and laboratory observations one usually conducts in human medicine can readily be applied to nonhuman primates. With the increasing number of chemotherapeutic agents coming under study, the nonhuman primate may become increasingly important as a model for definitive studies of the drugs used in infectious diseases prior to their ultimate application in man. A number of infectious disease studies in primates have been reported including tularemia (64, 251), influenza (249, 251), streptococcal, staphylococcal, and pneumococcal infections (274, 250, 251), rickettsia (250, 251, 287), and histoplasmosis (248, 251).

Study of Biochemical Changes in Infectious Disease Process

Alterations in plasma lipids have been described as incidental findings during infectious diseases (26) but have rarely been studied in a serial manner throughout the entire clinical course of an illness. Recently, Gallin et al. (95, 96) suggested that certain pathogenic microorganisms stimulate specific patterns of responses in the host's lipids. These investigators noted markedly increased concentrations in serum of total lipids, triglycerides (TG), and free fatty acids (FFA) in patients with gram-negative septicemia but not in patients with gram-positive infections or influenza. However, this concept has not been substantiated when tested during experimental infections in animals (76). Because of possible species-related differences, the responses of common laboratory animals may not be representative of changes in lipids during similar infections in humans. Fiser et al. (81) measured plasma lipids and lipoproteins in a sequential fashion in rhesus monkeys after experimental infection with either *D. pneumoniae* or *S. typhimurium*. Monkeys infected with the pneumococci generally had increased levels of triglycerides and pre-β-lipoproteins and decreases in cholesterol and β-lipoproteins in plasma. Concentrations of free fatty acids and phospholipids were either increased or decreased according to the stage of the infection. Monkeys infected with Salmonellae had increased levels of triglycerides, pre-β- and α-liproproteins and decreased levels of phospholipids, cholesterol, and β-lipoproteins in plasma. The free fatty acid responses as in the pneumococcal infection was variable. It was suggested that lipid metabolism varies with

Table 4-1. Use of Laboratory Animals for the Study of Induced Bacterial Infections

Animal Model	Species	Human Counterpart	Reference
Antiplague agents	Dog (Beagle)	Plague chemo-therapy	135
Bacterial endocarditis	Rabbit	Bacterial endo-carditis	62
Cholera	Chinchilla	Cholera	29
Chronic Staph-lococcal Osteo-myelitis	Rat	Chronic Osteo-myelitis	6
Entercoccal endocarditis	Dog	Bacterial endocarditis	22
Escherichia coli infection	Rat	Enteric Infection	39 134 191 231
Experimental pyelonephritis	Rat	Pyelonephritis	191
Hemobartonella sp., hemolytic anemia	Cat	Oroya fever	35 83 84
Hemophilus in-fluenza menin-gitis	Infant Rats	Meningitis	206
Johni's disease	Cattle	Whipple's disease	223
Leprosy	Armadillo Mouse	Leprosy	236 270
Mycobacteriosis	Rhesus monkey	Pulmonary tuberculosis	
Mycoplasma arthritis	Swine	Arthritis	255
Mycoplasma pneumonia	Young pigs	Mycoplasma pneumonia	36
Neisseria gonorrhea infection	Chimpanzee Rabbit	Gonorrhea	11 13 170

Table 4-1. (Continued)

Animal Model	Species	Human Counterpart	Reference
Non-spore forming anaerobic bacterial infection	Mouse	Liver Abscess	124
Pneumococcal Pneumonia	Rat Monkeys	Bacterial Pneumonia	81 163 196
Salmonella enterocolitis diarrhea	Rat	Enteritis	171
Scrub Typhus	Silver Leaf Monkey	Scrub Typhus	287
Staphylococcal Infection	Rat	Topical staphylococcal infection	106
Tularemia	Monkey	Tularemia	64 251
Urinary Tract Infection	Rat	Topical Staphylococcal infection	241

the nature of the invading microorganism as well as with the stage and duration of the infectious illness.

A number of other biochemical changes, including enzymes, proteins, carbohydrates, and the like, during the development of infectious disease in animals, have also been reported (194, 195, 196, 197, 198, 199, 200). These studies on experimental animals may serve as useful models for understanding the mechanisms of pathogenicity of infectious human diseases.

VIRAL DISEASES

Herpesvirus Infections

Infection of the newborn human infant with *Herpesvirus hominis* (HVH) may be manifested by a wide spectrum of diseases, including subclinical infection, encephalitis, and a disseminated visceral form with or without involvement of the central nervous system (CNS). Many experimental animal studies have been reported with herpesvirus infections (30, 57, 91, 92, 93, 165, 239). In the more severe forms of HVH infections, it has been estimated that more than 50 percent of untreated cases may terminate fatally or result in permanent neurological sequelae (151). The severity of neonatal herpesvirus infections clearly establishes the need for effective therapy. While various modes of treatment such as γ-globulin, cytosine arabinoside, and polyribonoside-polyribocytidylic acid (poly I.C.) have been attempted on a limited scale, the most commonly used chemotherapeutic agent has been 5-iodo-2'-deoxyuridine (IUDR). Although treatment of disseminated HVH infections of the newborn human infant has resulted in both success and failure, there have been no controlled studies to evaluate the effect of therapy on either the outcome or the pathogenesis of the disease. Infection of the newborn mouse with the genital type 2 strain of HVH by the intranasal route provides an experimental model of the disease in humans. After inoculation of the mouse, the virus multiplies in the lung and is disseminated through the blood to the liver and spleen and to the brain by both viremia and nerve route transmission (151). Therapy with IUDR did not affect mortality although a significant effect on pathogenesis was observed. Viral replication in the lung was reduced and viremia, as well as subsequent involvement of liver and spleen, was completely eliminated. In contrast, nerve route transmission to and replication of the virus in the central nervous system was not affected by therapy. Lack of inhibition of viral replication in the CNS appeared to result from inadequate levels of IUDR in brain tissue and is the likely explanation for the therapeutic failure of IUDR in this model infection. These observations suggest possible reasons for the variable results reported in the treatment with IUDR of disseminated herpesvirus infections of newborn human infants.

An animal model of encephalitis due to *Herpes simplex* (type I) was developed in the adult hooded rat by Marks and Carpenter (178). In this model, the disease is reproducible, restricted to the brain, and has histopathologic features similar to those produced by the disease in humans. Including the appropriate controls, 222 animals were studied. Infection can be characterized by a 2- to 4-day period of incubation followed by signs of infection of the central nervous system; the average survival time of the infected rats is 7 days. The model provided a clear end point of mortality with reproducible results and an inoculum dependent response.

Cho et al. (44) reported that marmosets inoculated with HVH readily developed encephalitis and/or disseminated infection. Thus, a nonhuman primate model is also available for the study of antiviral chemotherapy in severe herpesvirus infection. Cho et al. (45) tested idoxuridine (5-iodo-2'-deoxyuridine, IUDR) in encephalitis and disseminated infection due to herpes in marmosets. The disease in idoxuridine-treated marmosets was similar in clinical picture mortality and histopathologic changes to that in untreated controls. Slight modulation of the course of illness, as manifested by suppressed titres of virus in visceral organs, was observed when the drug was given before infection.

Herpesvirus has been associated with oncogenic properties in various experimental animal studies. Epstein-Barr virus (EBV), another herpesvirus isolated from human tissues is considered to be the likely etiological agent of Burkitt's lymphoma, although further studies are required to prove this. The fact that two herpesviruses, namely HVA and HVS, can induce oncogenic changes in nonhuman primates provide primate model systems for studying malignant lymphoma and leukemia (186, 202). *Herpesvirus saimiri,* the first oncogenic herpesvirus of primates, was shown to produce malignant lymphoma and leukemia in several species of nonhuman primates and rabbits. The recent demonstration by Melendez et al. (186) that the spider monkey develops malignant lymphoma after inoculation with the virus increases the list of susceptible primates to five: marmosets, owl monkeys, and spider monkeys develop malignant lymphoma, and the cebus and African green monkeys develop reticuloproliferative diseases.

Several other oncogenic viruses including herpesvirus have been studied in experimental animals which resemble human malignant diseases (1, 43, 54, 74, 125, 166, 167, 189, 208).

LCM Virus

Lymphocytic choriomeningitis (LCM) virus is a small, labile, pathogenic agent. When acquired congenitally or when injected at birth, mice develop persistent asymptomatic viremia. Intracerebral inoculations of LCM virus into older mice increase acute lethal meningoencephalitis in 6 to 9 days. Also, subcutaneous inoculation of older mice induces protective immunity against subsequent intracerebral challenge (224). Human infections have been associated with exposures to mouse-infested areas and to hamsters with experimentally propagated, LCM-contaminated tumors. The LCM mouse is a distinct strain with defined, predictable, virus-related disease characteristics.

1. All progeny are infected congenitally and the infection is lifelong.
2. The assay for LCM virus is rapid and accurate.
3. The development of chronic lesions is predictable.
4. The syndrome appears to result from an immunogenic dyscrasia.
5. The lesions respond to immunosuppressive therapy, but viremia persists.
6. The virus is associated with oncogenicity.

The chronic LCM syndrome resemble some human diseases in which etiology and mechanisms have not yet been determined.

Viral Hepatitis

The lack of a laboratory animal model or an apparently susceptible animal host to the human hepatitis viruses has hampered progress. Epidemiological and immunological evidence show that infectious Hepatitis A (infectious hepatitis) and Hepatitis B (serum hepatitis) are caused by at least two distinct infectious agents. Deinhardt et al. (53) reported biochemical and histological changes compatible with hepatitis in two species of marmosets, *Saguinus nigricollis* and *S. fuscicollis,* after inoculation of acute phase serum or plasma from patients with viral hepatitis. Parks and Melnick (220) found that inoculation of clinical human hepatitis material into cotton-topped tamarin marmosets (*S. oedipus oedimpomidas*) resulted in histological and biochemical evidence of liver damage, but such changes also developed spontaneously in control marmosets. Holmes et al. (129) have found that marmosets inoculated with well-documented, acute-phase infectious hepatitis plasma from three human volunteers developed hepatitis, whereas other marmosets injected with preinfection plasma from another volunteer showed no evidence of liver damage.

Lorenz et al. (168) reported that the biochemical evidence of liver damage and hepatic lesions was induced by injecting marmosets with samples from a pool of sera obtained from human volunteers who were suffering experimentally transmitted infectious hepatitis. Sera pooled from the marmosets injected with the human material were still effective in producing hepatic lesions even after five passages through marmosets. Two marmosets that had been previously inoculated with infectious marmoset serum did not develop evidence of hepatitis when challenged with human infectious hepatitis serum. No evidence of liver damage was found in marmosets injected with serum containing the Hepatitis B antigen, and this antigen was not found in any of the marmoset sera examined. Other experimental work on infectious hepatitis has been reported using various animal species (52, 55, 184, 201, 203, 243, 263).

Infantile Viral Diarrhea

A model of gastroentralia virus in piglets was studied by Kelly et al. (150). Piglets infected with transmissible gastroentralia virus, compared to matched-fed litter mates had massive diarrhea characterized by increased quantities and concentrations of sodium, potassium, and chloride ions. Determination of Na- and Ka- ATPase in mucosal homogenates from the small and large intestine revealed a decrease in the activity of this enzyme in the upper small bowel. It seems that a defect in active sodium transport in this region may be an important factor in the pathogenesis of viral diarrhea. Further studies using this model should help to define the mechanisms producing diarrhea in chronic infantile gastroenteritis.

Dengue Virus Infection

Dengue infections in monkeys were not generally accompanied by any of the more obvious clinical manifestations of human dengue fever, such as fever, rash, gastrointestinal symptoms, myalagia, or headache. However, other objective clinical or clinical laboratory abnormalities are well documented and serve to distinguish mild and severe dengue syndromes. Virological, serological, and clinical responses to infection were studied by Halstead et al. (113) in 122 *Macaca mulatta* monkeys. Seventeen monkeys of three other species were inoculated with dengue 1-4 viruses passaged in tissue culture. Susceptible rhesus monkeys inoculated with either high ($10^{3.7}$ to $10^{5.7}$ pfu) or low (8–50 pfu) doses of virus always developed antibodies frequently with dengue 2 infection but less frequently with dengue 1 infection; lymphadenomegaly, depression of leukocyte count, and lymphocytosis were noted. In approximately 90 percent of the infected animals, viremia began 2 to 6 days after inoculation; 90 percent of dengue 2 and 4 viremia lasted 6 days or less; the average duration of dengue 1 viremia was somewhat longer and the duration of dengue 3 viremia was shorter. The titers of hemagglutination inhibition (HAI) homologous antigen in convalescent sera were usually twofold higher than titers to heterologous dengue viruses; antibody response to dengue 4 infection was relatively specific. No abnormalities were observed in serial hematocrit, prothrombin time, and determination of total protein. Levels of complement in serum rose several days after the start of serial bleedings in both infected and control animals. The courses of infection due to dengue viruses were found to be similar to humans and monkeys.

Viral Infection of Upper Respiratory Tract

Acute Newcastle disease virus infection following intranasal inoculation of chicks with a misogenic strain of the virus produced a localized infection of the middle turbinate that was histologically demonstrable 18 hours after inoculation. There was destruction of mucous cells of individual acini in the under surface of the middle turbinate, and the infection rapidly spread to ciliated and goblet cells and to neighboring acini. By day 2, remodeling of the mucosa began, although there was continued destruc-

Table 4-2. Use of Laboratory Animals for the Study of Induced Viral Infections

Animal Model	Species	Human Counterpart	Reference
Blue tongue vaccine virus	Lamb	CNS Malformation, hydroencephaly	215
Dengue virus	Monkey	Dengue virus infection	114 174
Experimental Kuru	Chimpanzee	Kuru	94
Influenza virus	Murine Simian Primate	Rhino virus disease	28 71
Herpes Virus	Squirrel monkey Simian primate Guinea pig Monkeys nonhuman primates	Herpes, Herpes Encephalitis, Cutaneous Herpes Infection, Genital Herpes Type 2 infection, Neoplastic diseases	73 136 178 131 77,207 2, 67
LCM Virus	Mouse	Congenital & persistent LCM infection	229
Monkey Pox	Baboon	Pox Virus	121 279
Mouse cyto-megalovirus	Mouse	Viral Disease of Labyrinth	51
Rhinovirus infection	Vervet Monkey	Rhinovirus disease	181
Scrapie-like disease (slow virus infection of mink)	Mink	Kuru, Creuzfeldt-Jakob disease	180
Sendai virus	Rats	Acute respiratory infection	279
Avian Leukosis	Fowl	Lymphocytic leukemia	25

Table 4-2.   (Continued)

Animal Model	Species	Human Counterpart	Reference
Viral embryo-pathies	Hamster	Viral embryo-pathies	79
Viral gastro-enteritis	Piglets	Infantile viral diarrhea	150
Viral hemo-lytic anemia	Horse	Hemolytic anemia	123
Viral Hepa-titis	Dog	Viral hepatitis	229
	Turkey		275
	Nonhuman		
	primates		55,259,260
	Duck		69
	Rats		175
Viral Leukemia	Cat	Lymphocytic leukemia	141

tion, inflammatory infiltration, and frequent loss of cartilage basophilia. By day 3, polymorphonuclear cells almost disappeared, epithelial mitosis commenced, and lymphocyte infiltration intensified; the plasma cells normally present along the lateral nasal gland ducts were often destroyed, and very occasionally the glands themselves were destroyed. By days 5 and 6, inflammation greatly decreased, and by day 8, the mucociliated epithelium was essentially normal. The infection is sequentially comparable to acute and mild rhinitis in humans (20, 21). Other experimental virus infections include rubella virus in rats (16), influenza virus in mice (72, 140), Sendai virus in mice (139), Sindbis virus in mice (38, 146), Coxsackie virus in mice (37, 277), kuru virus in primates (15), Borna virus in hamsters, (9) and adenovirus infection in mice (128, 177). A number of other viral disease models in experimental animals are listed in Table 4-2.

PROTOZOAN DISEASES

Malaria

Before 1966, those concerned with developing more effective agents for treatment of human malarias were forced to use avian, simian, and rodent malaria as the models for human malarias. The demonstration that the splenectomized owl monkey, *Aotes trivirgatus* (also known as the night monkey and douroucouli) could be infected with *Plasmodium vivax* (292) and *P. falciparum* (282, 283, 293) presented investigators with their first opportunity to study the biological and chemotherapeutic features of infections with these human plasmodia in an experimental host that might be maintained easily in a laboratory environment. Efforts have been directed toward the development of a non-human primate model for the demonstration that the chemotherapeutic responses of owl monkeys infected with various strains of *P. falciparum* and *P. vivax* duplicated almost precisely the responses of naturally infected patients or human volunteers. Also, attention was directed toward the application of owl monkeys, *P. falciparum* and *P. vivax,* as models in the search for new and well-tolerated drugs effective against infections with both drug-susceptible and multiple-drug resistant strains. These models have been utilized for pilot studies in a remarkably economic manner. As a result of the experimental animal studies, a number of compounds with outstanding antimalaria activity— 4-quinoline metathanol, 9-phenathrene methanol, and 6-sulfur substituted, 2, 4, diaminoquinozoline derivatives—have been identified. Further studies have been made on (1) the breadth of activity of the most promising compounds against infections with drug-susceptible and drug-resistant strains; (2) the effectiveness of various dosages; (3) the liabilities of emergence of drug resistance; and (4) ways of delimiting such an event where there were prospects of its occurring.

Natural and induced malarias in the western hemisphere were reported by Young (294). *Plasmodium brasilianum* and *P. simium* are found naturally in western hemisphere monkeys. Geiman and Siddique (101) reported that *P. malariae* is transmitted readily from humans to the *Aotus* strain of monkeys. *Plasmodium malariae* strains have been experimentally induced into six species of monkeys that are not natural hosts. Seven species of monkeys have been experimentally infected with one or more of the human malarial parasites, *P. vivax, P. falciparum,* and *P. malariae.* The experimental infections differ among the monkey hosts and between these hosts and human hosts in various aspects. The vector of the naturally occurring monkey malarias is not known. Several mosquito species can be infected by human malarias in monkey hosts and can transmit the infections to other monkeys and to humans. These new host-parasite vector relationships present many problems for investigation. Nevertheless, human malaria induced in monkey hosts is a promising model for the study of the malarias of man.

Toxoplasmosis

Old-world primates are relatively resistant to infection with *Toxoplasma gondii* whereas new-world primate species are more susceptible. Some old-world species are readily infected with this protozoan but rarely show symptoms (261). Also *Toxoplasma* infection in humans rarely causes symptoms of disease. Araujo et al. (10, 11) developed a model for the study of *T. gondii* transmission through the placenta of *Macaca arctoides* (stump-tailed monkey). Using the Sabin-Feldman dye test, 20 percent were found to be positive. However, the serological responses varied with the size of the inoculum, the form of the parasite inoculated, and the route of inoculation. Skin tests were positive in all animals 15 weeks postinfection. Parasitemia, demonstrated by mouse inoculation, was found in four animals during the first 10 days postinfection but not thereafter. *Toxoplasma gondii* was demonstrated in mice inoculated with the brain and heart of one monkey that was sacrificed 10 weeks postinfection and from another that died of unrelated causes 30 weeks postinfection. *Macaca arctoides* is susceptible to infection with *Toxoplasma.* Organisms may persist in a latent form in tissues beyond the acute stage of the infection. No monkey died of infection despite large-size inocula. The "natural" resistance of this species to mouse virulent strains of *Toxoplasma* is similar to that observed in other mammalian species including humans.

Amebic Meningoencephalitis

The basic features of primary amebic meningoencephalitis (PAM) described in humans (46, 60, 59) have all been noted in experimental infections in the mouse (40, 46, 182). These features include the same incubation period and portal of entry; residence of amebas in the olfactory mucosa with invasion and migration through submucosal structures into nerve plexuses; passage of amebas through pores of cerebiform plate into the subarachnoid space; subsequent invasion of olfactory bulbs and lobes with spread to more distant areas of the brain; frequent aggregation of amebas in perivascular spaces; and a predominantly neutrophilic cellular response associated with and superimposed upon widespread areas of hemorrhagic necrosis.

In the natural disease in humans, there is a brief incubation period (5 to 6 days) and a short hospital course (72 hours) with rapid central nervous system deterioration, coma, and death. Using the same amebas isolated from fatal human cases, identical infections can be regularly induced in white Swiss (SW) mice by intranasal instillation of 10^3 to 10^4 viable trophozoites. About 45 percent of the mice develop disease and die in 5 to 7 days. Grey and white matter are both affected. There is an acute inflammatory reaction associated with hemorrhage, edema, disintegration of neural structures, and widespread invasion by amebas. Numerous organisms may be seen. This model may be used to clarify (a) the mechanisms of penetration by amebas through the nasal and olfactory epithelium; (b) factors regulating proliferation of amebas within the nasal mucosa; (c) host factors involved in invasion or spread of the organism; and (d) effects of chemotherapeutic agents and antibiotics on preventing or controlling infection.

Chagas' Disease

A new, highly susceptible laboratory animal model for Chagas' disease was described by Bafort et al. (17, 18). During their investigation on the virulence of the infection of trypanosomes from the nonhuman primate *Aotes trivirgatus* to different laboratory animals, they observed that *Chinchilla lanigira* was highly susceptible to a strain of *T. curzi*. The inoculation into mice (gib/TB), rat (Wistar), and chinchilla has shown that the susceptibility of the latter, at least for this particular strain, is considerably higher than that of the other animals. A continuously increasing parasitemia was observed and the infection proved to be consistently fatal. On histological examination, numerous pseudocysts with amastigote stages were found in the heart and to a lesser extent in other organs (e.g., skeletal muscle, gut wall). Mice and rats inoculated at the same time did not show trypanosomes in the peripheral blood during the period of investigation.

These findings suggested that the chinchilla may be an excellent experimental model for the study of Chagas' disease; the animal is not only highly susceptible to infection (either with small numbers of blood form or culture form trypanosomes) but also shows large numbers of trypanosomes in the blood as well as numerous intracellular forms in the organs. The outcome of infection is consistently fatal but is not too fulminating in nature. The animal is a relatively large mammal and easy to handle for experimental purposes.

Recently, many studies have been reported on experimental parasitic infections

Table 4-3. Use of Laboratory Animals for the Study of Induced Parasitic and Mycotic Infections

Animal Model	Species	Human Counterpart	Reference
I. PARASITIC INFECTIONS			
Amebic meningoencephalitis (CNS) Protozoan infection	Nonhuman primates	Amebic meningoencephalitis	183
Ascariasis	Mouse	Endemic Hookworm infection	130
Chagas's Disease	Nonhuman primates	Chagas's Disease	17
Fasciola hepatica infection	Cattle	Cholongiohepatitis	4
Filariasis	Nonhuman primates	Filariasis	253
Malaria	Owl monkey	Malaria	254,294
	Simian		98
	Penguin		110
	Avian		133
Nematospiroidis dubias infection	Mouse	Endemic Hookworm infection	24
Toxoplasma gondii infection	Nonhuman primates	Toxoplasmosis	14
Trichinosis	Guinea Pig	Trichinosis	41
II. MYCOTIC INFECTIONS			
Candidiasis	Mouse, Rat	Candidiasis	221
Dermatophytosis	Guinea pig	Dermatophytosis	171
Nocardiosis	Rhesus monkey	Pulmonary nocardiosis (nodular)	137
Aspergillosis	Rat	Pulmonary mycotic infection	89

including *Trypanosoma* infection in monkeys (209), *Coccidioides* (290), filariasis (253), *Trichospirurosis* (214), schistosomiasis (68, 70, 159, 230, 273, 284, 285), ascariosis (42), nosematosis (217), tristronglosis (162), and fascialiasis (120, 174). Other protozoa and helminthic disease models in laboratory animals are listed in Table 4-3.

SPONTANEOUS INFECTIOUS DISEASES OF EXPERIMENTAL ANIMALS AS MODELS OF HUMAN DISEASE

Many bacterial, viral, parasitic, and fungal diseases occur in laboratory animals spontaneously (Chapter 3). Some of these infectious disease may serve as excellent models for the study of human infections, as they are quite similar in nature to infective agents, hosts responses, and pathogenicity. Because of the vast literature available in this field, it is not feasible to discuss all the spontaneous infectious disease models in detail. Therefore, selected infectious diseases are listed in Table 4-4, with pertinent references for further information.

Table 4-4. Use of Laboratory Animals for the Study of Spontaneous Infections

Animal Model	Species	References
I. BACTERIAL INFECTION		
Bacillus piliformis focal liver necrosis	Mice	237
Corynebacterial infections-Pseudotuberculosis	Mice & Rats	89 223
Hemobartonella muris and Eperythrozoon coccidiosis	Mice & Rats	104 106 187 238 258
Leprosy	Rat	82
Leptospira vallum	Mice & Hamsters	261
Mycobacterial infections	Hamsters & Rhesus monkeys	87 255
Myocarditis & Pulmonary vasculitis related to a Rickettsia-like agent in neutrophil granulocytes	Mice	218
Mycoplasma infections	Mice & Rats	157 278

Table 4-4. (Continued)

Animal Model	Species	References
Paracoliform bacteria	Guinea pigs, Hamsters & Mice	58 75 157 256
Pasteurella pneumotropica, low-grade pulmonary infection, endocarditis	Mice & Rats	127 142
Pseudomonas infections	Rats & other animals	85 213
Salmonellosis	Mice	109
Staphylococcal infections	Rats treated with corticosteroids	90
Streptobacillus moniliformis	Mice & Rats	204
Streptococcal disease	Guinea Pig	6
Tetanus	Mice	173

II. VIRAL INFECTIONS

Dengue Virus	Monkey	114 174
Ectromelia Virus (Model for small pox)	Mice	34 78 115 143 289
Fetal and Neonatal lymphocytic choriomeningitis (LCM)	Mice	19 212 281 288
Machupo Virus (Model for Rubella)	Hamsters	144
Mammary tumors, viral oncogenesis in Humans	Mice	156
Mouse Hepatitis virus group (MHV, JHM, H74, 71)	Mice	5
Mouse Poliomyelitis	Mice	219
Salivary gland virus	Mice	185 265

168

Table 4-4. (Continued)

Animal Model	Species	References
III. PARASITIC INFECTIONS		
Protozoan:		
Pneumocystis infections	Rats with corti-costeroids, al-kylatings agents, (Cyclophosphamide) and with a depressed immune response	88
Nematodes:		
Dirofilaria immitis	Dogs	88
Litomosoides carinii	Rats	88
Syphacia obvelata	Mice	88
Trematodes:		
Schistosoma mansoni	Mice	100 247 286
Cestodes:		
Hymenoleposis nana	Mice & Rats	88
H. diminuta	Mice, Rats (Re-quires intermediate hosts)	88
Arthropods:		
Bartonellosis, Eperythozoon infection	Mice, Rats	104
Flea bite sensitivity	Guinea pigs	27 132
Myobia musculi, pruritic dermatitis and secondary amyloidiosis	C57, Black Mice	97
IV. MYCOTIC INFECTIONS		
Alternaria sp.	Rats treated with corticosteroids	89 161
Aspergillus flavus	Rats treated with corticosteroids	89 161
A. fumigatus	Rats treated with corticosteroids	89 161

Table 4-4. (*Continued*)

Animal Model	Species	References
A. glaucus	Rats treated with corticosteroids	89 161
A. niger	Rats treated with corticosteroids	89 161
Penicillum sp.	Rats treated with corticosteroids	89 161
Phycomycetes	Rats treated with corticosteroids	89 161

REFERENCES

1. Ablashi, D. V., and G. R. Pearson, *Cancer Res.,* **34**:1232 (1974).
2. Ablashi, D. V., and G. R. Pearson, *Clin. Orthop.,* **99**:1232 (1974).
3. Abrams, C. D., H. Schneider, S. C. Formal, and H. Spring, *Lab. Invest.,* **12**:1241 (1963).
4. Alicata, J. E., and L. E. Swanson, *Am. J. Vet. Res.,* **2**:417 (1941).
5. Andrews, C. H., and H. G. Pereira, *Viruses of Vertebrates,* Williams & Wilkins, Baltimore, 1967.
6. Andriole, V. T., D. A. Nagel, and W. O. Southwick, *Yale J. Biol. Med.,* **47**:33 (1974).
7. Angrist, A., and Mollor, *Am. J. Med. Sci.,* **215**:149 (1948).
8. Angrist, A., and M. Oka, in E. Bajusz and A. Josmin (eds.), *Experimental Pathology,* Vol. 3, S. Karger, New York, 1967.
9. Anzil, A. P., K. Blinzinger, and A. Mayr, *Arch. Gesamte Virusforsch,* **40**:52 (1973).
10. Araujo, F. G., M. M. Wong, J. Theis, and J. S. Remington, *Am. J. Trop. Med. Hyg.,* **22**:465 (1973).
11. Araujo, F. G. and J. S. Remington, *Proc. Soc. Exp. Biol. Med.,* **139**:254 (1972).
12. Arko, R. J., *Science,* **177**:1200 (1972).
13. Arko, R. J., *Lab. Anim. Sci.,* **23**:105 (1973).
14. Arko, R. J., *J. Infect. Dis.,* **129**:451 (1974).
15. Asher, D. M., C. J. Gibbs, Jr., D. E. Alpers, M. P. Gajdusek, and D. C. Gajdusek, in W. P. McNulty (ed.), *Symposia of the Fourth International Congress in Primatology,* Vol. 4, S. Karger, Basel, 1973.
16. Avila, L., W. E. Rawls, and P. B. Dent, *J. Infect. Dis.,* **126**:585 (1972).
17. Bafort, J. M., P. Kageruka, and G. T. Timperman, *Trans. R. Soc. Trop. Med. Hyg.* **67**:434 (1973).
18. Bafort, J. M., P. Kageruka, and G. Timperman, *Trans. R. Soc. Trop. Med. Hyg.,* **67**:435 (1973).
19. Baker, F. D., and J. Hotchins, *Science,* **158**:502 (1967).
20. Bang, B. G., and F. B. Bang, *J. Exp. Med.,* **125**:409 (1967).
21. Bang, B. G., and F. B. Bang, *Am. J. Pathol.,* **76**:333 (1974).
22. Barenfus, M., and W. L. Hewitt, *Am. J. Med. Sci.,* **263**:103 (1972).
23. Barker, H. J., and J. E. Lindsay, *Exp. Hematol.,* **14**:52 (1967).
24. Bartlett, A., and P. A. Ball, *Ann. Trop. Med. Parasitol.,* **66**:129 (1972).
25. Beard, J. W., *Ann. N.Y. Acad. Sci.,* **108**:1057 (1963).
26. Beise, W. R., and R. M. Fuser, *Am. J. Clin. Nutr.,* **23**:1069 (1970).

27. Benjamini, R., B. F. Feingold, B. F. Young, L. Kartman, and M. Shinizu, *Exp. Parasitol.,* **13:**143 (1963).

28. Berendt, R. F., *Infect. Immunol.,* **9:**101 (1974).

29. Blackman, U., S. J. Goss, and M. J. Pickett, *J. Infect. Dis.* **129:**376 (1974).

30. Blaskovic, D., J. Svobodova, K. Weidnerova, and B. Stastny, *Acta. Virol.,* **15:**522 (1971).

31. Blokhin, N. N., N. I. Kuzavova, T. V. Belysheva, and D. S. Momot, *Probl. Tuberk.,* **51:**86 (1973).

32. Bonventre, P. F., *Infect. Immunol.,* **7:**556 (1973).

33. Boros, D. L., and K. S. Warren, *Immunology,* **24:**511 (1973).

34. Briody, B. A., in *Viruses of Laboratory Rodents,* Natl. Cancer Inst. Monograph No. 20, 1966.

35. Brody, R. S., and W. Schalm, *J. Am. Vet. Med. Assoc.,* **143:**231 (1963).

36. Brunner, H., W. D. James, R. L. Horswood, and R. M. Chanock, *J. Infect. Dis.,* **127:**315 (1973).

37. Burch, G. E., and C. Y. Tsui, *Br. J. Pathol.,* **52:**360 (1971).

38. Burch, D. G., C. Y. Tsui, and J. M. Harb, *Lab. Invest.,* **26:**163 (1972).

39. Burrous, S. E., and J. B. Cower, *Appl. Microbiol.,* **18:**448 (1969).

40. Carter, R. F., *Trans. R. Soc. Med. Hyg.,* **66:**193 (1972).

41. Catty, D., *Monogr. Allergy,* **5:**1 (1969).

42. Chakravarty, A. A. K., S. K. Sharma and J. N. Ghose, *J. Commun. Dis.,* **5:**190 (1973).

43. Chino, F., *Acta Pathol. Jap.,* **23:**479 (1973).

44. Cho, C. T., M. Manhar, L. Muangmance, D. W. Voth, and C. Lin Fed. Proc., **30:**352 (1971).

45. Cho, C. T., C. Liu, D. W. Voth, and K. K. Feng, *J. Infect. Dis.,* **128:**718 (1973).

46. Colbertson, C. G., *Ann. Rev. Microbiol.,* **25:**231 (1971).

47. Clyde, W. A., Jr., *J. Infect. Dis. Suppl.,* **127:**S69 (1973).

48. Clyde, W. A. Jr., *J. Infect. Dis. Suppl.* **127:**587 (1973).

49. Couvit, J., and M. E. Pinardi, *Science,* **184:**1191 (1974).

50. Curlin, G. T., J. P. Craig, A. Subong, and C. C. Carpenter, *J. Infect. Dis.,* **121:**463 (1970).

51. Davis, G. L., and M. Strauss, *Ann. Otol. Rhinol. Larygnol.,* **84:**584 (1973).

52. Deinhardt, F., *Vox Sang.,* **19:**261 (1970).

53. Deinhardt, F., *Nature* (New Biol) **229:**130 (1971).

54. Deinhardt, F., L. Falk, B. Marczynska, G. Shramek, and L. Wolfe, *Haematology,* **39:**416 (1973).

55. De Smyter, J., W. T. Liu, P. De Somer, and J. Mcreilmans, *Vox Sang.,* **24:**17 (1973).

56. Dick, H. M., and J. E. Baird, *Br. J. Dermatol.,* **86:**Suppl. 8:28 (1972).

57. Dillard, S. H., W. J. Cheatham, and H. L. Moses, *Lab. Invest.,* **26:**391 (1972).

58. Dubos, R., R. W. Schaedler, R. Costello, and P. Holt, *J. Exp. Med.,* **122:**67 (1965).

59. Duma, R. J., W. C. Rosenblum, R. F. McGhee, M. M. Jones, and E. C. Nelson, *Ann. Intern. Med.,* **74:**923 (1971).

60. Duma, R. J., *CRC Critical Review Clin. Lab. Sci.,* **3:**163 (1972).

61. Durack, D. R., and P. B. Beeson, *Br. J. Exp. Pathol.,* **53:**44 (1972).

62. Durack, D. T., P. B. Beeson, and R. G. Petersdorf, *Br. J. Exp. Pathol.,* **54:**142 (1973).

63. Dzhikidze, E. K., in *Biology and Acclimatization of Monkeys,* Materials of a Symposium, Moscow, Nauka, (1973).

64. Eigelbach, H. T., S. Salow, J. T. Tules, and R. B. Hornick, in *Symposium of the Use of Non-Human Primates in Drug Evaluation,* University of Texas Press, 1968.

65. Eklund, C. M., and W. J. Hadlow, *Medicine* (Baltimore) **52:**357 (1973).

66. el-Garem, A. A., M. A. Rifaat, M. A. Madwar, and X. H. Mousa, *J. Egypt. Med. Assoc.,* **55:**491 (1972).

67. Ellenberg, T., and J. L. Sever, *Clin. Orthop.,* **99:**118 (1974).

68. Erickson, D. G., F. Von Lichtenberg, E. H. Sadun, H. L. Lucia, and R. L. Hickman, *J. Parasitol.,* **57:**543 (1971).

69. Fabricant, J., C. G. Richard, and P. D. Levine, *Avian Dis.,* **1:**256 (1957).

70. Fadl, A. M., *J. Helminthol.,* **45:**111 (1971).

71. Fairchild, G. A., and J. Roan, *Arch. Environ. Health,* **25:**51 (1972).

72. Fairchild, G. A., J. Roan, and J. McCarroll, *Arch. Environ. Health,* **25:**174 (1972).

73. Falk, L. A., L. G. Wolfe, and F. Deinhardt, *J. Natl. Cancer Inst.*, **51**:165 (1973).
74. Falk, L. A., *Lab. Anim. Sci.*, **24**:Part II:182 (1974).
75. Farrar, W. D., and T. M. Kent, *Am. J. Pathol.*, **47**:629 (1965).
76. Farshitchi, P., and V. J. Lewis, *J. Bact.*, **95**:1616 (1968).
77. Felsburg, P. J., R. L. Heberling, M. Brack, and S. S. Kalter, *J. Med. Primatol.*, **2**:50 (1973).
78. Fenner, F., *J. Immunol.*, **63**:341 (1949).
79. Fern, V., and L. Kilham, *Science*, **145**:510 (1964).
80. Fernald, G. W., and W. A. Clyde, Jr., *Infect. Immun.*, **1**:559 (1970).
81. Fiser, R. H., J. C. Dennison, and W. R. Beisel, *J. Infect. Dis.*, **125**:54 (1972).
82. Fite, G. L., *Natl. Inst. Health Bull.*, **173**:45 (1940).
83. Flint, J. C., M. M. Roepke, and R. Jensen, *Am. J. Vet. Res.*, **19**:164 (1958).
84. Flint, J. C., M. M. Roepke, and R. Jensen, *Am. J. Vet. Res.*, **20**:33 (1959).
85. Flynn, R. J., *Lab. Anim. Care*, **13**:69 (1963).
86. Freedman, L. R., and M. L. Johnson, *Yale J. Biol. Med.*, **45**:163 (1972).
87. Frenkel, J. K., *Am. J. Pathol.*, **34**:586 (1958).
88. Frenkel, J. K., in R. A. Marcial-Rogas (ed.), *Pathology of Protozoal and Helminthic Diseases*, Williams & Wilkins, Baltimore, 1969.
89. Frenkle, J. K., J. T. Good, and J. A. Shultz, *Lab. Invest.*, **15**:1559 (1966).
90. Frenkel, J. K., and M. Havenhill, *Lab. Invest.*, **12**:1204 (1963).
91. Frenkel, J. K., and H. R. Wilson, *J. Infect. Dis.*, **125**:216 (1972).
92. Gajdusek, D. C., *Am. J. Clin. Pathol.*, **56**:320 (1971).
93. Gajdusek, D. C., *Ann. Clin. Res.*, **5**:254 (1973).
94. Gajdusek, D. C., C. J. Gibbs, Jr., and D. M. Asher, *Science*, **162**:693 (1968).
95. Gallin, J. I., D. Kaye, and W. M. O'Leary, *N. Engl. J. Med.*, **281**:1081 (1969).
96. Gallin, J. I., W. M. O'Leary and D. Laye, *Proc. Soc. Exp. Biol. Med.*, **133**:309 (1970).
97. Galton, M., *Am. J. Pathol.*, **43**:855 (1963).
98. Garnham, P. C. C., *J. Parasitol.*, **49**:905 (1963).
99. Garrison, P. K. and L. R. Freedman, *Yale J. Biol. Med.*, **42**:394 (1970).
100. Gear, J. H. S., in I. K. Mostofi (ed.), *Bilharziasis*, Springer-Verlag, New York, 1967.
101. Geiman, Q. M., and W. A. Siddique, *Am. J. Trop. Med. Hyg.*, **18**:351 (1969).
102. Gemski, P., Jr., A. Takeuchi, O. Washington, and S. B. Formal, *J. Infect. Dis.*, **126**:523 (1972).
103. Gerone, P. J., *Lab. Anim. Sci.*, **24**:139 (1974).
104. Gledhill, A. W., *J. Gen. Microbiol.*, **15**:292 (1956).
105. Gledhill, A. W., J. S. F. Niven, and J. Seamer *J. Hyg.*, **63**:73 (1965).
106. Goldschmidt, F., *Appl. Microbiol.*, **23**:121 (1972).
107. Goldschmidt, F., *Appl. Microbiol.*, **23**:130 (1972).
108. Good, R. C., *Ann. N.Y. Acad. Sci.*, **154**:200 (1968).
109. Gowen, J. W., *Bact. Rev.*, **24**:192 (1960).
110. Griner, L. A. and B. W. Sheridan, *Am. J. Vet. Clin. Pathol.* **1**:7 (1967).
111. Guze, L. B., and P. B. Beeson, *J. Exp. Med.*, **104**:803 (1956).
112. Haberman, R. T., and F. P. Williams, *J. Natl. Cancer Inst.*, **20**:933 (1958).
113. Halstead, S. B., H. Shotwell, and J. Casals, *J. Infect. Dis.*, **128**:7 (1973).
114. Halstead, S. B., H. Shotwell, and J. Casals, *J. Infect. Dis.*, **128**:15 (1973).
115. Halstead, S. B., and C. Yamerat, *Am. J. Publ. Health*, **55**:1386 (1965).
116. Hamburger, M., E. A. Gall, and N. C. Scott, *Arch. Intern. Med.*, **129**:496 (1971).
117. Handl, R., and O. Kube, *Vet. Med.* (Praha) **17**:175 (1972).
118. Hannan, P. C. T., *Ann. Rheum. Dis.*, **30**:316 (1971).
119. Harrison, L. H., A. S. Cass, B. C. Bullock, and C. E. Cox, *J. Urol.*, **109**:163 (1973).
120. Hayes, R. J., J. Bailer, and M. Mitrovic, *J. Parasitol.*, **59**:314 (1973).
121. Heberling, R. L., and S. S. Kalter, *J. Infect. Dis.*, **124**:33 (1971).
122. Heberling, R. L., and S. S. Kalter, *Lab. Anim. Sci.*, **24**:Part II; 142 (1974).
123. Henson, J. B., J. A. Gorham, G. A. Padger, and N. C. Davis, *J. Am. Vet. Med. Assoc.*, **151**:1830 (1968).
124. Hill, G. B., S. Osterhout, and P. C. Pratt, *Infect. Immun.*, **9**:599 (1974).
125. Hirsch, M. S., P. H. Black, and M. R. Proffitt, *Fed. Proc.*, **30**:1852 (1971).
126. Hite, K. E., M. Locke, and H. C. Hesseltine, *J. Infect. Dis.*, **84**:1 (1942).

127. Hoag, W. G., P. W. Westmore, J. Rogers, and H. Meier, *J. Infect. Dis.,* **111:**135 (1962).
128. Hoenig, E. M., G. Margolis, and L. Kilham, *Am. J. Pathol.,* **75:**375 (1974).
129. Holmes, A. W. L. Wolfe, H. Rosenblate, and F. Dienhardt, *Science,* **165:**816 (1969).
130. Howes, H. L., Jr., *J. Parasitol.,* **57:**487 (1971).
131. Hubler, W. R., Jr., T. D. Felber, D. Troll, and M. Jarratt, *J. Invest. Dermatol.,* **62:**92 (1971).
132. Hudson, B. W., B. J. Feingold and L. Kartman, *Exp. Parasitol.,* **9:**18 (1960).
133. Huff, C. G., in *Advances in Parasitology,* Academic Press, New York, 1963.
134. Hughes, B. F., and D. Gomolka, *Invest. Urol.,* **11:**357 (1974).
135. Hull, P. S., R. M. Davies, and M. A. Lennon, *J. Peridont. Res. Suppl.* **10:**37 (1972).
136. Hunt, R. D., and L. V. Melendez, *Pathol. Vet. Basil.,* **3:**1 (1966).
137. Huppert, M., S. H. Sun, and A. J. Gross, *Antimicrob. Agents Chemother.,* **1:**367 (1972).
138. Ignarovich, V. F., *J. Hyg. Epidemiol. Microbiol. Immunol.* (Praha), **17:**176 (1973).
139. Iida, T., *J. Gen. Virol.,* **14:**69 (1972).
140. Iwasaki, Y., and H. Koprowski, *Lancet,* **1:**738 (1974).
141. Jarrett, W. F. H., W. B. Martin, and C. W. Crighton, *Nature* (London), **202:**566 (1964).
142. Jarvetz, E. J., *Infect. Dis.* **86:**172 (1950).
143. Johnson, K. M., S. B. Halstead, and S. N. Cohen, *Prog. Med. Virol.,* **9:**105 (1967).
144. Johnson, K. M., R. B. Mackenzie, P. A. Webb, and M. L. Kerns, *Science,* **150:**1618 (1965).
145. Johnson, R. T., and C. J. Gibbs, *Arch. Neurol.,* **30:**36 (1974).
146. Johnson, R. T., H. F. McFarland, and S. E. Levy, *J. Infect. Dis.,* **125:**257 (1972).
147. Jones, J. H., and D. Adams, *Br. J. Dermatol.,* **83:**670 (1970).
148. Kast, A., M. Herbst, and Stall, *Vet. Pathol.,* **8:**146 (1971).
149. Keane, W. F., and L. R. Freeman, *Yale J. Biol. Med.,* **40:**231 (1967).
150. Kelly, M., D. G. Butler, and J. R. Hamilton, *J. Pediatr.,* **80:**925 (1972).
151. Kern, E. R., J. C. Overall, Jr., and L. A. Glasgow, *J. Infect. Dis.,* **128:**290 (1973).
152. Keys, T. F., F. L. Sapico, R. Touchon, M. Barenfus, and W. L. Hewitt, *Am. J. Med. Sci.,* **263:**103 (1972).
153. Kirchheimer, W. F., *J. Med. Assoc. Thai.,* **55:**605 (1972).
154. Kirchheimer, W. F. and E. E. Stors, *Int. J. Lepr.,* **39:**693 (1971).
155. Kohn, D. F., *Lab. Anim. Sci.,* **24:**823 (1974).
156. Kaprowski, H., *Harvey Lectures,* **60:**173 (1966).
157. Kraemer, P. M., *Proc. Soc. Exp. Biol. Med.,* **117:**910 (1964).
158. Kuehne, R. W., *Appl. Microbiol.,* **26:**239 (1973).
159. Kuntz, R. E., A. W. Cheever, and B. J. Myers, *J. Natl. Cancer Inst.,* **48:**223 (1972).
160. Landy, M. and W. Braun, in *Bacterial Endotoxins,* Rutgers University Press, New Brunswick, N.J. 1964.
161. Louria, D. B., *N. Engl. J. Med.,* **277:**1065 (1967).
162. Leland, S. E., Jr., *J. Parasitol.,* **54:**437 (1968).
163. Leung, L. S., G. J. Szal and R. H. Drachman, *J. Infect. Dis.,* **126** (1972).
164. Liu, C., P. Jayanetra and D. W. Voth, *Ann. N.Y. Acad. Sci.,* **174:**828 (1970).
165. London, W. T., L. W. Catalano, Jr., A. J. Nahmias, D. A. Fuccillo, and J. L. Sever, *Obstet. Gynecol.,* **37:**501 (1971).
166. London, W. T., A. J. Nehmias, Z. M. Naib, D. A. Fuccillo, J. H. Ellenberg, and J. L. Sever, *Clin. Orthop.,* **99:**1118 (1974).
167. London, W. T., A. J. Nahmias, Z. M. Naib, D. A. Fuccillo, J. H. Ellenberg, and J. L. Sever, *Cancer Res.,* **34:**1118 (1974).
168. Lorenz, D., L. Barker, D. Stevens, M. Peterson and R. Kirschstein, *Proc. Soc. Exp. Biol. Med.,* **135:**348 (1970).
169. Lucas, C. T., F. Chandler, J. E. Martin, and J. D. Schmale, *J.A.M.A.,* **216:**1612 (1971).
170. Lucas, C. T., F. Chandler, J. E. Martin, and J. D. Schmale, *J.A.M.A.,* **218:**436 (1972).
171. Maenza, R. M., D. W. Powell, G. R. Plotkin, S. B. Formal, H. R. Jervis, and H. Sprinz, *J. Infect. Dis.,* **121:**475 (1970).
172. Maestrone, G., S. Sadek, and M. Mitrovic, *Am. J. Vet. Res.,* **34:**833 (1973).
173. Malgren, R. A., and C. C. Flanigan, *Cancer Res.,* **15:**473 (1955).
174. Mango, A. M., C. K. Mango, and D. Esamal, *J. Helminthol.,* **46:**381 (1972).
175. Marchette, N. J., S. B. Halstead, W. A. Falkler, Jr., A. Stenhouse, and D. Wash, *J. Infect. Dis.,* **128:**23 (1973).

176. Margolis, J. C., L. Kilham, and P. R. Ruffolo, *Exp. Molec. Pathol.,* **8:**1 (1968).
177. Margolis, G., L. Kilham, and E. M. Hoenig, *Am. J. Pathol.,* **75:**363 (1974).
178. Marks, M. I. and S. Carpenter, *J. Infect. Dis.,* **128:**331 (1973).
179. Marmion, B. P., and J. M. Goodburn, *Nature,* **189:**247 (1961).
180. Marsh, R. F., *Am. J. Pathol.,* **69:**209 (1972).
181. Martin, G. V., and R. B. Heath, *Br. J. Exp. Pathol.,* **50:**516 (1969).
182. Martinez, A. J., M. M. Jones, R. J. Duma, and W. I. Rosenblum, *Lab. Invest.,* **25:**465 (1971).
183. Martinez, A. J., E. C. Nelson, and R. J. Duma, *Am. J. Pathol.,* **73:**545 (1973).
184. McClure, H. M., and M. E. Keeling, *Lab. Anim. Sci.,* **21:**1002 (1971).
185. Medelaris, D. N., in K. Benerschle (ed.), *Comparative Aspects of Reproductive Failure,* Springer Verlag, New York, 1966.
186. Melendez, L. V., R. D. Hunt, M. D. Daniel, J. B. Blake, and F. G. Garcia, *Science,* **171:**1161 (1971).
187. Melendez, L. V., R. D. Hunt, M. D. Daniel, C. E. O. Fraur, H. H. Barahona, F. G. Garcia, and N. W. King, in P. M. Briggs, G. de-The, L. N. Payne (eds.), *Oncogenesis and Herpes Virus,* IARC Scientific Pub. No. 2, Lyon, France, 1972.
188. Mellins, R. B., O. R. Levine, H. J. Wigger, G. Leidy, and E. C. Curnen, *J. Appl. Physiol.,* **32:**309 (1972).
189. Melnick, J. L. (ed.), *Progress in Medical Virology,* Vol. 15, S. Karger, Basel, 1973.
190. Miller, T. E., *Am. J. Syph. Gonoc. Vener. Dis.,* **32:**437 (1948).
191. Miller, T. E., and K. B. Robinson, *J. Infect. Dis.,* **127:**307 (1973).
192. Mitruka, B. M., and A. M. Jonas, *Appl. Microbiol.,* **18:**1072 (1969).
193. Mitruka, B. M., L. E. Carmichael, and M. Alexander, *J. Infect. Dis.,* **119:**625 (1969).
194. Mitruka, B. M., and M. Alexander, *Appl. Microbial.* **17:**551 (1969).
195. Mitruka, B. M., A. M. Jonas, and Λ. Alèxander, *Inf. Imm.,* **2:**474 (1970).
196. Mitruka, B. M., *Yale J. Biol. Med.,* **44:**253 (1971).
197. Mitruka, B. M., *Yale J. Biol. Med.,* **45:**471 (1972).
198. Mitruka, B. M., in *Symposium on Rapid Methods of Identification of Microorganisms,* Stockholm, 1973.
199. Mitruka, B. M., *Proc. Int. Congress Bacteriol.,* **1:**173 (1974).
200. Mitruka, B. M. in C. G. Heden and I. Illeni (eds.), *New Approaches to the Identification of Microorganisms,* John Wiley and Sons, Inc., New York, 1975.
201. Monkey models for infectious hepatitis, *Nature* (New Biol.), **229:**130 (1971).
202. Morgan, D. C., M. A. Epstein, and B. G. Achong, *Nature,* **228:**170 (1970).
203. Morris, T. Q., and D. J. Gocke, *Proc. Soc. Exp. Biol. Med.,* **139:**32 (1972).
204. Morton, H. E., in R. J. Dubos and J. B. Hirsch (eds.), *Bacterial and Mycotic Infections of Man,* Lippincott, Philadelphia, 1965.
205. Moulder, J. W., *Perspect. Biol. Med.,* **14:**486 (1971).
206. Moxon, E. R., A. L. Smith, D. R. Averill, and D. H. Smith, *J. Infect. Dis.,* **129:**154 (1974).
207. Nahmias, A. J., W. T. London, L. W. Catalano, D. A. Fuccillo, J. L. Sever, and C. Graham, *Science,* **171:**297 (1971).
208. Nazerian, K., *Adv. Cancer Res.,* **17:**279 (1973).
209. Neal, R. A., W. H. Richards, and D. A. Farebrother, *Trans. R. Soc. Trop. Med. Hyg.,* **67:**277 (1973).
210. Nedzel, A. J., *Arch. Pathol.,* **24:**143 (1937).
211. O'Donoghue, J. M., A. I. Schweid, and H. N. Beaty, *Proc. Soc. Exp. Biol. Med.,* **146:**571 (1974).
212. Oldstone, M. B. A., and F. J. Dixon, *Science,* **158:**1193 (1967).
213. Order, S. E., A. D. Mason, M. L. Walker, R. F. Lundberg, W. E. Swotzer, and J. A. Mancrief, *Surgery,* **120:**983 (1965).
214. Orihel, T. C. and H. R. Seibold, *J. Parasitol.,* **57:**1366 (1971).
215. Osburn, B. I., and A. M. Silverstein, *Am. J. Pathol.,* **67:**211 (1972).
216. Osovskaia, A. M., and L. K. Surkova, *Probl. Tuberk.,* **51:**75 (1973).
217. Pakes, S. P., J. A. Shadduk, and R. G. Olsen, *Lab. Anim. Sci.,* **22:**870 (1972).
218. Pappenheimer, A. M., and J. B. Daniels, *J. Exp. Med.,* **98:**667 (1953).
219. Parker, J. C., R. W. Fennant, and F. G. Ward, in *Viruses of Laboratory Rodents,* Bethesda National Cancer Institute Monograph No. 20, 1966.

220. Parks, W. P., and J. L. Melnick, *J. Infect. Dis.,* **120:**539 (1969).
221. Pearsall, N. N., and D. Lagunoff, *Infect. Immun.,* **9:**999 (1974).
222. Peschanskaia, I. N., *Probl. Tuberk.,* **51:**65 (1973).
223. Pierce-Chase, C. H., R. M. Fauve, and R. Dubos, *J. Am. Med.,* **120:**267 (1964).
224. Pollard, M., *Am. J. Pathol.,* **67:**613 (1972).
225. Pollard, M., N. Sharon, and B. A. Teah., *Proc. Soc. Exp. Biol. Med.,* **127:**755 (1958).
226. Povysil, C., and L. Konickova, *Invest. Urol.,* **9:**313 (1972).
227. Powell, D. W., G. R. Plotkin, R. M. Maenza, L. I. Solberg, D. H. Catlin, and S. B. Formal, *Gastroenterology,* **60:**1053 (1971).
228. Prat, V., H. J. Mohr, M. Hatala, and L. Konickova, *Beitr. Pathol.,* **150:**55 (1973).
229. Preisig, R., K. D. Gocke, T. Morris, and S. E. Bradley, *Experientia,* **22:**701 (1966).
230. Preston, J. M., and C. James, *J. Helminthol.,* **46:**291 (1972).
231. Prohaszka, L., *Zentralbl. Bakteriol.* (Orig. A), **221:**314 (1972).
232. Quan, T. J., J. L. Meek, K. R. Tsuchiya, B. W. Hudson, and A. M. Barnes, *J. Infect. Dis.,* **129:**341 (1974).
233. Pankin, J. D., *Vet. Rec.,* **70:**693 (1958).
234. Rees, R. J., *Int. J. Lepr.,* **41:**320 (1973).
235. Rees, R. J., and A. G. Weddell, *Ann. N.Y. Acad. Sci.,* **154:**214 (1968).
236. Rees, R. J., and A. G. Weddell, *Trans. R. Soc. Trop. Med. Hyg.,* **64:**31 (1970).
237. Rights, F. L., E. B. Jackson, and J. E. Smadel, *Am. J. Pathol.,* **23:**627 (1947).
238. Riley, V., *Science,* **146:**921 (1964).
239. Robinson, T. W. E., and J. R. Dover, *Br. J. Dermatol.,* **86:**40 (1972).
240. Rocha, M., L. B. Guze, L. R. Freedman, and P. B. Beeson, *Yale J. Biol. Med.,* **30:**340 (1958).
241. Rosini, S., and D. Benetti, *Int. Urol. Nephrol.,* **4:**333 (1972).
242. Ross, R. F., and W. F. Switzer, *Med. Clin. N. Am.,* **52:**677 (1968).
243. Sabesin, S. M., and R. S. Koff, *N. Engl. J. Med.,* **290:**944 (1974).
244. Sabin, A. B., and J. Warren, *Science,* **92:**535 (1940).
245. Sack, R. B., and C. C. Carpenter, *J. Infect. Dis.,* **119:**138 (1969).
246. Sack, R. B., and C. C. Carpenter, *J. Infect. Dis.,* **119:**150 (1969).
247. Sadum, E. H., F. Von Lichtenberg, and J. I. Bruce, *Am. J. Trop. Med. Hyg.,* **15:**705 (1966).
248. Saslaw, S., H. N. Carlisle, and J. Sparke, *Proc. Soc. Exp. Biol. Med.,* **103:**342 (1960).
249. Saslaw, S., and H. N. Carlisle, *Proc. Soc. Exp. Biol. Med.,* **119:**838 (1965).
250. Saslaw, S., and H. N. Carlisle, *Br. Rev.,* **30:**636 (1966).
251. Saslaw, S., and H. N. Carlisle, *Ann. N.Y. Acad. Sci.,* **162:**568 (1969).
252. Saymen, D. G., P. Nathan, I. A. Holder, E. O. Hill, and B. G. Macmillan, *App. Microbiol.,* **23:**509 (1972).
253. Schacher, J. F., *S. E. Asian J. Trop. Med.,* **4:**336 (1973).
254. Schmidt, L. H., *Trans. R. Soc. Trop. Med. Hyg.,* **67:**446 (1973).
255. Schmidt, L. H., *Am. Rev. Tuberc. Pulmonary Dis., Suppl.,* **74:**138 (1956).
256. Schnierson, S. S., and E. Perlman, *Proc. Soc. Exp. Biol. Med.,* **91:**229 (1956).
257. Schutze, E., *Zbl. Bakt.* (Orig.), **208:**301 (1968).
258. Seamer, J., A. W. Gladhill, J. L. Barlow, and J. Hotchin, *J. Immunol.,* **86:**512 (1961).
259. Semtana, H. F., *Lab. Invest.,* **14:**1366 (1965).
260. Semtana, H. F., *Nature* (New Biol.), **229:**465 (1971).
261. Shadduck, J. A., and S. P. Pakes, *Am. J. Pathol.,* **64:**657 (1971).
262. Shier, D. R., and M. W. Long, *Am. Rev. Resp. Dis.,* **104:**206 (1971).
263. Shaw, E. D., A. P. McKee, M. Rancourt, and L. Hollenbeck, *J. Virol.,* **12:**1598 (1973).
264. Smith, D. W., *J. Infect. Dis.,* **128:**800 (1973).
265. Smith, M. G., *Progr. Med. Virol.,* **2:**171 (1959).
266. Sokoloff, L., *Am. J. Pathol.,* **73:**261 (1973).
267. Spira, W. M., and J. M. Goepfert, *Appl. Microbiol.,* **24:**34 (1972).
268. Stein, H., R. Yarom, S. Levin, T. Dishon, I. Ginsburg, and T. N. Harris, *Proc. Soc. Exp. Biol. Med.,* **143:**1106 (1973).
269. Stoenner, H. G., E. F. Grimes, F. Thraikall, and E. Davis, *Am. J. Trop. Med. Hyg.,* **7:**423 (1958).
270. Storrs, E. E., *Lepr. Rev.,* **45:**8 (1974).
271. Storrs, E. E., G. P. Walsh, H. P. Burchfield, and C. H. Binford, *Science,* **183:**851 (1974).

272. Storrs, E. E., *Int. J. Lepr.*, **39**:703 (1971).
273. Sturrock, R. F., *J. Helminthol.*, **45**:189 (1971).
274. Sulkin, S. E., *Bact. Rev.*, **25**:203 (1961).
275. Syoneyenbus, C. H. and H. Basch, *Avian Dis.*, **4**:477 (1960).
276. Thomas, L., *Fed. Proc.*, **32**:143 (1973).
277. Tsui, C-Y., G. E. Burch and J. M. Harb, *Arch. Pathol.*, **93**:379 (1972).
278. Tulley, J. J., and R. Ask-Nielson, *Am. N.Y. Acad. Sci.* **143**:345 (1967).
279. Tyrrell, D. A., and C. R. Coid, *Vet. Rec.*, **86**:164 (1970).
280. Ujiye, A., and K. Kobari, *J. Infect. Dis. Suppl.*, **121**:50 (1970).
281. Volkert, M., and J. H. Larsen, *Progr. Med. Virol.*, **7**:160 (1965).
282. Voller, A., D. R. Davies, and M. S. R. Hutt, *Br. J. Exp. Pathol.*, **54**:457 (1973).
283. Voller, A., C. C. Draper, T. Shwe, and M. S. R. Hutt, *Br. Med. J.*, **4**:208 (1971).
284. Von Lichtenberg, F., E. H. Sadun, A. W. Cheever, D. G. Erickson, A. J. Johnson, and H. W. Boyce, *Am. J. Trop. Med. Hyg.*, **20**:850 (1971).
285. Von Lichtenberg, F., T. M. Smith, H. L. Lucia, and B. L. Doughty, *Nature* (London), **229**:199 (1971).
286. Von Lichtenberg, F., E. H. Sadum, and J. I. Bruce, *Am. J. Trop. Med. Hyg.*, **11**:347 (1962).
287. Walker, J. S., F. C. Cadigan, R. A. Vosdingh, and C. T. Chye, *J. Infect. Dis.*, **128**:223 (1973).
288. Weigard, H., and J. Hotchin, *J. Immunol.*, **86**:401 (1961).
289. Wenner, H. A., and T. Y. Lou, *Progr. Med. Virol.*, **5**:219 (1963).
290. Wright, E. T., and L. H. Winer, *Int. J. Dermatol.*, **10**:17 (1961).
291. Yammanouchi, K., A. Shishido, and S. Honjo, *Exp. Anim.*, **22**:389 (1973).
292. Young, M. D., J. A. Porter, and C. M. Johnson, *Science,* **153**:1006 (1966).
293. Young, M. D., and J. A. Porter, *Trans. R. Soc. Trop. Med. Hyg.*, **63**:203 (1969).
294. Young, M. D., *Lab. Anim. Care,* **20**:361 (1970).

CHAPTER 5

LABORATORY ANIMALS IN THE STUDY OF NUTRITIONAL DEFICIENCY AND METABOLIC DISEASES

The study of animal nutrition has relevance to human dietary requirements in a very broad sense, but there are aspects of nutrition of each species (human or animal) that are peculiar to that species (21). By using an appropriate animal model for the study of nutritional deficiency disease, certain broad principles may be formulated that approximate the metabolic and nutritional behavior of humans. For example, early studies by Elvelhjem et al. (66) on the nutritional value of ascorbic acid (vitamin C) in guinea pigs led to the identification of the pellagra preventative factor (nicotinic acid) in humans. As with ascorbic acid studies, choice of the species of animal is important, for the animal analog of pellagra is black tongue in dogs; rats do not develop a classical pellagra under most conditions. Even so, despite the apparent anomaly posed by the rat, research on rat pellagra led to the recognition of the interrelationship between tryptophan and nicotinic acid and their involvement in the etiology of maize pellagra.

Rats, guinea pigs, dogs, swine, and nonhuman primates are commonly used animal species to study nutritional and metabolic characteristics of human beings. There have been many studies with experimental animals describing the effects of undernutrition (protein deficiency in particular) in early life on the development of the brain; the effects of vitamin, mineral, or amino acid deficiencies on growth and metabolism; and the effects of dietary changes on the metabolism of carbohydrates, lipids, and proteins. However, knowledge and study of specific genetic characteristics is essential for proper interpretation of the results and for establishing a baseline upon which the effect of diet on an individual response can be measured. Characteristics such as body size, food intake, body measurements (length, girth, skinfolds, etc.), organ weights, liver composition, voluntary activity, blood and body composition, and renal function vary with strains, age, and sex of animals. For example, Osborne and Mendel rats were reported to gain more body fat in certain types of diets than a strain of black rats (184). Marked differences in their responses to dietary carbohydrates in five strains of rats were reported by Marshall et al. (131). One of the more important needs in nutrition research is to determine why and to what extent individual variations in nutritional requirements exist and to learn whether the various requirements are the result of genetic determinants or of adaptations due to differences in the metabolic patterns of the individual resulting from long-standing dietary practices (16, 22, 76, 122, 128, 140, 179, 189, 196, 197, 231).

This chapter briefly describes selected animal models that may be applicable to understanding some problems of human nutrition and metabolism. A list of animal models used in the study of nutritional deficiency and metabolic disorders are given in Table 5-1. Inborn errors of metabolism and congenital diseases affecting metabolism and consequently nutritional adaptations are discussed in Chapter 6.

Table 5-1. Laboratory Animals in the Study of Nutritional Deficiency Diseases and Metabolic Diseases

Animal Model	Species	Human Counterpart	Reference
Adrenalamyloidosis	mouse	Adrenal insufficiency	98
Alcoholism	chimpanzee	Alcoholism	158
	rat	Pancreatitis	69, 182
Amyloidosis	mouse	Amyloidosis	59
	cattle		7
	horse		99
	dog		47, 87, 104
	hamster		230
	rat		82
Anemia from trichlorethylene extracted soybean meal	cattle	Hypoplastic anemia	167
Antidiuretic-hormone deficiency	mouse	Diabetes insipidus	100
	dog		160
Arteriosclerosis	cattle	Arteriosclerosis	46
	cat		119
	rabbit		78
	mouse		101
Arteriosclerosis, coronary	mouse	Coronary arteriosclerosis	101
Atherosclerosis	rabbit	Atherosclerosis	132
	pigeon		36, 180, 166
	swine		146, 114, 115
	dog		172
	Rhesus monkey		8, 927, 201
	rat, Obese rat		109, 203
Biotin deficiency	rat	Glycemic response	17
Bracken-fern poisoning	cattle	Hypoplastic anemia, thrombocytopenia, granulocytopenia	183
Cholesterol metabolism	dog	Hypercholesteremia	52
		Non-human primates	64
Cirrhosis	macaca	Cirrhosis	23
Cirrhosis, alcoholic	baboon	Alcoholic cirrhosis	49

Table 5-1. (*Continued*)

Animal Model	Species	Human Counterpart	Reference
Cushing's disease	dog	Cushing's disease	54,26,25
Essential fatty acid deficiency		Fat malabsorption	35
Factor VII deficiency	dog	Factor VII deficiency	149
Folic acid deficiency	mice	Folic acid deficiency	124
Gallstones	squirrel monkey	Gallstones	155
Gallstones, cholesterol	baboon	Gallstones	136
Glucagon treatment	pig dog	Pancreatitis	81,216
Glutathione deficiency	sheep	Glutathione deficiency	41
Glycogen-storage syndrome	dog	Von Gierke's syndrome	11,83
Growth of body, brain and kidney	rat & mice pig	Infant development	5,28,176, 187,15,73, 153
Hepatosis dietetica	swine	Massive hepatic necrosis	1,142
Hyperglycemia and obesity	rodent	Hyperglycemia and obesity	33,127,197
Hyperkeratosis	cattle	Hyperkeratosis	91
Hypocalcemia	rabbit	Acute hypoparathyroidism	229
Hypocalcemia, post-parturient	rabbit	Lactation hypocalcemia	177
Hypomagnesemic tetany	cattle	Hypomagnesemia	191
Hypothalmic, hyperpha-gic	rat	Hunger and appetite	84
Hypothyroidism	dog chicken	Hypothyroidism	125 37
Lantana camalia poisoning	sheep	Kwashiorkor	188
Lipidosis foamy macrophages	budgerigars	Lipid-storage disease	113

Table 5-1. (Continued)

Animal Model	Species	Human Counterpart	Reference
Lipotrope deficiency	rat	Fatty liver	174
Magnesium deficiency	rat	Magnesium deficiency	3,94
Malabsorption syndrome	dog	Nontropical sprue	107,173
Milk fever	cattle	Hypocalcemia	191
Muscle metabolism	rat	Muscle metabolism	199
Muscular dystrophy	mice rabbit sheep	Muscular dystrophy	143 129 150
Myelin degeneration phenylalanine defect	mouse	Myelin degeneration	169
Pericarditis,calcareuns	mouse	Calcification of heart	34
Porphyria	swine fox squirrel	Porphyria	34,51,208, 106 117
Protein-energy deficiency	Non-human primates	Protein deficiency	67
Protein-energy malnutrition	baby baboon	Kwashiorkor	43
Protein-losing gastropathy	dog	Protein-losing gastropathy	148
Purine metabolism	pig	Purine metabolism	24,58
Pyrrolizidine plant alkaloids	cattle rat	Veno-occlusive disease Hepatic megalocytosis	205 185
Scotty cramps	dog	Neurogenic muscular cramps	151
Siderosis and liver injury	Rhesus monkey dog	Siderosis and liver injury	152 123
Snell's dwarf	mouse	Thyrotropin deficiency Growth-hormone deficiency	218 14
Sodium Depletion	rat	Hypertension	105,205
Starvation	pig	Infant metabolism	84,95

Table 5-1.   (Continued)

Animal Model	Species	Human Counterpart	Reference
Starvation and refeeding	rat	Carbohydrate and lipid metabolism	96
Ulcer, duodenal or gastric	rat	Gastric ulcer	154
Undernutrition	rat	Malnutrition	159
Undernutrition and brain development	rodent	Nutrition and brain development	57
Vitamin A deficiency	rats & mice	Vitamine A deficiency	145
Vitamin B_{12} deficiency	mouse monkey	Vitamin B_{12} deficiency	75 157
Vitamine E, Selenium deficiency	cattle	Nutritional hepatic necrosis	206
Xanthomatosis	chicken	Xanthomatosis	204

ANIMAL MODELS OF STARVATION AND MALNUTRITION

A large number of human babies born at or near term are below the average birth weight and have, therefore, suffered intrauterine growth retardation. The overwhelming majority of such "small-for-dates" babies have suffered as fetuses from maternal malnutrition or undernutrition, as shown by the higher birth weights that can be achieved when mothers are properly fed. There are also many factors other than nutritional deficiency that may account for the full-term, "small-for-dates" human baby. These factors include restrictions of placental blood supply, maternal or placental pathology, and heavy smoking by mothers during pregnancy. The net effect on the fetus in such cases may well be similar to the much more common nutritional growth restriction. Adlard et al. (2) developed an animal model to study some of the nutritional implications. Rats were subjected to growth retardation either from conception to 5 days of postnatal age (FNR group) or during the period of 5 to 25 postnatal days (IR group). The 5-day-old rat had reached a stage of brain development comparable with the human baby at term; therefore, FNR group could be considered as a good model for the "small-for-dates" baby. Compared with well-nourished controls, both restricted groups showed deficits in neurological maturation. Both brain weight and body weight in adulthood were significantly reduced by earlier growth restriction in FNR group of animals. Cerebellar weight was consistently reduced more than that of the remainder of the brain. Brain stem acetylcholinesterase activity in adulthood was significantly higher than normal in IR, but not in FNR animals. A technique involving ligation of one uterine artery on the seventeenth day of gestation in rats have been used to develop a model for the human low birth weight baby (224). In this model, growth-retarded offspring have been studied, particularly in terms of hepatic carbohydrate metabolism. Such a model is clearly useful for examination of physiological events that are dependent on birth, for example, the rapid changes occurring in liver glycogen content and gluconeogenic capacity in the immediate postnatal period. Birth has, however, no comparable significance for brain development. A newborn rat brain is much less mature than that of a newborn human and corresponds to a human brain age of about 25 to 30 weeks gestation. Thus, in the techniques of Wiggleworth, the insult to the developing brain is applied for too short a time and at the wrong stage of brain growth to correspond to the insult to the brain of the intrauterine growth-retarded baby. The 5-day-old rat is probably a more appropriate model for the human newborn in terms of brain maturation.

Newborn pigs have sufficient similarities in their metabolism to indicate that they may be suitable experimental models for studies of the effects of starvation in the newborn human baby. An important difference between the two species is that the newborn human has approximately 16 percent fat whereas the newborn pig has only 1 percent. Thus, the pigs cannot draw on fat for energy reserve at birth, but by the age of 16 days, when its fat content is 15 percent, the metabolic adaptation of the pig to starvation may form a valid model for comparison to human beings at birth. Genz et al. (79) studied the effects of starvation for 4 or 5 days on normal pigs at birth or when aged 1, 3, 9, or 16 days. It was found that gluconeogenesis plays an important part in their response to starvation. Later, when the fat content of the piglet is similar to that of the newborn human, starvation does not provoke lipolysis comparable to that seen in the human infant. There is probably a quantitative difference between the two species in the sensitivity of the adipocyte to endocrine and other humoral stimuli of lipolysis. The difference is an important factor to consider when making a choice of an experimental animal to model the adaptation of the human baby to starvation.

Compared with the young of other commonly used animals in nutrition research, such as dog, cat, goat, and sheep, the baby pig appears in many ways to be most like the newborn human. Similarities in respiratory, renal, and hematologic systems have been discussed by Glauser (80). Because young pigs grow so rapidly, they have more stringent requirements than human infants; hence, a diet sufficient to sustain healthy growth in the baby pig will be more than adequate for the human baby. Piglets have been utilized to evaluate infant formulas and milk substitutes because such evaluations can be started on the newborn (163, 181). They have also been utilized in assessing diets for infants for rehabilitation from protein-caloric malnutrition (164). Probably the major utilization of pigs in nutrition research has been in the area of malnutrition. Piglets weaned on inadequate diets such as protein-deficient or calorie-deficient regimens have been used as models for the human conditions kwashiorkor and marasmus (162). Weanling miniature swine maintained on a protein-deficient diet for 3 months showed a number of blood and serum changes (208). The deficient group had lower mean serum total protein, albumin, β-globulin, calcium, and inorganic phosphorus concentrations and a decreased albumin/globulin ratio. The decreased packed cell volume, hemoglobin concentration, mean cell volume, and mean corpuscular hemoglobin values observed in the young pigs were also observed in piglets when their pregnant mothers were on a protein-deficient diet. However, the total protein albumin/globulin ratio, calcium and inorganic phosphorus were unaffected by deprivation during gestation (210). At the end of feeding, the protein-deprived group was about one-third the size of the control.

Selected organs were only about one-half (thyroid) to less than one-third the size of controls when pregnant sows were fed a restricted-protein diet; the brains of the offspring were less affected (209). The severity of symptoms on a low protein, high fat diet was reduced when the dietary fat was decreased. The anemia observed in young pigs on chronic deprivation has been reported not to be the result of a primary lack of required nutrients but an adaptation to reduced metabolic needs of the undernourished piglets (126).

Baby baboons were used to produce animal model of kwashiorkor-like disease seen in Ugandan children (43). In the initial studies, the baby baboons were weaned at 8 to 10 weeks of age and given a full-cream milk diet for 2 weeks. The milk intake was then gradually reduced and local staples with a low protein, high carbohydrate content were provided instead. Maintaining the baby baboons for as long as 100 days on this diet markedly impaired body growth but did not result in a kwashiorkor-like appearance. Subsequent stressing the animal by the introduction of periods of caloric restriction did apparently precipitate a clinical condition in many ways reminiscent of kwashiorkor. The baboons exhibited extreme mental apathy, had sparse hair, edema of the limbs and face, and skin lesions similar to the flaky paint rash found in severly malnourished children, but there was no gross accumulation of fat in the liver. The basal diet was subsequently modified by the addition of sucrose to bring the carbohydrate composition more into line with that in food eaten by local Ugandan children. This relatively minor change seemed to cause a more rapid clinical deterioration and the animals did develop fatty livers as well as other pathological changes. The final condition resembled marasmic kwashiorkor disease. This animal model is useful for investigation into the metabolic and structural changes that accompany the development of malnutrition and ultimately result in severe kwashiorkor.

The possibility that malnutrition in early life may permanently reduce the intellectual capacity of men and women has become increasingly recognized in recent years. Undernutrition has been considered for too long in adult terms as a series of deficiency diseases whose main consequences can be reversed on restoration of the deficient components. In the studies using animals, animal species should be selected whose brain growth characteristics are known and on whom nutritional or other restrictions at the supposed vulnerable time of development have beem imposed (57). Animals have then been allowed unrestricted nutrition in an attempt at rehabilitation and have been examined for residual deficiencies when they reached maturity. Permanent behavioral deficiencies as well as those of physical composition have long been sought (118, 223). Most experimental work has been with rats who have a conveniently timed brain "growth spurt" encompassed by the suckling period (86, 228).

Guinea pigs grow their brains in fetal life and the pig develops at perinatal period (53). The effects of malnutrition on the development of the brain have been best studied using the vulnerable period hypothesis which has been effectively tested and proved. Its corollaries have also stood the test, that a vulnerability related to rate of growth implied a comparative invulnerability before the brain growth spurt and an inviolate adult brain. There has been another finding in animals that could have considerable importance for extrapolation to human children. It can be shown that growth retardation does not alter the timing of the "growth spurt" in the brain. It is not true retardation of the delay of brain development. There is simply a reduction in the extent of process that nevertheless occurs at its proper, ordained time. The brain has a once-and-for-all opportunity to accomplish certain important processes of development. If this opportunity is lost, it can never be fully recovered. The same phenomenon has also been demonstrated for certain enzyme activities both in undernutrition and in phenyl-ketonuria (29, 30, 31). Metabolic reactions occurring in brain tissues are shown in Fig. 5-1.

One of the effects of starvation or undernutrition in humans or in animals is on neural development which leads to behavioral changes. Different techniques of undernu-trition may be used to produce the influence of nutrition on neural and behavioral developments of experimental animals (4, 6). Two techniques were used for producing undernutrition in infant rats and their effects on growth of the body and the brain were investigated.

In one procedure, the mothers were fed *ad libitum* and the size of their litters was varied (five, ten, and sixteen pups). In the other procedure, litter size was kept constant at eight pups per mother, but during lactation the mothers were fed either *ad libitum* or their food intake was restricted to 40 or 20 percent of normal consumption. The latter procedure was found to be the more reliable way of producing experimental retardation in the growth of the body and of the brain during preweaning period. Both of the

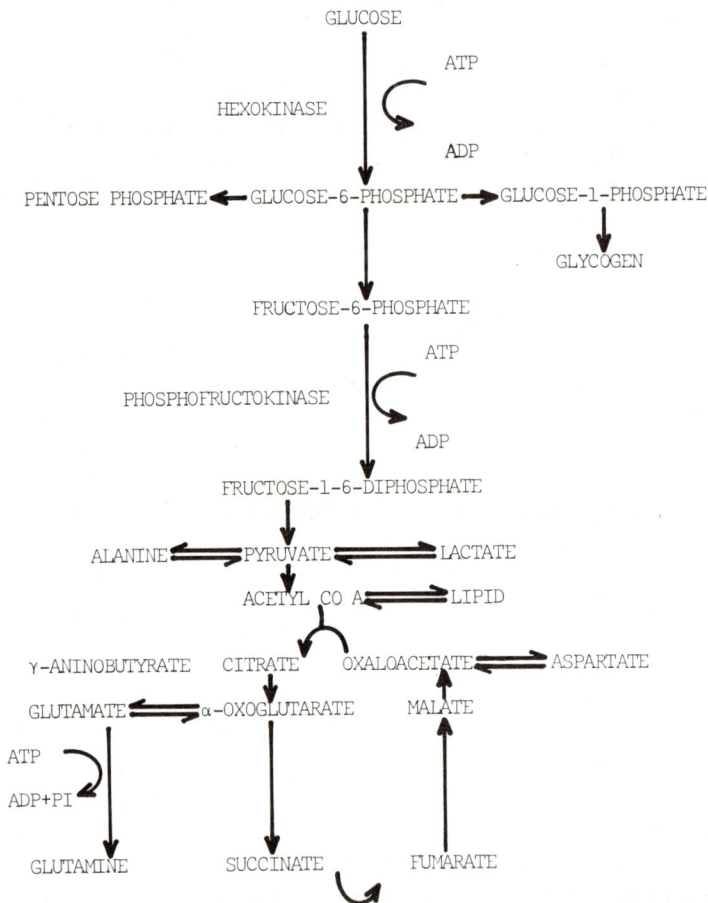

Fig. 5-1. Simplified scheme showing interrelationship between glucose me-tabolites. Adapted from *Applied Neurochemistry* (50) with permission.

procedures were used and produced appreciable retardation in the growth of the body. But the procedure of increasing a litter size to sixteen pups per mother (the usual upper limit) was found to be a less consistent and less drastic procedure, and it had little lasting effect on the growth of the brain. The body weight of "undernourished" young (i.e., pups from large litters) was 33–45 g. Apparently, the capacity of mothers for suckling large litters varies under different laboratory conditions. The irreversible retardation of brain growth produced by undernutrition is attributed to interference with cell proliferation and a reduced cell population in the brain. Because cell proliferation is very high in several brain regions during the first week after birth, a procedure of undernutrition that affects the pups from birth onward should produce a more severe retardation.

A great deal of information has been accumulated on the long-term effects of food deprivation or protein-caloric deficiency in humans and animals. It is important, however, to consider that most of the time it is very difficult to isolate malnutrition as the sole determinant of the observed abnormalities. It is necessary, therefore, to control as much as possible the experimental conditions so that the least number of determinants will interfere with the variable under study, in order to avoid unwanted and unnoticed factors that can affect the biochemistry, morphology, and the functions of the brain (12, 44, 45, 70, 71).

Exploratory behavior is clearly affected by the early experience of the animal. It has been reported that the restriction of food during early development, together with low protein diet after weaning, modified the behavior of rats. Lat et al. (111) found a positive correlation between size and weight of the rats and their spontaneous activity and a variation of the activity with the age. More recently, Frankova (72) reported data on the exploratory behavior and on the conditioned response of rats with over- and undernutrition during the preweaning period and with a low protein diet after weaning. The over- and undernutritioned rats showed both lower scores in conditioned response and less spontaneous activity compared to controls. Other data have called attention to the same poor results obtained in small and large litter animals by studying the visual and acoustic cortical evoked potentials. It seems that even a small variation from the normal could affect the brain development and in a long-term range could be highly damaging to its function, so that, to retain entirely its correct function, the brain may oscillate between very narrow limits. Furthermore, the unequivocal impairment of the biochemical and functional parameters in the two groups of animals suggests that the brain organization must be looked on in dynamic and quantitative rather than in static and qualitative terms, since, during development, a proper sequence of events must be followed that can be upset by overnutrition, which induces earlier and incorrect neural connections or by undernutrition, which causes them to develop later and also incorrectly.

The neurochemical data showed that important modifications have occurred in the amount and in the quantity of the cerebral glycoproteins and gangliosides. The presence of NANA-containing substances on the axonal and synaptosomal membranes strongly supports the hypothesis that the damage can be mediated by these structures. The importance of glial cells cannot be excluded, however, in contributing to the plasticity of the nervous tissue. The results of Dobbing (56), indicating two distinct periods of malnutrition in the human brain, can be applied also to the rat brain, the glial cells of which could be vulnerable during the stages of extrauterine development. Recent reports point out that even mild dietary restriction may induce severe chemical and functional deficits mainly in the first periods of life (21).

The studies with pigs and rats support the general conclusion that severe protein-caloric malnutrition in early life can have long-lasting, possibly permanent effects on learning behavior (45, 133, 134, 165, 221, 222). As the severity of the malnutrition decreases, the variety of behavioral abnormalities and possibly the degree to which they are affected decreases. Two types of protein-caloric malnutrition in the pig that might be likened to kwashiorkor and marasmus affect behavior in a manner that is interpreted as a retardation in the ability to learn. These animal models illustrate how the experimental animal may help in elucidating the causal relationship of early nutrition to behavior development in humans. They also offer some hope of providing information as to the manner and extent to which malnutrition may have specific effects on learning ability.

ANIMAL MODELS OF VITAMIN DEFICIENCY DISEASES

The production of experimental vitamin A deficiency in rats and mice have been reported by Moore and Holmes (145) by restricting dams and litters from parturition to a pelleted diet made mainly from white flour. Young rats usually developed clear signs of avitaminosis within 60 days from birth. Mice were more resistant, and some survived for periods up to 150 days from birth. Retention of traces of vitamin A in the liver was no more prolonged in mice than in rats. In mice, enlargement of the prostates and seminal vesicles and atrophy of the testes were usually the most prominent pathological features.

In rats, timely treatment with vitamin A acid (retinoic acid) cured xerophthalmia and restored growth. Signs of deficiency reappeared after its administration was stopped. The procedure allows supplies of animals to be kept in good general health but ready for the production of acute deficiency at a short notice. Retinoic acid was also effective in curing deficient mice.

These animal models are useful for (a) checking the presence of vitamin A in foods, particularly after its addition in stabilized form, and its efficient utilization, and (b) providing deficient minerals for use in research on the mode of action of the vitamin.

Vitamin A deficiency-associated central nervous system (CNS) disturbances have been reported in swine (97), chickens (138), dogs (137), rats (175), sheep (68), infants (42), and newborn rabbits (93).

Frenkel and White (75) reported a reproducible animal model of vitamin B_{12} deprivation. Weanling rats, protected against coprophagy, were maintained on B_{12}-deprived diets which resulted in limited growth and maturation. Specificity of the deprivation was documented by correction with only the addition of parenteral vitamin B_{12}. During deprivation serum B_{12} values declined first and then tissue level fell. Biochemical intermediates of disorders of B_{12} coenzyme metabolism (propionic acid and methylmalonic aciduria) were variable in excretion. Documentation of deranged B_{12} metabolism was obtained by evidence of defective propionate metabolism. The sequence of changes was characterized by the use of a more direct evaluation of coenzyme function. In vitro $^{14}CO_2$ liberation from radio-labeled propionate provides a form of functional coenzyme evaluation. The classical hallmarks of vitamin B_{12} deficiency in humans are megablastosis and neurological lesions. By contrast, hematological and neurological lesions are exceedingly rare in most animal models of B_{12} deprivation. The clearest exception is the rare neurological degeneration described in captive monkeys by Oxnard et al. (156). In other animal models, the absence of tissue injury has made the definition of the B_{12} deprived state difficult. Generally, serum and/or tissue B_{12} levels have been utilized as evidence of a significant alteration from the normal steady state to define deficiency. However, that such levels are not necessarily predictable criteria of deprivation has been best documented in humans where anatomical marks assist in the definition. Recently captive rhesus monkeys showed amounts of vitamin B_{12} in the serum that resembled those found in humans (150–600 pg/ml) (110). When allowed to become deficient, rhesus monkeys do not often show obvious changes; for instance, the macrocytic anemias that occur in humans do not appear. Individual animals sometimes have greater amounts of hemoglobin, higher red cell counts and higher serum iron levels than are found in recently captive animals, and obvious macrocytosis has not been noted. Anemias that have been found from time to time in isolated animals are microcytic and hyperchromic. Macrocytic anemias reported in monkey have been shown to respond to folic acid (and sometimes ascorbic acid) rather than to treatment with B_{12} (195). Monkeys fed vegetarian diets develop neurological and hematological abnormalities. Although actual signs may not be evident, paralysis can occur from the neural lesions (cerebral degeneration, posterior and lateral column degeneration of the spinal cord, segmental and Wallerian degeneration of the peripheral nerves) (157). Neuropathies are also well known in humans with vitamin B_{12} deficiency associated with pernicious anemia, malabsorption syndromes (39), and in true vegetarians; subacute combined degeneration of the cord may be the sole manifestation of the deficiency. Similarly, in a few monkeys that have been maintained on vegetarian diets and have been found to have low levels of the vitamin in the serum, paralysis of the hind limbs and tails can occur.

It has been known for some time that young growing rats fed a diet deficient in phosphorus and vitamin D develop rickets. In this condition, some linear bone growth continues, but the calcification of cartilage and bone matrices ceases or is subnormal. Although the cells in growth cartilage continue to proliferate and age, the naturally appearing hypertrophic cells are not reabsorbed and the growth plates become abnormally thick.

Simmons and Kunin (192) developed an animal model of rickets in weanling rats by maintaining the animals on a low phosphate vitamin D-free diet for 2 weeks. During the third week, subgroups were fed a basal rachitogenic ration supplemented with either inorganic phosphate or vitamin D_2, or both. The ability of these diets to produce and heal rickets was assayed histologically and histochemically by determinations of tibial ash/weight ratios and thymidine content. The efficiency with which the diets were able to support growth was also calculated by quantification of dietary intake versus weight gain. The width of proximal tibial epiphyseal cartilage of rats fed the rachitogenic diet for only 1 week was greater than that of animals fed commercial laboratory chow *ad libitum* due to unimpaired cellular proliferative activity and continued accumulation of hypertrophic cells. The increase in width persisted for the next 2 weeks despite a significant concomitant reduction in cellular proliferation. By the end of the second week, the number of cells in the proliferative zones of rachitic cartilage that could be labeled with ^{3}HTdr was not significantly different from that of the cartilages of pair-fed rats, but became significantly lower after 3 weeks. Dietary phosphate supplementation was more effective than vitamin D_2 alone in normalizing the DNA synthetic indices from the second through third weeks. By the end of the third week, there were no appreciable intergroup differences in growth cartilage thickness between pair-fed control rats and rats fed phosphorus or vitamin D supplements, or both. There was voluntary reduction of food intake by rats fed the rachitogenic ration and utilization of this diet for growth (feed efficiency) was much less than normal. Both inorganic phosphorus and vitamin D_2 dietary supplementation were able to restore feed efficiency to normal levels. The low tibial ash weights of the rachitic group were increased following either phosphorus or vitamin D_2 dietary supplementation, but the diet containing both nutrients provided the best overall improvement.

Jacob and Forbes (103) used weanling male albino rats in studies to investigate the relationship of vitamin D status and other parameters affecting citrate and mineral metabolism on calcification in magnesium deficiency. In experiment I, the basal diet contained 0.5 percent calcium, 0.5 percent phosphorus, 90 ppm of magnesium, and no known source of vitamin D. Magnesium (600 ppm), vitamin D (1–25 I.U./g), and galactose (15 percent) were included in diets to form a 2^3 factorial design. In experiment II, the effect of L-thyroxine on citrate excretion was studied in magnesium-deficient and normal animals. Kidney calcification was produced only in the presence of vitamin D and was prevented by thyroxine treatment, although these treatments each increased urinary citrate excretion. Serum citrate levels were not influenced by magnesium but were increased by vitamin D and by galactose. Thus urine citrate was not correlated with either serum citrate or kidney calcification. The mechanism of the effect of vitamin D in the development of rickets has been studied biochemically and biologically in experimental medicine (55, 65, 161).

ANIMAL MODELS OF INORGANIC ION DEFICIENCY

The baby pig like the human infant is subject to nutritional disorders involving iron. Piglets with no source of iron other than sow's milk develop anemia within 2 to 4 weeks of birth as a result of their high postpartum growth rate and poor store of iron at birth. The surviving animals begin to recover by 6 to 8 weeks of age, when they start to forage on their own. The sensitivity of the human infant to iron deficiency anemia is similar, usually occurring between 4 to 24 months of age. Anemia and neutropenia have been observed in copper-deficient piglets and infants. The similarities and differences in newborn pigs and human infants have been summarized by Mount (147). Both show a drop in body temperature at birth followed by a rise; both shiver and have a little thermal insulation; the metabolic rates of both increase in the first few days following parturation. However, the pig has a lower birth weight, more rapid growth rate, higher body temperature and metabolic rate, an appreciable inability to sweat, and a more limited fat reserve than the human baby. With regard to thermo regulation, keeping the species differences in mind, the newborn pig is an appropriate model for a nonsweating human neonate. The Belgrade laboratory rats (b-b rats) have also been used to study iron metabolism (193).

A consistent occurence of significant necrobiotic changes in muscles of rats severely depleted of magnesium by dietary means was reported by Heggtveit (94). The patterns of multifocal necroses frequently associated with calcification is in many respects similar to that seen in other spontaneous and experimentally induced muscular disorders of a dystrophic nature. This is perhaps not unusual, since skeletal muscle, like the myocardium, can only react to diverse types of injury in a limited number of ways. Ultrastructural studies of cardiac necrosis and calcification in experimental magnesium deficiency showed that early mitochondrial damage is fundamental to the pathogenesis of the lesions. Dietary restrictions of magnesium in the rat provokes a spectrum of dystrophic changes in skeletal muscle including multifocal hyaline, vacuolar, granular, and floccular necrosis and calcification of myofibers. Disturbances of blood and tissue magnesium levels may play a role in a variety of muscular disorders including alcohol myopathy and cardiomyopathy. The renal lesions of nephrocalcinosis associated with magnesium deficiency in the rat are characterized by the formation of intrarenal calculi in the thin limbs of Henli's loop as well as by metastatic intraluminal calcific deposits of debris accumulating in the thick ascending link of the loop. Alterations of renal functions observed in the deficiency state have been reviewed by Whang et al. (219) and Jacob and Forbes (102). It was found that although the effects of Mg deficiency are manifested both morphologically and functionally, the precise functional and anatomic correlation cannot be made at the present time.

Hypocalcemia is associated with magnesium depletion in humans, monkeys, sheep, dogs, and pigs. However, the magnesium-depleted rat appears to be unique in developing hypercalcemia when fed diets having the same calcium content as those that produce hypocalcemia in other species. A magnesium-deficient diet produced significant differences between rats and mice in symptomatology and in plasma calcium levels. Deficient young male rats developed the classical erythema, hyperirritability, and tonic-clonic convulsions. While there was a high mortality with the convulsions, a good proportion recovered. The deficient rats were either normocalcemic or hypercalcemic. In comparison, deficient male mice did not develop erythema or hyperirritability; they did convulse, but it was a single violent spasm with almost immediate death and rare survival. There was a positive correlation between the plasma magnesium and calcium in the deficient mice. Although growth of mice receiving 5 mg percent magnesium in their diet was close to that of controls with 40 percent, the plasma magnesium and calcium remained low for approximately 5 weeks. Renal glomerular dysfunction and calcification did not occur in depleted animals in either species.

Young et al. (229) reported that intravenous administration of *Escherichia coli*, L-asparaginase (Asnase) to rabbits at a dose of 1000 units/kg/day for 3 days resulted in an incidence of hypocalcemia of 100 percent and tetany in 70 percent. In addition, the majority of the rabbits had hyperphosphatemia, hypomagnesemia, hyperkalemia, and azotemia. Other rabbits given *E. coli* endotoxin failed to develop a significant hypocalcemia or parathyroid alteration. The majority of rabbits receiving prophylactic and therapeutic treatments designed to alleviate the fatal hypocalcemic tetany induced by Asnase responded favorably. Rabbits surgically thyroparathyroidectomized developed overt signs of and clinicopathologic alterations similar to Asnase-treated rabbits and died within 24 hours after surgery. The performance of bilateral ureterectomies in additional rabbits resulted in an increase in blood urea nitrogen levels up to fivefold greater than those occurring in Asnase-treated rabbits but failed to produce a hypocalcemia of a degree sufficient to induce tetany. The rabbit Asnase system presents us an experimental animal model of drug-induced hypoparathyroidism characterized by drug endocrine cell interactions somewhat analogous to the lethal selectivity of alloxan for pancreatic B-cells in the production of diabetes.

A number of other animal models of imbalance of inorganic ions have been reported, including magnesium in the rat (90), cobalt in sheep followed by discovery of vitamin B_{12} (194), chromium in the rat (186), zinc or sodium in many experimental species (202, 212), manganese in the rat (226, 227), cadmium in the rat (13), and others (77, 141).

ANIMAL MODELS OF DISORDERS OF PROTEIN, LIPID AND CARBOHYDRATE METABOLISM

New reports are published each year describing the occurence of human metabolic diseases due to specific protein, amino acid, and enzyme deficiencies (20, 92, 168, 169, 201, 225). The mouse as an animal model of hyperprolinemia was investigated by Blake (18). Deficiency of mouse proline oxidase activity was found to be the cause of hyperprolinemia. With regard to human hyperprolinemia, two variants of this metabolic disorder have been described by Efron (63). Shafter et al. (190) reported a familial hyperprolinemia associated with cerebral dysfunction and renal anomalies and deafness. A second case of familial hyperprolinemia was associated with congenital renal malformation, hereditary hematuria, and mental retardation (62). The metabolic disorder in the second case was further associated with a deficiency of liver proline oxidase activity. Hyperprolinemia caused by a deficiency of liver proline oxidase has been classified as type I. In type II hyperprolinemia, there was no evidence of renal disease as assessed by intravenous pyelography, blood urea, and nitrogen and creatinine clearance tests. A deficiency of the second enzyme in the degradation pathway of proline, 'pyrroline-5-carboxylate dehydrogenase would be expected to result in the accumulation of 'pyrroline-5-carboxylate as well as proline. The hyperprolinemia occurring in the highly inbred Pro/Re strain of mice is associated with a marked deficiency of proline oxidase activity in liver, kidney, and brain. F_1 hybrid mice exhibit an intermediate liver proline oxidase activity. The Pro/Re may serve as an animal model for biomedical investigation on type I hyperprolinemia.

An experimental model for type V hyperlipoprotinemia in rabbits was reported by Weber et al. (217). Combined treatment with oral cholesterol (1 g/day) and Tween 80 (three times per week) in rabbits induced a marked atherosclerosis with hyperchylomicronemia. This condition closely resembles the human type V hyperlipoprotinemia according to the classification of Fredrickson et al. (74). Similarities of these observations, hyperchylomicronemia without atherosclerosis, with the clinical findings of Fredrickson's type V indicate that this is a useful experimental model for the rare clinical disease.

The syndrome of cholestasis is associated with a variety of metabolic abnormalities. Some of these, such as jaundice, elevated bile salt concentration, and strectorehea, can be regarded as reflecting directly the failure of bilirubin and bile salt secretion into the bile ducts and duodenum and consequently their accumulation in plasma. The metabolic changes associated with cholestasis are diverse and may be related either directly or indirectly to failure of bile secretion. Thus, abnormalities in lipid and lipoprotein metabolism could result from the effects of (a) failure of lipids to be secreted into the bile, (b) qualitative or quantitative changes in the lipoproteins of plasma or intestinal lymph, or (c) altered liver cell function not directly attributable to the first two factors. Cooper et al. (40) described a method in which these factors were analyzed separately by comparing obstructed and unobstructed liver tissue from the same animal in regard to structure and function. In this model of selective biliary obstruction, light microscopic changes consistent with cholestasis were found in the obstructed median lobe but not in the unobstructed right lobe. Tissue concentrations of the total and free cholesterol and of phospholipids were similar in unobstructed and obstructed hepatic lobes. In serum, levels of cholesterol, phospholipid, and alkaline phosphatase were significantly elevated; an abnormal lipoprotein characteristic of biliary obstruction also appeared. It seems that selective biliary obstruction may be useful in the experimental approach to elucidate not only the abnormalities in lipid and lipoprotein metabolism in cholestasis but also the process of bile secretion in general.

Clark et al. (35) reported studies of fat absorption in the rat with essential fatty acid deficiency as an experimental approach to understanding the mechanism of fat malabsorption of unknown etiology. Male rats were made deficient in essential fatty acids by feeding them a fat-free diet supplemented with 4 percent tri-palmitin for 8 to 12 weeks from the time of weaning. After feeding 0.5 ml of ^{14}C-triolein on ^{3}H oleic acid, 72-hour stool recoveries of radioactivity were significantly less in deficient rats than in chow-fed controls. Essential fatty acid deficiency did not reduce the absorptive capacities for triolein or for a medium chain fat, measured after 3 and 2 hours of maximal rate duodenal infusion. In jejunal slices from essential fatty acid deficient rats, uptake of micellar ^{14}C-oleic acid at 0–1°C was similar to that of controls, but the rate of incorporation of fatty acid into triglyceride after rewarming to 37°C was significantly reduced. The specific activities of the microsomal esterifying enzymes, acyl CoA monoglyceride acyltransferase and fatty acid CoA ligase in jejunal mucosa were 30 percent lower in essential fatty acid-deficient rats. However, the total microsomal enzyme activity adjusted to constant weight did not differ significantly in deficient rats compared with controls. After intraduodenal perfusion of triolein, accumulation of lipid in the intestinal wall was increased in the deficient rats. Because over 90 percent of the absorbed mucosal lipid was present as triglyceride, essential fatty acid deficiency appears to affect the synthesis or release of chylomicron lipid from the intestine. Analysis of regions of intestine showed that this delay in transport was most marked in the midportion of the small intestine.

King et al. (108) demonstrated that intravascular fibrin deposition occurs in the lungs of the oleic acid canine model of fat embolism. Controversy has existed over whether intravascular coagulation occurs in experimental fat embolism. Baker et al. (10) showed that the presence of fibrin deposition predominantly in the lung, following injection of oleic acid into the right ventricle, completing the case for the occurrence of intravascular coagulation in this model. Both fibrin and fibrinogen can be demonstrated in the interstitium and alveoli of the lung 4 hours after the fatty acid assault. This model may be a good example of both intravascular and extravascular coagulation.

Fatty liver and cirrhosis have been recognized as sequelae of excessive intake of alcohol and its accompanying nutritional problems (48). In humans, fatty liver can result from either nutritional deficiency or alcohol intake alone, but the development of cirrhosis apparently requires both alcohol and a nutritionally damaged liver. Cirrhosis can be induced in rats by nutritional deficiency alone or by certain hepatotoxins but not by alcohol alone. The animal model of fatty liver and cirrhosis in lipotrope-deficient male rats was reported by Rogers and Newberne (174) for the study of this important disease. This animal model may be useful to investigators interested in its pathogenesis or that of related liver diseases. In this model, weanling rats are fed a diet deficient in choline, methionine, vitamin B_{12}, and folic acid, high in fat, and adequate in amino acids other than methionine. The severity of deficiency and the rapidity with which fatty liver and cirrhosis develop are governed by dietary levels of nutrients mentioned and by the rat's age. The fatty liver and cirrhosis induced in rats resemble closely the human disease in gross and microscopic lesions and in development of hepatic dysfunction. The variations in biochemical function and pathology that can be induced in rats by manipulation of the diet mimic the several patterns found in alcoholic patients, with the exception of alcoholic hepatitis and its associated hyaline cytoplasmic deposits. This facet of the disease may require exposure to alcohol or other hepatotoxins. The high incidence of esophageal and hepatic cancer in cirrhotic patients suggests that they have enhanced sensitivity to some environmental carcinogens. Aflatoxin B_1 and daily diethylnitrosamine are more effective hepatic carcinogens in lipotrope-deficient rats than in normal rats, an effect of diet which may be mediated through impairment of drug metabolism in the endoplasmic reticulum. The model has been produced in at least three strains of rat (Charles River C.D. strain, Wistar and Fischer strain).

The role of dietary factors in the genesis and evolution of atherosclerosis and its complications has received considerable attention in recent years (32, 89, 116, 121, 130, 178, 198, 213, 214, 215). Whereas the diet continues to be considered as the single most important etiological factor in atherosclerosis, the question still remains regarding the relative roles of its various components. Gupta et al. (85) studied the role of dietary fats, their quantity, and composition in the genesis of atherosclerosis using swine as an animal model. Two groups of pigs, each consisting of six animals, were fed for 18 months on isocaloric amounts of an experimental diet with a high fat content and cholesterol but with widely different levels of protein (5 percent versus 25 percent by weight of the diet). In addition, a third control group consisting of four animals was maintained on normal stock diet. Animals of the low protein group showed the maximal intimal surface area involvement with atherosclerotic lesions in the aorta and coronary arteries, and also the most severe changes among the three groups. No significant differences were noted in the extent and severity of lesions between the high protein-high fat-fed animals as compared with the high protein-low fat-fed controls. Lesions of the low protein group had a higher cholesterol content and a higher cholesterol : phospholipid ratio than those in the other two groups. Extremely low levels of dietary proteins seem to have had a promotive effect on the induction of atherosclerotic lesions by an atherogenic diet, whereas adequate levels of dietary proteins have had a protective influence. The precise mechanism by which varying levels of dietary proteins have such effects is not understood. It may possibly be related to the abberations in lipid metabolism induced by extremely low levels of dietary proteins.

Atherosclerosis, similar in many respects to that in humans, can be produced in a variety of experimental animals by feeding high fat diets. In general, primates develop distinct atherosclerotic plaques and show lipid deposition in the lesions more commonly than most of the lower animals. Lindsay and Chaikoff (120) reported that coronary and aortic atherosclerosis was encountered in seventeen nonhuman primates species and the disease had the same basic pathological features as in other mammals. In certain primates, the natural form of the arterial disease is similar if not identical to that in humans. Similarities between cholesterol-induced lipoprotein changes in chimpanzees and those obtained in human hyperbetalipoproteinemia were discussed by Blaton et al. (19). These similarities (i.e., tremendous increase of betalipoproteins and free and esterified cholesterol; no change in triglyceride levels; increase in phospholipids but less than that of cholesterol; the oleic to linoleic acid ratio of both lipoproteins increased mainly due to changes in the cholesterol esters, etc.) stress the usefulness of these nonhuman primates as a model for experimental atherosclerosis and for studying the molecular changes in lipoprotein patterns.

The effects of high carbohydrate diets on various parameters of lipid and carbohydrate metabolism were examined by Eaton and Kipnis (60, 61). Rat serum lipoproteins were characterized by paper electrophoresis, flotation, ultracentrifugation, and differential precipitation. Prealpha lipoprotein content was directly related to the serum-free fatty acids (FFA) concentration (Fig. 5-2). Pre-β-lipoprotein and triglyceride levels increased during long-term glucose supplementation, decreased following an oral gluose load, and were lowered maximally after 48 hours of fasting. A dissociation in the behavior of the serum triglyceride and prebeta lipoprotein levels was noted in the short-term glucose-fed rat. Basal insulin level and the insulin response to the oral glucose were significantly lowered by glucose feeding despite normal serum glucose levels and a normal to improved glucose tolerance. The glucose-fed rat exhibited a normal-to-exaggerated plasma glucose response following intravenous glucagon or theophylline and exhibited increased hepatic glycogen content, with normal total hepatic fatty acid ester concentration. Random serum FFA levels were markedly elevated in contrast to control

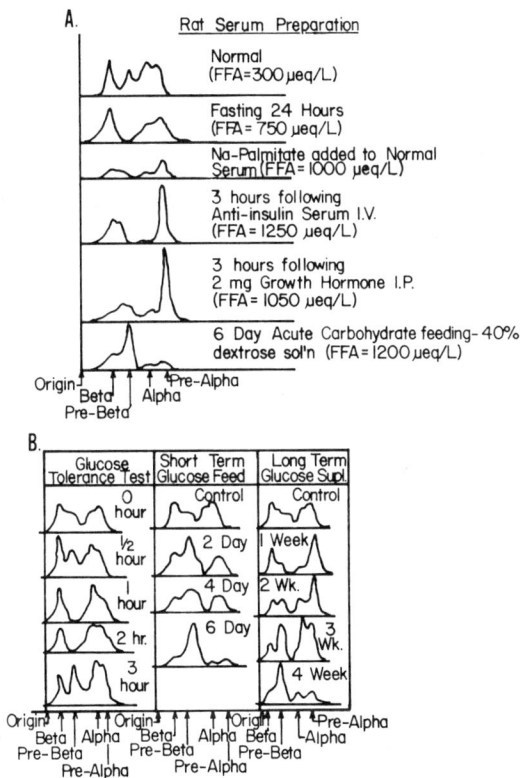

Fig. 5-2. Effects of high carbohydrate diets on lipid metabolism in the rat. (A) Effect of either in vitro or in vivo elevations of serum free fatty acids (FFA) on electrophoretic pattern of rat serum lipoproteins. (B) Effect of an acute glucose load, short-term glucose feeding, and long-term glucose supplementation on electrophoretic pattern of rat serum lipoproteins. Reproduced from *American Journal of Physiology* (60) with permission.

animals, showing no correlation with the corresponding serum insulin levels. Increased prominence of serum-β-lipoprotein fraction appeared within 48 hours on a diet consisting exclusively of 40 percent glucose dissolved in 0.5 percent saline. A comparable serum lipoprotein pattern associated with a marked increase in triglycerides developed in 3 to 4 weeks when rats were fed a regular chow diet supplemented with 10 percent glucose added to their drinking water. The effects of short-term glucose feeding (40 percent glucose in 0.5 percent saline exclusively) and long-term glucose supplementation (10 percent glucose in drinking water) on hepatic protein synthesis showed increased incorporation of leucine ^{14}C into all serum proteins in the glucose-fed rats as compared with those of control animals. The concentration of serum beta-plus prebeta (VLD-LD) lipoprotein was increased (control = 1.44 $\pm$ 0.09 mg/ml; glucose-fed rats = 2.14 $\pm$ 0.11 mg/ml) in short-term glucose-fed rats, whereas the serum albumin level remained unchanged. Leucine-^{14}C incorporation by liver slices into tissue VLD-LD lipoprotein was increased 200 percent after 48 hours of short-term glucose feeding and progressively increased to levels greater than 400 percent after 4 weeks of long-term glucose supplementation. These effects could not be attributed to changes in the plasma or hepatic-free amino acid pools. The data suggested that synchronous changes in hepatic lipoprotein-protein synthesis occur with changes in hepatic triglyceride synthesis in the carbohydrate-fed rat.

Since food intake and glucose metabolism are enmeshed in the energy exchange of mammals, it is attractive to consider glucose metabolism as a major determinant of food intake (38, 170, 171, 220). The glucostatic hypothesis for the control of food intake predicts that decreased glucose utilization leads to increased food intake. Three major characteristics of the glucostatic control of feeding have been specified: (a) it operates over a relatively short epoch of hours or a day; (b) the rate of cellular glucose utilization by brain tissue, especially the ventromedial area of the hypothalamus, is the critical parameter for food intake; and (c) food intake and glucose utilization are inversely related. Since the glucose analog, 2-deoxy-D-glucose(2-DG) produces decreased intracellular glucose utilization of most tissues and particularly of brain, the injection of 2-DG into five monkeys and twenty-eight rats provided an explicit test of the predicted relationship (195). Both monkeys and rats ate more after 2-DG administration. The most effective systemic doses of 2-DG were 300 mg/kg (IV) in monkeys and 750 mg/kg (ip) in rats. Larger doses (L1000 mg/kg) produced drowsiness, stupor, ataxia, or retching but not convulsions. When separate groups of rats were tested with either insulin (6U Iletin) or 2-DG (700 mg/kg), they increased their mean food intake to approximately the same amount, but the rats treated with 2-DG ate sooner. Shorter latency of feeding with 2-DG is interpreted as resulting from the fact that 2-DG produces decreased glucose utilization of brain directly, whereas insulin reduces cerebral glucose utilization indirectly as a consequence of hypoglycemia. Since 2-DG produced marked hyperglycemia during the period of increased feeding, these results show that the abrupt onset of decreased glucose utilization, not hypoglycemia, is a sufficient condition to induce food intake in mammals.

Marked changes occur in lipid and carbohydrates metabolism during starvation and refeeding. McGray et al. (135) examined the early time course of changes in gluconeogenic and ketogenic capacities of the intact animal and the perfused liver. Brief periods of starvation and refeeding caused marked changes in glucose and fatty acid metabolism in the rat or its perfused liver. During starvation there was no impairment in triglyceride synthesis. The enhanced ketogenesis resulted from activation of an early step such as acylcarnitine transferase. In considering the possible usefulness of the spontaneously occurring hyperglycemic syndromes of laboratory animals for our understanding of the pathogenesis of human diabetes, no single animal syndrome can yet be considered the exact or even the best animal model for the human disease (196, 170).

Undisputed similarities exist between the clinical course of the diabetic syndrome of the Chinese hamster and that typical of juvenile diabetes. Also, striking analogies can be established between the metabolic rate of animals transferred from desert regions to the laboratory and that of primitive populations suddenly faced with an abundance of food (188). However, current knowledge clearly does not allow for any conclusion regarding the identity of similarity between the underlying primary pathogenic defect(s) or to the nature of the endocrine and metabolic components of the system responsible for the diabetogenic effect of environment changes (see Chapter 8). In addition, neither the benign hyperglycemic animal syndromes nor those that may result in ketoacidosis can be considered to reflect in every respect the clinical picture and the metabolic or endocrine anomalies that we consider typical of adult onset diabetes in humans. More extensive study of animal strains and their more frequent use in testing hypotheses of not only carbohydrate metabolism but also protein and fat metabolism may prove to be extremely useful for understanding diabetes and other metabolic disorders of humans.

REFERENCES

1. Abel, A. L., *Acta. Pathol. Microbiol. Scand. Suppl.* **94:**1 (1953).
2. Adlard, B. P., J. Dobbing, and J. L. Smart, *Biol. Neonate,* **23:**95 (1973).
3. Alcock, N. W., and M. E. Shills, *Proc. Soc. Exp. Biol. Med.,* **146:**137 (1974).
4. Altman, J., G. D. Das, and K. Sudarshan, *Dev. Psychobiol.,* **3:**281 (1970).
5. Altman, J., G. D. Das, K. Sudarshan, and J. E. Anderson, *Dev. Psychobiol.,* **4:**55 (1971).
6. Altman, J., K. Sudarshan, G. D. Das, N. McCormick, and D. Barnes, *Dev. Psychobiol.,* **4:**97 (1971).
7. Anderson, A., *Skand. Vet. Tidsskr.,* **26:**241 (1936).
8. Armstrong, M. L., *Arch. Pathol.,* **92:**395 (1971).
9. Aubert, D., D. C. Ferrand, B. Lacaze, O. Pepin, E. Panak, and M. Padesta, *Artherosclerosis,* **20:**263, (1974).
10. Baker, P. L., M. C. Knenzog, and L. F. Peltier, *J. Trauma.* **9:**577 (1969).
11. Bardens, J. W., in R. W. Kirk, G. G. Rickard, and K. McEntee (eds.), *Small Animal Practice,* W. B. Saunders Co., Philadelphia, 1964.
12. Barnes, R. H., C. S. Neely, E. Kwong, B. A. Labadan, and S. Frankova, *J. Nutr.,* **96:**467 (1968).
13. Barr, M., *Teratology,* **9:**13 (1974).
14. Barthe, A. *Genetics,* **48:**882 (1963).
15. Batterham, E. S., *Br. J. Nutr.,* **31:**237 (1974).
16. Berg, B. N., *Proc. Soc. Exp. Biol. Med.,* **119:**417 (1965).
17. Bhagavan, H. N., D. B. Coursin, and C. N. Stewart, *Life Sci.,* **8:**1117 (1969).
18. Blake, R. L., *Biochem. J.,* **129:**987 (1972).
19. Blaton, V., R. Vercaemst, N. Vandecasteele, H. Caster, and H. Peeters, *Biochemistry,* **13:**1127 (1974).
20. Blaton, V., D. Vandamme, B. Declercq, M. Vastesaeger, J. Mortelmans and H. Peeters, *Exp. Mol. Pathol.,* **20:**132 (1974).
21. Blaxter, K. L., *J. Sc. Food. Agric.,* **23:**941 (1972).
22. Book, S. A. and L. K. Bustad, *J. Anim. Sci.,* **38:**997 (1974).
23. Bullock, B. C., R. Paula, and L. L. Rudel, *Fed. Proc.,* **33:**626 (1974).

24. Cameron, J. S., H. A. Simmonds, P. J. Hatfield, A. S. Jones, and A. Cadenhead, *Adv. Exp. Med. Biol.* **41:**691 (1974).
25. Capen, C. C., S. L. Martin and A. Koestner, *Pathol. Vet.* (Basel), **4:**301 (1967).
26. Capen, C. C., and A. Koestner, *Pathol. Vet.* (Basel), **4:**326 (1967).
27. Chakravarti, R. N., A. K. Sarker, G. Pal, and B. Mahata, *Indian J. Med. Res.,* **62:**413 (1974).
28. Chantler, C., E. Lieberman, and M. Holliday, *Pediatrics Res.,* **8:**109 (1974).
29. Chase, H. P., J. Dorsey, and G. M. McKahn, *Pediatrics,* **50:**551 (1967).
30. Chase, H. P., and D. O'Brien, *Pediat. Res.,* **4:**96 (1970).
31. Chase, H. P., R. A. Marlow, C. S. Dabiere, and N. N. Welch, *Pediatrics,* **52:**513 (1973).
32. Child, G. V., W. M. Thurlbeck, and B. I. Weigensberg, *Arch. Pathol.,* **98:**47 (1974).
33. Chlouverakis, C., and D. Hojnicki, *Metabolism,* **23:**133 (1974).
34. Clare, N. T. and E. H. Stephens, *Nature,* **153:**252 (1944).
35. Clark, S. B., T. E. Ekkers, A. Singh, J. A. Balint, P. R. Holt, and J. B. Rodgers, Jr., *J. Lipid Res.,* **14:**581 (1973).
36. Clarkson, T. B., W. R. Prichard, M. G. Netsky, and H. B. Lofland, *Arch. Pathol.* (Chicago) **68:**143 (1959).
37. Cole, R. K., *Genetics,* **53:**1021 (1966).
38. Conney, A. H., C. Coutinho, B. Koechlin, R. Swarm, J. A. Cheripko, C. Impellizzeri, and H. Baruth, *Clin. Pharmacol. Ther.,* **16:**176 (1974).
39. Cooke, W. T., and W. J. Smith, *Brain,* **89:**683 (1966).
40. Cooper, A. D., A. L. Jones, R. E. Koldinger, and R. K. Ockner, *Gastroenterology,* **66:**574 (1974).
41. Cornelius, C. E., and M. Ariasel, *Am. J. Med.,* **40:**165 (1966).
42. Cornfeld, D., and R. E. Cooke, *Pediatrics,* **10:**33 (1952).
43. Coward, D. G., and R. G. Whitehead, *Br. J. Nutr.,* **28:**223 (1972).
44. Cravioto, J., E. R. Delicardie, and H. G. Birch, *Pediatrics,* **38:**319 (1966).
45. Carvioto, J., in D. C. Glass (ed.), *Proceedings of a Conference Under the Auspices of Russell Sage Foundation and the Rockefeller University,* The Rockefeller University Press, New York, 1968.
46. Creech, G. T., *Am. J. Vet. Res.,* **2:**400 (1941).
47. Dahme, E., and N. Deutschlander, *Deut. Tieraerztl. Wochschr.,* **74:**134 (1967).
48. Davidson, D. S., *Am. J. Clin. Nutr.* **23:**427 (1970).
49. Davidson, C. S., and A. A. Mjhas, *N. Engl. J. Med.,* **291:**50 (1974).
50. Davison, A. N., and J. Dobbing, *Appl. Neurochem.,* F. A. Davis Co., Philadelphia, 1968.
51. De Matteis, F., *Enzyme,* **16:**266 (1973).
52. Den Besten, L., S. Safaie-Shirazi, W. E. Connor, and S. Bell, *Gastroenterology,* **66:**1036 (1974).
53. Dickerson, J. W. T., J. Dobbing, and R. A. McCance, *Proc. R. Soc.,* **166:**396 (1967).
54. Diener, R., and R. Langham, *Sm. Animal Clin.,* **1:**274 (1961).
55. Dixit, P. K., *J. Histochem. Cytochem.,* **17:**411 (1969).
56. Dobbing, J., *Nature* (London), **226:**639 (1970).
57. Dobbing, J., *Bibl. Nutr. Dieta,* **17:**35 (1972).
58. Duncan, H., *Henry Ford Hosp. Med. J.,* **19:**105 (1971).
59. Dunn, T. B., In, E. Cotchin, and F. J. C. Roe (eds.), *Pathology of Laboratory Rats and Mice,* Blackwell, Oxford England, 1967.
60. Eaton, R. P., and D. M. Kipnis, *Am. J. Physiol.,* **217:**1153 (1969).
61. Eaton, R. P., and D. M. Kipnis, *Am. J. Physiol.,* **217:**1160 (1969).
62. Efron, M. C., *N. Engl. J. Med.,* **272:**1243 (1965).
63. Efron, M. C. in J. B. Stanbury, J. B. Wyngaarden, and D. S. Frederickson (eds.), *The Metabolic Basis of Inherited Disease,* McGraw-Hill, New York, 1966.
64. Eggen, D. A., *J. Lipid Res.,* **15:**139 (1974).
65. Eisenstein, R., H. Ellis, and J. Rosato, *Proc. Soc. Exp. Biol. Med.,* **132:**58 (1969).
66. Eluelhjem, C. A., R. J. Madden, F. M. Strong, and D. W. J. Wooley, *J. Biol. Chem.,* **123:**137 (1938).
67. Enwonwu, C. O., R. V. Stambaugh, and K. L. Jacobson, *Am. J. Clin. Nutr.,* **26:**1287 (1973).
68. Eveleth, D. F., D. W. Ballin, and A. J. Goldsky, *Am. J. Vet. Res.,* **10:**250 (1949).

69. Falk, J. L., H. H. Samson, and G. Winger, *Science,* **177**:811 (1972).
70. Frankova, S., and R. H. Barnes, *J. Nutr.,* **96**:477 (1968).
71. Frankova, S., and R. H. Barnes, *J. Nutr.,* **96**:485 (1968).
72. Frankova, S., *Nutr. Metab.,* **12**:228 (1970).
73. Fraser, H. S., and G. A. L. Allyene, *Br. J. Nutr.,* **31**:113 (1974).
74. Frederickson, D. S., R. I. Levy, and R. S. Lees, *N. Engl. J. Med.* **271**:32 (1967).
75. Frenkel, E. P., and J. D. White, *Lab. Invest.,* **29**:614 (1973).
76. Fuller, J. L., *Adv. Psychosom. Med.,* **7**:2 (1972).
77. Gabbiani, G., M. C. Badonnel, and C. A. Baud, *Calcif. Tissue Res.,* **4**:224 (1969).
78. Gaman, E. M., A. S. Feigenbaum, and E. A. Schenk, *J. Ather. Res.,* **7**:131 (1967).
79. Genz, J., G. Bengtsson, J. Hakkarainen, R. Hellstrom, and B. Persson, *Am. J. Physiol.,* **218**:662 (1970).
80. Glauser, E. M., *Exp. Med. Surg.,* **24**:181 (1966).
81. Goodhead, B., *Arch. Surg.,* **103**:724 (1971).
82. Green, C. J., *Lab. Anim. Sci.,* **8**:99 (1974).
83. Gregoriadis, G., and R. A. Buckland, *Nature,* **244**:170 (1973).
84. Griminger, P., V. Villamil, and H. Fisher, *J. Nutr.,* **99**:368 (1969).
85. Gupta, P. P., H. D. Tandon, M. G. Karmarkar, and V. Ramalinjaswami, *Exp. Mol. Pathol.,* **20**:115 (1974).
86. Guthrie, H. A., and M. L. Brown, *J. Nutr.,* **94**:419 (1968).
87. Hadlow, W. J., and K. R. Reinhard, *Cornell Vet.,* **44**:475 (1954).
88. Hahn, H. J., E. Wappler, and H. Fiedler, *Z. Versuchstierkd,* **13**:275 (1971).
89. Hamm, T. E., C. R. Abee, T. A. Riggs, and T. B. Clarkson, *Fed. Proc.,* **33**:236 (1974).
90. Hann, S., M. Harrison, I. MacIntyre, and R. Fraser, *Lancet,* **II,** 1960 (1972).
91. Hansel, W., P. Olafson, and K. McEntee, *Cornell Vet.,* **44**:94 (1955).
92. Haris, G. W., and F. Naftalin, *Br. Med. Bull.,* **26**:3 (1970).
93. Harrington, D. D., and P. M. Newberne, *Lab. Anim. Care,* **20**:675 (1970).
94. Heggtveit, H. A., *Ann. N.Y. Acad. Sci.,* **162**:758 (1969).
95. Hegsted, D. M., *Nutr. Rev.,* **29**:45 (1971).
96. Hegsted, D. M., *Nutr. Rev.,* **31**:222 (1973).
97. Hentges, J. F., R. H. Gummer, P. H. Phillipa, G. Bohstedt, and D. K. Sorensen. *Am. J. Vet. Res.,* **120**:213 (1952).
98. Heston, W. E., C. D. Larsen, and M. K. Deringer, *J. Nat. Cancer Inst.,* **6**:41 (1945).
99. Hjarre, A., and I. Nordlund, *Skand. Vet. Tidsskr.,* **32**:385 (1942).
100. Hummel, K. P., and D. C. Chapman, *Roscoe B. Jackson Mem. Lab. Ann. Rep.,* **33**:31 (1962).
101. Hummel, K. P., *Roscoe B. Jackson Mem. Lab. Ann. Rep.,* **35**:96 (1964).
102. Jacob, M., and R. M. Forbes, *J. Nutr.,* **99**:152 (1969).
103. Jacob, M., and R. M. Forbes, *J. Nutr.,* **100**:228 (1970).
104. Jakob, W., *Vet. Path.,* **8**:292, (1971).
105. Jalowiec, J. E., and E. M. Stricker, *J. Comp. Physiol. Psychol.,* **70**:94 (1970).
106. Jorgensen, S. K., and T. K. With, *Ann. N.Y. Acad. Sci.,* **104**:701 (1963).
107. Kaneko, J. J., J. E. Moulton, R. S. Brodey, and V. D. Perryman, *J. Am. Vet. Med. Assoc.,* **146**:463 (1965).
108. King, E. G., P. K. Nakane, and D. G. Ashbaugh, *Surgery,* **69**:782 (1971).
109. Koletsky, S., *Exp. Mol. Pathol.,* **19**:53 (1973).
110. Krohn, P. L., C. E. Oxnard, and J. N. M. Chalmers, *Nature* (London), **197**:186 (1963).
111. Lat, J., E. M. Widdowson, and R. A. McCance, *Proc. R. Soc. B.,* **153**:347 (1961).
112. Leader, R. W. *Sci. Am.,* **216**:110 (1962).
113. Leav, I., A. C. Crocker, M. L. Petrak, and T. C. Jones, *Lab. Invest.,* **18**:433 (1968).
114. Lee, K. T., J. Jarmolych, D. N. Kim, C. Grant, J. A. Krasney, W. A. Thomas, and A. M. Bruno, *Exp. Mol. Path.,* **15**:170 (1971).
115. Lee, K. T., S. C. Nam, R. A. Florentin, and W. A. Thomas, *Med. Clin. North Am.,* **58**:281 (1974).
116. Lee, W. M., and K. T. Lee, *Fed. Proc.,* **23**:623 (1974).
117. Levin, E. Y., and V. Flyger, *Science,* **174**:59 (1971).
118. Levitsky, D. A., and R. M. Barnes, *Nature* (London), **225**:469 (1970).

119. Lindsay, S., and I. L. Chaikoff, *Arch. Pathol.*, (Chicago), **60:**29 (1955).
120. Lindsay, S., and T. L. Chaikoff, *J. Atheroscler.*, **6:**36 (1966).
121. Lindsay, S., and C. W. Nichols, *Exp. Med. Surg.*, **29:**42 (1971).
122. Lindstedt, K. J., *Comp. Biochem. Physiol.*, **39:**553 (1971).
123. Lisboa, P. E., *Gut*, **12:**363 (1971).
124. Lockner, D., and U. Ericson, *Acta Haematol.* (Basel), **49:**242 (1973).
125. Lombardi, M. H., C. L. Comar, and R. W. Kirk, *Am. J. Vet. Res.*, **23:**412 (1962).
126. Lopez, V. S. D. Davis, and N. J. Smith, *Pediatr. Res.* **6:**779 (1972).
127. Loten, E. G., A. Rabinovitch, and B. Jeanrenaud, *Diabetologia*, **10:**45 (1974).
128. Lozzio, B. B., A. I. Chernoff, E. R. Machado, and C. B. Lozzio, *Science*, **156:**1742 (1967).
129. Mackenzie, C. G., *Proc. Soc. Exp. Biol. Med.*, **40:**313 (1942).
130. Mansson, I., R. Norberg, B. Olhagen, and N. E. Bjorklund, *Clin. Exp. Immunol.*, **9:**677 (1971).
131. Marshall, M. W., A. M. AllenDurand, and M. Adams, in *Defining the Laboratory Animals*, National Academy of Sciences, Washington, 1971.
132. Mathur, K. S., V. K. Srivastava, R. Mathur, and R. D. Sharma, *Indian Heart J.*, **25:**200 (1973).
133. McCance, R. A., *Lancet* **2,** part 1:621 (1962).
134. McCance, R. A., *J. Pediatr.*, **65:**1008 (1964).
135. McGary, J. D., J. M. Meier, and D. W. Foster, *J. Biol. Chem.* **248:**270 (1973).
136. McSherry, C. K., F. Glenn, and N. B. Javitt, *Proc. Nat. Acad. Sci.*, **68:**1564 (1971).
137. Mellany, E., *J. Physiol.*, **99:**467 (1941).
138. Millen, J. W., and D. H. M. Woolam, *Br. J. Nutr.*, **10:**355 (1956).
139. Millen, J. W., and A. D. Dickson, *Br. J. Nutr.*, **11:**440 (1957).
140. Mitchell, H. H., in *Comparative Nutrition of Man and Domestic Animals*, Academic Press, New York, 1962.
141. Mohr, H. E., and L. L. Hopkins, Jr., *Lab. Anim. Sci.*, **22:**96 (1972).
142. Monier, D., and S. R. Wagle, *Proc. Soc. Exp. Biol. Med.*, **136:**377 (1971).
143. Montalbo, R. G., and J. J. Kabara, *Proc. Soc. Exp. Biol. Med.*, **145:**1225 (1974).
144. Moore, L. A., and J. F. Sykes, *Am. J. Physiol.*, **130:**684 (1940).
145. Moore, T., and P. D. Holmes, *Lab. Anim. Sci.*, **5:**239 (1971).
146. Moreland, A. F., T. B. Clarkson, and H. B. Lofland, *Arch. Pathol.* (Chicago), **76:**204 (1963).
147. Mount, L. E., in *The Climatic Physiology of the Pig*, Edward Arnold Ltd., London, 1968.
148. Munro, D. R., *Gastroenterology*, **66:**960 (1974).
149. Mustard, J. F., D. Secord, T. D. Hocksema, H. G. Downe, and H. C. Rowsell, *Br. J. Haematol.*, **8:**43 (1962).
150. Muth, O. H., *J. Am. Vet. Med. Aps.*, **126:**355 (1955).
151. Myers, K. M., J. E. Lund, and J. T. Boyce, *Fed. Proc.*, **27:**2221 (1968).
152. Nath, I., S. K. Sood, and N. C. Nayak, *J. Pathol.*, **106:**103 (1972).
153. Nitzan, M., B. E. Metzger, and J. F. Wilber, *Life Sci.*, **10:**671 (1971).
154. Okabe, S., and C. J. Pfeiffer, in C. J. Pfeiffer (ed.), *Peptic Ulcer*, Munksgaard, Copenhagen, 1971.
155. Osuga, T., and O. W. Portman, *Proc. Soc. Exp. Biol. Med.*, **136:**722 (1971).
156. Oxnard, C. E., and W. T. Smith, *Nature* (London) **210:**507 (1966).
157. Oxnard, C. E., W. T. Smith, and I. Torres, *Lab. Anim.*, **4:**1 (1970).
158. Pieper, W. A., M. J. Sheen, H. M. McClure, and P. G. Bourne, *Science*, **176:**71 (1972).
159. Plaut, S. M., *Dev. Psychobiol.*, **3:**157 (1970).
160. Pollock, S., *J. Am. Vet. Med. Assoc.*, **118:**12 (1951).
161. Ponchon, G., and H. F. DeLuca, *J. Nutr.*, **99:**157 (1969).
162. Pond, W. G., R. H. Barnes, I. Reid, L. Krook, K. Dworg, and A. V. Moore, in L. K. Bustad and R. O. McClellan (eds.), *Swine in Bio-Medical Research*, Pacific Northwest Laboratory, Richland, 1966.
163. Pond, W. G., W. Snyder, E. F. Walker, Jr., J. Stillings, and V. Sidewell, *J. Anim. Sci.*, **33:**587 (1971).
164. Pond, W. G., W. Snyder, J. T. Snook, E. F. Walker, Jr., D. A. Walker, and B. R. Stillings, *J. Nutr.*, **101:**1193 (1971).

165. Pratt, C. W. M., and R. A. McCance, *Brit. J. Nutr.,* **15:**121 (1961).
166. Prichard, R. W., T. B. Clarkson, H. O. Goodman, and H. B. Lofland, *Arch. Pathol.,* **77:**244 (1964).
167. Pritchard, W. R., C. E. Rehfeld, N. S. Misuno, J. H. Sautter, and M. O. Schultze, *Am. J. Vet. Res.,* **17:**425 (1956).
168. Ratnakar, K. S., M. Mathur, V. Ramalingaswami, and M. G. Deo, *J. Nutr.,* **102:**1233 (1972).
169. Rauch, H., and M. T. Yost, *Genetics,* **48:**1487 (1963).
170. Renold, A. E., D. P. Cameron, M. Amherdt, W. Stauffacher, E. Marliss, L. Orci, and C. Rouiller, *Isr. J. Med. Sci.,* **8:**189 (1972).
171. Resnick, S., *Vet. Med. Small Anim. Clin.,* **69:**585 (1974).
172. Roberts, J. C., Jr., and R. Straus, *Comparative Atherosclerosis,* Hoeber, New York, 1966.
173. Robinson, J. W., *Gut,* **13:**938 (1972).
174. Rogers, A. E., and P. M. Newberne, *Am. J. Pathol.,* **73:**817 (1973).
175. Rokkones, T., *Intern. Z. Vitaminoforsch.,* **26:**1 (1955).
176. Romsos, D. R., and G. A. Leveille, *Proc. Soc. Exp. Biol. Med.,* **145:**591 (1974).
177. Rutty, D. A., *Vet. Rec.,* **80:**28 (1967).
178. St. Clair, R. W., J. J. Toma, Jr., and H. B. Lofland, *Proc. Soc. Exp. Biol. Med.,* **146:**1 (1974).
179. Saito, S., and L. C. Fillies, *Am. J. Physiol.,* **207:**1277 (1964).
180. Santerre, R. F., T. N. Wright, S. C. Smith, and D. Brannigan, *Am. J. Pathol.,* **67:**1 (1972).
181. Sarett, H. P., in Saronoggi (ed.), *Nutrition and Technology for Growing Humans,* Basel, Karger, 1973.
182. Sarles, H., G. Lebreuil, F. Tasso, C. Figarella, F. Clemente, M. A. Devaux, B. Fagonde, and H. Payan, *Gut,* **12:**377 (1971).
183. Schalm, O. W., in *Veterinary Hematology,* 2nd ed., Lea and Febiger, Philadelphia, 1965.
184. Schemmiel, R. O., O. Mickelsen, G. Jersen, and S. Wegrizyn, *Fed. Proc.,* **27:**555 (1968).
185. Schoental, R., *J. Pathol. Bacteriol.,* **77:**485 (1959).
186. Schwartz, K., and W. Mertz, *Arch. Biochem. Biophys.,* **85:**292 (1959).
187. Sclafani, A., and L. Kluge, *J. Comp. Physiol. Psychol.,* **86:**28 (1974).
188. Seawright, A. A., *Pathol. Vet.* (Basel), **2:**175 (1965).
189. Sellers, A. L., S. Rosenfeld, and N. B. Friedman, *Proc. Soc. Exp. Biol. Med.,* **104:**512 (1960).
190. Shafter, L. A., C. R. Scriver, and M. L. Efron, *N. Engl. J. Med.* **267:**51 (1962).
191. Simensen, M. G., in C. E. Cornelius, and J. J. Kaneko (eds.) *Clinical Biochemistry of Domestic Animals,* Academic Press, New York, 1963.
192. Simmons, D. J., and A. S. Kunin, *Clin. Orthop.,* **68:**251 (1970).
193. Sladi c-Simi, D, N. Zivkovi, and D. Pavi, *Rev. Eur. Etud. Clin. Biol.,* **17:**197 (1972).
194. Smith, E. L., *Nature* (London), **162:**144 (1948).
195. Smith, E. L., in *Vitamin B*$_{12}$*,* Methuenen, London, 1965.
196. Smith, G. P., and A. N. Epstein, *Am. J. Physiol.,* **217:**1083 (1969).
197. Stauffacher, W., L. Orci, D. P. Cameron, I. M. Burr, and A. E. Renold, *Recent Prog. Horm. Res.,* **27:**41 (1971).
198. Stemerman, M. B. and R. Ross, *J. Exp. Med.,* **136:**769 (1972).
199. Strohfeldt, P., *Res. Exp. Med.,* **162:**7 (1973).
200. Stephens, T., S. Irving, P. Mutton, J. D. Gupta, and J. D. Harley, *Nature,* **248:**524 (1974).
201. Strong, J. P., *J. Med. Prim.,* **3:**89 (1974).
202. Swales, J. D., and J. D. Tange, *J. Lab. Clin. Med.,* **78:**369 (1971).
203. Taylor, C. B., K. J. Ho, and L. B. Liu, in W. P. McNulty, Jr. (ed.), *Recent Advances in Arteriosclerosis and Cholesterol Metabolism in Primates,* Symposia of the Fourth International Congress of Primatology, Vol. 4: Nonhuman primates and human diseases, Karger, Basel, 1973.
204. Thoonen, J., J. Hoorens, and E. Van Murhaeghe, *Arch. Geflnegelk.,* **23:**314 (1959).
205. Thrope, E., and E. J. H. Ford, *J. Comp. Pathol.,* **78:**195 (1968).
206. Todd, G. C., and L. Krook, *Pathol. Vet.* (Basel), **3:**379 (1966).
207. Tschudy, D. P., and H. L. Bonkowsky, *Fed. Proc.,* **31:**147 (1972).
208. Tumbleson, M. E., O. W. Tinsley, L. A. Corwin, Jr., R. E. Flatt, and M. A. Flynn, *J. Nutr.,* **99:**505 (1969).

209. Tumbleson, M. E., O. W. Tinsley, K. W. Hicklin, and J. B. Mulder, *Growth,* **36:**373 (1972).
210. Tumbleson, M. E., and D. P. Hutcheson, *Nutr. Rep. Internat.,* **6:**321 (1972).
211. Turchitto, E., and P. Barri, *Bibl. Nutr. Diet,* **11:**34 (1969).
212. Underwood, E. J., in *The Mineral Nutrition of Livestock,* F. A. O. Rome and C. A. B. Fornham Royal, 1966.
213. Vesselinovitch, D., G. S. Getz, R. H. Hughes, and R. W. Wissler, *Atherosclerosis,* **20:**303 (1974).
214. Ward, B. C. and J. N. Shively, *Arch. Pathol.,* **98:**90 (1974).
215. Waters, L. L., *Arch. Pathol.,* **93:**525 (1972).
216. Waterworth, M. W., G. O. Barbezat, R. Hickman, and J. Ferblanche, *Br. J. Surg.,* **61:**318 (1974).
217. Weber, G., L. Resi, E. Agradi, and C. R. Sirtori, *Res. Exp. Med.* (Berl), **161:**304 (1973).
218. Wegelius, D., *Proc. Soc. Exp. Biol. Med.,* **101:**225 (1959).
219. Whang, R., J. Oliver, L. G. Welt, and M. MacDowell, *Ann. N.Y. Acad. Sci.,* **162:**766 (1969).
220. Whittmore, C. T., and F. W. H. Elsley, *Vet. Rec.,* **94:**113 (1974).
221. Widdowson, E. M., J. W. T. Dickerson, and R. A. McCance, *Br. J. Nutr.,* **14:**475 (1960).
222. Widdowson, E. M., and R. A. McCance, *Proc. R. Soc. Biol.,* **158:**329 (1963).
223. Widdowson, E. M., in R. A. McCance and E. M. Widdowson (eds.), *Calorie Deficiencies and Protein Deficiencies,* Little, Brown and Co., Boston, 1971.
224. Wigglesworth, J. S., *J. Pathol. Bacteriol.,* **88:**1 (1964).
225. Wilfred, G., and T. N. Varma, *Biochim. Biophys. Acta.* **187:**442 (1969).
226. Witzleban, C. L., *Am. J. Pathol.,* **62:**181 (1971).
227. Witzleban, C. L., *Am. J. Pathol.,* **66:**577 (1972).
228. Yager, J. D., M. J. Lichtenstein, R. J. Bonney, H. D. Hopkins, P. R. Walker, C. G. Dorn, and V. R. Potter, *J. Nutr.,* **104:**273 (1974).
229. Young, D. M., H. M. Olson, D. J. Prieur, D. A. Cooney, and R. L. Reagen, *Lab. Invest.,* **29:**374 (1973).
230. Zoltowska, A., and T. Wrzolkowa, *J. Pathol.,* **109:**93 (1973).
231. Zucker, L. M., *Ann. N.Y. Acad. Sci.,* **131:**447 (1965).

CHAPTER 6
ANIMAL MODELS FOR THE STUDY OF HEREDITARY DISEASES

During the past few years a rather remarkable series of contributions have been made in the field of inherited animal diseases which are either exact models of or similar to their human counterpart (e.g., Chediak-Higashi syndrome (140, 173, 181) of cattle, mink, and mice; Dubin Johnson syndrome of sheep (6, 42); lupus erythematosis and Christmas disease of dogs (143); porphyria of cattle and swine (135); and many others listed in Table 6-1). Although the mechanisms involved in many of the hereditary disorders are not clear at present, the underlying factor in all these anomalies is the genetic transmissibility of the character(s). In some instances a certain enzyme is missing, thereby affecting metabolism of the host. In others, anatomic or physiological defect is the main characteristic of the congenital or inherited disorder in humans and animals. By using animal species, the hereditary disease processes may be studied, which can be very useful in the management and treatment of patients. This chapter briefly describes selected animal models of congenital disorders including inborn errors of metabolism, congenital disorders caused by viral agents, and other hereditary diseases. Numerous reviews and textbooks are available describing inherited diseases in human beings (106, 107, 113, 136, 209, 226, 231).

AMINOACIDOPATHIES

The primary cause of inborn errors of metabolism is usually the absence or inactivity of an enzyme system. This results in the accumulation of nonmetabolizable products which the affected organism must excrete unchanged or altered by biochemical reactions. Aminoacidopathies are metabolic diseases in which patients have abnormal accumulation of one or more amino acids in urine and/or blood and usually have the primary enzyme defect. A causal relationship exists between the metabolic defect and the clinical symptoms. It is interesting to note that, with the exception of the essential amino acids, the amino acids of proteins that are not related to known aminoacidopathy are all direct precursors of intermediates of the tricarboxylic acid cycle. A feasible explanation for this fact is that an enzyme defect related to one of these amino acids will not be compatible with life. The clinical symptoms in aminoacidopathies cover a wide spectrum, from no proven abnormalities at all, as in β-aminoisobutyricaciduria, hypersarcosinemia, and cystathionuria, to very early death in infancy as in ketotic hyperglycinemia and branched chain ketoaciduria. Most aminoacidopathies cause brain damage and mental retardation of varying degrees (3). It may be due to deficiency of biogenic amines and other neurotransmitters or impaired energy production (ATP deficiency) or impaired synthesis of enzymes and structural compounds (lipid, proteins). Important work with experimental hyperaminoacidemias, using methionine, glycine, histidine, and the branched chain amino acids, leucine, isoleucine, and valine, has been reported (77). Re-

Table 6-1. Animal Models for the Study of Hereditary Diseases

Animal Model	Species	Human Counterpart	References
Achondroplasia	Rabbit Cattle Mouse	Dwarfism	84 48, 127 54, 84
Adiposity, familial	Mouse	Obesity	18
Albinism	Mouse	Albinism	38
Alopecia, inherited symmetrical	Cattle	Inherited symmetrical alopecia	35
Aminoaciduria	Mouse	Aminoaciduria	14
Anemia, familial	Basenji dog	Hereditary spherocytosis	220
Anemia, hereditary	Mouse	Siderocytic anemia Thalassemia	62, 163
Anemia, inherited hypochromic	Mouse	Hypochromic anemia	57
Anemia, hereditary hemolytic	Mouse	Hemolytic anemia	210
Anemia, inherited	Alaskan Malamute Dog	Inherited Hemolytic anemia with stomato-cytosis congenital hemolytic anemia	67
Anemia, microcytic and jaundice	Mouse	Neonatal jaundice	211
Aniridia, hereditary	Horse	Aniridia	60
Ataxia, hereditary	Calf Mouse	Ataxia	115, 150
Atrial septal defect	Chimpanzee	Interatrial septal defect	87
Atrichosis and hypo-trichosis congenital	Cattle Dog	Congenital atricho-sis & Hypotrichosis	35
Autoimmune disease	Mouse	Systemic lupus erythematosus	

Table 6-1. (Continued)

Animal Model	Species	Human Counterpart	References
Cardiomyopathy	Syrian Hamster	Myocardial failure	78
Congenital hyper- bilirubinemia (Gilbert's syndrome)	Southdown Sheep	(Gilbert's syndrome)	45
Congenital hyper- bilirubinemia	Gunn rats	Crigler-Najjar syndrome	39
Congenital hyper- bilirubinemia	Sheep	Dubin-Johnson syndrome	43
Cyclic neutropenia	Dog	Cyclic neutropenia	149
Cataract, congenital	Dog	Congenital cataracts	207
Cataract, inherited	Dog Cattle Mouse	Cataract	28, 6,11
Cerebellar cord degeneration inherited	Horse	Cerebellar degenera- tion	137
Chrondrodystrophia fetalis	Dog	Achondroplasia	74
Clubfoot	Mouse	Clubfoot	193
Cystinuria	Dog Blotched genet	Cystinuria	44, 51
Deafness, hereditary	Mink Dog Cat Mouse	Deafness	201 108, 109 1, 21, 27, 50 53, 82, 83
Diabetes insipidus	Mouse	Diabetes insipidus	62
Dwarfism hypohyseal	Mouse	Dwarfism	208
Ehlers-Danlos syndrome	Dog	Ehlers-Danlos syndrome	93
Gangliosidosis GM_2	German Shorthaired Pointers	GM_2 Gangliosidosis	130
Glaucoma, hereditary	Rabbit	Glaucoma	89
Gross congenital anomalies	Mammals (many species)	Congenital malforma- tion due to vitamin A	203

Table 6-1. (*Continued*)

Animal Model	Species	Human Counterpart	References
Goiter, congenital	Cattle	Goiter	111
Hairlessness	Mouse	Alopecia	24
Heart disease, congenital	Dog Cattle Cat	Congenital heart disease	175 170 219
Hemophilia-A	Dog	Classical hemophilia (anti-hemophilic factor)	22
Hemophilia-B-like disease	Dog	Christmas disease (plasma thrombo-plastic component deficiency)	197
Hereditary diseases (Metabolic dysfunction)	Cattle	Hereditary diseases (Metabolic dysfunction)	233
Hydroencephaly & Porencephaly	Fetal lambs	Congenital malforma-tion	171
Hydronephrosis, inherited	Rat	Hydronephrosis	145
Hypoamylasemia-mucoid enteritis, neonatal	Rabbit	Hypoamylasemia	158
Hypotrichosis, genetic	Cattle	Hypotrichosis	91
Ichthyosis, congenital	Cattle	Ichthyosis, congenital	35, 224
Immunoproliferative disease	Mouse	Lymphocytic chorio-meningitis (LCM)	185
Insulin tolerance, inherited	Mouse	Insulin tolerance	31
Iridal heterochromia, hereditary	Cattle	Iridal heterochromia	142
Kidneys, Cystic or absent	Rat	Cystic kidneys	112
Legg-Perthes disease	Dog Cattle	Legg-Perthes disease	55 146

Table 6-1. (*Continued*)

Animal Model	Species	Human Counterpart	References
Leukodystrophy, Globoid cell	Dog	Krabbe's disease	64
Leukodystrophy, inherited	Mink	Familial meta-chromatic dystrophy	4
Leukomelanopathy, hereditary	Mink Cattle	Chediak-Higashi syndrome	141
Lipidosis, neonatal intestinal	Mouse Sheep Simian Primate	Intestinal degen-eration, hepatic disease, congenital	227 7 155
Lipodystrophy	Dog	Familial amaurotic	129
Lymphatic edema congenital	Swine	Lymphatic edema	165
Lymphedema, hereditary	Dog	Milroy's disease (congenital heredi-tary lymphedema)	176
Lymphoid tumors, abnormal lipid	Mouse	Niemann-Pick disease	151
Lysosomas, giant	Mouse	Chediak-Higashi syndrome	12
Mannosidosis (pseudolipidosis)	Angus cattle	Mannosidosis	117
Microphthalmis, congenital cystic	Swine	Microphthalmia	30
"Midget"	Mouse	Dwarfism	72
"Miniature"	Mouse	Dwarfism	13
Mongolism	Chimpanzee	Down's syndrome	157
Muscular dysgenesis	Mouse	Prenatal muscle degeneration	174
Muscular dystrophy	Mouse Chicken Duck	Muscular dystrophy	229, 164 8 191
Myoclonia, congenital	Swine	Myoclonia, congenital	205

Table 6-1. (Continued)

Animal Model	Species	Human Counterpart	References
Neuronal glyco-proteinosis	Dog	Progressive familial myoclonus epilepsy; Lafora's disease	97
Osteoporosis, hereditary	Rabbit	Osteoporosis	178
Osteoporosis, familial	Dog	Osteogenesis, imperfecti	27
Oxalate calculi	Cat	Oxaluria	110
Persistent viral infections (Aleutian Disease)	Mink	Immunologically mediated, glomerulo-nephritis & arteritis, Dysgammopathies	94
Phenylalanine hydroxylase deficiency	Mouse	Phenylketonuria	114
Polycythemia, familial	Jersey cattle Hereford cattle Dog	Polycythemia vera	221 73 56
Polymyopathy	Syrian Hamster	Muscular dystrophy	100
Porphyria, congenital (dominant)	Cat	Erythrocytic porphyria	79
Porphyria, congenital erythrocytic	Shorthorn cattle Holstein-Friesian cattle Swine Fox squirrel	Congenital erythro-cytic porphyria	142 128 123 225
Porphyria (dominant inherited)(defective porphyrin metabolism)	Cat	Porphyria (erythropoietic and non-erythropoietic hepatic type)	79, 222
Porphyria (recessive inherited)	Cattle	Porphyria	70, 124, 189, 228
"Pygmy"	Mouse	Dwarfism	152
Retinal dysplasia, congenital	Dog	Retinal dysplasia	198

Table 6-1. (Continued)

Animal Model	Species	Human Counterpart	References
Rheumatoid factor	Howler monkey	Rheumatoid factor	180
Sclerectasia, hereditary; retinal detachments	Dog	Hereditary **scler-** ectasia; retinal detachments	194
Sebaceous glands, lack of	Mouse	Hyperkeratosis	75
Sex chromosome anomaly (Tortoise-shell male cat)	Cat	Klinefelter's syndrome	119, 120
Short-Danforth syndrome	Mouse	Absence of kidney	199
Sickling of erythrocytes, in vitro	Mouse	Sickle-cell anemia	134
Spherocytosis, hereditary	Deer Mouse	Spherocytosis	5
Spina bifida	Mouse Rabbit	Spina bifida	86 49
Spina bifida, sacrococcygeal agenesis	Manx cat	Spina bifida, sacral dysgenesis	133
Teratoma, embryonal carcinoma, terato-carcinoma	Mice (Inbred strain)	Embryonal carcinoma, teratocarcinoma teratoma	214
Ventricular septal defect	Dog	Interventricular septal defects	88
Vitamin A deficiency. Hydrocephaly	Rabbit	Congenital Communi-cating Hydrocephalus	167
Vitiligo	Horse	Vitiligo	35
Waardenburg's syndrome	Cat	Waardenburg's syndrome	61

search on experimental hyperphenylalaninemia has progressed steadily. Criteria for developing an animal model for phenylketonuria (PKU) (or another aminoacidopathy) are (a) to obtain increased amino acid blood levels comparable to those found in the disease; (b) to cause formation and urinary excretion of the same metabolites that are excreted by the patient; and (c) to cause brain damage through long periods of increased blood amino acid levels at an early age, detectable by a deterioration of learning behavior in the animal. Investigations may also be made on detection of changes in different enzyme activities in the liver and brain and alterations in transport of essential compounds across the blood-brain barrier. When fed large amounts of phenylalanine, animals have increased blood phenylalanine levels and they excrete phenylketones. Deterioration of learning behavior is detected in these animals. The essential objections to this animal model are (a) the enzyme defect may not present in the animal; (b) the phenylalanine blood levels may be variable depending on dietary intake; (c) the tyrosine level in the blood which is below normal in a PKU patient is increased in the animal; and (d) the differences between species in enzyme patterns and responses to drugs and metabolic insults often make the application of animal data to human disease questionable.

The ideal animal PKU model may be a mutant of a laboratory animal with a deficiency of the enzyme, phenylalanine hydroxylase, but this genetic defect has not yet been found in animals.

There are two basic approaches to the use of animals as models for aminoacidopathies. (a) Long-term experiments during which chronic hyperaminoacidemia is maintained. This is usually done by feeding a standard diet, supplemented with a specific aminoacid. (b) Short-term experiments designed to elucidate the acute effects of hyperaminoacidemia. For this type of experiment, the amino acid is usually injected intraperitoneally. PKU is caused by deficiency of phenylalanine hydroxylase in the liver. It has been found that, of the normally used experimental animals, the rat has a specific liver phenylalanine hydroxylase activity of about forty times that of a human, and about ten times that found in the liver of the rhesus monkey (77). Feeding diets containing up to 7 percent L-phenylalanine to young rats up to 90 days of age causes an increase of phenylalanine blood levels ranging from 30 to 80 mg percent, which is comparable to levels found in PKU patients. The urine samples of these rats contain phenylpyruvate and other phenylalanine metabolites. There is growth depression compared to pair-fed controls and phenylalanine hydroxylase activity in these rats decreases to about 50 percent of the normal level. Horwitz and Waisman (104) reported that the changes in phenylalanine hydroxylase activity in hamsters were influenced by the protein content of the diet. In both rats and hamsters fed phenylalanine-supplemented diets, it was noted that tyrosine blood levels were considerably increased, sometimes to even higher levels than the phenylalanine.

Newborn monkeys could tolerate diets containing added amounts of phenylalanine or other amino acids sufficient to increase their blood levels comparable to those found in patients with PKU and other aminoacidopathies. The monkey is a much more expensive animal to maintain and breed than the rat. However, experiments can be performed with monkeys over much longer periods than with rats, and adequate blood and urine samples can be obtained easily without sacrificing animals. The development of the monkey is slower and more comparable to the human infant and the primate placenta functions like that of the human. Also, the monkey infant can be removed from the mother 6 hours after birth, and bottle-fed throughout infancy. Feeding monkeys at 4-hour intervals through the day may achieve a constant elevation of the desired aminoacid.

MANNOSIDOSIS

Animal model of mannosidosis (pseudolipidosis) in Angus cattle has been reported by Jolly (117, 118) who found similarities between the animal disease and mannosidosis of children.

Mannosidosis of Angus cattle is a nervous disease inherited as an autosomal recessive gene. It is characterized by ataxia, incoordination, head tremors, intention tremor, aggressive tendency, failure to thrive, and death, usually in the first year of life. An oligosaccharide composed of mannose and glucosamine can be extracted from brains and lymph nodes of affected calves, but not from normal animals, thus suggesting an anomaly of glycoprotein catabolism. Subsequent investigation showed that this is associated with a deficiency of L-mannosidase in the affected animals. It is an inherited and a storage disease. Probably, the storage material containing mannose and glucosamine is found in the lysome-like vacuoles of certain cells, since there is a deficiency of a lysomal enzyme which under normal circumstances would degrade the storage product. It is known that heterosaccharide fraction of glycoprotein contains a core of mannose and N-acetyl glucosamine with various side chains. Degradation of these heterosaccharides takes place by sequential splitting of sugars from the nonreducing end.

Mannosidosis of Angus cattle and children are similar in regard to pathology, nature of storage substance, and specific enzyme deficiency. As most inherited lysosomal storage diseases are biologically similar, mannosidosis of cattle has many features in common with diseases caused by deficiencies of other lysomal enzymes (118). Mannosidosis is a general model for studying the biology of lysosomal storage diseases particularly in regard to some of the problems concerning enzyme therapy. It is also a specific model and is being used as such for studying human mannosidosis. In addition, being an anomaly of glycoprotein catabolism, it may be useful for investigation of the structure and biology of glycoproteins.

AMYLOIDOSIS

Spontaneous amyloidosis has been reported in a variety of animal species, including dogs, cats, mice, and birds (34, 36, 154). Experimentally amyloidosis can be induced in mice, hamsters, guinea pigs, and rabbits, although susceptibility to the amyloid-inducing regimen varies among species and strains (186, 230). Cohen (37) described spontaneous and induced amyloidosis in various inbred and random-bred strains of mice. Although the clinical signs of the disease in mice are not specific, general weakness and loss of the body weight may occur. Animals may develop a nephrotic syndrome and associated illnesses; death from azotemia may occur. The major pathological feature of the disease is the presence of intercellular deposition of amyloid which by light microscopy has a hyalin eosinophilic appearance. It stains with Congo red and shows characteristic green birefringence under polarized light after such staining. It is PAS-positive and demonstrates crystal violet and methyl violet metachromasia. In the electron microscope, amyloid consists of characteristic fibrils approximately 100 Å wide, indeterminate in length, rigid, nonbranching, and often arranged in random array (34, 154). The deposition usually appears first in the perifollicular zone of the spleen, then in the space of Disse in the liver and in the glomerular mesangial region in the kidney. When the animal is kept for a long period of time on an amyloid-inducing-regimen (e.g., 5 to 10 percent casein solution) the deposition is widespread in a variety of organs, including the blood vessels (154).

Experimentally induced amyloidosis in animals is comparable in virtually all respects with human amyloidosis by histological and electromicroscopic standards (36). Therefore, the animal model can be used for the studies of the course, pathogenesis and treatment of amyloid disease in human beings.

CONGENITAL HYPERBILIRUBINEMIA
(Dubin-Johnson Syndrome)

Cornelius (40) reported Corriedale sheep as a model of Dubin-Johnson (Sprinz-Nelson) syndrome in human beings. It is characterized by a congenital hepatic excretory defect for organic anions, such as bilirubin (conjugated, direct), sulfobromophthalein sodium (BSP), phylloerythrin, metepinephrine glucuronide, and so forth. The condition may be inherited as a single autosomal recessive gene. The liver of sheep with this syndrome appears functionally and morphologically identical to those observed in the human counterpart of the disease. The syndrome in sheep, unlike that in humans, is lethal under field conditions due to the complication of photosensitivity from the ingestion of chlorophyll in green feed. Photosensitivity can be prevented by housing mutants indoors. Elevated levels of direct-reacting bilirubin in plasma is found in the affected animals but jaundice is not present. The major pathological finding is grossly dark brown to black liver due to accumulation of a melanin pigment in the hepatic lysosomes.

Congenital hyperbilirubinemia of unconjugated, "indirect-reacting" bilirubin due to bilirubin uptake defect of the liver in Southdown sheep was described by Cornelius (40). It is similar to Gilbert's syndrome (constitutional hepatic dysfunction) in humans. Gilbert's syndrome is characterized by a congenital hepatic uptake defect for many organic ions such as bilirubin, phylloerythrin, sulfobromophthlein sodium, indocyanine green, rose bengal, and so forth. In the sheep, it is a sublethal trait due to single autosomal recessive gene. A marked elevation (50 to 60 percent) in the unconjugated bilirubin without the signs of jaundice is generally present in this syndrome. Sheep are presented with acute photophobia and photosensitization at the time of weaning due to ingestion of chlorophyll in green feed. Chlorophyll is converted in the ruminant gastrointestinal tract to phylloerythrin and inadequately excreted from the portal blood by the mutant's liver. In the affected animal the liver is normal by microscopic examination and the major pathological finding a diffusely fibrotic kidney. Death usually results from renal failure.

In both mutant Southdown sheep and patients with Gilbert's syndrome, lower fractional removal rates for bilirubin from the early mixing pool (primarily plasma) to the storage pool (primarily liver) are found. The Southdown mutant, unlike Gilbert's syndrome in humans, has in addition to the bilirubin uptake defect, a marked defect in the uptake of many other organic ions such as BSP, indocyanine green, and rose bengal. Hepatic bilirubin glucuronyl transferase activities are normal in Southdown sheep mutants but in Gilbert's syndrome in humans, the enzyme activities may be lower than normal. The syndrome in humans is not lethal because of the absence of large amounts of intestinal phylloerythrin excreted from the portal blood by the liver.

It seems that the Southdown mutant sheep may provide a good model for the study of Gilbert's syndrome, particularly for understanding the mechanism of active transport of bilirubin and other organic anions by the liver.

Hereditary nonhemolytic, unconjugated hyperbilirubinemia in Gunn rats was reported by Cornelius (41). This condition in the rat results from the absence of hepatic UDP glucuronyl transferase activity. The affected animals are jaundiced, and bilirubin in the serum of these animals gives an indirect van den Bergh reaction, even after ligation of the common bile duct. The defect is transmitted as an autosomal recessive characteristic. Heterozygous Gunn rats appear normal, not jaundiced, and their livers exhibit glucuronide formation in vivo and in vitro that is intermediate between that observed in genetically normal and homozygous jaundiced litter mates. The enzymatic defect in the heterozygous rats apparently is not severe enough to result in the retention of bilirubin in the plasma. The deficiency in glucuronyl transferase activity in homozygous Gunn rats involves the intestine and kidneys as well as the liver. The animals produce bile pigment at a normal rate and quantities of unconjugated bilirubin in the serum and tissues are relatively constant despite the liver's inability to conjugate bilirubin. It seems that in the mutant Gunn rats, the major portion of bilirubin is catabolized to diazo-negative, polar bilirubin derivatives excreted in bile and urine. A small amount of unconjugated bilirubin is transferred across the mucosa into the intestine (41).

The animal model of hereditary nonhemolytic hyperbilirubinemia is useful in the study of Crigler-Najjar syndrome in human infants. The syndrome is a hereditary unconjugated hyperbilirubinemia in infants which produces nonhemolytic acholuric jaundice, often associated with kernicterus. Crigler-Najjar syndrome results from a deficiency in hepatic uridine diphosphate (UDP) glucuronyl transferase activity. Jaundice is first noticed in infants on or about the third day and persists throughout life. Most patients die in infancy from kernicterus (116). The characteristic features of the syndrome in infants are severe unconjugated hyperbilirubinemia, with serum bilirubin concentrations ranging from approximately 14 to 45 mg/100 ml. There is no evidence of hemolysis, the early labeled pigment is normal, the extrahepatic biliary system is patent, and light microscopic examinations shows that the liver is normal. There is no measurable bilirubin in urine. Glucuronyl transferase activity is absent. The congenital anomaly is transmitted by a gene with autosomal recessive characteristics. Nonicteric heterozygotes excrete glucuronide in vivo; this is intermediate between that observed in normal and jaundiced individuals in a family. There are two genotypes of the syndrome reported. The Crigler-Najjar syndrome corresponds to type I UDP-glucuronyl transferase deficiency in which blood bilirubin levels can be dramatically reduced by administering phenobarbital or other drugs. In type II UDP-glucuronyl transferase deficiency, the disorder is transmitted as an autosomal dominant characteristic and the bile contains small amounts of bilirubin glucuronide.

The model in Gunn rats may serve as a useful model for the study of bile metabolism, intestinal glucuronide formation, jaundice, and brain damage (kernicterus).

PORPHYRIA (Erythropoetic and Hepatic Types)

Animal model of porphyria in the domestic cat was reported by Glenn (80). In cats, it is characterized by biochemical features which overlap the various types of erythropoietic and nonerythropoietic (hepatic) types in man. Many of the clinical and biochemical features of feline porphyria are similar to those of congenital erythropoietic porphyria. Animals affected by porphyria exhibit brownish pigmentation of teeth and bones, dark brown to red discoloration of the urine, and varying degrees of anemia. Teeth, bone, and urine exhibit bright pink-red fluorescence when exposed to ultraviolet light due to excessive production of uroporphyrin and coproporphyrin, predominantly of the type I isomer which cannot be further utilized in the heme biosynthesis pathway. Consequently, these highly fluorescent pigments are deposited in osseus tissue and excreted in the urine. Feline porphyria also bears similarity to erythropoietic protoporphyria of humans in that protoporphyrin accumulates in excessive amounts in immature erythrocytes. In the cat, porphyria is further characterized by excessive urinary excretion of porphobilinogen, a feature that is characteristic of the hepatic porphyrias of humans.

Feline porphyria is inherited as an autosomal dominant trait (79). Porphyria among other domestic animal species occurs with relatively high frequency in cattle (2, 122) and rarely in swine (32, 121). The disease in cattle is similar to congenital erythropoietic porphyria of humans and is inherited as an autosomal recessive trait. Porphyria in swine, though less well characterized, is considered to be of the erythropoietic type but is inherited as an autosomal dominant trait like that in the cat.

Goldberg and Rimmington (81) have described hereditary diseases involving defective porphyrin metabolism in humans. Defective porphyrin biosynthesis occurs in the erythropoietic tissues (erythropoietic type) and disturbance in porphyrin metabolism occurs in nonerythropoietic tissues (hepatic type). Two forms of erythropoietic and three forms of hepatic porphyria occur in humans. However, the prophyrias of domestic animals, with the possible exception of that in the cat, are of the erythropoietic type. Therefore, feline porphyria may be a useful model for the study of inborn errors of porphyrin metabolism and for elucidation of control mechanisms of porphyrin metabolism.

INHERITED HEMOLYTIC ANEMIA

Inherited hemolytic anemia with stomatocytosis in the Alaskan malamute dog, described by Fletch and Pinkerton (67), is similar to congenital hemolytic anemia in humans. Short-limbed dwarfism (chondrodysplasia) is inherited as an autosomal recessive gene (*dan*) in the pure breed of Alaskan malamute dog. Chondroplasia is consistently associated with erythrocyte macrocytosis and mild anemia (184). A close parallel exists between the canine disease and at least two families described in humans (Table 6-2).

The animal model of the disease may be useful in the elucidation of metabolic processes whereby normal mammalian red cell's electrolyte balance is maintained (67, 162, 172, 234). Study of the canine disease, similar human disorders, spherocytosis in humans and mice, and elliptocytosis in humans may prove useful in understanding the red cell membrane as the inherited disorders of red cell enzyme function have in clarifying the glycolytic mechanisms of the red cell. Elucidation of the red cell defect caused by *dan* may also be expected to give useful information on normal bone growth (67).

Table 6-2. Comparison of Inherited Hemolytic Disease in Alaskan Malamute Dog and Humans

Feature	MCV	MCHC	Stomato-cytosis	Osmotic fragility	^{51}Cr t$_{\frac{1}{2}}$	Red cell sodium	Inheri-tance	Short Stature
Malamute (dan/dan)	inc	dec	+	inc	dec	inc	AR	+
Human Disease:								
Zarkowsky et al. (234)	inc	dec	+	inc	---	inc	AR	+
Oski et al. (172)	inc	dec	+	inc	dec	inc	AD	-
Miller et al. (162)	Normal	Normal	+	dec-normal	dec	inc	AD	-

inc = increased, dec = decreased, + = present, - = not reported, AR = autosomal recessive, AD = Autosomal dominant, MCV = mean corpuscular volume, MCHC = mean corpuscular hemoglobin concentration.

Reproduced from The American Journal of Pathology (67) with permission.

INHERITED HYPOCHROMIC ANEMIAS

Three types of inherited anemias, (a) sex-linked anemia of the mouse (gene symbol *sla*), (b) hereditary microcytic anemia of the mouse (gene symbol, *mk*), and (c) the anemia of the Belgrade laboratory rat (gene symbol *b*), were described by Edwards and Bannerman (57). These rodent anemias represent potential rather than definitive models for the study of normal and abnormal iron metabolism, hemoglobin synthesis, and related processes in humans. However, there appears to be no exact counterpart in humans to sex-linked anemia or hereditary hypochromic anemia of the mouse. In contrast to the situation in these two murine anemias, the tissue iron stores are increased in human

Table 6-3. Hematological and Biochemical Features of Inherited Hypochromic Anemias

Features	Disorders		
	sla*	mk**	b***
Serum iron concentration	Decreased (183)	Decreased (10)	Increased (206)
Serum total iron binding capacity	Increased (183)	Increased (10)	Increased (206)
Free erythrocyte proto-porphyrin concentration	Increased (138)	Increased (138)	Undetermined
Histochemical studies of tissue iron distribution			
Spleen	Decreased (182)	Decreased (10)	Decreased (206)
Duodenum	Increased (182)	Decreased (10)	Undetermined
Iron clearance	Rapid (183)	Unchanged (10)	Undetermined
Iron utilization	Increased (183)	Unchanged (10)	Undetermined
Intestinal iron absorption			
In vivo	Decreased (183)	Decreased (58)	Undetermined
In vitro	Impaired transfer from mucosa to serosa (59)	Impaired mucosal uptake (58)	Undetermined
Response to parenteral therapy	Complete (183)	Incomplete (10)	Incomplete
Fecal urobilinogen excretion	Increased (138)	Increased (138)	Undetermined

```
  * Sex-linked anemias
 ** Hereditary microcytic anemia
*** Anemia of the Belgrade laboratory rat
```

Adapted from <u>Comparative Pathology Bulletin</u> (57) with permission.

thalassemia, x-linked hypochromic anemia of humans, and the unusual familial hypochromic anemia. Furthermore, these conditions are associated with hypersideremia. Hereditary, hypochromic anemia of the rat is a potential model for human thalassemias; but although the diseases share common characteristics, such as the red cell morphological abnormalities and the hypersideremia, further studies are required before it can be determined whether the anemia of Belgrade rat is due to a primary disturbance of hemoglobin synthesis or of iron metabolism.

Red cell hypochromia is a common feature to all the three inherited rodent anemias. The main hematological and biochemical characteristics of rodent anemias are summarized in Table 6-3.

The inherited rodent anemias represent very useful models for the study of iron and hemoglobin metabolism. For example, the existence of animals with two separate and distinct genetically determined defects in the intestinal absorption of iron that is of the mucosal uptake in *mk* and of mucosal serosal transfer in *sla,* offers a powerful tool in the elucidation of iron absorption.

CYCLIC NEUTROPENIA

Cyclic neutropenia in the collie breed of dogs was described by Lund (148), with similar clinical manifestations of the disease in humans. Although the genesis of the disease cycles is not fully known, a genetic trait inherited as a simple autosomal recessive has been reported (71, 147), for example, in the collie breed, a specific dilution of the hair coat color ("grey collie"). However, it is not known whether the two conditions, that is, the blood disease and the dilution of the coat color are closely linked traits or if they are due to the same genetic defect.

Cyclic neutropenia in the dog results from a periodic failure of cell maturation in the bone marrow. Both red and white cell production is interrupted (148). However, because of the short interruption in the production in relation to the long erythrocyte life span in the peripheral blood, the defect in erythrocyte production is not clinically apparent. When bone marrow production is interrupted, there is a rapid depletion of the bone marrow reserve of the segmented neutrophile, followed by a severe neutropenia (0–400 neutrophils/mm). The neutropenia episodes occur at 11-day intervals (range of 9–13) and last from 2 to 5 days (average 3 days). The dogs have severe infections which occur during the neutropenic episodes. The affected animals usually do not survive the first year without careful medical treatment. Even with adequate medical treatment, most affected dogs die between the age of 1 and 2 years.

The interval between disease cycles is variable in humans; most have been 21 days in length, but cycles of 14 and 28 days have also been reported (188). Although the primary signs of the disease are similar in humans and in dogs, the secondary effects of the disease appear to be more severe in affected dogs than in human patients. Retardation of sexual development and growth has not been reported in humans. The difference may be due to the shorter cycles of neutropenic disorders in dogs which may therefore simulate a chronic disease process. It may, in part, also be attributed to the effects of amyloid deposition in the visceral organs. Renal amyloidosis in particular is present in all dogs well before maturity. Death at an early age, which characterizes the canine disease, does not appear to be a feature of the human disease.

Despite the differences between the animal model of neutropenia and the human disease, cyclic neutropenia in collie dogs may be a useful model to study causes and mechanism of the disease in humans.

HEMOPHILIA A AND B

Both types of hemophilia are sex-linked, recessive, bleeding diathese due to the lack of procoagulant trace plasma proteins, and symptoms are similar for both hemophilias. The missing protein in hemophilia A is antihemolytic factor (AHF) (Factor VIII), while plasma thromboplastin component (PTC) (Factor IX) is lacking in hemophilia B (23). The coagulation defects are temporarily corrected by transfusion of normal plasma or plasma concentrates; serum is not active in hemophilia A therapy but can be used for the treatment of hemophilia B.

Brinkhous and Gambill (22, 23) described hemophilia A in the Irish setter breed of dogs and hemophilia B in the cross-breeds between beagle and cairn terrier dogs. Hemophilic dogs exhibit localized hemorrhages in various tissues and organs of the body as well a hemarthrosis (spontaneous or trauma-induced). Common laboratory findings, identical in both canine and human hemophiliacs are shown in Table 6-4.

Canine hemophilia may be a good model to study severe hemophilia condition in human beings, since the hemophilias in humans and dogs are identical in symptomatology, inheritance, and coagulation defects.

Table 6-4. Common Laboratory Findings in Hemophilia A and B in Human and Canine Species

Test	Hemophilia A	Hemophilia B
Clotting Time	Prolonged	Prolonged
Bleeding Time	Normal	Normal
Secondary Bleeding Time	Prolonged	Prolonged
Prothrombin Time	Normal	Normal
Platelets	Normal	Normal
Partial Thromboplastin Time	Prolonged	Prolonged
Thromboplastin Generation Test	Prolonged	Prolonged
Correction with normal serum	No	Yes
Correction with $Ba(SO_4)$ adsorbed plasma	Yes	No
Factor VIII Assay	< 1%	Normal
Factor IX Assay	Normal	< 1%

Reproduced from <u>Comparative Pathology Bulletin</u> (22) with permission.

GROSS CONGENITAL ANOMALIES

Treatment of various experimental animal species with a large dose of vitamin A at known stages in pregnancy produce many types of gross structural malformations (203). The types and incidence of malformations depend on the stage of pregnancy and dosage and to a lesser extent on species and strain of the animal. High incidence of malformation has been observed by a single high dose of vitamin A to the rat, mouse, golden hamster, guinea pig, and rabbit and in limited experiments on monkeys, pigs, and dogs. Malformation produced in animal species by hypervitaminosis A mimic human malformations of genetic, environmental and unknown causes. More than seventy types of malformations are produced by hypervitaminosis A. In the hamster, eighteen types could be produced at more than 90 percent incidence, and twenty-eight types at 10 to 90 percent. Structures affected (and analogs of some human malformations) include: brain (anencephaly); spinal cord (spina bifida); face (cleft lip, cleft palate, micrognathia); eye (microphthalmia); all parts of the ear, teeth, salivary glands, arotic arch (malformation of several types); heart (ventricular septal defect); lungs; gastrointestinal tract (imperforate anus, omphalocele); liver and gallbladder; urinary system (renal "agenesis," hydronephrosis); genitalia; pituitary; thyroid; thymus; skull; vertebrae; ribs; extremities (phocomelia, digit malformations); muscles; and *situs inversus*.

The potential usefulness of animal models of hypervitaminosis is the study of sequential stages in the morphogenesis of malformations.

VITAMIN A DEFICIENCY HYDROCEPHALY

Maternal deficiency of vitamin A has practical implications in women, and the vitamin A-deficient rabbit provides a readily available model that is predictable and highly reproducible (167). Many animal species, including pigs, young dogs, chicks, rats, sheep, and newborn rabbits, show vitamin A deficiency-associated central nervous system (CNS) disturbances and increased cerebrospinal fluid (CSF). The young born of vitamin A-deficient mothers may be hydrocephalic at birth or they may develop it in a short time because of the associated elevated CSF pressure. Newberne (167) reported that young female Dutch Belted rabbits, 6 to 8 months of age, placed on the diet and bred when the serum vitamin A concentration stabilized at about 20–30 μg vitamin A/100 ml serum, produce hydrocephalic litters. Variations in severity of hydrocephaly can be achieved by modifying serum vitamin A levels at the time the females are bred. Permitting serum vitamin A concentration to decrease below about 20 μg/100 ml can result in failure to conceive or to carry to term. Serum concentrations, 30–50 μg/100 ml of vitamin A, usually permit clinically normal neonates, some of which develop hydrocephalus postnatally.

Signs and symptoms of vitamin A deficiency hydrocephalus in the newborn rabbit is essentially the same as those observed in the human infant. At birth, the condition is apparent by a distinct bulging over the frontal lobes of the brain. If the condition develops postnatally, the general anatomic alterations develop in a predictable pattern typical of postnatal hydrocephaly. Cerebrospinal fluid pressure is increased to a variable degree in any case and surviving offspring sometime develop opacity of the cornea and lens of the eye.

GM$_2$ GANGLIOSIDOSIS

Canine GM$_2$ gangliosidosis in German shorthair pointers as animal model of human GM$_2$ gangliosidosis was described by Karbe (130). The mode of inheritance seems to be autosomal recessive. Clinical signs of the disease in animals appear at about the age of 6 months, which consist of a decreased ability for training and increased nervousness. Before the animals reach 1 year of age, ataxia develops and becomes the most distinct sign of the general progressive neurological impairment. Vision is reduced, but most dogs never become totally blind. Seizures are seen occasionally; most animals die before they reach 2 years of age. The characteristic histoplasmic lesions affect practically all central neurons which contain in their cytoplasm granular material and are therefore enlarged to various degrees. Neurons of the retinal ganglion cell layer and of the spinal ganglia are severely involved, while vegetative neurons are only slightly affected. The ganglioside content of the cerebral cortex is increased about five fold mainly due to accumulation of GM$_2$ (16). The canine GM$_2$-ganglioside molecule is equivalent to that found in the human brain which is caused by hexosaminidase deficiency.

GM$_2$-gangliosidosis in German shorthair pointers is not associated with visceral histocytosis and therefore is not identical with type II GM$_2$ gangliosidosis in humans (Sandhoff's disease, infantile) (200). The canine disease takes an intermediate position between type I (Tay-Sach's disease, infantile) and type III (Bernheimer-Seitelberger's disease, late infantile) of the GM$_2$-gangliosidosis in humans with respect to the histoplasmic lesions particularly to the degree of neuronal involvement (160, 168, 169).

Canine GM$_2$-gangliosidosis offers a specific, genetically determined experimental research model for GM$_2$-gangliosidosis in humans.

NEURONAL GLYCOPROTEINOSIS

Canine neuronal glycoproteinosis (Lafora's disease) occurs spontaneously in at least two breeds, the bassett and the poodle (96, 97). The clinical signs in affected animals are vague signs of somnolence at onset, followed by incoordination, endocrine abnormalities, convulsions during sleep and progressive deterioration of neurological function. Diagnosis may be confirmed by cerebral biopsy and demonstration of the characteristic PAS positive inclusion bodies in neurons of the cerebral cortex. Lafora's disease is a rare familial neuronal metabolic disturbance associated with the accumulation of a complex insoluble glucoprotein material in motor neurons throughout the central nervous system and optic retina. The disease is progressive and fatal. In humans, the first symptoms become evident in the second decade of life usually presenting a history of incoordination, spasticity and seizures. Myoclonus frequently accompanies the syndrome in humans (202).

Accurate diagnosis depends on cerebral biopsy and demonstration of characteristic intraneuronal inclusion bodies. The inclusion bodies observed in humans and animals are similar in respect to histochemistry distribution and ultrastructure. They are intensely PAS positive, diatase resistant, and negative for glycogen, lipids, minerals, and nucelic acid. The inclusions are intracytoplasmic and predominantly associated with neurons or their processes.

The canine homolog of Lafora's syndrome is of importance, for it provides a valuable model for the study of the biochemical and genetic basis of this disease, and a means whereby corrective or palliative therapy can be developed.

GLOBOID CELL LEUKODYSTROPHY

Globoid cell leukodystrophy (GLD) in two breeds of dogs (cairn terriers and West Highland terriers) have been reported as animal model for Krabbe's disease in humans (69). Some dogs develop severe pelvic limb paralysis early in the disease; in other dogs, cerebellar signs dominate the early syndrome. The onset of the disease is marked by pelvic limb ataxia which progresses to paresis, the result of spinal cord and peripheral nerve involvement, followed by hypermetria of thoracic limb and head tremor—indicating cerebellar destruction. Finally, defects occur in placing reactions, vision, behavioral recognition, and alertness, due to cerebral damage. Terminally, dogs become prostrate, oblivious, anorexic, cachetic, and highly susceptible to secondary disease. Significant pathological changes are restricted to the nervous system (68). The dogs have an increased amount of protein in cerebrospinal fluid. Grossly, the involved regions of fixed white matter are grey and soft compared to normal white matter. Histopathologically, white matter of the central nervous system (CNS) has degenerate lesions, but changes in grey matter are minimal. The characteristic feature of GLD is the presence of PAS-positive, globoid-type macrophages which may occasionally be multinucleate. These foamy type, globoid macrophages accumulate perivascularly in regions of white matter destruction. GLD is the result of an inherited deficiency of the catabolic enzyme, galactocerebroside β-galactosidase (156). It appears to be inherited as a simple autosomal recessive trait.

In children with GLD, disturbances of temperament and mentation are more prominent, compared to GLD puppies where gait deficiency develops early and easily evaluated. In both species, GLD is similar histopathologically, ultrastructurally, and biochemically. Antimortem diagnosis is possible by assaying enzyme activity in leukocytes in both species (218).

Canine GLD is an excellent enzymatic-pathologic model for the human disease.

MONGOLISM (Down's Syndrome)

Down's syndrome in a female chimpanzee was described by McClure (157). The animal was of low normal birth weight and had a markedly decreased rate of growth as compared to other laboratory-reared chimpanzees. The animal also had bilateral partial syndactyly of the toes with clinodactyly, prominant epicanthal folds, hyperflexibility of the joints, muscle hypotonia, and a short neck with excess skin folds. Radiographic studies showed an abnormality of the left thoracic cavity at birth associated with the heart. Cytogenetic studies conducted on peripheral blood cell cultures revealed a model chromosome number of 49, as opposed to the normal diploid number of 48 for the chimpanzee. Karyotypes prepared from the aneuploid cells showed an additional small aerocentric chromosome. This autosomal trisomy was confirmed in bone marrow preparations and in fibroblast cultures obtained from skin cultures.

Frequent clinical episodes of anoxia, as manifested by cynosis of the lips and cold extremities was presented by the affected animal. Her clinical condition was characterized by retarded growth, delayed neurological and postural development, and general inactivity. The animal died at the age of 17 months.

In humans, Down's syndrome is associated with a specific chromosomal abnormality, trisomy of one of the small aerocentrics. The condition is reported to occur in the human population with a frequency of approximately 1 in 600 births, with increasing frequency associated with advancing maternal age. Physical and mental anomalies include oblique palpebral fissures with epicanthal folds; a short broad neck; cardiac anomalies; muscle hypotonia and hyperflexibility of the joints; dysplastic ears; short broad hands; dysplastic middle phalanx of the fifth finger; and mental retardation. Syndactyly of the toes is reported to occur with a frequency of 2 to 11 percent.

The occurrence of classical Down's syndrome in chimpanzees indicates that these animals may be used in investigations pertaining to the etiology of this condition.

HEREDITARY MUSCULAR DYSTROPHY

An animal model of hereditary muscular dystrophy in chickens was described by Julian (126). The condition in a commercial flock of New Hampshire chickens was inherited as an autosomal recessive trait. The first clinical sign of muscular dystrophy is the inability of the bird to rise after falling or being placed on its back. This disability is the interference of the pectoralis, the major depressor muscle of the wing, with the superacoracoideus, the major elevator muscle of the wing. Characteristics of myotonia are also present in the muscles involved which may contribute to the disability. Grossly discernible proximal white muscles are the first to be involved by the dystrophic process. Early microscopic changes in dystrophic muscles include an increase in the number of nuclei, fiber size variation, vacuolization destruction of muscle fibers, and the presence of ringed fibers. Fat deposition occurs at sites where muscle fibers have been lost. There is an apparent increase of connective tissue, although such an increase may be relative rather than absolute. The mouse, the chicken, and the hamster are affected with muscular dystrophy; the condition of the chicken is the mildest. The etiology of muscular dystrophy of chickens, as well as that of humans and other experimental animals, is unknown.

The condition in chickens has some features comparable to characteristics found in the myopathies in humans, although, there appears to be no single muscular dystrophic entity of humans directly comparable to that of the chicken. The dystrophic chicken is particularly useful for the study of such processes as hypertrophy and atrophy of skeletal muscle, fat deposition (and therefore pseudotrophy), "ring binden," and myotonia (101, 102, 125).

CARDIOMYOPATHY

Cardiomyopathy in the Syrian hamster as an animal model for human myocardial failure was described by Gertz (78). The cardiomyopathy was found to be hereditary and transmitted by an autosomal recessive gene (99). The disease involves both skeletal and cardiac muscle. Myocardial lesions occur in all animals in both sexes. They appear between 25 to 30 days of life in females and 10 days later in males (9). Acute lesions are widespread and occur with equal frequency in both sexes by 60 days. Two types of lesions, acute myolysis with primary dissolution of myofilaments in the absence of significant cellular infiltration and neurosis with significant round and inflammatory cells infiltrate, are found. The latter is more common in strains with the shortest life span. These lesions occur without vascular or valvular lesions (166). Pathological alterations are not usually found in early stages of cardiac involvement. In later stages, subcutaneous edema, ascites, hydrothorax, and hydropericardium are present. Liver, spleen, lungs, kidneys, and other visceral organs show congestive changes. Heart weight may increase as much as 50 to 70 percent above normal, and the heart may be extremely dilated.

The cardiomyopathic Syrian hamster offers a model for the study of muscle disease and for myocardial failure. However, the muscle disease in the hamster differs from the majority of cases of muscular dystrophy in humans where skeletal muscle disease is the cause of death with failure of the muscles of respiration. Congestive failure in these animals is clinically identical to the failure that occurs in the general category of human congestive heart failures.

HEREDITARY LYMPHEDEMA

Congenital hereditary lymphedema in the dog was described as an animal model to study Milroy's disease in human beings (177). The condition in the animal is congenital, nonpainful, pitting edema of the hind limbs, and it results from a malformation of the peripheral lymphatic system, which is inherited as an autosomal dominant trait. Pups with generalized edema usually die in the early neonatal period because they are unable to crawl and nurse effectively. Mildly affected individuals have pitting edema of the rear limbs at birth, which gradually disappears by the age of 3 months. An acute myopathy was observed in the nine pups that died with generalized lymphedema during the neonatal period but was not found in surviving dogs with lymphedema or normal littermate.

In Milroy's disease of humans, the edema is usually restricted to the lower limbs below the knee joints, sometimes involves the male genital organs, and is rarely found in the upper extremities. The cause of the edema in Milroy's disease in humans is unknown. Inheritance of Milroy's disease in humans, like the dog, fits an autosomal dominant pattern. In both species, the feature of edema varies greatly, some individuals being so mildly affected that they escape clinical detection.

Because of the clinical and genetic similarities between hereditary lymphedema in dogs and Milroy's disease, the former can be used as an animal model to elucidate pathological alterations in the latter condition. The present evidence suggests that whereas in the canine model there is a disturbance in morphogenesis of the peripheral lymphatic system at the level of the regional lymph nodes with failure to establish normal connections to the more central lymphatic system, the abnormality in Milroy's disease may involve hypoplasia or aplasia of the lymphatic system at a more peripheral location (177).

SEX CHROMOSOME ANOMALY
(Klinefelter's Syndrome)

Sex chromosome anomaly (tortoise shell) in male cats as animal models for the study of Klinefelter's syndrome was reported by Jones (120). Two sex-linked alleles in the domestic cat determine alternate colors in the hair coat orange (O) and black (O+). These allels are codominant, each is expressed in the heterozygote, called the tortoise shell ("tortie"). In the presence of white spotting ("piebald") (S, an autosomal dominant), the phenotype is often described as tortoise shell and white, tortie and white, tricolor, or "calico." Many modifying genes affect the intensity and pattern of the orange or black, necessitating care in identifying the phenotype. The homozygous male may be orange (O) or black (O+), the female may be black (O+O+) or orange (OO) in the homozygous state. The heterozygous female (O+O) is the tortoise shell. The rare occurrence of male tortoise shell cats has recently been explained on the basis of anomalies of sex chromosomes. One or more additional X chromosomes are present in male torties but more complex anomalies also occur. A diploid/triploid chimera chimera (38XX/57XXY) tortoise shell male and two chimeric torties (38XX/XY/39XXY/4OXXYY and 38XY/39XXY/4OXXYY) have been reported (47, 119). Most of these male cats are sterile because of aspermatogenesis. Klinefelter's syndrome in humans is characterized by infertility, enlargement of breasts, small testes, sex chromatin-positive cells, and usually an XXY karyotype or XX/XXY mosaics. The frequency in the newborn population is approximately 0.2 percent live male births. The clinical picture is somewhat varied, depending in particular on the degree and type of mosiacism. The testes degenerate progressively with age. The animal model may be useful to study varied expressions (e.g., reduced mental capacity, osteoporosis, endocrine changes, etc.) of the Klinefelter's syndrome in humans.

WAARDENBURG'S SYNDROME

Waardenburg's syndrome is characterized by anomalies of certain facioskeletal structures, congenital deafness, and pigmentary disorders. In the domestic cat (*Felis catus*) the autosomal dominant W gene causes a white, furred animal which may be deaf (15, 25, 223). Individual white animals may be blue-eyed, heterochromic, or yellow-eyed. The genes controlling eye color appear to be independent from gene W, but it is thought that the blue eye color is expressed only in the presence of the W gene. The blue eye color results from lack of pigment in the iris stroma; generally there is a concomitant lack of uveal trait pigmentation. The number of blue eyes is associated with the number of deaf ears within individual white cats, but no relationship exists between the side of deafness and the side of the blue iris.

Histopathologic changes in deaf animals include collapse of Reissner's membrane, atrophy of the hair cells, balling up at the tectorial membrane (i.e., atrophy and dysgenesis of the organ of Corti), hyalinization of the stria vascularis, degeneration or absence of the ontonconia, and collapse of the secular macula. Degenerative changes may be in only one part of one turn of the cochlea or occur as a partial defect in one ear.

In humans, like the cat, there is variability in the expression of different components of the syndrome. The general features of Waardenburg's syndrome in humans include lateral displacement of the medial canthi, broad prominent root of the nose, confluence of the eyebrows and hypertrichosis of their medial portions, white forelock, and leucism, iris heterochromia, and congenital deafness. Some individuals do not show the facial morphological characteristics but do express the other components of pigment and hearing defects. Both in cats and in humans, the role of the embryonal neural crest appears fundamental in the genesis of disease involving disorders of pigmentation and neural structure (187). Although abnormalities of facioskeletal structures have not been reported in deaf white cats, other pathological features are similar to Waardenburg's syndrome in humans. The disease in both cases is inherited as an autosomal dominant trait, leukoderma, heterochromia, and deafness with variability of expression are common features. The animal model of Waardenburg's syndrome should be very useful to study the genetic defect in man. An estimated 5 percent of humans with profound congenital deafness have this genetic disease.

TERATOMA, EMBRYONAL CARCINOMA, TERATOCARCINOMA

Teratomas are congenital malformations, but they are also neoplastic; they may be benign or highly malignant; their histopathological composition may be simple or may have many kinds of tissues; they can produce structures, embryoid bodies that resemble embryonic stages of the host tissues. Animal models of teratomas were described in inbred strains 129 A/He, F_1 hybrids between strains 129 and A/He and other inbred strains of mice by Stevens (214). Two new methods of experimentally producing teratomas have been described and new genetically defined mice have been developed to investigate teratocarcinogenesis (212, 213, 215, 216).

In human males, teratomas are rare and are usually highly malignant. In human females, teratomas (dermoid cysts of the ovary) are common and usually benign. The holologous tumor in male mice usually resembles the condition in human females; however, some of them progressively metastasize and kill the bearer. These animal models may be used to study genetic and environmental influences on the development of teratomas, embryonal carcinomas, and teratocarcinomas in humans.

AUTOIMMUNE DISEASE (Systemic LUPUS ERYTHEMATOSUS)

A complex immunologic disease resembling systemic lupus erythematosus (SLE) was described in certain inbred strains of New Zealand mice and their first generation (F_1) hybrids by Holmes and Burnet (98), Burnet and Holmes (26), and Lewis (144). New

Table 6-5. Major Immunological Lesions in New Zealand Mice

Strain	Lesion	
	Hemolytic anemia	Glomerulo-nephritis
New Zealand White (NZW)	−	−
New Zealand Black (NZB)	++	$\overset{+}{-}$
New Zealand Chocolate (NZC)	$\overset{+}{-}$	−
(NZB x NZW) F_1 Hybrid	−	++

Reproduced from Comparative Pathology Bulletin (144) with permission.

Zealand black mice (NZB) develop hemolytic anemia in the presence of a positive direct antiglobulin test; New Zealand black and New Zealand white (NZB × NZW) F_1 hybrids exhibit a high frequency of progressive glomerulonephritis leading to renal failure; the inbred strains of New Zealand mice have thymic lesions, and they develop antibodies to nuclear antigens including native DNA and exhibit positive LE cell tests. The major immunologic lesions in New Zealand mice are listed in Table 6-5. The NZB and F_1 hybrids are useful models in the study of environmental and genetic factors related to SLE.

CONGENITAL DISORDERS CAUSED BY VIRAL AGENTS

Recent reports indicate that most infectious agents, particularly viruses, produce congenital defects in specific organs through their cytopathic effects on the cells of the organs. Such effects kill the cells or alter them in such a way that they do not participate effectively in the normal differentiation of tissues. If extensive cell damage occurs during the early stages of gestation, when major organogenesis is taking place, the structural defects may be striking and serious. Although the major anatomical development is most rapid during early embryonic stages, other equally intricate aspects of differentiation, that is, the maturing of tissues for specialized biological functions, continue throughout the period of gestation and, in fact, into the postnatal period. It is reasonable to suppose that teratogenic influences may also act at later stages of development although the effects may be more subtle. Infection may also produce changes in fetal organs by destroying tissues that have already differentiated (Table 6-6). The congenital defect may include all defects that are present at birth and that cannot be corrected throughout normal processes of maturation as congenital defects. Although it appears unlikely that an animal model will be found that can accurately predict inherited disorders caused by a specific virus in humans, experimental studies in animal models are useful in analyzing mechanisms and studying influences that affect the outcome of infections acquired at various stages of development. There are also a number of animal viruses that are known to produce congenital anomalies in their natural host (Table 6-7). Some of these may be useful as spontaneous animal disease models to study similar human congenital disorders. Also, a number of viruses can be used to produce experimental animal models to study certain aspects of human congenital disorders (Table 6-8). From the data accumulated all over the world, both clinical and experimental, it is clear that viruses produce congenital defects and also account for perinatal infections in animals and in humans. Embryonic tissues are particularly susceptible to viral infections; it has been demonstrated in many studies utilizing the avian embryo. Of special interest are experiments indicating that strains of human viruses grow well in avian tissue, even though the hatched chicken is shown not to be susceptible to these same viruses (19).

A few animal models of congenital malformations of viral etiology are briefly described below.

Hydrencephaly and Porencephaly

Cerebral malformations, including hydrencephaly and porencephaly, have been reported in lambs and calves whose dams received a live attenuated blue tongue virus vaccine or contacted blue tongue infection during pregnancy (85, 190). Lambs infected with blue tongue vaccine virus at 50 to 55 days of gestation develop a severe necrotizing

Table 6-6. Microorganisms Associated with Congenital Defects of Intrauterine Infection in Humans

Microorganisms	Congenital Defect	Other Effects
Virus		
Togavirus		
Rubella	Heart defects, cataracts, microphthalmia, deafness microcephaly, mental retardation, immune defects	Abortion; expanded rubella syndrome consisting of hepatosplenomegaly, hepatitis, thrombocytopenia and purpura, long bone lesions, myocarditis, pneumonia, encephalitis
Western equine encephalomyelitis		Congenital encephalitis
Herpesvirus		
Herpesvirus hominis	Microcephaly, microphthalmia, retinopathy (2 cases reported, one associated with genital herpes)	Skin lesions, disseminated infection with neurologic signilae and/or chorioretinitis
Cytomegalovirus	Microcephaly, mental retardation, deafness, microphthalmia, retinopathy	Hepatosplenomegaly, jaundice, cerebral calcification, encephalitis, seizures, pneumonia, thrombocytopenia, anemia; Abortion, congenital chicken pox
Varicella-Zoster		
Picornavirus		
Coxsackievirus	Suspected congenital heart disease, myocarditis	Neonatal infection, diarrhea
Echovirus		Neonatal diarrhea;
Poliovirus		Abortion; congenital poliomyelitis
Ortho- and Paramyxovirus		
Influenza		Abortion (?)
Measles		Abortion; congenital measles
Mumps	(Suggested etiology in endocardial fibroelastosis is controversial)	Abortion

Table 6-6. (Continued)

Microorganisms	Congenital Defect	Other Effects
Poxvirus		
Vaccinia		Abortion; congenital vaccinia
Variola		Abortion; congenital smallpox
Other		
Hepatitis		Congenital hepatitis
Bacteria		
Brucella abortus		Abortion
Listeria monocytogenes		Abortion; meningoencephalitis, disseminated granulomatosis, neonatal death
Mycobacterium tuberculosis		Congenital tuberculosis
Treponema pallidum	Hutchinson's teeth; saddle nose, deafness, deformaties of nails, hydrocephalus	Congenital syphilis, hepatosplenomegaly, arthritis, bone fractures
Protozoa		
Toxoplasma gondii	Hydrocephalus, choriore-tinitis, more rarely micro-cephaly, mental retardation, micropthahalmia, nystagmus, squint, muscular paralysis	Congenital toxoplasmosis

Adapted from Progress in Medical Virology (19) with permission.

Table 6-7.   Viruses Causing Congenital Defects and Intrauterine Infections in Their Natural Hosts.

Virus and Animal Host	Congenital Defects	Other Effects
Hog cholera virus (vaccine strain)	Defects of limbs, ears, heart, snout, kidneys, tail, lungs, cerebellar hypo-plasia, hydro-cephalus	Neonatal illness, failure to thrive, death, asymptomatic infection
Bluetongue virus of sheep (vaccine strain)	Cerebral cysts, hydrencephaly	Acute, necrotizing meningoencephalitis
Bovine diarrheal-mucosal disease virus	Cataract, retinal atrophy, micro-phthalmia, cere-bellar hypoplasia	Abortion
Infectious bovine rhinotrachitis		Abortion; focal necrosis in liver and spleen
Lymphocytic choriomenin-gitis in mice		Congenital infection, neonatal illness, death
Japanese B encephalitis in swine		Abortion; stillbirth, encephalitis, death
Kilham rat virus	Cerebellar hypoplasia	Neonatal illness, hepatitis

From Progress in Medical Virology (19) with permission.

encephalopathy and retinopathy which, at 150 days of gestation, the time of birth, manifest as hydrencephaly and retinal dysplasia (171). Inoculation of lambs at 75 days results in a multifocal encephalitis and selective vacuolation of white matter, which manifest as porencephalic cysts in the newborn. No occular lesions are present in these animals. Lambs inoculated after 100 days of gestation show confined to mild focal me-ningoencephalitis. Intrafetal inoculation with blue tongue vaccine virus of time-dated pregnant sheep offers a rarely reproducible model for studying congenital hydren-cephaly and porencephaly. The morphologic features of hydrencephaly, porencephaly, and retinal dysplasia in fetal lambs are very similar to that in humans.

Immunoproliferative Disease

Animal models of congenital and persistent lymphocytic choriomeningitis (LCM) virus infection in mice was described by Pollard (185). LCM virus is a small pathogenic agent. When acquired congenitally or when inoculated at birth, mice develop persistent asymptomatic viremia (105). Intracerebral inoculations (IC) into older mice induce acute lethal meningocephalitis within 6 to 9 days thereafter and subcutaneous inocula-tions into older mice induces protective immunity against subsequent IC challenge.

The following are characteristics of LCM disease in the mouse: (a) all progeny are

Table 6-8. Experimental Animal Studies of Viruses Causing Congenital Defects

Virus	Animal Model	Defects	Other Effects	Reference
Rubella	Monkey (Macaca mulatta)	Lens changes	Abortion; otic, osseous and cutaneous lesions; retarded growth	52
	Baboon		Abortion; neonatal death	94
	Albino rat	Lens opacities; myocardial atrophy	Retarded growth; neonatal death	20, 46
Cytomegalovirus	Mouse		Abortion; stillbirth, no virus isolated from fetal tissues	159
Herpesvirus hominis	Rabbit		Transplancental transfer indicated by immunofluorescent studies and virus isolation	17
	Rabbit		Fetal death, death and resorption following high titers inoculated in mother	161
	Hamster		Transplacental passage	63, 66
	Chick embryo	Collapse of ventricles of brain; axial twists, defects of auditory anlage		92
Mumps	Chick embryo	Cataract, retarded feathering	General growth retardation	195
	Chick embryo	Endocardial fibroblastosis	Myocarditis, chronic infection until shortly after hatching	217

Table 6-8. (Continued)

Virus	Animal Model	Defects	Other Effects	Reference
Mumps	Chick embryo	Suppression of maturation of immunoglobulins		217
Coxsackie A	Chick embryo	Defects of skeletal muscle	Necrosis of epidermis skeletal muscle and feather follicle	179
Coxsackie B	Mouse	Heart defects	Ischemic necrosis; subendocardial hyalinization	139
Influenza A	Chick embryo	Collapsed ventricles of brain, 'microencephaly'; axis twists		103
	Chick embryo	Absence or defects of lens and auditory anlage; myloschisis		196
	Mouse		Transplacental passage	204
Vaccinia	Chick embryo	Collapsed ventricles, 'microencephaly'; defects of eye and auditory anlage		92
Reovirus	Mouse		Chronic neonatal infection; interstitial pneumonia, renal tubular necrosis, internal hydronephrosis	90
Newcastle disease virus	Chick embryo	Defects of neural tube, lens, auditory anlage	Necrosis of visceral arches, limb buds	232

229

Table 6-8. (Continued)

Virus	Animal Model	Defects	Other Effects	Reference
Lactic dehy-drogenase virus	Mouse		Transplacental infection	76
Parvovirus Group A				
Rat virus	Hamster		Transplacental infection	64
H-1 virus	Hamster	Microencephaly; exencephaly; spina bifida; ectopic heart; hernia; facial clefts	Fetal death and resorption	65
Feline pan-leucopenia	Cat	Cerebellar hypoplasia	Transplacental infection, death	131
	Ferret	Cerebellar hypoplasia	Transplacental infection	131
Minute virus	Mouse	Cerebellar hypoplasia		132
Porcine virus			Abortion; still-birth, neonatal death	29

Adapted from Progress in Medical Virology (19) with permission.

infected congenitally and the infection is lifelong; (b) the assay for LCM virus is rapid and accurate; (c) the development of chronic lesions is predictable; (d) the syndrome appears to result from an immunologic dyscrasia; (e) the lesions respond to immunosuppressive therapy, but viremia persists; and the virus is associated with oncogenicity. The chronic LCM syndrome manifests facets that resemble some human diseases in which etiology and mechanism have not yet been determined.

Persistent Viral Infections

Aleutian disease of minks (AD) as animal models for the study of immunologically mediated glomerulonephritis and arteritis (dysgammopathies) was reported by Henson and Gorham (95). A spontaneous mutation with dark blue coat color (resembling that of the Aleutian fox) is homozygous recessive for the Aleutian gene, *a*. All genotypes of mink were susceptible to the disease (AD), but a mink develops more severe disease with a more rapid course. The disease is characterized by gradual weight loss, anemia, uremic ulceration of the oral mucosa, and, rarely, nervous signs. The gross lesions include emaciation, hepatomegaly with small pinpoint pale foci scattered through the parenchyma, generalized lymphadenopathy with the spleen and lymph nodes two to four times normal size, nephritis characterized by enlarged kidneys with widespread petechiae early in the course, and shrunken pale kidneys with cortical cysts during the later stages. Histopathologically, there is widespread proliferation and infiltration of plasma cells in practically all organs. Gammaglobulin, C3, and viral antigen are deposited in the affected arteries, suggesting that deposition of antigen-antibody complexes is the causal factor of arteritis. The disease is readily transmitted from mink to mink by parenteral injection of blood or tissue homogenates.

Some human diseases are caused by or have been suggested to be the result of persistent viral infections. These include such conditions as infectious hepatitis, subacute sclerosing panencephalitis, kuru, Creutzfeldt-Jakob disease, multiple sclerosis, systemic lupus erythematosus (SLE), progressive multifocal leukoencephalopathy, and others. Similar mechanisms may occur in AD and in persistent viral infections of humans. The lesions in AD are immunologically mediated, as are those in such human disease as SLE. The arteritis that develops is also immunologically mediated and is probably similar to that occurring in some cases of polyarteritis nodosa in humans.

The AD in the mink model represents a readily reproducible viral disease with immunologically mediated lesions that can be used to elucidate mechanisms of host virus interaction in viral persistence and the induction of multisystem disease by immunologic means. Other animal models of virus diseases useful for the study of human congenital diseases of viral etiology are listed in Table 6-8.

REFERENCES

1. Altman, F., *Arch. Otolaryngiol.,* **51:**852 (1950).
2. Amorso, E. C., R. M. Loosmore, and C. Rimmington, *Nature,* **180:**230 (1957).
3. Andersen, A. E., and G. Guroff, *Proc. Natl. Acad. Sci.* (USA), **69:**863 (1972).
4. Andersen, H. A., *Acta Neuropath.* (Berlin) **7:**297 (1967).
5. Andersen, R., R. Huestis, and A. G. Motulsky, *Blood,* **15:**491 (1960).
6. Anderson, A. C., and F. T. Schultz, *Am. J. Pathol.,* **34:**967 (1958).
7. Arias, I., L. Bernstein, R. Toffler, C. Cornelius, A. B. Novikoff, and E. Epsner, *J. Clin. Invest.,* **43:**1249 (1964).
8. Asmundson, V. S., F. H. Kratzer, and L. M. Julian, *Ann. N.Y. Acad. Sci.,* **138:**49 (1966).
9. Bajusz, E., F. Hamburger, J. R. Baker, and L. H. Opie, *Ann. N.Y. Acad. Sci.,* **138:**213 (1966).
10. Bannerman, R. M., J. A. Edwards, M. Kreimer-Birnbaum, E. McFarland, and E. S. Russell, *Br. J. Hematol.* **23:**235 (1972).
11. Beasley, A. B., *J. Morphol.* **112:**1 (1963).
12. Bennet, J. M., R. S. Glume, and S. M. Wolff, *J. Lab. Clin. Med.,* **73:**235 (1969).
13. Bennett, D., *J. Hered.,* **52:**95 (1961).
14. Bennett, D., *Ann. Hum. Genet.,* **25:**1 (1961).
15. Bergsma, D. R., and K. S. Brown, *J. Hered.,* **62:**171 (1971).
16. Bernheimer, H., and E. Karbe, *Acta Neuropathol.,* **16:**243 (1970).
17. Biegelrisen, J., Jr., L. V. Scott, and W. Joel, *Am. J. Clin. Pathol.,* **37:**289 (1962).
18. Bielschowsky, M., and F. Bielschowsky, *Aust. J. Exp. Biol. Med. Sci.,* **34:**181 (1956).
19. Blattner, R. J., A. P. Williamson, and F. M. Heys, in H. L. Melnick (ed.), *Progress in Medical Virology,* Karger, Basel (1973).
20. Bohigian, G. N., J. Fox, and E. Cotlier, *Am. J. Ophthalmol.* **65:**196 (1968).
21. Bosher, S. K., and C. S. Hallpike, *Proc. R. Soc.* (London) **162:**147 (1965).
22. Brinkhous, K. M., and T. Gambill, *Comp. Pathol. Bull.* **3:**3 (1971).
23. Brinkhous, K. M., R. D. Langdell, G. D. Renick, J. B. Graham, and R. H. Wagner, *J.A.M.A.,* **154:**481 (1954).

24. Brooke, H. C., *J. Hered.,* **17:**173 (1926).
25. Brown, K. S., *Animal Models of Pigment Bearing Abnormalities in Man.* Second Conference on the Clinical Delineation of Birth Defects. The Johns Hopkins Medical Institution, 1969.
26. Burnet, F. M., and M. C. Holmes, Aust. Ann. Med., **14:**185 (1965).
27. Calkins, E., D. Kahn, and W. C. Diner, *Ann. N.Y. Acad. Sci.* **64:**410 (1956).
28. Carter, A. H., *Proc. N. Z. Soc. Animal Prod.,* **20:**108 (1960).
29. Cartwright, S. F., M. Lucas, and R. A. Huck, *J. Comp. Pathol.* 79:371 (1969).
30. Cella, F., *Nouva Vet.,* **24:**145 (1948).
31. Chase, H. B., M. S. Gunther, J. Miller, and D. Wolffson, *Science,* **107:**297 (1948).
32. Clare, N. T., and E. H. Stephens, *Nature,* **153:**252 (1944).
33. Clark, S. L., J. W. Ward, and I. S. Dribben, *J. Comp. Neurol.,* **74:**409 (1941).
34. Cohen, A. S., in G. W. Richter and M. A. Epstein (eds.), *The Constitution and Genesis of Amyloid. International Review of Experimental Pathology,* Vol. 4, Academic Press, Inc., New York, 1965.
35. Cohen, H. B., H. Beerman, and L. Nicholas, *Am. J. Med. Sci.,* **251:**475 (1966).
36. Cohen, A. S., *N. Engl. J. Med.,* **277:**522, 574, 628 (1967).
37. Cohen, A. S., and T. Shirahama, *Am. J. Pathol.,* **68:**441 (1972).
38. Coleman, D. L., *Arch. Biochem.,* **96:**562 (1962).
39. Cornelius, C. E., and I. M. Arias, *Am. J. Pathol.,* **69:**369 (1972).
40. Cornelius, C. E., *Comp. Pathol. Bull.,* **2:**3 (1970).
41. Cornelius, C. E., *Comp. Pathol. Bull.,* **2:**4 (1970).
42. Cornelius, C. E., and I. M. Arias, *Am. J. Med.,* **40:**165 (1966).
43. Cornelius, C. E., I. M. Arias, and B. I. Osburn, *J. Am. Vet. Med. Assoc.,* **146:**709 (1965).
44. Cornelius, C. E., J. A. Bishop, and M. H. Schaffer, *Cornell. Vet.,* **57:**177 (1967).
45. Cornelius, C. E., and R. R. Gronwall, *Am. J. Vet. Med.,* **29:**291 (1968).
46. Cotlier, E., J. Fox, G. N. Bohigian, C. Beaty, and A. DuPree, *Nature* (London) **217:**38 (1968).
47. Chu, E. H. Y., H. C. Thuline, and D. E. Norby, *Cytogenetics,* **3:**1 (1964).
48. Crary, D. D., and P. B. Sawin, *J. Hered.,* **43:**255 (1952).
49. Crary, D. D., R. R. Fox, and P. B. Sawin, *J. Hered.,* **57:**236 (1966).
50. Darwin, C., *The Variation of Animals and Plants under Domestication,* Murray, London, 1868.
51. Datta, S. P., and H. Harris, *Ann. Engen.* (London) **18:**107 (1953).
52. Delahunt, C. S., and N. Reiser, *Am. J. Obstet. Gynecol.,* **99:**550 (1967).
53. Deol, M. S., *J. Med. Genet.,* **5:**127 (1968).
54. Dickie, M. M., *Mouse Newsletter,* **25:**36 (1961).
55. Diploch, P. T., *J. Am. Vet. Med. Assoc.,* **141:**462 (1962).
56. Donovan, E. F., and W. F. Loeb, *J. Am. Vet. Med. Assoc.,* **134:**36 (1950).
57. Edwards, J. A., and R. M. Bannerman, *Comp. Pathol. Bull.,* **4:**3 (1972).
58. Edwards, J. A., and J. E. Hoke, *Proc. Soc. Exp. Biol. Med.,* **141:**81 (1972).
59. Edwards, J., and V. Jansen, *Br. J. Hematol.,* **19:**573 (1970).
60. Eriksson, K., *Nord. Vet. Med.,* **7:**773 (1955).
61. Faith, R. E., and J. C. Woodard, *Comp. Pathol. Bull.,* **5:**2 (1973).
62. Falconer, D. S., and J. H. Isaacson, *Genet. Res.,* **3:**248 (1962).
63. Falk, L. A., and F. Rapp, *Texas Rep. Biol. Med.,* **22:**649 (1964).
64. Ferm, V. H., and L. Kilham, *J. Embryol. Exp. Morph.,* **11:**659 (1963).
65. Ferm, V. H., and L. Kilham, *J. Embryol. Exp. Morph.,* **13:**151 (1965).
66. Ferm, V. H., and R. J. Low, *J. Pathol. Bact.* **89:**295 (1965).
67. Fletch, S. M., and P. H. Pinkerton, *Am. J. Pathol.,* **71:**477 (1973).
68. Fletcher, T. F., H. J. Kurtz, and D. G. Low, *J. Am. Vet. Med. Assoc.,* **149:**165 (1966).
69. Fletcher, T. F., and H. J. Kurtz, *Am. J. Pathol.,* **66:**375 (1972).
70. Foarie, P. J. J., *Onderstipoort J. Vet. Sci. Animal Ind.,* **2:**535 (1936).
71. Ford, L., *J. Hered.,* **60:**293 (1969).
72. Fowler, R. E., and R. G. Edwards, *Genet. Res.,* **2:**272 (1961).
73. Fowler, M. E., C. E. Cornelius, and N. F. Baker, *Cornell Vet.,* **54:**153 (1964).
74. Gardner, D. L., *J. Pathol. Bact.,* **77:**243 (1959).

75. Gates, A. H., and M. Karasek, *Science,* **148:**1471 (1965).
76. Georgie, A., L. Lenz, and H. Zobel, *Proc. Soc. Exp. Biol. Med.,* **117:**322 (1964).
77. Gerritsen, T., and F. L. Siegel, *Monogr. Hum. Genet.,* **6:**22 (1972).
78. Gertz, E. W., *Am. J. Pathol.,* **70:**151 (1973).
79. Glenn, B. L., H. G. Glenn and I. T. Omtvedt, *Am. J. Vet. Res.,* **29:**1653 (1968).
80. Glenn, B. L., *Comp. Pathol. Bull.,* **2:**6 (1970).
81. Goldberg, A. and C. Rimmington, *Diseases of Porphyrin Metabolism,* Charles C Thomas, Springfield, Ill., 1962.
82. Green, E. L. (ed.), *Biology of the Laboratory Mouse* 2nd ed., McGraw-Hill, New York, 1966.
83. Green, M. C. in E. L. Green (ed.), *Biology of the Laboratory Mouse,* McGraw-Hill, New York, 1966.
84. Greene, H. S. N., *J. Exp. Med.,* **71:**839 (1940).
85. Griner, L. A., B. R. McCrory, N. M. Foster, and H. Meyer, *J. Am. Vet. Med. Assoc.,* **145:**1013 (1964).
86. Gruneberg, H., *J. Genet.,* **52:**52 (1954).
87. Hackel, D. B., T. D. Kenney, and W. Wendt, *Lab. Invest.,* **2:**154 (1952).
88. Hamlin, R. L., D. L. Snulzer, and R. C. Smith, *J. Am. Vet. Med. Assoc.,* **145:**331 (1964).
89. Hanna, B. L., P. B. Sawin, and L. B. Sheppard, *Genetics,* **47:**512 (1962).
90. Hashimi, A., M. M. Carrutturs, P. Wolf, and A. M. Lerner, *J. Exp. Med.,* **124:**33 (1966).
91. Hatt, I. B., and L. Z. Saunders, *J. Hered.,* **44:**97 (1953).
92. Health, H. D., H. H. Sheer, D. T. Imagawa, M. H. Jones, and J. M. Adams, *Proc. Soc. Exp. Biol. Med.,* **92:**675 (1956).
93. Hegriberg, G. A., and G. A. Padgett, *Bull. Pathol.,* **8:**247 (1967).
94. Hendricks, A. G., *Conference on Nonhuman Primate Toxicology,* Arlie House, Virginia, 1966.
95. Henson, J. B., and J. R. Gorham, *Am. J. Pathol.,* **71:**345 (1973).
96. Holland, J. M., W. C. Davis, D. J. Prieur, and G. H. Collins, *Am. J. Pathol.,* **58:**509 (1970).
97. Holland, J. M., *Comp. Pathol. Bull.,* **4:**2 (1972).
98. Holmes, M. D., and F. M. Burnet, *Ann. Intern. Med.,* **59:**265 (1963).
99. Homburger, F., C. W. Nixon, J. Harrop, G. Wilgram, and J. R. Baker, *Fed. Proc.,* **22:**195 (1963).
100. Homburger, F., C. W. Nixon, M. Eppenberger, and J. R. Baker, *Ann. N.Y. Acad. Sci.,* **138:**28 (1966).
101. Holliday, T. A., J. R. VanMater, L. M. Julian, and V. S. Asmundson, *Am. J. Physiol.,* **209:**871 (1965).
102. Holliday, T. A., L. M. Julian, and V. S. Asmundson, *Anat. Rec.,* **160:**207 (1968).
103. Homburger, V., and K. Habel, *Proc. Soc. Exp. Biol. Med.,* **66:**608 (1947).
104. Horwitz, I., and H. A. Waisman, *Proc. Soc. Exp. Biol. Med.,* **122:**750 (1966).
105. Hotchin, J., and M. Cinits, *Canad. J. Microbiol.,* **4:**149 (1958).
106. Hsia, I. J., *Inborn Errors of Metabolism,* Vol. I, Year Book Medical Publishers, Chicago, 1966.
107. Hsia, I. J., *Inborn Errors of Metabolism,* Vol. II, Year Book Medical Publishers, Chicago, 1966 (b).
108. Hudson, W. R., and R. J. Ruben, *Arch. Otolaryngol.,* **75:**213 (1962).
109. Innes, J. R. M., and L. Z. Saunders, in *Comparative Neuropathology,* Academic Press, New York, 1962.
110. Jackson, O. F., *Vet. Rec.,* **83:**417 (1968).
111. Jamison, S., B. W. Simpson, and J. B. Russell, *Vet. Rec.* **57:**429 (1945).
112. Jay, G. E., Jr., in W. J. Burdette (ed.), *Methodology in Mammalian Genetics,* Holden-Day, Inc., San Francisco, 1963.
113. Jepson, J. B., and M. J. Spiro, in J. B. Stanbury, J. B. Wyngarden, and D. S. Frederickson (eds.), *The Metabolic Bases of Inherited Disease,* McGraw-Hill, New York, 1960.
114. Jervis, G. A., in *Phenylpyruvic Oligophrenia, Proceedings of the Association for Research in Nervous and Mental Disease,* Williams and Wilkins, Baltimore, 1954.
115. Johnson, K. R., D. L. Roart, R. H., Ross, and J. W. Bailey, *J. Dairy Sci.,* **41:**137 (1958).
116. Johnson, L., F. Sarmiento, W. A. Blanc, and R. Day, *Am. J. Dis. Child.,* **97:**591 (1959).

117. Jolly, R. D., *Am. J. Pathol.,* **74:**211 (1974).
118. Jolly, R. D., and W. F. Blakemore, *Vet. Rec.,* **92:**391 (1973).
119. Jones, T. C., in K. Benirschke (ed.), *Comparative Mammalian Cytogenetics,* Springer-Verlag, New York, 1969.
120. Jones, T. C., *Comp. Pathol. Bull.,* **1:**1 (1969).
121. Jorgensen, S. K., *Br. Vet. J.,* **115:**160 (1959).
122. Jorgensen, S. K., *Br. Vet. J.,* **117:**7 (1961).
123. Jorgensen, S. K., and T. K. With, *Ann. N.Y. Acad. Sci.,* **104:**701 (1963).
124. Jorgensen, S. K., T. K. With, in A. J. Rook and G. S. Walton (eds.), *Comparative Physiology and Pathology of the Skin,* Blackwell, Oxford, 1965.
125. Julian, L. M., and V. S. Asmundson, in G. H. Bourne and M. A. Golraz (eds.), *Muscular Dystrophy of the Chicken, Muscular Dystrophy in Man and Animals,* Karger, Basel, 1963.
126. Julian, L. M., *Am. J. Pathol.,* **70:**273 (1973).
127. Julian, L. W., W. S. Tyler, T. J. Hage, and P. W. Gregory, *Am. J. Anat.,* **100:**269 (1957).
128. Kaneko, J. J., *Ann. N.Y. Acad. Sci.,* **104:**701 (1963).
129. Karbe, E., and B. Schiefer, *Pathol. Vet.* (Basel) **4:**223 (1967).
130. Karbe, E., *Am. J. Pathol.,* **71:**151 (1973).
131. Kilham, L., G. Margolis, and E. M. Colby, *Lab. Invest.,* **17:**465 (1967).
132. Kilham, L., and G. Margolis, *Teratology,* **4:**43 (1971).
133. Kitchen, H., R. E. Murray, and B. Y. Cockrell, *Am. J. Pathol.,* **68:**203 (1972).
134. Kitchen, H., F. W. Putman, and W. J. Taylor, *Blood,* **29:**867 (1967).
135. Kitchen, H., *Pediat. Res.,* **2:**215 (1968).
136. Knox, W. E., in *Metabolic Basis of Inherited Diseases,* McGraw-Hill, New York, 1966.
137. Koch, P., and H. Fischer, *Tierarztl. Umschan.,* **5:**317 (1950), *Tierarztl. Umschan.,* **6:**158 (1951).
138. Kreimer-Birnbaum, M., R. M. Bannerman, E. S. Russell, and S. E. Bernstein, *Comp. Biochem. Physiol.* **43A:**21 (1972).
139. Krympotic, E., W. Thomas, Jr., and H. Lopey, *Excerpta, Med. Inv. Congr. Ser.,* **191:**56 (1969).
140. Leader, R. W., G. A. Padgett, and J. R. Gorham, *Blood,* **22:**477 (1963).
141. Leader, R. W., G. A. Padgett, and J. R. Gorham, in *The NINDB Monograph,* No. 2 (*Chediak-Higashi Syndrome of Man, Mink and Cattle*), 1965.
142. Leipold, H. W., and K. Huston, *J. Hered.,* **59:**3 (1968).
143. Lewis, R. M., *J. Am. Vet. Med. Assoc.,* **147:**939 (1965).
144. Lewis, R. M., *Comp. Pathol. Bull.,* **5:**396 (1973).
145. Lozzio, B. B., A. I. Chernoff, E. R. Machado, *Science,* **156:**1742 (1967).
146. Ljunggren, J., *Acta Orthopaed. Scand. Suppl.,* **95** (1967).
147. Lund, J. E., G. A. Padgett, and J. R. Gorham, *J. Hered.,* **61:**47 (1970).
148. Lund, J. E., in *Animal Models in Biomedical Research III,* Washington, National Academy of Sciences, 1970.
149. Lund, J. E., *Comp. Pathol. Bull.,* **5:**2 (1973).
150. Lyon, M. F., *J. Hered.,* **46:**77 (1955).
151. Lyon, M. F., E. V. Hulse, C. E. Rowe, *J. Med. Genet.,* **2:**99 (1965).
152. MacArthur, J. W., *Am. Naturalist,* **78:**142 (1944).
153. Madden, D. E., D. J. Ellis, R. D. Barner, I. Melcher, and J. M. Ortin, *J. Hered.,* **49:**125 (1958).
154. Mandema, E., L. Ruinen, J. H. Scholtin, and A. H. Cohen (eds.) *Amyloidosis,* Excerpta Medica Foundation, Amsterdam, 1968.
155. Maruffo, C. A., M. R. Malinow, J. R. Depaoli, and S. Katz, *Am. J. Pathol.,* **49:**445 (1966).
156. McGrath, J., H. Schutta, A. Yaseen, and S. Steinberg, *J. Neuropathol. Exp. Neurol.,* **28:**171 (1969).
157. McClure, H. M., *Am. J. Pathol.,* **67:**413 (1972).
158. McCuistian, W. R., *Vet. Med.,* **59:**315 (1964).
159. Medearis, D. N., *Johns Hopkins Hosp. Bull.,* **114:**181 (1964) (a).
160. Menkes, J. H., J. S. O'Brien, and S. Okada, *Arch. Neurol.,* **25:**14 (1971).
161. Middlekamp, J. N., C. A. Reed, and G. Patrizi, *Proc. Soc. Exp. Biol. Med.,* **125:**757 (1967).

162. Miller, D. R., F. R. Rickels, M. A. Lichtman, P. L. CaCelle, J. Bates, and R. I. Weed, *Blood,* **38:**184 (1971).
163. Mixter, R. and H. R. Hant, *Genetics,* **18:**367 (1933).
164. Montalbo, R. G., and J. J. Kabara, *Proc. Soc. Exp. Biol. Med.,* **145:**1225 (1974).
165. Morris, B., D. C. Blood, W. R. Sidman, J. D. Steel, and J. H. Whittem, *Aust. J. Exp. Biol. Med. Sci.,* **32:**265 (1954).
166. Nakao, K., M. Oka, C. Chen, E. Bajusz, and A. Angrist, *Pathol. Microbiol.,* **35:**118 (1970).
167. Newberne, P. M., *Comp. Pathol. Bull.,* **5:**4 (1973).
168. O'Brien, J. S., *N. Engl. J. Med.,* **284:**893 (1971).
169. O'Brien, J. S., D. L. Fillerup, M. L. Veath, and K. Adams, *Fed. Proc.,* **30:**956 (1971).
170. Olafson, P. J., *Tech. Meth. Bull. Intern Assoc. Med. Museums,* **19:**129 (1939).
171. Osburn, B. I., *Am. J. Pathol.,* **67:**211 (1972).
172. Oski, F. A., J. L. Naiman, S. F. Blum, H. S. Zarkowsky, J. Whaun, S. B. Shohet, A. Green, and S. G. Nathan, *N. Engl. J. Med.,* **280:**909 (1969).
173. Padgett, G. A., R. W. Leader, C. C. O'Mary, and J. R. Gorham, *Genetics,* **49:**505 (1964).
174. Pai, R. C., *Develop. Biol.,* **11:**93 (1965).
175. Patterson, D. F., *Ann. N.Y. Acad. Sci.,* **127:**541 (1965).
176. Patterson, D. F., W. Medway, H. Luginbühl, and S. Chacko, *J. Med. Genet.,* **4:**145 (1967).
177. Patterson, D. F., *Comp. Pathol. Bull.,* **3:**2 (1971).
178. Pearce, L., *J. Exp. Med.,* **92:**59 (1952).
179. Peers, J. H., S. E. Ranson, and R. J. Huebner, *J. Exp. Med.* **96:**17 (1952).
180. Persellin, R. H., and L. L. Wilson, in *Biology of Howler Monkey,* Karger, Basel, 1968.
181. Phillips, L. L., H. S. Kaplan, G. A. Padgett, and J. R. Gorham, *Am. J. Vet. Clin. Pathol.,* **1:**1 (1967).
182. Pinkerton, P. H., *J. Pathol. Bacteriol.,* **95:**155 (1968).
183. Pinkerton, P. H., R. M. Bannerman, T. D. Doeblin, B. M. Benisch, and J. A. Edwards, *Br. J. Hematol.,* **18:**211 (1970).
184. Pinkerton, P. H., and S. M. Fletch, *Blood,* **40:**963 (1972).
185. Pollard, M., *Am. J. Pathol.* **67:**613 (1972).
186. Ram, J. S., R. A. DeLellis, and C. G. Glenner, *Proc. Soc. Exp. Biol. Med.,* **130:**462 (1969).
187. Reed, W. B., V. M. Stone, E. Boder, and L. Ziprowski, *Arch. Derm.,* **95:**176 (1967).
188. Reiman, H. A., *Periodic Disease,* F. A. Davis, Philadelphia, 1963.
189. Rhode, E. A., and C. E. Cornelius, *J. Am. Vet. Med. Assoc.,* **132:**112 (1958).
190. Richards, W. P. C., C. L. Crenshaw, and R. B. Bushnell, *Cornell Vet.,* **61:**336 (1971).
191. Rigdon, R. H., *Ann. N.Y. Acad. Sci.,* **138:**28 (1966).
192. Rimington, C., *Onderstepoort. J. Vet. Res.,* **7:**567 (1936).
193. Roberts, S. R., *Vet. Scope,* **12:**2 (1967).
194. Robertson, G. G., A. P. Williamson, and R. J. Blattner, *Am. J. Anat.,* **115:**473 (1964).
195. Robertson, G. G., H. O. DeBandi, Jr., A. P. Williamson, and R. J. Blattner, *Anat. Rec.,* **158:**1 (1967).
196. Robins, M. W., *J. Hered.,* **50:**188 (1959).
197. Rowsell, H. C., H. C. Downie, J. F. Mustard, J. E. Leeson, and J. A. Archibald, *J. Am. Vet. Med. Assoc.,* **137:**247 (1960).
198. Rubin, L. F., *J. Am. Vet. Med. Assoc.,* **152:**260 (1968).
199. Russell, E. S., in W. J. Burdette (ed.), *Methodology in Mammalian Genetics,* Holden-Day, Inc., San Francisco, 1963.
200. Sandhoff, K., V. Andreae, and H. Jatzkewitz, *Life Sci.,* **7:**283 (1968).
201. Saunders, L. Z., *Pathol. Vet.* (Basel) **2:**256 (1965).
202. Schwartz, G. A., and M. Yanoff, *Arch. Neurol.,* **12:**172 (1965).
203. Shenefelt, R. E., *Am. J. Pathol.,* **66:**589 (1972).
204. Siem, R. A., H. Ly, D. T. Imagawn, and J. M. Adams, *J. Neuropath. Exp. Neurol.,* **19:**125 (1960).
205. Sink, J. D., M. D. Judge, R. G. Cassens, W. G. Hoekstra, R. H. Grummer, and E. J. Briskey, *Am. J. Vet. Res.,* **27:**1494 (1966).
206. Sladic-Simic, D., P. N. Martinovitch, N. Zivkovic, D. Pavic, J. Martinovic, M. Kahn, and H. M. Rauney, *Ann. N.Y. Acad. Sci.,* **165:**93 (1969).
207. Smith, H. A., and T. C. Jones, *Veterinary Patholology,* 3rd ed., Lea & Febiger,

Philadelphia, 1966.

208. Snell, G. D., *Proc. Natl. Acad. Sci.* (U.S.A.), **15**:733 (1929).
209. Stanbury, J. B., J. B. Wyngarden, and D. S. Frederickson, *The Metabolic Basis of Inherited Disease,* 3rd ed., McGraw-Hill, New York, 1972.
210. Stevens, L. C. and J. A. Mackinsen, *J. Hered.,* **49**:153 (1958).
211. Stevens, L. C., J. A. Mackinsen, and S. E. Bernstein, *J. Hered.,* **50**:35 (1959).
212. Stevens, L. C., *Proc. Natl. Acad. Sci.,* **52**:654 (1964).
213. Stevens, L. C., *Adv. Morphogen.,* **6**:1 (1967).
214. Stevens, L. C., *Comp. Pathol. Bull.,* **2**:4 (1970).
215. Stevens, L. C., *Dev. Biol.,* **21**:364 (1970).
216. Stevens, L. C., *J. Natl. Cancer Inst.,* **44**:923 (1970).
217. St. Game, J. W., Jr., C. W. L. Davis, M. D. Cooper, and G. R. Noun, *Fed. Proc.,* **29**:284 (1970).
218. Suzuki, Y. and K. Suzuki, *Science,* **171**:73 (1971).
219. Tashjian, R. J., M. Kushna, W. E. P. Das, R. L. Hamlin, and D. A. Yarn, *Ann. N.Y. Acad. Sci.,* **127**:581 (1965).
220. Tasker, J. B., G. A. Severin and S. Young, *J. Am. Vet. Med. Assoc.,* **154**:158 (1969).
221. Tennant, B., A. C. Asbury, R. C. Laben, W. P. C. Richards, J. J. Kaneko, and P. T. Cupps, *J. Am. Vet. Med. Assoc.,* **150**:1493 (1967).
222. Tobias, G., *J. Am. Vet. Med. Assoc.,* **145**:462 (1964).
223. Todd, M. B., T. C. Jones, and B. C. Zook, *Carnivore Genet. Newsletter,* **5**:100 (1968).
224. Tuff, P., and L. A. Gleditsch, *Nord. Vet. Med.,* **1**:619 (1949).
225. Turner, W. J., *J. Biol. Chem.,* **118**:519 (1937).
226. Wadman, S. K., F. J. Van Sprang, C. Van der Hedding, and D. Kettin, in H. Bickel, F. B. Hudson, and L. J. Wolf (eds.) *Phenyketonuria and Some Other Inborn Errors of Amino Acid Metabolism,* Georg Thine Verlag, Stuttgart, 1971.
227. Wallace, M. E., *Mouse Newsletter,* **32**:20 (1968).
228. Watson, C. J., and V. Perman, *AMA Arch. Intern. Med.,* **103**:436 (1959).
229. West, W. T., H, Meier, and W. G. Hoag, *Ann. N.Y. Acad. Sci.,* **138**:4 (1966).
230. Willerson, J. T., R. Asofsky, and W. F. Barth, *J. Immunol.* **103**:741 (1969).
231. Williams, H. E., and L. H. Smith, in W. L. Nyhan (ed.) *Amino Acid Metabolism in Gene Variation,* McGraw-Hill, New York, 1968.
232. Williamson, A. P., R. J. Blattner, and L. Simouseri, *J. Immunol.,* **76**:275 (1956).
233. Young, G. B., *Vet. Rec.,* **81**:606 (1967).
234. Zarkowsky, H. S., F. A. Oski, R. Shaafi, S. B. Shahet, and D. G. Nathan, *N. Engl. J. Med.,* **278**:573 (1968).

CHAPTER 7

USE OF ANIMALS IN BEHAVIORAL RESEARCH

Brij M. Mitruka

Although many psychiatrists are still unwilling to accept animal work as relevant to human disorders, animal models have become available in recent years for behavioral studies. Human behavior is extremely complex and influenced by many social, environmental, nutritional, and genetic factors. One way to study certain aspects of human behavior is to treat the animal as a human model. To do this we must be able to assume a reasonable match between the animal model and humans, at least in its response to the condition that happens to interest us. For this reason, using an animal as a human model works best when the process being studied is fairly well defined and when the research has a limited specific objective. These conditions are most often met in medical investigations in which the focus is on a particular disease category, but they are also found in certain kinds of behavioral research. This chapter briefly describes some processes of human behavior that can be studied by using experimental animals. A brief discussion on the usefulness of certain animal species in behavioral research is given with the emphasis on their applications to achieve a limited specific objective in the field of behavioral studies in humans.

USE OF RODENTS IN BEHAVIORAL RESEARCH

The white rat has been used more than any other animal for behavioral research because of its size, economy, cleanliness, and ease of housing. Other reasons include docility, adaptability to new surroundings, large litters, similarity to humans in nutritional, glandular, and neurological considerations, and a tendency to explore, which is obviously important in behavioral work. The emphasis in behavioral work using rats started with sensory capacity—the role of kinesthesis, vision, touch, smell, and the like and the appropriate cues for their arousal and control. Early studies on rodents were carried out on the reliability of mazes and training techniques, cerebral destruction, effects of reward or punishment, and drugs (alcohol, cocaine, opiates); kinesthetic, auditory, tactual, olfactory and visual cues; and age, diet, sex, transfer, heredity, instinct, and imitation (87). More recently, however, the emphasis has shifted to simpler behavioral situations which presumably involve simpler response systems (14, 19, 70, 72, 74, 84, 115).

USE OF CATS IN BEHAVIORAL RESEARCH

The domestic cat is a potential subject for experimentation in behavioral research. The cat is relatively uniform in size, yet small enough for laboratory maintenance and large enough for gross surgery, and tolerant of most anesthetics, antibiotics, or other pharmaceuticals implicated in satisfactory laboratory existence. It is adept in the use of its limbs, has frequent and multiple births, is reasonably docile, and is readily available in most urban areas.

The first behavioral experimentation with cats was probably directed by the evolutionary fervor created by Darwin's writings together with the orientation of psychology of that day toward broad theories and overall systematization. In recent years, experiments on cats have been directed toward:

.1. The neuroanatomical and neurophysiological researches which are given behavioral meaning; and

2. Some physiological or biochemical variable which may be related to certain behavioral sequelae.

The cat lends itself to physiological and biochemical experimentation, and much more information on cats exists in these areas than in the behavioral disciplines.

In behavioral research, the cat is utilized for studies of (a) behavioral development, (b) comparative evaluation, and (c) behavioral assessment in which questions are raised whose implications are much broader than conceivably encompassed by a single species (62).

The research on development, the promise offered by operant techniques applied to this species, and the favorable evaluation of the cat's behavior when compared to other animals assure the continued usage of the cat in behavioral research. However, problems of rearing, maintenance, and genetic control remain to be resolved if the species is to have a unique role in behavioral research.

USE OF DOGS IN BEHAVIORAL RESEARCH

Dogs are used in a wide range of experiments, including genetic, developmental, neurological, and psychopharmacological studies. There are many advantages in using dogs as research subjects, notably the ease of handling and management and the relatively large litters that mature slowly, thus facilitating either physiological or behavioral developmental studies. The dog is phylogenetically a highly evolved species and is a highly social animal; consequently it has a rich repertoire of behavior. The dog, unlike many captive laboratory species, adapts well to the laboratory environment. Its social bond with humans can be a useful attribute for training and adapting it to experimental test situations. In addition, this close bond between subject and experimenter has motivational or reward value which can be of use in certain tests much in the same way that food reward alone motivates a rat. As a result of domestication and selective breeding, a wide spectrum of various breeds has evolved. Differences between breeds, not only behavioral but also physiological and morphological, provide a relatively untouched source of research material for the comparative physiologist, anatomist, and immunologist, as well as for the geneticist and psychologist. (103).

USE OF PRIMATES IN BEHAVIORAL RESEARCH

The most important virtue of subhuman primates is that they constitute the group of animals phylogenetically, biochemically, physiologically, anatomically, and behaviorally closest to humans and therefore are often their best experimental substitutes. Research with nonhuman primates has already made many important, practical contributions to human health and welfare, and its promise for the future is impressive. Some of the areas in which research with nonhuman primates is carried out include the study of infectious diseases, degenerative diseases (especially of the cardiovascular system and cancer), neurological diseases, sensory physiology, and in hepatic coma therapy. Primate behavior of interest to neuroanatomy, neurophysiology, endocrinology, psychiatry, neurochemistry and pharmacology includes emotional behavior; motor activity and skill; reproductive activity; visual, auditory, and other sensory capacities studied through both innate and learned responses; various appetitive behaviors; and numerous learned behaviors (8, 18, 25, 43, 45, 86, 94, 97, 98).

Although primates as experimental subjects are expensive and more difficult to maintain and handle than small animals, primates are unique (either by reason of their genotype, phenotype, or treatment) and can be studied over many years in a setting that has a long-term commitment to primate research.

The capability of experimental work on nonhuman primates to contribute to the understanding of human schizophrenia is open to question. It is possible that certain basic factors, particularly genetic factors, which apparently underlie some schizophrenic disorders, may eventually lend themselves to study by the use of nonhuman animals, in the same sense that it has been possible to find parallels in laboratory monkeys for various inborn biochemical enzyme errors producing such disorders as phenylketonuria, galactosemia, and maple syrup disease in human children. Extensive research in nonhuman primates is being conducted to further the current understanding of the etiology of mental retardation, potential management of the individual retardate, and learning and related phenomena for understanding human cognitive development. The effects of partial social isolation on the social and sexual behavior of monkeys have been studied for many years.

The diversity of the available primates offer many opportunities to enrich the understanding of behavioral adaptation and the process of primate evolution. Studies of behavior, especially the field studies, strongly support the similarity between humans and apes. Man's inclination toward and control of aggression, his highly complex social behavior, his growth and development, and even his basic psychological processes of learning, forgetting, perception, and transfer skills also can be better understood through study of his homologs as manifested by the various nonhuman primates.

Various nonhuman primates have been used for prospective and longitudinal studies. Rhesus monkeys and other primates have greatly compressed life spans compared to human beings; thus longitudinal studies become feasible. Other major areas of interest for the behavioral scientist in primate models include:

1. The development of attachment behavior (115);
2. The effects of social isolation (99);
3. Separate studies (75);
4. Possible experimental simulation of learned helplessness (100);

5. Biological induction of abnormal behavior syndromes (76);

6. Biological studies of animals who have socially induced syndromes (71);

7. The development of biological rehabilitative methods (72).

USE OF OTHER ANIMALS SPECIES IN BEHAVIORAL RESEARCH

The four animal species just described have been used for the majority of studies in the approximation of various aspects of human behavior. However, there are some reports available in the vast literature dealing with this subject which indicate that animals such as horses, sheep, pigs, hamsters, guinea pigs, birds, reptiles, and fish may be used as models to study a well-defined process of human behavior (16, 35, 41, 57, 59, 73, 85). For example, the horse has been used to study neurotic and psychotic phenomena; sheep and birds have been used to study hypnosis; many animal species have been used to study the effects of toxicologic agents, drugs, and mutagenic agents on their behavior patterns (16, 17, 21, 22, 73, 83, 88). It is beyond the scope of this chapter to discuss these aspects of studies of human behavior in experimental animals. The reader is referred to many excellent textbooks and reviews on the subject for further information. (8, 34, 41, 44, 62, 64, 70, 71, 72, 100, 103, 114). The following section describes some studies on human behavior using rodents, cats, dogs, or primates as experimental animals.

NEUROSIS

Neuroses in humans refer to syndromes due to emotions (instead or organic disease) manifested by exaggerations of feelings. There are many causes of neurosis involving conflicts, often unconscious, deep seated, and rooted in childhood. Many varieties of neuroses include anxiety, hysteria, phobia, obsessive-compulsive behavior, depression, neurasthenia, depersonalization, hypochondriasis, and the like. In experimental animals, it is quite difficult to produce a human model of classical neurotic syndromes because of the problems in measuring the emotional conditions of animals. Broadhurst (12) observed that many of the reactions of "neurotic" animals described in earlier reports as symptomatic of "neurosis" were, in fact, a generalized conditioned emotional response (CER). The factors of fear and restraint in the conditioning apparatus of experimental subjects must be considered. With prior training, the subject becomes accustomed to the testing room and restraint in the conditioning harness, but as soon as the CER is established, the subject may be unwilling to enter the experimental room and vigorously attempt to escape from the apparatus. The CER cannot, therefore, be regarded as a "neurotic" reaction. Under natural conditions in which trauma is unavoidable and the animal has no means of escape, a state of vigilance may arise, and the combination of conditioned fear and anxiety, coupled with learning, may precipitate the CER in an animal showing abnormal or maladaptive motor and autonomic reactions. In experimental animals, there are considerable general disturbances of higher nervous activity. There are also the principles of autokinesis and schizokinesis, which imply that the splitting of body functions and the neurotic behavior can occur after the cessation of external stress, as a result of internal constitutional mechanisms. In human neurotics, experiments have shown various degrees of general impairment of excitatory and inhibitory processes and of the mobility of these processes.

As a general rule, the greater the general disturbances are, the longer the neurotic condition has lasted. It seems that human neuroses, like animal neuroses, are based on a variety of pathophysiological mechanisms. In a more simply structured animal neurosis, there is a short-lived disturbance of the equilibrium of excitatory and inhibitory processes, so that rest is sufficient for cure. The morbid foci (conditions associated with more localized disturbances) in animals and in humans are mostly associated with general impairment of higher nervous activity.

Many experimental studies have been reported on animals with various degrees of success in the development of models of human neurotic conditions. These studies include dogs (92), rats (23), dogs and cats (27, 28), chimpanzees (48), sheep (41, 64), and monkeys (24, 29, 72).

PSYCHOSIS

The common psychoneuroses in humans are due to anxiety reaction, conversion reaction, obsessive-compulsive reaction, and the like. The symptoms form a pattern or complex that is more characteristic of a particular neurosis, although these may develop with the symptoms of another neurosis. The symptom complex tends to be persistent in duration (months or years). In common functional psychoses (schizophrenic reactions, manic depressive reactions, etc.), the symptoms are hallucinations, delusions, ideas of reference, thought disturbances, overactivity and so forth. In the so-called reactive or psychogenic psychoses, it is presumed that stress plays a major etiological role. In the reactive depressions, conflict interferes with other functions, such as motor reflexes and autonomic reactivity. The patterns of functions are rather similar to those in neurosis, and from an experimental point of view, it is not clear what the main distinctions between depressive neurosis and depressive psychosis are. In experiments, confusional psychogenic reactions show severe impairment of higher mental processes. It appears that neurophysiological structures become overexcited by psychic traumatization and induce a "cortical blackout" demonstrated by primitive-response type subjects in association experiments.

The animal models of psychoses have been created in primates with drug-induced (e.g., LSD, amphetamine, etc.) mental states and also by the effects of social isolation to study certain aspects of psychotic states in humans. The effects of social isolation in early life (during 6 to 12 months of age) result in severe behavior defects manifested by the animals; most of the time, they huddled in a corner, rocked, self-clasped, and refused to enter into play or other normal social encounters. They may often show stereotyped behaviors and may also engage in self-aggression or in inappropriate and unpredictable aggression against, for example, a large dominant male. In adulthood, appropriate sexual behaviors are virtually absent among isolated-reared monkeys (9, 46, 84, 110). Sacket (100) reported that animals reared without physical contact with age mates and deprived of varied sensory experiences show abnormal levels of motor activity, emotionality, and maternal and sexual responsiveness. Similarities between abnormal humans and abnormal monkeys suggest that the isolation-reared monkey can serve as a model for the development of abnormal behavior in primates. However, there is no clear evidence that the underlying causes of abnormalities in the monkeys reflect processes identical with the causes of abnormal behavior in humans. The abnormalities appear to be uncorrelated with intellectual deficits. Attempts to reverse these effects of early experience were uniformly negative. After extensive postrearing social experience, animals deprived of physical peer contact during the first 3 to 6 months of life failed to perform socially at levels comparable to those achieved by animals reared with peer experience. Monkeys receiving visual social experience from pictures, but otherwise isolated from social contact, showed appropriate social behavior during rearing, but failed to develop normally after removal from isolation. Even when monkeys were gradually introduced to novel and complex social and nonsocial stimuli, their behavior was totally inadequate and failed to improve after repeated social experience. Thus, isolation per se from critical early social and nonsocial experience appears to produce permanent ananabasia within the animal, which persists regardless of the degree of adaptation to a postrearing test situation. Evidence has also been presented to show that early learning, perhaps within the first month of life, can produce social traits, such as hyperaggression and preference for specific social stimuli, that persist into adult life.

Suomi et al. (115) studied the effects of respective infant-infant separation of young monkeys. Previous studies have reported the effects of separation from an attachment object using a mother-infant model. In view of several methodological difficulties inherent in mother-infant separation, an alternative procedure, that of repetitive, short-term infant-infant separation, was proposed by Suomi et al. (115). Their results showed that each short-term infant-infant separation produced behavior patterns similar to those exhibited by infants separated from their mothers. A cumulative effect of repetitive infant-infant separation was an arrest of maturation of social development in the monkey. It may be a good model not only for the study of reaction to separation from an attachment object but also for other purposes, such as arrest of social maturation. The infant-infant separation of young rhesus monkeys produces behavior similar to that attributed to mother-infant separation of both human and nonhuman primates.

MANIC-DEPRESSIVE PSYCHOSIS

A manic-depressive psychosis provides a model of how changes in basic emotional processes alter responses to external environment. The melancholic patient, in contrast to the patient with reactive depression, shows only slight influences of psychological complex structures. Although unconditioned reflexes in patients with reactive depressions are essentially normal, many investigators have reported alteration of these responses in manic-depressive states (7). In particular, the defensive reflexes to pain tend to be inhibited. Seligman et al. (104) reported that dogs given inescapable shock in a Pavlovian harness later seem to "give up" and passively to accept traumatic shock in shuttlebox escape-avoidance training. The failure to escape was alleviated by repeatedly compelling the dog to make the response that terminated shock. This passive behavior in the face of trauma in dogs may be related to maladaptive passive behavior in humans. Other work with laboratory animals also suggests that lack of control (the independence of response and reinforcement) over the important events in an animal's environment produced abnormal behavior. For example, wild rats rapidly gave up swimming and drowned when placed in tanks of water from which there was no escape. (95). If, however, rats were repeatedly placed in and taken out of tanks or if the experimenter allowed them repeatedly to escape from his grasp, the animals swam for approximately 60 hours before drowning.

Maier (66) and Maier and Klee (67) reported that rats showed positional fixations when they were given insoluble discrimination problems (problems in which the response of the rat and the outcome are independent). Making the problem soluble alone did not break up these fixations. But the "therapeutic" technique of forcing the rats to jump to the nonfixed side when the problem was soluble eliminated the fixations. Liddell (64) reported that inescapable shocks produced experimental psychosis in lambs.

SCHIZOPHRENIA

Schizophrenia is the most common mental illness in humans. It is characterized by faulty reality sense, disharmony and inappropriateness of thinking, feeling "split," and often hallucinations and delusions. The tendency is hereditary, perhaps via biochemical (enzyme) abnormalities, but psychological factors are contributory. Common types are paranoid ("persecuted," pathologic conformity), catatonic (mute, stuporous, waxy or bizarre, excited, frenzied), hebephrenic (shallow, silly, withdrawn), and simple (apathetic, regressed). From the foregoing description, schizophrenia appears to be a very heterogenous group of psychoses clinically as well as experimentally. Nevertheless, these psychoses have as a common trait a great probability of a chronic, progressive course of illness. (6). Furthermore, schizophrenics have specific types of altered emotional responses incongruent with their perception of reality. In clinical terms, changes of instincts and drives along with dissociation of psychic functions stand in the foreground. One of the most characteristic features of the experimental performances of schizophrenics are inhibitions of defensive, unconditioned reflexes along with dissociation within and between the three levels of unconditioned reflexes and the two signal systems. Dissociation can be used as a parameter to measure schizophenic disturbances.

In experimental animals, schizophrenic models can be produced by isolation rearings (99) and through the use of certain drugs, such as amphetamines (110). At relatively low amphetamine doses, animals appear excited and will run around their cages at a furious rate, an effect that has close similarities to the central stimulant actions of amphetamines in humans. D-amphetamine was found ten times as potent as L-amphetamine in enhancing locomotor activity in rats, paralleling the tenfold difference between these isomers by inhibiting the reuptake of catacholamines by norepinephrine neurons. This finding suggests that amphetamine-induced locomotor stimulation in animals, and possibly central stimulation in humans involves brain norepinephrine. Drugs that block the conversion of dopamine to norepinephrine diminish the locomotor stimulant actions of amphetamines, which also suggests that norepinephrine rather than dopamine mediates the behavior (15). At somewhat higher doses, amphetamines elicit stereotyped compulsive behavior in animals, the exact pattern varying with different species. In rats, for example, whose major means of exploring their environment is olfactory, animals stay in one portion of the cage, sniff, lick, and gnaw. Motor components of this behavior resemble the involuntary movements elicited in Parkinsonian patients by large doses of L-dopa as well as the motor tics of Grilbes de la Tourette's disease (109).

In cats and monkeys, the stereotyped behavior produced by amphetamines takes on a searching quality. Chimpanzees intoxicated with amphetamines display patterns of looking from side to side, as well as self-picking and grooming behavior that are reminiscent of the stereotyped, purposeless searching and grooming behavior of human amphetamine addicts.

Ellenwood (30), who studied human amphetamine pyschosis as well as the effects of chronic amphetamine administration in rats, cats, and monkeys, has emphasized the formal similarities of the human and animal stereotyped behaviors.

DEPRESSIVE DISORDERS

Depression rivals anxiety as the most important and inclusive category in psychopathology. Depression is a neurotic symptom and is the salient feature of three psychoses: manic-depressive psychosis, involutional melancholia, and psychotic depressive reaction. Mild depression manifests itself largely by loss of pleasurable interest in the usual affairs of life. One does not feel physiologically ill but neither does one feel comfortable and well. Fatigue is excessive. In more severe depression, patients are frankly despondent or feel physically ill (or both). They are usually gloomy, hopeless, helpless, and bereft of self-esteem. Their thinking, speech, and movements may be slowed (psychomotor retardation) or may be tense, hypervigilant, and restless (agitation anxiety). Agitated depressed patients are likely to complain endlessly about aches and pains, fatigue, feelings of unworthiness, of guilt fears. If the depression is of psychotic severity, patients may actually believe things are as bad as they feel they are and they may have elaborate delusions, often of a hypochondrial nature. Of the numerous physical symptoms, insomnia is the most prominent. There are also neurotic symptoms in which hypersomnia is a problem. Anorexia and weight loss are also characteristic. Many somatic complaints commonly occur in depression, particularly obscure pain, gastronintestinal symptoms, menstrual irregularities, and the whole range of psychophysiologic disorders. Sexual disinterest or incapacity ("loss of libido") is described as a classical symptom.

Depressive disorders are perhaps the most distressful, and among the most common, maladies that afflict humans. In order to understand the basic mechanism involved, studies are being carried out in an attempt to create experimental animal models of depression, models that would simulate the central features of human depressions. Depressive-like states consisting of anorexia, weight loss, decrease in motor activity, slowing of responses, and loss of usual concerns have been observed in a variety of animals. As in human situations, object loss has been frequently associated with the onset of these states and reunion with the lost object has been noted to be curative. Senay (105) studied an animal model of depression by observing separation in dogs. Six dogs were raised in an experimental setting by the author who was their sole consistent object. The animals were ranked daily in an approach-avoidance test, and their temperament characterized in approach-avoidance terms. Eleven independent observers scored the animals on object seeking, object avoiding, and aggressive behavior in each of three conditions:

1. A preseparation period in which the animals' relationship with the author was undisturbed;

2. A separation condition in which the animals had no contact with the author; and

3. A reunion period.

Body weight and gross motor activity were also measured. Separation was associated with increases in object avoidances and aggressive behavior for animals of the avoidance temperament. Activity was significantly decreased in those animals for whom reliable data were obtained. Reunion was associated with further deviations from preseparation behavior patterns. These results tentatively indicated that models of separation and depression can be constructed in experimental animals.

Depressive behavior in young rhesus monkeys was described by McKinney et al. (80). An animal model of depression was defined as the production of behavioral changes in animals that resemble behaviors commonly associated with depression in humans. Previous work showed that mother-infant separation in rhesus monkeys leads to behavioral changes in the infant that are similar to those seen in anaclitic depression as described by Spitz (113) and as seen in Bowlby's depressed older child patients when they have been separated from their mothers (96). The experimental paradigm for monkey mother-infant separation is to have mother and infant live together for a preseparation period, during which essential baseline data are taken. The infants are then physically separated from their mothers for a certain length of time and finally returned to their mothers in a postseparation period. The reaction of the infant monkey occurs in two stages: an initial stage of protest followed by a stage of dispair and withdrawal. Bowlby described a third stage in humans, called detachment, but generally it has not been possible to duplicate it in monkeys. Rather, the infants show a great deal of positive mother-directed behavior upon reunion with their mothers, with increased ventral clinging and contact.

Male rhesus monkeys (*Macaca mulata*), 3 years of age, were studied socially before and after 10 weeks of confinement in a vertical chamber apparatus designed to facilitate production of psychopathological disturbances. The study represents an initial effort to move beyond the use of young monkeys in a depression research program. Chamber confinement results in a significant increase in contact clinging between animals and a decrease in locomotion following removal from the apparatus. These behavior patterns are very typical of laboratory research rhesus monkeys of that age and may represent a maturational regression induced by social means (1, 79, 105).

Although many factors influence the application of an animal model in depression studies, anaclitic depression and pharmacological depression studies have been made extensively using nonhuman primates. The animal models simulate some of the central features of clinical depression, for example, helplessness and objective loss, thereby allowing rigorous investigation of developmental, behavioral, and biochemical alterations. The objective loss model, as a concrete version of a metapsychological psychoanalytic concept, has enabled primatologists to study the disruption of an attachment bond. The behavioral model accommodates this concept to a broader generalization. Akiskal and McKinney (1) have reviewed the evidence that these processes involve the diencephalic centers of reward of reinforcement, thereby permitting integration of the psychoanalytical and behavioral formulations with biochemical hypotheses. The breaking of an attachment bond in the primate represents significant loss of reinforcement, which indicates helplessness and disrupts motivated behavior. The depressive syndrome could be caused by interactions of genetic, chemical, developmental, and interpersonal factors, all of which impinge on the diencephalic centers of reinforcement.

Certain drugs or metabolic disorders can suppress or interfere with neurotransmitters of this system, for example, reserpine and alphamethyldopa, reserpine and alphamethylparatyrosine in primates, and borderline hypothyroidism which could chemically interfere with the sensitivity of the monoaminergic receptor (39, 56, 78, 93,

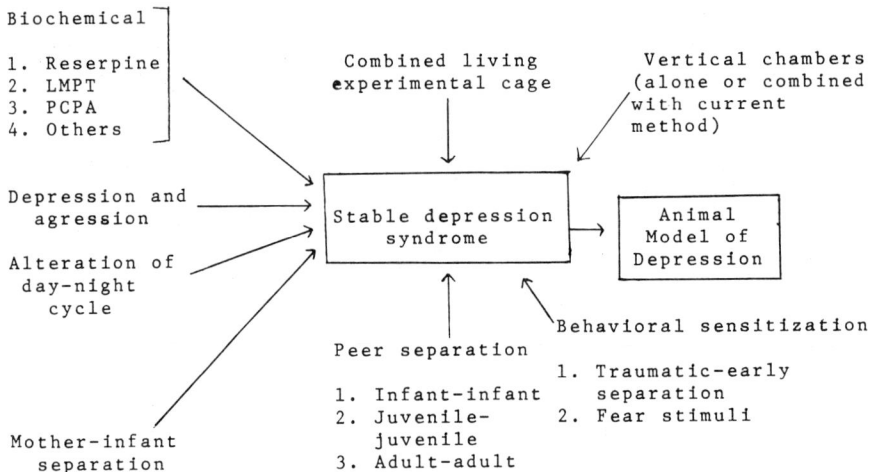

Fig. 1. Methods of inducing depression in rhesus monkeys. Adapted from *American Journal Psychiatry* (80) with permission.

112). Alterations in norepinephrine and/or serotonin metabolism are thought to be important as biological determinants for depression (13, 101, 102). Experimental production of depression in nonhuman primates by reserpine was reported by McKinney et al. (77). Reserpine was administered daily by injection for 81 days to three rhesus monkeys. Their behavior during the experimental period was compared to their behavior before and after the drug period as well as to that of a control group of three monkeys given water instead of reserpine. It was found that reserpine caused significant behavioral changes in the rhesus monkey, including decreases in visual exploration and locomotion and increases in self-huddling, posturing, and tremor. The behavioral effects of repeated daily dosage were not cumulative nor was a tolerance developed by the monkeys.

A number of methods can be used to induce depression in rhesus monkeys (Fig. 7-1). By using these methods, experimental investigation of many currently held concepts about depression can be investigated. Key variables can be experimentally manipulated or systematically controlled, and direct investigation of the parameters of depression becomes possible. Conclusive results yielded by an animal model cannot help but be of interest to therapists and theoreticians deeply concerned with the problems of human depression.

EPILEPSY

A major stumbling block for the understanding of the mechanisms of epilepsy in humans has been the difficulty in finding an experimental model that can serve as a surrogate for the phenomenon in nature. Epilepsy in humans is characterized by chronically recurrent, spontaneous, clinical seizures. While such techniques as electric shock, photogenic methods and chemical agents, cortical freezing, and intracerebral injections of tetanus toxin have been used in a variety of animals, the seizures induced have not fully satisfied these classical requirements (5, 47, 53, 69, 90, 119, 121). The Mongolian gerbil appears to be unique in that it is prone to chronically recurrent, spontaneous seizures that have been described as epileptiform in character (65, 116). Frequency of occurrence of seizures in a Mongolian gerbil population was shown to be a function of age but not of sex (55). Seizures first appeared at 2 months of age, and at 6 months their frequency was still increasing. Early stimulation in the form of weekly tests, from 1 week to 4 months of age, delayed appearance and sharply curtailed occurrence of seizures. These results were due not only to the environment but also to the early stimulation, which has been shown by others to affect emotional reactivity and general arousal level. It is suggested that gerbils might provide an effective model for the study of epilepsy in humans.

Acoustic priming is a technique whereby a few seconds of exposure to loud sound so alters the mouse or hamster that subsequent auditory exposure will produce audiogenic seizures (50, 51, 52, 121). Henry and Sabeh (49) examined C57 B1/6J mice for the absolute threshold of the auditory evoked potential (AEP) at the level of inferior colliculus. Acoustic priming at 16 days of age, which induces susceptibility to audiogenic seizures, elevated this threshold of individual frequencies from 5 kHz to 25 kHz in 21-day-old mice by 5–11 db. They concluded that tested mice show recruitment deafness, which is a good model for investigating recruitment deafness in humans.

Epileptic seizures have been produced experimentally in animals by photogenic and pharmacological methods to understand the mechanisms of human epileptic disorders (53, 63).

ALCOHOLISM

The study of alcoholism would be facilitated if an animal model possessing the major behavioral and physiological features of the human alcoholic could be developed. Recent works reported on this subject indicate that presently available experimental arrangements fall short of providing such a model (20, 26, 81, 82, 89, 120). In order to meet the behavioral requirements of the model, the following criteria should be fulfilled (60, 61).

1. The animal should orally ingest ethanol solutions excessively and chronically in a pattern that increases the concentration of blood ethanol analogous to that in the alcoholic.

2. Unequivocal physical dependence on ethanol must be demonstrated.

3. Food and ethanol should be available from sources physically separate so that factors determining ethanol intake are not inextricably bound to those primarily concerned with meeting nutritional requirements.

4. The experimental arrangement should retain an elective aspect to the ethanol ingestion by not programming extrinsic reinforcing events (e.g., shock avoidance, food pellet delivery) contingent upon drinking ethanol.

Falk (32) reported that volumes of water three to four times the normal 24-hour amounts were ingested when small food pellets were delivered intermittently to rats on a limited food regimen. Rats maintained on an intermittent food schedule with an available ethanol solution drink an excessive amount (13.1 g of ethanol per kilogram of body weight) of alcohol daily (33). Removal of ethanol produced symptoms of physical dependence, including death from tonic-clonic seizures. Overindulgence in oral self-administration of an aqueous ethanol solution (51 v/v), resulting in unequivocal physical dependence, is approximately a model of human alcoholism. Griffiths et al.

(40) developed a model of ethanol dependence in mice based on the inhalation of ethanol vapor.

This method consistently produced ethanol tolerance and dependence in mice. Groups of thirty male mice (18–22 g) were exposed to increasing concentrations of ethanol in the inspired air for 7 to 10 days. Brain ethanol levels were found to be closely related to blood levels which rose gradually throughout the experiment. A representative group of thirty mice had 150 ± 30 mg of ethanol/100 ml blood on day 2 and 400 ± 40 mg ethanol/100 ml blood on day 10. The treated mice showed locomotor depression and ataxia during the inhalation. They also ate and drank less than control mice. However, these differences were not large, and treated mice showed little change in weight compared to controls. The death rate of treated mice was between 10 and 15 percent. In mice treated for 10 days, there was a markedly increased rate of elimination of ethanol from the blood and also from the brain. These observations provide evidence for both metabolic and pharmacological tolerance during the administration of ethanol in this way. The dependence on ethanol was shown after ethanol concentration in inspired air was brought down rapidly to near zero levels. Two hours after ethanol was withdrawn, treated mice showed fine tremor and piloerection. A characteristic convulsion could be shown by holding a withdrawn mouse by the tail. The signs of withdrawal usually persisted for 10 to 20 hours after alcohol was withdrawn. From these observations, it may be concluded that the administration of ethanol in this way fulfills the condition for a valid model of ethanol dependence. However, it does not fulfill the other requirements of the alcoholic behavior model cited above.

A valid and useful model for alcohol physical dependence was produced in mice by Goldstein (37). It agrees with other animal models and human physical dependence in several aspects. The alcohol dose relationships and time course of the signs that develop after discontinuing the alcohol indicate a true alcohol withdrawal reaction. The model may be used to test several drugs that are used in the clinic to treat alcohol withdrawl syndromes because the mice behave much like patients. Meprobamate, paraldelhyde, or pentobarbitol had the same transient effect as ethanol. With the long-acting barbiturates or with the benzodiazepines, a single dose suppressed the seizures during the entire withdrawal period. A dose of 20 mg/Kg of diazepam or chloropromazine and promazine suppressed convulsions for several hours. Withdrawal syndromes in mice dependent upon alcohol, barbiturates, and bromide have been reported by Freund (36). The most important factor in the production of delirium tremors and related disorders is the withdrawal of alcohol following a period of chronic intoxication. Two factors in particular—hypomagnesemia and respiratory alkalosis—are consistently associated with all but the mildest withdrawal symptoms and are probably important in their genesis (118).

The behavior of rats manifesting psychogenic polydypsia is impressive because it is predictable, excessive, and atypical when the fluid consumed is ethanol. It is reminiscent of the compulsive drinking of the alcoholic human being. Lestor and Freed (61), using rat models, found that (a) alcohol intake, as a proportion of total fluid drunk, was positively related to food deprivation of the animals; (b) only one of the fluid choices was drunk at any clearly time-deferred period; and (c) the choice of alcohol was most likely to occur the longer the time since food was consumed. These results suggest the ability of the rat to detect alcohol caloric value and make this the basis for its selection of alcohol under conditions of food deprivation.

Many investigators have reported differences in ethanol preference between different strains of rats (10, 11, 31). There followed the cross-breeding of such strains in order to elucidate the genetic mechanism governing behavior.

The second approach, the selective breeding of new strains specifically for different degrees of preference or aversion for ethanol, has been demonstrated by Mardones et al. (68). Brewster (10) investigated preference for 5 percent ethanol in rats of the Mandsley reactive (MR) and nonreactive (MNR) strains, bidirectionally selected for emotional reactivity, in a paired-choice situation with plain water as the alternative choice. The MNR strain displayed significantly higher mean levels of both preference for an absolute intake of ethanol than the MR. A genetic analysis of preference for 5 percent ethanol over water was then performed by crossing rats of the two strains reciprocally to derive a hybrid generation. The genetic mechanism governing voluntary ethanol intake emerged as an additive system with a high heritability and complete dominance in the direction of high ethanol intake. Ethanol preference gave a similar picture but with no dominance. Prenatal maternal effects were excluded as a possible source of the strain difference.

Behavioral and physiological differences in rat strains specially selected for their alcohol consumption were described by Eriksson (31). With genetic selection, it was possible to demonstrate and estimate the part played by heritability in alcohol drinking behavior and to find the physiological and behavioral factors correlating with drinking. Estimates of heritability can be made by studying inbred mouse strains which differ greatly in alcohol intake. Mardones (69) has selected two inbred rat strains for their alcohol intake and shown that they differ in metabolic characters in respect of the unknown N-factor. However, the difference in alcohol consumption between these strains is rather small. Eriksson (31) noted a clear sex difference in alcohol intake in the drinker strain. The genetic analysis made with these strains showed that drinking behavior is a polygenically inherited, additive trait and that its estimate of heritability (h^2) is rather low. Rat strains tend to drink roughly the same absolute amount of alcohol per unit of body weight over a certain concentrate range, that is, 5 to 10 percent (v/v).

MENTAL RETARDATION

The genetic bases of many types of mental retardation are becoming increasingly clear. Classical family patterns of inheritance, enzyme deficiencies, and chromosomal aberrations have all been found to be associated with subnormal mental development. A number of disorders of amino acid, lipid, and carbohydrate metabolism were discussed in Chapter 6. Among the autosomal recessive diseases are phenylketonuria, tyrosinosis, maple syrup urine disease, Hartnup's disease, histidinemia, homocystinurea, arginosuccinic acidurea, and hyperprolinemia.

Experimental animal models of phenylketonuria have received much attention because this disorder is associated with a type of mental retardation whose origin is genetically and biochemically understood (3, 4). There is a chemical method of producing the model by giving phenylalanine alone in large amounts to experimental animals. This method produces hyperphenylalaninemia but fails to duplicate the high ratio of phenylalanine to tyrosine in blood or the absence of hepatic phenylalanine hydroxylase activity found in phenylketonuria. Some researchers have administered p-chlorophenylalanine supplemented with phenylalanine to produce a high ratio of phenylalanine to tyrosine in their blood. By this method, an experimental subject is produced whose be-

Table 7-1. Toxicity of Major Groups of Psychopharmacologic Drugs

	Antipsychotic Tranquilizers	Antidepressants	Sedatives	Sedative Withdrawal	Amphetamines	Narcotics	Narcotic Withdrawal
Autonomic Effects							
Blurred vision	+	+	-	-	+	-	-
Dry mouth	+	+	-	-	+	-	-
Urinary retention	+	+	-	-	+	-	-
Pupillary dilatation	+	+	-	+	+	-	+
Pupillary constriction	+	+	-	-	-	+	-
Paralysis of accommodation	+	+	-	-	-	-	-
Abdominal cramps, diarrhea	-	-	-	-	-	-	+
Constipation	+	+	-	-	-	+	-
Sweating	-	-	-	+	+	-	+
Cardiovascular Effects							
Bradycardia	+	+	-	-	-	+	-
Tachycardia	+	+	-	+	+	-	+
Hypotension, postural	+	+	-	-	-	+	-
Hypertension	-	-	-	+	+	-	+
Edema	+	+	-	-	-	-	-
CNS Effects							
Drowsiness	+	+	+	-	-	+	-
Hyperthermia	-	-	-	+	+	-	+
Insomnia	-	-	-	+	+	-	+
Excitement, hyperactivity, agitation	+	+	-	+	+	-	+
Convulsions	+	+	-	+	-	-	+
Euphoria	-	+	+	-	+	+	-
Depression	+	-	+	-	-	-	-
Ataxia, dysarthria, nystagmus	-	-	+	-	-	-	-
Unconsciousness, coma, anesthesia	+	+	+	-	-	+	-
Respiratory depression	+	+	+	-	-	+	-
Hallucinations	-	-	+	+	+	+	+
Tremors and rigidity	+	+	-	+	-	-	+
Dystonias and akathisia	+	+	-	-	-	-	-
Metabolic and Endocrine Effects							
Menstrual irregularity	+	+	+	-	-	+	-
Galactorrhea	+	+	-	-	-	-	-
Weight increase	+	+	-	-	-	-	-
Anorexia	-	-	-	+	+	-	+

*Variations in effect may be dose-related or may be due to setting or basic personality structure.
Adapted from Handbook of Psychiatry, P. Solomon and V.D. Patch (eds.), Lange Medical Publications, Los Altos, California, 1971, with permission.

+ = often present
- = usually absent

havioral, neurological, and biochemical characteristics are similar to those of patients with phenylketonuria. The treatment of rat pups with p-chlorophenylalanine and phenylalanine during the critical period of rapid brain growth between birth and 21 days of age results in a delay before weaning in neural development and enduring behavioral changes in activity and performance on learning tests in adult animals. Associated with these behavioral abnormalities is a lower than normal brain weight and a deficit in myelination.

EFFECTS OF DRUGS AND TOXIC COMPOUNDS ON BEHAVIOR

Numerous articles, monographs, and books are available on this subject. Many investigators have used experimental animals to study drug metabolism and the toxicity of major groups of psychopharmacologic drugs (Table 7-1). Silverman (108) reviewed laboratory models of behavioral manifestation as an indicator of toxicity. The reader is referred to Schildkraut (101), Shader and Di Mascio (106), Solomon (111), Usdin and Efron (117), Freund (36), Silverfeld and Goldberg et al. (107), Cheney et al. (19), Page and Glenner (91), Haler (42), Yamori and Okamoto (112), Goldstein et al. (38), Kumar and Stolerman (58), and Jaques (54) for further information on this subject.

REFERENCES

1. Akiskal, H. S., and W. T. McKinney Jr., *Science,* **182:**20 (1973).
2. Alexander, N., *Proc. Soc. Exp. Biol. Med.,* **146:**163 (1974).
3. Andersen, A. E., V. Rowe, and G. Guroff, *Proc. Natl. Acad. Sci.* (USA), **71:**21 (1974).
4. Andersen, A. E., and G. Guroff, *Proc. Nat. Acad. Sci.,* **69:**863 (1972).
5. Appleton, D. B., and D. C. DeVivo, *Proc. Aust. Assoc. Neurol.,* **10:**75 (1963).
6. Astrup, C., *Schizophrenia: Conditional Reflex Studies,* Charles C. Thomas, Springfield, Ill., 1962.
7. Astrup, C., *Pavlovian Psychiatry: A New Synthesis,* Charles C. Thomas, Springfield, Ill. (1965).
8. Benhar, E., and D. Samuel, *J. Med. Primatol.,* **2:**11 (1973).
9. Bowers, M. D., and D. X. Freedman, *Arch. Gen. Psychiatry,* **15:**240 (1966).
10. Brewster, D. J., *Ann. N.Y. Acad. Sci.,* **197:**49 (1972).
11. Brewster, D. J., *J. Gene. Psychol.,* **115:**217 (1969).
12. Broadhurst, P. L. in H. J. Eysenk (ed.), *Handbook of Abnormal Psychology. An Experimental Approach,* Basic Books, New York, 1961.
13. Bunney, W. E., and J. M. Davis, *Arch. Gen. Psychiat.,* **13:**483 (1965).
14. Butcher, R. E., *Am. J. Ment. Defic.,* **75:**755 (1971).
15. Carleson, A., in CE. Garattine (ed.), *Amphetamines and Brain Catecholamines in Amphetamines and Related Compounds,* Raven Press, New York, 1970.
16. Carson, T. L., G. A. VanGelder, G. G. Karas, and W. B. Buck, *Environ. Health Perspect.* **7:**233 (1974).
17. Caviness, Jr., V. S., D. D. K. So, and R. L. Sidman, *J. Hered.,* **63:**241 (1972).
18. Chance, M. R. A., E. Jones, and T. K. Pitcairn, *Lab. Anim.,* **8:**13 (1974).
19. Cheney, D. H., S. Slogoff, and G. W. Allen, *Anethesiology,* **40:**531 (1974).
20. Chey, W. Y., S. Kosay, H. Siplet, and S. H. Lorber, *Am. J. Dig. Dis.,* **16:**835 (1971).
21. Ciaranello, R. D., J. N. Dornbusch, and J. D. Barchas, *Science,* **175:**789 (1972).
22. Cohen, N., T. J. Kneip, D. H. Goldstein, and E. A. S. Muchmore, *J. Med. Prim.,* **1:**142 (1972).

23. Cook, S. W., *Psychosom. Med.,* **1:**293 (1939).
24. Cummings, J. F., *Am. J. Path.,* **66:**189 (1972).
25. Davenport, R. K., *Science,* **168:**279 (1970).
26. Deutsch, J. A., *Science,* **180:**880 (1973).
27. Dmitruk, V. M., *J. Abnorm. Psychol.,* **83:**97 (1974).
28. Dworkin, S., *Psychosom. Med.,* **1:**388 (1939).
29. Dzhalagoniya, S. L., in *Biology and Acclimitization of Monkeys. Materials of a Symposium,* Nauka, Moscow, 1973.
30. Ellinwood, E. H. Jr. *J. Nerv. Ment. Dis.,* **144:**273 (1967).
31. Eriksson, K., *Ann. N.Y. Acad. Sci.,* **197:**32 (1972).
32. Falk, J. L., *Science,* **133:**195 (1961).
33. Falk, J. L., *Science,* **177:**811 (1972).
34. Fox, M. W., *Abnormal Behavior in Animals,* W. B. Saunders Co., Philadelphia, 1968.
35. Fraser, A. F., *Farm Animal Behavior,* Williams and Wilkins Co., Baltimore, 1974.
36. Freund, G., *Ann. N.Y. Acad. Sci.,* **215:**224 (1973).
37. Goldstein, D. B., *Ann. N.Y. Acad. Sci.,* **215:**218 (1973).
38. Goldstein, M., A. F. Battista, S. Nakatani, and B. Anognoste, *Nature* (London), **224:**382 (1969).
39. Goodwin, F., and W. Bunney, *Semin. Psychiat.,* **3:**435 (1971).
40. Griffiths, P. J., J. M. Littleton, and A. Ortiz, *Br. J. Pharmacol.,* **47:**669 (1973).
41. Hafez, E. S. E. (ed.), *Behaviour of Domestic Animals,* 2nd ed., Williams and Wilkins, Baltimore, 1969.
42. Haler, D., *Int. J. Clin. Pharmacol.,* **9:**160 (1974).
43. Harlow, H. F., *Am. Sci.,* **59:**538 (1971).
44. Harlow, H. F., in, A. Abrams, H. H. Garner, and J. E. P. Tomal (eds.), *Tasks in the Behavioral Sciences,* Williams and Wilkins, Baltimore, 1964.
45. Harlow, H. F., P. E. Plubell, and C. M. Baysinger, *J. Autism. Child Schizo.,* **3:**299 (1973).
46. Harlow, H. F., W. D. Joslyn, M. Senko, and A. Dopp, *J. Anim. Sci.,* **25:**45 (1966).
47. Hartman, E. R., B. K. Colasanti, and C. R. Craig, *Epilepsia,* **15:**121 (1974).
48. Hebb, D. O., *Psychosom. Med.,* **9:**256 (1947).
49. Henry, K. R., and M. Saleh, *J. Comp. Physiol. Psychol.,* **84:**430 (1973).
50. Henry, K. R., *Science,* **158:**938 (1967).
51. Iturrian, W. B., and H. D. Johnson, *Experientia,* **27:**1193 (1971).
52. Iturrian, W. B., and G. B. Fink, *Fed. Proc.,* **26:**736 (1967).
53. Izquierdo, I., and A. G. Nasello, *Exp. Neurol.,* **27:**399 (1970).
54. Jaques, L. B., *New Istanbul Contrib. Clin. Sci.,* **10:**171 (1973).
55. Kaplan, H., and C. Miezejeski, *J. Comp. Physiol. Psychol.,* **81:**267 (1972).
56. Koe, B., and A. Weisman, *J. Pharmacol. Exp. Ther.,* **154:**499 (1966).
57. Kuenzel, W. J., and J. B. Rubenstein, *J. Exp. Zool.,* **187:**63 (1974).
58. Kumar, R., and I. P. Stolerman, *Psychol. Med.,* **3:**225 (1973).
59. Lehrich, J. R., M. Katz, L. B. Rorke, G. Barbanti-Brodano, and H. Koprowski, *Trans. Am. Neurol. Assn.,* **95:**51 (1970).
60. Lester, D., and E. X. Freed, *Pharmacol. Biochem. Behav.,* **1:**103 (1973).
61. Lester, D., and E. X. Freed, *Ann. N.Y. Acad. Sci.,* **197:**54 (1972).
62. Leyhausen, P., *Cat Behavior,* Van Nostrand Reinhold, New York, 1972.
63. Lezhava, G. G., *Neurosci. Behav. Physiol.,* **6:**122 (1973).
64. Liddell, H. S. *Emotional Hazards in Animals and Man,* Charles C. Thomas, Springfield, Ill., 1956.
65. Loskota, W. J., P. Lomaz, and S. T. Rich, *Epilepsia,* **15:**109 (1974).
66. Maier, N. R. F., *Frustration: The Study of Behavior Without Goal,* McGraw-Hill, New York, 1949.
67. Maier, N. R. F., and J. Klee, *J. Psychol.,* **19:**133 (1945).
68. Mardones, J. J., T. Segova, and F. Hederra, *Q. J. Stud. Alcohol,* **14:**1 (1953).
69. Mardones, J. *Int. J. Rev. Neurobiol.,* **2:**41 (1960).
70. Mason, W. A., in D. G. Glass (ed.), *Environmental Influences,* Russell Sage Foundation, New York, 1968.
71. Mason, W. A., in C. H. Southwick (ed.), *Primate Social Behavior,* Van Nostrand, Princeton, 1963.

72. Masserman, J. H., in *Experimental Psychopathology in Recent Research and Theory*, Academic Press, New York, 1971.
73. McGibbon, W. H., *J. Hered.*, **65:**124 (1974).
74. McKinney, W. T. Jr., and W. E. Bunney, Jr., *Arch. Gen. Psychiat.*, **21:**240 (1969).
75. McKinney, W. T., Jr., S. J. Suomi, and H. F. Harlow, *Arch. Gen. Psychiat.*, **26:**223 (1972).
76. McKinney, W. J., Jr., *Perspect. Biol. Med.*, **17:**529 (1974).
77. McKinney, W. J., Jr., R. C. Eising, E. C. Morgan, S. J. Suomi, and H. F. Harlow *J. Psychiat. Neurol.*, **32:**740 (1971).
78. McKinney, W. J., Jr., R. Eising, E. Morgan, S. Suomi, and H. Harlow, *Dis. Nerv. Syst.*, **32:**735 (1971).
79. McKinney, W. J., L. D. Young, S. J. Suomi, and J. M. Davis, *Arch. Gen. Psychiat.* **27:**490 (1973).
80. McKinney, W. J. Jr., S. J. Suomi, and H. F. Harlow, *Am. J. Psychiat.*, **127:**1313 (1971).
81. Mello, N. K., *Pharmacol. Biochem. Behav.*, **1:**89 (1973).
82. Mello, N. K., in D. H. Efron (ed.), *Psychopharmacology: A Review of Progress*, Government Printing Office, Washington, D.C., 1968.
83. Michaelson, I. A., and M. W. Sauerhoff, *Toxicol. Appl. Pharmacol.*, **28:**88 (1974).
84. Mitchell, G. D., and D. L. Clark, *J. Gen. Psychol.*, **113:**117 (1968).
85. Molony, V., *J. Physiol.*, **219:**12 (1971).
86. Muller-Calgan, E., *Activ. Nerv. Super.*, **16:**62 (1974).
87. Munn, N. L., *Handbook of Psychological Research on the Rat*, Haughton, Boston, 1950.
88. Murphey, D. L., and R. E. Dill, *Exp. Neurol.*, **34:**244 (1972).
89. Myers, R. D., and W. L. Veal, in H. Begleiter and B. Kissin (eds.), *Biology of Alcoholism*, Vol. 2, Plenum Press, New York, 1971, p. 131.
90. Naquet, R., *Arch. Ital. Biol.*, **111:**516 (1973).
91. Page, D. L., and G. G. Glenner, *Am. J. Pathol.*, **67:**555 (1972).
92. Pavlov, I. P., *Lectures on Conditioned Reflexes*, International, New York, 1928.
93. Redmond, D., J. Maas, A. Kling, C. Graham, and M. Dekirmenjian, *Science*, **174:**428 (1971).
94. Reite, M., J. D. Pauley, I. C. Kaufman, A. J. Stynes, and V. Marker, *Physiol. Behav.*, **12:**1021 (1974).
95. Richter, C. P., *Psychosomatic Med.*, **19:**191 (1957).
96. Robertson, T., and J. Bowlby, *Cours du Centre International de l'Enfance* **2:**131 (1952).
97. Rose, R. M., *Science,* **178:**643 (1972).
98. Rumbaugh, D. M., *J. Comp. Physiol. Psychol.*, **76:**250 (1971).
99. Sackett, G. P., *Neurosci., Res. Prog. Bull.*, **10:**388 (1972).
100. Sackett, G. P., in R. Porter (ed.), *CIBA Foundation Symposium on the Role of Learning in Psychotherapy*, J. A. Churchill, London, 1968.
101. Schildkraut, J. J., *Neuropsychopharmacology and the Affective Disorders*, Little, Brown, Boston, 1970.
102. Schildkraut, J. J., *Am. J. Psychiatry*, **122:**509, (1965).
103. Scott, J. P., and J. L. Fuller, (eds.), *Genetics and the Social Behavior of the Dog*, University of Chicago Press, Chicago, 1965.
104. Seligman, M. E. P., S. F. Mailer, and J. H. Greer, *J. Abnorm. Psychol.*, **73:**256 (1973).
105. Senay, E. C., *J. Psychiat. Res.*, **4:**65 (1966).
106. Shader, R. I., and A. DiMacio, *Psychotropic Drugs Side Effects*, Williams and Wilkins, Baltimore, 1970.
107. Silbergeld, E. K., and A. M. Goldberg, *Exp. Neurol.*, **42:**146 (1974).
108. Silverman, A. P., *Ciba Found. Study Group*, **35:**25 (1970).
109. Snyder, S. H., K. M. Taylor, and J. T. Coyle, *Am. J. Psychiatry*, **127:**199 (1970).
110. Snyder, S. H., *Am J. Psychiatry*, **130:**61 (1973).
111. Solomon, P. (ed.), *Psychiatric Drugs*, Grune & Stratton, New York, 1966.
112. Spector S., A. Sjoersma, S. Underfriend, *J. Pharmacol. Exp. Ther.*, **147:**86 (1965).
113. Spitz, R. A., *Psychoanal. Stud. Child*, **2:**313 (1946).
114. Stebbins, W. C., *Animal Psychophysics*, Appleton-Century-Crofts, New York, 1970.
115. Suomi, S. J., H. F. Harlow, and W. T. McKinney, *Am. J. Psychiatry*, **128:**927 (1972).
116. Thiessen, D. D., G. Lindzey, and H. C. Friend, *Psychonomic Sci.*, **11:**227 (1968).

117. Usdin, E., and D. Efron, *Psychotropic Drugs and Related Compounds,* U.S. Dept. of Health, Education and Welfare, Washington, D.C., 1967.
118. Victor, M., *Ann. N.Y. Acad. Sci.,* **215:**235 (1973).
119. Ward, A. A., Jr., in H. H. Jaspur, A. A. Ward, Jr., and A. Pope (eds.), *Basic Mechanisms of Epilepsy,* Little, Brown, Boston, 1969.
120. Woods, J. H., and G. D. Winger, *Prev. Med.,* **3:**49 (1974).
121. Wozny, J. R., and G. H. Wolfe, *Behav. Genet.,* **1:**87 (1970).
122. Yamori, Y., and K. Okamoto, *Singapore Med. J.,* **14:**393 (1973).

CHAPTER 8

ANIMAL MODELS OF ENDOCRINOPATHIES

The endocrine organs are a group of glands that secrete specific types of substances called hormones. The glands are specialized structurally and functionally, and their secretions are released directly into the bloodstream. For the efficient transport of the secreted material, these glands are highly vascularized. The endocrine organs are found in all parts of the body and have rather different embryologic origins. Several pairs (neurohypophysis and adenohypophysis, adrenal cortex and adrenal medulla, thyroid and parathyroid) are often closely associated anatomically. The level of secretory activity of most of these organs affects secretions by one or more of the others. The nature and actions of hormones are diverse, (Table 8-1). In general, the actions of hormones can be morphological (e.g., growth and differentiation responses involving the entire animal, an organ, a specific tissue, or an organelle in a cell); physiological (e.g., muscular contraction, cell permeability, movement of pigment granules, reflexes or behavioral responses, etc.), and metabolic (e.g., general and specific chemical alterations in proteins, fats, carbohydrates, and minerals.) The normal development and functioning of an organism is dependent on the balanced production of hormones and excessive or inadequate production of these substances may result in the development of pathological conditions (14, 25, 43, 60, 61, 106, 157, 159, 160).

Laboratory animals have been widely used for the study of comparative endocrinology, endocrinopathies, and for the assay of hormones. However, the development of newer radioimmunoassay procedures have considerably decreased the use of experimental animals for assay procedures in endocrine research. In recent years, however, there has been a steady increase in the use of experimental animal model systems for testing pathophysiological effects of new synthetic steroid hormones, for the preparation of hormones used in the therapy of endocrine disorders, and for understanding the mechanisms involved in endocrinopathies of humans. This chapter briefly describes selected animal models which may be useful in the study of certain endocrinopathies. Animal models used in endocrinological research are listed in Table 8-2.

PITUITARY GLAND

Secretions of the pituitary gland control most of the other endocrine glands and directly affect various somatic tissues. The pituitary has two morphologically and embryologically distinct parts: the anterior glandular part, or adenohypophysis; and the posterior part of neural origin, called the neurohypophysis. The two parts jointly occupy the sella turcica, but the function and control of each part are separate. Various segments of the

Table 8-1. The Hormones, Their Source, and A Brief Description of Their Action

Endocrine Gland and Hormone	Nature of Hormone	Site of Action	Principal Actions
HYPOTHALAMUS			
Various releasing factors	Polypeptides	Anterior pituitary	Release of trophic hormones
ANTERIOR PITUITARY			
Somatotrophin, growth hormone (STH,GH)	Protein	Body as a whole	Growth of bone and muscle
Adrenocorticotrophin (ACTH)	Polypeptide	Adrenal cortex	Stimulates formation and secretions of adrenocortical steroids
Melanophore-stimulating hormone (MSH, intermedin)	Polypeptide	Skin	Dispersion of pigment granules; darkening of skin
Thyrotrophin (TSH)	Glycoprotein	Thyroid	Stimulates formation and secretion of thyroid hormone
Follicle-stimulating hormone (FSH)	Glycoprotein	Ovary	Growth of follicles, with LH secretion of estrogens and ovulation
		Testis	Development of seminiferous tubules; spermatogenesis
Luteinizing or interstitial cell-stimulating hormone (LH or ICSH)	Glycoprotein	Ovary	Formation of corpora lutea, secretion of progesterone
		Testis	Stimulation of interstitial tissue-- secretion of androgen
Prolactin (lactogenic hormone, luteotrophin	Protein	Mammary gland	Proliferation of mammary gland and initiation of milk secretion.
		Corpus luteum	Maintenance of corpus luteum
POSTERIOR PITUITARY			
Vasopressin (ADH, anti-diuretic hormone)	Octapeptide	Arterioles Renal tubules	Elevates blood pressure Water reabsorption
Oxytocin	Octapeptide	Smooth muscle (uterus, mammary gland)	Contraction, action in parturition and in sperm transport; ejection of milk
THYROID			
Thyroxine and triiodothyronine	Iodoamino acids	General body tissue	Stimulates oxygen consumption and metabolic rate of tissues

Endocrine Gland and Hormone	Nature of Hormone	Site of Action	Principal Actions
Thyrocalcitonin (calcitonin)	Polypeptide	Skeleton	Inhibits calcium resorption; lowers plasma calcium and phosphate
PARATHYROID			
Parathyroid hormone (PTH, parathormone)	Polypeptide	Skeleton, kidney, gastrointestinal tract	Regulates calcium and phosphorus metabolism
ADRENAL CORTEX			
Adrenal cortical steroids--cortisol, aldosterone	Steroids	General body tissue	Carbohydrate, protein and fat metabolism; salt and water balance, inflammation, resistance to infection; hypersensitivity
ADRENAL MEDULLA			
Norepinephrine and epinephrine	Aromatic amines	Sympathetic receptor	Mimic sympathetic nervous system
		Liver and muscle	Glycogenolysis
		Adipose tissue	Release of lipid
OVARY			
Estrogens	Phenolic steroids	Female accessory sex organs	Development of secondary sex characteristics
Progesterone	Steroids	Female accessory reproductive structures	Preparation for ovum implantation; maintenance of pregnancy
Relaxin	Polypeptide	Symphysis pubis, uterus	Relaxation, aids in parturition
TESTIS			
Testosterone	Steroid	Male accessory sex organs	Development of secondary sex characteristics, maturation and normal function
PANCREAS			
Insulin	Polypeptide	Most cells	Regulation of carbohydrate metabolism; lipogenesis
Glucagon	Polypeptide	Liver	Glycogenolysis

Table 8-1. *(Continued)*

Endocrine Gland and Hormone	Nature of Hormone	Site of Action	Principal Actions
PLACENTA			
Estrogens, progesterone, gonadotrophins (HCG) Growth hormone-prolactin, relaxin	Same as above.	Same as above.	Same as above.
GASTROINTESTINAL TRACT			
Secretin and pancreozymin	Protein	Pancreas	Secretion of alkaline fluid and digestive enzymes
Cholecystokinin	Protein	Gallbladder	Contraction and emptying
Enterogastrone	Protein	Stomach	Inhibition of motility and secretion
Gastrin	Protein	Stomach	Secretion of acid

Adapted from Fundamentals of Clinical Chemistry (28) with permission.

258

Table 8-2. Animal Models in Endocrinology

Animal Model	Species	Human Counterpart	Reference
Accerlated Parathormone inactivation	mouse	Failure of secondary bone absorbtion	70
Acetonemia	cattle	Ketosis	91
Acromegaly	dog	Acromegaly	144
Addison's disease	dog	Addison's disease	62,129
Adenohypophyseal aplasia	cattle	Adenohypophyseal aplasia	86
Adenohypophyseal cysts	birds	Adenohypophyseal cysts	20
Adenomatosis	cattle	Adenomatosis	137,155
Adrenal amyloidosis	mouse	Adrenal insufficiency	69
Adrenal cortical hypertrophy	dog mouse	Hyperadrenocorticism	23,24 90
Adrenal induced polydypsia	mouse	Polydypsia	29
Antidiuretic hormone deficiency	mouse dog	Diabetes insipidus Diabetes insipidus	76 31,121
Cushing's disease	dog primates	Cushing's disease Cushing's disease	23,24,41 145
Diabetes insipidus	dog rat cat monkey	Diabetes insipidus Diabetes insipidus Diabetes insipidus Diabetes insipidus	15,31,102 136,152,153 50 50
Diabetes mellitus	dog Chinese hamster horse sand rat, mouse cat rat guinea pig primates	Diabetes mellitus Diabetes mellitus Diabetes mellitus Diabetes mellitus Diabetes mellitus Diabetes mellitus Diabetes mellitus Diabetes mellitus	58,126 114,124 148 17,33 78,107,166 53,54 11 105,109
Dwarfism	mouse	Human pygmy	87,127
Early senility	Syrian hamster	Senility	81

Table 8-2. (*Continued*)

Animal Model	Species	Human Counterpart	Reference
Familial adiposity	mouse	Obesity	12
Goiter	cattle	Goiter	80,110
	sheep	Goiter	79,163
Hyperinsulinism	dog	Hyperinsulinism	26
Hyperparathyroidism	dog	Hyperparathyroidism	92
Hypoparathyroidism	rat	Neonatal tetany	128
Hypertrophy of Island of Langerhan	mouse	Obesity	68
Hypophyseal Dwarfism	mouse	Dwarfism	141
Hypothyroidism	dog	Hypothyroidism	30,100,103,140
	rat		32
	chicken		46
Inherited insulin tolerance	rat	Insulin tolerance	27
Obese rats	rat	Hyperlipemia	175
Polyuria	Chinese hamster	Diabetes insipidus	82
Snell's Dwarf	mouse	Thyrotropin deficiency	164
		Dwarfism	141
		Growth hormone deficiency	8
	cattle	Growth hormone deficiency	2,5,63,85
Sleep related release of growth hormone	baboon	Growth hormone release	120,174
Spontaneous Hyperthyroidism	dog	Hyperthyroidism	92
Thyroiditis	Marmoset	Chronic thyroiditis	98
	rabbit	Chronic thyroiditis	169
Ultimobronchial neoplasm	bull	Medullary thyroid Carcinoma	22

pituitary have been named by the International Anatomic Nomenclature as follows:

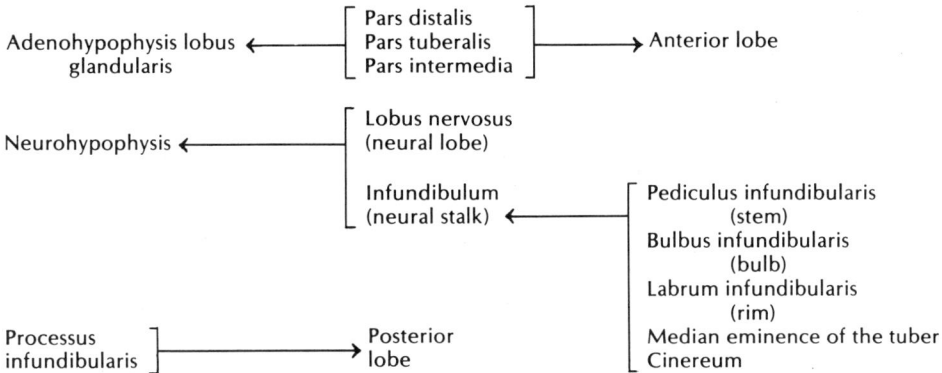

Adenohypophysis lobus ← ⎡ Pars distalis ⎤ → Anterior lobe
glandularis ⎢ Pars tuberalis ⎢
⎣ Pars intermedia ⎦

Neurohypophysis ← ⎡ Lobus nervosus
⎢ (neural lobe)

⎡ Infundibulum ⎡ Pediculus infundibularis
⎣ (neural stalk) ← (stem)
Bulbus infundibularis
(bulb)
Labrum infundibularis
(rim)
Processus ⎤ → Posterior Median eminence of the tuber
infundibularis ⎦ → lobe ⎣ Cinereum

Adenohypophysis

The adenohypophysis secretes at least six hormones: growth hormone (somatotropin); thyroid-stimulating hormone (TSH); luteinizing hormone (LH, interstitial cell-stimulating hormone); luteotropic hormone (LTH, prolactin); and adrenal cortical stimulating hormone (corticotropin, adrenal corticotropic hormone, ACTH). The functional cell types, staining characteristics, cell location, chemical nature, and hypothalamic controlling factors are summarized in Table 8-3.

Pituitary abnormalities may result from overproduction (e.g., from a tumor) or underproduction (e.g., due to hypophyseal destruction from a disease or surgical removal) of any or all of the adenohypophyseal hormones. Since the hypothalamus is so important to pituitary control, it is likely that isolated hormonal defects have their origin in individual loci of the hypothalamus.

Animal Models

The most dramatic pituitary dysfunctions are disorders of growth hormones. Pituitary dwarfism occurs when postnatal supplies of somatotropin are inadequate, since most such children had normal intrauterine development. Most of the earlier experimental studies were done using rats. In hypophysectomized rats, the administration of the growth hormone results in increased body proteins and decreased body fat content. In these experimental animals, the rate of conversion of amino acids into plasma and urinary urea is decreased since the entrance of amino acids into cells via the membrane is stimulated by growth hormones (GH). Hypophysectomy also causes a decrease in total hepatic RNA synthesis concomitant with a decrease in the number of ribosomes, mRNA and tRNA. The administration of GH in hypophysectomized animals improves organization of ribosomes and an increase in the amount of mRNA synthesis.

At the tissue level, GH affects bone, muscle, kidney, liver, and adipose tissue. In bone, the effects appear to be primarily on the cartilaginous portion as evidenced by increased mitotic activity at the epiphyseal line of growing bones and marked osteoblastic activity. The muscle mass and creatinine excretion increase in the presence of GH; the muscle also appears to resist the changes produced by insulin when it is under the influence of GH. The kidneys of hypophysectomized animals are small, with decreased glomerular filtration rate, diminished blood flow, and tubular secretion. The compensatory hypertrophy of the kidney following unilateral nephrectomy is possible evidence that GH is essential for adaptive changes in the organ. In adipose tissues, GH

Table 8-3. The Pituitary Cells and Their Hypothalamic Controlling Factors

Functional Cell Type	Hormone Secreted	Hypothalamic Controlling Factor
1. Somatotropic or STS cells	Growth hormone (HGH)	Somatotropin releasing factor (SRF)
2. Lactotropic or LTH cells	Prolactin	Prolactin inhibiting factor (PIF)
3. Thyrotropic or TSH cells	Thyrotropin (thyroid stimulating hormone--TSH)	Thyrotropin releasing hormone (TRH)
4. Gonadotropic or GTH cells	a. Follicular stimulating hormone (FSH) b. Luteotropic hormone (LH) or interstitial cell stimulating hormone (ICSH)	a. FSH releasing factor (FSH-RF) b. LH releasing factor (LH-RF) possibly identical to FSH-RF
5. Corticotropic or ACTH cells	Adrenal corticotropin hormone (ACTH)	Corticotropin releasing factor (CRF)
6. Melanotropic or MSH cells	Intermedin (melanocyte stimulating hormones--MSH) (a) alpha MSH (b) beta 1 MSH (c) beta 2 MSH	Melanotropin inhibiting factor (MIF)

Functional Cell Type	Granule Staining	Predominate Cell Location	Chemical Nature
1. Somatotropic or STS cells	Acidophilic	Lateral pars distalis	191 amino acid polypeptide, molecular weight 21,800
2. Lactotropic or LTH cells	Acidophilic		198 amino acid polypeptide
3. Thyrotropic or TSH cells	Basophilic	Pars distalis	Glycoprotein, molecular weight 28,000
4. Gonadotropic or GTH cells	Basophilic	Pars distalis	Glycoprotein, molecular weight approximately 30,000
5. Corticotropic or ACTH cells	Basophilic periodic acid-Schiff positive	Pars intermedia	A polypeptide containing 39 amino acids
6. Melanotropic or MSH cells		Pars intermedia	Polypeptides (a) alpha 13 amino acids (b) beta 32 amino acids

Three of the above hypothalamic regulating factors have been isolated: LH-RF, a decapeptide, and TRH and MIF, both tripeptides. The administration of TRH increases the pituitary elaboration of both TSH and prolactin; elaboration of the latter has been shown to occur even in a patient with isolated pituitary TSH deficiency.

Reproduced from Handbook of Endocrinology (43) with permission.

causes mobilization of fat with increased levels of nonesterified fatty acids (NEFA) in plasma. However, the level of free fatty acids does not increase if there is concomitant release of catecholamines (e.g., during sleep). The carbohydrate metabolism in muscle and liver is affected by GH as evidenced by hypoglycemia, decrease in glycogen levels, and abnormal sensitivity to insulin in hypophysectomized animals. These effects, except for abnormal sensitivity to insulin, can be counteracted by administration of adreno-cortical steroids, indicating that insulin sensitivity is under the control of GH. Insulin facilitates the entrance of glucose into the cells of muscle and adipose tissue; however, the effects of GH are somewhat transitory. The rate-limiting action in the utilization of glucose is phosphorylation, which is decreased by GH and adrenocorticosteroids.

The GH from nonprimates is ineffective in primates (88). In humans, administration of human growth hormone (hGH), 1–10 mg/day, produces major metabolic changes in hypo- and normal pituitary individuals. The changes include an increase in plasma levels of nitrogen, sodium, potassium, inorganic phosphorus, and magnesium, and the degree of the change is dose dependent (77). The diabetogenic effect of hGH has been demonstrated in normal and hypophysectomized subjects, and it is most pronounced in hypophysectomized diabetic patients. In other cases, the effect is not so pronounced because it is masked by the normal pancreatic reserves. The GH level in well-rested individuals is very low (3 ng/ml), however, it increases with exercise, ingestion of meal, and prolonged fasting. The secretion of GH shows a diurnal pattern with a sleep-related peak of about 40 ng/ml followed by gradual decrease and consistent levels during the day except for spikes after meals (147). The hormone has a half-life of 20 to 25 minutes, and it is rapidly metabolized and cleared from the plasma.

The secretion of GH is related to insulin as an important defense mechanism against hypoglycemia. Oral or intravenous administration of amino acids, particularly arginine, greatly stimulates GH production which is preceded by an increase in insulin levels.

Animal Model of Human Sleep-Related Release of Growth Hormone. Parker et al. (120) and Zir et al. (174) reported that an animal model of human sleep-related-release of GH is useful in studies on the anabolic effects of the hormone. Release of GH in the sleep of two adolescent male baboons was reported by Parker (119). Sleep was videomonitored for behavior, recorded polygraphically, and scored by human criteria. GH was measured by a human RIA since there is immunologic cross-reactivity of the baboon GH with human GH. The GH release was sleep related in reproducible patterns, and sleeping GH responses in fasting and during β-androgenic blockage were similar to results reported in humans. The results suggest that the baboon may be a useful model for the study of GH release in sleep.

Animal Model of the Human Pygmy. King (87) described a dwarfing gene in the house mouse for which he proposed the name pygmy (pg). This mutation results in proportionate dwarfism when present in the homozygous state, reducing adult body weight to approximately one-half that of normal litter mates. This occurs regardless of whether or not the gene is segregating (in large or small sized strain of mice). The mutant is not completely recessive, however, since +/pg heterozygotes are slightly smaller than their ±/+ litter mates. Any gross skeletal abnormalities or any defects in endocrine gland morphology (which could account for their small size) were not detected (in the pg/pg mice). Using pituitary gland implantation, King (87) demonstrated that their small size did not result from growth hormone (GH) deficiency; pg/pg mice did not grow following the implantation of normal rat pituitaries, a procedure that results in marked acceleration of growth of dw/dw hypopituitary mice. Rimoin and Richmond (127) reported that the growth rate of pg/pg mice was unresponsive to treatment with porcine growth hormone, whereas this same hormonal preparation resulted in marked growth acceleration in hypopituitary mice. These observations suggest that the pygmy mutant of the mouse is associated with a peripheral subresponsiveness to growth hormone, and thus may represent a model of the human African pygmy. Animal models of hypophyseal dwarfism in mice as a model of human dwarfism have been described by Snell (141). The growth hormone deficiency resulted in the Snell's dwarf mouse (8).

Neurohypophysis

The posterior lobe of the pituitary is the distal end of the neurosecretory system, including supraoptic and paraventricular nuclei of the hypothalamus and paraventricular tract. The hormones are synthesized in the ganglion cells of the hypothalamic nuclei and are transported down the axons of the neurohypophyseal tract to terminations in the median eminence, pituitary, and pars nervosa. Its release is influenced by impulses reaching the neurohypophyseal tract.

Two hypothalamic hormones, vasopressin and oxytocin, are present in an electrostatic complex form with protein neurophysin in the neurohypophyseal tract and posterior pituitary (44). The hormones are octapeptides with a five member S-bond ring and a tail of three amino acids.

Vasopressin

The major physiological role of this hormone relates to its antidiuretic activity. The release of the hormone is under the fine control of osmotic pressure of the blood in the carotid. The dilution of plasma or decrease of osmolarity suppresses the release of vasopressin, so there is a negative feedback control mechanism involving kidney and nervous system. However, the water loss also takes place through skin and lungs. In humans, there is a mechanism that alerts the individual for the need for water—thirst. Pioneering work has been done using the goat as an animal model. Andersson (4) reported that the antidiuretic and thirst centers were closely related in the hypothalamus and less than 0.01 ml of intrahypothalamus injection of 2 percent NaCl caused the goats to drink large quantities of water. Similar effect can be produced at will by electric stimulation of intrahypothalamic electrodes. Later the thirst, antidiuresis center was associated with milk secretion. Stimulation of the anterior part of the thirst center inhibited water diuresis and milk ejection simultaneously. In addition to neural controls, the hormone release is inhibited by alcohol and stimulated by the presence of nicotine, morphine, barbiturates, ferritin, and bradykinin. The plasma volume is another potent controlling factor for vasopressin. Hemorrhage almost immediately stimulates vasopressin release.

The antidiuretic action of vasopressin probably involves a countercurrent mechanism of urinary concentration. Using amphibian tissues and urinary bladders as models (135), it has been suggested that vasopressin acts by making the responsive epithelium more permeable to water and thus permitting transfer of large quantities of water with little solute. This action to accommodate this bulk movement of water is primarily at the luminal surface of the tubular cells. The nature and mechanism of changes in the membrane are not known. But cAMP mimics the action of vasopressin, and its presence increases adenylcyclase activity in toad bladder membranes (118).

Vasopressin has been shown to possess additional effects, such as release of corticotropin, adrenal steroidgenesis, release of radioactive iodine from thyroid, and contraction of smooth muscles. Its effect on blood pressure is varied. In the anesthetized animal, there is a rise in blood pressure with vasoconstriction in the peripheral and visceral blood vessels.

Vasopressin disappears quickly (varying with species); in the rat, it takes 1 minute; in humans, 10 to 20 minutes; and about 7 minutes in the dog. In rats, the liver and kidneys are considered the chief inactivating organs (71); however, in dogs, kidneys contribute 25 percent and liver 12 percent in the inactivation of vasopressin hormones (45).

Oxytocin

This hormone has a primary effect on the contraction of uterine musculature and lactation. The two hormones, oxytocin and vasopressin, share common properties. Vasopressin has the antidiuretic and vasopressor effects but also shares some oxytocic and milk ejection properties of oxytocin. Similarly oxytocin has some properties of vasopressin. The exact mechanism by which oxytocin acts on uterine musculature and myoepithelium surrounding mammary ducts is not known. Estrogens enhance the responsiveness to oxytocin. The sensitivity of the human uterus increases in late pregnancy, and at term, it is much more sensitive to oxytocin than to vasopressin. The nonpregnant uterus is much more sensitive to vasopressin.

Diseases of the Neurohypophysis

The neurohypophysis secretes only two hormones, and it appears that no diseases are associated with oxytocin. However, a deficiency disease called diabetes insipidus (DI) involves vasopressin. Clinically, DI is characterized by polyuria and polydipsia. The volume of urine ranges from 5–10/24 hr with a specific gravity of 1.004 to 1.005. Generally, no serious consequences of the disease are seen because patients have to drink more water. However, if the thirst center is also destroyed, clinical signs of dehydration and mental disturbances may be present.

The known etiology of the disease may be divided into traumatic, hereditary, inflammatory, and degenerative. However, a large number of cases appear idiopathic. In trauma, the common site of the lesion is in the stalk and may or may not be associated with fracture of the skull. Inflammation of the base of the brain may cause DI, and it generally results from chronic diseases like tuberculosis, meningitis, or syphilis. Other conditions, such as pertussis, sarcoidosis, and Hodgkin's disease, may cause DI. The disease may also follow surgical removal of pituitary tumors.

Animal Models of Diabetes Insipidus. Several animal models have been used to study the disease process. It has been shown that rats with familial DI lack arginine vasopressin and neurosecretory material in their hypothalamus and posterior pituitary (134). Sachs (133) has shown that labeled tyrosine and cystine can be incorporated in vitro by the hypothalamus and the rate increased under conditions of dehydration. Fisher et al. (50) showed that diabetes insipidus develops only by the complete degeneration or removal of neurohypophysis in cats and monkeys. Cats undergo a triphasic response to surgical removal: (a) immediate polyuria and polydipsia lasting for 4 to 5 days; (b) a period of intense antidiuresis for about 6 days; and (c) permanent polyuria and polydipsia. In humans, a similar pattern is observed following high section of the pituitary stalk (108).

The Brattleboro strain of rats with hereditary hypothalamic diabetes insipidus (DI) first described by Valtin and Schroeder (153) almost certainly do not secrete vasopressin although they do secrete oxytocin.

Sensitivity to antidiuretic action of vasopressin appears inversely related to the level of hydration in normal rats. The DI rats under ethanol anesthesia appeared twice as responsive to the acute antidiuretic action of intravenous arginine vasopressin as normal rats. The DI rats did not show typical prolonged oliguric responses to intravenous doses of nicotine that were effective in normal rats. Sawyer and Valtin (136) proposed the DI rat as an experimental animal when it is advantageous to avoid interference from endogenous vasopressin release.

The DI rats have normal antidiuretic sensitivity to oxytocin relative to their sensitivity to vasopressin; that is, they are also twice as sensitive to oxytocin as are normal rats. The authors could not rule out the possibility that oxytocin released in response to administered nicotine contributed to the production of antidiuresis.

Valtin and Harrington (151) suggested that the progressive increase in urinary concentration seen in DI rats with prolonged treatment with vasopressin may be due to a progressive rise in osmolal concentration of the renal medulla and papilla.

As an assay animal, the DI rat poses a number of problems: (a) they are not appreciably more consistent in their sensitivities; (b) hypertrophy of the urinary bladder makes it difficult to implant a catheter that will drain freely throughout the entire day; (c) dose response discrimination is parallel to normal rats. The DI rat may have its greatest use when very small concentrations of vasopressin must be measured, as suggested by Vierling et al. (156), or when one wishes to distinguish between antidiuresis resulting from a direct renal action and that resulting from neurohypophyseal release of vasopressin. This use is offset because DI rats become oliguric in response to anoxia, hypotension, or dehydration just as do normals. Moses and Miller (111) characterized the vasopressin synthesis and release abnormalities by studying the heterozygous and normal rats during dehydration and consequent rehydration. Heterozygous (for hypothalamic diabetes insipidus) and normal rats (both Brattleboro strain) were deprived of water for 4 days and rehydrated for 5 days. The heterozygous rats had 57 percent as rapid pituitary vasopressin depletion as the normals, even though the osmotic stimulus to vasopressin release was greater. The heterozygotes excreted a greater volume of more dilute urine throughout dehydration. Pituitary vasopressin reaccumulation in the heterozygotes during rehydration was only 38 percent of the normals' rate. At no time period was there dissociation between biologically and immunologically measurable pituitary vasopressin in the heterozygotes. The Brattleboro strain defect is indicated to be one involving both synthesis and release of vasopressin, while only the synthesis defect is apparent in the homozygous rat. The heterozygous rat partially expresses the abnormality and demonstrates the abnormal release mechanism in addition to the defective synthesis.

Lithium-induced diabetes insipidus was investigated in ninety-six patients and in a rat model by Forrest et al. (55). Polydipsia was reported by 40 percent and polyuria by 17 percent of the patients. Maximum concentrating ability after dehydration and vasopressin was significantly impaired in ten polyuric patients and was reduced in seven out of ten nonpolyuric patients studied before and after lithium therapy. Severe polyuria was not affected by vasopressin and chlorpropamide, but responded to chlorothiazide. Rats on lithium developed massive polyuria resistant to vasopressin as compared to rats with polyuria from drinking glucose. Renal tissue analysis in rats with lithium polyuria showed a progressive increase in the concentration of lithium from cortex to papilla. The normal gradient for sodium was not reduced by lithium treatment. Vasopressin given intravenously had no effect on polyuria nor did dibutyl cyclic AMP, which reverses water diuresis in normal and hypothalamic diabetes insipidus rats of the Brattleboro strain. These studies show that nephrogenic diabetes insipidus may be a common finding after lithium treatment and partly results from interference with the mediation of vasopressin at the same step in the formation of 3′, 5′ cyclic AMP.

ADRENAL GLAND

The adrenal gland, like the pituitary, consists of two functionally and embryologically distinct portions. The cortex is of mesodermal origin and secretes hormones essential to maintain life. The medulla derives from the ectoderm, and its secretions, though important for normal existence, are not vital. The cortical hormones have as their basic structure the cyclopenanoperhydrophenanthrene ring, a 17-carbon structure. The outer and middle zones of the adrenal cortex produce hormones with 21 carbon atoms. Those with a hydroxyl group at position 17 are 17-hydroxycorticosteroids or glucocorticoids, since one of their significant effects is on glucose metabolism. Mineralocorticoids, 21-carbon corticoids without the OH-group at C-17, principally affect mineral metabolism. The inner zone of the cortex produces steroids with nineteen carbon atoms and a detone group at C-17. These have androgenic activity and are produced in both males and females. All adrenal cortical hormones derive from cholesterol and acetate through a multitude of enzymatic reactions. The major regulator of adrenocortical growth and secretory functions is ACTH (anterior pituitary). This hormone attaches to the adrenal cortical cells and stimulates the production of cAMP (via adenyl cyclase), which is a major cofactor for the proteinkinases involved in the synthesis of hormones of this organ.

Biological Effects of Glucocorticoids

These steroids (cortisol, cortisone, and compound A) promote gluconeogensis, liver glycogen deposition, and elevation of glucose concentration (10). In addition, glucocorticoids increase protein breakdown and inhibit amino acid uptake by extrahepatic tissue but accelerate their uptake by liver. In chronic cases, they may affect the production of GH and inhibit somatic cell growth. Glucocorticoids suppress ACTH activity through corticotropin-releasing factor (CRF), and in large quantities they may inhibit inflammatory and allergic reactions. Other activities include stabilization of lysosomes, diapedsis of leukocytes across capillary walls, stimulation of hematopoiesis, promotion of fat deposition, promotion of uric acid excretion, promotion of appetite, reduction of circulating eosinophils, and maintenance of muscular work activity.

Adrenal Disorders

Hypercortisolism (Cushing's Syndrome)

The disease occurs in both sexes, but women in the age group 20 to 60 are most commonly affected. There is an increase in weight, primarily in adipose tissue. Protein wasting is manifested by a tendency to bruise easily in the beginning and eventually by pink or purple striae. The tissues have poor tensile strength, tear off easily; wounds heal slowly and are infected easily. General osteoporosis results from attrition of bone matrix.

Hypercalcuria with a value of 150–300 mg/day and renal stones are fairly common. Growth is arrested and short stature is inevitable. The patients show impaired glucose tolerance and high blood pressure associated with ventricular hypertrophy. The classical symptoms also include heightened color of face and neck, downy hirsutism, oligomenorrhea, mild erythrocytosis, lymphopenia, and eosinopenia.

Cushing's Disease

This condition differs from Cushing's syndrome in that the hypercortisolism is secondary to inappropriate secretion of ACTH. However, the clinical features of the two conditions are the same. The patients have relatively high plasma values of ACTH and urinary cortisol and 17-OHCS. The adrenal secretory activity is suppressed by dexamethasone and then has a vigorous response to metyrapone.

Animal Models of Cushings Disease. A primate model for the experimental production of Cushing's disease was reported by Strasberg et al. (145). Stimulation of the amygdala of the cerebellum of monkeys was achieved by a Grass-58 stimulator, two stimulus isolation units, and a constant current system. Mason (104) has shown that the amygdala stimulation in monkeys leads to increased 17-hydroxycorticosteroids exertion. It has been known that the median basal hypothalamus is the final common pathway through which the central nervous system influences pituitary adrenocorticotropic hormone (ACTH) release and subsequently the peripheral glucocorticoid level. Electrical stimulation of hypothalamus results in elevated concentrations of 17-OHCS together with a variety of autonomic responses. This method may provide a useful animal model for the study of Cushing's disease in humans.

Hyperaldosteronism

Aldosterone normally maintains fluid and electrolyte homeostasis but when produced in excess can cause potassium depletion and expansion of extracellular fluids, which may eventually result in edema or hypertension.

Primary Aldosteronism (Conn's Syndrome)

Excessive production of aldosterone from cortical adenomas results in this disease (34, 35). The usual symptoms of mild hypertension and impaired renal concentrating ability leads to polyuria, nocturia, and polydipsia. Hypokalemia may cause paresthesia, muscle weakness, and even paralysis.

Primary Adrenal Insufficiency

Irrespective of the cause, the main symptoms are due to insufficiencies of (1) aldosterone or (2) cortisol. The loss of adrenal androgen in women results in diminished growth of axillary hair. A combination of aldosterone and cortisol deficiency results in characteristic Addison's disease. Lack of aldosterone results in the inability to conserve sodium, thereby causing a decrease in the extracellular fluid volume, weight loss, hypovolemia, and hypotension; decrease in cardiac size and renal blood flow; prerenal azotemia, increase in renin production, decrease in pressor response to catecholamines, weakness, postural syncope, and shock. Lack of aldosterone also produces hyperkalemia, mild acidosis, and cardiac asystole. Salt intake may increase; however, the patient on a moderate or low salt diet could develop serious problems if anorexia, vomiting, diarrhea, or excessive sweating occur. Cortisol deficiency results in a much wider range of effects. The general symptoms include anorexia, abdominal pain, wasting of fat deposits, apathy, weakness, and diminished ability to excrete water. Other systems may also be effected.

- *Pituitary:* Unrestrained secretion of ACTH and MSH resulting in mucocutaneous hyperpigmentation.

- *Energy metabolism:* Impaired gluconeogenesis, impaired fat mobilization and utilization, liver glycogen depletion, fasting hypoglycemia.

- *Cardiovascular-renal:* Impaired ability to secrete free water, impaired pressor responses to catecholamines, and hypotension.

- *Mental:* Diminished vigor, lethargy, apathy, confusion, and psychosis.

Addison's Disease

This disease was described as early as 1855. The patient with Addison's disease has general poor health, is languid, indisposed to physical and mental exertion, and has an impaired appetite. The body wastes gradually, and the patient complains of pain in epigastrium with occasional vomiting. Areas of skin with a smoky appearance having various shades of deep amber or deep chestnut brown appear on the body. The main cause of Addison's disease has been tuberculosis. However, as the incidence of tuberculosis is decreasing, it appears that idiopathic atrophy is more common and it may be due to an autoimmune destruction of the adrenal cortex (13). The patients with this disease have a greater risk of developing other autoimmune diseases, such as hypothyroidism, primary ovarian failure, pernicious anemia, or hypoparathyroidism. Conversely, patients with these diseases tend to develop Addison's disease. Rare causes of this disease include amyloidosis, adrenal apoplexy, Waterhouse-Friderichsen syndrome with characteristic hemorrhage in the skin, and metastatic carcinoma. Surgical removal of adrenals and prolonged treatment with o, p'DDD, or heparinoids can result in Addison's disease (43).

Secondary Adrenal Insufficiency

This condition occurs as a result of ACTH deficiency and has the general symptoms of cortisol deficiency except for the cutaneous hyperpigmentation. Plasma cortisol concentration and urinary 17-OHCS and 17-ketosteroids are subnormal and rise only sluggishly in response to a test dose of ACTH (in comparison to the primary form when the values do not rise at all).

No suitable animal model for the above diseases have been reported except adrenal cortical hypertrophy in the dog (62, 129) and in the mouse (7) to study hyperadrenocorticism in humans.

Adrenal Medulla

The importance of the adrenal medulla lies in the fact that it synthesizes and stores two catecholamines, epinephrine and norepinephrine, both mediating the "fight or flight" reflexes. Epinephrine is the major medullary hormone. Of total secreted and stored hormone, approximately 80 percent is epinephrine. Norepinephrine, however, predominates in the urine, much of it deriving from postganglionic synapses of the autonomic nervous system. Catecholamine synthesis starts with the amino acid phenylalanine, which is oxidized to tyrosine. Subsequent oxidative and other reactions produce norepinephrine to which a methyl group is added in epinephrine synthesis. For both substances, the degradative end products are acids usually studied in terms of 3-methoxy-4-hydroxy mandelic acid, also called vanilmandelic acid (VMA). The complexities of physiological and pharmacological manipulations which influence the synthesis of catecholamine have been extensively reviewed by Austin (7); Gewirtz and Kopin (59); Mueller et al. (112); Roth et al. (131); Thonen and Mueller (149).

The two hormones have disparate cardiovascular effects: norepinephrine constricts vascular smooth muscle, slows the heart rate, and raises both systolic and diastolic blood pressure, while epinephrine increases heart rate and cardiac output, and produces only systolic hypertension. Both increase the metabolic rate, raise the blood sugar, and cause elevated plasma free fatty acid levels. Despite their widespread and dramatic physiological effects, the medullary catecholamines are not essential for life.

Diseases Associated with Abnormalities of Adrenal Medulla

No known diseases are associated with the insufficiency of the adrenal medulla. After bilateral adrenalectomy, urinary excretion of epinephrine falls rapidly but the excretion of norepinephrine remains the same (154). However, tumors at this site or in the extramedullary sites produce clinically significant syndromes. The most important tumor is pheochromocytoma (97, 143) and is found in about 1 percent of the hypertensive population. The generalized symptoms of nervousness, weight loss, palpitations, headache, and paroxysmal sweating may first bring the patient to medical attention. A suitable animal model of pheochromocytoma is not available at this time.

THYROID GLAND

Thyroid, the first endocrine gland to develop, is recognizable at about 1 month after conception. Its development begins as an invagination of the pharynx at the base of the tongue and at times may persist in the area as accessory thyroid tissue. As the fetus develops, this gland starts producing thyroxin at 15 weeks of age. In the adult, the encapsulated gland (10–20 g) is held around the anterior and lateral aspects of the trachea by loose connective tissue. The two lobes are connected by an isthmus which lies just below the cricoid cartilage. The unique function of the thyroid is to metabolize and incorporate iodine into a variety of organic compounds.

Hormone Chemistry and Biosynthesis

The steps involved in the biosynthesis of the thyroid hormone include (a) iodine trapping, (b) oxidation to iodine via a peroxidase, (c) iodination of tyrosine to monoiodotyrosine (MIT), (d) iodination of MIT to di-iodoform (DIT), and (e) condensation of two DIT to form thyroxin. It is possible that tri-iodothyroxine (T_3 or TRIT) may be produced from M, MIT, and DIT. Thyroid is most efficient in withdrawing and concentrating circulating iodine from plasma. The thyroid is able to concentrate iodine thirty to forty times that of plasma in the "thyroid-iodine trap." This trapping function is controlled by thyrotropin and blocked by perchlorate or thiocyanate.

The synthesis of thyroid hormones takes place in the follicles, and all the compounds described above are found as complexes with a special storage protein called thyroglobulin. The protein contains 10 percent carbohydrate and has a molecular weight of 660,000. The 195 molecules consist of four chains. Individual chains exist in monomeric (6–7S) or dimeric (12S) form. However, iodoproteins of higher sedimentation constants of 27S and 32S are also found. There are 120 tyrosyl residues of which only a few are naturally iodinated. Before thyroxin is secreted into the blood, the globulin-thyroxin complex is split by proteolytic and mucolytic enzymes. These steps are stimulated by thyrotropin. Thyroid hormone, T_4, is associated with inter-α-globulin (TBG) and T_4-binding prealbumin (TBPA), and to a lesser extent with albumin. T_3 on the other hand is mainly bound to TBG. In normal subjects, TBG is one-third saturated with thyroxin, and the extent of binding is an indication of thyroid function. TBPA has a higher binding capacity, but the binding is not as avid as TBG. The thyroxine bound to these proteins is not available to tissues, and the release of thyroxine is exceedingly slow; the free thyroxine is only about 0.06 percent of the total serum thyroxine. Thyroid secretes 90 μg of thyroxine, 50 μg of triiodothyroxine, one-third of which comes from peripheral conversion of T_4 to T_3.

Actions

The two thyroid hormones, T_3 and T_4, stimulate calorigenesis in eleven species, metamorphosis in amphibians, and growth in humans. The hormones also potentiate epinephrine and lower serum cholesterol. Thyroxin is necessary for the development of the central nervous system; for full brain development of the mammalian embryo, normal amounts of thyroxine must be present. Thyroxin promotes protein synthesis and increase in nitrogen retention, but at high thyroxin levels, protein breakdown is accelerated and there is a negative nitrogen balance. Excessive amounts of thyroxin impair the metabolism of creatine and phosphocreatinine and eventually leads to muscular weakness and increased O_2 requirements. The mycocardial contraction force and heart rate are also increased.

Thyroxin reduces organism sensitivity to insulin, and hyperthyroidism may aggravate or precipitate diabetes mellitus. Thyroid hormones affect energy metabolism, and at the cellular level, there is an increase in size (swelling) and number of mitochondria. Recently, it has been shown to increase the activity of membrane-bound sodium- and potassium-dependent ATPase.

Thyroid Disorders

Various clinical manifestations resulting from dysfunctions of the thyroid gland are basically due to hyper- or hypothyroid function. A number of thyroid function tests are available to diagnose the clinical cases.

Hyperthyroidism

Thyrotoxicosis is the term used to describe the complex biochemical and physiological changes in the presence of excessive amounts of thyroid hormones. The excess amount may result from overproduction by diffuse hyperplasia, from autonomously functioning adenomas, or from hormone production by ectopic thyroid tissue.

Animal Model of Hyperthyroidism. Sit and Kanagasuntheram (139) developed an experimental model for the study of hyperthyroidism in embryonic development in the local anuran larvae (*Bufo melanostictus*) by treatment with potassium perchlorate at a critical dose at the time of embryogenesis. Phocomely-hemimely and digital deformities of fore- and hind limbs were found associated with this treatment and also were produced by exogenous treatment with L-thyroxin sodium. Histological examination indicated incoordinate cellular differentiation in the malformed limbs. There was undifferentiated mesenchyme in the distal extremities although the overlying ectoderm had already formed mature epidermis with specialized skin glands. Deformities were aggravated if tadpoles had antithyroid treatment before critical stimulation with thyroxin in the sensitive phase of development.

In humans, various embryonic abnormalities are related to hypothyroidism (83) and gross digital and limb deformities have hitherto not been directly correlated with embryonic hyperthyroidism. Recently, however, a Negro family with familial hyperthyroidism was reported to be affected with digital clubbing and shortening acropathy (167).

Graves' Disease (Basedow's Disease)

This is a multisystem disease characterized by goiter, thyrotoxicosis, infiltrative ophthalmopathy, and infiltrative dermopathy. The symptoms may occur singly or in varying combinations and the full syndrome may never develop. The serum characteristically contains 7S immunoglobulin G, the long-acting thyroid stimulator (LATS), a material seen in significant amounts in Graves' disease.

Graves' disease generally is seen in patients in the third or fourth decades and is much more common in females than males, with a ratio of 7:1. These patients generally have lymphoid hyperplasia, lymphocytic infiltration of thyroid, and retroorbital tissues and circulating thyroid autoantibodies. In fact, it appears that Graves' disease may have an autoimmune mechanism with genetic elements as evidenced by the occurrence of the disease in members of the same family and in several generations of the same family.

The role of TSH and pituitary in Graves' disease is questionable as LATS is capable of producing hyperplasia of thyroid in hypophysectomized animals. The biological activity of LATS can be prevented by antibodies directed against this class of proteins (7S IgG). Recent evidence indicates that a LATS-protector may have a direct stimulatory effect of thyroid tissue and that the cell-mediated immunity may be much more important than humoral antibodies (1, 95). The role of TSH (β subunit) in producing ophthalmopathy has been suggested; however, the β-subunit cannot be detected by radioimmunoassay in thyrotoxic ophthalmopathic patients (89). The exact mechanism of production of Graves' disease is not clear. Clinical symptoms of Graves' disease are similar to hyperthyroidism, but the mechanism of origin of hyperactivity is different.

Hypothyroidism

Inadequate secretion of thyroid may develop at any time in life, including in utero. The inadequate production may be due to a disorder of the thyroid or to anterior pituitary insufficiency. The clinical picture depends on the degree of hormone deficiency rather than the cause of deficiency. Several organs and systems may be involved. The general symptoms include deficient growth, low BMR, mental retardation, and mental and physical sluggishness. The affected individuals are sensitive to cold, somnolent, and generally show hypercholesterolemia and myxedema.

Animal Model of Hypothyroidism. Cole (32) reported that by selective breeding, the frequency of hypothyroidism in a strain of white Leghorn fowls was increased to a level exceeding 80 percent in females and 75 percent in males. The trait is recognized at 6 to 8 weeks of age. It is characterized by obesity, silky and elongated feathers, a subsequent marked reduction in rate of growth, and often absent or delayed sexual maturity. The pathological changes in the thyroid gland consist of infiltration of lymphoid cells and proliferation of cords of epithelial-like cells and are recognizable at 2 weeks of age. They are progressive, although some recovery in both the structure and function of the gland may occur later.

Inheritance of hypothyroidism is polygenic with some of the causative genes dominant. After 6 generations of selection, there may be a considerable variation in the expression of the syndrome.

Supplemental thyroid-stimulating hormone does not maintain the integrity of the thyroid gland. Supplemental thyroxine, administered in the diet as iodinated casein (Protamine), does compensate for the low level or absence of endogenous thyroxine.

Thyroxine-supplemented birds lay eggs of good size and reproduce well. This may be a useful model for the study of hypothyroidism in humans.

Simple or Nontoxic Goiter (Diffuse and Multinodular)

This condition occurs sporadically in nonendemic areas. The thyroid enlargement is not associated with thyrotoxicosis or hypothyroidism and does not result from an inflammatory or neoplastic process.

Thyroid Neoplasms

Thyroid nodules can be seen in as many as 4 percent of the population on physical examination; probably its incidence is even higher. Adenomas (benign tumors) of thyroid glands are encapsulated with few mitoses. Several types of neoplasms can be distinguished on histopathological examinations.

1. Embryonal (trabecular) adenoma: histological by the tissue; resembles embryonic thyroid prior to the development of follicles;

2. Fetal adenoma;

3. Microfollicular adenoma;

4. Macrofollicular adenoma;

5. Papilliary cystadenoma;

6. H urthle cell adenoma (large, pale, acidophilic cells that are arranged in a trabecular pattern).

Malignant Neoplasm. It is usually of follicular epithelium origin or it may arise from parafollicular (C-cell) elements. There are three categories of carcinoma of follicular epithelium: (1) papilliary, (2) follicular, and (3) anaplastic. The most common types of carcinoma of parafollicular origin are medullary carcinoma and solid carcinoma with amyloid stroma.

Animal Model of Medullary Thyroid Carcinoma. A high incidence of naturally occurring thyroid neoplasm has been reported in old rats (99). The neoplasm appears to be of parafollicular origin similar to human medullary thyroid carcinoma. Boorman et al. (16) reported necropsy findings of 334 older WAG/Rij rats (84 percent were older than 2 years). Medullary thyroid carcinoma accounted for 123 neoplasms and one follicular carcinoma was found. Multiple endocrine neoplasia was common, but no increased association with thyroid neoplasia could be found. Many of the neoplastic cells showed a considerable similarity in their fine structure to normal parafollicular cells. Amyloid deposits, a characteristic of medullary cell carcinoma of the thyroid in humans, was present. The high spontaneous incidence, comparable ultrastructure, and similar biological behavior seem to make the rat an ideal animal model for the study of human medullary thyroid carcinoma.

A syndrome of ultimobranchial thyroid neoplasms in aged bulls, which shares many similarities with human medullary thyroid carcinoma and multiple endocrine neoplasia (Sipple's syndrome) was described by Capen (22). Jubb and McEntee (84) reported that approximately 30 percent of the bulls had neoplasms and an additional 15 to 20 percent had hyperplasia of ultimobranchial derivatives. These growth disturbances have been found only in the thyroid glands of bulls, not cows. The most characteristic cell type in ultimobranchial neoplasms had large aggregations of concentric or interwoven microfilaments which often partially indented the nucleus. Secretion granules were membrane limited, composed of fine dense particles, and appeared similar to those in normal parafollicular cells of control bulls. Ultrabranchial tumors were firm, and large aggregations of fine amyloid fibrils often were observed between bundles of collagen fibers. Bioassay of ultimobranchial adenomas and carcinomas demonstrated the presence of calcitonin activity (466 $\pm$ 84 and 409 $\pm$ 93 MRC mU/g respectively). Parathyroid glands from bulls with ultimobranchial neoplasms had ultrastructural evidence of secretory inactivity and atrophy of chief cells. Cytoplasmic organelles were poorly developed, and secretion granules were infrequent. The large cytoplasmic area had numerous lipofusion granules. Parathyroid hyperplasia and adenomas, reported in patients with familial medullary thyroid carcinoma, have not been observed in bulls. Prominent aggregations of amyloid fibrils occasionally were observed around the inactive chief cells.

Medullary thyroid carcinoma in humans and ultimobranchial neoplasms of bulls both appear to be derived from parafollicular cells, have an amyloid stroma, contain calcium which can be released by calcium infusion, and are associated with a syndrome of multiple endocrine neoplasia. The cells comprising medullary thyroid carcinoma are more differentiated and have well-developed organelles and numerous secretion granules.

Since the animal model shares many important characteristics with the human disease, it offers a unique opportunity to investigate the long-term effects of excessive calcitonin secretion on calcium homeostatic mechanisms and bone metabolism. Investigations with this animal model may help to further define the physiological role of calcitonin and its interaction with parathyroid hormone in the pathogenesis of certain metabolic diseases.

Thyroiditis

Hashimoto's disease (lymphocytic thyroiditis, struma lymphomatosa), first described by Hashimoto (66), is an exaggeration of normal lobular pattern. Intrafollicular infiltration by lymphocytes and plasma cells, as well as granular, oxyphilic changes in the cytoplasm of follicular epithelium, is a common feature of thyroiditis (95).

Riedels Thyroditis. This is a rare condition characterized by extensive fibrosis of the thyroid gland and adjacent structures. It may be associated with retroperitonial fibrosis (150).

Acute Pyogenic Thyroiditis. This is due to infection by pyogenic organisms and characterized by severe pain and tenderness in the affected area.

Subacute Thyroditis (Granulomatous, Giant Cell, or de Quervain's Thyroiditis). This may have a viral etiology since it often follows a respiratory infection.

Animal Models of Thyroiditis. Human chronic thyroiditis is characterized histologically by varying degrees of lymphocytic infiltration of the thyroid glands. Thyroiditis also occurs spontaneously in the dog (113) and in the obese strain of the white Leghorn chicken (168). Autoimmune chronic thyroiditis has been induced in guinea pigs, dogs, rats, mice, goats, rabbits, and rhesus monkeys by injections of autologous or homologous thyroid extract. Levy et al. (98) reported a chronic thyroiditis in marmosets which is histologically similar to the natural and experimental forms of the disease. In these small South American primates the disease seemed to be species related and may be associated with estrus; of 494 animals examined 40 had chronic thyroiditis. Approximately 60 percent of the colony-born females, 28 percent of the colony-born males, 12 percent of the wild caught females, and 9 percent of the wild caught males of the genus had chronic thyroiditis. The high incidence of thyroiditis in marmosets of the genus *Callithrix* provides a new primate model for the study of human chronic thyroiditis.

Witebsky and Rose (169) induced thyroiditis in rabbits by an immunizing procedure. The induced allergic thyroiditis has been duplicated in mice. It is now accepted that the induced thyroiditis depends on the presence of the thymus, especially thymus-derived lymphoid cells or T-cells. Neonatal thymectomy in chickens and rats made them less responsive to the induction of thyroiditis.

However, the spontaneous disease has been reported in dogs, chickens, and rats. Here, neonatal thymectomy increased the severity of the spontaneous thyroiditis in chickens (165), while neonatal bursectomy decreased the frequency and severity of the disease. The bursa-dependent lymphocytes are equivalent B-cells in mammals. Since these cells play a major role in thyroiditis in birds, there may be pathologically different processes of the diseases in different species.

In mice, Nishizuka et al. (115) found that neonatal thymectomy induces ovarian dysgenesis in long-term studies and that these dysgenetic ovaries are often infiltrated by lymphoid cells. Administration of human chorionic gonadotropin produces lymphocytic oophoritis in predysgenetic ovaries of these mice. Neonatal thymectomy alone induces lymphocytic thyroditis in some hybrid mice. Different strains of mice show varying sensitivities to development of the disease.

Another study involved the subcutaneous grafting of a neonatal whole thymus or an intraperitoneal injection of 7-day thymus cells into 7-day-old neonatally thymetomized mice. This treatment prevented infertility with ovarian dysgenesis and lymphocytic thyroiditis. The assumption is made that the disease may depend on the sustained imbalance between T- and B-cells and that an intact thymus or enough T-cells would prevent the disease.

PARATHYROID GLANDS

The parathyroid glands, which may vary from two to ten, produce only one hormone. Parathyroid hormone or parathormone (PHT) regulates calcium and phosphorus metabolism. Through its effect on plasma calcium concentration, parathormone influences neuromuscular function in all organs and affects a wide spectrum of body functions.

The secretion of PHT is regulated by calcium levels in plasma. Chelatin agents (EDTA) bind calcium and the resulting hypocalcemia stimulates PHT secretion. Infusion of calcium decreases PHT secretion by the parathyroid glands. Hypocalcemia, in the presence of severe hypomagnesemia does not increase PHT levels, indicating that magnesium is required for the release of PHT. Calcitonin (CT) and vitamin D also affect the secretion of PHT. The presence of CT releases PHT; however, the concentration of CT needed for this effect in organ cultures is much higher than normally present in plasma. Vitamin D and its metabolites increase the level of PHT. Vitamin D acts as a set point around which the plasma calcium level controls PHT secretion (3, 6, 21, 36, 38, 40, 67). Major function of PHT are to:

1. Increase plasma calcium concentration and decrease phosphate concentration;

2. Decrease urinary excretion of calcium and increase the excretion of phosphate and hydroxyproteine containing peptides;

3. Regulate the renal synthesis of $1,25(OH)_2D_3$;

4. Activate adenylcyclase in the cells of target tissues;

5. Increase the rate of skeletal remodeling and net rate of bone reabsorption;

6. Increase the extent of osteocytic osteolysis in bone;

7. Increase the number of oestoclasts and osteoblasts upon bone surfaces;

8. Cause the initial increase in calcium entry into cells of target tissues;

9. Alter the acid base balance of the body;

10. Increase the gastrointestinal absorption of calcium.

The effect of PHT on soft tissue has not been fully established. However, there is evidence that PHT effects the liver and thymus (122, 162). The PHT action can be modified by concentration of phosphate, pyrophosphate, and CT. The effect of PHT is also related to cAMP as the latter mimics the action of PHT.

Calcitonin

Calcitonin is produced by specialized "C" cells of ultimobranchial origin. The hormone is usually present in the human thyroid, although in some cases it is also found in parathyroid and thymus tissues. The rate of CT secretion is directly controlled by calcium concentration in plasma. Magnesium at nonphysiologically high concentrations also induces the secretion of CT. Increase in CT level is related to injection of gastrin and pancreozymin.

Functions of CT

Calcitonin decreases the resorptive activity of osteoclasts, osteocytes, and the rate of activation of osteoprogenitor cells to preosteoclasts and osteoclasts while the modulation of osteoclasts to osteoblasts is increased. No definitive function of CT in intestinal absorption of calcium can be assigned. Urinary electrolyte excretion is affected by CT, but it does not stimulate renal adenylcyclase or alter the exertion of cAMP.

The mode of action of CT is not yet well established but it appears that it causes immediate release of calcium from bone (bone cells) followed by profound inhibition. It also causes a retention of calcium by isolated cell and inhibition of bone citrate production.

Animal Models of Hyperparathyroidism

Krook (92) described primary hyperparathyroidism, renal secondary hyperparathyroidism in the dog, and the relationship between the chondrodystrophoid dogs and hyperparathyroidism.

The functional capacity of an endocrine organ is related to the size of the organ. In all dogs, regardless of body size, the normal weight of the parathyroids was the same (1.25–1.32 mg/kg body weight). There was no difference in the weight of the parathyroid due to age or sex of the animals. The canine parathyroid did not show a uniform cell picture as previously indicated in the literature. The dog shows some epithelial cells in the parathyroid similar to those seen in humans, that is, chief cells (predominate), water clear cells (only in adenomas and secondary hyperplasia), and oxyphil cells (appearing with increasing frequency with age).

In humans, primary hyperparathyroidism is caused by (a) primary hyperplasia of the parathyroids, (b) parathyroid adenoma, or (c) parathyroid carcinoma. Primary hyperparathyroidism in the dog was described for the first time by Krook (92). The three cases described were all in very old dogs—two boxers and one German shepherd. The dogs showed a variable picture similar to that seen in the human.

Comparison of chondrodystrophoid and nonchondrodystrophoid breeds showed that the parathyroid weights of chondrodystrophoid dogs are significantly greater than those of nonchondrodystrophoid dogs. Morphologically, the parathyroids were identical. The parathyroid hyperplasia in chondrodystrophoid dogs showing ricketic skeletal changes may be due to alterations in calcium metabolism associated with a defect in calcification of the bone matrix. Thus, every chondrodystrophoid dog is considered to be an example of secondary hyperparathyroidism and is therefore predisposed to nephrocalcinosis and urolithiasis.

Animal Model of Hypercalcemia

Rice et al. (125) reported that a transplantable Leydig cell tumor of the Fishcher rat produces hypercalcemia and hypercalciuria. An increase in the serum calcium concentration is accompanied by an increase in the excretion of calcium in the urine and by a decrease in serum inorganic phosphorus concentration. Continued serum hypercalcemia leads to azotemia and an eventual decrease in the serum calcium and calcium excretion. The greatest change in serum calcium and calcium excretion was found in castrated and/or thyroparathyroidectomized animals, suggesting a regulatory role for gonadotrophins and thyrocalcitonin. A VX-2 papilloma has also been reported to maintain hypercalcemia in the acute thyroparathyroidectomized rabbit (158). However, this tumor metastasizes in the rabbit, while the testicular tumor (Leydig cell tumor in the Fischer rat) remains localized. It seems that the tumor-induced hypercalcemia may serve as a suitable model for the study of hypercalcemia and neoplasia in humans.

Geary and Cousins (57) described a nonfamilial nephrocalcinosis unrelated to hypercalcemia and occurring only in female rats. Ovariectomized rats did not develop this condition; therefore estrogens were found to be essential. When the authors instituted estrogen replacement therapy following gonadectomy in animals of either sex, kidney calcification resulted. An unidentified dietary factor may be concerned in the initiation of nephrocalcinosis. Calcification commenced at the corticomedullary junction, spreading to involve the medulla. Electron microscopy revealed the presence of both intratubular and intracellular lamellated deposits in the nephron.

The association of estrogens with renal mineralization is unexpected, since in humans the incidence of renal calculi in females is less than half that in males.

Androgen and Enhancement of Hypocalcemic Response to Thyrocalcitonin in Rats

Thyrocalcitonin is a hypocalcemic principle extracted from the thyroid. It exerts its effects by inhibiting the resorption of bone. The balance between accretion and resorption plays a major role in maintaining constancy of the serum calcium level. Copp and Kuczerpa (37) reported the effects of growth hormone on phosphate levels in rats. Since thyrocalcitonin is a relatively specific inhibitor of a process in the bone metabolism, it may be used as a tool for examining metabolic activities in bone. Ogata et al. (116) showed that endogenous androgens, as well as pharmacological doses of testosterone propionate, markedly enhanced hypocalcemia and suggested that androgens considerably influence the metabolic activities in bone.

The serum calcium level is not influenced by either castration or spaying of experimental rats. Porcine thyrocalcitonin (0.02–0.08 MRC units) injected subcutaneously, produced, in 60 minutes, a more pronounced hypocalcemia in the sham-operated male rats than in the castrated male rat. The response was enhanced by prior treatment with 0.1–1.0 mg testosterone propionate. The same response, though lesser in magnitude, was seen in female spayed and sham-operated animals.

Animal Model of Diet-Induced Hypoparathyroidism:
A Model for Neonatal Tetany

Tetany in newborns has been attributed to transient physiological hypoparathyroidism caused by immaturity of the parathyroid glands, renal tubules, or both. Fetal calcium-phosphorus homeostasis is regulated by maternal and placental mechanisms, and in utero parathyroid stimulation is thought to be minimal during normal pregnancies.

Rogers and Bergstrom (128) induced transient hypoparathyroidism in mature male albino rats by dietary means. Rats were kept on a low-phosphate diet with an elevated Ca/P ratio and were challenged with parenteral sodium phosphate. Hyperphosphatemia, hypocalcemia, and fatal tetany were seen in experimental animals but not in controls fed a standard ration. Parathyroid extract was found partially protective, and prefeeding standard rations reversed susceptibility to phosphate loading within a week. It seems that the capacity to defend serum calcium levels against exogenous phosphate depends on prior stimulation of the mechanism for calcium-phosphorus homeostasis rather than on "maturity" per se.

PANCREAS

The islets of Langerhans produce two different hormones, insulin and glucagon. The α cells of the islets secrete glucagon and the β cells secrete insulin. The β cells contain secretory granules. Insulin is stored in the cells until a stimulus such as glucose, glucagon, or tolbutamine is applied. Insulin directly or indirectly affects each organ and almost all biochemical constituents. The main function of insulin is to stimulate anabolic reactions involving carbohydrates, fats, proteins, and nucleic acids and their biosynthesis from monomeric compounds. It is important that plasma glucose levels (60–100 mg/dl) be maintained in humans. Integration of the metabolic processes of carbohydrates, proteins, and lipids is necessary to keep plasma glucose at a fairly constant level. Intermediary metabolism of carbohydrates, proteins, and lipids is regulated by insulin, glucagon, catecholamines, glucosteroids, and GH (19, 48, 49). Several noninsulin factors can simulate some of the effects of insulin. These factors include GH, anoxia, chymotrypsin, and a plasma protein containing insulin-like activity.

Diabetes Mellitus

The diabetic syndrome in humans and in animals consists of a variety of metabolic lesions that are not necessarily present all at once. Hyperglycemia is the most consistent sign of diabetes, but it is not a sensitive indicator at the onset of the disease. Hyperglycemia may result from inadequate insulin production, production of abnormal insulin products, presence of insulin antagonists, or defects in peripheral glucose utilization. Full-blown diabetes mellitus is marked by the obvious and well-known symptoms of polyuria, polydipsia, and weight loss despite increased appetite. A considerable amount of work on the etiology, diagnosis, pathogenesis, and management of diabetes

Table 8-4. Types of Inherited "Diabetes" in *Mus musculus* and in Other Small Laboratory Rodents

A. Mus musculus
Single gene mutations

Gene symbol**	Gene name	Existing stocks	Synonyms previously used but not now recommended
A^Y	Yellow or lethal yellow	Many	Obese yellow, yellow obese
A^{vy}	Viable yellow	C57BL/6J-A^{vy}	
A^{iy}	Intermediate yellow	C57BL/6J-A^{iy}	
ob	Obese	C57BL/6J-ob	AO, obese hyperglycemic, North American obese hyperglycemic, etc.
ad	Adipose		Adipose--Edinburgh
db	Diabetes	C57BL/ksj-db	

Inbred strains and F_1 hybrids

Recommended name	Synonym	Synonym previously used but not now recommended
NZO	New Zealand obese	
KK	KK mouse	Japanese obese
C3Hf F_1	C3f1 F_1	Wellesley mouse

B. Other species

Recommended name	Synonym	Synonym previously used but not now recommended
Acomys cahirinus	Spiny mouse	Acomys dimidiatus
Psammomys obesus	Sand rat, desert rat	
Cricetulus griseus	Chinese hamster	
"Fatty": single-mutant gene in the rat	Fatties	

*Diabetes has not been established in all these animals. It is suspected in all, with the possible exception of "fatties", where obesity and hyperlipemia are the predominant features.
**Mouse News Letter, semiannual bulletin produced and distributed by the International Committee on Laboratory Animals and the Laboratory Animals Centre, MRC Laboratories, Woodmansterne Road, Carshalton, England. Reproduced from Diabetes Mellitus: Theory and Practice (47) with permission.

280

has been reported in the literature. The theory and practice of diabetes have been described in detail in a recent book by Ellenberg and Rifkin (47).

Animal Models for the Study of Diabetes Mellitus

Spontaneous Diabetes in Animals. The spontaneous occurrence of diabetes in dogs and cats has been known for a long time (124, 126), but it was not until the development of inbred strains of rodents that a systematic study of diabetes could be carried out. At least thirteen diabetic strains of rodents have been developed (Table 8-4).

The Chinese Hamster. Extensive metabolic studies during the early phases of development of diabetes mellitus have been conducted using the Chinese hamster. Although, the exact genetic factors in the development of diabetes during inbreeding is not clear, a recessive diabetic gene *dd* (173) and polygenic system of four genes have been suggested (18). The severity of diabetes and the amount of glucose excreted in the urine varies greatly. The glucose level increases by a factor of three or four, while the free fatty acids that are relatively high (2000 μeq/l) remain the same in the diabetic animal. Other manifestations of diabetes in the Chinese hamster include changes in the islet cells, a low incidence (about 3 percent) of arteriolar and capillary aneurysms, mesangial cell alterations, cystic dilation, and fusion of capillary loops.

Renold et al. (123) reported that the Chinese hamster is unique in that obesity has never been observed in the clinical course of diabetes. The dynamics of immunoreactive insulin (IRI) release in some species and strains predispose to inappropriate hyperglycemia. Early IRI response to several stimuli is slightly or greatly decreased in spiny mice.

The Mouse. Many strains of mice, especially the males, develop natural diabetes characterized by obesity and high blood sugar levels. Diabetes in mice is generally seen in old animals. It is mild and does not require insulin treatment for survival.

The mutation, diabetes (db), occurred in the C57 BL/Ks strain (Jackson Laboratory) and is a unit autosomal recessive gene with full penetrance. The disease causes metabolic disturbances in homozygous mice resembling those of maturity-onset diabetes mellitus in humans (75). The sequence is abnormal deposition of fat at 3 to 4 weeks of age followed by hyperglycemia, polyuria, and glycosuria. First, there is a marked increase in levels of plasma insulin, rates of lipogenesis, gluconeogenesis, and glucose oxidation, and a reduction of β-cell granules in the islets of Langerhans. Second, as a late stage, there is a near-normal level of circulating insulin, a marked decrease in glucose utilization, with a continued high rate of gluconeogenesis. These findings suggested a defect in the peripheral utilization of insulin rather than in the synthesis and release of the hormone from the pancreas.

There are several advantages of this mutant mouse: (a) the mutation arises and is maintained in an inbred strain, which facilitates studies involving tissue and organ transplantations; (b) inheritance is under control of a recessive gene with complete penetrance; (c) offspring exhibiting diabetic syndrome can be distinguished at 3 weeks.

The disadvantages are (a) diabetic homozygotes do not breed; heterozygotes cannot be distinguished from normals except by progeny testing; (b) the early onset of obesity and diabetes makes studies of the preclincal stages difficult. Outcrossing to another inbred strain may introduce genetic modifiers to offset the severity of the disease. Also, ovary transplantation is used to produce diabetic offspring. This, coupled with artificial insemination, should yield litters composed entirely of diabetics and should greatly facilitate detailed studies of preclinical stages.

Renold et al. (123) reported endocrine metabolic anomalies in rodents with hyperglycemic syndromes of hereditary and/or environmental origin. Twelve hereditary components of abnormal hyperglycemia in small rodents were characterized (Table 8-5).

Ultrastructural observations suggested coexistence in *Acomys cahirimes* (spiny mouse) of active insulin biosynthesis and inhibited insulin release. They also exhibit (Basel-Geneva strain) a remarkable rate of increase and thickening of glomerular capillary basement membranes with advancing age, even in the absence of hyperglycemia which enhances or accelerates the vascular aging process (117).

The KK and A^+ mouse both exhibit many symptoms of diabetes; the most satisfactory test for this condition is glycosuria (18). Glucosuria is inherited as a dominant character in both strains, but factors such as diet, fighting, breeding, and other stress can reduce its penetrance. In the KK strain, the penetrance can be increased from 23 to 62 percent by crossing with C57BL/10J and then selecting the most effective background modifiers. Since the response is rapid and fixation is fast, the number of modifiers must be small. The cross KK X A^+ gives equal numbers of yellow and black offspring. The yellows all have dominant glucosuria genes and the males all become glu-

Table 8-5.   Genetic Transmission of "Predisposition to Inappropriate Hyperglycemia" Characters in Small Rodents

Animal Strain	Mode of Inheritance
A. Single gene mutants: Mice	
Yellow (A^y + variants)	Autosomal dominant
Obese (ob)	Autosomal recessive
Adipose (ad)	Autosomal recessive
Diabetes (db)	Autosomal recessive
Rats	
"Fatty"	Autosomal recessive
B. Inbred Strains and Hybrids:	
New Zealand Obese mice (NZO)	Inbred probably polygenic
KK mice (Japan)	Inbred one component dominant, with varying penetrance due to other genes.
C3Hf X.1 F_1 (Wellesley mice)	Hybrid of 2 inbred strains, polygenic
Chinese hamster (cricetulus griseus)	Inbred several recessive genes
C. Distinct species with presumably polygenic hereditary component strongly influenced by environment:	
Acomys cahirinus (spiny mouse)	As yet unknown
Psammomys obesus (sand rat)	As yet unknown
Genomys tolarum (tuco-tuco)	As yet unknown

cosuric by 8 months of age, while 88 percent of the yellow females are glucosuric by 1 year. The black males without the gene from the A$^+$ but from the KK have no glucosuria at 8 months and only 2 percent have it at 1 year of age. The F$_1$ of the KK X A$^+$ cross may provide a good animal model for further studies on glucosuria.

The Sand Rat. Diet-induced diabetes in the sand rat has been reported (64). It is not clear if a genetic mechanism is involved or if it is due to the effects of a high caloric diet on metabolism. The animal model mimics the human diabetic condition.

The Dog and the Cat. In dogs, the incidence is 1 in 200 and old females are generally affected. The most common cause is pancreatitis; however, idiopathic lesions of β-cells are also seen. Retinal lesions, typical of human conditions, are also seen in diabetic dogs, including degeneration of mural pericytes, widening of retinal capillary shunts, and microaneurysms. Changes in the kidney are not a characteristic finding but, when present, 90 percent of glomeruli may be affected. Other findings include fatty liver and adenomas of pituitary.

The incidence of diabetes is much lower in cats (1 : 800), and it is most frequently seen in males. Typical hydropic changes are usually seen in islets; however, occasionally no pathological changes may be present.

Nonhuman Primates. Clinical features of spontaneous diabetes in monkeys is very similar to the human disease (82); therefore, the monkey is a useful model for future studies on abnormal carbohydrate metabolism. Sokolovera (142) first described in detail diabetes mellitus in the sacred baboon. Rabb et al. (132) reported diabetes in the tree shrew. Jones (82) described three cases of diabetes in monkeys maintained on a high carbohydrate diet, with severe peridontal disease. These three cases were among a colony of 300 placed on a high sucrose diet for a dental study. Although the incidence is small, it does show a correlation between diet and the disease, and it also shows the usefulness of monkeys for the study of diabetes.

DiGiacomo (42) suggested that spontaneous diabetes mellitus may be more common in nonhuman primates than previously thought. He described diabetes in the male rhesus monkey, controlled by insulin and diet, although there were problems from insulin shock after injection. Further investigations into the occurrence of the spontaneous disease showed abnormal glucose tolerance in females giving birth to large infants as seen in human patients. Other reports indicated that the rhesus monkey may be a good model for human diabetes mellitus.

Stullman et al. (146) reported serum glucose values of 1423 blood samples from 620 *Mastromys albicandatus.* Serum glucose was not affected by sex, inbreeding, or age in these animals. However, weight affected the serum glucose values. It was indicated that spontaneous diabetes mellitus in Mastromys is very similar to the disease in humans.

Experimentally Induced Diabetes in Laboratory Animals. Experimental production of diabetes dates back to 1889 when Von Mering and Minkowski (161) showed that the extirpation of the pancreas of dogs resulted in the diabetic condition. Most of the studies on diabetes in animals have been carried out on metabolic alterations resulting from pancreatectomy, damage to the islet tissues by alloxan or other chemical agents, interference with the release of insulin, or interference with the effects of insulin on target organs.

Diabetes induced in laboratory animals by pancreatectomy or by chemical agents also cause severe secondary effects. Howard (74) developed a nonhuman primate model of diabetes by infusing streptozotocin directly into the pancreas through the femoral artery or through the celia artery. Since a direct application was used, a relatively low dose of the drug was needed to induce diabetis free from side effects. The drug produced a loss of β-cells or at least loss of granules in β-cells. This model can be used for the study of long-range effects of metabolic alterations associated with diabetes and diabetes-related atherosclerosis and microangiopathies.

Hamilton (65) induced electrolytic lesions in ventromedial hypothalamus in the rhesus monkey to produce a model for experimental study of obesity and diabetes mellitus. The animal was observed for 14½ years. Rapid increase of weight followed the hypothalamic lesions. Glucose was first detected 7 months after the operation. High fasting levels of glucose and abnormal tolerance were observed. Insulin treatment was begun 3 years after the operation and the dosage was gradually increased. Although the animal continued to show glycosuria, weight gain was reinstated with lower food intake. After large doses of insulin, the animal began to lose weight, although food intake and glucose output were unchanged. When insulin treatment was continued and food intake halved, glucose was absent in urine. As food was increased, glucose levels remained low but gradually increased to prerestriction levels. This work demonstrates the feasibility of long-term studies on diabetes using nonhuman primates as models.

Lehner (96) found that the rate of disappearance of intravenously administered glucose was inversely related to serum cholesterol concentration and the indices of atherosclerosis. Squirrel monkeys were fed a diet containing 1 mg of cholesterol per calorie for 3 years. Hypothyroid and insulin-deficient monkeys had significantly greater concentrations of serum cholesterol and β-lipoprotein than those of control animals. However, controls and hypertensive monkeys did not differ. The three groups developed extensive atherosclerosis of the coronary arteries and the aorta as compared to controls. Most severe atherosclerosis developed in the insulin-deficient monkeys. Systolic blood pressure, serum cholesterol, and serum-β-lipoprotein were significantly and positively correlated with coronary arterial and aortic atherosclerosis. This suggested that the squirrel monkey was a good experimental animal in which to study the mechanisms of these disorders (insulin deficiency, hypothyroidism, and hypertension) because of their similarity to those in humans. The insulin deficiency was induced in the squirrel monkey by intravenous injections of 125 mg alloxan monhydrate per kilogram of body weight. Electrophoresis of sera of insulin-deficient monkeys resembled type II hyperproteinemia with an increased β-lipoprotein band and without a prebeta band. Sera were clear, without an increase in chylomicrons indicated by unchanged triglyceride concentrations. These findings differ from type IV or V hyperlipoproteinemia which may be present in humans with diabetes (56). The high concentration of serum cholesterol in the insulin-deficient monkey is consistent with the finding in diabetic humans and animals (130). Hypercholesteromia is a recognized sequelae of human hypothyroidism. Prolonged elevated concentration of blood glucose in hypothyroid monkeys is similar to that found in hypothyroid humans (94) and is apparently unrelated to the availability of insulin.

In the rat, removal of 95 percent of the pancreas is normally followed by the development of diabetes. The time of the onset of diabetes can be shortened or lengthened by treatment with hormones. Female sex hormones have been shown to have a protective effect, while male hormones have a deleterious effect on the development of diabetes (52). Testosterone propionate was subcutaneously injected into 3-day-old rats. The animals, which were found to be in permanent estrus by the second month, were partially pancreactomized. Female rats injected with testosterone showed a significantly higher glycemia than the control group within the first and second month postpancreactomy. However, there was no significant difference between the injected and control males, both being hyperglycemic at the same time.

The presence or absence of diabetes in rats is dependent on the hormonal balance between diabetogenic agents and insulin. The control of diabetes by the removal of diabetogenic hormones was first reported by Houssay and Biassotti (72) through hypophysectomy and by Long and Lukens (101) through adrenalectomy. Bates and Garrison (9) explored the use of rats for studies of the diabetogenic agents to prevent the production of permanent diabetes by diabetogenic hormones. Rats bearing transplantable, hormone-producing (growth hormone, ACTH, and prolactin) pituitary tumors with 80 percent of the pancreas removed were used for the studies. All such animals developed severe irreversible glycosuria. Proper prophylactic insulin treatment prevented glycosuria and also prevented permanent diabetes. Treatment for 20 days with metyrapone and aminoglutethimide (0.5 and 0.1 percent of diet respectively) delayed diabetes and prevented its permanent development during the 20-day treatment in nine of the seventeen rats, a result not unlike adrenalectomy. Metyrapone (0.5 percent) given alone was not as effective as 0.1 percent aminoglutethimide in preventing corticosterone production and diabetes. Permanent development of diabetes produced basal insulin levels of 0.16 I.U./g of pancreas and < 13 μU/ml of plasma. The level was reached as soon as the glucosuria was more than 3.5 g/day. Aminoglutethimide alone and the combined drugs reduced the mean corticosterone levels in the plasma and adrenals, but there was no reduction in the size of the liver, kidneys, and heart. Adrenalectomy caused atrophy of these organs.

GONADS

The ovaries and the testes have certain endocrine functions influenced by pituitary and other hormones. The cortical cells of the ovary differentiate into the granulosa and theca cells of the ovarian follicle, which produce estrogens and progestins, respectively. Critical levels of estrogen induce secretion of follicle-stimulating hormone (FSH) from the pituitary which, in turn, promotes estrogen production. Follicle-stimulating hormone and luteinizing hormone (LH) together induce ovulation, after which FSH and estrogen levels decline while the LH and progesterone secretion increase. Under the influence of LH, the ruptured follicle becomes the corpus luteum, and luteotropic hormone promotes continued elaboration of progesterone. Shortly before menstruation, progesterone secretion declines and there is a second, but lower, peak of estrogen production. Ovarian hypofunction usually means estrogen deficiency, since an isolated deficiency of progesterone is rare. Combined deficiency of estrogen and progesterone results in menstrual irregularities and difficulty in conception.

The Leydig or interstitial cells of the testes secrete testosterone under the influence of the interstitial cell-stimulating hormone (ICSH), which is apparently the same as the luteinizing hormone in females. An upsurge of pituitary gonadotropins heralds the onset of puberty and promotes the secretion of testosterone, which initiates the male sex characteristics.

The anatomical and physiological characteristics of the gonads in animal species are described in detail in Chapter 9. Among the endocrinopathies, ovarian and testicular tumors are rare in laboratory animals except in certain inbred strains of rodents (Chapter 11).

For greater details on the endocrinopathies of the gonads, the reader is referred to Fleischmann (51), Bloodworth (14), and Dillon (43). The influence of ovarian hormones on the metabolism of carbohydrate and lipids have been extensively reported.

Animals Model of Estrogen-Induced Hyperlipidemia

The chick as a suitable animal for the study of estrogen-induced hyperlipidemia was reported by Kudzma et al. (93). Plasma levels of triglyceride, cholesterol, and free fatty acids of the chick are similar to those in humans. Also, the lipid composition of high and low density lipoproteins, very low density lipoproteins, and chylomicrons resemble those of humans.

The estrogen component of the oral contraceptives is known to induce hyperlipidemia in premenopausal women (171). The widespread and long-term use of oral contraceptives and the association between hypercholesterolemia, hypertryglycridemia, and coronary atherosclerosis have aroused interest in the elucidation of the mechanism underlying estrogen-induced hyperlipidemia. An analogous situation exists in the chicken. Laying hens have higher blood lipid levels than cockrels or hens that have not come into lay. Striking hyperlipidemia results from administration of diethylstilbesterol (DES) (0.1, 1.0, and 5.0 mg/day for 18 days). This is dose related and consists predominantly of triglycerides. Although chylomicrons and low and high density lipoproteins contribute to the hypertriglyceridemia, the very low density, endogenous lipoproteins predominate as in estrogen-treated women. Livers from DES-treated birds incorporate four and one-half times more acetate into triglyceride than livers from untreated birds, suggesting that estrogen-induced hyperlipidemia may partly be due to enhanced hepatic lipogenesis. Levels of plasma free fatty acids are also elevated by estrogen treatment. The rise does not occur until after hypertriglyceridemia is well established, indicating that the elevation of free fatty acid is not part of the mechanism of the disorder. From the qualitative similarities between hyperlipoproteinemia in the chick and that in estrogen-treated women, it appears that the chick is a suitable laboratory animal model for the study of the mechanism and control of hyperlipidemic conditions in humans.

Effect of Progesterone on Blood Glucose

The influences of many hormones such as growth hormone, glucocorticoids, and estrogens on the carbohydrate metabolism in mammals have been widely studied (39, 73). Very large doses of progesterone (50–100 mg/rat/day) may cause exacerbation of glycosuria in the partially depancreatized, force-fed rats) while small doses (0.5 mg/day) of progesterone do not significantly affect the blood glucose levels. In humans, administration of progesterone in a single dose of 250 mg to women produced a favorable affect on their glucose tolerance, while a combination of progestogens and estrogens used as contraceptives caused an abnormal glucose tolerance in some healthy women (170). Yang (172) reported that a single dose of progesterone (6 mg/100 g) in virgin rats produced hyperglycemia in 30 to 60 minutes and 5 hours, respectively, after injection; adrenergic blockage with ergotamine tartarate abolished the hyperglycemic effect. In adrenalectomized rats, a single dose of progesterone (2–4 mg/100 g) caused hypoglycemia unaccompanied by a rise in plasma insulin level, whereas progesterone did not affect the blood glucose level in the adreno-demedullated rats. This model may be used for the study of the effect of progesterone on the adrenal medulla and carbohydrate metabolism.

REFERENCES

1. Adams, D. D., and T. H. Kennedy, *J. Clin. Endocrinol.,* **33**:47 (1971).
2. Amstutz, H. E., *Vet. Scope,* **92**:2–9 (1964); Ibid, **92**:23–2 (1964).
3. Anast, C. S., J. M. Mohs, S. L. Kaplan, and T. W. Burns, *Science,* **177**:606 (1972).
4. Andersson, B., and S. M. McCann, *Acta Physiol. Scand.,* **35**:191 (1955).
5. Andrews, E. N., and J. M. Fransen, *Am. J. Vet. Res.,* **10**:822 (1958).
6. Arnaud, C., F. Glorieux, and C. Scriver, *Science,* **173**:845 (1971).
7. Austin, L., B. G. Livett, and I. W. Chubb, *Life Sciences,* **6**:97 (1967).
8. Bartke, A., *Genetics,* **48**:882 (1963).
9. Bates, R. W. and M. M. Garrison, *Endocrinology,* **86**:107 (1970).
10. Baxter, J. D., and P. H. Forsham, *Am. J. Med.,* **53**:573 (1973).
11. Benson, B., *Acta Endocrinol.* (Kbh), **53**:663 (1966).
12. Bielschowsky, M. and F. Bielchowsky, *Anst. J. Exp. Biol. Med. Sci.,* **34**:181 (1956).
13. Blizzard, R. M., and M. Kyle, *J. Clin. Invest.,* **42**:1653 (1963).
14. Bloodworth, J. M. B., Jr., *Endocrine Pathology,* Williams & Wilkins Co., Baltimore, 1968.
15. Bloom, F., *North Am. Vet.,* **23**:727 (1942).
16. Boorman, G. A., M. J. Van Noord, and C. F. Hollander, *Arch. Pathol.,* **94**:35 (1972).
17. Brodoff, B. N., *Metabol.,* **16**:744 (1947).
18. Butler, L., *Can. J. Genet. Cytol.,* **14**:265 (1972).
19. Cahill, G., *Diabetes,* **20**:785 (1971).
20. Campbell, J. G., *Res. Vet. Sci.,* **3**:50 (1962).
21. Cannigia, A., C. Gennari, P. Piantelli, and A. Vattimo, *Clin. Sci.,* **43**:171 (1972).
22. Capen, C., and H. E. Black, *Am. J. Pathol.,* **74**:377 (1974).
23. Capen, C. C. and A. Koestner, *Pathol. Vet.* (Basel), **4**:326 (1967).
24. Capen, C. C., S. L. Martin, and A. Koestner, *Pathol. Vet.* (Basel), **4**:301 (1967).
25. Catt, K. J., *An ABC of Endocrinology,* Little, Brown & Co., Boston, 1971.
26. Cello, R. M., and P. C. Kennedy, *Cornell Vet.,* **47**:538 (1957).
27. Chase, H. B., M. S. Gunther, J. Miller, et al., *Science,* **107**:279 (1948).
28. Chattoraj, S. C., in N. W. Tietz (ed.), *Fundamentals of Clinical Chemistry,* W. B. Saunders Co., Philadelphia, 1970.
29. Chi, C. K. and M. M. Dickie, in E. Green (ed.), *Biology of the Laboratory Mouse,* 2nd ed., McGraw-Hill Book Co., Inc., New York 1966, p. 387.
30. Coffin, D. L., and T. G. Munson, *J. Am. Vet. Med.,* **123**:403 (1953).
31. Coffin, D. L., and A. Thordal-Christensen, *Vet. Med.* **48**:193 (1953).
32. Cole, R. K., *Genetics,* **53**:1021 (1966).
33. Coleman, D. L., and K. P. Hummel, *Diabetologia,* **3**:238 (1967).
34. Conn, J. W., *J. Lab. Clin. Med.,* **45**:3 (1955).
35. Conn, J. W., R. F. Knopf and R. M. Nesbit, *Am. J. Surg.,* **107**:159 (1964).
36. Copp, D. H., *Am. Rev. Pharm.,* **9**:327 (1969).
37. Copp, D. H., and A. V. Kuczerpa, *Fed. Proc.,* **26**:368 (1967).
38. Cuthbertson, W. F. J., (ed.), *Calcium and Cellular Function,* McMillan & Co. Ltd., London, 1970.
39. DeBodo, R. C., and N. Altszuler, *Physiol. Rev.,* **38**:389 (1958).
40. De Luca, H. F., *Recent Progr. Hormone Res.,* **27**:479 (1971).
41. Diener, R., and R. Langham, *Small Anim. Clin.,* **1**:274 (1961).
42. Di Giacomo, R. F., R. E. Myers, and L. R. Baez, *Lab. Anim. Sci.,* **21**:572 (1971).
43. Dillon, R. S., *Handbook of Endocrinology,* Lea & Febiger, Philadelphia, 1973.
44. Du Vigneaud, V., *Science,* **123**:967 (1956).
45. Du Vigeneaud, V., *Harvey Society Lectures,* Academic Press, New York, 1956.
46. Eayrs, J. T., *J. Anat.,* **88**:164 (1954).

47. Ellenberg, M., and H. Rifkin (eds.), *Diabetes Mellitus: Theory and Practice,* McGraw-Hill, Inc., New York, 1970.
48. Exton, J. H., N. Friedman, E. H. Wong, J. A. Brineux, J. D. Corbin, and C. R. Park, *J. Biol. Chem.,* **247:**3579 (1972).
49. Felig, P., *Metabolism,* **22:**179 (1973).
50. Fisher, C., W. R. Ingram, and S. W. Ranson, *Diabetes Inspidus and the Neuro-Hormonal Control of Water Balance,* Edward Bros., Ann Arbor, 1938.
51. Fleischmann, W., in W. E. Ribelin and J. R. McCoy (eds.), *Pathology of Laboratory Animals,* Charles C. Thomas, Springfield, Ill., 1965.
52. Foglia, V. G., J. C. Basabe, and R. A. Chieri, *Diabetologia,* **5:**258 (1969).
53. Foglia, V. G., R. E. Mancinsi, and A. F. Cardeya, *Arch. Pathol.,* **50:**75 (1950).
54. Foglia, V. G., N. Schuster, and R. R. Rodriguez, *Endocrinology,* **41:**428 (1947).
55. Forrest, J. N., A. D. Cohen, J. Torretti, J. M. Himmelhock, and F. H. Epstein, *J. Clin. Invest.,* **53:**1115 (1974).
56. Fredrickson, D. S., R. I. Levy, and R. S. Lees, *N. Engl. J. Med.,* **276:**34 (1967); Ibid, p. 94, 148, 215, and 273.
57. Geary, C. P., and F. B. Cousins, *Br. J. Exp. Pathol.,* **50:**507 (1969).
58. Gepts, W., and D. Toussaint, *Diabetologia,* **3:**249 (1967).
59. Gerwirtz, G. P., and I. J. Kopin, *J. Pharm. Exp. Ther.,* **175:**514 (1970).
60. Gorbman, A. (ed.), *Comparative Endocrinology,* John Wiley & Sons, Inc., New York, 1959.
61. Gorbman, A., and H. A. Bern, *Textbook of Comparative Endocrinology,* John Wiley & Sons, Inc., New York, 1962.
62. Hadlow, W. J., *Am. J. Pathol.,* **20:**353 (1953).
63. Hafey, E. F. E., C. C. O'Mary, and M. E. Ensminger, *J. Hered.,* **40:**111 (1958).
64. Haines, H., D. B. Hackel, and K. Schmidt-Nielsen, *Am. J. Physiol.,* **208:**297 (1965).
65. Hamilton, C. L., *J. Med. Prim.,* **1:**247 (1972).
66. Hashimoto, A., *Arch. Klin. Chir.,* **97:**219 (1912).
67. Haussler, M. R., D. W. Boyce, and E. Littlekike, *Proc. Natl. Acad. Sci. (USA),* **68:**177 (1971).
68. Hallman, B. and C. Hellerstrom, *Z. Zellforsch,* **56:**97 (1962).
69. Heston, W. E., C. D. Larsen and M. K. Deringer, *J. Natl. Cancer. Inst.,* **6:**41 (1945).
70. Hirsh, M. S., *Bull. Johns Hopkins Hosp.,* **110:**257 (1959).
71. Holmes, R. L., and J. N. Ball, *Pituitary Gland: A Comparative Account,* Cambridge University Press, Cambridge, 1974.
72. Houssay, B. A., and A. Biasotti, *Endocrinology,* **15:**511 (1931).
73. Houssay, B. A., V. A. Foglia, and R. R. Rodriguez, *Acta Endocrinol. (Kbh)* **17:**146 (1954).
74. Howard, C. F., *Primate News,* **9:**3 (1971).
75. Hummel, K. P., M. M. Jickie, and D. L. Coleman, *Science,* **153:**1127 (1966).
76. Hummel, K. P., *R. B. Jackson Memorial Lab. Ann. Rep.,* **63–64:**35 (1964).
77. Hutchings, J. J., *J. Clin. Endocrinol,* **19:**759 (1959).
78. James, O. P., *Vet. Rec.,* **72:**630 (1960).
79. Jamieson, S., and H. E. Harbour, *Vet. Rec.,* **59:**102 (1947).
80. Jamieson, S., B. W. Simpson, and J. B. Russie, *Vet. Rec.,* **57:**429 (1945).
81. Jay, G. E., Jr., in W. J. Burdette (ed.), *Methodology in Mammalian Genetics,* Holden Day, Inc., San Francisco, 1963.
82. Jones, S. M., *Lab. Anim.,* **8:**161 (1974).
83. Jost, A., *Proc. Soc. Study Drug Toxicity,* **1:**35 (1963).
84. Jubb, K. V., and K. McEntee, *Cornell Vet.,* **49:**41 (1959).
85. Julian, L. M., W. S. Tyler, and P. W. Gregory, *J. Am. Vet. Med. Assoc.,* **135:**106 (1959).
86. Kennedy, P. C., J. W. Kendrick, and C. Stormont, *Cornell Vet.,* **47:**160 (1957).
87. King, J. W. B., *J. Hered.,* **41:**249 (1950).
88. Knobil, E. and R. O. Greep, *Recent Progr. Horm. Res.,* **15:**1 (1959).
89. Kourides, I. A., B. D. Weintraub, M. A. Levko, and F. Maloof, *Clin. Res.,* **21:**496 (1973).
90. Krischbaum, A., M. H. Franty, and W. L. Williams, *Cancer Res.,* **6:**707 (1946).
91. Kronfeld, D. S., *Ann. N.Y. Acad. Sci.,* **104:**799 (1963).
92. Krook, L., *Acta Pathol. Microbiol. Scand.,* **41:**1 (1957).

93. Kudzma, D. J., P. M. Hegstad, and R. E. Stoll, *Metabolism,* **22:**423 (1973).
94. Lemberg, B. A., *Acta Med. Scand.,* **178:**351 (1965).
95. Lamki, L., V. V. Row and R. Volpe, *J. Clin. Endocrinol.,* **36:**358 (1973).
96. Lehner, N. D. M., *Exp. Mol. Pathol.,* **15:**230 (1971).
97. Levit, S. A., and C. G. Sheps, *N. Engl. J. Med.,* **281:**805 (1969).
98. Levy, B. M., *J. Comp. Pathol.,* **82:**99 (1972).
99. Lindsay, S., C. W. Nichols, Jr., and I. L. Chaikoff, *Arch. Pathol.* **86:**353 (1968).
100. Lombardi, S., C. L. Comar, and R. W. Kirk, *Am. J. Vet. Res.,* **23:**412 (1962).
101. Long, C. N. H., and F. D. W. Lukens, *J. Exp. Med.,* **63:**465 (1936).
102. Mahoney, W., and D. Sheehan, *Am. J. Physiol.,* **112:**250 (1935).
103. Marine, D., and C. H. Lenhart, *Arch. Intern. Med.,* **4:**253 (1909).
104. Mason, J. W., *Am. J. Physiol.,* **196:**44 (1959).
105. Matuso, T., K. Furuno, and K. Shimakawa, *J. Takeda Res. Lab.,* **30:**307 (1971).
106. Mazzaferri, E. L., *Endocrinology,* Medical Examination Publishing Co., Inc., 1974.
107. Miller, M., *J. Am. Vet. Med. Assoc.,* **118:**31 (1951).
108. Miller, M., T. Dalakos, A. M. Moses, H. Felluman, and D. M. P. Streeten, *Ann. Intern. Med.,* **73:**721 (1970).
109. Miersky, I. A., N. Nelson and S. Eligart, *Science,* **93:**586 (1941).
110. Marrison, S. H., and C. K. Whitehair, *Nutrition in Diseases of Cattle,* 2nd ed., American Veterinary Publication Inc., Santa Barbara, Calif., 1963.
111. Moses, A. M., and M. Miller, *Endocrinology,* **86:**34 (1970).
112. Mueller, R. A., H. Thoenen, and J. Axelrod, *Science,* **163:**468 (1969).
113. Musser, E., and W. R. Graham, *Lab. Anim. Care,* **18:**58 (1968).
114. Meir, H., and G. A. Yerganian, *Pro. Soc. Exp. Biol. Med.,* **100:**810 (1959).
115. Nishizuka, Y., Y. Tanaka, T. Sakakura, and A. Kojima, *Experientia,* **29:**1396 (1973).
116. Ogata, E., E. Shimazawa, H. Suzuki, Y. Yoshitoshi, H. Asano, and H. Ando, *Endocrinology,* **87:**421 (1970).
117. Orci, L., W. Stauffacher, M. Amhudt, R. Pictet, A. E. Renold, and C. Rouiller, *Diabetologia,* **6:**343 (1970).
118. Orloff, J., and J. Handler, *Am. J. Med.,* **42:**757 (1967).
119. Parker, D. C., *Endocrinology,* **91:**1462 (1972).
120. Parker, D. C., J. F. Sassin, J. W. Mace, R. W. Gotlin, and L. G. Rossman, *J. Clin. Endocrinol. Metab.,* **29:**871 (1969).
121. Pollock, S., *J. Am. Vet. Med. Assoc.,* **118:**12 (1951).
122. Rasmessen, H., *Am. J. Med.,* **50:**567 (1971).
123. Renold, A. E., D. P. Cameron, M. Amgerdt, W. Stauffacher, E. Marliss, L. Orci, and C. Rouiller, *Isr. J. Med. Sci.,* **8:**189 (1972).
124. Renold, A. E., and W. E. Dulin, *Diabetologia,* **3:**63 (1967).
125. Rice, B. F., R. L. Ponthier, Jr., and M. C. Miller, 3rd., *Endocrinology,* **88:**1210 (1971).
126. Ricketts, H. A., E. S. Peterson, P. E. Steiner, and N. Tupekova, *Diabetes,* **2:**288 (1953).
127. Rimoin, D. L., and L. Richmond, *J. Clin. Endocrinol. Metab.,* **35:**467 (1972).
128. Rogers, M. C. and W. H. Bergstron, *Pedriatrics,* **47:**207 (1971).
129. Rogoff, J. M., *Arch. Pathol.,* **38:**392 (1944).
130. Root, H. F., and R. F. Bradley, in E. P. Joslin, H. F. Root, P. White, and A. Marble (eds.), *Treatment of Diabetes,* Lea & Febiger, Philadelphia, 1959.
131. Roth, R. H., L. Stzame, R. J. Lurine, and N. J. Giarman, *J. Lab. Clin. Med.,* **72:**397 (1968).
132. Rubb, G. B., R. E. Getty, W. H. Williamson, and L. S. Lombard, *Diabetes,* **15:**327 (1966).
133. Sachs, H., L. Share, J. Osinchak, and A. Carpi, *Endocrinology,* **81:**755 (1967).
134. Sawyer, W. H., *Parmacol. Rev.,* **13:**225 (1961).
135. Sawyer, W. H., *Am. J. Med.,* **42:**678 (1967).
136. Sawyer, W. H., and H. Valtin, *Endocrinology,* **80:**207 (1967).
137. Seaton, V. A., *Am. J. Vet. Res.,* **19:**600 (1958).
138. Sirek, O. V. and A. Sirek, in M. Ellenberg, and H. Rifkin (eds.), *Diabetes Mellitus: Theory and Practice,* McGraw-Hill Book Co., Inc., New York, 1970.
139. Sit, K. H., and R. Kangasuntheram, *J. Embryol. Exp. Morphol.,* **28:**223 (1972).
140. Smith, H. A., and T. C. Jones, *Veterinary Pathology,* 3rd ed., Lea & Febiger, Philadelphia, 1966.

141. Snell, G. D., *Proc. Natl. Acad. Sci.* (USA), **15:**733 (1929).
142. Sokolovera, I. M., in I. A. Utkin (ed.), *Theoretical and Practical Problems of Medicine and Biology in Experiment on Monkeys,* Pergamon, New York, 1960.
143. Spitzer, R., R. Borrison, and R. A. Castillino, *Radiology,* **98:**577 (1971).
144. Stockard, C. R., *Am. Anat. Memoir 19,* Wistar Institute of Anatomy and Biology, Philadelphia, 1941.
145. Strasberg, S. M., *Surg. Forum,* **21:**87 (1970).
146. Stuhlman, R. A., J. T. Packer, and R. E. Doyle, *Diabetes,* **21:**715 (1972).
147. Takashashi, Y., and D. M. Kipins, *J. Clin. Invest.,* **47:**2 (1968).
148. Tasker, J. B., C. E. Whiteman, and B. R. Martin, *J. Am. Vet. Med. Assoc.,* **149:**393 (1966).
149. Thoenen, N., and R. A. Mueller, *PNAS* (USA), **65:**58 (1970).
150. Turner-Warwick, R., and J. D. N. Nabarro, *Proc. R. Soc. Med.,* **59:**596 (1966).
151. Valtin, H., and A. R. Harrington, *Fed. Proc.,* **25:**204 (1966).
152. Valtin, H., W. H. Sawyer, and H. W. Sockol, *Endocrinology,* **77:**701 (1965).
153. Valtin, H., and H. A. Schroeder, *Am. J. Physiol.,* **206:**425 (1964).
154. Vassalle, M., and J. H. Stuckey, *Am. J. Physiol.,* **217:**930 (1969).
155. Vickens, C. L., W. T. Car, B. W. Bierer, J. B. Thomas, and H. D. Valentrine, *J. Am. Vet. Med. Assoc.,* **137:**507 (1960).
156. Vierling, A. F., J. B. Little, and E. P. Radford, Jr., *Endocrinology,* **80:**211 (1967).
157. Villee, D. B., *Human Endocrinology: A Developmental Approach,* W. B. Saunders Co., Philadelphia, 1975.
158. Vogel, S. B., W. F., Enncking and W. C. Thomas, Jr., *Endocrinology,* **80:**404 (1967).
159. Von Evler, U. S., and H. Heller (eds.), *Comparative Endocrinology,* Vol. 1, Academic Press, New York, 1963.
160. Von Evler, U. S., and H. Heller (eds.), *Comparative Endocrinology,* Vol. 2, Academic Press, New York, 1963.
161. Von Mering, J., and O. Minkowski, *Arch. Exp. Pathol. Pharmacol.,* **26:**371 (1889–90).
162. Vallach, S., A. B. Chausner, *Clin. Orthop.,* **78:**40 (1971).
163. Watson, W. A., G. D. Breadhead, and R. Kilpatrick, *Vet. Rec.,* **74:**506 (1962).
164. Wegelius, O., *Proc. Soc. Exp. Biol. Med.,* **101:**225 (1959).
165. Wick, G., J. H. Kite, Jr., R. K. Cole and E. Witebsky, *J. Immunol.,* **104:**45 (1970).
166. Wildinson, J. S., C. F. Sehotthaur, and J. A. S. Miller, *J. Am. Vet. Med. Assoc.,* **118:**31 (1951).
167. Wilroy, R. S., Jr., and J. N. Etteldorf, *J. Pediat.,* **79:**625 (1971).
168. Witebsky, E., J. H. Kite, Jr., G. Wick and R. K. Cole, *J. Immunol.,* **103:**708 (1969).
169. Witebsky, E., and N. R. Rose, *J. Immunol.,* **83:**41 (1959).
170. Wynn, V., and J. W. H. Doar, *Lancet,* **2:**715 (1966).
171. Wynn, V., et al., *Lancet,* **2:**750 (1969).
172. Yang, M. M. P., *Endocrinology,* **86:**924 (1970).
173. Yerganian, G., in B. S. Leibel and G. A. Wrenshall (eds.), *On the Nature and Treatment of Diabetes,* Excerpta Medica Foundation, Amsterdam, 1965.
174. Zir, L. M., D. C. Parker, R. A. Smith, and L. G. Rossman, *J. Clin. Endocrinol. Metab.,* **34:**1 (1972).
175. Zucker, T. F., and L. M. Zucker, *Proc. Soc. Exp. Biol. Med.,* **110:**165 (1962).

CHAPTER 9

ANIMAL MODELS FOR RESEARCH IN REPRODUCTIVE PHYSIOLOGY

A variety of animal species has been used to obtain a better understanding of the pathophysiological aspects of human reproduction. However, the choice of an animal model in reproductive physiology is limited because of some of the unique reproductive characteristics in humans, for example, menstrual cycle, long gestation period, upright posture, and long postreproductive life. The nonhuman primates come close to expressing these characteristics, although there are significant differences in the physiology of different monkey genera. Mice, rats, guinea pigs, rabbits, cats, dogs, and many other domestic animals may be suitable for the study of certain aspects of human reproduction, such as assay subjects for gonadal hormones, pregnancy tests, study of normal and teratogenic embryology, and the like.

This chapter presents a description of male and female reproductive organs of experimental mammals from the comparative point of view. Special morphological, anatomical, and functional characteristics of the reproductive organs of various animal species are pointed out which may be applicable for the study of problems in human reproduction. Nonhuman primate models of certain reproductive processes in humans (such as sex cycles, gametogenesis, ovulation, fertilization, fertility regulation, and pregnancy), are briefly discussed. Finally, the uses of experimental animals for the study of metabolism and effects of contraceptive steroids are described. For an extensive review of the whole field of reproduction, the reader is referred to Hafez (119), Diczfalusy and Standley (67), Asdell (12), Kaplan (153), Kent (161), Cole and Cupps (55), Lamning and Amorso (180), and Parkes (242). A summary of reproduction in several species of experimental animals is given in Table 9-1.

COMPARATIVE ANATOMY AND PHYSIOLOGY OF THE MALE REPRODUCTIVE ORGANS

The mammalian male reproductive system consists of the primary sex organs (testes) which produce the sperm; accessory sex glands (the prostate, vesicular glands, bulbourethal glands, and lesser structures) which produce fluids of semen; tubular system (epididymis, ductus deferens, and urethra) for the transportation of spermatozoa and fluids; and the external genitalia (the penis and, in some forms, a scrotum). Certain anatomical characteristics of the male reproductive structures are presented in Table 9-2.

Table 9-1. Summary of Reproduction in Laboratory Mammals

	Cat	Chinchilla	Dog	Gerbil
Weight for adult male	3.5-4.0 kg	400-500 gm	14-18 kg	46-131 gm
Weight for adult female	2.25-2.8 kg	400-600 gm	14-16 kg	53-133 gm
Birth weight	95-140 gm	35-50 gm	.36-.45 kg	2.5-3.5 gm
Male breeding age (weight)	9-12 mos. (3.5 kg)	6 mos. (350 gm)	10-12 mos.	10-12 wks.
Female breeding age (weight)	5-7 mos. (2.5 kg)	6-8 1/2 mos. (400 gm)	9-12 mos.	10-12 wks.
Estrous cycle length (days)	regular 14 days	41 (Nov.-May)	bi-annual	4-6
Duration of estrus	3-6 days	2 days	7-9 days	
Time of ovulation	25-27 hrs. post-coitum (induced)	? (spontaneous)	1-3 days after onset of estrus (spontaneous)	
Gestation (days)	65±4	111 (105-115)	63	24-26
Weaning Age (weeks) (weight)	7-8 (800 gm)	6-8 (130 gm)	6-8	3
Litter size at birth	4 (1-6)	2 (1-5)	4-8 (1-12)	4.5 (1-12)
First estrus after parturition	next heat 4-6 wks after lactation	immediately	next heat period	immediately
Reproductive life span of female (years)	8-14	>10	6-10	2 (6-10 litters)
Reproductive life span of male (years)	5-7	>10	6-14	2
Mating system	harem or hand bred	polygamous	hand bred	pair
Number of breeding females/1 male	15-30	4-12	60	1

	Guinea Pig	Hamster (Golden)	Mink	Monkey (Rhesus)
Weight for adult male	1000–1200 gm	90–120 gm	1.8 kg	12 kg
Weight for adult female	850–900 gm	95–140 gm	0.9 kg	10 kg
Birth weight	90 gm	2 gm		500–700 gm
Male breeding age (weight)	3–5 mos (550 gm)	2 mos (85–100 gm)	10 mos	6 yrs
Female breeding age (weight)	1–5 mos (500 gm)	2 mos (95–100 gm)	10 mos	5 yrs
Estrous cycle length (days)	16.5	4–5	7–10	28
Duration of estrus	6–11 hrs	20 hrs	2 days	none
Time of ovulation	10 hrs after onset of estrus (spontaneous)	8–12 hrs after onset of estrus (spontaneous)	36–48 hrs post-coitum (induced)	11–14 days after onset of menses (spontaneous)
Gestation (days)	68 (65–71)	16 (15–17)	51 (45–70) according to date of mating	165 (150–180)
Weaning age (weeks) (weight)	3.5 (250 gm)	3 (35 gm)	8	12–27 (0.9 kg)
Litter size at birth	4 (1–6)	6–10	4.4 (1–12)	1
First estrus after parturition	immediately	1–8 days		after weaning previous young
Reproductive life span of female (years)	4–5	1 (4–5 litters)	7	12–15
Reproductive life span of male (years)	>5	2	7	12–15
Mating system	colony	individual pairs, family	pair	pair or colony
Number of breeding females/1 male	3–10	8	5	10

Table 9-1. (*Continued*)

	Mouse (Laboratory)	Rabbit	Rat (Laboratory)
Weight for adult male	20–40 gm	4–5 kg	200–400 gm
Weight for adult female	25–90 gm	4–6 kg	250–300 gm
Birth weight	1.5 gm	60 gm	5–6 gm
Male breeding age	50 days	6–7 mos	100 days
(weight)	(20–35 gm)	(4 kg)	(300 gm)
Female breeding age	35–60 days	5–6 mos	100 days
(weight)	(20–30 gm)	(4 kg)	(200 gm)
Estrous cycle length (days)	4–5	—	5
Duration of estrus	10 hrs	continuous	13–15 hrs
Time of ovulation	2–3 hrs after onset of estrus (spontaneous)	10–11 hrs post-coitum (induced)	8–10 hrs after onset of estrus (spontaneous)
Gestation (days)	19 (17–21)	31 (30–32)	21 (20–22)
Weaning age (weeks)	3	8	3
(weight)	(7–15 gm)	(1.5 kg)	(40–50 gm)
Litter size at birth	11	8 (1–18)	8–12
First estrus after parturition		35 days	immediately
Reproductive life span of female (years)	(6–10 litters)	1–3	1
Reproductive life span of male (years)	1.5	1–3	1
Mating system	pair or colony	pair-mate in buck's cage	pair or colony
Number of breeding females/ 1 male	3	9	4

From: Reproduction and Breeding Techniques for Laboratory Animals (119) with permission.

Table 9-2. Comparative Anatomy of the Male Reproductive Structures

Species	Testis	Ampullary Glands	Seminal Vesicles	Prostate Gland	Bulbo-urethral Glands
Bat	3 X 2	4 X 15	A	5 X 3	1 X 1
Cat	14 X 8	A	A	5 X 2	4 X 3
Dog	40 X 30	F	A	25 X 16	A
Gerbil	14 X 9	6 X 5	20 X 12	9 X 7	6 X 4
Guinea pig	25 X 15	A	115 X 7	15 X 8	8 X 5
Hamster	14 X 11	3 X 3	11 X 6	8 X 6	4 X 3
Mink	11 X 8	F	A	10 X 5	A
Mouse	6 X 4	2 X 1	13 X 4	4 X 4	3 X 2
Opossum	15 X 11	A	A	40 X 9	19 X 15
Rabbit-Euopean	35 X 15	15 X 4	19 X 7	19 X 6	6 X 3
Rabbit-cottontail	35 X 17	F	A	25 X 6	6 X 2
Rat	20 X 12	4 X 4	20 X 10	13 X 10	5 X 3
Shrew	3 X 2	3 X 1	A	3 X 1.5	1 X 1
Squirrel-tree	30 X 12	F	7 X 4	28 X 8	13 X 13
Squirrel-ground	14 X 8	F	8 X 7	18 X 18	8 X 6
Subhuman primate	10 X 7	A	14 X 7	6 X 5	A
Ferret	F	F	A	F	A

Table 9-2. (Continued)

Species	Paraprostate Glands	Baculum	Urethral Glands	Bulbar Gland	Preputial Glands	Inguinal Glands
Bat	A	F	F	A	A	A
Cat	A	F	A	A	A	A
Dog	A	F	A	A	A	A
Gerbil	A	F	F	A	A	A
Guinea pig	A	F	A	A	PD	A
Hamster	A	F	F	A	3 X 1	A
Mink	A	F	A	A	A	A
Mouse	A	F	F	A	6 X 5	A
Opossum	A	A	A	A	A	A
Rabbit-European	6 X 2	A	A	A	A	15 X 7
Rabbit cottontail	A	A	A	A	A	10 X 4
Rat	A	F	F	A	16 X 4	A
Shrew	A	A	A	A	Diffuse	A
Squirrel-tree	A	F	A	10 X 7	A	A
Squirrel ground	A	F	A	10 X 8	F	A
Subhuman primate	A	F	A	A	Diffuse	A
Ferret	A	F	A	A	A	A

F indicates that a functional gland is present, but was not measured; A indicates that the structure does not occur; PD indicates that the gland is poorly developed.

* Measurements are in millimeters.

Reproduced from Reproduction and Breeding Techniques for Laboratory Animals (119) with permission.

Testes and Scrotum

The testis is an oval or pear-shaped mass which enlarges at puberty. Primates, dogs, cats, and opossums maintain constant size testes throughout their lives. In these species the male is capable of breeding at any time. In other species, such as shrews, ground squirrels, and some strains of mice, the testes undergo alternate regression and enlargement, and the male is capable of breeding only at certain periods.

The position of the testes is highly variable among the animal species. In primates including humans, carnivores (dogs), and omnivores (opossums), they descend into the scrotum and remain there permanently. In rats, rabbits, and bats, the testes are carried within the abdominal cavity part of the year. During the breeding season the scrotum of bats and squirrels enlarges and the testes descend through the inguinal canals and enlarge. In other rodents and rabbits, no true scrotum is present and the testes lie in the inguinal pouch.

Accessory Tubules

The epididymis consists of caput epididymis, corpus epididymis, and cauda epididymis. The caput epididymis forms the ductuli efferentes and ductus epididymis. The epithelium of ductus efferentes in nutria (*Moyocastor coypus*) consists of alternating tall and short groups of cells (275) resembling very closely those in humans. In cats, the epithelium is simple cuboidal with sterocilia on the surface. In the opossum, the lining is a mixture of ciliated and nonciliated cells (179). In rat epididymis, two types of cells (holocrine and principal) are present (204). The main function of ductus epididymis and ductuli efferentes is the movement of spermatozoa.

The epithelium of ductus deferens differs from species to species in animals; for example, it is stratified squamous in cats, simple squamous to columnar with basal cells in guinea pigs, high columnar with incomplete basal cells in the opossum, and folded or straight simple columnar in rabbits.

Accessory Sex Glands

The major accessory glands vary considerably in their anatomical characteristics in various laboratory animals. (Table 9-3). Other accessory glands that are restricted to certain experimental animals include bulbar gland in *Sciuridae* and *Aplondontidae* (227), paraprostate as evagination of the urethra in European rabbits (31), and inguinal glands (sebaceous and odor producing) in socially dominant European rabbits (229); urethral glands in bats and hamsters and smegma-scenting preputial glands in rats are located on either side of the distal end of the penis, while in primates they are located near the base of the glands (26). Secretory activity of male accessory sex glands and urethra of laboratory animals is summarized in Table 9-4.

Penis

In most laboratory animals except the opossum, the penis is located anterior to scrotum. Functionally, this organ transfers semen from male to female. Structurally, it consists of three cavernous (erectile) bodies. Some animals (bats, dogs, mustelids, squirrels, and some subhuman primates) have a bone (os penis) in the penile structure. The size and shape of the bone is characteristic of each species and changes with age (45, 134).

Table 9-3.   Anatomical Characteristics of Male Accessory Sex Glands of Laboratory Animals

Organ	Anatomical Characteristics	Species
Ampullary glands	Spindle-shaped enlargement composed of branched tubular diverticulae which open by ducts on entire circumference of the ductus deferens	Dog Mink Rabbit
	A distinct, abrupt enlargement composed of branched, tubular diverticulae, the proximal ends of which form the lumen of the ductus deferens	Bats Shrews
	Distinct lobes projecting from the ductus deferens	Mice Rats
	Shallow diverticulae in wall of ductus deferens	Guinea pig
Vesicular glands	Simple or branched tubule which coils to form compact glands	Squirrels
	Simple, straight or curved tubules	Guinea pig
	Straight or curved saccules	
	single and semibilobed	European rabbit
	bilateral structures	Mice Rat
Prostate gland	Disseminate	Opossum
	semi-disseminate, bilobed	Cat
	non-disseminate, bilobed	Bat
	mass subdivided by septa	Dog Mink Some primates
	Two or three pairs of distinct lobes	Mice Rats
Bulbo-urethral glands	Three lobed	Opossum
	Single lobed	
	ducts opening into distal spongy urethra via the bulbar gland and penile duct	Squirrels
	ducts opening into the proximal spongy urethra	All others

Reproduced from <u>Reproduction and Breeding Techniques for Laboratory Animals</u> (119) with permission.

Semen

The semen, representing the complete male secretions during normal ejaculation, consists of spermatozoa and seminal plasma. Some laboratory animals, such as rabbits, ejaculate a jelly-like copulation plug to assure retention of spermatozoa. The consistency of semen varies greatly. It is of low viscosity in dogs and rabbits, high viscosity in cats, and jelly-like in rodents and primates. A comparison of ejaculate volume and semen density is given in Table 9-5. The time required for spermatozoa to travel from testes to the distal end of the epididymis varies with species and the sexual activity of the male (253). In rabbits, this period is 1 day, in rats and mice 6 to 8 days, in hamsters 9 to 10 days, and in guinea pigs 14 to 18 days. The sperms are immotile, and their metabolic activity is low, due to low oxygen pressure, high levels of bicarbonate, mechanical crowding, and metabolic inhibitors. In the rabbit, the sperms are coated with a polysaccharide material which has to be removed by the female reproductive tract before fertilization can take place (299). The survival time of sperm for most species is between

Table 9-4. Histology and Secretory Activity of Male Accessory Sex Glands and Urethra of Laboratory Animals

Organ	Epithelium	Properties of Secretion
Ampullary glands	Simple, cuboidal, or columnar	Yellowish liquid containing ergothianeine, fructose, sialic acid, and phosphorus
Vesicular glands	Simple or pseudostratified columnar	White, yellowish or bluish fluid or gelatinous substance containing fructose, protein, citric acid, and ascorbic acid
Prostate gland	Simple or pseudostratified columnar	Colorless liquid containing fibrinolysin, fibrinogenase, diastase, carbonic anhydrase, amylase, vesiculase, fructose, citric acid, free amino acids, and zinc
Paraprostate glands	Pseudostratified cuboidal or low columnar	No data
Bulbo-urethral glands	Simple columnar	Viscous fluid containing sialoprotein
Bulbar gland	Simple high columnar	Liquid which facilitates passage of bulbo-urethral secretion
Urethral glands	Simple cuboidal	Clear, watery fluid containing mucoprotein
Preputial glands	Stratified squamous	Contains 7-dehydrocholesterol
Prostatic urethra	Transitional	None
Membranous and proximal spongy urethra	Stratified or pseudostratified columnar	None
Distal spongy urethra	Stratified squamous	None

Reproduced from Reproduction and Breeding Techniques for Laboratory Animals (119) with permission.

Table 9-5. Minimum Age and Weight to Begin Breeding Male Laboratory Animals

Species	Begin Breeding Life		Ejaculate Volume (ml)		Density (sperm/ml×10^8)	
	Age	Weight	Range	Average	Range	Average
Cat	9 mos	3.5 kg	0.01-0.3	0.04	1.5-28	14
Dog	10-12 mos	-	2-25	9.0	0.6-5.4	1.3
Guinea pig	3-5 mos	450 gm	0.4-0.8	0.6	0.05-0.2	0.1
Hamster	2 mos	90 gm	-	-	-	-
Monkey (M. mulatta)	6 yrs	9 kg	1.0-4.5	1.8	0.9-8.1	4.0
Mouse	45 days	25 gm	-	-	-	-
Rabbit	4-12 mos	-	0.4-6	1.0	0.5-3.5	1.5
Rat	45-50 days	225 gm	-	-	-	-

Reproduced from Reproduction and Breeding Techniques for Laboratory Animals (119) with permission.

50 to 70 days; however, they may lose their ability to produce viable embryos much earlier. In the guinea pig, the incidence of nonviable embryos increases from a normal of 4 percent to 20 percent if the sperms are retained for more than 20 days (309).

The total length (tail and head) of sperms from man, rabbits, dogs, and domestic animals range from 0.55–0.65 mm, while the spermatozoa of mice, rats, and hamsters are several times longer. The sperm head shows morphological variation. In fowl, it is elongated and cylindrical; in rats and hamsters, it is elongated and hooked; in rabbits, cats, dogs, and most other animals, the head is ovoid; the guinea pig has flat headed sperms with a large wrap-around acrosome; and the mouse has a flat headed sperm with a spicule-like hook at the anterior end. In man, the sperm head is almond-shaped.

The seminal plasma varies considerably in volume and composition (201, 301, 302). It appears that this fluid is not essential for fertilization. However, it may play an important role in natural breeding by providing diluent effect, buffer effect and pharmacodynamic effect due to prostaglandins. Chemically, fructose and lactic acid are most commonly present but sorbitol, phospholipids, amino acids, and fatty acids may be catabolized to a limited extent. The plasma of dogs consists of three fractions and has been thoroughly studied (22) (Table 9-6). Rabbit seminal plasma contains a high content of catalase, which makes rabbit sperm relatively resistant to the action of hydrogen peroxide (201).

Table 9-6.   Chemical Components of Dog Seminal Plasma

Components	Unfractionated ejaculate	1st Fraction	2nd Fraction	3rd Fraction
Dry weight (%)	2.5	2.1	3.1	2.5
pH	6.4	6.2	6.3	6.5
Na (mEq/1)	114	–	192	172
K (mEq/1)	8.1	5.8	12.4	7.5
Ca (mEq/1)	0.7	1.4	0.4	0.5
Mg (mEq/1)	0.7	1.5	0.8	0.3
Cl (mEq/1)	152.3	–	206.3	160.1
Lactic acid (mg/dl)	17.5	9.0	27.2	16.3

[1]Average values given for unfractionated ejaculate and seminal fractions in the order of appearence at ejaculation.
From Reproduction and Breeding Techniques for Laboratory Animals (119) with permission.

The deposition of semen in the female tract varies with the species. In dogs and pigs, the insemination of a large amount of semen is uterine with temporary retention of the penis. In guinea pigs, rats, and mice, small amounts of gelled semen are discharged into the uterus, while vaginal insemination with or without gel plug is seen in cats, rabbits, and primates.

The sperms may be ready for fertilization, or prefertilization (capacitation) may be necessary in certain species (the rat, rabbit, hamster, ferret, and sheep). The time required for capacitation varies from 1½ hour in sheep to about 6 hours in rabbits. The physiological requirements, including removal of polysaccharide layers and loosening of achromsomes, have been investigated (3, 124).

FEMALE REPRODUCTIVE SYSTEM OF LABORATORY ANIMALS

The female reproductive organs include ovaries, oviduct, uterus, vagina, and external genitalia. The development of these organs is controlled by hormones. A summary of female reproductive anatomy in experimental mammals is given in Table 9-7.

Ovaries

Ovaries, like the testes, are paired organs, but unlike testes, they remain in the abdominal cavity. These glands have dual exocrine (ovum producing) and endocrine (estrogen, progesterone, and androgen secreting) functions.

The ovaries of the rabbit are located in the ovarian bursae, whereas in the rat, ovaries are enclosed in a periovarial sac. Injury or rupture of the periovarial sac could cause fertility problems probably resulting from the escape of eggs. The germinal epithelium of the ovary usually has cuboidal or low columnar cells. The medulla contains large polyhedral interstitial cells which are more prominent in the ovaries of rodents and carnivores than in primates and ungulates (126). The Graafian follicles differentiate from the germinal epithelium which lies on the surface of the ovary, and often has the appearance of a blister. The egg-containing follicles reach maximum size just before ovulation. The number of follicles depends on the species and environmental conditions. In primates, including humans, one or two eggs are released, while rodents release between four and fourteen. Polyovular follicles occur in a number of species.

After the release of ova the follicular cavity is filled with blood and lymph (corpus hemorrhagicum). Later, the corpus luteum develops and grows rapidly for about half the estrous cycle. If fertilization does not occur, it regresses and corpus albicans is formed, which disappears after two to three estrous cycles. The mammalian corpus luteum is a short-lived progesterone-secreting endocrine gland. The exact regulatory mechanisms are not known, but it appears that the nongravid uterus plays an important role in guinea pigs, rats, and ungulates.

Oviduct

The oviduct is an important organ because its fluids provide the environment for capacitation, fertilization, and early stages of embryonic development. The oviduct is divided into the infundibulum, the ampulla, the isthmus, and the uterotubal junction. Length, coiling, and other features vary with species; in mouse it is coiled into ten loops, whereas it is almost straight in rabbits. The development of fimbriae around the infun-

Table 9-7. Summary of Female Reproductive Anatomy in Laboratory Mammals

Organ	Main Anatomical Types	Species
Ovarian bursa	Infundibulum funnel-shaped and lies close to ovary forming an open ovarian bursa	Rabbit, primates
	Infundibulum and ovary enclosed by a fold of mesosalpinx which forms a periovarial sac	Rat, mouse
Fimbriae	Extensively developed fimbriae enclose the ovarian surface	Rabbit, cat
	Fimbriae moderately developed	Primates
	Poorly developed fimbriae make very limited contact with ovarian surface	Rat, mouse
Uterotubal junction	Very complex; glandular mucosal lip; projecting papilla	Guinea pig
	Rosette type projections; no intramural portion	Rabbit
	Single papilla; short intramural portion of oviduct	Rat
	Single mound; short intramural portion of oviduct	Dog
	Simple fold which projects into a pocket in uterine wall; no mound	Mink
	No projections; extended intramural portion	Primates
Uterus, cervix and vagina	Two uteri; two cervices; two vaginas	Opossum, kangaroo
	Two uteri; two cervices; one vagina	Rabbit, rat (Norwegian), chinchilla
	Two well developed horns, small uterine body; one cervix; one vagina	Cat, dog
	One uterine body without compartments; one cervix; one vagina	Primates

Reproduced from Reproduction and Breeding Techniques for Laboratory Animals (119) with permission.

303

dibulum is inversly related to the presence of the ovarian bursa. Minks, mice, rats, and dogs have poorly developed fimbriae, while rabbits, primates, and ungulates, with no bursae, have well-developed fimbriae (36).

The isthmus has a narrow lumen and apparently acts as a sphincter in several species (rabbit, cat, dog, rat, and humans). Epithelial folding, which is nonexistent in the mouse and rat, is very extensive in primates and rabbits. Ciliated cells, which are rare in mice and rats, are present in the guinea pigs but most prominent in rabbits and primates.

The uterotubular junction in primates has extramural and intramural portions. The junction is in the form of an elevated slit in dogs, a rosette-like structure in rabbits, papilla in rats, but has no projections in primates (120).

Mammalian Egg

The complex structure of the mammalian egg has been described by Sotelo and Porter (273). The structure of the egg, particularly its membranes, varies among species. Fish and amphibians have gelatinous envelopes, monotremes and marsupials have several layers of albumin, and lagomorph eggs have mucopolysaccharides.

Egg transport takes place rapidly to the ampullary-isthmic junction where it stays up to 50 hours in most mammals; fertilization occurs at this location. Transport of the ovum depends on kinocilia, smooth muscle of the uterus, and ovarian hormones. Fertilized or unfertilized eggs reach the uterus in 2 to 3½ days after ovulation, and the time is related to the activation of the corpus luteum. The fertilizable life of the egg varies with species, but it is usually less than 24 hours.

Uterus

Several types of uteri occur in animals and the gross anatomical features are illustrated in Fig. 9-1. Uterine activity is under complex endocrine control. The uterus has several important functions. At copulation, its contractile action facilitates the transport of sperm into the oviduct. Prior to implantation, it produces fluids that sustain the blastocyst. After implantation, it participates in the formation of the placenta and is the site of fetal development. At parturition, it plays the major role in delivery (fetal expulsion). The uterus undergoes tremendous changes in size, structure, and position in order to accommodate the growing conceptus. Nonetheless, it returns to nearly its former size and condition after parturition, by a process called involution.

Cervix

This muscular sphincter projects from the posterior end of the uterus into the vagina. It has a small lumen and thick wall containing annular rings. The cervix seals off the uterine lumen from the outside environment except during estrus, when it is relaxed for the entry of sperms. During pregnancy, it has a mucous plug which dissolves at parturition. The mucus consists of sialoglycoproteins, free sugar, glycogen, cholesterol, lipids, amino acids, and minerals (296). In primates, ruminants, and rabbits, the uterine cervix acts as a sperm reservoir. In horses, rats, and mice, the sperms are deposited in the uterus and the uterotubal junction is the sperm reservoir.

Vagina

The vagina acts as a copulatory canal and its exact location varies with species. In the guinea pig, it is sealed off by a membrane, which ruptures spontaneously during estrus; in the rat and rabbit, the vaginal opening appears at the time of first ovulation.

Duplex

Bicornuate

Two cervices

Uterine horns
completely separated

Prominent uterine horns

Rabbit
chinchilla

Guinea
pig

Swine
insectivores

Bipartite

Simplex

A septum
separates horns

Prominent
uterine body

Prominent uterine body
no horns

Cat, dog
cattle, sheep

Horse

Primates

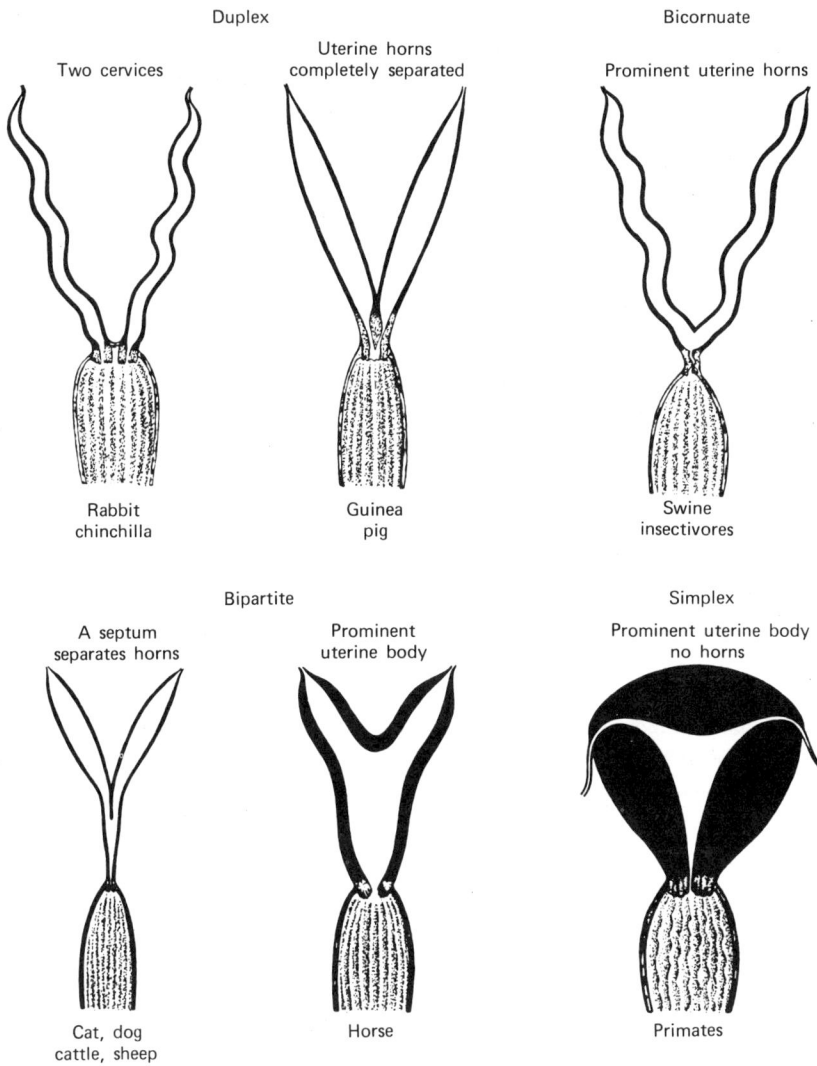

Fig. 9-1. The relative length of the uterine horns and the uterine body in various laboratory mammals. Reproduced from *Reproduction and Breeding Techniques for Laboratory Animals* (119) with permission.

External Genitalia

The anatomical feature of external genitalia consisting of vaginal vestibule, labia majora and minora, and clitoris vary considerably in animal species. True labia are present only in human females.

Some primates (baboon, chimpanzee, mangabey, mandrill, celebes, and black ape) have well-developed sexual skin. The sexual skin undergoes changes in color and general conformation during the menstrual cycle. These changes are most pronounced in the pig-tailed monkey and the baboon but are absent from the rhesus monkey.

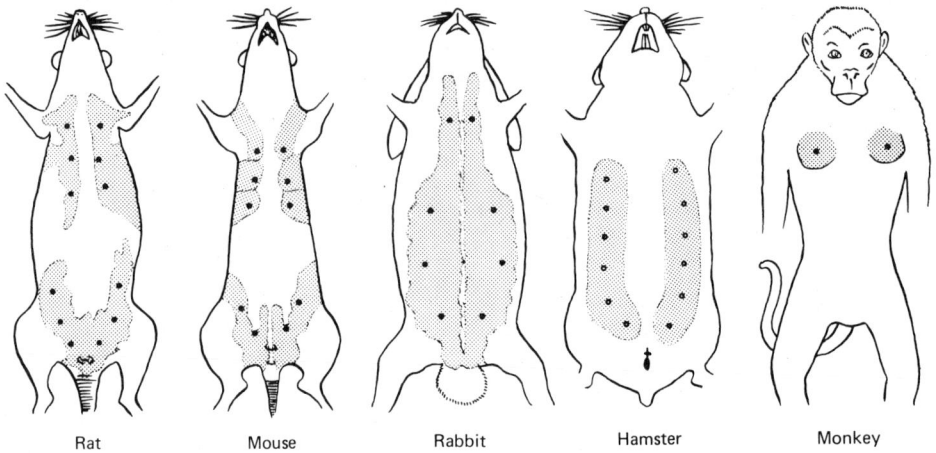

Fig. 9-2. Location of the nipples and size of the mammary glands in common laboratory mammals during lactation. Reproduced from *Reproduction and Breeding Techniques for Laboratory Animals* (119) with permission.

Mammary Glands

The number and position of mammary glands are characteristic of each species (Fig. 9-2). The abdominal glands are usually the largest and produce the most milk. Each mammary gland is a streak canal. The latter is lined by stratified squamous epithelium. The mammary tissue itself is divisible into lobules, each of which is composed of many alveoli. The alveolus is the secretory unit of the gland. It is lined by a single layer of cuboidal to columnar epithelial cells and surrounded by branched myoepithelial cells. The latter contract under the influence of oxytocin and cause the discharge of accumulated milk within the alveolus.

CHARACTERISTIC FEATURES OF REPRODUCTIVE ORGANS IN ANIMAL SPECIES

Rats, Mice, Hamsters, and Gerbils

Male Reproductive Organs

1. The penis is flexed posteriorly.

2. The glans lies within a nonprotruding prepuce.

3. A pair of well-developed preputial glands lies between the skin and ventral body wall and lateral to the penis (except in the gerbil).

4. A pair of large seminal vesicles is present. They consist of sac-like structures which are flexed ventrally at the anterior end.

5. Ampullary glands form two distinct lobes from each ductus deferens.

6. The prostate is well developed and consists of three pairs of lobes. The anterior lobe is known as the coagulating gland and lies along and is attached to the ventral curvature of the seminal vesicle.

7. A urethral sinus is located proximal to the bulb of the penis.

8. Paired bulbourethral glands are located dorsolateral to the bulb of the penis.

9. The testes lie in a scrotal sac, the posterior end of which may be partially or completely devoid of hair.

10. In the rat, mouse, and gerbil, an enlongated, prominent fat body extends from the testis along the spermatic cord. Its anterior end is free of the cord and extends well into the coelom. In hamsters the fat body is less developed.

Female Reproductive Organs

1. In rats, the ovary lies within the ovarian bursa, whereas in mice it lies just below (ventral) the kidneys within transparent ovarian capsules. In hamsters, the ovaries are compact and encapsulated.

2. In the rat, periovarial space opens into the peritoneal cavity through a slit on the antimesometrial side of the bursa at the tip of each uterine horn. In the mouse, a narrow, tunnel-like passage connects the periovarial space with the peritoneal cavity.

3. In the hamster, two cervical canals remain separate for about two-thirds of the length of the cervix and then fuse.

4. The vagina of the hamster has a muciferous type of epithelium and its wall contains urethral glands similar to those in the prostate.

5. The rat has twelve mammary glands made up of two ventrolateral series along the thoracic and inguinal regions; the mouse has ten mammary glands (six thoracic and four abdominoinguinal); the hamster has twelve or fourteen mammary glands, which are located in the thoracic and abdominal regions.

Guinea Pigs

Male Reproductive Organs

1. The penis of the guinea pig curves caudally, and the glans is covered with cornified epithelial spines or papillae, which occur singly or are arranged in incomplete circular rows.

2. The preputial opening lies between posterior pads, which are covered with hairless skin of the perineum.

3. The seminal vesicles consist of a pair of greatly enlongated, uncoiled tubes that are circular in cross section and taper toward their free end. The ductus deferens joins the seminal vesicle; the seminal vesicles then unite and all four structures open into the urethra by a single wide canal.

4. The prostate consists of three pairs of lobes.

5. The testes lie in a scrotal pouch and their posterior end and cauda epididymis produce a slight bulge in the perineum.

6. A large fat body covers the anterior end of each testis; it lies along the spermatic cord and projects through a large inguinal canal into the coelom. The testes may be withdrawn into the coelom through this canal at any time.

Female Reproductive Organs

1. The guinea pig has two internal cervical openings, but only one common external os.

2. Intestinal and urinary tracts open into a groove (the fossa anovagi-nourethralis).

3. The lower end of the vagina is closed by an epithelial membrane but opens periodically at estrus and during parturition.

4. The guinea pig has two mammary glands in the inguinal region.

Rabbits

Male Reproductive Organs

1. The rabbit has a posteriorly directed penis that lies in a prepuce which hangs in a fold of skin below the body wall and opens into a single, short, ejaculatory duct along with the excretory duct of the seminal vesicle.

2. A vesicular gland is present in the European rabbit (*Orctolagus cuniculus*) but not in the North American cottontail (*Syvilagus floridanus*). It lies dorsal and medial to the ampullary glands and dorsal to the anterior urethra.

3. The prostate consists of two distinct lobes, anterior and posterior, in both the European and cottontail rabbit.

4. Bulbourethral glands are present in both species, but they are better developed in the European rabbit than in the cottontail.

5. Testes are relatively large and flaccid in sexually active animals. They lie in inguinal pouches that have a large canal through which the testes can be withdrawn into the coelom even when they are enlarged.

Female Reproductive Organs

1. The rabbit has complete duplication of the uterine segment with two long uterine horns and two entirely separate cervical canals, each of which has an internal and external os.

2. The endometrium is arranged in numerous transverse and longitudinal folds which are particularly prominent along the mesometrial borders.

3. Cervical canals have a narrower lumen and more extensively folded mucous membrane than the uterine horns.

4. Vaginal portions of the cervical segments are surrounded by a complete ring of the fornices.

5. The rabbit has eight mammary glands arranged in ventrolateral series.

Cats

Male Reproductive Organs

1. The domestic cat has a posteriorly directed penis encased in a free prepuce which lies just ventral to the scrotum.

2. Spines cover the penis and are directed proximally.

3. A baculum is present.

4. The two lobes of the prostate are separated by a connective tissue septum.

5. Numerous prostrate-collecting ducts open on a well-developed colliculus seminalis just posterior to the ductus deferens openings.

6. A pair of bulbourethral glands lies anterolateral to the base of the penis.

7. The testes are permanently located in the scrotum. They do not undergo seasonal atrophy.

Female Reproductive Organs

1. The ovary is ovoid and lies within a peritoneal fold.

2. The cervix is remarkably short and it lacks a true internal os; it is directly continuous with dorsal wall of vagina.

3. External os is V-shaped.

4. The cat has eight mammary glands; all or some may function at one time.

Dogs

Male Reproductive Organs

1. The penis lies in a fold of skin beneath the ventral body wall.

2. A well-developed baculum lies in the distal two-thirds of the organ; the bone curves ventrally at its distal end.

3. The glans extends along the entire length of the baculum.

4. During erection, a distinct *urethral process* with the external urethral orifice at its apex is present at the free extremity of the *pars longa glandis*. A *corona glandis* is present on the distal end of the pars longa.

5. The prepuce retracts onto the glans about midway to the bulbus.

6. A retractor penis muscle, composed mostly of smooth muscle, extends from near the preputial fornix to the first and second coccygeal vertebrae.

7. Dogs (and foxes) have no accessory sex glands except the prostate and poorly developed ampullary glands.

8. The prostate is a well-developed gland which completely surrounds the urethra.

9. Numerous collecting ducts open on the entire urethral circumference.

10. The testes lie within the pendulous scrotum at all times, and they do not undergo seasonal atrophy.

Female Reproductive Organs

1. The ovary is flattened and completely enclosed in a roomy peritoneal pouch.

2. Slender uterine horns are long and straight.

3. The cervix is a short, thick-walled segment.

4. The vagina is wider above (cranially) than below.

5. The dog has ten mammary glands arranged in two ventrolateral series.

Nonhuman Primates

The reproductive organs of nonhuman primates vary among families and among genera of the same family (83). The main characteristics of reproductive organs of two families *Ceboidea* or New World monkeys (the squirrel, the spider, and the howler monkeys) and *Cercopithecoidea* or Old World monkeys (the rhesus monkey and baboon) are described below.

Male Reproductive Organs of Monkeys

New World Monkeys

1. These monkeys have a short, stout, cylinderical penis; the glans projects from the prepuce; and the entire organ projects from the anterior scrotum.

2. Squirrel monkeys have cornified epithelial spines on the glans and distal stalk; spider monkeys have black spines.

3. A small baculum is present in squirrel monkeys but is absent in spider monkeys.

4. Seminal vesicles are well developed and, in some species, their ducts join with the ductus deferens to form ejaculatory ducts.

5. No ampullae are present.

6. Bulbourethral glands are either lacking or very small.

7. The testes are suspended in a scrotum which varies from a shallow pouch to a pendulous structure.

8. The epididymis is large in proportion to the size of the testis and, in squirrel monkeys, the caput epididymis is approximately one-third the size of the testis.

Old World Monkeys

1. In these monkeys, most of the penis lies in the anterior wall of the scrotum.

2. A bulb-like glans lies at the end of the neck; the latter is surrounded by the prepuce.

3. A well-developed baculum is present.

4. Ampullae are lacking or poorly developed.

5. Seminal vesicles are large in the rhesus monkey and join the ductus deferens to form ejaculatory ducts.

6. The ducts and lower seminal vesicles are surrounded by the lobulated cranial lobe of the prostate which is comparable in function to the coagulating glands of the rodents.

7. A smooth, dark caudal lobe is also present.

8. Small bulbourethral glands are located just below the caudal lobes of the prostate.

9. Testes of the rhesus monkey lie within a well-developed scrotum at birth, but ascend to near the inguinal canal shortly thereafter. They descend again and remain permanently in the scrotum after sexual maturity is attained.

10. The caput epididymis is well developed, as it is in New World monkeys.

Female Reproductive Organs of Monkeys

1. The uterus of the monkey is divided into an upper segment, consisting of the body and fundus, and a lower segment, formed by the thick-walled cervix. The fundus is irregularly pear-shaped and has a shallow depression.

2. Sexual skin does not usually swell but undergoes variations in color which are not always clearly related in the menstrual cycle.

3. The monkey has two mammary glands which are poorly developed except during pregnancy.

4. Supernumerary nipples are sometimes present below normal nipples.

SEXUAL CYCLES

The sexual cycle of the female is a complex phenomenon involving neuronal and hormonal factors that influence fertilization, implantation, and developmental processes. The reproductive cycle may be defined according to the effect of season on breeding and the number of periods of reproductivity occurring within a season. The reproductive pattern may be classified further by type of ovulation, reflex or spontaneous, and by type of cycle, estrous or menstrual (Table 9-8).

Puberty

The age of puberty varies among different species and within species between the male and female. Growth toward puberty is under hormonal control. The major gonadotropins are follicular stimulating hormone (FSH), luteinizing hormone (LH), and lactogenic or prolactin hormone. Besides the specific hormones, other pituitary gonadotropic hormones also affect prepuberal and puberal growth. The pituitary of immature animals contains a considerable amount of gonadotropins, but their release is either inhibited or unintegrated, thus resulting in a nonfunctional reproductive system (92). In immature rats, rabbits, pigs, and cattle, the prepuberal pituitary levels of FSH and LH exceed those of postpuberal animals. In immature males, the concentration of gonadotropins is also high, and FSH is required for testicular growth. Unlike in females, FSH administration in males does not elicit gonadal growth beyond adult size. Apparently, FSH is not involved in spermatogenesis. Fetal gonads, with the possible exception of those of male rabbits, are not under the influence of fetal pituitary gonadotropins (147). Steroid sensitivity of the pituitary appears to be another variable dependent on sex hormones; in rats, it has been shown that administration of androgens or estrogens to neonatal rats during a critical period (1 to 5 days of age) blocks the integration of the central nervous system control of pituitary functions (21).

Age and Weight at Puberty

Puberty occurs prior to the attainment of mature body weight. It is characterized by an abrupt onset concomitant with a shift in hormonal balance. There are considerable differences in the age of onset of puberty in laboratory animals. The differences are species and strain dependent, particularly in rabbits, mice, and dogs (11, 232). The season of birth affects puberty particularly in rabbits and gerbils. The ambient temperature also has an important influence. The onset of puberty may be advanced or delayed by administration of hormones such as TSH (thyroid-stimulating hormone) and triiodothyroxin. Inbreeding for selection of a pool of genetic traits often delays puberty. By selective breeding, strains can be produced with a particular trait, like hypogonadism in rabbits, for experimental purposes.

In contrast to humans, the photoperiod regulates the reproductive function in animals. The exact mechanism is not clear, but the effect is mediated through light receptors in the retina, pineal gland, and hypothalmus (61). In rats, continuous light induces persistent estrus, prolonged vaginal cornification, and cystic follicular ovaries devoid of corpus luteum. However, light does not affect the estrous cycle or the vagina in hamsters. In ferrets, light affects the pineal gland (71). The ovarian weights of rats are dependent on the age of the animal and its exposure to light. This has been related to levels of FSH and LH. Other features affected by light include adrenal cortical steroid levels in mice, rats, and monkeys; ovarian ascorbic acid levels in psuedopregnant rats; ovulation-inducing hormones in chickens; time of the day that the guinea pig comes into estrus; prolactin content of the pituitary, blood, and pituitary FSH in rats (61, 304).

Table 9-8.   Effect of Species on Type, Season, and Selected

Species	Type and Season of Cycle	Length of Sexual Cycle (days)	Duration of Heat
Cat	Polyestrus seasonal	15-28	4-10 days ♂ depend.
Chinchilla	Polyestrus	~24-28	2 days
Dog	Monestrus (spring and fall)	22	7-13 days
Ferret	Polyestrus (April-August)		In absence of ♂ 5 months
Gerbil	Polyestrus	4-6	12-18 hrs
Guinea pig	Polyestrus	16-19	6-15 hrs
Hamster	Polyestrus	4	4-23 hrs
Mink	Polyestrus	8-9	
Mouse	Polyestrus any time	4-5	9-20 hrs
Opossum	Polyestrus seasonal	22-38	1-2 days
Rabbit	Polyestrus any time	15-16	1 month
Rat	Polyestrus any time	4-6	9-20 hrs
Squirrel	Diestrus	10	
Subhuman Primates			
Marmoset	Menstrual?		
Pigtail macaque	Menstrual all year	32	
Rhesus	Menstrual all year	23-33	
Squirrel monkey	Menstrual seasonal	12	range 6-18 days recept. 7-8 days
Chimpanzee	Menstrual	37	menstruation 3 days

The nutritional status of the animal is very important. Decrease of 15 to 30 percent in food intake in rats results in retarded growth, loss of body weight, cessation of estrus, atrophy of testes, loss of libido, and decrease in androgen secretion (183). Both the amount and type of dietary proteins affect the reproductive capacity of rats (184). Protein deficiency results in decrease in testicular size, decrease in gametogenesis, decrease in tubular RNA, and increase in lipids, while high protein content causes increase in pituitary gonadotropin, retarded ovarian growth, and delayed initiation of estrous cycles (185). Essential fatty acid deficiency in rabbits results in diminished secretion of anterior pituitary hormones (6).

Reflex Ovulation

Ovulation is a complex mechanism under neurochemical control. A releasing factor (RF) is released by the hypothalamus in response to mechanical and physical stimuli. The RF reaches the pars nervosa of the pituitary, releasing the gonadotropins, which, in turn, reach the target organ by the circulatory system and cause changes in the Graafian follicle for the release of the ovum (285). In mammalian systems, LH plays an important role. Rabbits have been extensively used as models for these studies. Ovulation takes place ten hours after LH stimulation in rabbits. The presence of the vagina or intromission is not necessary for threshold stimulation of the central nervous system (276).

Characteristics of the Sexual Cycle in Laboratory Mammals

Ovulatory Mechanism	Time of Ovulation	Viability of Ova	Number of Eggs Shed
Reflex	24-36 hrs postcopulation		4-6*
			4
Spontaneous	Near beginning of estrus	5-8 days	8-10*
Reflex	30 hrs postcopulation	30 hrs	5-13
Spontaneous			1-7
Spontaneous	10 hrs from onset of estrus	20 hrs	2-4
Spontaneous	Early estrus	10 hrs	1-12
Reflex	42-50 hrs postcopulation		8-10
Spontaneous	2-3 hrs from onset of estrus	10-12 hrs	6*
Spontaneous	Early in heat		22
Reflex	10 hrs postcopulation	8 hrs	10*
Spontaneous	8-11 hrs from onset of estrus	10-12 hrs	10*
Reflex			1-4
Many cycles, not necessarily ovulatory			1-3
Spontaneous	Not known		
Spontaneous	11-15 days from start of cycle	24 hrs	1**
Ovulation not necessarily spontaneous			1**
Spontaneous	22-28 days from start of menses		1**

*Vary with strain or breed;   ** Ovaries alternate ovulation
Reproduced from Reproduction and Breeding Techniques for Laboratory Animals (119) with permission.

In laboratory rabbits, copulation is not always sufficient for ovulation. In addition, a specific threshold of stimulation is necessary (99). The pathways involved in rabbits are shown in Fig. 9-3.

Estrous Cycle

In mature, nonpregnant females, a species-specific rhythmic estrous cycle occurs. The onset, duration, and other relevant data are shown in Table 9-8. The cycle in most mammalian species consists of estrus, metestrus, diestrus, and proestrus. In animals, ovulation is synchronized with estrus to ensure a high degree of fertility. Estrus in animals is under the control of photoperiodic environment and pheromones. The hormonal control is illustrated in Fig. 9-4.

Menstrual Cycle

The menstrual cycle, in principle, is similar to the estrous cycle. However, the receptivity of the female is over a wider period of time or throughout the cycle and the sloughing of the superficial layer of endometrium differentiates it from the estrous cycle. The four phases of the menstrual cycle include menstrual period, proliferative period (follicular), ovulatory period, and progestational period (luteal period).

The changes in the endometrium are under the influence of progesterone, LH, and

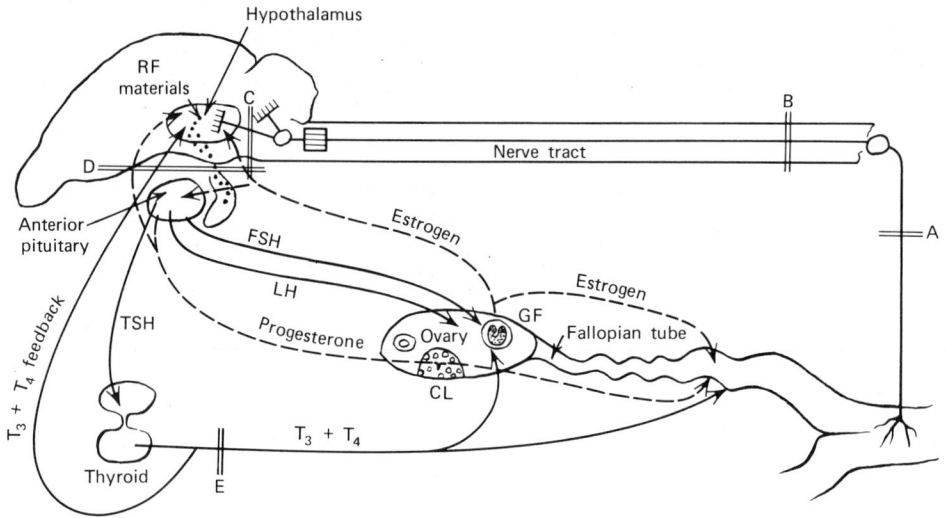

Fig. 9-3. Diagrammatic illustration of the pathways involved in reflex ovulation in the rabbit. Points A, B, C represent sites in the central nervous system where ovulation may be blocked. D represents an ovulatory block due to hypophysectomy, while E indicates an ovulatory block due to hypothyroidism. CL = corpus lutea, GF = Graafian follicle, RF = releasing factor, T_3 and T_4 = tri- and tetraiodothyronine, and TSH = thyroid-stimulating hormone. Reproduced from *Reproduction and Breeding Techniques for Laboratory Animals* (119) with permission.

FSH, and the hormonal changes involved in the menstrual and estrous cycles are similar. The length of the menstrual cycle varies in primates; it is 12 days in the squirrel monkey and 37 days in the chimpanzee.

Fertilization

Most of the mammals produce an egg covered with cells of cumulus and with gelatinous, residual antral fluid. The eggs have to be denuded while the sperms have to be capacitated before fertilization can take place. In the rabbit, most of the achrosome is lost during the penetration of the cumulus mass (27). Hyaluronidase mediates this process in experimental rabbits, but it is less effective in dogs and foxes (89). The first sperm that enters the zona pellucida enters the inside of the ovum after a lag period of a few minutes. The midregion of the sperm head is first partly incorporated into the egg. Normally one sperm enters the egg. However, in some species, for example, rats, mice, and rabbits, occasionally an additional sperm may be taken into the cytoplasm, particularly in the older ova of the rabbit. At any rate, fusion of the genetic material of one sperm with each egg takes place and this excludes the incorporation of more than one sperm into an egg (16, 17). In most laboratory animals, the ova are in the metaphase of the second meiotic division at the time of penetration, with the exception of the dog, in which the first polar body is released after fertilization. After penetration, the two nuclei become adjacent to one another, the pronuclei decrease in volume, and the nucleoli coalesce and then disappear. The chromosomes from both gametes are organized on a single spindle, thus marking the end of syngamy and fertilization. Replication of DNA occur in 3 to 6 hours after fertilization (286). Despite the presence of induced ovulation in certain species where sperm are in the oviduct for some time (10 hours in rabbits and 48 hours in minks) the fertilizable life of ova is between 12 to 20 hours with the exception of the ferret (30 to 36 hours). However, older ova are more prone to abnormalities such as polyspermia, gynogenesis (failure of sperm head to participate in the first cleavage), androgenesis (failure of female pronucleus to participate in the first-cleavage), and polypoidy (failure of the second meiotic division of the ovum) (214). Androgenesis has been noted in mice (203).

Several attempts have been made to effect in vitro fertilization in laboratory animals. In vitro fertilization has been achieved with satisfactory results in the hamster and rabbit. In rabbits, viable newborn rabbits have been obtained from in vitro fertilized eggs (53). Several attempts have been made in primates but with limited success.

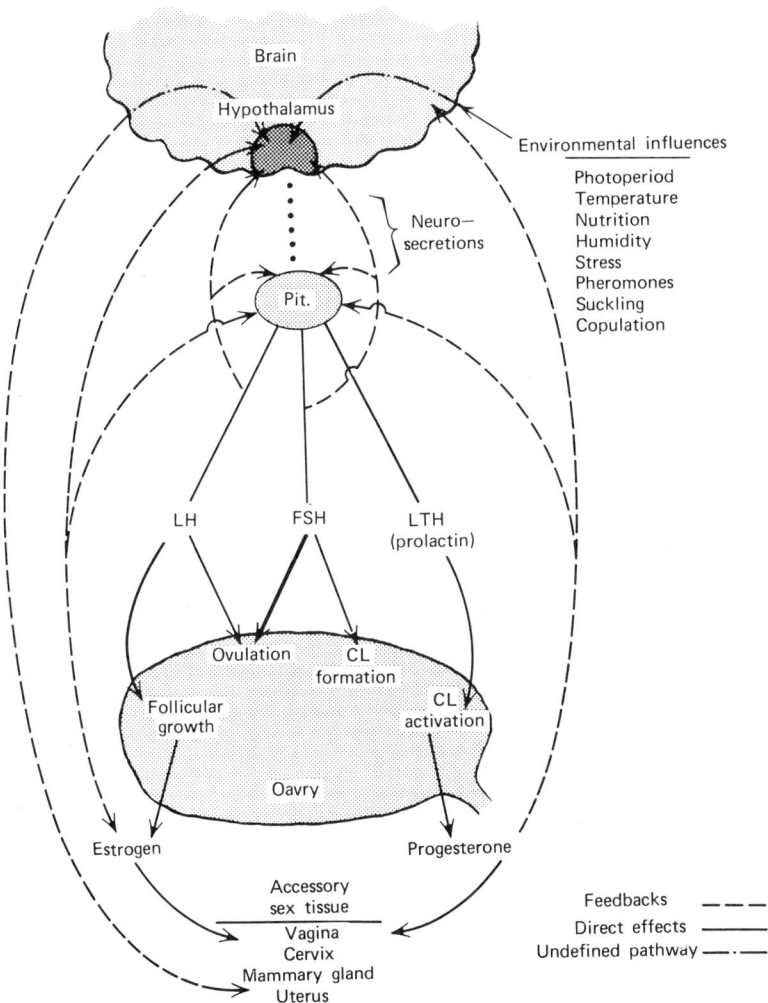

Fig. 9-4. Diagram illustrating the hypothalamic-pituitary-gonad axis hormone relationships regulating the mammalian estrous cycle. Individual species differences are not shown. Reproduced from *Reproduction and Breeding Techniques for Laboratory Animals* (119) with permission.

Implantation

A uniform feature of the fertilized ovum after it enters the uterine cavity in humans and animals is its transformation into the blastocyst. The morula forms a cavity that enlarges and expands rapidly to line the zona pellucida. Certain cells gather at one pole and form the inner cell mass from which the embryo develops. The layer of epithelial cells on the zona pellucida comprises the trophoblast. When primitive mosoderm lines the trophoblast, it becomes the chorion. In most laboratory animals, the morula or early blastocyst stage has been reached in the few days preceding its entry into the uterus from the oviducts. The blastocyst remains free in the uterine lumen for a variable period of time (rat and mouse, 36 to 48 hours; guinea pig, 6 days; macaque, 9 days; baboon, 7 to 8 days; cat, 13 days; and human, at least 5 to 6 days). In cases of delayed implantation, nidation may be deferred for days and has been reported delayed for even as long as 10 months (European badger, *Meles meles*). Under conditions of delayed implantation, the appearance of the blastocyst remains largely unchanged.

The development and appearance of the blastocysts has been studied in the baboon (128), in the lesser bush baby (47), and in the rhesus monkey (130). The blastocyst of each species is oriented in utero in a specific direction for implantation, and this orientation is correlated with the later development of fetal membrane (226). In the guinea pig, mouse, and rat, the embryonal pole of the trophoblast is directed toward the endometrium for attachment. Butler (48) and Hendricks (128) have described blastocyst orientation and its attachment and invasion into the maternal endometrium, in nonhuman primates. Unfortunately, very little is known concerning nidation in nonhuman primates. The major episodes of implantation are usually described as (a) the development of the decidual reaction of the endometrium or the preparation of the endometrium to establish a proper environment for implanting the blastocyst; (b) orientation of the blastocyst so that its trophoblast may make firm attachment and invade the endometrium or at least make sufficiently intimate contact to obtain nourishment; and (c) the invasive process itself in which the trophoblast either actively penetrates the endometrium or is passively engulfed by the nidus (278). The induction of decidualization and its rapid development constitute the major uterine contribution to the whole process of blastocyst attachment and implantation in the rat (62, 195). Shelesnyak et al. (267, 278) have demonstrated that in normal pregnancy in the rat, decidualization involves various histaminergic stimuli through the release of histamine. These findings have been confirmed by the uterine freezing technique of Nalbandoy (231) and Zipper et al. (307) in which the mast cells are selectively destroyed. Thus histamine cannot be mobilized and decidualization does not occur. Decidualization in the guinea pig and possibly in primates is related to the initial active invasion of the blastocyst into the maternal endometrium. However, decidualization in the guinea pig is not initiated unless there is an active invasion of endometrial stroma by the trophoblast. Some kind of injury to the endometrium, induced either artificially or by the blastocysts is the event that leads to nidus formation in the guinea pig (37). The extent to which stromal cells react is different in different groups of animal species; for example, in most of the carnivores, there is no true decidual reaction in that the stromal cells do not respond to trauma at the time of implantation. On the other hand, there is a marked epithelial proliferation in the carnivores, and in response to trauma, irregular giant cells and symplasmic masses are formed from the proliferating epithelial cells.

EARLY DEVELOPMENT OF THE EMBRYO

Mammals develop the same basic extraembryonic membranes as reptiles and birds. In most marsupials only the yolk sac vascularizes the chorion; the allantois remains a relatively simple sac. The region of the trophoblast that is vascularized by the maternal endometrium is called a choriovitelline placenta. In eutherian mammals, the vessels of the allantois vascularize the trophoblast to form a chorioallantoic placenta. Both the extent to which this trophoblast invades the endometrium and final arrangement of the regions of the chorioallantoic placenta are different in different animals. In the carnivores, the trophoblast comes to surround the individual maternal vessels in an arrangement known as endotheliochorial or vasochorial condition. In this type of placenta, the maternal endothelium, a thick basement membrane, the trophoblast, and fetal endothelium intervene between the two blood streams. In most other animal species, the trophoblast of the chorion is bathed directly by maternal blood. In monkeys, two main features are that the amnion and the extraembryonic mosoderm, coelom, and the mesodermal primordium of the body stalk form precociously. In conjunction with the formation of a body stalk, primary attachment of the blastocyst to the uterine wall is at the embryonic pole in which case the placenta is bidiscoidal. These characteristics are strikingly similar to that seen in humans, although the temporal aspect may differ slightly.

SELECTED ANIMAL MODELS

Nonhuman primates represent the best animal models for the study of pathophysiology of human reproduction. This section briefly describes selected animal models, which may be used to further understand certain aspects of human reproductive processes. A list of animal models of human reproduction is given in Table 9-9.

Animal Models of Gametogenesis

The primordial germ cells in mammals arise in the extraembryonic tissues and migrate to the gonads by ameboid movements or via the bloodstream (19, 101). Several studies have shown that the early embryonic development in rhesus monkeys, baboons, and humans is similar (128, 281, 295). Germ cells can be recognized in the gonads of humans and macaques between the thirty-third and thirty-fifth day post conception, and the number is increased by mitosis (125). The wide range in the number of such cells (300–1300 in humans) may in part reflect the sex of the individual, since Beaumont and Mandl (25) have shown that male rat embryos consistently possess many more primordial germ cells than their coeval sisters.

The sexual differentiation of the gonad in embryos of humans and rhesus monkeys occurs between the thirty-sixth and forty-second day after fertilization. However, the ovary in the rhesus monkey cannot be recognized until about day 46 post conception, which is some 4 to 6 days later than in humans (295). The presence or absence of x and y chromosomes determines the sex; however, some autosomal genes can override this determinant (239). Intersexuality in nonhuman primates is extremely rare; the only case in the literature relates to a rhesus monkey with bilateral ovotestes (282).

Although many aspects of gametogenesis have been studied in nonhuman primates, they are incompletely understood (20). Nonetheless, nonhuman primates serve as ex-

Table 9-9. Animal Models in Reproductive Physiology

Animal Model	Species	Human Counterpart	References
Aplasia segmentalis ductus wolffi	cattle	Aplasia segmentalis ductus wolffi	13
Bacterial mastitis	mouse	Bacterial mastitis	52
Cervical cancer	several	Cervical cancer	266
Cervical dysplasia	rhesus-monkey	Cervical dysplasia	151
Chimerism	cattle	Possible gonadal sterilization	240
Club foot	guinea pig	Club foot	86
Cryptorchidism	swine dog,sheep & cattle horse	Cryptorchidism	215 271 108
Embryonic & reproductive physiology & teratology	Japanese quail	Embryonic & reproductive physiology & teratology	181
Endometrial-physiology	Macaca arctoides	Endometrial physiology	65
Estrogen evaluation	rhesus-monkey	Estrogen evaluation	262
Estrogen synthesis	baboon	Estrogen synthesis	291
Fetal growth retardation	rhesus-monkey	Fetal growth retardation	168, 228
Full term small for dates human baby	Rat	Full term small for dates human baby	4
Genital Herpes virus hominis type 2 infections	Cebus monkey	Genital Herpes virus type 2 infections	230
Gonorrhea	Several rhesus-monkey	Gonorrhea	10 44
Gynecological pathology	Macaca mulatta	Gynecological pathology	68
Hermaphroditism	Mink	Hermaphroditism	233
Hydramnion	rats	Hydramnion	90,245

Table 9-9. (Continued)

Animal Model	Species	Human Counterpart	References
Hydronephrosis of pregnancy	Macaca mulatta	Hydronephrosis of pregnancy	254
Hypogonadia	rabbits	Hypogonadia	100
Identical multiple offspring	nine banded armadillo	Heterokaryotic monozygous twins	28
Intrauterine growth retardation	sheep	Intrauterine growth retardation	60
Malignant ovarian tumors	rats	Malignant ovarian tumors	131
Mammary turmors	DDD mouse	Mammary tumors	209
	C-3H Avyfb. 90% trans- mission by either parent		35
	BL/LyDe & SWR/Lyde		66
	GR		293
	C57 Bl/M		129
	Feline		138
	Mice SPFCeH/A[ay]		288
Maternal-fetal model	sheep	Maternal-fetal model	272
Meconium aspiration	many species	Meconium aspiration	113
Multiple ovulations	squirrel- monkey	Multiple ovulations	78
Orchitis	guinea pig	Orchitis	43
	swine		162
Ovarian tumors	dog		142
	mouse		105,106, 107,194, 258
	rat		34
Ovum implantation	several	Ovum implantation	191

Table 9-9. (Continued)

Animal Model	Species	Human Counterpart	References
Persistent corpora lutea	mouse	Persistent corpora lutea	93
Polycystic ovary infertility	rat	Polycystic ovary infertility	30,42
Post-parturient hypocalcemia	rabbit	Lactation hypocalcemia hypocalcemia	259
Prolonged gestation	cattle	Prolonged gestation	133
Prostatic carcinoma	golden hamster	Prostatic carcinoma	1
Prostatic growth	many species	Prostatic growth	283
Prostatic hyperplasia	Canine	Prostatitis	29
Regular fraternal twinning	marmosets	Twinning	277
Sex chromosome anomalies	cat	Sex chromosome anomalies	54, 200
	mouse		236
	cattle		156,237
	swine		199,211
	mink		233
	marmosets		28
Spontaneous miscarriage	guinea pig	Spontaneous miscarriage	81
Testicular teratomas	mice (129/ter SV)	Testicular teratomas	278
Testicular teratomas (zinc induced)	Japanese quail	Testicular teratomas	118
Toxemia of pregnancy	rat	Toxemia of pregnancy	9,59,74, 123,148,197, 198,205,207, 238,241,274, 284
	monkeys		50
	guinea pig		143
	sheep		14
	rabbit		30,115,116, 135,206,298

Table 9-9. *(Continued)*

Animal Model	Species	Human Counterpart	References
	dog		79
Triploidy & tetraploidy in blastocysts	swine	Abortion	211
Trophoblastic tumors of placenta	Armadillo rat	Trophoblastic tumors of placenta	202 269
Twinning	Non-human primate	Human twinning	303
Uterine tumors	various animals	Uterine tumors	57,258
	rabbits		116,117,140, 216
	mouse		141,189,261, 287
	guinea pig		194
Vasectomy	rhesus-monkey	Vasectomy	7

cellent models to elucidate many aspects of reproduction in humans, particularly germ cell degeneration; the mechanism by which Sertoli cell activity influences the regulation of spermatogenesis; origin of germ cells; aging of gametes in the male and female genital tracts; early products of fertilization; and cleaving embryos with respect to factors leading to trisomy and polyploidy. Marmoset monkeys, always blood chimeric fraternal twins, could serve as a useful model not only for the study of chimerism but also to determine whether or not germ cells are dispersed in the circulation of primates and what their fate may be.

Animal Models of Implantation and Ovum Development

Implantation has been described in great detail by Blandau (37) and Lindner (192). This subject is of particular interest for problems of teratogenesis, which have received a great deal of attention following the thalidomide episode. Four species of macaque have been used for implantation studies; staging of the time scale of embryogenesis is somewhat faster in these species than in humans.

The galago and African green monkey embryos develop at a rate more like that of humans, even though the gestation period for the galago is 133 ± 4 days and for the African green 155 + 7, compared to 165 for macaques and 175 for baboons. The embryo development evidently must be accelerated in both galago and green.

Nonhuman primate teratological studies have largely been directed toward developing a model using thalidomide. Posuillo et al. (248) reported that the marmoset was sensitive to thalidomide, while the galago was insensitive. The sensitive period for the marmoset is poorly defined; for all other species, it falls in the interval between 24 and

30 days, except in the green monkey, whose slower developmental rate moves the sensitive period up to 30 to 33 days. Other studies have shown that solanine, the toxin in spoiled potatoes, produces malformations in the marmoset which are similar to those in humans (248). Corticosteroids have been associated with cleft palate in humans, and triamcinolone has now been shown to produce similar orofacial defects in the baboon and the bonnet monkey as well as in experimental rodents.

Animal Models of Sexual Cycles

Studies of the integrated mechanism that controls reproductive cycle in rats (264) and humans (292) have been reported. Investigation of the mechanism in nonhuman primates has been made possible by radioimmunoassay techniques for measurement of gonadal steroids in plasma and for assay of releasing factor and gonadotropic activities.

Neurophysiological control of the ovarian cycle have been studied in rodents, for example, the functions of the preoptic nucleus as a "biological clock" regulating the tonic output of the ventromedial-arcuate region. Studies in the baboon indicate substantial differences in the effect of electrolytes and differentiation. In contrast to the rat, injury to the area of the preoptic nucleus of the baboon produces a transient or a delayed effect on the menstrual cycle and rhythm. Another important difference between these two species is the inability of constant illumination to alter ovarian cycles in the baboon, whereas it produces an ovulatory persistent estrus in the rat (121). On the other hand, a "critical period" during rat proestrus (the time when ovulation can be blocked by pentobarbital) has now been shown to exist in the baboon as well. Synthetic (but not natural) corticosteroids can block induced ovulation in the immature rat; a similar observation has been made in nonhuman primates (122). In the baboon, the blocking effect of triamcinolone acetonide can be overridden by administration of gonadotropin-releasing factor. Synthetic gonadotropin-releasing factor has been shown to be active in a wide variety of animal species including humans (87).

The pattern of plasma levels of FSH, LH, estrogen, and progestin has been examined throughout the cycle in numerous species, and remarkable similarities may be observed in the rat, macaque, baboon, chimpanzee, and human. The similarity of corpus luteum function during pregnancy between human and rhesus has also been commented upon (171). The dynamics of positive and negative feedback effects of estrogens and progesterone or synthetic progestins have been extensively investigated in the rat, in the human, and in the rhesus (154, 188, 234).

Nixon et al. (235) examined the similarity of antigenic determinants in pituitary and chorionic gonadotropins from primates. A marked similarity between chorionic and pituitary gonatotropins exists both in chimpanzees and in humans. Pituitary gonadotropins from rhesus, chimpanzee, baboon, and human gave similar responses in radioimmunoassay (RIA) and these were similar to the response to HCG. The RIA slope of chimpanzee chorionic gonadotropin was similar to that of HCG, while the responses of baboon and rhesus chorionic gonadotropin were different. These data suggest a large degree of similarity in antigenic sites and in primary structure of pituitary gonadotropins in several primate species. However, substantial differences appear to exist in the gonadotropins of chorionic origin.

Animal Models of Sexual Behavior in Humans

The endocrine factors play a significant role in the sexual behavior of nonhuman primates (114, 220). In lower primates, the endocrine control of sexual behavior is somewhat similar to that in nonprimate mammals; for example, in *Lemur catta,* which has an ovarian cycle length of 33 to 41 days, receptivity to the male is limited to about 12 hours (91). In *Galago crassicaudatus,* in which estrus persists for about 12 days, the onset of sexual behavior occurs after 3 to 4 days of endogenous estrogen stimulation. The intensity of behavior reaches its peak promptly and remains elevated for about 1 to 2 days, after which there is a rapid decline. In contrast, many monkeys and apes copulate throughout the menstrual cycle, although completed copulation increases at midcycle and diminishes shortly after ovulation. The establishment of corpus luteum was promptly followed by a rapid decline in the incidence of full copulatory activity. However, the inhibitory effect of the corpus on sexual behavior was incomplete. Some pig-tailed macaques continue to copulate for prolonged periods even after bilateral oophorectomy. The effect of pregnancy on primate sexual behavior is generally one of suppression of ordinary sexual activity with occasional passive cooperation with male. The copulatory patterns of various species of experimental animals are summarized in Table 9-10.

Michael et al. (220) have reported various behavioral parameters to study the effects of endogenous and exogenous hormones on sexual behavior. Since the number and frequency of female invitational gestures depend not only on the attitude of the female to the male but also on the male's attitude toward the female several behavioral characteristics of the pair need to be observed in order to assess hormonal effects on female behavior. Ovariectomy markedly reduces the interactions between males and females. Administration of estradiol to females restores female receptivity and attractiveness; progesterone treatment counteracts the effects of estrogen. The behavior of the male partner is influenced by estrogenizing the deeper, phylogenetically older parts of the female brain. Testosterone propionate, administered intracerebrally (hypothalamus, amygdala, and thalamus) elicited a well-marked increase in female invitations but only a moderate increase in male mounting behavior. However, the intracerebral testosterone implants raised plasma testosterone levels so that interpretation of the results is equivocal (247).

The effects of estrogen on female sexual motivation was shown by a marked increase in operant behavior by the female during a period around ovulation which correlated with increased excretion of urinary estrone and increase in serum estradiol. A similar dose dependent behavioral response was produced in ovariectomized females by graded amounts of estradiol.

The influence of sexual behavior by olfactory communication via pheromones has been recognized in the breeding of farm animals. Prosimians are believed to possess specialized apocrine scent glands which are used for self-marking, territorial marking, and for marking each other. Under both natural and laboratory conditions, macaque display scenting activity related to estrogen-induced stimuli aroused by vaginal secretions. It has been possible to obtain behavioral effects in the rhesus monkey by using vaginal secretions from baboons (163).

In humans, sexual behavior is the result of complex interactions of endocrine, sociocultural, and psychological factors. Women are reported to display a premenstrual peak in sexual desire when levels of endogenous ovarian hormone are low and falling rapidly. Ovariectomy or spontaneous menopause do not significantly alter sexual response (208). On the other hand, adrenalectomy or hypophysectomy decrease libido and orgasm, suggesting that adrenal androgens may be endocrinologically important. The influence of estrogens and progestins on the level of human sexuality is minimal,

Table 9-10. Summary of Copulatory Patterns of Laboratory Mammals

Species	Characteristics of Intromissions	Pre-ejaculatory Intromissions	Number of ejaculations Per Episode	Other Characteristics	References
Hamster	Fairly brief with cessation of thrusting	Average = 10	Average = 9	Females hold lordoses between intromissions	24
Gerbil	Brief with cessation of thrusting	Exceptionally high (average = 51)	Average greater than 3	Males show pattern of "foot-stomping"	173
Mouse	Average 15 to 20 sec with 1 per sec thrust rate	Average 5 to 20	Usually cease after first ejaculation	Large, reliable strain differences	212
Guinea pig	Average about 5 sec with thrusting	Up to 13; 1/4 ejaculate on first	Usually cease after first ejaculation	Female resistance after intromission common	305
Chinchilla	Average less than 10 sec with continued thrusting	Up to 18; 2/3 ejaculate on first	Usually cease after first ejaculation	Female resistance after intromission common	32
Rabbit	Average about 2 sec with cessation of thrusting	Ejaculate on first	Multiple ejaculation pattern	Pre-intromission thrusts average 13 per sec	256
Cat	Average about 7 sec with cessation of thrusting	Ejaculate on first	Average about 7	Distinctive post-ejaculation display in female	300
Dog	Locks with thrusting average 20 min	None	Frequently multiple locks	Male may dismount during lock	104
Rhesus Macaque	Average 11 mounts, most with intromission	Average 11 Intromissions, with thrust rate of 3 per sec	1 to 3 in 1 hr	Male mounting varies with menstrual cycle	219

From *Reproduction and Breeding Techniques for Laboratory Animals* (119) with permission.

certainly less important than a basal level of androgens in the female. Environmental, psychosocial, and other factors are probably of overriding importance.

Animal Models of Hemodynamic and Metabolic Changes During Pregnancy and Early Postpartum

The developing mammalian fetus is completely dependent on the maternal organism to meet its metabolic needs for growth and development. Reversible changes in maternal vascular system of pregnant women, sheep, and monkeys have been studied (46, 139, 218, 244). Inability of the maternal cardiovascular system to make adequate adjustments for proper nourishment of placental and fetal mass may result in fetal death and resorption, premature delivery, stillbirth, small-for-age fetuses, or increased neonatal mortality. The pygmy goat has been used as the animal model for long-term studies on reproductive physiology and perinatal physiology (217). Hoversland et al (137) studied hemodynamic adjustments at 3-week intervals beginning prior to conception and continuing 6 weeks postpartum. Significant changes occurred during pregnancy: body weight increased 15 percent at term; heart rate was elevated 32 percent at term; arterial systolic pressure dropped 7 percent; arterial diastolic pressure dropped 9 percent, and mean arterial pressure dropped 8 percent, all being lowest at 16 to 18 weeks of pregnancy. Other changes included 13 percent elevated plasma volume, 11 percent elevated blood volume, 48 percent increased cardiac output, 29 percent increased cardiac output per kilogram, 10 percent increased stroke volume, 37 percent decreased peripheral vascular resistance, and 31 percent elevated hemoglobin flow per kilogram of body weight at 16 to 18 weeks of gestation. Mean pulmonary artery pressure, systemic arterial pulse pressure, hematocrit, hemoglobin concentration, red cell volume, and blood volume per kilogram did not change significantly during pregnancy as compared to those values at postpartum.

Hoversland et al. (137) also reported that during pregnancy, maternal oxygen consumption rose until term when it was 36 percent higher than the mean postpartum value. The CO_2 concentration declined 13 percent in arterial blood and 12 percent in mixed venous blood, with lowest values near term. Concurrently pCO_2 values declined 13 percent in arterial blood and 12 percent in mixed venous blood, and plasma bicarbonate declined 12 percent in arterial blood and 14 percent in mixed venous blood. The maternal vascular and cardiodynamic adjustments during pregnancy were adequate to meet the increased O_2 requirement of the pregnant doe without drawing on blood O_2 reserves, as evidenced by stable values for mixed venous pO_2, arteriovenous O_2 concentration difference, coefficient of O_2 utilization, and plasma pH.

The study of respiratory and cardiodynamic adjustments in pygmy goats during late pregnancy and postpartum may be useful models for understanding the effects of hormonal (estrogens and progesterone) level changes in blood during pregnancy.

Toxemia of Pregnancy

Toxemia of pregnancy is the most common single cause of maternal death in late pregnancy. It is a leading cause of perinatal deaths and, in addition, has been linked to intrauterine growth retardation and to behavioral disorders in surviving children (243). Toxemic states of pregnancy are known to occur in various animal species, although none has a presentation precisely similar to human eclampsia in which the principal triggering factors are of uterine or placental origin. A number of experiments on toxemia of pregnancy have been carried out in rodents, dogs, and monkeys (30, 132,

178, 197). The trigger mechanism in the pathogenesis of toxemia seems to be uteroplacental ischemia which probably results in placental degeneration with the liberation of thromboplastin and the development of intravascular coagulation. This may be an acute or chronic process and may be localized or generalized, resulting in characteristic morphological toxemic lesions on the kidneys and perhaps elsewhere. Then, as a result of aberrations in the renin-angiotensin-aldosterone system and the increased vascular response to endogenous vasoconstrictors, the generalized vasoplasm, hypertension, edema, and proteinuria so characteristic of the disease develop (51). An experimental model in the subhuman primate (baboon) was developed by Cavanagh et al. (51) in an effort to clarify the pathophysiology of the eclamptic toxemia. Hemoclips were placed around the uterine arteries so as to produce chronic, experimental uterine artery stenosis. The ovarian arteries were transsected to eliminate this source of collateral blood supply. When the baboons became pregnant, hypertension and proteinuria developed during the third trimester. The renal artery flow was also markedly reduced. Also, heavy deposits of fibrin or fibrinogen were present in the glomeruli of the hypertensive baboons. It seems that when uterine blood flow becomes insufficient during pregnancy and there is no significant collateral blood supply available to the placenta and developing fetus, changes similar to eclampsia will develop in the subhuman primate. The authors concluded that the baboon is a suitable experimental model to study eclamptic toxemia of pregnancy in the human.

Animal Models in Contraceptive Research

Steroidal Contraceptives

Two estrogens have been used in the various contraceptive preparations: ethynylestradiol and 3-methyl ether of ethynylestradiol mestranol. The synthetic progestogens used in these preparations are related to (a) 19-nortestosterone and (b) 17α-hydroxyprogesterone.

The main action of estrogens is to suppress ovulation; 50 μg/day of ethynylestradiol produces infertility in women and 80 μg/day totally suppresses ovulation. The nortestosterone derivatives are to some extent metabolized to estrogenic compounds and hence have an antiovulatory effect more pronounced than 17α-hydroxyprogesterone. Progestogens have an effect on the cervix, the tubes, and the endometrium, but their contraceptive effect is considered to be mainly due to an alteration in a cervical mucus resulting in a cervical bar to fertilization. Steroids used as contraceptives markedly affect the balance of endogenous hormones, particularly the gonadal steroids, and expose the woman to a quite different hormonal environment from that which normally occurs during the menstrual cycle. Doll and Vessey (69) and Klopper (169) have written excellent critical reviews on an evaluation of the adverse effects of systemic contraceptives.

The major obstacle to the development of improved fertility-regulating agents is the lack of suitable pharmacological and toxicological animal models. Clinical studies have shown rare adverse reactions, including jaundice, venous thrombosis and pulmonary embolism, ischemia, cerebrovascular effects, hypertension, amenorrhea, breakthrough bleeding, anxiety, depression, weight gain, and changes in libido. Preclinical studies in rodents dosed with oral contraceptives did not reveal more side effects (186). A carcinogenic effect can be produced in certain strains of mice and rats when estrogens, progestogens, and combinations of these hormones are given in high doses throughout their life span. However, the susceptibility to tumor induction by hormonal contraceptives is not consistent in different strains of mice and rats. Ideally, carcinogenic testing of oral contraceptives should be conducted in animal species that have a metabolic disposition of the drugs and feedback mechanisms which control ovulation and menstruation similar to that of women; nonhuman primates and guinea pigs possess some of these characteristics. Responses to endogenous and exogenous sex steroids in the dog are characteristic for that species, differing in many important aspects from other laboratory animals and humans. Long-term administration to dogs of progesterone and synthetic compounds structurally related to this natural gestagen has resulted in precocious development of mammary gland tumors and other changes, suggesting stimulation of anterior pituitary activity. Progestogens structurally related to testosterone have not induced similar effects (8, 72, 73). Long-term administration of either progesterone or chlormadinone acetate induces mammary gland hyperplasia in the female beagle and enhance the appearance of mammary gland neo-plasia (49, 94). Results obtained to date indicate a lack of relevance of data obtained in the dog to the prediction of clinical potential of sex steroids for mammary gland carcinogenesis.

The fate of estrogenic and progestational steroids has been examined to some extent in various primate species (Table 9-11). The urinary metabolites are mostly glucuronides in all species tested. However, the pattern of individual urinary metabolites differs considerably. In humans, approximately 20 percent of radioactivity is estrone; 20 percent, estriol; 20 percent, 2-hydroxyestrone; and 5 percent, 2-methoxyestrone. The gorilla appears to be the only species which, like humans, has a substantial ability to hydroxylate estrone forming urinary estriol. The increase in urinary estriol during preg-

Table 9-11.   Metabolic Fate of Some Estrogens and Progestins in Various Primates

| Species | Steroid Administered | Excretion | | Comment | Reference |
		Per cent in urine (days)	Per cent in faeces (days)		
Homo sap.	Oestrone, Oestradiol	50–80 (4–6)	15–20 (4–6)	MCR 1000–1400 L/d; plasma chiefly as E_1-SO_4	196, 257
Pan satyrus	Oestrone	72(3)	–	Little urinary E_3	145
M. mulatta	Oestradiol	48–100(5)	5(4)	At first E_1 later epiE_2 important	182
Papio sp.	"	50–70(3)	–	Mostly E_1	175, 176
Homo sap.	Ethynyloestradiol	23–60	9(5)	MCR=1350+220 L/d; EE_2-SO_4 chief plasma constituent	33, 112
Papio sp.	"	77–83(6)	–		175
Homo sap.	Mestranol	10–52(5–8)	–	MCR=1740+390 L/d; EE_2-SO_4 chief plasma constituent	33,112
Homo sap.	Progesterone	50–60(5–7)	10	MCR–1740+140 L/d; mostly pregnanediols in urine	170, 190
P. satyrus	"	33–50(3)	–	Mostly urinary pregnanediols	255
Papio sp.	"	33–41(4)	14–17(4)	Urinary androsterone important. Pregnanediols insignificant	110
M. mulatta	"	28–46(9)	41–57(9)	"	246,250
M. nemestrina	"	44–60(6)	17(5)	"	144
M. mulatta	Chlormadinone acetate	35–36(7)	26–28(7)		–
M. mulatta	Megestrol acetate	14–33(6)	36–44(6)		–
Papio sp.	" "	33–43(5)	0.4–2.8(5)		–
Homo sap.	Trengestone	50 in ♀, 10 in ♂		20α and β reduction; in human only 20α; in rat, 16α-hydroxylation only	172

Reproduced from <u>Pharmacological Models in Contraceptive Development</u> (111) with permission.

nancy in the gorilla is similar to that in humans; it is not seen in pregnancy in the other primates. After administration of estrone to chimpanzees, Jirku and Layne (145) found 72 percent of the dose in the urine within 3 days and Breckwoldt et al. (40) recovered 44 to 48 percent over 5 days in the rhesus monkey. In the latter species, 65 percent of a dose of estradiol was recovered over a 3-day period by Flickinger and Wu (95) and 44 to 48 percent over 5 days by Breckwoldt et al. (40), whereas in the baboons 53 to 71 percent was excreted in the urine with 4 days (174, 177). As in humans, much of the radioactivity excreted in urine is excreted within 24 hours of administration of the dose. In rhesus monkeys, the major metabolites of the estradiol are estrone and estradiol itself together with small amounts of estriol and 16-epiestriol (95). In the baboon, estrone with small amounts of 17 β-estradiol and estriol are the major metabolites of estradiol (174). Leung et al. (187) also reported a number of other metabolites, in the baboons, including 15-α-hydroxy estradiol, 16-oxo-estradiol, 16-α-hydroxyestrone, and 2-methoxyestrone. Only small amounts (less than 5 percent of the dose) appear to be excreted in the feces (174).

Most of the information regarding the metabolism of synthetic estrogens in baboons has been obtained by Goldzieher and Kraemer (112). Metabolism of estrogens have been also studied estensively in the dog (18, 56), rabbit (2, 260), guinea pig (70, 249, 270), and rat (38, 39, 160).

Other reviews related to the metabolism of synthetic progestogens and estrogens have appeared (98, 251, 290). The interaction between naturally occuring and synthetic sex steroids and the liver has been described by Adlercreutz and Tenhuen (5). The similarity of progesterone metabolism in rabbits and humans, considered by Fotherby (97), suggests a close resemblance in the metabolism of the synthetic progestins in the two species. The effects of estrogens and estrogen-progestin combinations on various aspects of lipid metabolism have been described in humans, canines, and baboons (64, 111). All three species show a slight increase in cholesterol concentrations; however, there is a marked difference between human triglyceride response and that of dogs and baboons. A considerable rise of triglyceride in the chylomicron, α_2-lipoprotein and β-lipoprotein fractions is evident in humans but no such changes are observed in the two other species. In women, there is a tendency for the phospholipid level in the chylomicron and α_2-lipoprotein fractions to rise; in the β-lipoprotein they do not appear to change within 3 months of estrogen administration. In the other two species, however, there is a profound fall in the phospholipid fractions. Depending on the parameter selected, the two animal species may or may not be suitable models for the study of effects of steroid contraceptives on human lipid metabolism.

Data on contraceptive steroids derived from properly conducted and interpreted animal studies can serve to predict effects that may be expected to occur in women (75, 76, 77, 213, 252, 263). The absence of certain effects of the steroids in both animals and humans is also important. Since the drugs are for human use, it is necessary that adequate clinical studies be conducted to confirm the predictions made regarding effectiveness and safety.

Nonsteroidal Contraceptives

The rhesus monkey has been widely used for the study of nonsteroidal fertility regulating agents (Tables 9-12 and 9-13). Species differences in response to potential antifertility agents exist between laboratory rodents and rhesus monkeys on one hand and between rhesus monkeys and human subjects on the other. The estrogen pattern of

Table 9-12. Antifertility Effectiveness of Some Nonsteroidal Compounds in the Female Rhesus Monkey

Status		Compound	References
Anti-ovulatory	(+)	Reserpine	63
Anti-implantation	(+)	ORF-3858 (2-methyl-3-ethyl-4-phenyl- Δ4-cyclohexanecarboxilic acid)	225
Anti-implantation	(+)	Diethylstilboestrol	222
Anti-implantation	($\pm$)	U-11100A (1-(2-(p-(3,4-dihydro-6 methoxy-2-phenyl-1-napthyl) phenoxy) ethyl)-pyrrolidine hydrochloride)	225
Anti-implantation	(-)	Clomiphene, MRL41 (1-(p-(B-diethylamino-ethoxy) phenyl) 1,2-diphenyl-2-chloroethylene	225
Anti-implantation	(+)	66 179 (2-phenyl-3-p(B-pyrrolidinoethoxy)-phenyl-(2:1.b) napthofuran	149
Anti-implantation	(+)	dl-cis-bisdehydrodoisynolic acid methyl ether	223
Anti-implantation	(-)	Prostaglandin F_{2a}	80
Abortifacient	(-)	Colcemide (desacetylmethyl colchicine)	224
Abortifacient	(-)	BW57-323H (2-amino-6-(1'-methyl-4'-nitro-5' imidazolyl) mercaptopurine	224
Abortifacient	(+)	Prostaglandins E_2 and F_{2a}	166, 167
Abortifacient	($\pm$)	Triacetyl-6-azauridine	294
Anti-implantation	(-)	H1067	-
Abortifacient	($\pm$)	H1067	-
Abortifacient	(-)	N-desacetyl thiocolchicine	-
Luteolytic	(-)	Amphenone (U-7256)	-
Anti-implantation	(+)	Centchroman (3,4-trans-2,2-dimethyl-3-phenyl-4-(p- β -pyrrolidinoethoxy)-phenyl)-7-methoxychroman)	150
Anti-ovulatory	(+)	Dithiocarbamoylhydrazine	210

+ = positive effect, - = negative effect, $\pm$ = marginal effect.
From <u>The Use of Non-Human Primates in Research on Human Reproduction</u> (67) with permission.

early pregnancy in the rhesus monkey is similar to that of the human (15, 136) so that testing of hormonal agents, administered postcoitally in this species may be relatively more applicable to the human than the use of observations derived from rodent studies. However, Glasser et al. (109) and Lipner and Greep (193) have found that aminoglutethimide, which inhibits steroid synthesis in the rat and human adrenal cortex, also prevents implantation in the rat. Since this affect is reversed by progesterone, its antifertility effect is attributable to impairment of progesterone synthesis by the corpus luteum. Amphenone, a congener of aminoglutethimide with similar thyroid and adrenal blocking properties, has been tested for its effect on the menstrual cycle in the rhesus monkey; the treated animals showed normal bleeding patterns. Another attempt to acheive luteolysis in the rhesus monkey involves the use of prostaglandins. In one study, prostaglandins $F_{2\alpha}$ was administered on days 11 to 15 postovulatory (30 mg/day × 5, subcutaneously); one pregnancy was achieved in six matings, whereas the usual pregnancy rate is six of ten matings. Progestin levels were depressed in all treated animals when compared to nontreated controls, (164). Subcutaneous injections of prostaglandins E_2 or $F_{2\alpha}$ ended pregnancy in eight of seventeen animals when administered after day 30. Intravenous infusions caused abortions in three of seven animals. Progestin concentrations fell within 24 to 48 hours of administration at days 30 to 40 (164). In another study, the administration of prostaglandins $F_{2\alpha}$ caused uterine contractions in

Table 9-13. Antifertility Effectiveness of Some Nonsteroidal Compounds in the Male Rhesus Monkey

Status		Compound	References
Anti-spermatogenic	(+)*	Cadmium chloride	155
Anti-spermatogenic	(+)	Ferric chloride, ferrous sulphate	159
Anti-spermatogenic	(+)	WIN 13,099 N,N'-bis (dichloroacetyl)-N,N¹-diethyl-1,4-xylylenediamine	58
Anti-spermatogenic	(−)	ORF 1616 (1-(N,N diethylcarbamoyl-methyl)-2,4-dinitropyrrole)	156
Anti-spermatogenic**	(+)	CIBA 32644-Ba (4-(5-nitro-2-thiozolyl)-2-imidazolidinone)	270
Anti-spermatogenic	(+)	Busulphan (1,4-dimethane-sulphonoxy-butane)	158
Anti-spermatogenic	(+)	Tretamine (2,4,6-triethylen eimino-triazine)	158
Anti-spermatogenic	(−)	ICI 33828 (1-a-methylallalythiocarbamoyl-2-methyl-thiocarba-moyl-hydrazine)	158
»Functional Sterility«	(+)	U-5897 a-chlorohydrin (3-chloro-1,2-propanediol)	165
Anti-spermatogenic	(−)	Histamine	157
Anti-spermatogenic	(−)	Serotonin (5-hydroxytryptamine)	157
Anti-spermatogenic	(−)	Fluoroacetamide	–

*+ = positive effect, − = negative effect.
** Tested in Cynocephalus sp.
From The Use of Non-human Primates in Research on Human Reproduction (67) with permission.

midterm and 150-day pregnancies, but the uteri of only 50 percent of the animals were not evacuated until 48 hours after termination of the infusion (103).

From many studies reported in literature, it is reasonable to assume that the rhesus monkey, or any other nonhuman primate, is the species of choice to study antifertility activity applicable to the human. There are striking similarities in response to some drugs, in endocrinological function, and menstrual pattern between the human and the rhesus monkey as well as other nonhuman primates. These parallels serve to emphasize the desirability of more extended exploration of the species for the study of fertility regulation (265).

Intrauterine Devices (IUD)

Eckstein (84) and El Sahwi and Moyer (88) reported that the contraceptive effects of IUDs in women and nonhuman primates may result from interference with implantation, impairment or destruction of the ovum, premature expulsion of the ovum, or possibly a combination of these. Although not a great deal of experimental work on this subject has been carried out in laboratory primates during the past few years, it is significant that increasing use is being made of baboons (Papio species). Baboons have several advantages over rhesus monkeys in this type of research, for example, the larger size of their reproductive organs; an almost straight, human-like cervix which permits direct access to the uterine cavity without laporotomy; and the occurrence of well-marked sexual skin cycle from which the approximate time of ovulation can be inferred (127, 128). From the animal studies with IUDs, it is generally agreed that the endometrial response consists chiefly of polymorphonuclear leukocytes and mononuclear elements. A comparative study of the effects of an IUD on the cellular and chemical composition of uterine flushings from rats, baboons, and women showed significant increases in the intrauterine concentrations of white blood cells and lysosomal enzymes in rats but not in either women or baboons (146). Moreover, there was no change in the protein content of human and baboon uterine fluid. However, in another study reported by Breed et al. (41), uterine fluid was sampled by injecting physiological saline through a special flushing device into the cavity of the uterus and then aspirating the washing. Flushings were carried out from approximately 2 weeks before to about 2 weeks after the estimated time of ovulation. The concentrations of total proteins, reducing sugars, and proteinase activity (measured at acid pH) were consistently higher when an IUD was present. The difference was statistically significant during the early secretory stage (3 to 7 days after ovulation) when the devices are thought to exert their contraceptive activity. The levels of amylase, lysozyme, and proteinase were not significantly altered in the presence of an IUD at any of the cycle stages examined (41). Eckstein (85) reported that the devices (IUD) do not interfere with the life span and function of the corpus luteum in the baboon and monkey, and hence the possible systemic or extrauterine effects of IUDs may be through interference with the hormonal control of the cycle as a whole.

Various IUDs consisting of synthetic organic polymers, inorganic macromolecules such as metals in solid form, and silicone-organic-inorganic hybrids have been studied in experimental animals for their adverse effects. There have been a number of alternatives (of oral contraceptives) involving uniform release of steroid hormones over comparatively long periods from subcutaneous or intrauterine implants of a polymer depot containing the steroid. The intrauterine depot has the theoretical advantage that the amount of required hormone is small enough that extrauterine increases of this hormone need not occur. However, the development of the fibrous capsule around the implant is a problem. In dogs, a silicone rubber capsule containing tri-iodothyronine and isoproterenol implanted in the myocardium demonstrated a pacemaker's effect for only a week. Implantation to another area after removal from the fibrous capsule restored the activity (96). In sheep, intramuscularly implanted silicone capsules containing melengestrol acetate were effective for 2 years in preventing ovulation (82). In a preliminary short-term study in human females implanted with subcutaneous silicone capsules containing a synthetic progestin to be released in micro quantities, no serious local or systemic complications were noted (289).

In rats, the subcutaneous implant must remain in situ during a critical period, about one-fifth the life span, for sarcomas to develop. Removal of the implant after this period does not prevent the development of a sarcoma in the fibrous capsule many months later. The mean latent period is 50 percent of the life span. In humans, mesotheliomas arising at the site of asbestosis (297) and pulmonary scar carcinomas related to embedded foreign bodies have long latent periods of about 30 years. At the human sites, foreign body-induced cancers had latent periods in the 10-year range. An implantation period of 1 year or less for contraceptive devices involving plastics would likely be too short for solid-state cancer to develop. However, repetitive implantations may have a cumulative effect.

Chemical methods of fertility control by introducing chemicals into the Fallopian tubes, cervix, and vagina (23), a progestin released from a vaginal ring (221), and injection of liquid silicone into the Fallopian tubes to block passage of sperm by forming a rubbery solid are being investigated. The latter have been tested on rabbits and monkeys and await clinical trial (102).

REFERENCES

1. Abdalla, A. M., and J. A. Oliver, *J. Urology,* **106:**590 (1971).
2. Abdel-Azia, M. T., and D. I. H. Williams, *Steroids,* **15:**695 (1970).
3. Adams, C. E., and W. L. Williams, *Bibl. Reprod.* **10:**177 (1967).
4. Adlard, B. P. F., J. Dobbing, and J. L. Smart, *Biol. Neonate,* **23:**95 (1973).
5. Adlercreutz, H., and R. Tenhuen, *Am. J. Med.,* **49:**630 (1970).
6. Ahluwalia, B., G. Pincus, and R. T. Holman, *J. Nutr.,* **92:**205 (1967).
7. Alexander, N. J., *J. Reprod. Fertil.,* **31:**399 (1972).
8. Anderson, A. C., *J. Am. Vet. Med. Assoc.,* **147:**1653 (1965).
9. Arhelger, R. B., B. H. Douglas, and H. G. Longford, *Arch. Pathol.,* **84:**393 (1967).
10. Arko, R. J., *Science,* **177:**1200 (1972).
11. Asdell, S. A., *Patterns of Mammalian Reproduction,* Cornell University Press, Ithaca, N.Y., 1964.
12. Asdell, S. A., *Patterns of Mammalian Reproduction,* 2nd ed., Cornstock, New York, 1964.
13. Ashdown, R. R., *Vet. Rec.,* **70:**467 (1958).
14. Assali, N. S., L. W. Holm, and D. L. Hutchinson, *Circ. Res.,* **6:**468 (1958).
15. Atkinson, L., J. Hotchkiss, G. R. Fritz, A. S. Surve, and E. Knobil, (Abstr.) Proceedings of the 4th Annual Meeting of the Society for the Study of Reproduction, Boston University, Boston, 1971, p. 14.
16. Austin, C. R., *Mammalian Egg,* Charles C. Thomas, Springfield, Ill., 1961.
17. Austin, C. R., *Fertilization,* Prentice-Hall, Englewood Cliffs, N.J., 1965.
18. Bachelor, A., E. T. Bell, and D. W. Christie, *J. Endocrinol.,* **51:**26 (1971).

19. Baker, T. G., *Mutation Res.,* **11**:9 (1971).
20. Baker, T. G., in E. Diczfalusy and C. C. Standley (eds.), *Use of Non-Human Primates in Research on Human Reproduction,* Karolinska Institute, Stockholm, 1972.
21. Barraclough, C. A., *Endocrinology,* **68**:62 (1961).
22. Bartlett, D. J., *J. Reprod. Fertil.,* **3**:190 (1962).
23. Battelle Memorial Institute, cited by Seattle UPI, *Santa Barbara News Press,* March 11, 1973, p. H14.
24. Beach, F. A., and R. G. Rabedeau, *J. Comp. Physiol. Psychol.,* **52**:56 (1959).
25. Beaumont, H. M., and A. M. Mandl, *J. Embryol. Exp. Morphol.,* **11**:715 (1963).
26. Beaver, D. L., *Anat. Rec.,* **146**:47 (1963).
27. Bedford, J. M., *Am. J. Anat.,* **123**:329 (1968).
28. Benirschke, K., M. M. Sullivan, and M. Marin-Padilla, *Obstet. Gynecol.,* **24**:819 (1964).
29. Berg, O. A., *Acta Endocrinol.* (Kopenhaven) **27**:140 (1958).
30. Berger, M., and D. Cavanaugh, *Am. J. Obstet. Gynecol.,* **85**:293 (1963).
31. Bern, H. A., and B. Krichesky, *Univ. Calif. Publ. Zool.,* **47**:175 (1943).
32. Bignami, G., and F. A. Beach, *Anim. Behav.,* **16**:45 (1968).
33. Bird, C. E., and A. F. Clark, *J. Clin. Endocrinol.,* **36**:296 (1973).
34. Biskind, G. R., and M. S. Biskind, *Am. J. Clin. Pathol.,* **19**:50 (1949).
35. Blahakis, G., W. E. Heston, and G. H. Smith, *Science,* **170**:185 (1970).
36. Blandau, R. J., in E. S. E. Hafez and R. J. Blandau (eds.), *Mammalian Oviduct,* University of Chicago Press, Chicago, 1969.
37. Blandau, R. J., in E. Diczfalusy and C. C. Standley (eds.), *Use of Non-Human Primates in Research on Human Reproduction,* Karolinska Institute, Stockholm, 1972.
38. Bolt, H. M., and H. Remmer, *Xenobiotica,* **2**:77 (1972).
39. Bolt, H. M., and H. Remmer, *Xenobiotica,* **2**:77 (1972).
40. Breckwoldt, M., G. L. Flickinger, T. Muravec, and J. C. Touchstone, *Endocrinology,* **91**:1 (1972).
41. Breed, W. G., A. Fraser, P. Eckstein and P. V. Peplow, *J. Reprod. Fertil.,* **30**:143 (1972).
42. Browman, L. G., *J. Exp. Zool.,* **75**:375 (1937).
43. Brown, P. C., J. Darling, and L. E. Glynn, *J. Pathol.,* **106**:229 (1972).
44. Brown, W. J., C. T. Lucas, and U. S. G. Kuhn, *Br. J. Vener. Dis.,* **48**:177 (1972).
45. Burt, W. H., *Univ. Mich. Mus. Zool. Misc. Publ.,* **113**:1 (1960).
46. Burwell C. S., and J. Metcalfe, *Heart Disease and Pregnancy-Physiology and Management,* Little, Brown & Co., Boston, 1958.
47. Butler, H., *J. Anat.,* **93**:257 (1959).
48. Butler, H., in D. Starck, R. Schneider, and H. J. Kuhn (eds.), *Neue Ergebnisse der Primatologie,* Gustav Fischer Verlag, Stuttgart, 1967.
49. Capel-Edwards, K., D. E. Hall, K. P. Fellows, D. K. Vallance, M. J. Davies, D. Lamb, and W. B. Robertson, *Tox. Appl. Pharmacol.,* **24**:474 (1973).
50. Cavanagh, D., P. S. Rao, K. Tung, and L. W. Gaston, *Obstet. Gynecol.,* **39**:637 (1972).
51. Cavanagh, D., P. S. Rao, K. S. K. Tung, and L. Gaston, *Am. J. Obstet. Gynec.,* **120**:183 (1974).
52. Chandler, R. K., *J. Med. Micro.,* **3**:273 (1970).
53. Chang, M. C., *J. Anim. Sci.,* **27**:15 (1968).
54. Chu, E. H. Y., H. C. Thuline, and D. E. Norby, *Cytogenetics,* **3**:1 (1964).
55. Cole, H. H., and P. T. Cupps, (eds.), *Reproduction in Domestic Animals,* Academic Press, New York, 1969.
56. Collins, D. C., H. D. Robinson, C. N. Howard, and J. R. K. Preedy, *J. Clin Invest.,* **49**:2324 (1970).
57. Cotchin, E., *Br. J. Cancer,* **18**:209 (1964).
58. Coulson, F., A. L. Beyler, and H. P. Drobeck, *J. Toxicol. Appl. Pharmacol.,* **2**:715 (1960).
59. Craig, J. M., *Fed. Proc.,* **26**:486 (1967).
60. Creasy, R. K., C. T. Barrett, M. de Sweit, K. Kahanpaa, and A. M. Rudolph, *Am. J. Obstet. Gynecol.,* **112**:566 (1972).
61. Critchlow, V., In A. V. Nalbandov (ed.), *Advances in Neuroendocrinology,* University of Illinois Press, Urbana, Ill., 1963.

62. De Feo, V. J., in R. M. Wynn (ed.), *Cellular Biology of the Uterus,* Appleton-Century-Crofts, New York, 1967.

63. De Feo, V. G., and S. R. M. Reynolds, *Science,* **124:**726 (1956).

64. de la Pena, A., C. Matthijissen, and J. W. Goldzieher, *Lab. Anim. Sci.,* **22:**249 (1972).

65. Demors, L. H., G. J. MacDonald, A. T. Hertig, N. W. King, and J. J. Mackey, *Fertil. Steril.,* **23:**529 (1972).

66. Deringer, M. K., *J. Natl. Cancer Inst.,* **45:**215 (1970).

67. Diczfalusy, E. and C. C. Standley (eds.), *Use of Non-Human Primates in Research on Human Reproduction* (Acta Endocinologica), Karolinska Institute, Stockholm, 1972.

68. Di Giacomo, R. F., and T. O. McCann, *Am. J. Obstet. Gynecol.,* **108:**538 (1970).

69. Doll, R., and M. P. Vessey, *Br. Med. Bull.,* **26:**33 (1970).

70. Dolly, J. O., C. G. Curtis, K. S. Dodgson, and F. A. Rose, *Biochem. J.,* **128:**347 (1972).

71. Donovan, B. T., and J. J. Van der Werft ten Bosch, *J. Physiol.* (London) **132:**123 (1956).

72. Dorn, C. R., D. O. N. Taylor, F. L. Frye, and H. H. Hibbard, *J. Natl. Cancer Inst.,* **40:**295 (1968).

73. Dorn, C. R., D. O. N. Taylor, R. Schneider, and H. H. Hibbard, *J. Natl. Cancer Inst.,* **40:**307 (1968).

74. Douglas, B. H., and H. G. Langford, *Proc. Soc. Exp. Biol. Med.,* **120:**238 (1965).

75. Drill, V. A., *Oral Contraceptives,* McGraw-Hill, New York, 1966.

76. Drill, V. A., *Proceedings of European Society for the Study of Drug Toxicity* (1973).

77. Drill, V. A., *Proceedings of European Society for the Study of Drug Toxicity* (1973).

78. Dukelow, W. R., *J. Reprod. Fert.,* **22:**303 (1970).

79. Dull, L. V., and C. C. Erickson, *Proc. Soc. Exp. Biol. Med.,* **39:**362 (1938).

80. Duncan, G. W., *Prostaglandins in Fertility Control,* World Health Organization Research & Training Center on Human Reproduction, Karolinska Institute, Stockholm, 1971.

81. Durieux, Rad Bret, A. J., *Rev. Fr. Gynecol. Obstet.,* **65:**225 (1970).

82. Dziuk, P. J. and B. Cook, *Endocrinology,* **78:**208 (1966).

83. Eckstein, P., In H. Hofer, A. H. Schultz, and D. Starck (eds.), *Primatologia,* Vol. III, S. Karger, New York, 1958.

84. Eckstein, P., *Br. Med. Bull.,* **26:**52 (1970).

85. Eckstein, P. *In* E. Diczfalusy, and C. C. Standley (eds.), *Use of Non-Human Primates in Research on Human Reproduction,* Karolinska Institute, Stockholm, 1972.

86. Edwards, M. J., *J. Pathol.,* **103:**49 (1971).

87. Ehara, Y., K. J. Ryan, and S. S. C. Yen, *Contraception,* **6:**465 (1972).

88. El Sahwi, S., and D. L. Moyer, *Contraception,* **2:**1 (1970).

89. Enders, A. C., *Am. J. Anat.,* **116:**29 (1965).

90. Eneroth, G. U., U. Forsberg, and C. A. Grant, *Acta Paed. Intr. Scand.,* Suppl., **206:**43 (1970).

91. Evans, C. S. and R. H. Goy, *J. Zool.,* **156:**181 (1968).

92. Everett, J. W., *Physiol. Rev.,* **44:**373 (1964).

93. Fekete, E., *Anat. Rec.,* **117:**93 (1953).

94. Finkel, M. J., and V. R. Berliner, International Academy of Pathology Meet., 1973.

95. Flickinger, G. L. and C. H. Wu, *Proc. Soc. Exp. Biol.,* **124:**1310 (1967).

96. Folkman, J., and D. M. Long, Jr., *J. Surg. Res.,* **4:**139 (1964).

97. Fotherby, K., in G. Raspe (ed.), *Advances in the Biosciences,* Pergamon Press, Oxford, 1969, p. 43.

98. Fotherby, K., and F. James, *Advances in Steroid Biochemistry and Pharmacology,* Academic Press, New York, 1972.

99. Fox, R. R., *Proc. Soc. Exp. Biol. Med.,* **128:**639 (1968).

100. Fox, R. R., and D. D. Crary, *J. Hered.,* **62:**163 (1971).

101. Franchi, L. L., A. M. Mandl, and S. Zuckerman, in S. Zukerman, A. M. Mandl, and P. Eckstein, (eds.), *Ovary,* Vol. I, Academic Press, London, 1962.

102. Franklin Institute, cited by *Los Angeles Times,* April 15, 1973, "The Nation," p. 2.

103. Fuchs, F. M., Prieto, and S. Marcus, *Ann. N.Y. Acad. Sci.,* **180:**531 (1971).

104. Fuller, J. L., and M. W. Fox, in E. S. E. Hafex (ed.), *Behavior of Domestic Animals,* 2nd ed., Balliere, Tindall & Cassell, London, 1969.

105. Furth, J., and O. B. Furth, *Am. J. Cancer,* **28:**54 (1936).

106. Furth, J., and H. Sobel, *J. Natl. Cancer Inst.,* **8:**7 (1947).

107. Gardner, W. U., *Cancer Res.,* **15:**109 (1955).

108. Garlick, N. L., *J. Am. Vet. Med. Assoc.,* **121:**101 (1952).

109. Glasser, S. R., R. C. Northcutt, and F. Chytil, Program of the 53rd Annual Meeting of the Endocrine Society, Abst. 200, 1971.

110. Goldzieher, J. W. and L. R. Axelord, *Gen. Comp. Endocrinol.,* **13:**201 (1969).

111. Goldzieher, J. W., S. Joshi, and D. C. Kraemer, in M. H. Briggs, and E. Diczfalusy (eds.), *Pharmacological Models in Contraceptive Development,* Karolinska Institute, Stockholm, 1974.

112. Goldzieher, J. W., and D. C. Kraemer, *Acta Endocrinol.,* Suppl. 166, **71:**389 (1972).

113. Gooding, C. A., G. A. Gregory, P. Taber, and R. R. Wright, *Radiobiology,* **100:**137 (1971).

114. Goy, R. W., and J. A. Resko, *Recent Prog. Hormone Res.,* **28:**707 (1972).

115. Greene, H. S. N., *Proc. Soc. Exp. Biol. Med.,* **40:**606 (1939).

116. Greene, H. S. N., and B. L. Newton, *Cancer,* **2:**673 (1948).

117. Griffiths, T. C., M. Tomic, J. M. Craig, and R. W. Kistner, *Surg. Forum,* **14:**399 (1963).

118. Guther, J., *Br. J. Cancer,* **25:**311 (1971).

119. Hafez, E. S. E., *Reproduction and Breeding Techniques for Laboratory Animals,* Lea & Febiger, Philadelphia, 1970.

120. Hafez, E. S. E., and D. L. Black, in E. S. E. Hafez and R. J. Blandau (eds.), *Mammalian Oviduct,* University of Chicago Press, Chicago, 1969.

121. Hagino, N. *Endocrinology,* **89:**1322 (1971).

122. Hagino, N., *J. Clin. Endocrinol.,* **35:**717 (1972).

123. Hall, C. E. and O. Hall, *Texas Rept. Biol. Med.,* **21:**16 (1963).

124. Hammer, C. E., in *Endocrinologia de la Reproduction,* D. F. La Prensa Medica Mexicana, Mexico, 1969.

125. Hardisty, M. W., *Biol. Rev.,* **42:**265 (1967).

126. Harrison, R. J. In S. Zuckerman (ed.), *Ovary,* Vol. I, Academic Press, New York, 1962.

127. Hendricks, A. G., and D. C. Kraemer, *J. Reprod. Fertil.,* Suppl., **6:**119 (1969).

128. Hendricks, A. G., and D. C. Kraemer, in A. G. Hendricks (ed.), *Embryology of the Baboon,* University of Chicago Press, Chicago, 1971, p. 3.

129. Heston, N. E., G. Vlahakis, and G. H. Smith, *J. Natl. Cancer Inst.,* **49:**805 (1972).

130. Heuser, C. H., and G. L. Streeter, *Contr. Embryol. Carneg. Inst.,* **29:**16 (1941).

131. Hilfrich, J., *Br. J. Cancer,* **28:**46 (1973).

132. Hodgkinson, A., P. M. Zarumbski, and B. E. Nordin, *Nature,* (London) **214:**1045 (1967).

133. Hohn, L. W. *Adv. Vet. Sci.,* **11:**159 (1967).

134. Hooper, E. T., and B. S. Hart, *Univ. Mich. Mus. Zool. Misc. Publ.,* **120:**1 (1962).

135. Horrobin, D. F., *Lancet,* **1:**170 (1968).

136. Hotchkiss, J., L. E. Atkins, and E. Knobil, *Endocrinology,* **89:**177 (1971).

137. Hoversland, A. S., J. Metcalfe, and J. T. Parer, *Biol. Reprod.,* **10:**589 (1974).

138. Hyden, D. W., and S. W. Nielsen, *J. Small Anim. Practice,* **12:**687 (1971).

139. Hytteen, F. E., and I. Lietch, *Physiology of Human Pregnancy,* Blackwell Scientific Publications, Oxford, 1964.

140. Ingalls, T. H., W. M. Adams, M. B. Lurie, and J. Iisen, *J. Natl. Cancer Inst.,* **33:**799 (1964).

141. Ijima, H., K. Nasu, and I. Taki, *Am. J. Obstet. Gynecol.,* **89:**946 (1964).

142. Jabora, A. G., *Australian J. Exp. Biol. Med. Sci.,* **40:**139 (1962).

143. Jay, G. E., Jr., in W. J. Burdette (ed.), *Methodology in Mammalian Genetics,* Holden-Day, Inc., San Francisco, 1963.

144. Jeffery, J., *J. Endocrinol.,* **34:**387 (1966).

145. Jirku, H. and D. S. Layne, *Steroids,* **5:**37 (1965).

146. Joshi, S. G., D. C. Kraemer and C. B. Chenault, *Contraception,* **2:**339 (1970).

147. Jost, A., *Recent Prog. Hormon Res.,* **8:**379 (1953).

148. Kaley, G., H. Demopoulos, and B. W. Zweifach, *Proc. Soc. Exp. Biol. Med.,* **109:**456 (1962).

149. Kamboj, V. P., H. Chandra, B. S. Setty, and A. B. Kar, *Contraception,* **1:**29 (1970).

150. Kamboj, V. P., A. B. Kar, S. Ray, P. K. Grover, and N. Anand, *Indian J. Exp. Biol.,* **9:**103 (1971).

151. Kaminetzky, H. A., *Obstet. Gynecol.,* **38:**232 (1971).

152. Kanagawa, H., K. Kawata, and T. Ishikawa, *Jap. J. Vet. Res.,* **13**:43 (1965).
153. Kaplan, H. M., *Rabbit in Experimental Physiology,* Scholars Library, New York, 1962.
154. Karsch, F. J., R. F. Weick, J. Hochkiss, D. J. Dierrschke, and E. Knobil, *Endocrinology,* **93**:206 (1973).
155. Kar, A. B., *Endocrinology,* **69**:116 (1961).
156. Kar, A. B., and H. Chandra, *Indian J. Exp. Biol.,* **4**:174 (1966).
157. Kar, A. B., and H. Chandra, *Current Science,* **35**:365 (1966).
158. Kar, A. B., V. P. Kamboj, and H. Chandra, *J. Reprod. Fertil.,* **16**:165 (1968).
159. Kar, A. B., V. P. Kamboj, and A. Goswami, *J. Reprod. Fertil.,* **9**:115 (1965).
160. Keith, W. N., and K. I. H. Williams, *Biochim. Biophys. Acta,* (Amst.) **210**:328 (1970).
161. Kent, G. C., Jr., in R. A. Hoffman, P. F. Robinson, and H. Magalhaes (eds.), *Golden Hamster: Its Biology and Use in Medical Research,* Iowa State University Press, Ames, Iowa, 1968.
162. Kernkamp, H. C. H., M. H. Roepke, and D. E. Jasper, *J. Am. Vet. Med. Assoc.,* **108**:215 (1946).
163. Keverne, E. B., and R. P. Michael, *J. Endocrinol.,* **51**:313 (1971).
164. Kirton, K. T., R. J. Ericsson, J. A. Ray, and A. D. Forbes, *J. Reprod. Fertil.,* **5**:275 (1970).
165. Kirton, K. T., R. J. Ericsson, J. A. Ray, and A. D. Forbes, *J. Reprod. Fertil.,* **21**:275 (1970).
166. Kirton, K. T., B. B. Pharris, and A. D. Forbes, *Proc. Soc. Exp. Biol.,* **133**:314 (1970).
167. Kirton, K. T., B. B. Pharris, and A. D. Forbes, *Biol. Reprod.* **3**:163 (1970).
168. Klionsky, B., and J. S. Wigglesworth, *Br. J. Exp. Pathol.,* **51**:361 (1970).
169. Klopper, A., *Br. Med. Bull.,* **26**:39 (1970).
170. Klopper, A., F. A. Michie and J. E. Brown, *J. Endocrinol.,* **12**:209 (1955).
171. Knobil, E., *Biol. Reprod.,* **8**:246 (1973).
172. Knuppen, R., and H. Breuer, *Excerpta Med. Int. Congr. Ser.,* **210**:234 (1970).
173. Kuchn, R. E., and I. Zucker, *J. Comp. Physiol. Psychol.,* **66**:747 (1968).
174. Kulkarni, B. D., *Int. J. Biochem.,* **1**:532 (1970).
175. Kulkarni, B. D., *J. Endocrinol.,* **48**:91 (1970).
176. Kulkarni, B. D., C. S. Kammer, and J. W. Goldzieher, *Gen. Comp. Endocrinol.,* **14**:68 (1968).
177. Kulkarni, B. D., C. S. Kammer, and J. W. Goldzieher, *Gen. Comp. Endocrinol.,* **14**:68 (1970).
178. Kumar, T. C. A., in E. Diczfalusy and C. C. Standley (eds.), *The Use of Non-Human Primates in Research on Human Reproduction,* WHO Symposium, Karolenska Institutat, Stockholm.
179. Ladman, A. J., *Anat. Rec.,* **157**:559 (1967).
180. Lamming, G. E. and E. C. Amorso (eds.), *Reproduction in the Female Mammal,* Plenum Press, New York, 1967.
181. Lansdown, A. B. G., S. J. Crees, and R. G. Wilder, *J. Inst. Anim. Tech.,* **21**:71 (1970).
182. Laumas, K., *Gen. Comp. Endocrinol.,* **2**:141 (1969).
183. Leathem, J. H., in J. H. Leathem (ed.), *15th Ann. Conf. on Protein Metabolism,* Rutgers University Press, New Brunswick, N.J., 1959.
184. Leathem, J. H., in W. C. Young (ed.), *Sex and Internal Secretion,* Williams & Wilkins, Baltimore, 1961.
185. Leathem, J. H., *J. Anim. Sci.,* **25**:68 (1966).
186. Leonard, B. J., in M. H. Briggs and E. Diczfalusy (eds.), *Pharmacological Models in Contraceptive Development,* Karolinska Institute, Stockholm, 1974.
187. Leung, K., I. Mernatz, and S. Solomon, *Endocrinology,* **91**:523 (1972).
188. Leyendecker, G., S. Wardlaw, and W. Nocke, *Acta Endocrinol.,* **71**:160 (1972).
189. Li, M. H., and H. Y. Tsai, *Acta Unio Intern. Contra Cancrum,* **20**:1514 (1964).
190. Lin, T. J., R. B. Billiar, and B. Little, *J. Clin. Endocrinol.,* **35**:879 (1972).
191. Lindner, H. R., *Acta Endocrinol.,* (Suppl.) **155**:93 (1971).
192. Lindner, H. R., in E. Diczfalusy and C. C. Standley (eds.), *Use of Non-Human Primates in Research on Human Reproduction,* Karolinska Institute, Stockholm, 1972.
193. Lipner, H., and R. O. Greep, *Endocrinology,* **88**:602 (1971).
194. Lipschutz, A., *Steroid Hormones and Tumors,* Williams & Wilkins, Baltimore, 1950, p. 6.
195. Loebel, B. L., E. Levy, and M. C. Shelesnyak, *Acta Endocrinol.* (Kbh.) **123**:47 (1967).

196. Longscope, C., D. S. Layne, and J. F. Tait, *J. Clin. Invest.*, **51**:93 (1968).
197. MacKay, D. G., V. Goldenberg, H. Kauntz, and I. C. Vossy, *Arch. Pathol.*, **84**:557 (1967).
198. Mackay, D. G., and T. C. Wong, *Am. J. Pathol.*, **42**:387 (1963).
199. Makino, S., M. S. Sasakti, T. Sofuni, and T. Ishiawa, *Proc. Jap. Acad.*, **38**:686 (1962).
200. Malouf, N., K. Benirschke, and D. Hoefnagel, *Cytogenetics*, **6**:228 (1967).
201. Mann, T., *Biochemistry of Semen and of the Male Reproductive Tract*, Methuen, London, 1964.
202. Marin-Padilla, M., and K. Benirschke, *Am. J. Pathol.*, **43**:999 (1963).
203. Marston, J. H., and M. C. Chang, *J. Exp. Zool.*, **115**:237 (1964).
204. Martan, J., and P. L. Risley, *Anat. Rec.*, **146**:173 (1963).
205. Masson, G. M. C., A. C. Corcoran, and I. H. Page, *Arch Pathol.*, **56**:217 (1952).
206. Masson, G. M. C., L. A. Lewis, A. C. Corcoran, and I. H. Page, *J. Clin. Endocrinol.*, **13**:300 (1953).
207. Masson, G. M. C., A. Mikasa, and H. Yosuada, *Endocrinology*, **71**:505 (1962).
208. Masters, R., and V. Johnson, *Human Sexual Response*, Little, Brown & Co., Boston, 1967.
209. Matsuyawa, A., T. Yamamoto, and K. Suzuki, *Jap. J. Exp. Med.*, **40**:159 (1970).
210. McArthur, J. W., and J. Ovadia, *Biol. Reprod.*, **5**:183 (1971).
211. McFeely, R. A., *J. Reprod. Fertil.*, **13**:579 (1967).
212. McGill, T. E., in F. A. Beach (ed.), *Sex and Behaviour*, John Wiley, New York, 1965.
213. McKinney, G. R., and J. H. Weikel, in H. Vagtborg (ed.), *Use of Non-Human Primates in Drug Evaluation*, University of Texas, Austin, 1968.
214. McLaren, A., in E. S. E. Hafez (ed.), *Reproduction in Farm Animals*, 2nd ed., Lea & Febiger, Philadelphia, 1968.
215. McPhee, H. C., and S. S. Buckley, *J. Hered.*, **25**:297 (1934).
216. Meissner, W. A., S. C. Sommers, and G. Sherman, *Cancer*, **10**:500 (1957).
217. Metcalfe, J., A. S. Hoversland, L. F. Erickson, A. L. Rogers, and P. L. Clary, in *Animal Models for Biomedical Research*, National Academy of Sciences, Publ. 1594, Washington, 1968.
218. Metcalfe, J., and J. T. Parer, *Am. J. Physiol.*, **210**:821 (1966).
219. Michael, R. P., and G. Slayman, *J. Comp. Physiol. Psychol.*, **64**:223 (1967).
220. Michael, R. P., D. Zumpe, E. B. Keverne, and R. W. Bonsall, *Recent Progr. Hormone Res.*, **28**:665 (1972).
221. Mishell, D. R., Jr., M. Lumkin, and S. Stone, *Am. J. Obstet. Gynecol.*, **113**:276 (1972).
222. Morris, J. M., and G. Van Wagenen, *Am. J. Obstet. Gynec.*, **96**:804 (1966).
223. Morris, J. M., G. Van Wagenen, and R. I. Dorfman, *Contraception*, **4**:15 (1971).
224. Morris, J. M., G. Van Wagenen, G. D. Hurten, D. W. Johnston, and R. A. Carlsen, *Fertil. Steril.*, **18**:7 (1967).
225. Morris, J. M., G. Van Wagenen, T. McCann, and D. Jacob, *Fertil. Steril.*, **18**:18 (1967).
226. Mossman, H. W., in R. J. Blandan (ed.), *Biology of the Blastocyst*, University of Chicago Press, Chicago, 1971.
227. Mossman, H. W., R. A. Hoffman, and C. M. Kirpatrick, *Am. J. Anat.*, **97**:257 (1955).
228. Myers, R. E., A. B. Holt, R. E. Scott, E. D. Mellits, and D. B. Cheek, *Bio. Neonate*, **18**:379 (1971).
229. Mykytowycz, R., and M. L. Dudzinksi, *CSIRO Wildl. Res.*, **11**:31 (1966).
230. Nahmias, A. J., W. T. London, L. W. Catalano, and D. A. Fucillo, *Science*, **171**:297 (1971).
231. Nalbaudov, A. V., in R. J. Blandan (ed.), *Biology of the Blastocyst*, University of Chicago Press, Chicago, 1971.
232. Napier, J. R., and P. H. Napier, *Handbook of Living Primates*, Academic Press, New York, 1967.
233. Nes, N., *Heriditas*, **56**:159 (1960).
234. Nillius, S. J., and L. Wide, *J. Obst. Gynaecol. Br. Commonw.*, **79**:865 (1971).
235. Nixon, W. E., W. W. Tullner, P. L. Rayford, and G. T. Ross, *Endocrinology*, **88**:702 (1971)
236. Ohno, S., and M. F. Lyon, *Chromosoma*, **16**:90 (1965).
237. Ohno, S., J. M. Trijillo, C. Stenius, L. C. Christian, and R. L. Teplitz, *Cytogenetics*, **1**:258 (1962).
238. Okuuda, T., and A. Grollman, *Arch. Pathol.*, **82**:246 (1966).

239. Overzier, C. (ed.), *Intersexuality,* Academic Press, London, 1963.
240. Owen, R. D., *Science,* **102:**400 (1945).
241. Page, E. W., and N. B. Glendening, *Am. J. Obstet. Gynecol.,* **69:**666 (1955).
242. Parkes, A. S. (ed.), *Marshalls Physiology of Reproduction,* Longmans, London, 1966.
243. Pasamanick, B., in *Conference on Obstetrics: Factors in Child Development,* Emory University, School of Medicine, Atlanta, 1965.
244. Peterson, E. N., and R. E. Behrman, *Am. J. Obstet. Gynecol.,* **104:**988 (1969).
245. Petter, G., and R. K. Creasy, *C. R. Soc. Biol.,* **164:**939 (1970).
246. Plant, T. M., V. James, and R. P. Michael, *J. Endocrinol.,* **51:**751 (1971).
247. Plant, T. M., and R. P. Michael, *Acta Endocrinol.* (Kbh.), **155:**69 (1971.
248. Poswillo, D. E., W. J. Hamilton, and D. Sopher, *Nature* (London) **239:**460 (1972).
249. Zuamme, G. A., D. S. Layne, and D. G. Williamson, *Can. J. Physiol. Pharmacol.,* **50:**45 (1972).
250. Reddy, S. V., H. Balin, and W. R. Nes, *Steroids,* **17:**493 (1971).
251. Reed, M. J., K. Fotherby, J. E. Peck, and Y. Gordon, *J. Endocrinol.,* **59:**569 (1973).
252. Rieche, K., *Arch Geschwulstforsch,* **29:**158 (1967).
253. Risley, P. L., in C. G. Hartman (ed.), *Mechanisms Concerned with Conception,* Macmillan Co., New York, 1963.
254. Roberts, J. A., *Invest. Urol.,* **8:**610 (1970).
255. Romanoff, L. P., M. P. Grace, E. M. Sugarman, and G. Pincus, *Gen. Comp. Endocrinol.,* **3:**649 (1963).
256. Rubin, H. B., and N. H. Azrion, *J. Exp. Anal. Behav.,* **10:**219 (1967).
257. Ruder, H. J., L. Loriaux, and M. B. Lipsett, *J. Clin. Invest.,* **51:**1020 (1972).
258. Russfield, A. B., *Tumors of the Endocrine Glands and Secondary Sex Organs,* Public Health Service No. 1332, Washington, D.C., 1966, (a) p. 45, (b) 81.
259. Rutty, D. A., *Vet. Rec.,* **80:**28 (1967).
260. Sandberg, A. A., R. Y. Kirdani, N. Back, P. Weyman, and W. R. Slaunwhite, *Am. J. Physiol.,* **213:**1138 (1967).
261. Scarpelli, D. G., and E. Von Haam, *Progr. Exp. Tumor Res.,* **1:**179 (1960).
262. Schane, H. D., A. J. Anzalone, and G. O. Potts, *Fertil. Steril.,* **23:**745 (1972).
263. Schardein, J. L., D. H. Kaump, E. T. Woosley, and M. M. Jellema, *Toxicol. Appl. Pharmacol.,* **16:**10 (1970).
264. Schwartz, N. B., *Recent Progr. Hormone Res.,* **25:**1 (1969).
265. Segal, S. J., L. Atkinson, A. Brinson, R. Hertz, W. Hood, A. B. Kar, L. Southhann, and K. Sunderam, in E. Diczfalusy, and C. C. Standley, (eds.), *Use of Non-Human Primates in Research on Human Reproduction,* WHO, Karolinska Institute, Stockholm, 1972.
266. Sever, J. L., *Cancer Res.,* **33:**1509 (1973).
267. Shellesnyak, M. C., *Endeavour,* **19:**81 (1960).
268. Shelesnyak, M. C., in C. R. Austin and J. S. Perry (eds.) *Biological Council Symposium on Agents Affecting Fertility,* Churchill, London, 1965.
269. Shintani, S., L. E. Glass, and E. W. Page, *Am. J. Obstet. Gynecol.,* **95:**550 (1966).
270. Sinari, V. S. P., *Acta Tropica* (Basel) Suppl., **9:**289 (1966).
271. Smith, H. A., and T. C. Jones, *Veterinary Pathology,* 3rd ed., Lea & Febiger, Philadelphia, 1966.
272. Soma, L. R., R. J. White, and P. B. Kane, *J. Surg. Res.,* **11:**85 (1971).
273. Sotelo, J. R., and K. R. Porter, *J. Biophys. Biochem. Cytol.,* **5:**327 (1959).
274. Stamler, F. W., *Fed. Proc.,* **20:**408 (1961).
275. Stanley, H. P., H. H. Hilleman, *J. Morphol.,* **106:**277 (1960).
276. Staples, R. E., *J. Rep. Fertil.,* **13:**429 (1967).
277. Stellar, E., *J. Comp. Physiol. Psychol.,* **53:**1 (1960).
278. Steven, L. C., *J. Natl., Cancer Inst.,* **50:**235 (1973).
279. Stoa, K. F., and B. W. Borjesson, *Biochim. Biophys. Acta,* (Amst.) **239:**337 (1971).
280. Strauss, F. H., E. Dordal, and A. Kappas, *Arch. Pathol.,* **76:**693 (1963).
281. Streeter, G. L., *Contr. Embryol. Carneg. Inst.,* **34:**167 (1951).
282. Sullivan, D. J., and H. P. Drobeck, *Folia Primat.,* **4:**309 (1966).
283. Sufrin, G., and D. S. Coffey, *Invest. Urol.,* **11:**45 (1973).
284. Symeonidis, A. J., *J. Natl. Cancer Inst.,* **10:**711 (1939).

285. Szentagothai, J., B. Flerko, B. Mess, and B. Halasz, *Hypothalmie Control of the Anterior Pituitary,* Akademiai Kiado, Budapest, 1968.
286. Szollosi, D., *Anat. Rec.,* **154:**209 (1966).
287. Taki, I., and H. Iijima, *Am. J. Obstet. Gynecol.,* **87:**926 (1963).
288. Tan, W. C., and A. Sabesteny, *Lab. Anim.,* **1:**1 (1972).
289. Tatum, H. J., *Contraception,* **1:**253 (1970).
290. Thijissen, J. H. H., in M. Tausk (ed.), *International Encyclopedia of Pharmacology and Therapeutics 2,* Pergamon Press, Oxford, 1972, p. 217.
291. Townsley, J. D., *Acta Endocrinol.,* **166:**191 (1972).
292. Vandewiele, R. L., J. Bogumil, J. Dyrenfurth, M. Ferin, R. Jewelewicz, M. P. Warren, T. Rizkallah, and G. Mikhail, *Recent Progr. Hormone Res.,* **26:**63 (1970).
293. Van Nie, R., and A. Dux, *J. Natl. Cancer Inst.,* **46:**885 (1971).
294. Van Wagenen, G., R. C. De Conti, R. E. Handschumacher, and M. E. Wade, *Am. J. Obstet. Gynecol.,* **108:**272 (1970).
295. Van Wagenen, G., and M. E. Simpson, *Embryology of the Ovary and Testis; Homo Sapiens and Macaca Mulatta,* Yale University Press, New Haven, Conn. 1965.
296. Vickery, B. H., and J. P. Bennett, *Physiol. Rev.,* **48:**135 (1968).
297. Wagner, J. C., *Some Pathological Aspects of Asbestosis in the Union of South Africa,* Little, Brown & Co., Boston, 1960.
298. Wardle, E. N., and A. N. Wright, *Am. J. Obstet. Gynecol.,* **115:**17 (1973).
299. Weinman, D. E., and W. L. Williams, *Nature* (London) **203:**423 (1964).
300. Whalen, R. E., C. Battle, and W. Luttge, *Behav. Biol.* **7:**311 (1972).
301. White, I. G., in E. S. E. Hafez (ed.), *Reproduction in Farm Animals,* Lea & Febiger, Philadelphia, 1968.
302. White, I. G. and J. Macleod, in C. G. Hartman (ed.), *Mechanisms Concerned with Conception,* Pergamon Press, New York, 1963.
303. Wildt, D. E., and W. R. Dukelow, *Lab. Primate Newsletter* **13:**15 (1974).
304. Wurtman, R. J., in L. Martini and W. F. Ganong (eds.), *Neuroendocrinology,* Academic Press, New York, 1967.
305. Young, W. C., and J. A. Grunt, *J. Comp. Physiol. Psychol.,* **44:**492 (1951).
306. Young, W. D., *J. Exp. Biol.,* **8:**151 (1931).
307. Zipper, J. G., G. Ferrando, E. Guiloff, G. Saez, and A. Tchernitchin, *Am. J. Obstet. Gynecol.,* **93:**510 (1965).

CHAPTER 10

ANIMAL MODELS IN TOXICOLOGY AND DRUG METABOLISM RESEARCH

A new drug when administered in humans produces desired and useful effects in most cases. However, at times it may produce harmful or toxic effects. The basic aim of toxicological studies, therefore, is to determine the degree and kind of harmful effect before a drug is considered safe for therapeutic purposes in humans. According to federal government regulations, toxicological studies should involve four mammalian species, of which one should be a nonrodent. Evaluation of a new drug in animal species is usually made by the determination of dose responses; absorption of the drug; systemic distribution, metabolism, and excretion; site and mechanism of action; activity of degradation products of the drug; acute, subacute, or chronic toxicity and allergic reactions. Occasionally, special toxicity studies such as teratogenic and carcinogenic properties of the test drug are also determined. Such tests may be performed in vitro with smooth muscle or microorganisms if the drugs are antibiotics. However, the biological activity of the drug should always be tested in animals before human trials are initiated. Animal studies are fairly reliable to provide evidence of the therapeutic values and safety of the drug. The World Health Organization (426) has also recommended that biochemical and toxicological studies of drugs and chemicals should be carried out in animals prior to their use in humans. Although the predictive value of data from animals and their interpretation with regard to human application is far from absolute, the pharmacological actions of the drugs are, by and large, similar to those in humans. The toxicological studies in animals pose other problems, for example, variations due to species or strain differences and translation of information from healthy animals under controlled conditions to pathological conditions in human beings.

This chapter describes the uses of experimental animals for toxicological studies, particularly for the evaluation of toxicity and metabolism of "new" drugs. Uses of animals in testing certain chemical compounds and toxins (microbial and plant origin) are briefly discussed. Evaluation of drugs for their potential carcinogenic or teratogenic activities have been described elsewhere in this book.

CHOICE OF EXPERIMENTAL ANIMALS FOR TOXICOLOGICAL STUDIES

Schmidt-Nielsen (337) reported that the dog is the favorite animal used in toxicological studies (27 percent of all experimental animals used in this field), followed by the rat (22 percent), the monkey and human (21 percent), the cat (8 percent), the rabbit (4 percent), and the frog (4 percent). Invertebrates, paramecia, planaria, lobsters, and insects are also used for special applications. Fish, alligators, tadpoles, lizards, and frogs are used in studies including drug metabolism, behavior and intestinal absorption, and determination of neurophysiological effects. Other mammals such as squirrels, beavers, desert rodents, pigs, and hamsters are also used for hibernation studies, renal function, blood pressure, and blood-brain barrier determinations. Certain unique anatomical or physiological properties of an animal may make it a favorite subject for particular studies. For example, the goose fish is used for renal studies because of the tubular structure of the kidney which is devoid of glomeruli and arterial supply.

The ideal species for toxicity studies should possess the following criteria:

1. Under 1 kg in body weight;
2. Easy to bleed and large enough to supply a reasonable amount of blood;
3. Easy to breed and maintain in the laboratory;
4. Easy to handle and to administer drugs by various routes;
5. Should have a short life span;
6. Physiology should approximate that of humans.

The weight of the experimental animal is important, because during early stages of development of drugs, only small quantities of it may be available. When selecting nonhuman primates because of their close relationship to humans, choice of species of nonhuman primate is important (17). For example, a completely vegetarian species may not be as useful because of the differences in the microflora of the intestine, which may affect drug metabolism.

Among the various species of animals used in toxicological studies, mice, rats, dogs, and nonhuman primates are most common.

Mice

Mice are particularly suitable when many animals are needed, such as in the evaluations of acute toxicity and carcinogenic activities. Mice may give a large number of false-positive results with carcinogenic agents. Also, the occurrence of spontaneous tumors is fairly common in mice, particularly after 18 months of age (386). Mice have a few idiosyncrasies, such as the susceptibility of the male mouse to chloroform (102) and formation of new bone within the medullary cavity. Despite these difficulties and the difficulty of obtaining reasonable amounts of blood for hematological studies, the mouse meets most of the criteria for a good animal model for toxicological work. The SPF mice should be used, and inbred mice should be avoided in toxicology.

Rats

The rat is probably the ideal species for toxicological work because it grows up to 500 g, can be easily handled and bled, and drugs can be easily administered. The rat generally behaves in a manner similar to mice, dogs, and monkeys with regard to toxicity studies.

Dogs

Various breeds of dogs are available for selection, but animals with short hair and adult weight of 12 kg are considered ideal for toxicological studies. Beagles are the animals of choice but others like spaniels and Pembroke Corgis have also been used. It is best to use 14- to 16-week-old dogs and allow about 4 weeks time for adjustment to the new environment, in which time unsuitable animals can be replaced. The dogs should be examined by a veterinarian, reinoculated against distemper, canine hepatitis, and leptospirosis, and anthelmintic drugs should be administered. Prothrombin time is determined to remove animals with Factor VII deficiency. During the drug administration period, the data from the study should be continuously compared to controls. General parameters such as body weight, food consumption, ophthalmoscopy, electrocardiography, and laboratory investigations (hematological and biochemical) are recorded and compared to normal animals.

Nonhuman Primates

Species difference in the incidence of tuberculosis, virus B, and foaming agents should be considered when using the nonhuman primate for toxicological studies (388). The animal handler should be warned about the zoonoses and should use protective clothing all the time. Monkeys have certain advantages over dogs, in weight, noise (barking), exercise (requiring large houses, etc). The upright posture of the monkey is similar to that of humans, making it possible to record meaningful observations, especially when peripheral neuropathy is a toxic manifestation. The brain and eyes approximate similarities to those of humans. However, it is almost impossible to administer drugs to monkeys parenterally, and a gum elastic stomach tube has to be used. Large number of animals are needed for fertility studies because of the low productivity.

INTERPRETATIONS OF TOXICOLOGICAL STUDIES IN EXPERIMENTAL ANIMALS

The United States federal government has established certain guidelines for animal toxicity studies (Table 10-1) (166). According to the Food and Drug Administration regulations, toxicological study of a drug should be made in four mammalian species, one of which should be a nonrodent. An outline of routine toxicological examination of drugs in animal species is given in Table 10-2.

Acute toxicity is determined in mice, rats, guinea pigs, and dogs (432, 433). However, the mouse is often the first test animal used to determine LD_{50}. The animals are observed for the effects of the drug on locomotion, behavior, circulation, and respiration 48 to 72 hours posttreatment (Table 10-3). Clinical symptoms are correlated with pathological and histological examination. The subacute toxicity studies are performed in both sexes of the experimental animal. Three dosage levels are usually tested for about 13 weeks. During this period, body weight, feed intake, and hematological and biochemical values in blood samples are routinely determined. These studies should provide information on (a) the maximum dose level at which a drug can be administered to the experimental animal without producing any clinical signs and (b) the rate of excretion of the drug or the tendency of the drug to accumulate.

The chronic toxicity of a drug is usually determined in rats and dogs. The animals are given the test material at three dosage levels for 18 months to 1 year. Table 10-4 summarizes the clinical tests and histopathological examination that should be made during the chronic toxicity studies.

Drug metabolism studies include absorption, distribution, excretion, and metabolism of a drug administered orally in a single dose. The distribution in various organs, tissues, and body fluids is determined by the most sensitive methods available. The extent of absorption would depend on surface area, blood flow, solubility of the

Table 10-1. Synopsis of Food and Drug Administration Guidelines for Animal Studies

Category	Duration of Human Administration	Phase[1]	Subacute or Chronic Toxicity[2]
Oral or Parenteral	Several Days	I,II,III,NDA	2 species; 2 weeks
	Up to 2 Weeks	I	2 species; 2 weeks
		II	2 species; up to 4 weeks
		III,NDA	2 species; up to 3 months
	Up to 3 Months	I,II	2 species; 4 weeks
		III	2 species; 4 months
		NDA	2 species; 6 months
	6 Months to Unlimited	I,II	2 species; 3 months
		III	2 species; 6 months or longer
		NDA	2 species 12 months (nonrodent) 18 months (rodent)
Inhalation (General Anesthetics)		I,II,III,NDA	4 species; 5 days (3 hours/day)
Dermal	Single Application	I	1 species; single 24-hour exposure followed by 2 week observation
	Single or Short-term Application	II	1 species; 20 day repeated exposure (intact and abraded skin)
	Short-term Application	III	As above
	Unlimited Application	NDA	As above, but intact skin study extended up to 6 months
Ophthalmic	Single Application	I	(Eye irritating tests; graded doses)

344

Table 10-1. (Continued)

Category	Duration of Human Administration	Phase[1]	Subacute or Chronic Toxicity[2]
Ophthalmic	Multiple Application	I,II,III	1 species; 3 weeks daily application, as in clinical use
		NDA	1 species; duration commensurate with period of drug administration
Vaginal or Rectal	Single Application	I	(2 species; local & systemic toxicity)
	Multiple Application	I,II,III,NDA	2 species, duration and number of applications determined by proposed use
Drug Combinations[3]		I	(LD50 by appropriate route)
		II,III,NDA	2 species, up to 3 months

[1]Phases I, II, and III are defined in 130.3 of the New Drug Regulations.
[2]Acute toxicity should be determined in 3 to 4 species; subacute or chronic studies should be by route to be used clinically.
[3]Where toxicity data are available on each drug individually.
Adapted from FDA Papers (166) with permission.

345

Table 10-2. Toxicological Procedures in Experimental Animals

I. Acute
1. LD_{50} determination in rats or mice.
2. Pyramiding single-dose studies in dogs.
3. Local effects (topical or parenteral agents).

II. Subacute
1. 6–13 weeks' administration to 40 rats (3 dose levels).
2. 4–13 weeks' administration to 6 dogs (3 dose levels).

III. Chronic
1. 1 year's administration to rats (3 dose levels).
2. 6 months' administration to dogs (3 dose levels).
(3. 6 months' administration to a third species.)
4. Reproduction experiments in rats and rabbits.

IV. Special studies
1. Metabolism: absorption, blood and tissue levels, transportation across membranes, excretion.
2. Effects on physiologic functions: blood pressure, cardiac output, respiration, renal function, CNS activity, hormonal effects, influences on appetite.
3. Histochemical studies when indicated.

Table 10-3. Symptoms Recorded During Acute Toxicity Test

Increased motor activity	Anesthesia
Tremors	Arching and rolling
Clonic convulsions	Ptosis
Tonic extensor	Lacrimation
Straub reaction	Exophthalmos
Piloerection	Salivation
Muscle spasm	a) viscid
Catatonia	b) watery
Spasticity	Diarrhea
Opisthotonos	Writhing
Hyperesthesia	Respiration
Loss of righting reflex	a) depression
Decreased motor activity	b) stimulation
Ataxia	c) respiratory failure
Sedation	Skin color
Muscle relaxation	a) blanching
Hypnosis	b) cyanosis
Analgesia	c) vasodilatation

Table 10-4. Guidelines for the Performance of Chronic Toxicity Studies

Study records	Autopsy record
Hematological Tests:	Complete gross pathology
	Histopathological examination of:
1. Hematocrit	
2. Hemoglobin	Heart
3. Total RBC count	Lungs
4. Erythrocyte sedimentation rate	Liver
5. Total leucocyte count	Pancreas
6. Differential leucocyte count	Kidneys
7. Reticulocyte count	Spleen
	Adrenals
Liver Function Tests:	Small and large intestines
	Stomach
1. Bromsulphalein retention	Mesenteric lymph nodes
2. Direct and indirect serum bilirubin	Gonads
3. Thymol turbidity	Brain
4. Serum alkaline phosphatase	Spinal Cord
5. Serum glutamic oxaloactic transaminase	Peripheral nerves
6. Serum glutamic pyruvic transaminase	Pituitary
7. Lactic dehydrogenase	Thyroid
8. Serum isocitric dehydrogenase	Bone marrow

Kidney Function Tests:

1. Blood urea nitrogen
2. Urine examination (color, transparency, pH, specific gravity, red blood cells, WBC, phosphate crystals)
3. Albumen, glucose, bilirubin, acetone, calcium oxalate crystals, epithelial cells and bacteria

drug, and its lipophilic and hydrophilic properties (101, 330). The binding of the drug to proteins and other macromolecules is critical in the competition of drugs and other metabolites for the same binding site on the molecule (37). The relationship of systemic activity and the concentration in plasma should be determined, although a complex mechanism controls the plasma concentration of the drug and its metabolites (Fig. 10-1) (261). Fleischi and Cohen (138) have developed a mathematical model to simulate the distribution of drugs in various fluids. The drug may act as an enzyme inducer, which may modify its action (32, 81, 82). The dose-response curve in an animal species is an important indicator of the use of the drug; it may be sigmoid, logarithmic, or a straight line (168, 286). The shape of the dose-response curve may be influenced by competetive, noncompetetive, and metaphasic antagonists (259, 309, 403).

A number of factors influence the drug response in experimental animals; some of these factors are briefly discussed in the following section.

Species Variation

Interspecies variation is primarily due to rate and pattern of metabolism, renal and biliary excretion, binding properties of the plasma proteins, tissue distribution, and response at "receptor site." Gillette (160) reported that carisoprodol (muscle relaxant) was effective at 10, 5, 1.5, and 0.2 hours postinoculations in cats, rabbits, rats, and mice, respectively. However, these apparent differences were related to plasma levels

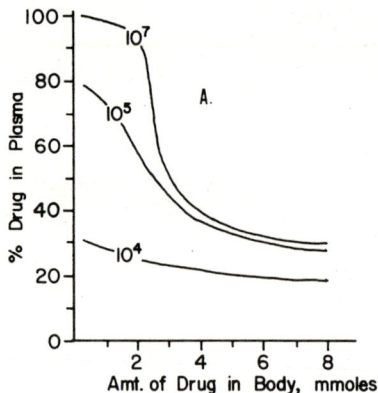

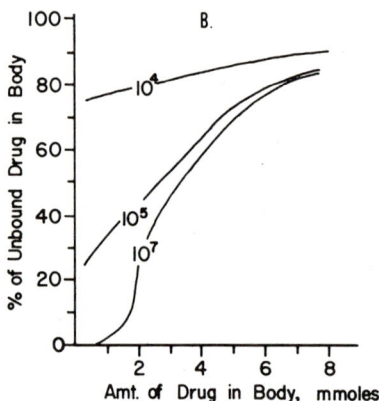

Fig. 10-1. The binding of drugs by plasma proteins. Plots are based on theoretical considerations; each curve represents a different value of k. (A) Illustrates the potential influence of plasma binding on the distribution of drug between plasma and other aqueous compartments in the body. (B) Illustrates the potential influence of plasma binding on the distribution of drug between bound and unbound forms in the body. Reproduced from *Journal of Pharmaceutical Sciences* (261) with permission.

which, at recovery, ranged between 125–130 μg/100 ml in all animals. Phenylbutazone (anti-inflammatory drug) has a half-life of over 72 hours in humans, whereas in most animals including monkeys it is between 3 and 8 hours (57). Similarly, sulfadimethoxine is readily metabolized in the rabbit, guinea pig, and the rat; the metabolism is slow but similar in humans and monkeys (43). Tables 10-5 and 10-6 illustrate the species differences in response to two drugs, methylfluoracetate and aminoxytriphene. Besides the species differences, drug metabolism is affected by other chemicals, disease state of the animal, dosage level, and route of drug administration (58, 97).

The metabolic differences in animal species are also related to the inherent ability of the test species to induce drug-metabolizing enzyme in liver (80, 316, 317). The extent of induction depends on the dosage level administered, age of the animal, and effective concentration in the liver (318, 319). Most of the work on enzyme induction has been reported with phenobarbital. It is not clear whether the same factors will apply to other drugs, and the applicability of these data from laboratory animal studies to human condition is possible.

Species differences are also noticed in the excretion of drugs (416), for example, N-acetyl-p-aminophenol is excreted as a glucuronide in dogs and humans, but it is not excreted in the urine of the cat because of the lack of conjugating enzymes in the liver (414). In addition to the excretion of a drug by kidneys, the biliary excretion is equally

Table 10-5. Comparison of LD_{50} Values of Methylfluoroacetate (mg/kg) in Various Animal Species

Species	Route of Injection				Lag Period	Terminal Symptoms
	I.V.	I.P.	S.C.	P.O.		
Mouse			6.8	6.7		
Rat			3.8	3.5	1 or more hours	convulsive (tonic)
Rabbit	0.24			4.0	minutes	cardiac arrest
Guinea pig		0.35	0.2	0.4		
Cat	0.5		0.3	0.3		
Dog	0.06		0.15	0.15		convulsive (tonic and clonic)
Pig			0.4			
Monkey			11.0	11.0		mixed
Man				3.85 (estimated)		mixed

Adapted from Coulston, F., In, Conference on Non Human Primate Toxicology, Dept. of Health, Education and Welfare FDA, Warrenton, Virginia, 1966, with permission.

Table 10-6. Toxicity of Aminoxytriphene in Various Animal Species

Species	Daily Dose (mg/kg)	P.O. No. of days	Symptoms of toxicity	Mortality
Mice	126	7	respiratory and general depression	0/10
	200	7	convulsions	1/10
	316	7	convulsions	2/10
	500	7		8/10
	LD_{50} 385±80			
Rats	15	21	None	0/10
	30	21	None	0/10
	60	21	Slight weight gain depression	0/10
Monkeys	6.25	365		0/3
	12.5	365	Estrogenic effect in male monkeys	0/3
	25.0	365		0/3
	50.0	8	Tremor, ptosis	2/3 (4–8 days)
	160.0		Slight ataxia	0/2
	320	Single dose	Slight ataxia, salivation, ptosis	0/2
	640		Convulsions, prostration, ptosis	2/2 (1 hour)
Dogs	1.56	30	None	0/4
	3.125	180	None	0/4
	6.25	180	None	0/4
	12.5	180	Slight ataxia, fine tremors	4/4 (1–3 days)
	25.0	180	Slight ataxia, fine tremors	4/4 (1–5 days)
	50.0	1	Clonic convulsions, prostration	4/3 (1 day)

Adapted from Coulston, F., In, Conference on Non Human Primate Toxicology, Dept. of Health, Education and Welfare FDA, Warrenton, Virginia, 1966, with permission.

important because it not only provides a mode of elimination but also because the excreted material is further degraded in the intestines (335, 352). Biliary excretion of simple low molecular weight compounds is small in most of the laboratory animal species. However, there is considerable species differences in the biliary excretion of strongly polar compounds of medium molecular weight (300–500). The excretion rate tends to be higher in rats, dogs, and hens than in rabbits, guinea pigs, and rhesus monkeys. The excretion rate is intermediate in the cat and sheep. The high molecular weight aromatic compounds (500–1000) are excreted at a similar rate in most animal species (353).

Strain Variations
The strains within a species of animal used to be a major source of differences in drug toxicity and metabolism. However, this variability has been considerably decreased in recent years due to the development of defined strains by inbreeding methods. However, mutation and other genetic factors can cause variability (7, 172, 211, 263, 270, 406).

Sex Differences
The importance of differences in the reactivity to compounds in male and female is well documented, for example, norepinephrine in cats (31, 311), tubocurarine in rats (424), lithium chloride in rats (6), reserpine in rats (313), histamine in guinea pigs (399), creatinine in rats (182).

Environmental Conditions
In mice, the type of bedding (corncob or red cedar chip) affects the action of sodium hexobarbital (136). Stress due to the isolation of animals affect the calcium, magnesium, sodium, and potassium levels in myocardium of rats (306). Exogenous noise may greatly increase the toxicity of sympathomimetic amines in mice (69). Crowding of animals may increase the effect on motor activity of amphetamine in mice (73, 112, 179). Temperature also affected the action of reserpine in rats (313) and caffeine and procaine in mice (272).

Diet
Although most animal feeds are available in standardized preparations, the extent of feeding and the feeding intervals affect drug action and toxicity (118, 148, 410a,b).

Age
Particular attention should be given to the age of the test animals when studying the drugs that affect the central nervous system (61, 213, 288, 313, 384).

Seasonal and Circadian Rhythms
The time of year and the time of day not only affect the endogenous substances in the plasma of animal (229, 307, 427) but also modify the actions of many drugs, for example, histamine in rats (89, 130), phenobarbital in mice (262), and biogenic amines (314, 336, 372).

Route of Drug Administration
In addition to animal-associated factors, drug toxicity depends on the route of administration (87, 93, 226, 275), the solvent used as a carrier (78, 147, 157, 358), and the chemical form of the drug (375, 411).

Table 10-7. Human Clinical Drug Schedules and Their Correlation with the Dose Schedules Used in Animal Toxicology Studies

Drug	Clinical schedule	Dose (mg/sq m)	Schedule and route	Dog	Monkey	Mouse
			Human Clinical Data	**Correlation of schedules for animal toxicology and schedules used in humans***		
Alkylating agents						
Cyclophosphamide	a	1,000	Single dose, i.v.	+	0	+
	b	375	d x 10, i.v.	–	–	+
Iphosphamide		5,000	Single dose, i.v.	+	0	+
Yoshi-864		75	d x 5, i.v.	+	+	–
Dibromodulcitol		131	d x 42, p.o.	0	–	–
TIC-mustard		900	d x 5, i.v.	+	+	+
Estradiol mustard**		80	d x 14, p.o.	+	–	–
Phenesterin**	a	28	d x 28, p.o.	+	0	0
	b	75	Biweekly (8 weeks), p.o.	+	0	0
Antimetabolites						
Cytosine arabinoside***	a	200	d x 5(continuous infusion),i.v.	+	–	–
	b	131	d x 10, i.v.	–	–	+
5-Azacytidine	a	250	d x 5, i.v.	+	0	+
	b	200	Biweekly, i.v.	+	0	+
6-MP-riboside		94	d x 28, i.v. or p.o.	–	–	–
5-FUdR		1,500	d x 5, i.v.	+	0	–
Tubercidin	a	56	Every week for 2 weeks, i.v.	–	0	+
	b	9.3	d x 10, i.v.	–	0	+
Guanazole		25,000	d x 5(continuous infusion), i.v.	–	0	–
β-TGdR**		300	d x 5, i.v.	+	–	+
Dichloromethotrexate	a	5.6	d x 42, p.o.	–	0	–
	b	75	d x 14, i.m.	+	0	–
Nitrosoureas						
BCNU		250	Single dose, i.v.	–	–	+
CCNU		130	Single dose, p.o.	+	+	+
MeCCNU		225	Single dose, p.o.	+	+	+
Streptozotocin		500	d x 5, i.v.	–	0	+
Antibiotics						
mithramycin	a	0.94	d x 7, i.v.	–	0	–
	b	1.88	Every other day for 8 days, i.v.	–	0	–
Porfiromycin	a	75	Single dose, i.v.	+	0	+
	b	19	Biweekly, i.v.	+	0	+
	c	8.4	d x 21, i.v.	–	0	–
Bleomycin		15	Biweekly, i.v.	+	+	+
Daunomycin	a	180	Single dose, i.v.	+	0	+
	b	45	d x 5, i.v.	–	–	–
Miscellaneous						
DTIC	a	169	d x 10, i.v.	–	–	+
	b	250	d x 5, i.v.	–	–	+
	c	1,150	Single dose, i.v.	–	–	+
Methyl-GAG	a	1,400	Single dose, i.m.	+a	–	–
	b	188	d x 20, i.m.	–	0	+
Pyran copolymer**		450	d x 14, i.v.	–	–	–
ICRF-159	a	3,000	weekly, p.o.	+	0	+
	b	1,000	d x 3, p.o.	–	–	–
Pseudourea		150	d x 5, i.v.	–	–	–
Emetine		56	weekly, i.v.	0	+	+
Camptothecin	a	120	Single dose, i.v.	+	0	+
	b	20	d x 5, i.v.	–	–	–

* Animal schedule and human schedule; +, identical; –, different; 0, no data.
** Human clinical data on file in Cancer Therapy Evaluation Program, National Cancer Institute.
*** Human clinical data for Schedule b on file in Cancer Therapy Evaluation Program.
a Different route.
Adapted from Cancer Research (167) with permission.

Because of these factors and many others, there has been increased concern over the predictability of experimental animal variations (187). Interpretation of data from animal studies are based on statistical evaluation of accuracy (i.e., how close the measured value approximates "true" value), precision (i.e., how well the repeated tests agree with one another), reproducibility (how well the results match when the tests are performed by different people), and repeatability (how well the results agree when the same text is performed on the same animal or patient by the same investigator). These statistical evaluations may reduce errors and increase the chances of predictability. Many reports are presented in the literature supporting as well as criticizing the applicability of animal studies to humans (108, 210, 365, 409). A recent editorial (254) pointed out some important considerations, such as the cost-benefit relationship, existence of threshold, appropriate experimental design, and ways to extrapolate data from experimental animal studies to humans. Freireich et al. (143), Homan (190), and Schein et al. (333) have compared human and animal toxicological studies, particularly interspecies difference in toxicities to organs, in mice, rats, hamsters, dogs, and monkeys. It was found that there was about 5.9 percent probability of exceeding the maximum tolerated dose (MTD) in humans if the clinical trials were started using the recommended one-third MTD found in the most sensitive large animal species (the dog or the monkey). In view of these findings, the protocol for large animal toxicological studies has been revised by the National Cancer Institute (NCI) toxicology laboratory (305). There has been considerable debate as to how the human dosage can be predicted on the basis of experimental animal studies. It has been emphasized that the calculation of the body surface area is a much better method than using body weight (143,300). However, Owen (287) supported his earlier belief that the clinical practice of starting human doses at one-tenth of the MTD (milligrams per kilogram) in the sensitive large animal is more suitable. Goldsmith et al. (167) reported that if the phase I clinical trials were based on one-third of the dose, then at least 17 percent of the thirty compounds compared would have shown human toxicity (Table 10-7). Therefore, the starting dose should not be strictly based on the one-third rule with the dog and the monkey studies. It seems that animal studies have a definite value in predicting human drug toxicity, although the predicting model may vary with the test drug.

ANIMAL MODELS FOR THE EVALUATION OF DRUG TOXICITY

Drugs Affecting the Neuromuscular Junction

Muscle relaxants have been widely used in surgery, orthopedics, and electric shock therapy. The effect of these drugs can be determined using nerve muscle preparations in vitro (303, 315) or in vivo by monitoring rate and depth of respiration (191) and by induction of tetany and fibrillation (236, 237). The most commonly used animal species are cats, rabbits, chicken, mice, and frogs.

The sciatic-gastrocnemius and femoral-gracilis nerve-muscle complex are often used in decerebrate and immobilized cats (42, 47, 56). In rabbits, the "head drop" procedure is most commonly used to test tubocurarine and other competitive antagonists (191, 397). In chickens, the depolarizing agents cause opisthotonos (rigid extension of limbs and retraction of head), whereas competitive agents cause flaccid paralysis (59, 377). In mice, a rotating cylinder producer, in which the time required for fall from a rotating rod, is determined following injection of a drug (113, 208, 404). The original animal trials using ligated legs and electrical stimulation in frogs (29) are not commonly used at the present time.

Drugs Affecting Parasympathetic-Neuroeffector Junction

Agents affecting the autonomic nervous system can be separated into (a) muscarinic (agents causing stimulation blocked by atropine) and (b) nicotinic (agents that are blocked by nicotine). The primary site of action of muscarinic results in characteristic SLUD syndrome (salivation, lacrimation, urination, and defecation). The mode of action may be by antagonists per se or by indirect inhibition of the enzyme, cholinesterase. Some of the experimental animals used for these studies include dogs, cats, rabbits, rats, and mice. In dogs, the cannulated femoral vein is used for test purposes. The ratio of response of blood pressure, both before and after administration of the test compound, is compared to that of acetylcholine (240, 430). Other procedures involve electrical stimulation of the vagus nerve in an anesthestized dog (110), electrical stimulation of the chorda tympani nerve and collection of saliva from Wharton's duct (243), collection of secretion from gastric pouches (328), and recording of intestinal activity by intraluminal balloons (240, 430). In cats, intramuscular injection produces mydrias, and in large doses, crying, prolonged whining, restlessness, gagging, vomiting, tremors, and ataxia. For quantitative tests, the diameter of the pupil in response to light (243), blood pressure changes (247), and balloon technique (296) are used. Rabbits have been used for screening tests. The inhibition of pilocarpine-induced salivation is used for test purposes (240, 243). Guinea pigs are very sensitive to histamine-induced gastric ulcers, and the test compound should prevent their development (5, 109). Other procedures used include lung overflow (79) and spasm of the bile duct (88).

Rats are probably most useful because both the stimulatory and inhibitory activities of agents on the parasympathetic-neuroeffector junction can be tested (202). The peripheral actions are salivation, lacrimation, and miosis, whereas the central actions include scratching, hunching of back, tremors, and compulsory chewing movement. Quantitative procedures for lacrimation and chromodacryorrhea (red tears) have been developed using (a) filter paper soaking technique (251) or color changes in silver or mercury salts complexed to iodide (349); (b) the pouch technique (380); (c) perfusion with 0.25 mm NaOH (158). Restriction of movement occurs in ulcer formation (45) and results from the effects of other agents like aspirin, reserpine (188); 5-hydroxytryptophane (326); and indomethacin (356). The urinary bladder of charcoal-fed rats have also been used (240, 396) to test the effects of these drugs.

The mydriatic effect has been tested in mice (198, 240). The heart of the frog, particularly during winter, is very sensitive to acetylcholine and has been used for assay procedure (431).

Drug Affecting Sympathetic Neuro Junctions

Activation of the sympathetic nervous system is characterized by increase in the rate and force of myocardial contractility, increase in blood pressure, dilation of pupil, and stimulation of vasoconstriction of fibers in the skin. Metabolic changes include increase in circulating glucose, fatty acids, and overall stimulation of metabolism for the production of energy. These actions result from stimulation of α or β receptors. Stimulation of α receptors (e.g., norepinephrine) results in vasoconstriction in the viscera, skin, contraction of nictating membrane, decrease in intestinal motility, and mydriasis. Stimulation of β receptors (e.g., isoproterenol) results in vasodilation in skeletal muscles and coronary vessels, contractility of myocardium, and dilation of bronchi. The agents for α and β receptors are phentolamine and propranolol, respectively. The actions of these drugs in animals are measured by gross symptoms, such as piloerection, increase in motor activity, excitability, and clonic and tonic convulsions. Cardiovascular effects are also used as test indicators. Sometimes indirect tests using animals pretreated with reserpine (which releases stored catecholamines) are also used.

Animal Models

Dogs. The electrocardiogram or blood pressure in the carotid artery is used as a test indicator in unanesthetized dogs (2, 235, 281). Other methods include measurements of bronchodilation and cardiovascular changes in anesthetized dogs (105), observation of ileal loops containing water-filled balloons, (3) and recording the movement of unrestrained dogs (299).

Cats. Similar cardiovascular measurements are used in cats (23, 31, 407). Other workers have tested drugs on systems involving the nictating membrane, submaxillary duct, hypogastric nerve-uterus junction (128), splenic volume (54), and chronic denervation of specific organs (55).

Guinea Pigs. The ability of the drug to control the histamine-induced bronchial spasm is used as a test system (235). The effect on vas deferens and urinary bladder in response to stimulation of hypogastric nerve is also used for testing new drugs (253 a,b).

Rats and Mice. The most commonly used model uses pithed rats with cannulated trachea, femoral vein, and carotid artery (159, 342). The effect of the drug on the blood glucose concentration in 24-hour-fasted rats is also used (224). In mice, the increased amount of motor activity is recorded by electrical or mechanical systems (310, 341).

A unique phenomenon of twitter up to 250 times per minute was noticed in chickens after injection of 2–5 mg of amphetamine. Additional symptoms include tilt of the body, drooping wing, vigorous head movement, and ruffled feathers.

Drugs Effecting Ganglionic Block

The drugs were used for treatment of hypertension. Because the agents affect both the sympathetic and parasympathetic nervous systems, they produced profound effects on other systems.

The cat is the animal of choice. Cervical ganglions are most often used as both the pre- and postganglions and the blood supply can be easily isolated (294, 383, 421). Other workers have used ciliary ganglions (297) or autonomic nerves of the heart (298). The hypogastric nerve-urinary bladder preparation in the guinea pig is often used (253 a,b). Dogs have been used but the experiments should be carried out with unanesthetized animals (153).

Drugs Acting on Smooth Muscles

The activity of Smooth muscle is controlled by neurotransmitters which are biologically active compounds (angiotensin, bradykinin, caerulein, darmstoff, eledoisin, and others). Except for the perfusion of organs (395) in dogs or cats, no generalized testing procedures are available because of the unique nature of action of each compound.

Angiotensin

A decapeptide with pressor activity is produced by action of renin on angiotensinogen. Most of the methods used are in vitro procedures.

Bradykinin

A polypeptide with γ-amino acids is released from α-globulin by limited proteolytic hydrolysis. It affects nonvascular smooth muscles, the cardiovascular system, and cell permeability. The in vivo procedure uses perfusion of the hindquarters of the animal by connecting the carotid and femoral arteries. The drug is generally injected into the jugular.

Histamine

A beta imidazolyl ethylamine is a decarboxylation product of histidine and is produced in the granules of mast cells and basophils in large quantities. There are few direct therapeutic applications of histamines; most of the test procedures are to evaluate antihistamines.

In dogs and cats, the decrease in blood pressure, collection of gastric juice, and production of ulcers are used as test indicators (188). The guinea pig is probably the best model. The lung overflow procedure with cannulated trachea (79) and capillary permeability changes using dyes is often used (134).

Serotonin (5-Hydroxytryptamine)

It is widely distributed in the body and is present in high concentration in most cells and the central nervous system. Serotonin is tested mostly by in vitro studies.

Oxytocin

This hormone is most active on the uterus and mammary glands. Animal models used for its action include the uterus of nonpregnant cats (25). This model is much more sensitive than the milk ejection from mammillary glands of rabbits.

Prostaglandins

These are a group of naturally occurring long chain hydroxy fatty acids that stimulate the contraction of the uterus and other smooth muscles, lower blood pressure, and affect the action of other hormones.

Administration of prostaglandins to anesthetized dogs causes a decrease of blood pressure by direct action on smooth muscle of vascular system (350, 408). Prostaglandins can also be tested by collecting gastric secretion in fundic pouches either innervated or denervated (322). The actions on adrenal medulla (215) and nasal mucosa (366) have also been used as test systems. In cats, sedation and catatonia (193) and changes in pupillary size and intraocular pressure are used as test systems. The oviduct of the rabbit is a very good system (194). Miosis in the profused arteries chamber of the rabbit eye has also been used (401).

Vasopressin

This is an octapeptide that causes an increase in blood pressure and has an antidiuretic effect. The blood pressure changes in the rooster (379), antidiuretic activity (140, 374), and blood pressure changes in rats are often used as test systems (140, 374).

ANIMAL MODELS FOR THE EVALUATION OF ANALGESICS

Analgesics are agents that relieve pain. There are two classical groups of analgesics: (a) narcotics (e.g., morphine) and (b) nonnarcotics (e.g., aspirin). Using biological systems, it is easier to evaluate the affects in the narcotic type than in the nonnarcotic type. The use of analgesic stimuli, such as heat, electricity, pressure, and chemicals, have been found useful in the evaluation of analgesic drugs. Chemicals used to produce pain invariably cause inflammation and hence analgesics may be judged on the basis of their anti-inflammatory activity. The test procedure for analgesics must permit quantitative determination of threshold values of stimuli; provide information on the least discernible difference between the intensities of two stimuli; be applicable to humans and animals; and the method should quantitatively determine different qualities of pain.

Heat Stimulation

In vivo methods have been developed with several animals.

Guinea Pigs

The use of radiant heat to produce flinching of the skin in a guinea pig has been used to evaluate analgesics (417). The threshold of pain is determined by the time of exposure, and it is compared with threshold in the presence of analgesic. Guinea pigs have also been used to study anti-inflammatory activity (271, 418).

Rats and Mice

The time required for the production of twitching of tail on exposure to heat and its decrease on the application of analgesic compound has been used as a test system (21, 91). Similar techniques have been developed for mice (176). However, the movement of hind paws using the hot-plate technique (117) and ultrasonics (277) are more often used.

Electrical Stimulus

Monkeys with electrodes implanted in the gasserian ganglia are often used to evaluate analgesic activity (413). The tooth pulp nerve of dogs and cats (38, 163) is very sensitive to electrical stimulus and is often used in the evaluation of analgesic drugs. Tooth pulp of rabbit (429) and movement of rabbit ear have also been used (234). The behavioral effects to electrical stimulation and its relief by analgesic agents in rats has been used by several workers (124, 412). Similarly, electrical stimulation of tails in rats and mice (63, 276, 278) and scrotum in rats (246) has been used.

Pressure Stimulation

The patellar reflex, ipsilateral flexor, crossed extensor reflex along with pupillary diameter, heart rate, respiration rate, and body temperature are used as tests in dogs (256). Mechanical pressure applied to the tails of rats and mice has been used for the evaluation of analgesic drugs (171, 373).

Chemical Stimulation

The modification of pseudoaffective response (vocalization, hyperpnea, and increase in systemic blood pressure) by narcotic and nonnarcotic agents in the dog is used for test purposes (177, 178). Inflammatory responses in the joints of dogs have also been used for evaluating anti-inflammatory drugs (260). In rats, inflammatory responses to the hind paw by 0.1 ml of 20 percent yeast cells and the pain threshold to air pressure or the volume of edema have been used as models (152, 308, 420).

ANIMAL MODELS FOR THE EVALUATION OF ANTICONVULSANT DRUGS

In animal studies, convlusion conditions are produced by electroshock seizures, and the protective anticonvulsion agent abolishes the extensor tonic component of the seizure. Such experiments have been performed with cats (122, 382), rabbits (382), rats (425), and mice (341). Several modifications of the basic test involving a particular area have been suggested (144, 151). In rats, rapid eye movement (REM) during sleep has also been used for testing (77).

In addition to electrical stimulation, intermittent light has been used to cause seizure in baboons (218), and the accompanying muscular activation is observed in triceps muscle. Sound (noise) seizure has been used in rats and mice (370). Another study using freezing of restricted area of brains of cats, dogs, and monkeys resulted in seizure that could be relieved by anticonvulsant drugs (279). Sudden alcohol withdrawal in dogs results in convulsion, and this procedure has also been used to test anticonvulsant drugs (123).

ANIMAL MODELS FOR THE EVALUATION OF ANESTHETICS

Local Anesthetic

The primary action comes from the cessation of nerve impulse conduction at the dendrite, cell body, or axion level. In vivo experiments using exposed sciatic or phrenic nerve of dogs or injection into lumbar spinal region are often carried out to test the activity (20, 98). The wink response to the blunted end of lead pencil, hair, and similar objects is evaluated after the test drug is placed in the cornea of rabbits (20, 164). The rabbit is also a favorite model for testing spinal anesthesia because of a large intervertebral opening at the lumbar sacral junction. The guinea pig is a most sensitive animal model and several locations, such as cornea abrasion, denuded back, skin burns, and sciatic nerve, are used (60). Mice (208), frogs, and earthworms (33, 368) are also used.

General Anesthesia

The best model for determining the general anesthetic activity of a compound is the mouse. Several closed systems have been developed for administration of volatile compounds (245, 347). The experiments with mice may provide LD_{50} information and dose response data (146, 221). Dogs are very useful in the determination of induction time, potency, and duration of action of anesthetics (139, 302a, b, 343). Several authors, realizing the difficulties of testing volatile compounds, have developed special emulsions for intravenous injections (67, 222).

ANIMAL MODELS FOR THE EVALUATION OF ANTITUSSIVE AGENTS

Antitussive agents suppress the cough response, which involves sensory receptors in respiratory tract, afferent nerves to medullae, and efferent nerves to the musculature.

The artificial cough response in experimental animals can be evoked by mechanical, electrical, and chemical means to test the antitussive agents. Dogs are most commonly used. Mechanical devices such as pig bristles are used to cause tracheal irritation (212, 376). The electrical stimulation of the vagus causes coughing in dogs and cats (34). Electrodes placed in the trachea (362) and stimulation of discrete regions of the medulla, particularly in cats, have also been used as the test system (39). The inhalation of acrotein in guinea pigs (345); ammonia (7 percent in air) in cats, rabbits, and guinea pigs (363, 419); and sulfur dioxide in cats, dogs, and guinea pigs (170) are effective agents of the coughing response. Other less commonly used agents are citric acid in the guinea pig (199) and dimethylpentylpiperazinium in cats (331).

ANIMAL MODELS FOR THE EVALUATION OF ANTITREMOR AGENTS

Involuntary tremors, a cardinal symptom of parkinsonism, has been the main target for antitremor agents. The oxymetabolite of termorine is the most commonly used agent for causing tremors (107, 285, 357, 428). Mice are often used. Rabbits and rats are also affected but cholinesterase inhibitors are usually used in these animals (92, 415). Repeated administration of iron (9) and damage to mesencephalic and pentile tegmentum also cause tremors in monkeys (204, 398). Antiparkinsonism drugs reduce the reserpine-induced rigidity in rats; this has been also used as a test system (209).

ANIMALS MODELS FOR THE EVALUATION OF SPINAL DEPRESSANTS

Spinal cord depressants, selectively or preferentially, depress the polysynaptic reflex and are clinically used to control pain associated with spastic muscular contraction.

Spinal cord depressants cause inhibition of posttetanic fibrillation in the musculature of dogs (237). Other procedures include hyperventilation of lungs in dogs with thyroid and parathyroid removed (236). In cats, the patellar and linguo-mandibular reflex in the presence of electrical or mechanical stimulation is used for testing spinal cord depressants (184, 220). In 1- to 7-day-old chicks the patellar and crossed extensor reflexes are often used (284). In mice, corneal and pinna reflexes give the best results (422).

ANIMAL MODELS FOR THE EVALUATION OF PSYCHOTROPIC DRUGS

Psychotropic drugs affect the activities of the mind. The animal methods involve observation of normal and drug-induced behavior. The central nervous system (CNS) stimulants cause hallucinations and sham rage (inappropriate, purposeless, aggressive behavior). The symptoms can be controlled by tranquilizers and CNS depressants. In animal behavior experiments, a number of factors such as environment (temperature, humidity, range area, noise), presence or absence of observer, and presence of other animals affect the results. Since the tests are subjective, it is important that "blind" observers (not knowing the treatment) are used.

Cats are often used in these studies and their behavior is categorized into four classes: sociability, contentment, excitement, and hostility. These are paired in two groups with opposite types of behavior: (a) sociability versus hostility and (b) contentment versus excitement. Further subclassification is also made for evaluation; for example, sociability is divided into: jump up, meowing, approach, tail up, and stand up (282). Psychotropic drugs administered orally or into the cerebroventricular system cause improvement in sociability (132, 200, 206, 320).

The rat is the most popular species for the evaluation of psychotropic drugs based on observations in the presence of drug-conditioned behavior for a reward or avoidance of noxious stimuli. Observation parameters include exploration, investigation and mating, aggression, flight submission, flight escape, maintenance, and residual (344). Other workers have used open field behavior (44), entrance into holes (223), bar press activity of untrained and trained rats in Y-boxes (255), and several types of maze tests (111, 127). Chemicals such as reserpine and apomorphine create an abnormal behavior, and antagonism of this behavior by the test drug is a measure of its effectiveness (120, 367). The double alley frustration test (141) and the ability of rats to work for a reward or conditioned avoidance have also been used (106, 207, 264).

In pigeons, reserpine-induced emesis (104) and observation of pecking activity (103) are used for testing psychotropic drugs. The twitch response in mice (84) produced by hallucinatory agents and the fighting behavior induced by chemical and electrical stimuli are used for testing purposes (35, 94, 229). The conditioned tail reflex in response to light stimulus in crabs (85) and the web-building ability of the spider (423) have been also used for testing psychotropic drugs.

ANIMAL MODELS FOR THE EVALUATION OF DRUGS ACTING ON BLOOD VESSELS

The effect of drugs on the vascular system can be determined by measurement of blood flow to a restricted area or peripheral resistance to an isolated portion of the cardiovascular system. Several reviews detail the procedures (145, 338).

Dogs have been most extensively used to produce chronic hypertension. Earlier experiments (165, 289) used bilateral renal ischemia. Other procedures, including neurogenic methods (174, 378), compression of the renal parenchyma by umbilical cord (175), chronic constriction of carotid sinus and internal carotid (402), injection of microspheres in renal artery (252), and a combination of several of these basic techniques, have been used (228). Cross-circulation procedures have been designed for studying the effect on heart, blood vessels, and CNS (173, 186). The venous portion of the peripheral vasculature has also been studied (1, 385).

In monkeys, chronic catheterization of superior vena cava or cannulization of jugular and carotid artery (16, 19) are used as the test system. In cats, perfusion of the hind limb by connecting abdominal aorta and femoral veins (169) or perfusion involving cerebral circulation (156) or protal vein (205) are used. The size of the central artery of a rabbit ear is determined under a binocular microscope for drug evaluation (10). Radioactive rubidium has also been used (26). Mice (26), 6- to 14-day-old chicks (359), and frogs (361) have been used occasionally.

ANIMAL MODELS FOR THE EVALUATION OF DRUGS ACTING ON THE HEART

Drugs affecting smooth muscle and blood vessels also have an effect on the heart, but in this section only drugs directly affecting the myocardium are considered.

Dogs have been most widely used in these studies. Several measurements in dogs have been made to evaluate the effects of drugs on the heart: conduction velocity, ventricular excitability (266), heart rate (129), and dimensional changes in the left ventricle during systole (328). Other techniques have used the strain gauge (360), vagal stimulation (90), and heart-lung preparations (46, 370). Other workers have used complete heart block (304) and chronic heart block (364) as test procedures. The use of chemicals for production of various conditions have also been tried, for example, barium chloride for ventricular extrasystoles (348), halothane-induced sensitization to epinephrine (355), and propiophenone sensitization to catecholamine (268, 291, 321).

ANIMAL MODELS FOR THE EVALUATION OF DRUGS AFFECTING KIDNEYS

A clinical sign of kidney dysfunction is edema. The agents that increase the amount of urine production, thus relieving edema, are called diuretics.

Trained (able to lie in supine position for 4 to 7 hours) female dogs make useful models for these studies (351). Glomerular filtration rate (GFR) is measured by insulin or creatinine clearance and phenolsulfonphthalein and paraaminohippuric acid are used for renal plasma flow (RPF) measurements. Both unanesthetized (15, 339) and anesthetized dogs (325, 339) may be used. The stop-flow technique to determine the site of action of drug within the nephron has been used by a number of authors (292, 339, 393).

Rats are the most widely used species for routine testing. Basically, the amount of urine collected during a 5-hour period was used as criterion of drug activity (13, 290, 389). Several other procedures with modified collection methods have been used (265, 334). Perfusion techniques using isolated tubules (28) and Henle's loop (273) have been reported. A few workers have also used mice (269, 301) and rabbits (278).

In addition to the animal models described above, a number of in vitro systems have been widely used for the evaluation of drug toxicity and metabolism. Table 10-8 summarizes the in vitro procedures using experimental animal organs as model systems.

Table 10-8. In Vitro Procedures for the Evaluation of Drug Toxicity Using Animal Organs

Drug action on	Animal species	In vitro system	Reference
Neuromuscular junction	Rat	1. Phrenic nerve-diaphragm 2. Obturator nerve-anterior gracilis muscle	51,142 225
	Chicken	1. Biventer cervicis 2. Semispinalis cervicis 3. Ichiatic nerve-tibialis muscle	161 71 392
	Frog	1. Rectus abdominis 2. Sciatic-gastrocnemius 3. Sciatic sartorius	390 329 274
	Guinea pig	Esophagus preparations	18
	Rabbit	Lumbrical muscle	203
	Cat	Lumbrical muscle	203
Parasympathetic-neuromuscular junction	Rabbit	Ileum and circular muscle	96,387
	Guinea pig	Circular muscle	295
	Cat	Circular muscle using Auerbach's nerve complex	72
	Guinea pig	Circular muscle strips and isolated trachea	22,83,180
	Rat	Stomach-Duodenum Jejunum and ileum	214,11

Table 10-8. (Continued)

Drug action on	Animal species	In vitro system	Reference
Parasympathetic-neuromuscular junction	Rat	Uterus	326
		Profused lung	232
	Chicken	Crop	125
	Frog	Perfused heart	41
	Squid	Giant axon	27
Sympathetic neuroeffectors	Dogs & Cats	Isolated perfused spleen	231
	Rabbits	Rabbit papillary muscle	2
		Rabbit isolated ear	99,267
		Spiral strip of aorta	149
		Isolated uterus	243
	Guinea pig	Isolated circular or longitudinal smooth muscle of esophagus	14
		Superperfusion of ileum	150
		Periarterial-nerve-circular muscle preparation	4
		Bronchial tree	2
		Trachial chain preparation	381
		Isolated uterus	185
	Chicken	Rectal cecum	126
		Expansor secundariorium of the wing and esophagus	50
	Rat	Rat fundus	12,201,283,394

363

Table 10-8. (Continued)

Drug action on	Animal species	In vitro system	Reference
Ganglionic Block	Rabbit	Ileum	133
		Isolated sympathetic ganglion	116,119
	Guinea pig	Urinary bladder	253a,b
	Rat	Cervical ganglion	323
	Marine gastropod aplysia	Pericarya of the abdominal ganglia	76
Smooth Muscles			
a) Angiotensin	Guinea pig	Ileum	162
	Rat	Ascending colon	312
	Hamster	Gastric fundic strip	262
b) Bradykinin	Cat	Isolated jugular strips and perfused with blood	
	Guinea pig	Ileum	216
		Bronchial musculature	346
	Rat	Isolated virgin uterus	332
		Duodenum	192
	Hamster	Fundic strips	262
c) Histamine	Cat	Isolated skin flap from hind legs;Ileum	135

Table 10-8. (*Continued*)

Drug action on	Animal species	In vitro system	Reference
d) Serotonin	Rat	Stomach fundus	394
	Chicken	Crop	125
		Rectal cecum	183
e) Oxytocin	Guinea pig	Isolated virgin uterus	53
	Rat	Uterus	25,183
		Myoepithelium of the rat	354
		Strips of isolated mammary glands	257
f) Prostaglandins	Cat	Tracheal chain preparation	250
		Isolated carotid artery	217
	Dog	Perfused spleen	95
	Rabbit	Uterus	217
		Ileum, jejunum and aorta	195
	Rat	Gastrointestinal tract	24,75,217
	Guinea pig	Tracheal rings	24
		Ileum	408
g) Vasopressin	Guinea pig	Proximal end of colon	40
h) Heat stimulant	Dog	Red blood cells hemolysis	48,49
Spinal cord depressants	Frog	Isolated perfused spinal cord	258

Table 10-8. (Continued)

Drug action on	Animal species	In vitro system	Reference
Drugs affecting blood vessels	Cow	Isolated facial artery	155
	Dog	Isolated femoral arterial vessels	64,65
		Nerve-arterial segment	324
		Isolated lateral saphenous, jugular and mesenteric veins	74
	Rabbit	Strips of aorta	149,293
		Helical strips from thorasic aorta	30
		Central artery of ear	391
		Isolated nerve-arterial	100
		Smooth muscle preparation	154
		External jugular and posterior venacava	369
		Isolated portal vein	197,205
	Chicken	Dually innervated longitudinal arterial muscle strip	36
	Rat	Isolated perfused rat lung	181
		Perfused hind quarters	249
		Isolated renal artery	196
		Isolated mesenteric arteries	
		Superior and inferior venacava	248
Drugs affecting the heart	Toad	Aorta and abdominal veins	219
	Dog	Isolated heart muscle	189
	Cat	Langendorff heart preparation	340
		Papillary muscle of heart	68,86

366

Table 10-8. (Continued)

Drug action on	Animal species	In vitro system	Reference
<u>Drugs affecting the heart</u>	Cat	Isolated sinoatrial node	227
	Rabbit	Langendorff heart	242
		Isolated auricles	238
	Chicken	Isolated auricles	238
	Rat	Vagus nerve and right atrium	70
		Isolated ventricle	131
<u>Drugs affecting the kidney</u>	Dog	Isolated perfused kidney	121,280,405
	Cat	Kidney with heart–lung preparation for perfusion	239
	Rabbit	Single nephrons	52
	Guinea pig	Isolated kidney	8

REFERENCES

1. Abboud, F., and J. W. Eckstein, *J. Clin. Invest.,* **47**:10 (1968).
2. Ablad, B., M. Brogard, and L. Ek, *Acta Pharmac., (Suppl. 2)* **25**:9 (1967).
3. Ahlquist, R. P., and B. Levy, *J. Pharmacol. Exp. Ther.,* **127**:146 (1959).
4. Akubue, P. I., *J. Pharm. Pharmacol.,* **18**:390 (1966).
5. Anderson, W., and P. D. Soman, *J. Pharm. Pharmacol.,* **17**:92 (1965).
6. Andredi, V. M., *Experientia,* **24**:1155 (1968).
7. Angel, C. R., D. T. Makin, R. D. Farris, K. T. Woodward, D. M. Yuhas, and J. B. Storer, *Science,* **156**:529 (1967).
8. Apte, B. K., and R. R. Chaudhary, *Arch. Int. Pharmacodyn.,* **160**:14 (1966).
9. Aranda, L. C., and A. Asenjo, *J. Neurosurg.,* **30**:35 (1969).
10. Armin, J., and R. T. Grant, *J. Physiol.,* **121**:593 (1953).
11. Armitage, A. K., and A. C. B. Dean, *J. Physiol.,* **182**:42 (1966).
12. Armitage, A. K., and J. R. Vane, *Br. J. Pharmacol.,* **22**:204 (1964).
13. Aston, R., *Toxic. Appl. Pharmacol.,* **1**:277 (1959).
14. Bailey, D. M., *J. Pharm. Pharmacol.,* **17**:782 (1965).
15. Baer, J. E., J. K. Michaelson, D. N. Mckinstry, and K. H. Beyer, *Proc. Soc. Exp. Biol. Med.,* **115**:87 (1964).
16. Baker, M. A., E. Burrell, J. Penkhus, and J. N. Hayward, *J. Appl. Physiol.,* **24**:577 (1968).
17. Baker, S. B., and C. De, *Proc. Eur. Soc. Study Drug Toxic.,* **11**:52 (1970).
18. Barlet, A. L., *Q. J. Exp. Physiol.,* **53**:170 (1968).
19. Barnstein, N. J., R. S. Grilfillan, N. Pace, and D. F. Rahlmann, *J. Surg. Res.,* **6**:511 (1966).
20. Bass, W. B., L. A. Schroeder, and M. J. Vander Brook, *Curr. Res. Anesth. Analog.,* **33**:234 (1954).
21. Bass, W. B., and M. J. Vander Brook, *J. Am. Pharm. Assoc.,* **41**:569 (1952).
22. Beaver, W. T., and W. F. Riker, *J. Pharmacol. Exp. Ther.,* **138**:48 (1962).
23. Benfey, B. G., K. Greeff, and E. Heeg, *Br. J. Pharmacol.,* **30**:23 (1967).
24. Bennett, A., K. G. Eley, and G. B. Scholes, *Br. J. Pharmacol.,* **34**:630 (1968).
25. Berde, B., W. Doepfner, and H. Konzett, *Br. J. Pharmacol.,* **12**:209 (1957).
26. Berde, B., W. R. Schalch, and W. Doepfner, *Helv. Physiol. Pharmacol. Acta,* **22**:110 (1964).
27. Berger, W., and L. Barr, *J. Appl. Physiol.,* **26**:378 (1969).
28. Bergeron, M., and F. Morel, *Am. J. Physiol.,* **216**:1139 (1969).
29. Bernard, C., *C. R. Acad. Sci.* (Paris), **43**:825 (1856).
30. Bevan, J. A., and M. A. Verity, *J. Pharmacol. Exp. Ther.,* **157**:117 (1967).
31. Bhargava, K. P., K. N. Dhawan, and R. C. Saxena, *Br. J. Pharmacol.,* **31**:26 (1967).
32. Bickel, M. H. and J. H. Weder, *Arch. Int. Pharmacodyn.,* **173**:433 (1968).
33. Block, B. P., D. J. Patts, and R. H. S. Finney, *J. Pharm. Pharmacol.,* **16**:(Supp):85 (1964).
34. Bobb, J. R. R., and S. Ellis, *Am. J. Physiol.,* **167**:768 (1951).
35. Boissier, J. R., and C. A. Aron, *Eur. J. Pharmacol.,* **4**:145 (1968).
36. Bolton, T. B., *J. Physiol.,* **186**:129 (1966).
37. Bordie, B. B., *Proc. R. Soc. Med.,* **58**:946 (1965).
38. Boreus, L. O. and F. A. Sandberg, *Acta Pharmacol.,* **11**:198 (1955).
39. Borison, H. L., *Am. J. Physiol.,* **154**:55 (1948).
40. Botting, J. H., *Br. J. Pharmacol. Chemother.,* **24**:156 (1965).
41. Boyd, I. A., and C. L. Potnak, *J. Physiol.,* **176**:191 (1965).
42. Breckenridge, B. McL., J. H. Burns, and F. M. Matachinsky, *PNAS(US),* **57**:1893 (1967).
43. Bridges, J. W., M. R. Kibby, S. R. Walker, and R. T. Williams, *Biochem. J.,* **109**:851 (1968).
44. Brimblecombe, R. W., *Psychopharmacologia,* **4**:139 (1963).
45. Brodie, D. A., and H. M. Hansen, *Gastroenterology,* **38**:353 (1960).
46. Brooks, C. McC., H. H. Lu, G. Lange, R. Mangi, R. B. Shaw, and K. Geoly, *Am. J. Physiol.,* **211**:1197 (1966).
47. Brown, G. L., and B. D. Burns, *J. Physiol.,* **108**:54 (1949).
48. Brown, J. H., and H. K. Mackey, *Proc. Soc. Exp. Biol. Med.,* **128**:504 (1968).
49. Brown, J. H., H. K. Mackey, and D. A. Riggilo, *Proc. Soc. Exp. Biol. Med.,* **125**:837 (1967).

50. Buckley, G. A., and L. E. Wheater, *J. Pharm. Pharmacol.,* **20:**1145 (1968).
51. Bulbring, E., *Br. J. Pharmacol. Chemother.,* **1:**38 (1946).
52. Burg, M., J. Grantham, M. Abramow and J. Orloff, *Am. J. Physiol.,* **210:**1293 (1966).
53. Burn, J. H., in *Methods of Biological Assay,* Oxford University Press, London, 1928, pp. 29–52.
54. Burn, J. H., in *Practical Pharmacology,* Oxford Scientific Publications, New York, 1952.
55. Burn, J. H., and M. J. Rand, *J. Physiol.,* **147:**135 (1959).
56. Burns, B. D., and W. D. M. Paton, *J. Physiol.,* **115:**41 (1951).
57. Burns, J. J., *Ann. N.Y. Acad. Sci.,* **151:**959 (1968).
58. Burns, J. J., *Proc. Eur. Soc. Study Drug Toxic.,* **12:**9 (1970).
59. Buttle, G. A. H., and E. J. Zaimis, *J. Pharm. Pharmacol.,* **1:**991 (1949).
60. Campbell, A. H., J. A. Stasse, G. H. Lord, and J. E. Wilson, *J. Pharm. Sci.,* **57:**2045 (1968).
61. Campbell, B. A., L. D. Lytle, and H. C. Fibiger, *Science,* **166:**635 (1969).
62. Campbell, D. E. S., and W. Richter, *Acta Pharmacol.,* (Kobenhaven) **25:**345 (1967).
63. Caroll, M. N., Jr., and R. K. S. Lim, *Arch. Int. Pharmacodyn.,* **125:**383 (1960).
64. Carrier, O., Jr., M. Cowsert, J. Hancock, and A. C. Guyton, *Am. J. Physiol.,* **207:**169 (1964).
65. Carrier, O., Jr., and W. C. Holland, *J. Pharmacol. Exp. Ther.,* **149:**212 (1965).
66. Carter, R. L., W. H. Perciral, and F. J. C. Roe, *Br. J. Cancer,* **22:**116 (1968).
67. Cascorb, H. F., and F. G. Rudo, *Anesth. Analog.,* **43:**333 (1964).
68. Cottell, M., and H. Gold, *J. Pharmacol. Exp. Ther.,* **62:**116 (1938).
69. Chance, M. R. A., *J. Pharmacol. Exp. Ther.,* **89:**289 (1947).
70. Chiang, T. S., and F. E. Leadears, *J. Pharmacol. Exp. Ther.,* **149:**225 (1965).
71. Child, K. J., and E. Zaimis, *Br. J. Pharmacol.,* **15:**412 (1960).
72. Christensen, J., and E. E. Daniel, *J. Pharmacol. Exp. Ther.,* **159:**243 (1968).
73. Clark, W. C., H. J. Blackman, and J. E. Preston, *Arch. Int. Pharmacodyn.,* **170:**350 (1967).
74. Clement, D., and P. Vanhoutte, *Arch. Int. Pharmacodyn.,* **166:**181 (1967).
75. Cocegni, F., and L. S. Wolfe, *Can. J. Physiol. Pharmac.,* **43:**445 (1965).
76. Coggeshell, R. E., *Aplysia Calif. J. Neurophysiol.,* **30:**1263 (1967).
77. Cohen, H. B., and W. C. Dement, *Science,* **150:**1318 (1965).
78. Coldman, M. F., B. J. Paulsen and T. Higuchi, *J. Pharm. Sci.,* **58:**1098 (1969).
79. Collier, H. O. J., J. A. Holgate, M. Schachter, and P. G. Shorley, *Br. J. Pharmacol.,* **15:**290 (1960).
80. Conney, A. H., in *Proc. II International Pharmacology Meeting,* Prague, Vol. 4, Pergamon, New York, 1965.
81. Conney, A. H., *Pharmacol. Rev.,* **19:**317 (1967).
82. Conney, A. H., E. C. Miller, and J. A. Miller, *J. Biol. Chem.,* **228:**753 (1957).
83. Constantine, J. W., *J. Pharm. Pharmacol.,* **17:**384 (1965).
84. Corne, S. J., and R. W. Pickering, *Psychopharmacologia,* **11:**65 (1967).
85. Corning, W. C., D. A. Feinstein, and J. R. Haight, *Science,* **148:**394 (1965).
86. Covino, B. G., and H. E. D'Amato, *Circ. Res.,* **10:**148 (1962).
87. Crawford, J. S., *Br. J. Anaesthesiol.,* **38:**628 (1966).
88. Crema, A., G. Benzi, G. M. Frigo, and F. Berte, *J. Pharm. Pharmacol.,* **17:**405 (1965).
89. Dage, R. C., and H. F. Hardman, *Eur. J. Pharmacol.,* **4:**231 (1968).
90. Daggett, W. M., G. C. Nugent, P. W. Carr, P. C. Powers, and Y. Harda, *Am. J. Physiol.,* **212:**8 (1967).
91. D'Amour, F. E. and D. L. Smith, *J. Pharmacol. Exp. Ther.,* **72:**74 (1941).
92. Dandiya, P. C., and L. P. Bhargava, *Arch. Int. Pharmacodyn.,* **176:**157 (1968).
93. Daugherty, R. M., Jr., J. B. Scott, T. E. Emerson, Jr., and F. J. Haddy, *Am. J. Physiol.,* **214:**611 (1968).
94. DaVanzo, J. P., M. Daugherty, R. Ruchart, and L. Kang, *Psychopharmacologia,* **9:**210 (1966).
95. Davies, B. N., and P. G. Withrington, *Br. J. Pharmacol.,* **32:**136 (1969).
96. Day, M. D., and P. R. Warren, *Br. J. Pharmacol.,* **32:**227 (1968).
97. Dayton, P. G., S. A. Cneinell, M. Weiss, and J. M. Perel, *J. Pharmacol. Exp. Ther.,* **158:**305 (1967).
98. Defalque, R. J., and V. K. Stoeling, *Anesth. Analog.* **45:**106 (1966).

99. De la Lande, I. S., and J. A. Harvey, *J. Pharm. Pharmacol.,* **17:**589 (1965).
100. De la Lande, I. S., and M. J. Rand, *Aust. J. Exp. Biol. Med. Sci.,* **43:**639 (1965).
101. DeMarco, T. J., and R. R. Levine, *J. Pharmacol. Exp. Ther.,* **169:**142 (1969).
102. Deringer, M. K., T. B. Dunn, and W. E. Heston, *Proc. Soc. Exp. Biol. Med.,* (N.Y.), **83:**474 (1953).
103. Dews, D. B., *J. Pharmacol. Exp. Ther.,* **122:**137 (1958).
104. Dhawan, K. N., G. P. Gupta and B. P. Raju, *Br. J. Pharmacol.,* **34:**248 (1968).
105. Diamond, L., *J. Pharm. Sci.,* 57:971 (1968).
106. Dicara, L. V., and N. E. Miller, *Science,* **159:**1485 (1968).
107. Dill, R. E., H. L. Dorman, and W. M. Nickey, *J. Appl. Physiol.,* **24:**598 (1968).
108. Dinman, B. D., *Science,* **175:**495 (1972).
109. Djahanguiri, B., D. Sadeghi and S. Hemmati, *Toxic. Appl. Pharmacol.,* **12:**568 (1968).
110. Domer, E. R., and F. W. Schueler, *Arch. Int. Pharmacodyn.,* **114:**217 (1958).
111. Domer, F. R., and F. W. Schueler, *Arch. Int. Pharmacodyn.,* **127:**449 (1960).
112. Doss, D., and F. K. Ohnesorge, *Pfluegers Arch. Ges. Physiol.,* **289:**91 (1966).
113. Dunham, N. W., and T. S. Miya, *J. Am. Pharm. Assoc.,* **46:**208 (1957).
114. Dunson, W. A., and E. G. Buss, *Science,* **161:**167 (1968).
115. Durbin, C. G. and J. F. Robens, *Ann. N.Y. Acad. Sci.,* **111:**696 (1963).
116. Eccles, R. M., *J. Physiol.,* **117:**181 (1952).
117. Eddy, N. B. and D. Leimbach, *J. Pharmacol. Exp. Ther.,* **107:**385 (1953).
118. Eichenwald, H. F., and D. C. Fry, *Science,* **163:**644 (1969).
119. Elliott, R. C. and J. P. Quilliam, *Br. J. Pharmacol.,* **23:**222 (1964).
120. Ernst, A. M., and D. G. Smelik, *Experientia,* **22:**837 (1966).
121. Erslev, A. J., T. W. Solit, R. C. Camishion, S. Amsel, J. Ilda, and W. F. Ballinger, *Am. J. Physiol.,* **208:**1153 (1965).
122. Essig, C. F., *Arch. Int. Pharmacodyn,* **143:**189 (1968).
123. Essig, C. F., and R. C. Lam, *Arch. Neurol.,* **18:**626 (1968).
124. Evans, W. O., *Psychopharmacologia,* **2:**318 (1961).
125. Everett, S. D., *J. Pharm. Pharmacol.,* **16:**767 (1964).
126. Everett, S. D., *Br. J. Pharmacol.,* **33:**342 (1968).
127. Ewing, P. L., B. M. Moore, and W. T. Moore, *J. Pharmacol. Exp. Ther.,* **105:**343 (1952).
128. Exley, K. A., *Br. J. Pharmacol.,* **12:**297 (1957).
129. Farmer, J. B., and G. P. Levy, *Br. J. Pharmacol.,* **32:**193 (1968).
130. Fearn, H. J., S. Karady, and G. B. West, *J. Pharm. Pharmacol.,* **18:**408 (1966).
131. Feigen, G. A., P. T. Masooka, C. H. Theines, P. R. Saunders, and G. B. Sutherland, *Stanford Med. Bull.,* **10:**27 (1952).
132. Feldberg, W., in *A Pharmacological Approach to the Brain from its Inner and Outer Surfaces,* Williams & Wilkins, Baltimore, 1963.
133. Feldberg, W., and R. C. Y. Lin, *Br. J. Pharmacol.,* **4:**33 (1949).
134. Feldberg, W., and A. A. Miles, *J. Physiol.,* **120:**205 (1953).
135. Feldberg, W., and W. D. M. Paton, *J. Physiol.,* **114:**490 (1951).
136. Ferguson, H. C., *J. Pharm. Sci.,* **55:**1142 (1966).
137. Finney, A. J., in *Protest Analysis. A Statistical Treatment of Sigmoid Response Curve,* Cambridge University Press, Cambridge, 1952.
138. Fleischli, G., and E. N. Cohen, *Anesthesiology,* **27:**64 (1966).
139. Folkman, J., S. Winsey, and T. Moghul, *Anesthesiology,* **29:**410 (1968).
140. Forshing, M. L., J. J. Jones, and J. Lees, *J. Physiol.,* **196:**495 (1968).
141. Freedman, P. E., and A. J. Rosen, *Psychopharmacologia,* **15:**39 (1969).
142. Freeman, S. E., *J. Pharmacol. Exp. Ther.,* **162:**10 (1968).
143. Freireich, E. J., E. A. Gehan, D. P. Rall, L. H. Schmidt, and H. E. Skipper, *Cancer Chemother. Repts.,* **50:**219 (1966).
144. Fromm, G. H., and J. M. Killian, *Neurology,* **17:**275 (1967).
145. Fry, D. L., and J. Rose, Jr., *Meth. Med. Res.,* **11:**50 (1966).
146. Fugijmori, H., *Psychopharmacologia,* **1:**374 (1965).
147. Fujiik, K., H. Jaffe, and S. S. Epstein, *Toxic. Appl. Pharmacol.,* **13:**431 (1968).
148. Fujimoto, J. M., and R. A. Donnelly, *Clin. Tox.,* **1:**297 (1968).
149. Furchgott, R. F., *Meth. Med. Res.,* **8:**177 (1961).

150. Gaddum, J. H., *Br. J. Pharmacol.*, **8**:321 (1955).
151. Gangloff, H., and M. Monnier, *Electroenceph. Clin. Neurophysiol.*, **9**:43 (1957).
152. Garland, L. G., S. J. Smith, and M. F. Sims, *J. Pharm. Pharmacol.*, **20**:236 (1968).
153. Garrett, J., *Arch. Int. Pharmacodyn.*, **144**:381 (1963).
154. Gay, W. S., M. J. Rand, and P. Ross, *J. Pharm. Pharmacol.*, **21**:374 (1969).
155. Gebert, G., P. Konold, and H. Seboldt, *Pfluegers Arch. Ges. Physiol.*, **299**:285 (1968).
156. Geiger, A. and J. Magnes, *Am. J. Physiol.*, **149**:517 (1947).
157. Gerarde, H. W., and D. B. Ahlstram, *Toxic. Appl. Pharmacol.*, **9**:185 (1966).
158. Ghosh, M. N., and H. O. Schild., *Br. J. Pharmacol.*, **13**:54 (1958).
159. Gillespie, J. S., and T. C. Muir, *Br. J. Pharmacol.*, **30**:78 (1967).
160. Gillette, J. R., *Am. N.Y. Acad. Sci.*, **123**:42 (1965).
161. Ginsberg, B. L. and J. Warriner, *Br. J. Pharmacol.*, **15**:410 (1960).
162. Godfrained, T., A. Kaba, and P. Polster, *Br. J. Pharmacol. Chemother.*, **28**:93 (1966).
163. Goetzl, F. R., D. Y. Burrill, and A. C. Ivy, *Bull. Northwest Med. School*, **17**:280 (1943).
164. Goldberg, L., *Acta Physiol. Scand.*, **18**:1 (1949).
165. Goldblatt, M. W., J. Lynch, and R. F. Hanzal, *Am. J. Pathol.*, **9**:942 (1933).
166. Goldenthal, E. I., *FDA Papers*, **2**:13 (1968).
167. Goldsmith, M. A., M. Slavik, and S. K. Carter, *Cancer Res.*, **35**:1354 (1975).
168. Goldstein, A., L. Arnow, and S. M. Kalman, in *Principles of Drug Action. The basis of Pharmacology*, Hoeber Medical Division, Harper & Row, New York 1968.
169. Green, A. F., and R. D. Robson, *Br. J. Pharmacol.*, **25**:497 (1965).
170. Green, A. F., and N. B. Ward, *Br. J. Pharmacol.*, **10**:418 (1955).
171. Green, A. F., and D. A. Young, *Br. J. Pharmacol.*, **6**:572 (1951).
172. Green, G. C., *Anesth. Analg.*, **47**:509 (1968).
173. Grewal, R. S., C. L. Kaul, and J. David, *J. Pharmacol. Exp. Ther.*, **160**:268 (1968).
174. Grimson, K. S., *Arch. Surg.*, **43**:284 (1941).
175. Grollman, A., *Proc. Soc. Exp. Biol. Med.*, **57**:102 (1944).
176. Grotto, M., and F. G. Sulman, *Arch. Int. Pharmacodyn.*, **165**:152 (1967).
177. Guzman, F., C. Braun, and R. K. S. Lim, *Arch. Int. Pharmacodyn.*, **136**:353 (1962).
178. Guzman, F., C. Brown, R. K. S. Lim, G. D. Potter, and D. W. Rodgers, *Arch. Int. Pharmacodyn*, **149**:353 (1964).
179. Hall, J. A., R. B. Nelson, and A. I. Edlin, *J. Pharm. Sci.*, **56**:298 (1967).
180. Harry, J., *Br. J. Pharmacol.*, **20**:399 (1963).
181. Hauge, A., *Acta Physiol. Scand.*, **72**:33 (1968).
182. Harvey, A. M., and R. L. Malvin, *Am. J. Physiol.*, **209**:849 (1965).
183. Hawker, R. W., W. G. North, and B. Zerner, *Br. J. Pharmacol.*, **35**:175 (1969).
184. Henneman, E., A. Kaplan, and K. Unna, *J. Pharmacol. Exp. Ther.*, **97**:331 (1949).
185. Hermansen, K., *Br. J. Pharmacol.*, **16**:116 (1961).
186. Heymans, C., G. R. DeVleeschhouwer, and A. F. DeSchaepdryver, *Arch. Int. Pharmacodyn.*, **157**:216 (1965).
187. Hill, B. C. F., *Charles River Dig.*, **7**:4 (1968).
188. Hillyard, I. W., and R. P. Grandy, *J. Pharmacol. Exp. Ther.*, **142**:358 (1963).
189. Hoffman, B. F., and E. E. Suckling, *Am. J. Physiol.*, **173**:312 (1953).
190. Homan, E. R., *Cancer Chemother. Repts.*, **3**:13 (1972).
191. Hoppe, J. O., *Curr. Res. Anesth. Analg.*, **30**:262 (1951).
192. Horton, E. W., *Br. J. Pharmacol.*, **14**:125 (1959).
193. Horton, E. W., *Br. J. Pharmacol.*, **22**:189 (1964).
194. Horton, E. W., I. H. M. Main, and C. J. Thompson, *J. Physiol.*, **180**:514 (1965).
195. Horton, E. W. and C. J. Thompson, *Br. J. Pharmacol.*, **22**:183 (1964).
196. Hrdina, P., A. Bongccors, and S. Garattini, *Eur. J. Pharmacol.*, **1**:99 (1967).
197. J. Hughes, and J. R. Vane, *Br. J. Pharmacol.*, **30**:46 (1967).
198. Ing, H. R., G. S. Dawes, and I. Wajda, *J. Pharmacol. Exp. Ther.*, **85**:85 (1945).
199. Iolanpaan-Hekkila, J. E., K. Jalonen, and A. Vartiainen, *Acta Pharmacol.*, **25**:333 (1967).
200. Irwin, S., *Psychopharmacologia*, **9**:259 (1966).
201. Jansen, J. A., *Acta Pharmacol.*, (Kobenhavn), **25**:329 (1967).
202. Janssen, P. A. J., and C. J. E. Niemegeers, *Psychopharmacologia*, **11**:231 (1967).
203. Jenden, D. J., K. Kamijo, and B. D. Taylor, *J. Pharmacol. Exp. Ther.*, **111**:229 (1954).

204. Jenkner, F. L., and A. Ward, Jr., *Neuro. Psychiat.*, **70**:489 (1953).
205. Johannson, B., and B. Ljung, *Acta. Physiol. Scand.*, **70**:299 (1967).
206. John, E. R., B. M. Wenzel and R. D. Tschirgi, *J. Pharmacol. Exp. Ther.*, **123**:193 (1958).
207. Johnson, H. E., and M. E. Goldberg, *J. Pharm. Pharmacol.*, **17**:54 (1965).
208. Jones, G., and D. J. Roberts, *J. Pharm. Pharmacol.*, **20**:302 (1968).
209. Jurna, J., *Arch. Exp. Pathol. Pharmakol.*, **259**:181 (1968).
210. Jusko, W. J., *Proc. Eur. Soc. Study Drug Toxic.*, **14**:9 (1973).
211. Kalow, W., *Fed. Proc.*, **24**:1259 (1965).
212. Kase, Y., *Jap. J. Pharmacol.*, **2**:7 (1952).
213. Kato, R., and A. Takanaka, *Jap. J. Pharmacol.*, **18**:381 (1968).
214. Kavin, H., N. W. Levine, and M. W. Stanley, *J. Appl. Physiol.*, **22**:604 (1967).
215. Kayaalp, S. O., and R. K. Turker, *Eur. J. Pharmacol.*, **2**:175 (1967).
216. Khairallah, P. A., and I. H. Page, *Ann. N.Y. Acad. Sci.*, **104**:212 (1963).
217. Khairallah, P. A., I. H. Page, and R. K. Turker, *Arch. Int. Pharmacodyn.*, **169**:328 (1967).
218. Killam, K. F., E. K. Killam, and R. Nagnet, *Electroenceph. Clin. Neurophysiol.*, **22**:497 (1967).
219. Kimoro, Y., and M. Goto, *Jap. J. Physiol.*, **17**:365 (1967).
220. King, E. E., and K. R. Unna, *J. Pharmacol. Exp. Ther.*, **111**:293 (1954).
221. Krantz, J. C., Jr., C. J. Carr, S. E. Forman, W. E. Evans, and H. Wollenweber, *J. Pharmacol. Exp. Ther.*, **72**:233 (1941).
222. Krantz, J. C., Jr., H. F. Cascorbi, M. Helrich, R. M. Burgison, M. I. Gold, and F. A. Rudo, *Anesthesiology*, **22**:491 (1961).
223. Krnjevic, H., and M. Videk, *Psychopharmacologia*, **10**:308 (1967).
224. Kvam, D. C., D. A. Riggilo, and P. M. Lish, *J. Pharmacol. Exp. Ther.*, **149**:183 (1965).
225. Laity, J. L. H., *J. Pharm. Pharmacol.*, **19**:265 (1967).
226. Lamanna, C., and C. J. Carr, *Clin. Pharmacol. Ther.*, **8**:307 (1967).
227. Lange, G., H. H. Lu, A. Chang, and C. McC. Brooks, *Am. J. Physiol.*, **211**:1192 (1966).
228. Lape, H. E., D. J. Fort, and J. O. Hoppe, *Arch. Int. Pharmacodyn.*, **160**:342 (1966).
229. Lapin, I. P., *Psychopharmacologia*, **11**:79 (1967).
230. Lawton, I. E., and N. B. Schwartz, *Am. J. Physiol.*, **214**:213 (1968).
231. Leaders, F. E., and C. Dayrit, *J. Pharmacol. Exp. Ther.*, **147**:152 (1965).
232. Levey, S., and R. Gast, *J. Appl. Physiol.*, **21**:313 (1966).
233. Lindsay, H. A., and V. S. Kullman, *Science*, **151**:576 (1966).
234. Lippert, T. H., and F. Hellersberg, *Arch. Int. Pharmacodyn.*, **165**:337 (1967).
235. Lish, P. M., J. Weikel, and K. W. Dungan, *J. Pharmacol. Exp. Ther.*, **149**:161 (1965).
236. Liu, C. T., *Archs. Int. Pharmacodyn.*, **174**:1 (1968).
237. Liu, C. T., K. Anderson and B. Norman, *Arch. Int. Pharmacodyn.*, **169**:288 (1967).
238. Lock, J. A., *Br. J. Pharmacol.*, **21**:393 (1963).
239. Lockett, M. F., *Proc. Soc. Exp. Biol. Med.*, **121**:937 (1966).
240. Long, J. P., and M. F. Armaly, *J. Pharmacol. Exp. Ther.*, **150**:389 (1965).
241. Long, J. P., and H. H. Keasling, *J. Am. Pharm. Assoc.*, **43**:616 (1954).
242. Lucchesi, B. R., and H. F. Hardman, *J. Pharmacol. Exp. Ther.*, **132**:372 (1961).
243. Luduena, F. P., E. Anenenko, O. H. Siegmund, and L. C. Miller, *J. Pharmacol. Exp. Ther.*, **95**:155 (1949).
244. Luduena, F. P., and L. M. Lands, *J. Pharmacol. Exp. Ther.*, **110**:282 (1954).
245. Luschei, E. S., and J. J. Mehaffey, *J. Appl. Physiol.*, **22**:595 (1967).
246. Macht, D. I., and M. B. Macht, *J. Am. Pharm. Assoc.*, **29**:193 (1940).
247. MacIntosh, F. C., and W. L. M. Perry, *Meth. Med. Res.*, **3**:78 (1950).
248. Macleod, P. P., and E. G. Hunter, *Can. J. Physiol. Pharmacol.*, **45**:463 (1967).
249. Mahler, R. J., O. Szabo, and J. C. Penhose, *Diabetes*, **17**:1 (1968).
250. Main, I. H. M., *Br. J. Pharmacol.*, **22**:511 (1964).
251. Malone, M. H., and R. C. Robichand, *Lloydia*, **24**:204 (1961).
252. Malvin, R. L., *Nature* (London) **206**:938 (1965).
253a. Mantegazza, P., and K. M. Naimzada, *Eur. J. Pharmacol.*, **1**:396 (1967).
253b. Mantegazza, P., and K. M. Naimzada, *Eur. J. Pharmacol.*, **1**:402 (1967).
254. Mantel, N., and M. A. Schneiderman, *Cancer Res.*, **35**:1379 (1975).
255. Marriott, A. S., and P. S. J. Spenser, *Br. J. Pharmacol.*, **25**:432 (1965).

256. Martin, W. R., C. G. Eades, H. F. Fraser, and A. Wikler, *J. Pharmacol. Exp. Ther.*, **144**:8 (1964).
257. Martin, P. J. and H. O. Schild, *Br. J. Pharmacol. Chemother.*, **25**:418 (1965).
258. Matsuura, S., S. Kawaguchi, M. Ichiki, M. Sorimachi, K. Katgoka, and A. Inouye, *Eur. J. Pharmacol.*, **6**:13 (1969).
259. Mautner, H. G., *Pharmac. Rev.*, **19**:107 (1967).
260. McCarty, D. J., Jr., P. Phelps, and J. Pyenson, *J. Exp. Med.*, **124**:99 (1966).
261. Meyer, M. C., and D. E. Guttman, *J. Pharm. Sci.*, **57**:895 (1968).
262. Mikos, E., *J. Pharm. Pharmacol.*, **18**:684 (1966).
263. Miller, F. P., R. H. Cox, Jr., and R. P. Maickel, *Science*, **162**:463 (1968).
264. Miller, N. E., *Science*, **163**:434 (1969).
265. Modi, K. N., N. N. Shah, and U. K. Sheth, *Arch. Int. Pharmacodyn.*, **144**:61 (1963).
266. Moe, G. K. and R. Mendez, *Circulation*, **4**:729 (1951).
267. Moller, K. O., *Arch. Int. Pharmacodyn*, **57**:51 (1937).
268. Moore, J. I., and H. H. Swain, *J. Pharmacol. Exp. Ther.*, **128**:243 (1960).
269. Moppert, J., and K. O. Fresen, *Virchows Arch. Pathol. Anat.*, **342**:304 (1967).
270. Motulksy, A. G., *Ann. N.Y. Acad. Sci.*, **123**:167 (1965).
271. Mule, S. J., T. H. Clements, R. C. Layson, and C. A. Haertzen, *Arch. Int. Pharmacodyn*, **173**:210 (1968).
272. Muller, P. J. and J. Vernikos-Danellis, *Proc. West Pharmacol. Soc.*, **11**:52 (1968).
273. Murayama, Y., A. Suzuki, M. Tadakoro, and F. Sakai, *Jap. J. Pharmacol.*, **18**:518 (1968).
274. Mastuk, W. L., *J. Cell. Comp. Physiol.*, **42**:429 (1953).
275. Matoff, I. L., *J. Pharm. Pharmacol.*, **19**:612 (1967).
276. Neal, M. J., and J. M. Robson, *Br. J. Pharmacol.*, **22**:590 (1964).
277. Nicak, A., *Arch. Int. Pharmacodyn.*, **153**:214 (1965).
278. Nielsen, O. E., *Acta Pharmacol.*, **18**:23 (1961).
279. Nims, L. F., C. Marshall, and A. Nielsen, *Yale J. Biol. Med.*, **13**:477 (1941).
280. Nizer, A., Y. Cuypers, P. Deetjen, and K. Kramer, *Pfluegers Arch. Ges. Physiol.*, **296**:179 (1967).
281. Noel, R. H., *J. Pharmacol. Exp. Ther.*, **84**:278 (1945).
282. Norton, S., and E. J. deBeer, *Ann. N.Y. Acad. Sci.*, **65**:249 (1965).
283. Offermeier, J., and E. J. Ariens, *Arch. Int. Pharmacodyn.*, **164**:192 (1966).
284. Osnide, G., *Eur. J. Pharmacol.*, **3**:283 (1968).
285. Otis, L. S., J. A. Willis, and G. T. Pryor, *J. Appl. Physiol.*, **26**:640 (1969).
286. Ott, R. L., and R. H. Myers, *Clin. Pharmacol. Ther.*, **10**:207 (1969).
287. Owens, A. H., *J. Chronic Dis.*, **15**:233 (1963).
288. Oyvin, I. A., P. Y. Gaponuik, and V. I. Oyvin, *Experimentia*, **23**:925 (1967).
289. Page, I. H., *J.A.M.A.*, **113**:2046 (1939).
290. Paila, J. J., J. W. Poutsiaka, C. I. Smith, J. C. Burke, and B. N. Chaver, *J. Pharmacol. Exp. Ther.*, **134**:273 (1961).
291. Paintal, A. S., *J. Physiol.*, **124**:166 (1954).
292. Papadopoulou, Z. L., L. M. Slotkoff, G. M. Eisner, and L. S. Lilienfield, *Proc. Soc. Exp. Biol. Med.*, **130**:1206 (1969).
293. Pardo, E. G., E. Hong, and J. LeLorier, *J. Pharmacol. Exp. Ther.*, **157**:303 (1967).
294. Paton, W. D. M., and W. L. M. Perry, *J. Physiol.*, **119**:43 (1953).
295. Paton, W. D. M., and M. A. Zai, *J. Physiol.*, **194**:13 (1968).
296. Perret, G. E. and F. H. Hesser, *Gastroenterology*, **38**:219 (1960).
297. Perry, W. L., and J. Talesnik, *J. Physiol.*, **119**:455 (1953).
298. Perry, W. L. M., and C. W. M. Wilson, *Br. J. Pharmacol.*, **11**:81 (1956).
299. Peterson, A. E., *J. Appl. Physiol.*, **25**:103 (1968).
300. Pinkel, D., *Cancer Res.*, **18**:853 (1956).
301. Pope, R. S., *J. Appl. Physiol.*, **22**:1024 (1967).
302a. Poznak, A. V., and J. F. Artusio, *Toxic. Appl. Pharmacol.*, **2**:263 (1960).
302b. Poznak, A. V., and J. F. Artusio, *Toxic. Appl. Pharmacol.*, **2**:374 (1960).
303. Preston, J. B., and E. E. VanMaanen, *J. Pharmacol. Exp. Ther.*, **107**:165 (1953).
304. Preutt, J. K., and E. F. Woods, *J. Appl. Physiol.*, **22**:830 (1967).

305. Prieur, D. J., D. M. Young, R. D. Davis, D. A. Cooney, E. Homan, R. L. Dixon, and A. M. Guarino, *Cancer Chemother. Rep.*, **4**:1 (1973).
306. Raab, W., E. Bajusz, H. Kimura, and H. C. Herrlich, *Proc. Soc. Exp. Biol. Med.*, **127**:142 (1968).
307. Radzialowski, F. M. and W. F. Bousquet, *Pharmacologist*, **9**:190 (1967).
308. Randall, L. O., and J. J. Sellitto, *Arch. Int. Pharmacodyn.*, **111**:409 (1957).
309. Rang, H. P., and J. M. Ritter, *Molec. Pharmacol.*, **5**:394 (1969).
310. Redefzki, H. M., *J. Pharmacol. Exp. Ther.*, **147**:232 (1965).
311. Rees, J. M. H., *Br. J. Pharmacol.*, **32**:253 (1968).
312. Regoli, D., and J. R. Vane, *Br. J. Pharmacol. Chemother.*, **23**:351 (1964).
313. Reilly, J. F., A. P. Ahlstrom, J. S. Watts, P. S. Cassidy, and L. M. Lusky, *Toxic. Appl. Pharmacol.*, **15**:97 (1969).
314. Reis, D. J., and R. J. Wurtman, *Life Sci.*, **7**:91 (1968).
315. Reitzel, N. L., and J. P. Long, *Arch. Int. Pharmacodyn.*, **119**:20 (1959).
316. Remmer, H., *Proc. 1st Int. Pharmacol. Meet.*, **6**:235 (1962).
317. Remmer, H., *Proc. Eur. Soc. Study Drug Toxic.*, **4**:57 (1964).
318. Remmer, H., *Proc. Eur. Soc. Study Drug Toxic.*, **11**:14 (1970).
319. Remmer, H., and G. Sigert, *Naunyn. Schmiedeberg's Arch. Exp. Path. Pharmak.*, **247**:522 (1962).
320. Rice, W. B. and J. D. McColl, *Arch. Int. Pharmacodyn.*, **127**:249 (1960).
321. Riker, W. F., F. Depierre, J. Roberts, B. B. Roy, and J. Reilly, *J. Pharmacol. Exp. Ther.*, **114**:1 (1955).
322. Robert, A., J. E. Nezamis, and J. P. Phillips, *Am. J. Dig. Dis.*, **12**:1073 (1967).
323. Roch-Ramel, F., *Helv. Physiol. Pharmacol. Acta (Suppl)*, **13**:1 (1962).
324. Rogers, L. A., R. A. Atkinson, and J. P. Long, *Am. J. Physiol.*, **209**:376 (1965).
325. Ross, C. R., and E. J. Cafruny, *J. Pharmacol. Exp. Ther.*, **140**:125 (1963).
326. Rubin, H. J., J. J. Piala, J. C. Burke, and B. N. Craver, *Arch. Int. Pharmacodyn.*, **152**:132 (1964).
327. Rudick, J., L. S. Semb, and L. M. Nyhus, *J. Surg. Res.*, **7**:383 (1967).
328. Rushmer, R. F., D. K. Crystal, C. Wagner, R. M. Ellis, and A. A. Nash, *Circ. Res.*, **2**:142 (1954).
329. Sachdev, K. S., M. H. Panjwani, and A. D. Joseph, *Arch. Int. Pharmacodyn.*, **146**:36 (1963).
330. Sample, R. G., G. V. Rossi, and E. W. Pachman, *J. Pharm. Sci.*, **57**:795 (1968).
331. Sanart, N. D., F. B. Fainman, and J. F. Emele, *J. Pharmacol. Exp. Ther.*, **162**:190 (1968).
332. Sardesai, V. M., *Can. J. Physiol. Pharmacol.*, **46**:77 (1968).
333. Schein, P. S., R. O. Davis, S. K. Carter, J. Newman, D. R. Schein, and D. P. Rall, *Clin. Pharmacol. Ther.*, **14**:3 (1970).
334. Scheline, R. R., *J. Pharm. Pharmacol.*, **17**:52 (1965).
335. Scheline, R. R., *J. Pharm. Sci.*, **57**:2021 (1968).
336. Scheving, L. E., W. H. Harrison, P. Gordon, and J. E. Pauly, *Am. J. Physiol.*, **214**:166 (1968).
337. Schmidt-Neilsen, B., *Fed. Proc.*, **20**:902 (1961).
338. Schnek, W. G., Jr., and D. Race, *J. Surg. Res.*, **6**:361 (1966).
339. Scriabine, A., S. Y. P'An, D. Rowland, and B. Bertrand, *J. Pharmacol. Exp. Ther.*, **133**:351 (1961).
340. Senturia, J. B., and M. Menaker, *J. Appl. Physiol.*, **21**:1869 (1966).
341. Shillito, E. E., *Br. J. Pharmacol.*, **26**:248 (1966).
342. Shipley, R. E. and J. H. Tilden, *Proc. Soc. Exp. Biol. Med.*, **64**:453 (1947).
343. Shulman, A., and M. S. Sadove, *Toxic. Appl. Pharmacol.*, **7**:473 (1965).
344. Silverman, A. P., *Br. J. Pharmacol.*, **24**:579 (1965).
345. Silverstrini, B., and C. Pozzatti, *Arch. Int. Pharmacodyn.*, **129**:249 (1960).
346. Simke, J., M. L. Graeme, and E. B. Sigg, *Arch. Int. Pharmacodyn.*, **165**:291 (1967).
347. Simmons, M. L., and L. H. Smith, *J. Appl. Physiol.*, **25**:324 (1968).
348. Singh, K. P., *Arch. Int. Pharmacodyn.*, **172**:475 (1968).
349. Sivadjian, J., M. Vautrin, and H. Matgae, *Arch. Int. Pharmacodyn.*, **153**:359 (1965).
350. Smith, E. R., J. V. McMarrow, Jr., B. G. Covino, and J. B. Lee, *Clin. Res.*, **15**:222 (1967).

351. Smith, H. W. in *Principles of Renal Physiology,* Oxford University Press, New York, 1956.

352. Smith, R. L., in E. Jucker (ed.), *Progress in Drug Research,* Vol. 9, Birkhauser Verlag, Basel, 1966.

353. Smith, R. L., *Proc. Eur. Soc. Study Drug Toxic.,* **12:**19 (1970).

354. Smith, W. M., *Nature,* **190:**541 (1961).

355. Somani, P., and B. K. B. Lum, *J. Pharmacol. Exp. Ther.,* **152:**235 (1966).

356. Somogyi, A., K. Kovacs, and H. Selye, *J. Pharm. Pharmacol.,* **21:**122 (1969).

357. Spencer, D. S. J., *Br. J. Pharmacol.,* **25:**442 (1965).

358. Spiegel, A. J., and M. M. Naseworthy, *J. Pharm. Sci.,* **52:**917 (1963).

359. Spooner, C. E., and W. D. Winters, *Arch. Int. Pharmacodyn.,* **161:**1 (1966).

360. Stanton, H. C., T. Kirchgessner, and K. Pormenter, *J. Pharmacol. Exp. Ther.,* **149:**174 (1965).

361. Steedman, B. L., R. A. Jones, D. E. Rector, and J. Seigel, *Toxic. Appl. Pharmacol.,* **9:**160 (1966).

362. Stefko, P. L. and W. M. Benson, *J. Pharmacol. Exp. Ther.,* **108:**217 (1953).

363. Stekfo, P. L., J. Denzel, and I. Hickey, *J. Pharm. Sci.,* **50:**216 (1961).

364. Steiner, C., and A. T. W. Kovalik, *J. Appl. Physiol.,* **25:**631 (1968).

365. Stirling, T. D., *Science,* **174:**1358 (1971).

366. Stovall, R., and R. T. Jackson, *Ann. Otol.,* **76:**1051 (1967).

367. Sulser, F., J. Watts, and B. B. Brodie, *Ann. N.Y. Acad. Sci.,* **96:**279 (1962).

368. Suskevich, J., A. H. Campbell, and G. H. Lord, *J. Pharm. Pharmacol.,* **19:**456 (1967).

369. Sutter, M. C., *Br. J. Pharmacol.,* **24:**742 (1965).

370. Swain, H. H., and G. P. Curtis, *Pharmacologist,* **7:**176 (1965).

371. Swinyard, E. A., W. C. Brown, and L. S. Goodman, *J. Pharmacol. Exp. Ther.,* **106:**319 (1952).

372. Snyder, S. H., J. Exelrod, and M. Zweig, *J. Pharmacol. Exp. Ther.,* **158:**206 (1967).

373. Takagi, H., T. Inukai, and M. A. Nakama, *Jap. J. Pharmacol.,* **16:**287 (1966).

374. Tata, P. S. and D. H. Gauer, *Pfluegers Arch. Ges. Physiol.,* **290:**279 (1966).

375. Tawaski, R., *Science,* **160:**76 (1968).

376. Tedeschi, D. H., R. J. Fowler, W. H. Cromley, J. F. Pauls, R. Z. Eby, and E. J. Fellows, *J. Pharm. Sci.,* **53:**1046 (1964).

377. Thampi, S. N., F. R. Domer, V. B. Haarstad, and F. W. Schueler, *J. Pharm. Sci.,* **55:**381 (1966).

378. Thomas, C. B., *Bull. Johns Hopkins Hosp.,* **74:**335 (1944).

379. Thompson, R. E., *J. Pharmacol. Exp. Ther.,* **80:**373 (1944).

380. Thorbjarnarson, B., R. Rees, S. Bjornsson, G. M. Watkins, and K. A. Martin, *J. Surg. Res.,* **7:**449 (1967).

381. Timmerman, H. and N. G. Sheffer, *J. Pharm. Pharmacol.,* **20:**78 (1968).

382. Toman, J. E. P., E. A. Swinyard, and L. S. Goodman, *J. Neurophysiol.,* **9:**231 (1946).

383. Trendelenburg, U., *Egrehn. Physiol.,* **59:**1 (1967).

384. Tschetter, T. H., A. C. Klassen, and J. R. Resch, *Proc. Soc. Exp. Biol. Med.,* **131:**1244 (1969).

385. Tsoter, T. and H. Tom, *Br. J. Pharmacol. Chemother.,* **31:**407 (1967).

386. Tucker, M. J. and S. B. Baker, in E. Cotchin and F. J. C. Roe (eds.), *Diseases of Specific Pathogenic Free Mice in Pathology of Laboratory Rats and Mice,* Blackwell Scientific Publishing Co., Oxford, 1967.

387. Tweeddale, M. G., *J. Pharm. Pharmacol.,* **15:**846 (1963).

388. Udall, V., *Proc. Eur. Soc. Study Drug Toxic.,* **11:**70 (1970).

389. VanArman, C. G., *J. Pharmacol. Exp. Ther.,* **111:**285 (1954).

390. VanMaanen, E. F., *J. Pharmacol. Exp. Ther.,* **99:**255 (1950).

391. VanNueten, J. M., *Eur. J. Pharmacol.,* **6:**286 (1969).

392. VanReizen, H., *J. Pharm. Pharmacol.,* **18:**688 (1966).

393. Vander, A. J., R. L. Malvin, W. S. Wilde, and L. P. Sullivan, *J. Pharmacol. Exp. Ther.,* **125:**19 (1959).

394. Vane, J. R., *Br. J. Pharmacol.,* **12:**344 (1957).

395. Vane, J. R., *Br. J. Pharmacol.,* **35:**209 (1969).

396. Vanou, S., *Br. J. Pharmacol.,* **24:**591 (1965).

397. Varney, R. G., C. R. Linegar, and H. A. Holaday, *J. Pharmacol. Exp. Ther.,* **97:**72 (1949).
398. Vernier, V. G., and K. R. Unna, *Arch. Int. Pharmacodyn.,* **141:**30 (1963).
399. Voith, K. and B. A. Kovacs, *Can. J. Physiol. Pharmacol.,* **44:**515 (1966).
400. Wada, J. A., and T. Asakura, *Exp. Neurol.,* **24:**19 (1969).
401. Waitzman, M. B., and C. D. King, *Am. J. Physiol.,* **212:**329 (1967).
402. Wakerlin, G. E., E. Crandall, M. H. Frank, D. Johnson, L. Pomper, and H. E. Schmid, *Circ. Res.,* **2:**416 (1954).
403. Wand, D. R., *Pharmacol. Rev.,* **20:**49 (1968).
404. Watzman, N., H. Barry, W. J. Kinnard, Jr., and J. P. Buckley, *Arch. Int. Pharmacodyn.,* **169:**362 (1968).
405. Waugh, W. H., and T. Kubo, *Am. J. Physiol.,* **217:**277 (1969).
406. Webb, R. E., and E. Horsfald, Jr., *Science,* **156:**1762 (1967).
407. Webster, R. A., *Br. J. Pharmacol.,* **25:**566 (1965).
408. Weeks, J. R., N. C. Sekhar, and D. W. Ducharme, *J. Pharm. Pharmacol.,* **21:**103 (1969).
409. Weil, C. S., *Toxic. Appl. Pharmacol.,* **22:**318 (1972).
410a. Weiner, M., and D. Moses, *J. Appl. Physiol.,* **23:**601 (1967).
410b. Weiner, M., and D. Moses, *Proc. Soc. Exp. Biol. Med.,* **125:**277 (1967).
411. Weiperre, J., Y. Cohen, and G. Valette, *Eur. J. Pharmacol.,* **3:**47 (1968).
412. Weiss, B., and V. G. Laties, *J. Pharmacol. Exp. Ther.,* **131:**120 (1961).
413. Weitzman, E. D., and G. S. Ross, *Neurology,* **12:**264 (1962).
414. Welch, R. M., A. H. Conney, and J. J. Burns, *Biochem. Pharmacol.,* **15:**521 (1966).
415. White, R. P. and E. J. Westerbeke, *Biochem. Pharmacol.,* **8:**92 (1961).
416. Williams, R. T., *Fed. Proc.,* **26:**1029 (1967).
417. Winder, C. V., C. C. Pfeiffer, and G. L. Maison, *Arch. Int. Pharmacodyn.,* **72:**329 (1946).
418. Winder, C. V., J. Wax, V. Burr, M. Been, and C. E. Rosiere, *Arch. Int. Pharmacodyn.,* **116:**261 (1958).
419. Winter, C. A. and L. Flactaker, *J. Pharmacol. Exp. Ther.,* **112:**99 (1954).
420. Winter, C. A., and L. Flactaker, *J. Pharmacol. Exp. Ther.,* **150:**165 (1965).
421. Winters, A. D., III and R. L. Volle, *Eur. J. Pharmacol.,* **2:**347 (1968).
422. Witkin, L. B., P. Spitaletta, and A. J. Plummer, *J. Pharmacol. Exp. Ther.,* **126:**330 (1959).
423. Witt, P. N., and C. F. Reed, *Science,* **149:**1190 (1965).
424. Wolf, S. M., R. L. Simmons, and W. L. Nastirk, *Proc. Soc. Exp. Biol. Med.,* **117:**1 (1964).
425. Woodbury, L. A., and V. D. Davenport, *Arch. Int. Pharmacodyn.,* **92:**97 (1952).
426. World Health Organization Technical Report Series 341, *Principles for Preclinical Testing of Drug Safety,* WHO, Geneva, 1966.
427. Wurtman, R. J., and J. Axelrod, *PNAS(USA),* **57:**1594 (1967).
428. Yen, H. C. Y., and C. A. Day, *Arch. Int. Pharmacodyn.,* **155:**69 (1965).
429. Yim, G. K. W., H. H. Keasling, E. G. Gross, and C. W. Mitchell, *J. Pharmacol. Exp. Ther.,* **115:**96 (1955).
430. Yusuf, S. M., and O. D. Pirddle, Jr., *Arch. Int. Pharmacodyn.,* **152:**198 (1964).
431. Zapata, P., and C. Eyzaquirre, *Can. J. Physiol. Pharmacol.,* **45:**1021 (1967).
432. Zbinden, G., *Adv. Pharmacol.,* **2:**1 (1963).
433. Zbinden, G., *Clin. Pharmacol. Ther.,* **5:**537 (1964).

CHAPTER 11

ANIMAL MODELS FOR CANCER RESEARCH

Cancer is a generic term for a wide variety of malignant neoplasms which may arise from any human body tissues. Neoplastic growths result in deleterious effects on the host due to their invasive and metastasizing character. Cancer is not unique to human beings but affects almost all animal systems as well as plants. Animal models provide a unique opportunity to study the etiology, development, and treatment of cancer in humans, since spontaneous and induced neoplasms can be established in a variety of animal species. Among the animal species, mice and rats are more frequently used in cancer research because of the availability of well-established genetic strains, their susceptibility to neoplasms, and relatively low cost of the animals (Table 11-1). Animal experiments, properly designed and carried out, are meaningful in acquiring information on conditions leading to cancer in humans. For example, all the chemical agents known to be carcinogenic to humans also produce tumors in animal species. A vast amount of literature exists in the area of cancer research with experimental animals. This chapter briefly describes the comparative aspects of neoplasia and the use of animals for the study of carcinogens and anticarcinogens. Selected animal models for specific neoplasms of humans are described to illustrate the usefulness of animal species in cancer studies.

CHOICE OF EXPERIMENTAL ANIMALS IN CANCER RESEARCH

Primates

Nonhuman primates, because of their phylogenetic relationship with humans, develop morphologically similar neoplasms to human neoplasms. However, the total number of tumors in nonhuman primates is low (probably in the hundreds) and varies with species, sex, age, and length of captivity. During 1957–1960, almost 800,000 primates were used for the production of polio vaccine, but no good records on tumorogenesis in these animals are available except on the central nervous system (CNS) tumors. Spontaneous neoplasms are most frequently seen in the rhesus monkey kept under laboratory conditions (231, 257). The incidence in the zoo animals is about 1 per 100 animals. The most frequent tumors in older animals, as in human beings, are epithelial neoplasms and malignant lymphomas; brain tumors are rare in the animals. Granulocytic leukemia in rhesus monkeys has been induced by procarbazine (328), as well as neutron and proton irradiation (403, 480). Lymphoid leukemia in owl monkeys (287) and marmosets have been reported (471). Recent reports suggest that primitive primates (prosimians) are more sensitive, particularly to chemical carcinogens, than the anthropoid species (6).

Neoplastic growth in nonhuman primates bears resemblance to that in humans in terms of metastasis, invasion, lethality to host, aneuploidy, cytological atypicalism, and preservation of phenotypic properties of the cells of origin.

Table 11-1. Comparison of Various Animal Models Used in Cancer Research

Animal Species	Number of Times Cited
Monkey	7
Mice	284
Rats	298
Hamster	61
Guinea pig	5
Chicken	7
Chick embryos	8
Cattle, cow	5, 3
Cat	3
Dog	17
Pig	1
Quail	1
Rabbits	14
Rainbow trout	3
Sheep	3

* Data collected from cumulative index Cancer Research 1966–1970.

Domestic Animals

The study of neoplasms in domestic animals has contributed a great deal toward the understanding of human cancer (81, 86, 222, 263, 305, 412). There is a considerable variance in the cytological and histological characteristics and in the frequency of occurrence of neoplasms among the domesticated animal species. For example, mammary tumors are a rarity in cows but are fairly common in bitches. However, the tumors in bitches are of mixed type (parenchymal, epithelial, chondroid, and osteoid). Nephrobastomas are relatively common in pigs but not in dogs, cats, horses, sheep, or oxen. Carcinoma of the liver has been reported in sheep and cattle and carcinoma of the penis frequently occurs in horses. Melanomas are more common in horses, cattle, and dogs than in sheep, pigs, or cats. Most cell tumors of the skin, aorta, and carotid are particularly seen in the boxer and Boston terrier dogs (371). Cancer of the conjunctiva occurs in Hereford cattle (15, 383, 436). A list of commonly occurring tumors (374) in some animal species is presented in Table 11-2 (86). The neoplasms of domestic animals with human counterparts represent good animal models for the study of the effect of environment; and epidemiological studies can be made when good breeding records are kept. Great amounts of information on the incidence of tumors can be obtained from the animals used for meat purposes by keeping ante- and postmortem records at slaughter houses.

Laboratory Rodents

The use of mice in the study of cancer was first reported by a geneticist (406) who emphasized the similarities in affected tissues, clinical course of the disease, and histological similarities. Inbred strains have been most useful in cancer research. At

Table 11-2. Comparative Distribution of Neoplasm in Several Domestic Animals

System of Origin	Horse	Ox	Sheep	Pig	Dog	Cat
Skin	Papilloma	Papilloma			Basal cell and glandular tumours	Basal cell glandular tumours
Mammary gland	Melanoma	Melanoma			Melanoma "Mixed" and carcinoma	Carcinoma
Skeletal					Bone Sarcoma	Bone Sarcoma
Respiratory						
Cardiovascular				Congenital rhabdomyoma of heart		
Lymphatic		Leukosis		Leukosis	Leukosis	Visceral lymphosarcoma
Alimentary		Papilloma of oesophagus Liver carcinoma	Liver carcinoma		Carcinoma of tonsil Intestinal lymphosarcoma	Carcinoma of tongue and oesophagus
Urinary				Nephro-blastoma		Lymphosarcoma of kidney
Male genital	Carcinoma of penis	Fibro-papilloma of external genital organs			Testis tumours	
Female genital		Fibro-leiomyoma			Fibro-leiomyoma	
Endocrine		Adrenal tumours			Thyroid tumour	
Nervous system						
Eye	Carcinoma	Carcinoma				
Ear						Ceruminous carcinoma

Adapted from Neoplasms of the Domestic Mammals (86) with permission.

present, 244 strains or substrains of mice are available (417). Frequently used inbred strains of mice for studies of neoplasms are listed in Table 11-3a (92) and the incidence of spontaneously occurring tumors in inbred strains is given in Table 11-3b.

Rats have been widely used in cancer research. The strains of rats available and the incidence of spontaneous and induced tumors has been reviewed (218, 362, 373, 409). A list of commonly used strains to induce neoplasms is given in Table 11-4. Hepatic neoplasms can be easily induced by chemical carcinogens in rats (301), but the incidence of natural hepatomas is rather low. The occurrence of leukemia is not as common in rats as in mice (strain C58). When present, it is of the granulocytic type in the rat as compared to the lymphocytic type in the mouse.

Golden hamsters are used for studies on tumors caused by DNA viruses (107, 154). They have been used particularly for xeno- and allograft because of evertible cheek pouches and also in carcinogenic studies requiring hairless epithelium surface and movement of neoplastic cell through capillary beds (124, 159, 195).

Guinea pigs have been used in cancer research only recently. It was believed that these animals did not develop neoplasms easily. However, several authors (51, 450) have shown that guinea pigs develop a variety of spontaneous tumors (92) (Table 11-5). A serially transplanted leukemia of virus origin has been established in certain strains (313, 332). Chemically induced tumors by urethane and anthracine derivatives have also been developed (450). Naturally occurring spontaneous tumors rarely develop before 3 years of age but have a greater tendency for metastasis as compared to those in rats and mice (51).

Mastomys show high frequency of male specific tumors of glandular stomach with typical argyrophilic carcinoids (329, 410, 411). These tumors can be transplanted and those 1.0 cm or larger are often accompanied by ulcers of the stomach, duodenum, and upper jejunum. These tumors represent excellent experimental models for the foregut tumors of humans (411). Spontaneous clear cell renal adenocarcinoma and thymomas in mastomys resemble very closely the human condition (410, 422).

Wild rats and mice do not develop as many neoplasms as laboratory species possibly because these animals do not reach the same age as laboratory species. This was confirmed when wild mice were bred in the laboratory (17). The animals (25 to 36 months of age) showed alveolar lung tumors, type B reticulum cell carcinoma, granulosa cell tumor of the ovary, hepatomas, hemangioendotheliomas, and mammary tumors (only in breeding females).

Wild rabbits show papillomas, fibromatosis, and myxomatosis. All these tumors may be of viral origin (120, 437).

The literature on spontaneous and induced tumors in rodents is vast. The reader is referred to Green (150), Cornelius (81), Mühlbock and Nomura (307), Nettesheim et al. (317), and reports of Penrose Research Lab (370) for further details in this area.

Birds

The naturally occurring neoplasms of the leukosis-sarcoma complex type have been extensively studied and are definitely of viral origin (65, 105). Marek's disease caused by herpesvirus is of interest not only from the standpoint of fowl lymphomatosis but as a model for Burkitt lymphoma of humans, herpes-induced lymphoma in owls and monkeys, and Lucke renal adenocarcinoma in frogs. The discovery of reticuloendotheliosis virus (REV) adds a new dimension to the use of fowl as models for human conditions (331, 445).

Table 11-3a. Inbred Strains of Mice for Studies of Neoplasms

Neoplasm	Strains	Incidence of neoplasm
Mammary gland tumors	C3H	Almost 100% in ♀♀
	A	High in breeding ♀♀; low in virgin
	DBA	High in breeding ♀♀
	RIII	High in breeding ♀♀
	BALB/c	Incidence low, but high following introduction of mammary tumor agent
	CC57BR	Do not develop spontaneous mammary tumors:
	CC57W	55% after introduction of mammary-tumor agent
Pulmonary tumors	A	90% in mice living to 18 months
	SWR	80% in mice living to 18 months
	BALB/c	26% in ♀♀, 29% in ♂♂
	BL	26% in mice of all ages
Hepatomas	C3H	85% in males 14 months old
	C3Hf	72% in males 14 months old
	C3He	78% in males 14 months old
		91% in breeding ♂♂; 59% in virgin ♀♀; 30% in breeding ♀♀; and 38% in force-bred ♀♀
Leukemia and other reticular cell neoplasms	C58	High leukemia
	AKR	High leukemia
	C57L	Hodgkin's-like lesion, reticular cell neoplasm Type B, approximately 25% at 18 months
Papilloma and carcinoma	HR	Papillomas occurred in all mice, both haired and hairless, painted with methylcholanthrene, and the carcinomatous transformation occurred in most of the animals
	I	Most susceptible of 5 strains tested with methylcholanthrene
Harderian gland tumors	C3H	Occur in C3H and hybrids resulting from outcrossing C3H

Table 11-3a. (Continued)

Neoplasm	Strains	Incidence of neoplasm
Subcutaneous sarcomas	C3H	Occurred spontaneously in 57 of 1774 C3H and C3Hf ♀♀
	CBA	Most susceptible of 8 strains tested with carcinogenic hydrocarbons High occurrence after subcutaneous injection of methylcholanthrene
Stomach lesion	I	Occurs in practically all mice of this strain
	BRS	Occurs following injection with methylcholanthrene and spontaneously
Adrenal cortical tumors	CE	High occurrence following castration
	NH	High occurrence of spontaneous adenoma High occurrence of carcinoma following castration
Teratomas, testicular	129	1% congenital
Interstitial cell tumors of testis	A	Readily induced with estrogen (see Gardner)
	BALB/c	High incidence following treatment with stilbestrol (Males tolerate stilbestrol better than do those of other strains)
Pituitary adenoma	C57BL	Occurred in almost all mice treated with estrogens
Hemargioendotheliomas	HR	19 to 33% in untreated mice 54 to 76% in mice injected with 4-o-tolylazo-o-toluidine
	BALB/c	High occurence particularly in interscapular fat pad and lung in mice treated with o-aminoazotoluene
Ovarian tumors	C3He	47% in virgin ♀♀; 37% in breeding ♀♀; 29% in force-bred ♀♀
Myoepitheliomas	A&BALB	Occur in region of salivary gland and clitoral gland

Adapted from <u>Cancer Medicine</u> (92) with permission.

Table 11-3b. Percentage of Spontaneous Tumors in Inbred Strains of Mice

Tumor	AKR ♀	AKR ♂	C3H ♀	C3H ♂	A/He ♀	A/He ♂	A/JAX ♀	A/JAX ♂	DBA/1 ♀	DBA/1 ♂	DBA/2 ♀	DBA/2 ♂	BALB/c ♀	BALB/c ♂	C57BR/cd ♀	C57BR/cd ♂	C57L/He ♀	C57L/He ♂	C57BL/6 ♀	C57BL/6 ♂
Mammary	3		95		74		28		80		43		8				3			
Lungs		2		4	30	51	41	49	2	2	2	5	26	29		5	4			
Skin			2				5		3	2					36	18	38	35	11	4
Liver				21						6						14	5	7	4	3
Leukemia	85	77				3					2	4	2	2			7		6	3
Reticulum Cell Sarcoma								3									2	3	9	5
Kidney						2	2	2												

Adapted from Cancer Medicine (92) with permission.

Table 11-4.   Commonly Used Strains of Rats in the Study of Human Neoplasms

Strain of Rat	Type of Tumor
Furth Wistar	Pituitary with mammotropic function and leukemia
Sprague-Dawly	Mammary tumors, spontaneous or x-irradiation induced
Osborne Mendel	Mammary, adrenal cortical ovarian
Marshall 520	Adrenal medullary tumors

Among wild birds, the budgerigar was suggested as a model for Wolman syndrome in humans because of the similarity in the development of lipomas and xanthomas of the pectoral subcutaneous tissue and lipid storage disease (250).

Amphibians

The Lucke's adenocarcinoma of the kidney in frogs is well known. The viral etiology was suggested (253); at least four different viruses have been associated with this tumor (148, 149). The plasma cell tumor (395) in *Rana pipiens* is the only example of this neoplasm outside the mammalian system. Special features of the cellular and humoral immunity (77, 354) make the amphibian species particularly useful for studies of the role of immunosuppression in the development of tumors such as lymphoreticuloma (241). The smooth nature of the skin provides another important model in understanding human skin tumors.

Fish

Spontaneous neoplasms in this group are widely spread except for cartilaginous fish (466). Although viral etiology is not important, the enzootic and epizootic nature of tumors suggest an infectious causative agent.

Fish, particularly rainbow trout, are sensitive to chemical carcinogens (aflatoxin, dimethylnitrosamine) which produce hepatomas and nephroblastomas (21, 23, 158, 418). The fish represent ideal models for testing environmental carcinogens. Lymphomas of northern pike could be used as a model for disease in dogs and humans.

ANIMALS FOR THE STUDY OF CANCER ETIOLOGY

The etiology of cancer has baffled scientists for a long time. Although no clear-cut answers are currently available, the etiological agents can be arbitrarily classifed into (a) genetic factors, (b) viruses, (c) chemicals, (d) radiation, (e) immunological aspects, (f)

Table 11-5. Spontaneous Neoplasms of the Guinea Pig

Tissue or organ	Neoplasm	
Bronchus	Papillary adenoma	64
	Adenocarcinoma	1
Subcutaneous	Fibrolipoma	2
and mesentry	Fibrosarcoma	2
	Fibroliposarcoma	7
	Neurogenic sarcoma	3
Reticuloendothelial	Lymphosarcoma (spleen, lymph nodes)	3
	Leukemia	10
Mammary gland	Adenoma and papillary cystadenoma	3
	Adenocarcinoma	8
	Lipofibroma	1
Uterus	Fibro and Leiomyoma	4
	Adenomyoma	1
	Fibro-leiomyosarcoma	3
	Mesenchymal mixed tumors	1
Gastrointestinal	Stomach-Fibromyoma & Lipoma	5
	Intestine (Liposarcoma)	1
	Liver (adenoma, hemangioma)	2
	Gallbladder (Papilloma)	1
Ovary	Teratoma	3
Endocrine	Adrenal-adenoma, carcinoma	2
	Thyroid (adenoma)	1
Bone	Osteo- and chondrosarcoma	2
Heart	Fibro- and round cell sarcoma	2
Kidney	Osteo- and round cell sarcoma	2
Skin	Epithelioma adenoidescyshcum	1
Eye	Dermoid of cornea	1
Brain	Teratoma of pons	1
Testis	Embryonal carcinoma	1

Reproduced from <u>Cancer</u> <u>Medicine</u> (92) with permission.

trauma and inflammation, (g) hormones and internal environments, and (h) parasites. Various animal species have been used to study most of these etiological agents.

Genetics

Animal models have been widely used for genetic studies of tumors. Among the invertebrates, hydra (55), planaria (247), mollusks (342), and earthworms (81) develop neoplasms. Tumors in drosophila have been reported (165, 382). However, these tumors are nonmalignant and have no value as models in human cancer research. Burdette (58) correlated metagenic and carcinogenic activity of tumors in invertebrates with those in humans.

The vertebrates are more prone to tumors. The genetic basis of neurolemmonas in a colony of goldfish has been established (392). Similarly, genetic influences have been implicated in hepatomas of fish (157, 158) and Lucké adenocarcinoma of the leopard frog (294). In fowl, Marek's disease and lymphomas show strain predilection (344).

Genetic influences in mammals have been reported, for example, ocular squamous carcinoma of cattle (14), mammary tumors in mice (182), and the relative resistance of the guinea pig to tumors (375). Hybridization of the species (261) and strains (184) generally increase the rate of tumor induction. The inbred strains, on the other hand, can be developed as high tumor risk strains or low tumor strains. These inbred strains of mice have been useful in the development of selection of host genomeprovirus hypothesis (35) and oncogene hypothesis (201). The role of genetics extends beyond the spontaneous tumors, since chemically induced cancers are considered to be under genetic control (57, 58, 181, 442). The interaction of genetic factors with other types of tumors have been reported in such tumors as mammary tumors in mice (210), hypophyseal tumors of mice (185), and adrenocortical tumors (185, 473). Transmission of tumors induced by MuLV (379, 380, 381) in AKR mice and mouse MTV (34, 35, 306) have also been shown under genetic control.

Hereditary neoplasms due to human genetic disorders may be localized or may occur as a part of other neoplasms or development anomalies. A list of hereditary human neoplasms is given in Table 11-6. In humans, familial cancer risk is rather small.

Table 11-6. Hereditary Neoplastic and Preneoplastic States in Humans

Heredity Neoplasms	Pre-Neoplastic State
1. Retinoblastoma	I. Hamartomatous syndromes
2. Trichoepithelioma	a. Neurofibromatosis

3. Multiple endocrine adeno-
   matosis
	b. Tuberous sclerosis
	  c. Von Hippel-Lindau syndrome

4. Thyroid carcinoma (medullary
   and pheochromocytoma)
	d. Multiple exostoses
	  e. Pentz-Jeghers syndrome

5. Chemodectomas	II. Genodermatoses
6. Polyposis coli	a. Xeroderma pigmentosum
7. Gardner's syndrome	b. Albinism

8. Tylosis with esophageal
   carcinoma
	c. Werner's syndrome
	  d. Epidermody aplasia

9. Nevoid basal cell carcinoma	verruciformis
	e. Polydysplastic epidermoly- sis bullosa
	f. Dyskeratosis congenita
	III. Chromosome Breaking disorder
	a. Bloom's syndrome
	b. Fanconi's syndrome
	IV. Immune Deficiency syndrome
	a. Ataxia-telangiectasia
	b. Wiskott-Aldrich syndrome
	c. Late-onset immunologic deficiency
	d. X-linked agamma globulinemia

However, site specific risks such as cancer of the breast, stomach, lung, colon, endometrium and prostate are three times as great in close relatives compared to the general population (266). In children, the site-specific risk among siblings is at least fourfold greater for leukemia, brain tumors, and soft-tissue sarcoma (293). Common environment, genes, or both could account for familial occurrences of cancer (244, 268, 449). These factors may also be important in racial and ethnic differences. The role of genetics in cancer in animals (183) and humans (243) has been reviewed recently (128).

Radiation

Many species of experimental animals have been used to study the mechanisms of tumors produced by radiation (454). Reviews of radiation effects on humans are also available (454).

Skin Cancer

Cutaneous carcinogenesis has been reported in experimental animals using rats (7), mice (205), and other animals (141, 249). The type of tumors produced vary with the host factors and conditions of irradiation. High yield of tumors required large doses causing ulceration and/or permanent damage to the hair follicles (141). In rats, the ulcers generally appeared near the margins of damaged follicles (62). Nonulcerative doses are not as carcinogenic and may not produce dermatitis (205, 249). The effect of radiation is enhanced by croton oil (402) and by chemical carcinogens (72) in mice. In rats the damage to the hair follicles parallels tumor yield and is maximum at 2000 to 4000 rems (62). In humans, the occupational hazards of radiologists have almost been eliminated by using better equipment safety measures. In general, a dosage of four rads over long periods of exposure may be harmful (205).

Whole animal irradiation can produce neoplastic transformations in various tissues, such as thymic lymphosarcoma and granulocytic leukemia in mice, and tumors of the endocrines, particularly females, in rats (454). Exposure of lung to x-ray irradiation may result in bronchial adenocarcinoma in rats (155), intestinal neoplasms in mice (82), and ovarian neoplasms in dogs (16). None of the tumors of the central nervous system have been reported to rise from adult neuronal tissue. Internal emitters may be carcinogenic (48, 63, 278) depending on the distribution of radionuclide and radiation energy. In princepal, cells affected are endosteal preosteoblasts or osteoblasts (200, 262).

Leukemia

As early as 1911, leukemogenic effect of ionizing radiation was noticed. Since then, there have been several reports confirming this relationship (87, 187). The type of myelo- and lympho-proliferative changes depend on the age of the cells at the time of exposure, dosage levels, and distribution factors. Generally the acute and myelocytic changes are produced; however the incidence of chronic lymphocytic leukemia is not related to irradiation. Animal studies have also contributed greatly to the mechanism of the occurence of leukemias (82). Mice with a low spontaneous rate show an increase in the incidence, whereas high leukemia strains of mice fail to show increased occurrences (213, 264). Nonthymic lymphoma can be induced by ^{90}Sr in some strains of mice (214). Animal studies have been useful to provide dose response data; however, a complex relationship has been observed in mice. It appears that fractions of the total dosage given at appropriate times are more damaging than the total dosage as such (225). The induction period is inversely related to the dosage level, but the peak mortality is found between

150 to 400 days. Susceptibility to induction varies with strain, age, and sex. Females are more sensitive (82). Myeloid leukemias and other myeloid proliferative disorders have been reported to occur in dogs (103), rats (221), and swine (199).

Osteosarcoma

A number of bone tumors have been reported in experimental animals caused by the tumerogenic activity of radium, plutonium, uranium, strontium, and calcium (121). The differences are based on quantity of radiation emitted, its uptake, distribution, and retention. The incidence of bone tumor varies proportionally to the square of the final concentration of radium in the skeleton (275). Similar relationship has been observed in humans; however, there is no evidence of tumor induction with less than 0.5 microcuries. The incidence increases to 20 percent level at 5 microcuries (116, 272). The sensitivity of endosteum to tumor induction seen in animals has been reported in humans (296).

Thyroid Tumors

When subjected to radioiodine therapy, rats, mice, or sheep show papillary and follicular growth. The lesions are similar to those reported in humans; however, the whole gland may be destroyed if the dosage level is not controlled. Children exposed to x-ray therapy (and those who were exposed to atomic bomb radiation) had a high incidence of thyroid tumors (carcinomas, adenomas, hyperplastic nodules) (76, 176). The latent period in these cases was 10 to 20 years.

Pulmonary Tumors

Uranium mine workers develop pulmonary tumors from irradiation. The average weekly dosage is estimated at ¼–1 rad (454). In animals, alveolar and bronchial neoplasms have been reported with external and internal sources of irradiation (67).

Trauma and Inflammation

The role of trauma and inflammation in cancer etiology is controversial (303, 421, 428, 459). The main controversy concerns the time interval between the trauma and the appearance of neoplasm. The time, of course, has to be flexible, because thyroid tumors may occur after 8 to 9 years of irradiation (468) and 18 to 24 years after chemical (e.g. stilbesterol) treatment (180). Similarly, thermal injury has been implicated in causation of cancer in Indian (318) and Japanese populations (452). Other trauma-associated tumors have been reported, including bone tumors (459), brain tumors (461), and testicular (303, 470) and uterine cervix tumors (54). Most of the data have been collected from epidemiological studies. Animal models should provide some definitive answers as to the importance of trauma and inflammation in cancer etiology.

Parasites

The role of parasites in human cancer has been suggested but not unequivocally confirmed, for example, schistosomiasis and bladder cancer in Egypt (3, 104, 171), Ghana (312), and East Africa (111, 358). However, it is not clear if the etiology is due to the parasite or parasite-associated carcinogenic metabolite or a virus.

Experiments with mice models have not resulted in the induction of bladder carcinoma (111, 170, 462).

Rats have been used as animal models for the study of sarcomas associated with *Cystcercus fasciolaris* infection (104). The relationship of parasites and neoplastic response has been reviewed by Audy (24).

Virus

Virus etiology of cancer was first reported in chickens in 1908 by Ellerman and Bang (112). In 1911, Rous (378) isolated chicken sarcoma virus, and Bittner (43) reported mammary adenocarcinoma in mice. Other earlier examples of virus etiology of cancer in animals include leukemias in murine mammals (153) and multiple tumors in mice (424). A total of 110 viruses that may induce neoplasms have been isolated up to 1970 (Table 11-7) (365), and more have been added to the list recently. The viruses usually have a large host range and some of these viruses not only replicate in tissue cultures but also may cause cancerous transformations of the cells (Table 11-8) (365). It shows that humans are susceptible to viral oncogenesis, and experiments with subhumans hold great relevance to human cancer studies. It is pointed out that there is no real danger to humans from accidental contact with the viruses through vaccinations (SV40 from polio, avian adenoleukosis with yellow fever and measles vaccines), exposure to cats, dogs, chickens, and cows, or through food such as eggs, milk, or meat. Nonetheless, there is a need for caution because of new or mutant isolates and the interactions due to other disease. It has been shown that cell cultures from patients with diseases such as Fanconi's anemia and Down's syndrome are more susceptible (50 to 100 times) to SV_{40} and adenovines 12 (447). Similarly a combination of viral agents in the presence of environmental influences (e.g., smog) may cause neoplasms (129). Another point is the development of the same type of neoplasm by widely different types of viruses, Table 11-8 (365).

Oncogenic RNA viruses

The RNA viruses that cause tumors have been given different names such as leuko viruses (119), thylaxoviridiae (90), and oncorna viruses (326). Other terms, C-type, B-type, H-type, and A-type, have also been used based on morphological, enzymatic, and antigenic differences. Many exhaustive reviews describe the oncogenic properties of RNA viruses (154, 443, 457), avian viruses (458), and murine viruses (46, 53, 372). Certain oncorna viruses and their properties are listed in Table 11-9 (36). Most of these viruses produce leukemia-sarcoma types of tumors. The leukemias are vertically (congenitally) transmitted.

Avian Leukemia-Sarcoma Complex. The infectious nature of avian leukemia (112) and avian (Rous) sarcoma (378) were discovered at the beginning of this century. Several pure and mixed strains of ALV and RSV have now been recognized (154, 458). Four major (A-D) and one minor (E) group (154, 346, 465), as well as nontransforming Rous associated virus (RAV), are known (Table 11-10) (36). A dominant cell gene governs the susceptibility of such cells to avian oncorna viruses (345).

Murine Leukemia-Sarcoma Complex. Four well-defined strains, G-MuLV (153), F-MuLV (130), M-MuLV (298), and R-MuLV (363), are known. On the basis of antigenic properties, these have been grouped into (a) F-M-R viruses and (b) G-virus (136, 330, 341). A new grouping of these agents into N-tropic (NIH Swiss mice), B-tropic (BAL B/c cell line), and NB-tropic, which replicates in both cell lines, have been reported. The latter classification does not agree with antigenic classification since group

Table 11-7. Viruses that Induce Neoplasms in Animals

Common name of virus	No. of major isolates	Host of origin	Produces neoplasia in:	Tumour type in animals
Mouse leukemia (MLV)*	16	Mouse	Mouse, rats, hamsters	Leukemia, lymphoma
Mouse sarcoma (MSV)	6	Mouse	Mouse, rats, hamsters, cats, tissue culture	Sarcomas
Polyoma (Py)	2	Mouse	Mouse, hamsters, tissue cultures	All types except leukemia
Mammary tumor (MTV)	2	Mouse	Mouse	Carcinoma
Chicken leukemia (ALV)	4	Chicken	Chicken, tissue culture	Leukemia
Twiehaus	1	Chicken	Chicken, quail, hamster	Reticuloendotheliosis
Chicken sarcoma (Rous et al.) (RSV)	9	Chicken	Chicken, quail, turkey, duck, hamster, monkey, snake, tissue culture	Sarcoma
Marek's (MHV)	1	Chicken	Chicken	Lymphoma
CELO	1	Chicken	Hamster	Sarcoma
Cat leukemia (FLV)	4	Cat	Cat	Leukemia, lymphoma
Cat sarcoma (FSV)	3	Cat	Cat, rat, dog, monkey, tissue culture	Sarcoma
G. pig leukemia	1	G. pig	G. pig	Leukemia
G. pig herpes	1	G. pig	G. pig	Sarcoma
Deer fibroma	1	Deer	Deer	Fibroma
Squirrel fibroma	1	Squirrel	Squirrel	Fibroma
Shope fibroma	1	Rabbit	Rabbit	Fibroma
Shope papilloma	1	Rabbit	Rabbit	Papilloma
Dog sarcoma	1	Dog	Dog, tissue culture	Sarcoma
Dog mast cell	1	Dog	Dog	Carcinoma
Lucke	1	Frog	Frog	Carcinoma
Human adeno	31**	Man	Hamster, mouse, tissue culture	Sarcomalymphoma
Wart	1	Man	Man	Papilloma

Table 11-7. (Continued)

Common name of virus	No. of major isolates	Host of origin	Produces neoplasia in:	Tumour type in animals
Hybrids#	7	Monkey Man Cat Mouse	Hamster, cat, tissue culture	Sarcoma, lymphoma
Yaba	1	Monkey	Monkey, man	Histiocytoma
H. saimiri	1	Monkey	Monkey	Lymphoma
Simian adeno	6	Monkey	Hamster, tissue culture	Sarcoma, lymphoma
SV40##	1	Monkey	Hamster, mouse, tissue culture	Lymphosarcoma
Graffi hamster	2	Hamster	Hamster	Lymphoma, papilloma
Bovine papilloma	1	Cow	Cow, horse, mouse, hamster	Papilloma, fibroma, sarcoma
Bullhead papilloma	1	Fish	Fish (bullhead catfish)	Papilloma

*Strains of mouse leukemia virus (MLV) are usually designated by names of original investigator: GLV=Gross leukemia virus, FLV=Friend leukemia virus, RLV=Rauscher leukemia virus.
**As of May 1970 approximately 12 of 31 human adenoviruses induce malignancies in hamsters. These 12 and the remaining 19 induce discrete foci of transformed (apparently cancerous) cells in tissue cultures.
#Hybrid=Genotypic recombinates of 2 different viruses, e.g. SV_{40}^+ adeno; cat leukemia + mouse sarcoma.
##Simian virus 40.
Adapted from Cancer Medicine (365) with permission.

Table 11-8. Tumor Viruses that Replicate and/or Transform Human Cells

Common name of virus	No. of major isolates	Type	Host of origin	Replication/ transformation*
A. "Known Tumor Viruses"				
Chicken sarcoma	3	RNA	Chicken	T
CELO	1	DNA	Chicken	T
Mouse sarcoma	2	RNA	Mouse	R,T
Mouse leukemia	1	RNA	Mouse	R
H. saimiri	1	DNA	Monkey	R
Yaba	1	DNA	Monkey	T
SV40	1	DNA	Monkey	R,T
"Hybrids"**	7	DNA	Monkey-man	T
Cat sarcoma	2	RNA	Cat	R,T
Cat leukemia	1	RNA	Cat	R
Human adeno (type 12)	1	DNA	Man	R,T
EBV	2	DNA	Man	R,T
B. "Suspect Tumor Viruses"#				
Influenza	2	RNA	Man	R
Sarcoma particle	2	("C")	Man	R
Herpes -2	1	DNA	Man	R
Herpes -1	1	DNA	Man	R
Human breast particle	3	"RNA"	Man	R
C. Other				
Shope papilloma##	1	DNA	Rabbit	R
Monkey breast particle	1	"RNA"	Monkey	R

*R=Replication of intact detectable virion; T=transformation as represented by nonlytic morphologic alteration and prolonged life.

**Genotypic recombinates of 2 different viruses, e.g. SV40+Adenovirus 12; cat leukemia + mouse sarcoma viruses.

#Viruses or virus-like particles reported in "consistent" association with cancers or which enhance (or are enhanced by) physical or chemical agents.

##This virus may replicate in man as judged by significant antibody levels and low serum arginine levels in some exposed laboratory personnel.

Adapted from Cancer Medicine (365) with permission.

Table 11-9. Some Properties of RNA Containing Tumor Viruses (Oncorna viruses)

Virus	Abbreviations used	Host of origin	Natural tumors (host of origin)*	Experimental host range in Vivo tumor*	Experimental host range in Vitro cell transformation	Virion morphology (C- or B-type particles)##
Avian complex						
Leukemia	ALV**	Chicken	Yes	Chicken, turkey	Chicken#	C
Sarcoma (Rous)	RSV	Chicken	Yes	Avian, rodent, monkey	Avian, rodent, bovine, monkey, man	C
Murine complex x						
Leukemia	MuLV	Mouse	Yes	Mouse, rat, hamster		C
Sarcoma	MSV	Mouse	No	Mouse, rat, hamster	Mouse, rat, hamster	C
Murine mammary tumor (Bittner)	MTV	Mouse	Yes	Mouse		B
Feline complex						
Leukemia	FeLV	Cat	Yes	Cat		C
Sarcoma	FeSV	Cat	Yes	Cat, dog, rabbit, monkey	Cat, dog, monkey man	C
Other						
Viper		Viper	Yes			C
Hamster, leukemia?	HaLV	Hamster				C
Rat leukemia?	RaLV	Rat				C
Guinea pig, leukemia		Guinea pig	Yes			C
Bovine, lymphoma		Cow	Yes			C
Primate						
Sarcoma (Wooly monkey, gibbon)		Monkey, ape	Yes	Monkey	Monkey	C
Mammary carcinoma (Mason-Pfizer)		Monkey		Monkey		B

**The term RAV has been used for ALV strains associated with the defective Bryan strain of RSV.
*Usually there is a persistence of infectious virus in the tumor.
##Size ranges from 70 to 100 nm: virus matures at cell membrane by a process of budding.
#Attained with avian myeloblastosis virus only.
Adapted from Molecular Biology of Cancer (36) with permission.

Table 11-10. Classification of Avian Oncorna viruses by Host Range

Antigenic subgroup	Representative viruses* Leukemia strains	Representative viruses* Rous sarcoma strains	Ability of viruses within antigenic subgroup to grow in: Genetically defined chick embryo cells						Japanese quail cells
			C/O	C/A	C/B	C/AB	C/BC	C/E	
A	RAV-1, AMV-1, RIF-1, RAV-2, RIF-2	SR-RSV-A, MH-RSV, PR-RSV-A	+**	0	+	0	+	+	+
B	AMV-2, RIF-2	SR-RSV-B, HA-RSV	+	+	0	0	0	+	0
C	RAV-7, RAV-49	PR-RSV-C#, B77#	+	+	+	+	0	+	±
D	RAV-50	SR-RSV-D#, CZ-RSV-D#	+	+	±	±	±	+	0
E	RAV-0, RAV-60, ILV	None	+	+	0	0	0	0	+

*Abbreviations: RSV = Rous sarcoma virus; RAV = Rous-associated virus; AMV = Avian myeloblastosis virus; RIF (resistance-inducing factor) = field strains of avian leukemia viruses that interfere with the focus-forming capacity of RSV; ILV = induced leukemia viruses ; SR = Schmidt Ruppin strain; MH=Mill Hill strain; PR = Prague strain; HA = Harris strain; B77 = Bratislava 77 strain of RSV; and CZ = Carr-Zilber strain.

**+ = cells fully susceptible; ± = intermediate cell susceptibility; 0 = cells resistant.

#Strains of RSV capable of inducing tumors in rodents and of transforming rodent cells. Adapted from _Molecular Biology of Cancer_ (36) with permission.

394

G can be either N or B tropic and FMR and G are NB tropic (164). A dominant cell gene governs the resistance to MuLV (352, 353).

Five strains of murine sarcoma virus (MSV) have been recognized as FJB-MSV (121), GZ-MSV, (169), Ki-MSV (234), and M-MSV (298). However, these strains have antigenic association with MuLV.

Murine mammary tumor virus (MTV). This virus is also called the Bittner milk agent (43). All strains identified cause mammary tumors in special strains of mice and are transmitted by milk (46, 154). All MTV strains share some common antigens but differ from other murine oncoviruses (47, 326).

Feline Leukemia-Sarcoma Complex. This is a new group of viruses FeLV which cause leukemias in cats and FeSV causing sarcomas in cats (135, 413), dogs (134), and marmoset monkeys (96).

Hamster Leukemia-Sarcoma Complex. This is also a new group, and HaLV are probably indigenous to hamsters (146, 419). In tissue cultures, these are nontransforming agents present in hamster-specific sarcoma (HaSV) viruses. The HaSV have been derived from MSV and are oncogenic to hamsters but not to mice (32, 228, 229, 348). They appear to contain MSV in the envelope of helper HaLV which is responsible for altered host and antigenic specificity (228, 229, 238). HaLV and HaSV, however, appear different from C-type murine viruses (228, 229, 334, 335).

Primate Oncorna Viruses. Three viruses have been isolated from nonhuman primates: (a) B-type Mason-Pfizer virus from rhesus monkey mammary tumors (69, 219); (b) C-type from woolly monkey fibrosarcoma (444, 471); and (c) C-type from gibbon ape (tymphisarcoma) (227). These agents are antigentically distinct from other simian viruses.

Oncorna Viruses of Other Animals. C-type particles have been detected from malignant tumors of snakes (481), rats (68), guinea pigs (154, 209), and PHA-stimulated lymphocytes of cattle with lymphosarcoma (292, 427). Cross-reaction properties between various groups of viruses are given in Table 11-11 (36).

Oncogenic DNA Viruses

The DNA viruses, which cause tumors in animals and humans, are a diverse group varying in their size, structure, and biological properties (36).

Papovaviruses. Papilloma viruses have the largest host range among the DNA viruses. However, the studies of this group have been hampered because of their inability to grow in tissue cultures in vitro. The polyoma virus first isolated from mice can induce tumors in newborn mice, rats, and hamsters (108, 109, 426). SV40 originally isolated from rhesus monkeys (434) have shown no cytotoxic effects.

Adenoviruses. This group contains a large number of viruses; at least thirty-one distinct antigenic forms are present in humans, sixteen in monkeys, eight in chickens, seven in cattle, four in swine, two in mice, and two in dogs (325). The adenoviruses of humans are associated with several human conditions such as respiratory diseases and conjunctivitis (18). They have been shown to be oncogenic to newborn hamsters (151, 451). Some of the types (c) can transform rat embryos and some of the simian, bovine, and avian adenoviruses are tumorigenic to newborn hamsters.

Herpesviruses. This group has definite oncogenic properties. Herpes simplex type I (HSV) is the causative agent of human fever blisters, acute stomatitis, eczema,

Table 11-11. Cross-Reaction Between Group Specific Antigens of Oncorna viruses

Viruses	Species-specific antigens (gs-1)									Interspecies-specific antigens (gs-3)								
	C-type viruses							B-type viruses		C-type viruses							B-type viruses	
	A	M	F	R	H	P	V	M	P	A	M	F	R	H	P	V	M	P
C-type																		
Avian	+	–	–	–	–	–	–	–	–	–	–	–	–	–	–	–	–	–
Murine	–	+	–	–	–	–	–	–	–	–	+	+	+	+	+	–	–	–
Feline	–	–	+	–	–	–	–	–	–	–	+	+	+	+	+	–	–	–
Rat	–	–	–	+	–	–	–	–	–	–	+	+	+	+	+	–	–	–
Hamster	–	–	–	–	+	–	–	–	–	–	+	+	+	+	+	–	–	–
Primate	–	–	–	–	–	+	–	–	–	–	+	+	+	+	+	–	–	–
Viper	–	–	–	–	–	–	+	–	–	–	–	–	–	–	–	–	–	–
B-type																		
Murine	–	–	–	–	–	–	–	+	–	–	–	–	–	–	–	–	–	–
Primate	–	–	–	–	–	–	–	–	+	–	–	–	–	–	–	–	–	–

*A = avian, M = murine, F = feline, R = rat, H = hamster, P = Primate, and V = viper.

Adapted from Molecular Biology of Cancer (36) with permission.

keratoconjunctivitis, and meningioencephalitis. It does not possess oncogenic properties but can transform hamster cells in tissue cultures (102). However, the HSV-2 induces genital infections and is transmitted venereally (314, 368).

It may play a role in the development of cervical carcinoma (25, 315, 367). The UV-irradiated HSV can induce malignant transformation in hamster cells in vitro (100). Epstein-Barr (EB) virus is intimately connected with Burkitts lymphoma but an absolute etiological relationship has not been established. However, EB as a causative agent of infectious mononucleosus is well established (178).

Oncogenic herpesviruses (HVS & HVA) have been isolated from monkeys (286, 287). The etiology of Marek's disease, a lymphoproliferative disease of chickens, is definitely a herpesvirus. Similarly, the role of herpesvirus in tumors in frogs (265) and wild rabbits is well known.

Pox Viruses. These viruses cause tumors in rabbits, monkeys, and humans (26, 93). The tumors in rabbits (Shope-fibroma) and monkeys (Yaba viruses) show spontaneous regression.

Viruses and Human Cancer

Evidence presented in the preceding section indicates viral etiology of certain cancers in animals. This has led to speculation of viral etiology of certain human cancers (9, 45, 114, 154, 279, 364).

Animal studies for the presence of C-particles in malignancies resulted in the recognition of similar particles in leukemia, Hodgkin's disease, lymphoma, and lymphosarcoma (154). Similar particles have been detected in cell lines derived from human malignancies (154, 302, 448); however, no clear evidence has been obtained with tumor tissues or cultivated cells (37, 279, 320). Activation experiments with cells from human sarcoma and electron microscopic examination show C-type particles (425, 426). However, other reports using embryonal human rhabdomyosarcoma suggest the absence of C-type particles (280, 281). In conclusion, the status of the role of C-particles in human cells is still uncertain.

B-type particles, characteristics of murine mammary tumor viruses, have been detected in human mammary cancer and in the milk of Parsi (Indian) women and American women with a familial history of breast cancer (300). However, later work by Sarkan and Moore (388) questions the validity of these microscopic particles. The physical studies indicate that they contain high molecular weight RNA (70 S) and reverse transcriptase which is characteristic of oncorna viruses (391, 392). Relationship between the two types of particles based on immunological similarities has been suggested (308). Also, hybridization (DNA-RNA) studies show similar possibilities (26, 27). In other studies RNA complimentary to R-MuLV has been reported in nonmammary tumors such as leukemias, lymphomas, and sarcomas. It hybridizes with DNA synthesized by MuLV (173a, 173b, 242). Other workers have shown the presence of reverse transcriptase in cells in acute human lymphoblastic leukemia (133, 389).

Models of DNA tumorigenic viruses in animals

As indicated earlier, papova and adenoviruses in animals produce solid tumors but no infectious agent can be isolated from the tumors. Similar results have been obtained with human tumors (154, 279). In animals, tumor virus specific mRNA can be detected; however, in human tumors specific-mRNA has not been isolated (279, 280). It appears that studies with herpesvirus, which is known to exist in latent form in humans, should produce interesting results. The isolation of new agents unrelated to papovavirus have been isolated from leukoencephalopathy (337, 463) and ureteric obstruction (135).

Immunological Studies

The results of animal studies on tantigens and tantibodies do not correlate with similar studies in humans (140).

Herpesvirus and human malignancies

There is some evidence relating herpesvirus to oncogenics in humans. A new isolate, herpes simplex type 2, is venereally transmitted in humans (314, 366, 367), and a relationship between cervical carcinoma and infection with this agent in women has been reported (4, 5, 25, 315, 368). Support for this concept has come from studies with hamster embryo cells (100, 101). A strong association has been observed between EB (Epstein-Barr) virus and two human malignancies: Burkitts lymphoma (60, 61) and post nasal carcinoma (194). The etiological relationship has been based on high EB antibodies (177, 179, 236). However, EB has also been found in normal populations (321, 356). These studies, however, have led to etiological association of EB virus with benign immunoproliferative infectious mononucleosis (115, 178, 322) as well as lupus erythematosus (117), sarcoidosis (192), Hodgkin's disease (220), and lepromatous leprosy (339).

Chemical Carcinogens

The nonspontaneous nature of cancer was evidenced by observations that nasal cancer was common in snuff users (190), scrotal cancer was common in chimney sweeps (357), and bladder cancer occurred in workers in dye manufacturing industries. The chemical carcinogens are structurally complex organic and inorganic compounds with various activities and selectivities (71, 166, 167, 168, 446). The majority of the chemicals are low molecular weight (about 500) organic compounds (Table 11-12) (19).

It is clearly established that certain chemicals induce neoplastic transformations in animals and cell cultures in reproducible manner (40, 91, 98, 174, 245, 467). Many other chemicals such as diethylstilbestrol (180), aflatoxins (211, 239, 399), sterigmaticystin (361), and nitrosamines (31, 255, 271, 472) are currently being tested for their carcinogenic properties. The list of chemicals that have established carcincogenic properties is small at the present time. Most of the chemical agents that cause cancers are incidental and considered occupational hazards with the exceptions of lung cancer due to smoking and oral cancer due to betel nut or tobacco chewing.

Immunocompetency and malignancy

The importance of the immune system in the etiology of cancer is well established (8, 175, 235, 237, 288). Laboratory animals with impaired immune mechanisms have been utilized to study tumorogenesis.

Polyoma virus, which infects laboratory and wild mice, does not produce tumors under natural conditions because of the elimination of transformed cells by thymus dependent (T) lymphocytes. It has been shown that antilymphocytic serum, ALS (246) or thymectomy permits the appearance of tumors in rats and mice (10, 456), and the reconstitution with thymus, spleen, or lymph nodes prevents the development of neoplasms (11, 249, 405). Mice and rat models have also been used to demonstrate that immunosuppressing agents such as irradiation (431) or cortisones (398) can increase susceptibility to a variety of tumors caused by C-type cells. In thymectomized rats and

Table 11-12. Dosing Schedules Used in the Administration of Some Known Types of Carcinogens

Type of carcinogen	Dose and vehicle	Route and frequency	Species
Polycyclic hydrocarbons and heteroaromatic compounds	⎧ 0.1 ml 0.3-0.4% in solvent ⎨ 0.5-2 mg (occasionally up to ⎩ 10 mg in oil or by trocar)	Skin painting 2 X weekly Single subcutaneous administration	Mice Mice, rats
4-Aminobiphenyl and 2-acetyl-aminofluorene: derivatives and related compounds	1.62 millimoles/kg diet	Mixed in the diet	Rats
2-Acetylaminofluorene	0.1 ml 0.5% in acetone ⎰ 300 mg/day for 50 days	Skin painting 2 X weekly Mixed in the diet	Rats Dogs
2-Naphthylamine	160 mg/kg diet	Mixed in the diet	Mice
2-Anthramine	1% in acetone, starting 0.02 ml and gradually increasing to 0.3 ml at 300 days	Skin painting 2 X weekly	Rats
Amino azo dyes	2.40 millimoles/kg diet (occa-sionally 2.00 or 2.67 milli-moles/kg)	Mixed in the diet	Rats
o-Aminoazotoluene	10-200 mg by trocar	3-10 repeated subcutaneous implanta-tions	Mice
4-Dimethylaminostilbene and derivatives	⎰ 0.22 millimoles/kg diet ⎱ 1 or 5 mg/week in oil	Mixed in the diet 8-10 repeated subcutaneous injections	Rats Mice, rats
Trypan Blue	1 ml 1% aqueous solution	Subcutaneous injection every 2nd week	Rats
Estrone	50-200 ug/week in oil	Weekly subcutaneous injections	Mice, rats
Methyl bis(β-chloroethyl) amine hydrochloride and related nitrogen mustards	0.10-0.25 mg in 0.25 ml water	4-8 repeated subcutaneous injections	Mice

399

Table 11-12. (Continued)

Type of carcinogen	Dose and vehicle	Route and frequency	Species
Ethyl carbamate and deriva-tives	⎰ 0.15-0.75% ⎱ 0.10%	Mixed in the diet In the drinking water	Mice Mice
Dialkylnitrosamines	1 mg/week in water	13 weekly intraperitioneal injections	Rats
Ethionine	5.4 μmoles/day in water	Stomach tube	Rats
	19 millimoles/kg diet	Mixed in the diet	Rats
Carbon tetrachloride	0.1 ml of 40% solution in oil	Stomach tube 2 X weekly	Mice
Dioxane	1%	In the drinking water	Rats
Tannic acid	1-2% aqueous solution	Subcutaneous injection every 5 days	Rats
Acetamide	2-5%	Mixed in the diet	Rats

Adapted from <u>Chemical Induction of Cancer</u> (19) with permission.

400

Table 11-13.   Relationship Between Immunodeficiencies and Neoplasms

Condition	Tumor/patients	Reference
Combined immunodeficiency	0/70	193
Wiskott–Aldrich syndrome	1/18	79
Wiskott–Aldrich syndrome	4/16	460
Ataxia-telangiectasia	1/20	460
Hypogammaglobulinemia	8/176	110
Selective IgA deficiency	5/102	13
Intestinal lymphangiectasia	3/50	460

mice, the incidence of lymphatic leukemia after administration of chemical carcinogen is drastically reduced (74). However, the treatment methyl nitrosourea administered to rats with ALS did not increase the incidence of tumors (97). The effect of immunosuppression on the increased development of neoplasms is not very clear with chemical carcinogens.

The most important contributions in this field have been made by the development of special athymic nude mice (122, 338). The mice do not reject skin grafts from other species such as birds, reptiles, or amphibians (273, 369, 385, 475). It is interesting to note that the nude mice do not develop neoplasms easily (89), which is in contrast to the immuno-surveillance theory. Nevertheless, it supports the concept that immunity is required for development of neoplasms (359).

The relationship between high incidence tumor strains of mice (AKR, NZB & SFL/J) and immunocompetence has been studied. The results vary with the strain. It cannot be concluded that the high incidence of tumors is related to decreased immunocompetence (145, 162, 163, 191, 289).

The applicability of these data to human conditions is clearly seen by the fact that inconsistencies observed in animals are also found in humans (Table 11-13) (13, 79, 110, 193, 460). However, there is no doubt that the relationship between immunocompetence and occurrence of neoplasms exists. The exact nature may vary with internal and external environmental conditions. From limited data, it is suggested that incidences of malignancy may be 200 times the expected rate.

SCREENING AND EVALUATION OF ANTITUMOR AGENTS

The cure of cancer, and its etiology, is not fully understood at the present time. However, good progress has been made in this field in the last decade as a result of finding that certain cancers are curable by specific drugs (482). In general, rapidly growing tumors are more responsive to treatment than slowly growing ones; both the rate of growth and drug responsiveness are related to the growth fraction and the percentage of cells undergoing cell division at one time. It should be evident that normal tissues with high growth fractions (e.g., bone marrow) should also be sensitive to these drugs. As the hope for cure of certain cancers from chemotherapeutic agents is real, at least in some cases, it is important that proper screening of potential drugs be carried out in experimental animals prior to administration in humans. The primary aim of the screening of drugs used in cancer therapy is to develop and to evaluate effective antitumor agents. The objectives can be divided into four groups. (a) Screening and evaluation in experimental animals. The test system should be sufficiently sensitive but not too sensitive to select many false-positives. Also, it should not be too insensitive to provide false-negatives. In addition, the individual tests should be reproducible. (b) Study of toxicological and pharmacological effects in animals. (c) Development of new improved test models. (d) Study of biochemical mechanism and drug-structure relationship to develop new agents with a rational approach.

Preclinical Evaluation

Primary Screening

Most cytotoxic effects in vitro are determined in the KB system. This is generally followed by in vivo testing in experimental tumor systems and sequence of dosage levels to get a dose-response curve (20, 252, 377). The next step is activity confirmation, which involves repeating the experiments to determine if the observed activity is reproducible with sufficient magnitude to warrant additional testing.

Secondary Evaluation

The detailed study based on preliminary screening data is carried out to obtain schedule dependency, optional routes of administration, activity against intracranial or other types of sequestered tumor cells, determination of appropriate vehicles, and dose-response relationships. Drug activity has been characterized in several types of tumor systems. This approach is valid if the data are going to approximate human conditions where the tumors vary in their cytokinetic, metastatic, and biochemical properties.

In Vitro Systems

A number of in vitro systems (seventy-four in all) have been tested (137) including mouse-adapted virus, bacteriophage, bacteria, fungi, slime mold, frog embryo, chick embryo, and drosophila (Table 11-14) (138). Other in vitro systems include tissue cultures such as hamster adenocarcinoma of duodenum, HEP-2 (human epidermoid carcinoma), HEP-2/755, HEP-2/740, lysogenic induction, Rous sarcoma, and more recently KB system (human epidermoid carcinoma of nasopharynx) (CCNSC) (143, 251, 440). The in vitro test systems are primarily based on cytotoxicity testing and are not very quantitative. There is no basis for the measurement of relative drug toxicity for abnormal cells as compared to normal cells. The utilization of cell cultures for cancer chemotherapy has been reviewed by Foley and Epstein (123).

In Vivo Systems

The in vivo system is more important but more complex since the three important parameters, host, tumor, and the drug, interreact with one another. The drug can be tested against the host (Table 11-15) (19) or the tumor, or a combination of the two can be developed.

The Host

The appropriate strain and sex should be selected for the experiment.

Table 11-14. Biological Systems

Experimental tumors	Microbiological	Differentiation and development	Biochemical synthesis
Sarcoma 180	Feline pneumonitis	E. coli mutation SD4	Tumor 755 deoxyribonucleic acid
Flexner-Jobling	Vaccinia	E. coli mutation WP14	Ehrlich protein
Walker 256	Influenza A	Dictyostelium aggregation	Ehrlich adenine
Carcinoma E0771	Western equine encephalitis	Dictyostelium culmination	Ehrlich guanine
Sarcoma T241	Poliomyelitis	Frog embryo differentiation	Gardner protein
Carcinoma 1025	Bacteriophages	Frog embryo drug sensitivity	Gardner adenine
Mecca lymphosarcoma	Escherichia coli	Chick embryo abnormalities	Gardner guanine
Carcinoma RC	Bacillus subtilis	Chick embryo growth inhibition	Flexner-Jobling Q_{d2}
Recently isolated mammary carcinoma	Serratia marcescens	Drosophila larva	Flexner-Jobling CO_2
Carcinoma 755	Staphylococcus aureus	Drosophila puparium	Flexner-Jobling protein in vitro
Brown-Pearce carcinoma	Mycobacterium smegmatis	Drosophila imago	Flexner-Jobling protein in vivo
Glioma 26	M. phlei	Drosophila larva/imago	Flexner-Jobling nucleic acids
Leukemia L1210	Lactobacillus casei	Drosophila phenotype abnormalities	Flexner-Jobling adenine
Ehrlich ascites EF	K. brevis	Drosophila chromosome breakage	Flexner-Jobling guanine
Ehrlich ascites ELD	T. utilis	Drosophila crossing-over	Spleen Q_{d2}
	Saccharomyces cerevisiae 358	Drosophila variegation W	Spleen CO_2
	S. cerevisiae 376	Drosophila variegation Y	Spleen protein in vitro
	Penicillium notatum		Spleen protein in vivo
	Actinomyces fumigatus		Spleen nucleic acids
	Streptomyces griseus		Spleen adenine
	S. antibioticus		Spleen guanine

Adapted from Cancer Research (138) with permission.

Table 11-15. Test Systems Which Have Been Used as Primary Screens by CCNSC

Mouse tumors	**Rat tumors**
Adenocarcinoma 755	Dunning leukemia ascites
Cloudman Melanoma (S91)	Dunning leukemia solid
Ehrlich ascites	Human Sarcoma HS1
Hepatoma 129	Murphy-Sturm lymphosarcoma
Lewis lung carcinoma	Walker 256 (intramuscular)
Lymphoid Leukemia L1210	Walker 256 (subcutaneous)
Osteogenic Sarcoma HE10734	
Sarcoma 180	**Human tumor**
	Human Sarcoma HS1
Hamster tumors	
Adenocarcinoma of duodenum	
Adenocarcinoma of endometrium	
Adenocarcinoma of small bowel	
Melanotic melanoma	

Adapted from <u>Chemical Induction of Cancer</u> (19) with permission.

Propagation of Tumor Lines

If a tumor system is to be established, it is important that tumors are examined histologically and the samples are maintained frozen in tumor banks for renewal of sublines and comparisons with existing sublines. Several tumor lines are available at CCNSC (Table 11-16) (142). Table 11-16 also indicates transfer time and strains used for testing purposes.

Tumor Transplantation

The tumor should be transplanted as quickly as possible after removal. The size should be standardized (2–6 mm), and the tumor fragment should be kept cold. It is important that the first four fragments from each tumor be tested for bacteriological contamination.

Tumor beri can also be transplanted. The donor tumor is excised on the day it is needed and about 6 g are minced with scissors or tissue grinder and held in normal saline and at a cold temperature. It should be quickly transplanted (30 minutes) in approximately 0.2 ml quantities such that each animal gets about 0.1 g of the tissue. As usual, the first 0.4 ml is used to test bacterial contamination.

Ascites tumor preparations are also used, particularly for leukemic cells. The ascites fluid is withdrawn aseptically and tested for cell morphology and cell count. The fluid is diluted with sterilized balanced fluid and 0.1 ml inoculum is used.

The inoculated animals are observed for the development of tumors and the number of animals without tumors, "no takes," should be less than three for fifteen to twenty-four mice, four for twenty-five to thirty-four mice, five for thirty-five to forty-four hamsters. Otherwise, it is considered excessive "no takes" and the underlying problem should be discovered.

After the establishment of tumors the testing of the drug can start with the following considerations.

Table 11-16.   Test Systems Employed in Screening Programs of CCNSC

Mouse	<u>Transplantable</u>
	Lymphoid leukemia L1210
	Adenocarcinoma 755
	Cloudman melanoma (S91)
	Ehrlich ascites
	Hepatoma 129
	Lewis Lung carcinoma
	Osteogenic sarcoma HE10734
	Sarcoma 180
	P388 leukemia
	L5178Y leukemia
	LPCI plasma cell
	<u>Primary and transplantable</u>
	AKR virus leukemia
	Moloney virus leukemia
	Rauscher virus leukemia
	Friend virus leukemia
	C3H mammary tumor
Rat	Dunning Leukemia
	Murphy-Sturm lymphosarcoma
	Walker 256
Hamster	Adenocarcinoma of duodenum, endometrium and small bowel
	Melanotic melanoma
Chicken	Rous Sarcoma
Carcinogen induced tumors	Mammary carcinoma
	1) 3 methyl cholanthrene
	2) Dimethyl benzanthrene
	Fibrosarcoma-3,4,9,10-Dibenzpyrene
Heterologous host	HS-1 in conditioned rats
	HEP-3 in conditioned rats
	DBA/2 mouse lymphatic leukemia in conditioned hamsters
	Human amelanotic melanoma in conditioned hamster

Adapted from <u>Methods in Cancer Research</u> (142) with permission.

Solvent for the drug

The solvents employed are water, saline, 0.1N HCl NaHCO₃, Na₂CO₃, 95 percent alcohol or acetone diluted with distilled water, 0.5 percent carboxymethylcellulose in saline, olive oil, or peanut oil.

Administration of Material and Dosage

Intraperitonial and subcutaneous routes are most commnly used. Other routes can be used as well. The drug may be given as a single dose a day, several doses a day, or continuous infusion. In the first trials, if no information is available, 500 mg/kg is generally used. If 34 percent or fewer deaths occur and antitumor effects are seen, no further tests are made. If more than 34 percent deaths occur, the dosage is reduced until the 34 percent death level is reached. When toxicity data are available, the dose is usually one-third of the acute LD_{50} dosage.

Test Evaluation

Most commonly used test criteria include recording of survival time, measurement of tumor diameters, and excision and weight of local tumors. Details of the protocols to be followed for testing of antitumor activity have been previously described (139).

ANIMAL MODELS FOR SELECTED NEOPLASMS (TABLE 11-17)

Cancer of the Stomach

Gastric cancer in humans may occur in the form of a carcinoma (adenocarcinoma, adenocanthoma, squamous cell carcinoma, and carcinoid tumor), lymphoma, leiomyosarcoma, and other sarcomas.

Gastric cancers, though decreasing (12), still claimed 15,000 lives in 1971 and were rated fifth on the list of cancers causing mortalities. The incidence of this cancer occurs more often in men than in women (3:2). There are strong indications of environmental influences, such as smoked meats and saki in Japan, molded soy beans in Korea, smoked fish in Ireland, and cabbage, rye bread, and purgatives in the United States (52, 147). Several benign diseases of the stomach, gastric ulcers, gastric adenomas, chronic gastritis, and intestinal metaplasia, are not related to gastric carcinoma with the possible exception of pernicious anemia (196).

Pathologically, gastric carcinoma frequently involves pylorus and antrum (51 percent), followed by fundus (21 percent), lesser curvature (18 percent), cardia (7 percent), and greater curvature (3 percent) (39). Four main types of lesions are found: (a) ulcerative (75 percent of the time), (b) polypoid (10 percent), (c) scirrhous (10 percent), and (d) superficial (5 percent). Histologically, microscopic multicentricity is seen most (22 percent) and the tendency increases with pernicious anemia (75).

Gastric adenocarthomas and squamous cell carinomas are rare, making up 0.04 to 0.07 percent. These may arise from (a) totipotential cells capable of giving tumors of any cell type (b) ectopic squamous epithelium, and (c) squamous metaplasia of preexisting cells (430).

Gastric lymphomas are found in 50 percent of necropsy series, which is a result of systemic lymphomatous disease (376); however, the primary gastric involvement is between 0.5 to 8.0 percent. Pathologically, they are grossly indistinguishable from adenocarcinomas and are generally bulky, lobulated, and vascular, with some demarcation for the stomach wall. Microscopic picture is characteristic of lymphoma. The usual route of spread is lymph nodes, paraaortic nodes, omentum, peritonium, mesentery, extraabdominal nodes, and liver.

Gastric leiomyosarcoma accounts for 1 to 3 percent of all gastric malignant tumors (336, 349, 386). These are primarily large tumors of anterior and posterior walls of the corpus of the stomach and are generally over 7 cm in diameter with weight reaching up to 20 lb. The tumor may be intra- or extraluminal; it is usually well supplied with small blood vessels and bleeds quickly.

Table 11-17. Animal Models in Cancer Research

Animal Model	Species	Human Counterpart	Reference
Abnormal lipid in lymphoid tumors	Mouse	Niemann-Pick Disease	267
Adenocarcinoma	Dog		309
	Cat	Adenocarcinoma	83
	Rabbit,aged	Endometrial adeno-carcinoma	28
Adenocarcinoma of the colon, DMH induced	Rat	Adenocarcinoma of colon	319b
Basal cell carcinoma	Dog	Basal cell carcinoma	305
Brown-Pearce tumor	Rabbit		423
Carcinoma	Dog	Carcinoma	126
Embryonal nephroma	Pig	Embryonal nephroma	432
Epulis	Dog	Embryonal nephroma	144
Fibroma	Horse	Fibroma	216
Granuloma pyogenicum	Cattle	Granuloma pyogenicum	73
Granulosa cell tumor	Cow	Granulosa cell tumor	248
Hemangioma	Dog	Hemangioma	396
Hepatocarcinoma	Monkey	Carcinoma	230
Hepatocellular carcinoma (Aflatoxin carcinogenesis)	Rat	Primary hepatocellular carcinoma	319a
Hepatoma	Fish		414
(Liver cell carcinoma)	Trout	Hepatoma	22
Herpesvirus induced malignant lymphoma	Squirrel-monkey	Burkitt's lymphoma, lymphocytic leukemia	207
Interstitial cell adenoma	Dog	Interstitial cell adenoma	203
Leukemia L2C/N-B	Guinea pig	Leukemia	313
Leukosis	Cattle	Lymphocytic leukemia	106
Lymphomatosis	Chicken	Lymphomatosis	197
Lymphatic tumors of ganglions	Mouse	Hodgkins disease	311
Lymphocytic Leukemia	Dog	Lymphocytic leukemia	217
Lymphoma	Gerbil	Lymphoma	160
Lympho sarcoma	Cattle	Lymphoma	277
	Chicken		355
Lymphosarcoma, (Leukosis, malignant lymphoma, leukemia)	Cattle Sheep Horse Pig	Malignant lymphoma	291
Malignant lymphoma	Cattle	Lymphoma	276,407, 416
	Dog		49,309, 323,333, 407,416
	Cat		324,407, 416
	Pig		299

Table 11-17.   (Continued)

Animal Model	Species	Human Counterpart	Reference
Mammary neoplasia	Dog	Breast cancer	429
Melanoma	Horse	Melanoma	304
	Dog		408
Melanotic melanomas	Hamster	Melanoma	125
Mixed mammary gland tumor	Dog	Breast cancer	41,309
Multiple myeloma	Horse	Multiple myeloma	80
	Cattle		347
	Swine		113
	Dog		284
	Cat		198
Mycetoma	Dog	Mycetoma	56
Neurofibroma	Dog	Solitary neurofi-broma	396
Osteosarcoma	Dog	Osteosarcoma	84
Ovarian tumors	19-Progesterone mouse	Ovarian tumor	260
	Stilbesterol Dog		215
	Rat		42
Papilloma	Cattle		30
	Dog	Papilloma	408
	Rabbit		340
Perianal gland adenoma	Dog	Adenoma	172
Schwannoma	Dog	Schwannoma granular cell	476
Seminoma	Dog	Seminoma	202
Sertoli's cell tumor	Dog	Sertoli's cell tumor	202,203
Squamous cell carcinoma	Cat		305,390
	Cattle	Squamous cell carcinoma	186,383, 435
	Dog		85,172
	Horse		408
Transmissible venereal tumor	Dog	Transmissible venereal	50,226, 384
Trichoepithelioma	Dog	Trichoepithelioma	396
Trophoblastic tumors of placenta	Armadillo		274
Ultimobranchial thyroid neoplasm	Bull	Medullary thyroid carcinoma	66
Uterine tumor	Rabbit	Uterine tumor	152,212
	Endometrium chemically induced		438
	Myometrium hormonally induced guinea pig		259
Viral avian leukosis	Fowl	Lymphocytic leukemia	33
Viral leukemia	Cat	Lymphocytic leukemia	310

Laboratory animals, with the exception of rats, are mostly refractile. However, among farm animals, the horse sometimes develops stomach tumors. No good animal models are available for this condition.

Cancer of the Liver

Primary hepatic cancers are mostly hepatomas (90 percent), cholengiomas (5 to 10 percent), mixed liver cell and ductulue tumors (2 to 5 percent), and hepatoblastomas, which occur frequently in children. These neoplasms account for 0.75 percent with approximately 2500 deaths a year in the United States. However, the cancer is more common in adult males in Africa and Asia (189).

No definite agent can be incriminated in human conditions, although several hepatocarcinogenic agents have been demonstrated in animals, for example, aflatoxins and senecio alkaloids (290, 469).

In South China and Thailand, parasitic infection with *Clonorchis sinensis* or other parasites is a probable cause of cholangiocellular carcinoma. Certain nutritional deficiencies such as kwashiorkor have also been shown to predispose the patient to hepatomas. In addition, the association between cirrhosis, viral hepatitis, and hepatoma is a likely possibility (188, 433).

Nodular, massive, and diffuse types of hepatomas have been reported (188), but the nodular form is most common (66 percent) followed by massive (30 percent). Diffuse form (5 percent) is mostly seen with cirrhosis. Microscopically well-differentiated tumors may resemble very closely the normal cells in structure and function; however, nuclear irregularities, cytoplasmic peculiarities, and absence of bile duct residual and Kupffer's cells are helpful. Frequently, death results from failure of liver function, and tumor invasion seldom extends to adjacent organs. Metastasis to periportal lymph nodes and lungs is most frequently seen (35 to 50 percent). Because of rich blood supplies, minor or massive hemorrhages are often present.

Cholangiomas may arise from bile ducts or intrahepatic ducts. They are present either in massive or nodular form with a firm or cut surface and with little or no tendency to necrosis or hemorrhage (188).

Vascular carcinoma present no unusual features because of their location.

Experimental animals, including the laboratory (mice, rats, rabbits) and farm animals (chicken, turkey, duck, cattle, horses) as well as other animals, such as trout and shrimp, have been used to develop experimental hepatomas. Spontaneous occurrence of hepatomas have been reported in mice (C3H, C3Hf), rats, and guinea pigs. The list of hepatocarcinogens includes urethane, A_{30} dyes, aromatic amines, nitrosamines, polycyclic hydrocarbons, mycotoxins, and toxins from plants (59). Studies in mice (1) showed the presence of a distinct α-globulin in mice with hepatoma and a similar protein was also found in the sera of human hepatoma patients (441). These proteins are similar to α-feto protein (α-FP) which is found in the fetus and newborn. However, when the test is applied to humans, the presence of α-FP is considered specific for hematoma which occurs in 36 percent of the cases (204). However, in Russia, Africa, and Asia, the positive reaction is found in 50 to 90 percent of the cases (327).

Cancer of the Large Bowel

The cancer of the large bowel may be a carcinoma (adeno or squamous cell and carcinoid) or sarcoma (lymphoma, leiomyosarcoma, and mixed).

Adenocarcinoma

Except for epitheliomas of skin, it is the most frequently occurring cancer in this country, affecting 75,000 persons during 1971. Worldwide, the incidence is highest in the United States and Europe and lowest in Asia, Mexico, and Central and South America. There is a negative correlation between the incidence of stomach and large bowel cancer (404, 479).

The etiology is not clear; however, environmental influences, particularly food, tobacco, alcohol, purgatives, as well as the biliary and bacterial population of the gut, have been implicated (59, 404, 479). It is now agreed that chronic ulcerative colitis and multiple polyposis predispose an individual to cancer of the large bowel (38). However, there are conflicting reports about the premalignant nature of adenomatous polyps (295, 415, 453).

The segmental distribution in decreasing order of frequency is rectum, sigmoid, descending colon, cecum, ascending colon, transverse colon, splenic flexure, and hepatic flexure. The adenocarcinomas are well demarcated from normal mucosa with the classical appearance of a nodular, fungating mass with areas of ulceration surrounded by inflammating reaction of paracolonic tissues. Of the several forms, scirrhous type produce firm tumors, while mucinous or colloid forms produce soft, bulky, and friable tumors. Papillary forms, often seen in the rectum, are soft, friable fronds of epithelium. Multicentricity is seen in above 3 percent of the cases and is exaggerated in multiple polyposis and chronic ulcerative colitis.

Histologically, the cells may resemble normal mucosa to cells in sheets and cords.

The development of adenocarcinoma goes through an intermediate premalignant stage when the cells continue to divide and produce new DNA even when the cells move through the midpositions of the crypts. In addition, they produce more protein and have higher thymidine kinase activity. In carcinoma, it appears that the increased mass is due to the inability of cells to slough off rather than higher multiplication rate (258). The spread is circumferential and eventually causes clinical obstruction. Direct invasion may involve stomach, duodenum, liver, pancreas, small bowel, kidney, spleen, peritoneal area, and abdominal wall. Other organs may also be affected, producing devastating conditions. The metastasis of regional lymph nodes often takes place; arterial walls resist invasion but venous involvement is often seen (95). Death in 50 percent of the cases results from distant metastasis (cerebral, liver, etc.).

Carcinoids of the large bowel are found mostly at the rectal and cecal extremities. The lesions are like those of the small intestine except that the amount of serotonin is similar to normal cells as compared to carcinoids of the small intestine, where the serotonin content is very high.

Squamous cell carcinoma of the large bowel is very rare. When present, the tumors cluster around rectum, sigmoid, and left colon (29, 59, 95).

Lymphonia is found in about 10 percent of the cases, but the primary involvement is less than 0.5 percent (29, 59, 400). Microscopically, these are reticulum cell sarcomas and very rarely seen in Hodgkin's disease (88, 477). It is most commonly seen in males (70 percent), and it is often not associated with ulcerative collitis (477).

Leiomyosarcomas are evenly distributed through the bowel. In spite of the abundance of smooth muscle, it is seen in 0.1 percent of the cases (127). Clinically, obstruction and bleeding are most commonly seen.

The occurrence of large bowel cancer in laboratory or domestic animals is not common. Among the domestic animals, dog, horse, and fowl are the only species showing tumor lesions in the intestine. Laboratory animals such as mice and rats have been widely used for cancer studies and do not show spontaneous or induced tumors of intestines.

Cancer of the Male Genitalia

Testes

Testicular tumors are rare (2.2/100,000 males) and are often seen in undescended or cryptorchid testes, usually in the older males (2, 64, 161). Germinal as well as non-germinal tumors are seen but seminoma are most commonly found. Serotal invasion is generally not present, probably due to barriers of tunica albuginea and vaginalis. Metastasis takes place by vascular or lymphatic routes and lungs or liver are most often involved.

Among laboratory species, mice and dogs have been used to study the human condition. Embryonal carcinoma has been widely studied in mice strain 129 (351). It has been observed that tetratocarcinomas are produced in the testicular tubules of fetal mice (420). Ultrastructurally, the primordial germ cell and embryonal cancer cells are related (420). Transplantation studies show that these tumors arise from primordial germ cells (350). In cases of seminomas, spermatogonia or spermatocytes are involved (351). Teratocarcinomas arise from multipotent stems (99, 357). The human tumors resemble closely those in mice.

Cancer of the Prostate

Prostatic cancer is almost always adenocarcinoma and often presents a heterogenous histological picture. The tumor is fairly common; a death rate of 18.9/1000 has been reported. Etiology is not known, though hormones and viruses have been implicated (439). Carcinoma is often multicentric (254). Histologically, the cells are cuboidal with acidophilic cytoplasm and hyperchromatic nuclei and nucleoli are present. Histochemically, acid phosphatase, beta glucoronidase, and mucin are often found (232, 233, 474) and aminopeptidase is absent.

Carcinoma of the prostate is rarely present in laboratory or farm animals. However, a virus induced (SV40) anaplastic carcinoma has been reported as a model in the hamster (343). However, because of its inability to metastasize to bone and its anaplastic appearance it is not a good model.

Breast Cancer

Cancer of the breast has an incidence of 65/100,000 and a mortality rate of 25/100,000. The incidence is highest in North America and very low in Japan and the Far East. The incidence increases with age (118) and it decreases with multiple pregnancies, prolonged lactation, and pregnancy before the age of 20 years (269, 270, 478). Other factors such as race and socioeconomic status have also been implicated. Virus etiology has also been suggested (see animal models).

Pathologically, the proliferation starts with the lining cells of small ducts and it appears that acini are less frequently involved (132, 455). Histologically, infiltrating duct carcinoma with productive fibrosis (scirrhous carcinoma) is most frequently (72 percent) followed by infiltrating lobular carcinoma, medullary, colloid, comedo, and pillary types (282). The regional axillary lymph nodes are most frequently involved. In general, internal mammary nodes are not involved unless numerous axillary metastasis are found

(156). Distant metastases are found in mediastinal lymph nodes, lungs, liver, bone, adrenal, skin, ovary, spleen, skin, pancreas, kidney, brain, thyroid, and heart (285).

Among laboratory animals, mice and rats have been widely studied as models for the human condition, but direct application of these studies to the human condition is not possible, except the involvement of genetic factors, that is, strain specificity. However, the discovery of B-type particles similar to murine mammary tumors has increased the investigations using the mouse model (300). Pathologically persistant hyperplasia in terminal ducts and alveolar epithelium precedes neoplastic growth (94). Similar epithelial involvement has been reported in human oncogenesis (387). The murine model should mimic human conditions such that it produces morphological changes in adolescence, is influenced by endocrine and reproductive activities (269), and stimulated by drug-induced mammary stimulation. Exposure to radiation, chemical carcinogens, immunosuppression, and pregnancy accelerates tumor production (283, 397, 464). Although the murine model does not satisfy several of these conditions, testing of the virus etiology in the animal offers a very promising model (see section on virology, this chapter).

Multiple Endocrine Neoplasia (*Sipples Syndrome*)

Medullary thyroid carcinoma in humans is derived from parfollicular C-type cells and has amyloid stroma containing calcitonin. These syndromes are associated with multiple endocrine neoplasia. The animal model that mimics this condition is the ultimobranchial neoplasm of the bull (44, 66). It has been reported that 30 percent of bulls may have this type of neoplasm (223). Bulls given long-term excessive calcium (three to six times normal) show a high incidence of the neoplasms (240). The low incidence under similar conditions in cows is related to high physiological requirements of calcium for lactation. This model may help to explain the physiological role of calcitonin and calcium in human diet.

Malignant Lymphoma (*Lymphosarcoma*)

The lymphosarcoma of cattle, sheep, pigs, and horses closely resemble the lymphosarcoma of humans both clinically and pathologically (256). Other comparisons include the histological picture and the increase in incidence with age. The chronic lymphocytic leukemia (CLL) of humans and enzootic bovine lymphosarcoma resemble each other in the course of the disease which is protracted. The model in cattle and other animals presents a good model for the study of viral, genetic, and environmental factors (291).

Canine Mammary Neoplasia (*Breast Cancer*)

The mammary neoplasia of dogs bears close resemblance to that of humans (360, 394) and epidemiological data have been reported. However, the breast tumors are two to three times less common in women than breast cancers in bitches when the age group in dogs (8 to 11 years) is compared to humans ($\sim$50 to 58 years). Other similarities include the role of hormones, and the high incidence in nulliparous females. The histological picture and the infiltration from the duct system is also common. However, mixed tumors in bitches constitute about 50 percent of the mammary carcinoma, whereas it is not common in human females. It appears that the naturally occurring carcinoma represents an important model for human breast cancer (429).

Burkitts Lymphoma, Malignant Lymphoma, and Lymphocytic Leukemia

Several models for Burkitts lymphoma have been described. Recently a herpesvirus-induced malignant lymphoma model using marmosets and spider and owl monkeys has been reported (207, 208). Spontaneous tumors in owl monkeys have also been reported (206). In humans, the Epstein-Barr virus (EBV) is carried through life without any ill effects, and only primary infection produced mononucleosis. A condition similar to this is seen in monkeys. EBV causes Burkitts lymphoma and naspharyngeal carcinoma under certain conditions. But such a situation does not exist in monkeys. Burkitts lymphoma-like syndrome is produced by *H. saimiri* (squirrel monkey) and *H. ateles* (spider monkey) in the species of monkeys that are not natural hosts for these viruses. The pathological features of the human and monkey conditions are compatible, although the histopathology is not the same.

REFERENCES

1. Abelev, G. I., S. D. Perova, N. Kharmkova, Z. A. Postnikova, and I. S. Irlin, *Transplantation*, **1**:174 (1963).
2. Abell, M. R., and F. Holtz, *Cancer*, **21**:852 (1968).
3. Aboul Nasr, A. L., M. E. Gazayerli, R. M. Fawzi, and I. El-Sibai, *Acta Unio Int. Contra Cancrum*, **18**:528 (1962).
4. Adam, E., A. H. Levy, W. E. Rawls, and J. L. Melnick, *J. Natl. Cancer Inst.*, **47**:941 (1971).
5. Adam, E., S. L. Sharma, O. Zeigler, K. Iwamoto, J. L. Melnick, A. H. Levy, and W. E. Rawls, *J. Natl. Cancer Inst.*, **48**:65 (1972).
6. Adamson, R. H., R. W. Cooper, and R. W. O'Gara, *J. Natl. Cancer Inst.*, **45**:455 (1970).
7. Albert, R. E., W. Newman, and B. Altshuler, *Rad. Res.*, **15**:410 (1961).
8. Alexander, P., *Nature* (London) **235**:137 (1972).
9. Allen, D. W., and P. Cole, *N. Engl. J. Med.*, **286**:70 (1972).
10. Allison, A. C., and L. W. Law, *Proc. Soc. Exp. Biol. Med.*, **127**:207 (1968).
11. Allison, A. C., and R. B. Taylor, *Cancer Res.*, **27**:703 (1967).
12. American Cancer Society, *Cancer Facts and Figures*, 1971.
13. Ammann, A. J., and R. Hong, *Medicine*, **50**:223 (1971).
14. Anderson, D. E., in Staff of M. D. Anderson Hospital & Tumor Institute (eds.), *Genetics and Cancer*, University of Texas Press, Austin, 1959.
15. Anderson, D. E., *J. Hered.*, **51**:51 (1960).
16. Anderson, A. C., and M. E. Simpson, *Ann. Rep. UCD-472-116*, University of California, Davis Radiobiol. Lab., 1969.
17. Andervont, H. B., and T. B. Dunn, *J. Natl. Cancer Inst.*, **28**:1153 (1962).
18. Andrewes, C., and H. G. Pereira, *Viruses of Vertebrates*, 2nd ed., Williams & Wilkins, Baltimore, 1967.
19. Arcos, J. C., M. F. Argus, and G. Wolf, *Chemical Induction of Cancer*, Vol. I, Academic Press, New York, 1968.
20. Armitage, P., and M. A. Schneiderman, *Ann. N.Y. Acad. Sci.*, **76**:896 (1958).
21. Ashley, L. M., *Bull. Wildlife Dis. Assoc.*, **3**:86 (1967).
22. Ashley, L. M., *Am. J. Pathol.*, **72**:345 (1973).
23. Ashley, L. M., and J. E. Halver, *J. Natl. Cancer Inst.*, **41**:531 (1968).
24. Audy, J. R., *Trop. Med. Hyg. News*, **19**:15 (1970).
25. Aurelian, L., I. Royston, and H. J. Davis, *J. Natl. Cancer Inst.*, **45**:455 (1970).
26. Axel, R., J. Schlom, and S. Spiegelman, *Nature* (London) **235**:32 1972.
27. Axel, R., J. Schlom, and S. Spiegelman, *PNAS* (USA) **69**:535 (1972).
28. Baba, N., and E. Van Haam, *Am. J. Pathol.*, **68**:653 (1972).

29. Bacon, H. E., *Cancer of the Colon, Rectum and Anal Canal,* J. B. Lippincott Co., Philadelphia, 1964.
30. Bagdonas, V., and C. Olson, *J. Am. Vet. Med. Assoc.,* **122:**393 (1953).
31. Barnes, J. M., in W. J. Hayes, Jr. (ed.), *Essays in Toxicology,* Vol. 4, Academic Press, New York, 1974.
32. Bassin, R. H., P. J. Simons, F. C. Chesterman, and J. J. Harvey, *Int. J. Cancer,* **3:**265 (1968).
33. Beard, J. W., *Ann. N.Y. Acad. Sci.,* **108:**1057 (1963).
34. Bentvelzen, P., *Genetical Control of the Vertical Transmission of the Mühlbock Mammary Tumor Virus in the GR Mouse Strain,* Hollandia, Amsterdam, 1968.
35. Bentvelzen, P., in P. Emmelot and P. Bentvelzen (eds.), *RNA Viruses and Host Genome in Oncogenesis,* Amsterdam, North Holland, 1972.
36. Benyesh-Melnick, M., and J. S. Butel, in H. Busch (ed.), *Molecular Biology of Cancer,* Academic Press, New York, 1974.
37. Benysh-Melnick, M., K. O. Smith, and D. J. Fernbach, *J. Natl. Cancer Inst.,* **33:**571 (1964).
38. Berk, J. E., and W. S. Haubrich, in H. L. Bockus (ed.), *Gastroeuterology,* Vol. II, W. B. Saunders, Philadelphia, 1964.
39. Bakson, J., in W. H. Remine, J. T. Priestley, and J. Berkson, *Cancer of the Stomach,* W. B. Saunders, Philadelphia, 1964.
40. Berwald, Y., and L. Sachs, *J. Natl. Cancer Inst.,* **35:**641 (1965).
41. Biggs, R., *J. Pathol. Bacteriol.* **59:**437 (1947).
42. Biskind, G. R., and M. S. Biskind, *Am. J. Clin. Pathol.,* **19:**50 (1949).
43. Bittner, J. J., *Science,* **84:**162 (1936).
44. Black, H. E., C. C. Capen, and D. M. Young, *Cancer* **32:**865 (1973).
45. Black, P. H., W. H. Burns, and M. S. Hirsch, in M. Sanders and M. Schaeffer (eds.), *Viruses Affecting Man and Animals,* Green Publishing, St. Louis, 1971.
46. Blair, P. B., *Current Topics Microbiol. Immol.,* **45:**1 (1968).
47. Blair, P. B., *Israel J. Med. Sci.,* **7:**161 (1971).
48. Blair, W. J., in M. G. Hanna, Jr., P. Netteshein, and J. R. Gilbert (eds.), *Inhalation Carcinogensis,* U.S. AFEC, Oak Ridge, Tenn., 1970.
49. Bloom, F., *Arch Pathol.,* **33:**661 (1942).
50. Bloom, F., G. H. Paff, and C. R. Norback, *Am. J. Pathol.,* **27:**119 (1951).
51. Blumenthal, H. T., and J. B. Rogers, In W. E. Ribelin and J. R. McCoy (eds.), *Pathology of Laboratory Animals,* Charles C Thomas, Springfield, Ill., 1965.
52. Bockus, H. L., *Gastroenterology,* Vol. I, 2nd ed., W. B. Saunders, Philadelphia, 1963.
53. Boiron, M., J. P. Levy, and J. Peries, *Prog. Med. Virol.,* **9:**341 (1967).
54. Boyd, J. T., and R. Doll, *Br. J. Cancer,* **18:**419 (1964).
55. Brien, P., *Bull. Biol. France Belg.,* **95:**301 (1961).
56. Brodey, R. S., H. F. Schryver, M. J. Deubler, et al., *J. Am. Vet. Med. Assoc.,* **151:**442 (1967).
57. Burdette, W. J., *Cancer Res.,* **3:**318 (1943).
58. Burdette, W. J., *Cancer Res.,* **15:**201 (1955).
59. Burdette, W. J., *Carcinoma of the Colon and Antecedent Epithelium,* Charles C Thomas, Springfield, Ill., 1970.
60. Burkitt, D., *Br. Med. J.,* **2:**1019 (1962).
61. Burkitt, D. P., *J. Natl. Cancer Inst.,* **43:**19 (1969).
62. Burns, F. J., R. E. Albert, and R. D. Heimbach, *Rad. Res.,* **36:**225 (1968).
63. Bustad, L. K., C. W. Mays, M. Goldman, L. S. Rosenblatt, W. H. Hetherington, W. J. Blair, R. O. McClellan, C. R. Richmond, and R. E. Rowland, *Proc. Int. Conf.* (4th), **11:**125 (1972).
64. Campbell, H. E., *J. Urol.,* **81:**663 (1959).
65. Campbell, J. H., *Tumors of the Fowl,* W. Heinemann, London, 1969.
66. Capen, C. C., and H. E. Klack, *Am. J. Pathol.,* **74:**377 (1974).
67. Cember, H., in L. Severi (ed.), *Lung Tumors in Animals,* University of Perugia, Italy, 1966.
68. Chopra, H. C., and R. M. Dutcher, in R. M. Dutcher (ed.), *Comparative Leukemia Research,* S. Karger, Basel, 1970.
69. Chopra, H. C., and M. M. Mason, *Cancer Res.,* **30:**2081 (1970).

70. Chopra, H. C., and D. H. Moore, *Nature* (London), **229:**627 (1971).

71. Clayson, D. B., *Chemical Carcinogenesis,* Little, Brown, Boston, 1962.

72. Cloudman, A. M., K. A. Hamilton, R. S. Clayton, and A. M. Brues, *J. Natl. Cancer Inst.,* **14:**1077 (1955).

73. Cohen, H. B., H. Beerman, and L. Nicholas, *Am. J. Med. Sci.,* **251:**475 (1966).

74. Cohen, S. M., D. B. Headley, and G. T. Bryan, *Cancer Res.,* **33:**637 (1973).

75. Collins, W. T., and E. A. Gall, *Cancer,* **5:**62 (1952).

76. Conard, R. A., B. M. Dobyns, and W. W. Sutow, *J.A.M.A.,* **214:**316 (1970).

77. Cooper, E. L., in M. Mizell (ed.), *Biology of Amphibian Tumors,* Springer-Verlag, New York, 1969.

78. Cooper, E. L., *J. Natl. Cancer Inst. Monogr.,* **31:**655 (1969).

79. Cooper, M. D., H. P. Chase, J. T. Lowman, W. Krurt, and R. A. Good, *Am. J. Med.,* **44:**499 (1968).

80. Cornelius, C. E., R. F. Goodbary, and P. C. Kennedy, *Cornell Vet.* **49:**478 (1959).

81. Cornelius, E. A., *N. Engl. J. Med.,* **281:**934 (1969).

82. Cosgrove, G. E., H. E. Waburg, Jr., and A. C. Upton, *Excerpte Med. Monog. Nucl. Med. Biol.,* **1:**303 (1968).

83. Cotchin, E., *Proc. R. Soc. Med.,* **45:**671 (1952).

84. Cotchin, E., *Br. Vet. J.,* **109:**248 (1953).

85. Cotchin, E., *Vet. Rec.,* **66:**879 (1954).

86. Cotchin, E., *Neoplasms of the Domestic Mammals,* Lamport Gilbert & Co., Reading, England, 1958.

87. Cronkite, E. P., W. Moloney, and V. P. Bond, *Am. J. Med.,* **5:**673 (1960).

88. Culp, C. E. and J. R. Hill, *Dis. Colon Rectum,* **5:**426 (1962).

89. Custer, R. D., H. C. Outzen, G. J. Eaton, and R. J. Prehn, *J. Natl. Cancer Inst.,* **51:**707 (1973).

90. Dalton, A. J., E. de Harven, L. Domochowski, D. Feldman, F. Hagnenau, W. W. Harris, et al., *J. Natl. Cancer Inst.,* **37:**395 (1966).

91. Dao, T. L., and D. Sinha, *J. Natl. Cancer Inst.,* **49:**591 (1972).

92. Dawe, C. J., in J. F. Holland and J. Frei, III, (ed.), *Cancer Medicine,* Lea & Febiger, Philadelphia, 1973.

93. De Monbreun, W. A., and E. W. Goodpasture, *Am. J. Pathol.,* **8:**43 (1932).

94. De Ome, K. B., and D. A. Media, *Cancer,* **24:**1255 (1969).

95. De Peyster, F. A., and R. K. Gilchrist, in R. Turell (ed.), *Diseases of the Colon and Ano-rectum,* W. B. Saunders, Philadelphia, 1969.

96. Deinhardt, F., L. G. Wolfe, G. H. Theilen, and S. P. Snyder, *Science,* **167:**881 (1970).

97. Denlinger, R. H., J. A. Swenberg, A. Koestner, and W. Wechsler, *J. Natl. Cancer Inst.,* **50:**87 (1973).

98. Di Paolo, J. A., P. Donovan, and R. Nelson, *J. Natl. Cancer Inst.,* **42:**867 (1969).

99. Dixon, F. J., Jr., and R. A. Moore, *Tumors of the Male Sex Organs,* Armed Force Institute of Pathology, Section VIII, Washington, D.C., 1952.

100. Duff, R., and F. Rapp, *J. Virol.,* **8:**469 (1971).

101. Duff, R., and F. Rapp, *Nature* (New Biol.), **233:**48 (1971).

102. Duff, R., and F. Rapp, *Proc. Am. Assoc. Cancer Res.,* **14:**147 (1973).

103. Dungworth, D. L., M. Goldman, and D. M. McKelvie, in W. J. Clarke, E. B. Howard, and P. L. Hackett (eds.), *Myleoproliferative Disorders of Animals and Man,* USAEC Symposium Series, No. 19, 1970.

104. Dunning, W. F., and M. R. Curtis, *Cancer Res.,* **6:**668 (1946).

105. Dutcher, R. M., *Comparative Leukemia Research,* S. Karger, Basel, 1971–1972.

106. Dutcher, R. M., E. P. Larkin, and J. J. Tumibowicz, *Recent Studies on Bovine Leukemia,* Pergamon Press, New York, 1966.

107. Eddy, B. E., *Virol. Monog.,* **7:**1 (1969).

108. Eddy, B. E., S. E. Stewart, M. F. Stanton, and J. M. Marcotte, *J. Natl. Cancer Inst.,* **22:**161 (1959).

109. Eddy, B. E., S. E. Stewart, R. Young, and G. B. Mider, *J. Natl. Cancer Inst.,* **20:**747 (1958).

110. Editorial, *Lancet,* **1:**163 (1969).

111. El-Ghaffar, Y. A., *Cancer,* **19:**1225 (1966).

112. Ellerman, V., and O. Bang, *Zentralbl. Bakteriol. Parasitenk,* **46:**595 (1908).
113. Englert, H. K. *Zbl Veterinaermed,* **2:**607 (1955).
114. Epstein, M. A., *Lancet,* **1:**1344 (1971).
115. Evans, A. S., *J. Infec. Dis.,* **124:**330 (1971).
116. Evans, R. D., *Br. J. Radiol.,* **39:**881 (1966).
117. Evans, A. S., N. F. Rothfield, and J. C. Niederman, *Lancet,* **1:**167 (1971).
118. Feinleib, M., and R. J. Garrison, *Cancer* **24:**1109 (1969).
119. Fenner, F., *Biology of Animal Viruses,* Academic Press, New York, 1968.
120. Fenner, F. J., and F. N. Ratcliffe, *Myxomatosis,* Cambridge University Press, New York, 1965.
121. Finkel, M. P., B. O. Biskis, and P. B. Jinkins, *Science,* **151:**698 (1966).
122. Flanagan, S. P., *Genet. Res.,* **8:**295 (1966).
123. Foley, G. E., and S. S. Epstein, in A. Golden and F. Hawking, (eds.), *Advances in Chemotherapy,* Vol. I, Academic Press, New York 1964, p. 175.
124. Fortner, J., *Cancer,* **10:**1153 (1957).
125. Fortner, J. G., A. G. Maley, and G. R. Schrodt, *Cancer Res.,* **21:**198 (1961).
126. Fowler, E. H., L. Kasza, and A. Koestner, *Cancer Res.,* **26:**2409 (1966).
127. Franklin, R., and B. McSwain, *Ann. Surg.,* **171:**811 (1970).
128. Fraumeni, J. F., Jr., in J. F. Holland and E. Frei, III (eds.), *Cancer Medicine,* Lea & Febiger, Philadelphia, 1973.
129. Freeman, A. E., P. J. Price, R. J. Bryan, R. J. Gordon, R. V. Gilden, G. J. Kelloff, and R. J. Huebner, *PNAS* (USA), **68:**445 (1971).
130. Friend, C., *J. Exp. Med.,* **105:**307 (1957).
131. Gall, E. A., and T. B. Mallory, *Am. J. Pathol.,* **18:**381 (1942).
132. Gallager, H. S., and J. E. Martin, *Cancer* **24:**1170 (1969).
133. Gallo, R. C., S. S. Yang, and R. C. Ting, *Nature* (London) **228:**927 (1970).
134. Gardner, M. B., R. W. Rongey, P. Arnstein, J. D. Estes, P. Sarma, R. J. Hueber, and C. G. Ricard, *Nature* (London), **226:**807 (1970).
135. Gardner, S. D., A. M. Field, D. V. Coleman, and B. Hulme, *Lancet,* **1:**1253 (1971).
136. Geering, G., L. J. Old, and E. A. Boyse, *J. Exp. Med.,* **124:**753 (1966).
137. Gellhorn, A., and E. Hirschberg, *Cancer Res.,* **3:**1 (1955).
138. Gellhorn, A., and E. Hirschberg, *Cancer Res.,* **3:**125 (1955).
139. Geran, R. I., N. H. Greenberg, M. M. MacDonald, A. M. Schumacher, and B. J. Abbott, *Cancer Chemotherapy Reports,* Part 3, **3:**1 (1972).
140. Gilden, R. V., Y. K. Lee, S. Oroszlan, J. L. Walker, and R. J. Huebner, *Virology,* **41:**187 (1970).
141. Glucksmann, A., L. F. Lamerton, and W. V. Mayneord, in R. W. Raven (ed.), *Cancer,* Vol. I, Butterworth & Co., London, 1957.
142. Goldin, A., *Methods in Cancer Res.,* **4:**193 (1968).
143. Goldin, A., A. A. Serpick, and M. Mantel, *Cancer Chemother. Repts.,* **50:**173 (1966).
144. Gorlin, R. J., C. N. Barron, A. P. Chandhry, and J. J. Clark, *Am. J. Vet. Res.,* **20:**1032 (1959).
145. Gottlieb, C. F., E. H. Perkins, and T. Makinodan, *J. Immunol.,* **109:**974 (1972).
146. Graffi, A., T. Schramm, E. Bender, I. Graffi, K. H. Horn, and D. Bierwolf, *Br. J. Cancer,* **22:**577 (1968).
147. Graham, S., A. M. Lilienfeld, and J. E. Tidings, *Cancer,* **20:**2224 (1967).
148. Granoff, A., P. E. Came, and K. A. Rafferty, *Ann. N.Y. Acad. Sci.,* **126:**237 (1965).
149. Granoff, A., M. Gravell, and R. W. Darlington, in M. Mizell (ed.), *Biology of Amphibian Tumors,* Springer-Verlag, New York, 1969.
150. Green, E. L., *Biology of the Laboratory Mouse,* McGraw-Hill, New York, 1966.
151. Green, M., *Ann. Rev. Biochem.,* **39:**701 (1970).
152. Greene, H. S. N. and B. L. Newton, *Cancer,* **2:**673 (1948).
153. Gross, L., *Proc. Soc. Exp. Biol. Med.,* **76:**277 (1951).
154. Gross, L., *Oncogenic Viruses,* 2nd ed., Pergamon Press, New York, 1970.
155. Gross, P., E. A. Pfitzer, J. Watson, R. T. P. de Treville, M. Kaschak, E. B. Tolker, and M. A. Babyak, *Cancer,* **23:**1046 (1969).
156. Haagensen, C. D., *Diseases of the Breast,* 2nd ed., W. B. Saunders, Philadelphia, 1971.

157. Halver, J. E., in L. A. Goldblatt (ed.), *Aflatoxin-Scientific Background, Control & Implications,* Academic Press, New York, 1969.
158. Halver, J. E., L. M. Ashley, and R. R. Smith, *Natl. Cancer Inst. Monog.,* **31:**141 (1969).
159. Homburger, F., *Prog. Exp. Tumor Res.,* **16:**152 (1972).
160. Handler, A. H., S. I. Magalini, and D. Pav, *Cancer Res.,* **26:**844 (1966).
161. Hansen, J. L., *J.A.M.A.,* **199:**944 (1967).
162. Haran-Ghera, N., M. Benyaakov, A. Peled, and Z. Bentwich, *J. Natl. Cancer Inst.,* **50:**1227 (1973).
163. Hargis, B. J., and S. Malkiel, *Cancer Res.,* **32:**291 (1972).
164. Hartley, J. W., W. P. Rowe, and R. J. Huebner, *J. Virol.,* **5:**221 (1970).
165. Hartung, E. W., *J. Hered.,* **41:**269 (1950).
166. Hartwell, J. L., *Survey of Compounds Which Have been Tested for Carcinogenic Activity,* US P. H. S. Publ. No. 149, 2nd ed., Washington, D.C., 1951.
167. Hartwell, J. L., *Survey of Compounds Which Have Been Tested for Carcinogenic Activity,* US P. H. S. Publ. No. 149, 2nd ed. Suppl. 1, Washington, D.C., 1957.
168. Hartwell, J. L., *Survey of Compounds Which Have Been Tested for Carcinogenic Activity,* US P. H. S., Publ. No. 149, 2nd ed., Suppl. 2, Washington, D.C., 1959.
169. Harvey, J. J., *Nature* (London), **204:**1104 (1964).
170. Hashem, M., and K. Boufros, *J. Egypt. Med. Assoc.,* **44:**598 (1961).
171. Hashem, M., S. A. Zaki, and M. Hussein, *J. Egypt. Med. Assoc.,* **44:**579 (1961).
172. Head, K. W., *Vet. Rec.,* **65:**926 (1953).
173. Hehlmann, R., D. Kufe, and S. Spiegelman, (a) *PNAS* (USA) **69:**1727 (1972); (b) *PNAS* (USA) **69:**435 (1972).
174. Heidelberger, C., *Eur. J. Cancer,* **6:**161 (1970).
175. Heilstrom, K. E., and I. Hellstrom, *Adv. Cancer Res.,* **12:**167 (1969).
176. Hempelmann, L. H., *Science,* **160:**159 (1969).
177. Henle, G., W. Henle, P. Clifford, V. Diehl, G. W. Kafuko, B. G. Kirya, et al., *J. Natl. Cancer Inst.,* **43:**1147 (1969).
178. Henle, G., W. Henle, and V. Diehl, *PNAS* (USA), **59:**94 (1968).
179. Henle, W., G. Henle, H. C. Ho, D. Burtin, Y. Cachin, P. Clifford, A. de Schryver, G. de The, V. Diehl, and G. Klein, *J. Natl. Cancer Inst.,* **44:**225 (1970).
180. Herbst, A. L., H. Ulfelder, and D. C. Pozkanzer, *N. Engl. J. Med.,* **284:**878 (1971).
181. Heston, W. E., *J. Natl. Cancer Inst.,* **3:**69 (1942).
182. Heston, W. E., *Proc. of 4th National Cancer Conference,* Lippincott, Philadelphia, 1960.
183. Heston, W. E., In F. F. Becker (ed.), *Cancer I,* Plenum Press, New York, 1975.
184. Heston, W. E., and G. Vlahakis, *J. Natl. Cancer Inst.,* **26:**969 (1961).
185. Heston, W. E., G. Vlahakis, and B. Desmukes, *J. Natl. Cancer Inst.,* **51:**209 (1973).
186. Hewlett, K., *J. Comp. Pathol. Ther.,* **18:**161 (1905).
187. Heyssel, R., and A. B. Brill, in G. R. Meneely (ed.), *Radioactivity in Man,* Charles C Thomas, Springfield, Ill., 1961.
188. Higgins, G. K., *Recent Results Cancer Res.,* **26:**15 (1970).
189. Higginson, J. *Recent Results Cancer Res.,* **26:**38 (1970).
190. Hill, M. and J. Hillova, *Nature* (New Biol.), **237:**35 (1972).
191. Hirsch, M. S., P. H. Black, M. L. Wood, and A. P. Monaco, *J. Immunol.,* **111:**91 (1973).
192. Hirshaut, Y., P. Glade, L. O. B. D. Vieira, E. Ainbender, B. Dvorak, and L. E. Siltzbach, *N. Engl. J. Med.,* **283:**502 (1970).
193. Hitzig, W. G., in R. A. Good and D. Bergsma (eds.), *Immunological Deficiency Diseases in Man,* National Foundation, New York, 1968.
194. Ho, H. C., *Un. Int. Contre Cancer Mongr.,* **10:**58 (1968).
195. Hoag, W. G., *Ann. N.Y. Acad. Med.,* **108:**805 (1963).
196. Hoffman, N. R., *Geriatries,* **25:**90 (1970).
197. Hoffman, H. A., and D. E. Stoner, *Bull. Calif. Dept. Agri.,* **31:**7 (1942).
198. Holzworth, J., and J. Meier, *Cornell Vet.,* **47:**302 (1957).
199. Howard, E. B., and W. J. Clarke, in W. J. Clarke, E. B. Howard, and P. L. Hackett (eds.), *Myleoproliferative Disorders of Animals and Man,* USAEC, 1970.
200. Howard, E. B., W. J. Clarke, M. T. Karogianes, and R. F. Palmer, *Radiat. Res.,* **39:**594 (1969).

201. Huebner, R. J., and R. V. Gilden, in P. Emmelot and P. Bentvelzen (eds.), *RNA Viruses and Host Genome in Oncogenesis,* North Holland, Amsterdam, 1972.

202. Huggins, C., and P. V. Moulder, *Cancer Res.,* **5:**510 (1945).

203. Huggins, C., and R. Pazos, *Am. J. Pathol.,* **21:**299 (1945).

204. Hull, E. W., C. F. Smith, C. G. Moertel, R. W. O'Gara, and P. P. Carbone, in W. J. Burdette (ed.), *Experimental Studies of Fetoprotein in Man & Monkey,* Charles C Thomas, Springfield, Ill., 1970.

205. Hulse, E. V., *Br. J. Cancer,* **21:**531 (1967).

206. Hunt, R. D., F. G. Garcia, H. H. Barahona, N. W. King, C. E. O. Fraser, and L. V. Melendez, *J. Infect. Dis.,* **127:**723 (1973).

207. Hunt, R. D., and L. V. Melendez, *Am. J. Pathol.,* **76:**416 (1974).

208. Hunt, R. D., L. V. Melendez, F. G. Garcia, and B. F. Trum, *J. Natl. Cancer Inst.,* **49:**1631 (1972).

209. Hurwitz, J., and J. P. Leis, *J. Virol.,* **9:**116 (1972).

210. Husbey, R. A., and J. J. Bittner, *Acta Unio. Int. Contra. Cancrum,* **6:**197 (1948).

211. Hutt, M. S. R., *Liver Cancer,* IARC Sci. Publ. No. 1, Int. Agency for Research on Cancer, Lyon, France (1973).

212. Ingalls, T. H., Wm. Adams, M. B. Lurie, and J. Ilsen, *J. Natl. Cancer Inst.,* **33:**799 (1964).

213. International Commission on Radiological Protection, *ICRP* Publ. No. 11, Pergamon, Oxford, 1968.

214. Ito, T., K. Yokoro, A. Ito, and E. Nishihara, *PNC Soc. Exp. Biol. Med.,* **130:**345 (1969).

215. Jabora, A. G., *Austral. J. Exp. Biol. Med. Sci.* **40:**139 (1962).

216. Jackson, C. *J. Vet. Sci. Anim. Indust.,* **6:**1 (1936).

217. Jarrett, W. F. H., W. B. Martin, G. W. Crighton, R. G. Dalton, and M. F. Stewart, *Nature* (London) **202:**566 (1964).

218. Jay, G. E., In J. W. Burdette (ed.), *Methodology in Mammalian Genetics,* Holden-Day, San Francisco, 1963.

219. Jensen, E. M., I. Zelljadt, H. C. Chopra, and M. M. Mason, *Cancer Res.,* **30:**2388 (1970).

220. Johansson, B., G. Klein, W. Henle, and G. Henle, *Int. J. Cancer,* **6:**450 (1970).

221. Jones, R. K., A. L. Brooks, and A. C. Ferris, in W. J. Clarke, E. B. Howard, and P. L. Hackett (eds.), *Myleoproliferative Disorders of Animals & Man,* USAEC Symposium Series No. 19, 1970.

222. Jones, T. C., *Fed. Proc.,* **28:**162 (1962).

223. Jubb, K. V., and K. McEntee, *Cornell Vet.,* **49:**41 (1959).

224. Jungherr, E., *Ann. N.Y. Acad. Sci.,* **108:**77 (1963).

225. Kaplan, H. S., and M. B. Brown, *J. Natl. Cancer Inst.,* **13:**185 (1952).

226. Karlson, A. G., and F. C. Mann, *Ann. N.Y. Acad. Sci.,* **54:**1197 (1952).

227. Kawakami, T. G., S. D. Huff, P. M. Buckley, D. L. Dungworth, S. P. Snyder, and R. V. Gilden, *Nature (New Biol.),* **235:**170 (1972).

228. Kelloff, G., R. J. Huebner, N. H. Chang, Y. K. Lee, and R. V. Gilden, *J. Gen. Virol.,* **9:**19 (1970).

229. Kelloff, G., R. J. Huebner, S. Oroszlan, R. Toni, and R. V. Gilden, *J. Gen. Virol.,* **9:**27 (1970).

230. Kelly, M. A., R. W. O'Gara, M. D. Walker, C. J. Dave, et al., *J. Natl. Cancer Inst.,* **39:**153 (1968).

231. Kent, S. P. and J. E. Pickering, *Cancer,* **11:**138 (1958).

232. Kirchheim, D., and C. V. Hodges, *Surg. Forum,* **16:**502 (1965).

233. Kirchheim, D., N. R. Niles, E. Franks, and C. V. Hodges, *Cancer,* **19:**1683 (1965).

234. Kirsten, W. H., and L. A. Mayer, *J. Nat. Cancer Inst.,* **39:**311 (1967).

235. Klein, E., *Ann. Inst. Pasteur,* **122:**593 (1972).

236. Klein, G., *Adv. Immurol.,* **14:**187 (1971).

237. Klein, G., *Transpl. Proc.,* **5:**31 (1973).

238. Klement, V., J. W. Hartley, W. P. Rowe, and R. J. Huebner, *J. Natl. Cancer Inst.,* **43:**925 (1969).

239. Krishnamachari, K. A. V. R., R. V. Bhah, V. Nagamjan, and T. B. G. Tilak, *Lancet,* **1:**1061 (1975).

240. Krook, L., L. Lutwak, K. McEntee, P. Hendrickson, K. Braun, and S. Roberts, *Cornell Vet.,* **61:**625 (1971).
241. Kruger, G. R. F., R. A. Malmgren, and C. W. Berard, *Transplantation,* **11:**138 (1971).
242. Kufe, D., R. Hehlmann, and S. Spiegelman, *Science,* **175:**182 (1972).
243. Kundson, A. G., Jr., in F. F. Becker (ed.), *Cancer I,* Plenum Press, New York 1975.
244. Kurita, S., *Acta Haemat. Jap.,* **31:**748 (1968).
245. Kuroki, T., and H. Sato, *J. Nat. Inst.,* **41:**53 (1971).
246. Lance, E. M., P. B. Medawar, and R. N. Taut, *Adv. Immunrol.,* **17:**2 (1973).
247. Lange, C. S., *J. Embryol. Exp. Morphol.,* **15:**125 (1966).
248. Laugham, R. F., and C. F. Clark, *Am. J. Vet. Res.,* **6:**81 (1945).
249. Law, L. W. and R. C. Ting, *Proc. Soc. Exp. Biol. Med.,* **119:**823 (1965).
250. Leav, I., A. C. Crocker, M. L. Petrak, and T. C. Jones, *Lab. Invest.,* **18:**433 (1968).
251. Leiter, J., B. J. Abbot, and S. A. Schepartz, *Cancer Res.,* **25:**1626 (1965).
252. Leiter, J., and M. A. Schneiderman, *Cancer Res.,* **19:**31 (1959).
253. Leuke, B., *J. Exp. Med.,* **74:**397 (1941).
254. Liavag, I., *Scand. J. Urol. Nephrol.,* **2:**65 (1968).
255. Lijinsky, W., and S. S. Epstein, *Nature* (London), **225:**21 (1970).
256. Lingeman, C. H., *Natl. Cancer Inst. Monog.,* **32:**177 (1969).
257. Lingeman, C. H., R. E. Reed, and F. M., *Natl. Cancer Inst. Monog.,* **32:**157 (1969).
258. Lipkin, M., E. Descher, and F. Troncale, *Gastroenterology,* **59:**303 (1970).
259. Lipschutz, A., *Steroid Hormones and Tumors,* Williams & Wilkins, Baltimore, 1950, p. 6.
260. Lipschutz, A., R. Iglesias, and S. Latinus, *Nature,* **198:**946 (1962).
261. Little, C. C., *Proc. Natl. Acad. Sci.,* **25:**452 (1939).
262. Litvinov, N. N., *Radiation Injury of the Bony System,* Medgiz, Moscow, 1964.
263. Lombard, L. S., and E. J. Witte, *Cancer Res.,* **19:**127 (1959).
264. Loutit, J. F., *Sci. Basic Med. Ann. Rev.,* 340 (1967).
265. Lucke, B., *Am. J. Cancer,* **20:**352 (1934).
266. Lynch, H. T., *Rec. Results Cancer Res.,* **12:**186 (1967).
267. Lyon, M. F., E. V. Hulse, and C. E. Rowe, *J. Med. Genet.,* **2:**99 (1965).
268. MacMohan, B., *Cancer Res.,* **26:**1189 (1966).
269. MacMohan, B., and P. Cole, *Cancer,* **24:**1146 (1969).
270. MacMohan, B., T. M. Lin, C. R. Lowe, H. P. Mirra, B. Ravinhar, E. V. Salber, D. Trichopoulos, V. G. Valaoras, and S. Yuasa, *Bull. WHO,* **42:**185 (1970).
271. Magee, D. N., *Food Cosmet Toxiol.,* **9:**207 (1971).
272. Maletskos, C. J., A. G. Braun, M. M. Shanahan, and R. D. Evans, *Proc. Symp. Assesment of Radioactive Body Burdens in Man,* Int. Atomic Energy Agency, Vienna, 1964.
273. Manning, D. D., N. D. Reed, and C. F. Shaffer, *J. Exp. Med.,* **138:**488 (1973).
274. Marin-Padilla, M. and K. Benirschke, *Am. J. Pathol.,* **43:**999 (1963).
275. Marinelli, L. D., *Am. J. Roentgen.,* **80:**729 (1958).
276. Marshak, R., L. L. Coriell, W. C. Lawrence, J. E. Croshaw, Jr., H. F. Schrgner, K. P. Altera, and W. W. Nichols, *Cancer Res.,* **22:**202 (1962).
277. Marshak, R. R., W. C. D. Hare, D. C. Dodd, R. A. McFeely, J. E. Martin, and R. M. Datcher, *Cancer Res.,* **27:**498 (1967).
278. Mays, C. W., W. S. S. Jee, R. D. Lloyd, B. J. Stoner, J. H. Doughtery, and G. N. Taylor, *Delayed Effects of Bone-Seeking Radionuclides,* University of Utah Press, Salt Lake City, 1969.
279. McAllister, R. M., *Prog. Med. Virol.,* **16:**48 (1973).
280. McAllister, R. M., R. V. Gilden, and M. Green, *Lancet,* **1:**831 (1972).
281. McAllister, R. M., W. A. Nelson-Rees, E. Y. Johnson, R. W. Rongey, and M. B. Gardner, *J. Natl. Cancer Inst.,* **47:**603 (1971).
282. McDivitt, R. W., F. W. Stewart, and J. W. Berg, *Atlas of Tumor Pathology,* 2nd Series, Fascicle 2, Armed Forces Institute of Pathology, Washington, D.C., 1967.
283. Media, D., *J. Natl. Cancer Inst.,* **46:**900 (1971).
284. Medway, W., W. T. Weber, and J. A. O'Brien, *J. Am. Vet. Med. Assoc.,* **150:**386 (1967).
285. Meissner, W. A., and S. Warren, in W. A. D. Anderson (ed.), *Pathology,* Vol. I, C. V. Mosby, St. Louis, 1971.

286. Melendez, L. V., M. D. Daniel, R. D. Hunt, and F. G. Garcia, *Lab. Anim. Care,* **18:**374 (1968).
287. Melendez, L. V., R. D. Hunt, M. D. Daniel, C. E. O. Fraser, H. H. Barahona, F. G. Garcia, and N. W. King, in P. M. Biggs, G. de The, and L. N. Payne (eds.), *Oncogenesis and Herpesviruses,* Int. Agency for Research on Cancer, Lyon, 1972.
288. Melief, C. J. M., and R. S. Schwartz, in F. F. Becker, (ed.), *Cancer I,* Plenum Press, New York 1975.
289. Metcalf, D., and R. Moulds, *Int. J. Cancer,* **2:**53 (1967).
290. Miller, J. A., in W. J. Burdette (ed.), *Primary Hepatoma,* University of Utah Press, Salt Lake City, 1965.
291. Miller, J. M., *Am. J. Pathol.,* **75:**417 (1974).
292. Miller, J. M., L. D. Miller, C. Olson, and K. G. Gillette, *J. Natl. Cancer Inst.,* **43:**1297 (1969).
293. Miller, R. W., *J. Natl. Cancer Inst.,* **46:**203 (1971).
294. Minzell, M., *Recent Results in Cancer Research* (Sp. Suppl.), Springer, New York, 1969.
295. Moertel, C. G., J. R. Hill, and M. A. Adson, *Arch. Surg.,* **100:**521 (1970).
296. Mole, R. H., in C. M. Mays, W. S. S. Jee, R. D. Lloyd, B. J. Stover, J. H. Dougherty, and G. N. Taylor (eds.), *Delayed Effects of Bone-Seeking Radionuclides,* University of Utah Press, Salt Lake City, 1969.
297. Moloney, J. B., *J. Natl. Cancer Inst.,* **24:**933 (1960).
298. Moloney, J. B., *Natl. Cancer Inst. Monog.,* **22:**139 (1966).
299. Monlux, A. W., W. A. Anderson, and C. L. Davis, *Am. J. Vet. Res.,* **17:**646 (1956).
300. Moore, D. H., C. Charney, B. Kramarsky, E. Y. Lasfargnes, N. H. Sarker, M. J. Bernan, J. H. Burrows, S. M. Sirsat, J. C. Paymaster, and A. D. Vidya, *Nature,* **229:**611 (1971).
301. Morris, H. P., *Prog. Exp. Tumor Res.,* **3:**370 (1963).
302. Morton, D. L., R. A. Malngren, W. T. Hall, and G. Schidlovsky, *Surgery,* **66:**152 (1969).
303. Mosinger, M., J. P. Glaunes, H. Fiorentini, and H. Bandler, *Ann. Med. Leg.,* **41:**472 (1961).
304. Mostafa, M. S. E., *Br. Vet. J.,* **109:**201 (1953).
305. Moulton, J. E., *Tumors in Domestic Animals,* University of California, Berkeley, 1961.
306. Mühlbock, O., *Eur. J. Cancer,* **1:**123 (1965).
307. Mühlbock, O., and T. Nomura, *Gann Monog.,* **5:**127 (1968).
308. Muller, M. and H. Grossman, *Nature (New Biol.),* **237:**116 (1972).
309. Mulligan, R. M., *Neoplasms of the Dog,* Williams & Wilkins, Baltimore, 1949.
310. Mulligan, R. M., *Ann. N.Y. Acad. Sci.,* **108:**642 (1963).
311. Murphy, E. D., and B. Roscoe, *Jackson Mem. Lab. Ann. Report,* **35:**37 (1965).
312. Mustacchi, P., and M. B. Shimkin, *J. Natl. Cancer Res.,* **20:**825 (1958).
313. Nadel, E., W. Banfield, S. Burstein, and A. J. Tousimis, *J. Natl. Cancer Inst.,* **38:**979 (1967).
314. Nahmias, A. J., W. R. Powdle, J. M. Naib, W. E. Josey, D. McLone, and G. Domescik, *Br. J. Vener. Dis.,* **45:**294 (1969).
315. Nahmias, A. J., W. E. Josey, Z. M. Naib, C. F. Luce, and A. Duffey, *Am. J. Epidemiol.,* **91:**539 (1970).
316. Nahmias, A. J., W. E. Josey, Z. M. Naib, C. F. Luce, and B. A. Gues, *Am. J. Epidemiol.,* **91:**547 (1970).
317. Nettesheim, P., M. G. Hanna, and J. W. Deatherage, Jr., *Morphology of Experimental Respiratory Carcinogenesis,* National Tec. Information Service, U.S. Dept. of Commerce, Springfield, Va., 1970.
318. Neves, E. F., *Br. Med. J.,* **2:**1255 (1923).
319. Newberne, P. M. and A. E. Rogers, *Am. J. Pathol.,* **72:**137 (1973).
320. Newell, G. R., W. W. Harris, K. O. Bowman, C. W. Boone, and N. G. Anderson, *N. Engl. J. Med.,* **278:**1185 (1968).
321. Niederman, J. C., A. S. Evans, L. Subrahmanyan, and R. W. McCollum, *N. Engl. J. Med.,* **282:**361 (1970).
322. Niederman, J. C., R. W. McCollum, and G. Henle, *J.A.M.A.,* **203:**205 (1968).
323. Nielsen, S. W., and C. R. Cole, *Am. J. Vet. Res.,* **19:**417 (1958).
324. Nielsen, S. W., and J. Holzworth, *J. Am. Vet. Med. Assoc.,* **122:**189 (1953).

325. Norrby, E., in K. Maramorosch and E. Kurstak (eds.), *Comparative Virology*, Academic Press, New York, 1971.
326. Nowinski, R. C., and N. H. Sarkar, *J. Natl. Cancer Inst.*, **48**:1169 (1972).
327. Oconer, G. T., Y. S. Tartarinov, G. I. Abelev and J. Uriel, *Cancer*, **25**:1091 (1970).
328. O'Gara, R. W., R. H. Adamson, M. G. Kelly, and D. W. Dalgard, *J. Natl. Cancer Inst.*, **46**:1121 (1971).
329. Oettle, A. G., *Br. J. Cancer*, **11**:415 (1957).
330. Old, L. J., E. A. Boyse, and E. Stockert, *Cancer Res.*, **25**:813 (1965).
331. Olson, L. D., *Am. J. Vet. Res.*, **28**:1501 (1967).
332. Opler, S. R., *J. Natl. Cancer Inst.*, **38**:797 (1967).
333. Orkin, M. and R. M. Schwartzman, *J. Invest. Derm.*, **32**:451 (1959).
334. Oroszlan, S., C. Foreman, G. Kelloff and R. V. Gilden, *Virology*, **43**:665 (1971).
335. Oroszlan, S., R. J. Huebner, and R. V. Gilden, *PNAS (US)* **68**:901 (1971).
336. Pack, G. T., *Ann. N.Y. Acad. Sci.*, **114**:985 (1964).
337. Padgett, B. L., D. L. Walker, G. M. ZuRhein, R. J. Eckroade, and B. H. Dessel, *Lancet*, **1**:1257 (1971).
338. Pantelouris, F. M., *Nature* (London) **217**:370 (1968).
339. Papageorgiou, P. S., C. Sorokin, K. Kouzoutzakoglou, and P. R. Glade, *Nature* (London) **231**:47 (1971).
340. Parsons, R. J., and J. G. Kidd, *J. Exp. Med.*, **77**:233 (1943).
341. Pasternak, G., *Adv. Cancer Res.*, **12**:1 (1969).
342. Pauley, G. B., *Nat. Cancer Inst. Monogr.*, **31**:509 (1969).
343. Paulson, D. F., E. E. Fraley, A. S. Robson, and A. J. Ketcham, *Science*, **159**:200 (1968).
344. Payne, L. N., in P. E. Emmelot, and P. Bentvelzen (eds.), *RNA Viruses and Host Genome in Oncogenesis*, North Holland, Amsterdam, 1972.
345. Payne, L. N., L. B. Crittenden, and W. Okazaki, *J. Natl. Cancer Int.*, **40**:907 (1968).
346. Payne, L. N., P. K. Pani, and R. A. Weiss, *J. Gen. Virol.*, **13**:455 (1971).
347. Pedini, B., and V. I. Romanelli, *Arch. Vet. Ital.*, **6**:193 (1955).
348. Perk, K., M. V. Viola, K. L. Smith, N. A. Wivel, and J. B. Moloney, *Cancer Res.*, **29**:1089 (1969).
349. Phillips, J. C., J. W. Linsay, and J. A. Kendall, *Am. J. Digest Dis.*, **15**:239 (1970).
350. Pierce, G. B., *J. Natl. Cancer Inst.*, **28**:247 (1962).
351. Pierce, G. B., *Cancer*, **19**:1963 (1968).
352. Pincus, T., J. W. Hartley, and W. P. Rowe,*J. Exp. Med.*, **133**:1219 (1971).
353. Pincus, T., W. P. Rowe, and F. Lilly, *J. Exp. Med.*, **133**:1234 (1971).
354. Pollara, B., W. A. Cain, J. Finstad, and R. A. Good, in M. Mizell (ed.), *Biology of Amphibian Tumors*, Springer-Verlag, New York, 1969.
355. Ponten, J., and B. R. Burmester, *J. Natl. Cancer Inst.*, **38**:505 (1967).
356. Porter, D. D., I. Wimberly, and M. Benyesh-Melnick, *J. Am. Med. Assoc.*, **208**:1675 (1969).
357. Pott, P., *Natl. Cancer Inst. Monog.*, **10**:7 (1962).
358. Prates, M. D., and J. Gillman, *S. Afr. J. Med. Sci.*, **24**:13 (1959).
359. Prehn, R. T., *Science*, **176**:170 (1972).
360. Prier, J. E., and R. S. Brodey, *Bull. WHO*, **29**:331 (1963).
361. Purchase, I. F. H., and J. J. Van der Watt, *Food Cosmete. Toxicol.*, **8**:289 (1970).
362. Ratcliffe, H. L., in E. J. Farris and J. O. Giriffith, Jr., (eds.), *Rat in Laboratory Investigation*, Lippincott, Philadelphia, 1949.
363. Rauscher, F. J., Jr.,*J. Natl. Cancer Inst.*, **29**:515 (1962).
364. Rauscher, F. J., Jr., *Proc. Natl. Cancer Conf.*, **6**:93 (1970).
365. Rauscher, F. J., and T. E. O'Conner, in J. F. Hollan and J. Frei, III, (eds.), *Cancer Medicine*, Lea & Febiger, Philadelphia, 1973.
366. Rawls, W. E., in A. S. Kaplan (ed.), *Herpes Viruses*, Academic Press, New York, 1973.
367. Rawls, W. E., H. L. Gardner, R. W. Flander, S. P. Lowry, R. H. Kaufman, and J. L. Melnik, *Am. J. Obstet. Gynecol.*, **110**:682 (1971).
368. Rawls, W. E., W. A. F. Tompkins, M. E. Figueroa, and J. L. Melnick, *Science*, **161**:1255 (1968).

369. Reed, N. D., and J. W. Jutila, *Proc. Soc. Exp. Biol. Med.,* **139:**1234 (1972).
370. Reports of the Penrose Research Lab. of the Zoological Society of Philadelphia, 1924–25; 1926–27; 1936–37; 1940–41.
371. Rhim, J. S., and B. Creasy, *Proc. Nat. Acad. Sci.,* **68:**2212 (1971).
372. Rich, M. A., and R. Siegler, *Ann. Rev. Microbiol.,* **21:**529 (1967).
373. Robinson, R., *Genetics of Norway Rat,* Pergamon Press, Oxford, 1965.
374. Robinson, F. R., R. J. Brown, H. Casey, and S. Rozof, *Registery of Veterinary Pathology. References on Naturally Occuring Neoplasms In Animals,* Armed Forces Institute of Pathology, Washington, D.C., 1974.
375. Rogers, J. B., *J. Geront.,* **6:**142 (1951).
376. Rosenberg, S. A., H. D. Diamond, B. Jaslowitz, and L. F. Craver, *Medicine,* **40:**31 (1961).
377. Rosenoer, V. M., in R. J. Schnitzer, and F. Hawkins (eds.), *Experimental Chemotherapy,* **4:**9 (1966).
378. Rous, P., *J. Exp. Med.,* **13:**397 (1911).
379. Rowe, W. P., *J. Exp. Med.,* **136:**1272 (1972).
380. Rowe, W. P., and J. W. Hartley, *J. Exp. Med.,* **136:**1286 (1972).
381. Rowe, W. P., J. W. Hartley, and T. Bremuar, *Science,* **178:**860 (1972).
382. Russel, E. S., *Genetics,* **27:**612 (1942).
383. Russel, W. O., E. S. Wynne, and G. S. Loquvam, *Cancer,* **9:**1 (1956).
384. Rust, J. H., *J. Am. Vet. Med. Assoc.,* **114:**10 (1949).
385. Rygaard, J., *Acta Pathol. Microbiol. Scand.,* **77:**761 (1969).
386. Salmela, H., *Acta Chir. Scan.,* **134:**384 (1968).
387. Sandison, A. T., *Nat. Cancer Inst. Monog.,* Public Health Service, Washington, D.C., 1962.
388. Sarkar, N. H., and D. H. Moore, *Nature* (London), **236:**103 (1972).
389. Sarngadharan, M. G., P. S. Sarin, P. S. Reitz and R. C. Gallo, *Nature (New Biol.),* **240:**67 (1972).
390. Schaffner, M. H., in R. W. Kirk (ed.), *Current Veterinary Therapy,* W. B. Saunders, Philadelphia, 1964.
391. Schlom, J., and S. Spiegelman, *Science,* **174:**840 (1971).
392. Schlom, J., S. Spiegelman, and D. H. Moore, *Science,* **175:**542 (1972).
393. Schlumberger, H. G., *Cancer Res.,* **12:**890 (1952).
394. Schneider, R., *Cancer,* **26:**419 (1970).
395. Schochet, S. S., and P. W. Lampert, in M. Mizell (ed.), *Biology of Amphibian Tumors,* Springer-Verlag, New York, 1969.
396. Schwartzman, R. M., and M. A. Orkin, *Comparative Study of Skin Diseases of Dog and Man,* Charles C Thomas, Springfield, Ill., 1962.
397. Segaloff, A., and W. S. Mazfield, *Cancer Res.,* **31:**166 (1971).
398. Shachat, D. A., A. Fefer, and J. B. Moloney, *Cancer Res.,* **28:**517 (1968).
399. Shank, R. C., N. Bhamazapravati, J. E. Gordon, and G. N. Wogan, *Food Cosmete Toxicol.,* **10:**171 (1972).
400. Sherlock, P., S. G. Winawer, M. J. Goldstein, and D. G. Bragg, in J. B. Class (ed.), *Progress in Gastroenterology,* Vol. 2, Grune & Stratton, New York, 1970.
401. Shimkin, M. B., and V. A. Triolo, *Prog. Exp. Tumor Research,* **11:**1 (1969).
402. Shubik, P., A. R. Goldfarb, A. C. Ritchie, and H. Lisco, *Nature,* **171:**934 (1953).
403. Siegl, A. M., H. W. Casey, R. W. Bowman, et al., *Blood,* **32:**989 (1968).
404. Silverberg, E., *Cancer of the Colon and Rectum,* American Cancer Society, New York, 1970.
405. Sjogren, H. O., and K. Borum, *Cancer Res.,* **31:**890 (1971).
406. Slye, M., *J. Cancer Res.,* **7:**107 (1922).
407. Smith, H. A., *Am. J. Clin. Pathol.,* **38:**75 (1962).
408. Smith, H. A., and T. C. Jones, *Veterinary Pathology,* Lea & Febiger, Philadelphia, 1961.
409. Snell, K. C., in W. E. Ribelin and J. R. McCoy (eds.), *Pathology of Laboratory Animals,* Charles C Thomas, Springfield, Ill., 1965.
410. Snell, K. C., and H. L. Stewart, *J. Natl. Cancer Inst.,* **39:**95 (1967).
411. Snell, K. C., and H. L. Stewart, *Gann Mong.,* **8:**39 (1969).
412. Snyder, R. L., and H. L. Ratcliffe, *Ann. N.Y. Acad. Sci.,* **108:**793 (1963).

413. Snyder, S. P., and G. H. Theilen, *Nature* (London), **221:**1074 (1969).
414. Solomon, G., R. Jensen, and H. Tanner, *Am. J. Vet. Res.* **26:**764 (1965).
415. Spratt, J. S., Jr., L. V. Ackerman, and C. A. Moyer, *Ann. Surg.,* **148:**682 (1958).
416. Squire, R. A., *Am. J. Vet. Res.,* **26:**97 (1965).
417. Staats, J., *Cancer Res.,* **32:**1609 (1972).
418. Stanton, M. F., *J. Natl. Cancer Inst.,* **34:**117 (1965).
419. Stenback, W. A., G. L. J. Van Hoosier, and J. J. Trentin, *J. Virol.,* **2:**1115 (1968).
420. Stevens, L. C., *J. Natl. Cancer Inst.,* **32:**1249 (1959).
421. Stewart, F. W., *Bull. N.Y. Acad. Med.,* **22:**145 (1947).
422. Stewart, H. L. and K. C. Snell, *J. Natl. Cancer Inst.,* **40:**1135 (1968).
423. Stewart, H. L., K. C. Snell, L. J. Dunham, and S. M. Schlyen, *Transplantable and Transmissible Tumors of Animal,* Am. Reg. of Path. AFIP Fascile, Washington, D.C., 1959, p. 378.
424. Stewart, S. E., B. E. Eddy, A. M. Gochenour, N. G. Borgese, and G. E. Grubbs, *Virology,* **3:**380 (1957).
425. Stewart, S. E., G. Kasnic, Jr., C. Draycott, W. Feller, A. Golden, E. Mitchell, and T. Ben, *J. Natl. Cancer Inst.,* **48:**273 (1972).
426. Stewart, S. E., G. Kasnic, Jr., C. Draycott, and T. Ben, *Science,* **175:**198 (1972).
427. Stock, N. D., and J. F. Ferrer, *J. Natl. Cancer Inst.,* **48:**985 (1972).
428. Stoll, H. L., and J. T. Crissey, *N.Y. J. Med.,* **62:**496 (1962).
429. Strandberg, J. D., and D. G. Goodman, *Am. J. Pathol.,* **75:**225 (1974).
430. Straus, R., S. Heschel, and D. J. Fortman, *Cancer,* **24:**985 (1969).
431. Stutman, O., and J. M. Dupny, *J. Natl. Cancer Inst.,* **49:**1283 (1972).
432. Sullivan, D. J. and W. A. Anderson, *Am. J. Vet. Res.,* **20:**324 (1959).
433. Sutnik, A. I., W. T. London, and B. S. Blumberg, *Ann. Int. Med.,* **74:**442 (1971).
434. Sweet, B. H., and M. R. Hilleman, *Proc. Soc. Exp. Biol. Med.,* **105:**420 (1960).
435. Sykes, J. A., L. Dmochowski, E. S. Wynne, and W. O. Russell, *J. Natl. Cancer Inst.,* **26:**445 (1961).
436. Sykes, J. A., M. Scanlon, W. D. Russel, et al., *Tex. Rep. Biol. Med.,* **22:**741 (1964).
437. Syverton, J. T., *Ann. N.Y. Acad. Sci.,* **54:**1126 (1952).
438. Taki, I., and H. Iijima, *Ann. J. Obstet. Gynecol.,* **87:**926 (1963).
439. Tannenbaum, M. and J. K. Lattimer, *J. Urology,* **103:**471 (1970).
440. Tarnowski, G. S., F. A. Schmid, J. G. Cappuccino, and C. C. Stock, *Cancer Chemother.,* **43:**181 (1966).
441. Tartarinov, Y. S., *Vop. Med. Khim.,* **11:**20 (1965).
442. Tatchell, J. A. H., *Nature* (London), **190:**837 (1961).
443. Temin, H. M., *Ann. Rev. Microbiol.,* **25:**609 (1971).
444. Theilen, G. H., D. Gould, M. Fowler, and D. L. Dungworth, *J. Natl. Cancer Inst.,* **47:**881 (1971).
445. Theilen, G. H., R. E. Zeigel, and M. Twiehans, *J. Natl. Cancer Inst.,* **37:**731 (1966).
446. Thompson, J. I. & Co., PHS Publ. No. 149, Washington, D.C., 1972.
447. Todoro, G., H. Green, and M. R. Swift, *Science,* **153:**1252 (1966).
448. Todaro, G. J., Zeve, V. and S. A. Aaronson, *In Vitro,* **6:**355 (1971).
449. Tokuhata, G. K. and A. M. Lilienfeld, *J. Natl. Cancer Inst.,* **30:**289 (1963).
450. Toth, B., *Cancer Res.,* **30:**2583 (1970).
451. Trentin, J. J., Y. Yabe, and G. Taylor, *Science,* **137:**835 (1962).
452. Treves, N., and G. T. Pack, *Surg. Gynecol. Obstet.,* **51:**749 (1930).
453. Turell, R., and J. D. Haller, in R. Turell (ed.), *Diseases of the Colon and Anorectum,* W. B. Saunders, Philadelphia, 1969.
454. Upton, A. C., *Methods Cancer Res.,* **4:**53 (1968).
455. Urban, J. A., *Surg. Clin. N. Am.,* **49:**291 (1969).
456. Vandeputte, M., *Transpl. Proc.,* **1:**100 (1969).
457. Vigier, P., *Prog. Med. Virol.,* **12:**240 (1970).
458. Vogt, P. K., *Adv. Virus Res.,* **11:**293 (1965).
459. Voutilainen, A., H. Teir, and A. Kivivouri, *Ann. Chir. Gynaec. Fenn. Supp.,* **152:**1 (1967).
460. Waldman, T. A., W. Strober, and R. M. Blaese, *Ann. Int. Med.,* **77:**605 (1972).
461. Walshe, F., *Lancet,* **2:**993 (1961).

462. Warren, K. S., *Trans. R. Soc. Trop. Med. Hyg.,* **61:**795 (1967).
463. Weiner, L. P., R. M. Herndon, O. Narayan, R. T. Johnson, K. Shah, L. J. Rubinstein, T. J. Prejiosi, and F. K. Conley, *N. Engl. J. Med.,* **286:**385 (1972).
464. Weiss, D. W., *Cancer Res.,* **29:**2368 (1969).
465. Weiss, R. A., R. R. Friis, E. Katz, and P. K. Vogt, *Virology,* **46:**920 (1971).
466. Wellings, S. R., *Natl. Cancer Inst. Monog.,* **31:**59 (1969).
467. Williams, G. M., J. M. Elliott and J. H. Weisburger, *Cancer Res.,* **33:**606 (1973).
468. Winship, T., and R. V. Rosvoll, *Cancer,* **14:**734 (1961).
469. Wogan, G. N., in F. F. Becker (ed.), *Cancer I,* Plenum Press, New York 1975.
470. Wojewski, A., *J. Urol.,* **89:**709 (1963).
471. Wolfe, L. G., F. Deinhardt, G. H. Theilen, H. Rabin, T. Kawakami, and L. K. Bustad, *J. Natl. Cancer Res.,* **47:**1115 (1971).
472. Wolff, I. A., and A. E. Wasserman, *Science,* **177:**15 (1972).
473. Wolley, G. W., M. M. Dickie, and C. C. Little, *Cancer Res.,* **13:**231 (1953).
474. Woodard, H. Q., *Cancer Res.,* **2:**497 (1942).
475. Wortis, H. H., *Clin. Exo. Immunol.,* **8:**305 (1971).
476. Wyand, D. S. and R. E. Wolke, *Am. J. Vet. Res.,* **29:**1309 (1968).
477. Wychullis, A. R., O. H. Beahrs and L. B. Woolner, *Arch. Surg.,* **93:**215 (1966).
478. Wynder, E. L., *Cancer,* **24:**1235 (1959).
479. Wynder, E. L. and T. Shigematsu, *Cancer,* **20:**1520 (1967).
480. Zalusky, R., J. J. Ghidoni, and J. McKinley, *Rad. Res.,* **25:**410 (1965).
481. Zeigel, R. F. and H. F. Clark, *J. Natl. Cancer Inst.,* **43:**1097 (1969).
482. Zubrod, C. G., in J. F. Holland and E. Frei, III (eds.), *Cancer Medicine,* Lea & Febiger, Philadelphia, 1973.

CHAPTER 12

ANIMAL MODELS IN GERONTOLOGY

Gerontology is the scientific study of problems of the aging processes. Aging may be defined as the sum total of progressive changes in clinical, biological, historical, and sociological processes with the passage of time. The aging process in an individual is influenced by a number of factors, such as heredity, sex, medical history, nutrition, and environmental conditions. Despite the universality of the aging process, not enough attention has been focused on the determination of the causes and mechanism of the phenomenon or on the elimination of the causes in order to extend the life span.

Gerontological research involves long-term studies that can be ideally carried out in experimental animals. Several animal species have been used for gerontological studies to understand morphological, physiological, pathological, and biochemical changes occurring in them with the passage of time. This chapter describes physiopathological and biochemical aspects of aging reported by investigators from their studies with a variety of experimental animals. Emphasis is placed throughout this chapter on the comparative changes in aging humans and animals to illustrate the usefulness of experimental animals as suitable models for human gerontological studies.

CHOICE OF EXPERIMENTAL ANIMAL FOR GERONTOLOGICAL RESEARCH

Among the more frequently used animal species for the study of aging processes are mice, rats, rabbits, guinea pigs, miniature pigs, dogs, and chickens. Other animals, such as pigeons, cats, hamsters, chipmunks, fishes, gerbils, goats, opossums, squirrels, turtles, and snakes, have also been used for experimental purposes in this field. Sulkin (257) reported from a survey that the annual need for experimental animals in fundamental research was approximately 90,000 mice and 37,000 rats. He also reported that the research was hampered because of nonavailability of old animals. However, old animals are now available in increasing numbers, which should accelerate research activities in the study of the aging process. The following criteria should be observed in the test species for long-term studies:

1. A well defined life span, with minimum of animal to animal variation, should be known (Table 12-1).

2. The experimental animal should be resistant to infectious diseases, particularly to those causing high mortality or morbidity.

3. There should be anatomical and physiological similarities between the animal species and humans with particular reference to similarity in diseases.

4. Dietary and nutritional requirements should resemble those of humans.

5. The information on karyotype, stem cells, and immune system should be well known.

Table 12-1.   Life Span of Commonly Used Laboratory Animals

Animal	Lifespan (years)	
	Max.	Average
Horse	35	20–25 (depending on the size)
Cow	30	20–25
Dog	34	10–12
Chicken		20
Sheep	20	10–15
Goat	19	12–15
Pig	27	16–18
Monkey (Rhesus)	29	15
Guinea pig	7	2
Rat	4	3
Rabbit	8	2

Adapted from Lifespan and Factors Affecting It (114) with permission.

6. The animals should be easy to handle and house and economical to maintain.

7. The data obtained should be extrapolated to humans.

On the basis of these criteria the following animal species are most frequently used for the study of the human aging process.

Rats and Mice

Rats and mice are the most commonly used animal species as models for gerontological research because they have a short life span and a low maintenance cost. A detailed study of SPF mice and rats showed that the life span varied considerably (312 to 802 days) between different strains and sexes (Table 12-2) (73). The longevity in SPF strains of mice appears to be longer, compared to that of conventional mice. However, there was an increased incidence of neoplasms in the SPF strains of animals (Table 12-3). In another study reported by Pollard and Kajinie (196), a decrease in the incidence of tumors was found in germfree Wistar rats surveyed during 1962–1969 period. The tumors were not malignant and no virus was isolated from the tumors. From these studies, it appears that the germfree rat may be an excellent model for the study of the aging process.

Miniature Swine

Life span data of miniature swine are not clearly established because the animal model has only been developed recently. However, from the limited data available, it appears that the average life span is about 10 years as compared to 16 years for the conventional swine. Similar diseases occur in the miniature pig and the standard swine, except that the incidence of arthritis is lower in the miniature pig (35, 65). The pig has many anatomical, physiological, and biochemical similarities to humans (Chapter 1). The weight of various organs in humans and swine are compared in Table 12-4. Cardiovascular systems of swine particularly have been used for atherosclerosis studies, because the animals produce spontaneous disease and because disease can be induced by dietary manipulations (85, 246). Atherosclerosis and heart disease (165, 166) are related to the aging process.

Domestic Pig as Model of Aging

Histomorphological uterine changes in postmenopausal women, primarily involving thickening of arterial walls of the endometrium, have been reported by several

Table 12-2. Mean Life Span (days) in Various Strains of Mice

Strain	Females		Males	
	Number	Longevity, mean (S.E.)	Number	Longevity, mean (S.E.)
A	68	558 (19.7)	65	512 (21.1)
AKR	79	312 (9.7)	79	350 (10.8)
A2G	51	644 (19.4)	49	640 (21.8)
BALB/c	33	561 (30.3)	35	509 (26.3)
CBA	38	825 (32.5)	37	486 (39.0)
CE	23	703 (37.3)	20	498 (48.5)
C3H	193	676 (9.8)	147	590 (18.6)
C57BL	29	580 (35.8)	31	645 (34.2)
C57BR/cd	46	660 (22.7)	45	577 (29.8)
C57L	26	604 (27.6)	22	473 (30.9)
DBA/1	39	686 (33.3)	35	487 (35.9)
DBA/2	23	719 (35.4)	22	629 (42.1)
NZB	111	441 (12.1)	110	459 (13.3)
NZW	20	733 (42.8)	28	802 (34.0)
129/Rrj	36	666 (23.2)	35	699 (29.8)
LACA	38	664 (29.9)	41	660 (38.5)
LACG	40	617 (26.2)	36	536 (38.9)
WA	22	749 (40.1)	25	645 (29.9)
P	10	782 (51.9)	14	729 (42.9)

Adapted from *Laboratory Animals* (73) with permission.

Table 12-3.   Percentage Incidence of Pathological Conditions in 17 Strains of SPF Mice

				Number of mice					
Pathological condition	Combined 574	A 30	A2G 18	AKR 67	BALB/c 21	CBA 21	C3H 131	CE 17	C57BL 13
Liver tumour	4-8	0-12	0-26	0-5	0-17	1-26	9-23	11-57	0-26
Lung tumour	5-9	4-31	17-65	0-5	2-32	0-17	2-10	0-20	0-26
Mammary tumour	6-11	0-12	0-17	0-5	0-17	3-37	21-36	0-30	0-26
All neoplasia (excluding lymphatic leukaemia)	23-31	6-35	26-74	0-5	2-32	14-57	48-76	17-77	0-26
Lymphatic leukaemia	9-15	10-43	0-17	66-87	0-17	0-17	0-4	0-20	1-36
Cardiac calcinosis	3-7	0-12	0-17	0-5	2-32	1-26	8-2	0-20	0-26
All heart defects (including calcinosis)	4-8	0-12	0-26	0-5	17-62	3-37	13-26	0-20	0-26
Cystic ovaries/uterus*	3-7	0-12	0-17	0-5	0-17	3-37	13-26	0-20	0-26
Haemothorax	2-4	0-12	0-17	0-5	0-17	3-37	0-7	0-30	0-26
Bacterial infections	3-7	15-50	0-17	0-5	0-17	0-17	0-3	0-20	0-26
Miscellaneous	9-15	4-31	7-46	0-12	6-43	3-37	6-16	2-37	2-44
No significant lesions	26-34	6-35	0-26	8-26	14-57	19-63	23-37	1-57	46-95

Table 12-3. (Continued)

Pathological condition	Number of mice								
	C57BR 13	C57L 14	DBA/1 23	DBA/2 30	NZW 27	NZB 84	129 16	LACA 38	LACG 12
Liver tumour	1-36	0-23	0-15	6-35	0-13	1-8	0-21	0-9	0-26
Lung tumour	1-36	0-23	2-27	1-23	2-24	0-5	4-46	12-39	0-26
Mammary tumour	0-26	0-23	0-15	0-11	0-13	0-5	0-21	0-9	0-26
All neoplasia (excluding lymphatic leukaemia)	2-44	2-42	5-37	20-56	3-29	1-8	11-57	29-61	6-57
Lymphatic leukaemia	0-26	0-23	0-23	0-17	3-29	1-8	0-21	0-9	1-37
Cardiac calcinosis	0-26	0-23	0-15	0-11	0-13	1-10	0-21	0-9	0-38
All heart defects (including calcinosis)	0-26	0-34	0-15	0-11	2-24	3-15	0-21	2-17	9-65
Cystic ovaries/uterus*	0-26	0-34	0-15	0-17	0-13	0-5	1-32	0-9	0-26
Haemothorax	0-26	0-23	3-33	4-31	1-19	0-5	0-21	0-9	0-38
Bacterial infections	1-36	0-23	2-27	0-11	1-19	2-13	0-21	0-11	9-65
Miscellaneous	14-67	2-42	7-41	10-42	4-33	22-43	19-69	6-30	0-26
No Significant lesions	14-67	28-82	16-57	10-42	23-62	33-58	7-50	13-42	0-38

95% confidence interval for the percentage of affected animals--i.e. there is a 95% probability that the true percent incidence of the condition in the colony lies between the 2 percentage points given in the table.

*Percent of all animals, not females only (roughly equal numbers males and females).

Adapted from Laboratory Animals (73) with permission.

Table 12-4. Comparison of Organ Weights of Humans and Swine

Organs	Man (70 kg)	Swine (55 kg)
Muscle and fat	57.0	56.0
Skin and subcutaneous tissue	8.7	16.5
Skeleton	10.0	6.8
Liver	2.4	2.0
Brain	2.1	0.16
Lungs	1.4	0.5
Kidneys	0.43	0.43
Heart	0.43	0.43
Spleen	0.21	0.15
Pancreas	0.10	0.12
Testes	0.057	0.65
Eye	0.043	0.027
Thyroid	0.029	0.018
Adrenals	0.029	0.0006
Others	9.4	8.3

Reproduced from Laboratory Animals in Gerontological Research (36) with permission.

authors (151, 192, 288). The uteri of rats (234) golden hamsters (219), mice (160), dogs (254), cows (179, 291), and swine, ranging from 1 to 10 years (10, 11), have been studied. The endometrial thickening of the pig uterus starts at 1½ years of age and progressively increases with age. At about 6 years of age, the endometrial gland becomes cystic, a condition similar to that in women. It appears that the pig uterus duplicates the aging conditions of the human female and therefore should serve as an excellent model.

Dogs

On the basis of life span, dogs may not be considered ideal models; however, they have been used for several studies involving long-range experiments. Natural aging changes can be observed with the pets in veterinary clinics.

Coturnix

The Japanese quail has become an important experimental model during the last two decades. In many studies it has replaced domestic fowl because of its smaller space requirement, faster maturity (35 to 45 days), and larger egg volume per unit body weight. For gerontological research, they offer other advantages such as a short life span and higher metabolic rate (122). The age-related changes in the reproductive system are also seen in quail.

Insects

Insects possess a number of properties that make them unique for the study of the aging process: (a) diversity of habits; (b) large numbers due to fecundity; (c) availability of highly inbred, homogenous, and genetically pure strains; (d) small size; (e) low cost of breeding and maintenance; (f) short life span—maximum life span for most flies is about 2 months (42, 212, 213).

The most obvious external manifestation of aging is the senile skin and the gradually increasing failure of motor ability. Studies with houseflies indicate that males have a shorter mean life span than females (17.4 and 54 days). The male flies start losing their wings around 1 week and practically all males lose their wings by 3 weeks, whereas the females show no loss of wing in this time (216, 217). Concomitant with maturation, there is an increase in enzymes in the brain tissues and acetylcholine esterase. Intramitochondrial magnesium activated ATPase activity (42) continues to rise until adult life (4 to 5 days) and then declines sharply with old age. The number of flight mitochondria in males reaches a maximum peak by the eighth day, levels off in 11 to 12 days, and then begins to decline (215). Alpha glycerophosphate dehydrogenase, (the counterpart of LDH in vertebrates), increases rapidly during early life, reaching a peak at 5 to 6 days and then declines with age. However, if the wings are removed at day 1, the α-glycerophosphate dehydrogenase activity remains at the plateau without declining (42, 216). Trehalose concentration, which is related to flight, decreases with aging (47) and thiamine concentration in the thoraces of houseflies also changes with age (214).

MORPHOLOGICAL AND PHYSIOLOGICAL CHANGES DURING THE AGING PROCESS

Morphological and physiological changes during aging appear to occur in all eukaryotic cells. The general cellular changes in mammals follow a set pattern of processes, for example, puberty, cessation of skeleton enlongation, menopause, increased incidence of tumors, and so forth (76, 77). The study dealing with the aging process in cells, manifested by structural and functional changes, is termed cytogerontology (110). It has been known that normal cell cultures in vitro have limited life spans (112). Human cells, derived from fetuses, have a life span of 10 months which represents about 50 times doubling of the cell population. The doubling of embryonic cells varies among different animal species but is somewhat related to the life span of the animal species (111) (Table 12-5). Extensive morphological changes of in vivo cells occur during the aging process (Table 12-6), resulting in significant changes in the weights of organs (Table 12-7).

Functional changes occur concomitantly with the structural and chemical changes in cells during the aging process (Table 12-8). In general, functional capacities of human organs decline progressively during the aging process (Fig. 12-1).

Biochemical Changes

One of the concepts of aging is based on alterations in tissue metabolism which eventually causes the death of the organism (56). It is not yet clear whether the somatic mutation (56) or stem cell mutation (32) is responsible for these alterations.

Connective Tissue

Connective tissue serves as an extracellular support and binds other tissues of the body. It is found in tendons, ligaments, skin, blood vessels, bone, teeth, cartilage, lining of the gastrointestinal tract, and lungs. It surrounds or is located in close proximity to

Table 12-5. In Vitro Multiplication and Life Span of Normal Fibroblasts from Humans and Animal Species

Species	Range of population doublings	Maximum lifespan
Galapagos tortoise	90–125	175
Man	40–65	110
Chicken	15–35	30
Mouse	14–28	3.5
Mink	30–34	10

Adapted from Federation Proceedings (111) with permission.

Table 12-6. Cellular Alterations in Aging

Cell Structure	Alteration
Nucleus	Clumping, shrinkage, fragmentation, dissolution of chromatin Increased staining Intranuclear inclusions Nuclear enlargement Amitotic division Invagination of nuclear membrane
Nucleolus	Increased size and number
Cytoplasm	Accumulation of pigments, fat, glycogen, hyaline droplets and formation of vacuoles
Mitochondria	Decrease in number and alteration in shape
Golgi apparatus	Fragmentation
Endoplasmic Reticulum	Loss of Nissel substance

Reproduced from Aging Life Processes (8) with permission.

Table 12-7.   Comparison of Average Relative Weights of Human Organs

Organ	% change at age 70 (age 21 to 30 = 100)
Heart	+11.1
Lungs	+10.8
Brain	+4.5
Skeletal muscle	−403
Spleen	−28
Liver	−20.1
Kidneys	−8.9

Adapted from <u>Physiological and Pathological Aging</u>. (139) with permission.

all cells of the body. The connective tissue chemically contains collagen, elastin, and mucopolysaccharides. Profound chemical changes take place in these tissues during the aging process.

Collagen. Collagen constitutes about 30 percent of all proteins of the body. The complex tropocollagen molecule is 260 nm long and 1.4 nm wide. Three separate chains, approximately 100,000 daltons each, are twisted together to form a super helix (202). Using rats as models, it has been reported that newly synthesized collagen is held together by hydrogen bonds, but as the tissue matures, the chains become cross-linked to produce dimers or trimers (194, 273). As the collagen matures, the proportion of dimers and higher polymers increases. In mouse skin, a decrease in α component does not increase in β component (113). However, these types of changes are not seen in humans (9). Characteristic changes are found in collagen of chickens with age (Fig. 12-2) (130). The load extension curves derived from rabbit tendons show differences with age (274). In humans, the stiffness factor shows a definite shift at age 45 (208). Several studies in humans and animals have been performed to determine the effect of collagenase. Post-mortem tissues, as well as in vivo skin, bone, and disc collagen, have been studied (138). Bacterial collagenase has been used in most of the earlier studies. However, with the discovery of mammalian collagenase (66, 154) the role of the enzyme in the production of structural changes in collagen has become much more important. Studies in animals show that the degree of attack by collagenase decreases with increasing age (103). In other conditions, which may or may not be age oriented, excessive collagenic activity occurs, for example, involution of the postpartum uterus (387), remodeling of wound scar

Table 12-8. Functional Changes with Age

Organ System	Function	% Function at Age 80 (Age 30 = 100)	Reference
Multi-component systems			
	Basal metabolic rate	85	242
	Standard cell water	80	242
	Blood pressure		152
	Systolic	115	
	Diastolic	104	
Cardiovascular			
	Cardiac output	65	242
	Total peripheral resistance	150	150
	Velocity of pulse wave (aorta)	200	149
Respiratory			
	Vital capacity	60	242
	Maximum breathing rate	40	242
Urinary			
	Renal plasma flow	45	242
	Glomerular filtration rate	60	242
Nervous			
	Nervous conduction velocity	90	242
Muscle			
	Hand grip strength	65	186
	Creatinine excretion, mg/24 hrs	65	185

Adapted from Aging Life Processes (8) with permission.

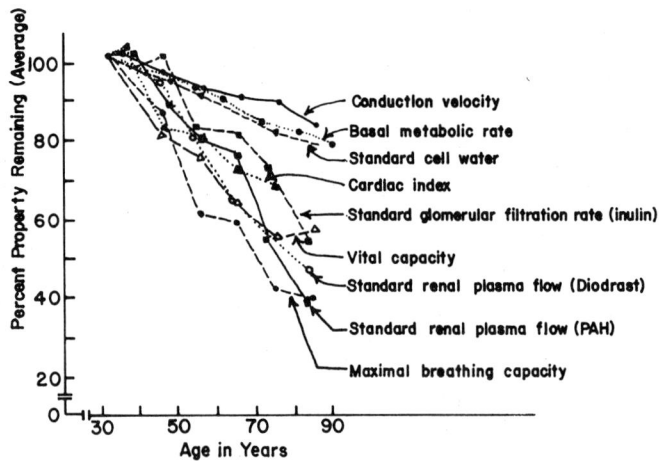

Fig. 12-1. Decline in various human functional capacities. Reproduced from *Biology of Aging* (213) with permission.

434

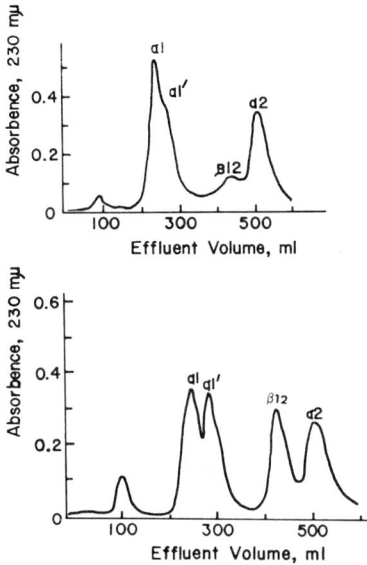

Fig. 12-2. Chromatographic patterns of primary structure of collagen in denatured chick skin. The upper portion of the figure is a pattern of younger (salt-extracted) collagen, showing almost identical α_1 chains and differences in α_2 chains. The lower curve is a pattern of older (acid extractable) collagen, showing another species of α_1 chain ($\alpha_{1'}$) and a considerably increased amount of the $\beta12$ component. Reproduced from *Biochemistry* (130) with permission.

tissue (120), and the like. However, changes in bone matrix are enzyme and age related (108).

Elastin. Elastin, a component of the arterial wall, has been thoroughly studied in connection with its role in atherosclerotic lesions. The most commonly used models are guinea pig and ox (101). The studies indicate that old elastin is frayed, fragmented, worn brittle, and yellow in color. The true elastic content decreases with age and a new protein, pseudoelastin, which has amino acid composition between collagen and elastin, appears. These changes found in oxen (101) have also been observed in humans (148). However, hydroxyproline generally is not excreted in urine. Hall (103) reported that the annual excretion of hydroxyproline was not over 300 mg and on a daily basis it was 30–40 μg percent. Plasma elastase is considered as an important factor in producing changes in collagen (102).

Experimental Lathyrism: A Model of Collagen Aging

Lathyrism is a disease of connective tissue which affects farm animals. It can be produced in laboratory animals by feeding lathrogens. The biochemical characteristics in rats, mice, and chicks include lack in degree of maturation of newly synthesized collagen and elastin (156, 170, 187). The formation of sufficient numbers of cross-linkages is inhibited, which eventually affects the mechanical strength of the molecule (21, 90, 190). It has been shown in rabbits that a defect in the synthesis collagen and elastic synthesis takes place which is related to structural changes in aorta, articular cartilage and Achilles tendon (22). In the adult animal, no alterations were observed. However, in young animals, accumulation of glycosaminoglycan took place. Age related changes have also been noticed in humans (22).

Effect of Food Intake on Aging of Collagen

The prolongation of life span by reducing food intake in rats was shown by McCay et al. (172). Later it was shown that restriction of diet intake retards the aging of collagen in the rat tail tendons (41, 87). At feeding level of 70 percent *ad libitum* or below, the aging of collagen is independent of food intake (69); however, increased amount of food intake or thyroxin injection accelerated collagen aging (70). Restricted diet, in addition to affecting aging of collagen, may affect the synthesis of collagen as shown in guinea pig skins (99) and malnourished rats (285). The chemical composition of the tail tendon of rats at different ages has been reported by Robert et al. (209) (Table 12-9).

Changes in Enzyme Patterns

Age-related enzyme changes in laboratory animals and humans have been extensively studied. Many workers have reported transcription and translation studies of enzymes. However, it has been pointed out that these mechanisms are under complex biosynthetic and degradation controls and the measured levels may not be a true representation of the effect of age (235). The enzymes from liver, kidney, heart, brain, erythrocytes, and prostate have been studied using mice and rats as animal models (Table 12-10 a–d). Comparison of the data on liver and kidney indicate that 80 to 88 percent of enzymes are not altered, 8 to 13 percent of enzymes show increase in activities (25 percent or more) and 3 to 7 percent are decreased. It would appear from these animal experiments that the gene function is generally not altered inasfar as the enzyme production is concerned during the aging process (76, 77). In old rats and mice, liver regeneration capacity is not lost (24, 197). Similar results were observed with RNA-DNA hybridization experiments in mice (40). As early as 1939, it was shown that certain types of gene activity may be increased, for example, gonadotropins in rats (153). Similar results have been obtained in women (128). Hairy pinna is another example in human males and in aging mice (53). In mice, many specific changes occur in intestinal epithelial cells in the duodenum (155) but not in the ileum (83). The induction in the hepatic tyrosine aminotransferase (TAT) is generally delayed in aged rats under cold stress. The effect is due to neuronal and hormonal changes rather than changes in liver (76, 100). Similar results have been obtained with other enzymes (1). The role of enzymes in aging has been recently reviewed by Wilson (286). It has been conclusively proved that age-related changes in endocrine function occur in experimental animals as well as in humans. Thyroid-dependent function declines in rats and other mammals, (29, 59), for example, reduced urinary output of cortisal in old men (221) and gonad function changes (71, 195). It appears that aging involves selective but not random changes in the activities of genes. In the study of enzyme differences, it should be emphasized that the usual animal factors are important. Even environmental factors like air pollution can effect the results (77). Other factors such as diurnal variations and disease condition both of normal and senescent animals should also be kept in mind in these studies.

Catecholamine Metabolism in the Brains of Aging Male Mice

In C57B1/6J male mice several catecholamine related changes have been reported, including reduced level of striatal dopamine, reduced conversion of tyrosine and L-DOPA to catecholamine in various regions of the brain, and slowed catabolism of total norepinephrine in the hypothalamus and total dopamine in the striatum (79).

The gross weight and cellular changes are generally not observed in old (28 to 30 months) mice and rats (79, 116). The loss of neurons in well defined nerve tracts such as the spinal cord is less than 15 percent in mice (289), rats (232), and cats (182). The experimental animal data are comparable to those in humans (51). However, a recent study (26) reported large-scale neuronal loss during the aging process.

Table 12-9. Chemical Composition of the Rat Tail Tendons at Various Ages

Age	Collagen content			Hydro content	Hexose	Hexosamine content in NaCl extract	Hydroxylysine	Arginine
	CTC	TCA	Urea	(mg/g)	(mg/g)	(mg/g)	(mg/g)	(mg/g)
	(mg protein/g tissue)							
1 month	607.5	328.1	48.7	30.3	0.96	0.82	0.30	1.63
3.5 months	125.6	660.9	31.2	3.36	0.53	0.57	0.32	1.60
10 months	115.5	567.0	22.0	3.89	0.57	0.54	0.31	1.63
17 months	60.1	533.5	23.1	3.38	0.55	0.62	0.29	1.64
24 months	40.9	613.1	30.9	10.09	0.58	0.62	0.32	1.64

Adapted from <u>Gerontologia</u> (209) with permission.

Table 12-10a. Age-Related Profiles of Enzyme Activities in the Liver

Enzyme Species	Strain	Sex	Compared in months	% change	Change given per unit	Reference
Acid DNAse (E.C.3.1.4.5.)						
mouse	CF-1	M	12 vs. 20	0	DNA	146
Acid phosphatase (E.C.3.1.3.2)						
mouse	N.M.R.I.		8 vs. 20	+10	weight	67
mouse	C57B1/6J	M & F	12 vs. 24	-20	protein	298
rat	Wistar	F	12 vs. 24	0	protein	81
Adenyl cyclase (E.C.3.6.1.3)						
rat	Sprague-Dawley	M	2 vs. 24	-20	weight	1
Alkaline DNA-se (E.C.3.1.4.5)						
mouse	CF-1	M	12 vs. 20	-20	DNA	146
Alkaline phosphatase (E.C.3.1.3.1)						
mouse	C57B1/6J	M & F	12 vs. 24	+30	protein	298
rat	wild	F	14 vs. 34	0	DNA	15
rat	Wistar	M	6 vs. 21	-40	DNA	225
rat	Wistar	M	13 vs. 24	-30	cell	224
D-amino acid oxidase (E.C.1.4.3.2)						
rat	wild	F	14 vs. 34	0	DNA	15
rat	Wistar	M	6 vs. 21	+20	RNA	225
Arginase (E.C.3.5.3.1)						
rat	Wistar	M	8 vs. 24	0	weight	243
Cathepsin (E.C.3.4.4.9)						
rat	wild	F	14 vs. 34	+50	DNA	15
rat	Wistar	F	12 vs. 24	+40	DNA	16
rat	Wistar	F	13 vs. 27	+10	DNA	18
rat	Wistar	M	13 vs. 25	+25	DNA	18
rat	Sprague-Dawley	M	13 vs. 27	+10	DNA	18
rat	Sprague-Dawley	F	12 vs. 24	+50	DNA	18
rat	Wistar	M	6 vs. 21	-20	DNA	225

Table 12-10a. (Continued)

Enzyme Species Strain	Sex	Compared in months	% change	Change given per unit	Reference
Cytochrome oxidase (E.C.1.9.3.1)					
mouse N.M.R.I.		6 vs. 24	0	weight	67
rat Wistar		13 vs. 26	0	weight	89
Fructose-1, 6-diphosphatase (E.C.3.1.3.11)					
rat Wistar	M	12 vs. 15	-20	DNA	244a,b
Glucokinase (E.C.2.7.1.2)					
rat Sprague-Dawley	M	2 vs. 24	+20	weight	1
Glucose-6-phosphatase (E.C.3.1.3.9)					
mouse N.M.R.I.		6 vs. 24	0	weight	67
rat Wistar	M	12 vs. 15	-30	DNA	244a
rat C57Bl/6J	M & F	12 vs. 24	0	DNA	299
α-glycerolphosphate dehydrogenase (E.C.1.1.99.5)					
rat Wistar	F	6 vs. 24	-30	protein	29
Histidase (E.C.4.3.1.3)					
rat Wistar	M	6 vs. 21	+30	DNA	225
rat Wistar	M	13 vs. 27	-25	cell	224
Lactic dehydrogenase (E.C.1.1.1.27)					
rat Wistar	F	12 vs. 24	0	protein	236
rat Wistar	M	6 vs. 21	+10	DNA	225
Malate dehydrogenase (E.C.1.1.1.37)					
rat Wistar	F	12 vs. 24	-10	protein	236
rat Wistar	M	6 vs. 21	+10	DNA	225
Pyrophosphatase (E.C.3.6.1.1)					
rat Wistar	M	6 vs. 21	0	DNA	225
RNAse (E.C.2.7.7.1 7)					
rat Wistar	F	13 vs. 27	0	DNA	18

Table 12-10a. (Continued)

Species	Strain	Sex	Compared in months	% change	Change given per unit	Reference
Succinic dehydrogenase (E.C.1.3.99.1)						
rat	Wistar	F	12 vs. 24	-10	DNA	15
rat	wild	F	14 vs. 34	0	DNA	15
rat	McCollum	M	13 vs. 26	0	DNA	13
rat	Sprague-Dawley	F & M	12 vs. 24	0	DNA	14
rat	Wistar	M	6 vs. 21	0	DNA	225
Succinic dehydrogenase (E.C.1.3.99.1) in mitochondria						
rat	McCollum	M & F	13 vs. 26	0	protein	13
rat	Sprague-Dawley	M & F	12 vs. 24	0	protein	14
Tryptophan pyrrolase (E.C.1.13.1.12)						
rat	Wistar	F	13 vs. 26	0	dry wt.	96
Tyrosine aminotransferase (E.C.2.6.1.5)						
rat	Wistar	F	13 vs. 26	0	dry wt.	96
rat	Wistar	F	9 vs. 16	-20	DNA	80
mouse	C57Bl/6J	M	16 vs. 26	0	DNA	80

Table 12-10b. Age-Related Profiles of Enzyme Activities in the Kidney

Enzyme Species	Strain	Sex	Compared in months	% change	Change given per unit	Reference
Acid DNAse (E.C.3.1.4.5)						
mouse	CF1	M	12 vs. 20	+10	DNA	146
rat	Wistar	F	6 vs. 24	-25	protein	81
Acid phosphatase (E.C.3.1.3.2)						
rat	Wistar	F	6 vs. 24	-10	protein	81
rat	Wistar	M	6 vs. 21	0	DNA	225
Alkaline phosphatase (E.C.3.1.3.1)						
rat	Sprague-Dawley	F	12 vs. 24	0	DNA	14
rat	Wistar	M	6 vs. 21	-30	DNA	225
Arginase (E.C.3.5.3.1)						
rat	Wistar	M	8 vs. 24	0	weight	243
Arylsulphatase B(E.C.3.1.6.1)						
rat	Wistar	F	6 vs. 24	0	protein	81
ATPase (Na-k activated) (E.C.3.6.1.3) in mitochondria						
rat	Wistar (outer layer)	F	13 vs. 20	0	protein	17
rat	Wistar (inner layer)	F	13 vs. 20	-20	protein	17
Cathepsin (E.C.3.4.4.9)						
rat	Sprague-Dawley	F	12 vs. 24	+50	DNA	14
rat	wild	F	14 vs. 34	+50	DNA	15,16
Cytochrome exidase (E. C1.9.3.1)						
rat	Wistar		13 vs. 26	-10	weight	89
Fructose 1, 6-diphosphatase (E.C.3.1.3.11)						
mouse	C57B1/6J	M & F	6 vs. 23	0	protein	300
Fumarase (E.C.4.2.1.2)						
mouse	C57B1/6J	M & F	6 vs. 23	-10	protein	300
Glucose-6-phosphatase (E.C.3.1.3.9)						
mouse	C57B1/6J	M	6 vs. 23	-20	protein	300

Table 12-10b. (Continued)

Enzyme Species	Strain	Sex	Compared in months	% change	Change given per unit	Reference
Lactic dehydrogenase (E.C.1.1.1.27)						
mouse	C57Bl/6J	M	6 vs. 23	-10	protein	300
mouse	C57Bl/6J	F	6 vs. 23	-10	protein	300
rat	Wistar	F	12 vs. 24	0	protein	236
rat	McCollum	F	13 vs. 26	+10	protein	237
Lysozyme (E.C.3.2.1.17)						
mouse	B10-Lp		14 vs. 25	+500-1200	protein	271
mouse	(B10-LPxC57B1/10)	F	9 vs. 25	+250	protein	271
Malate dehydrogenase (E.C.1.1.1.38)						
rat	Wistar	F	12 vs. 24	-10	protein	236
Phosphoenolpyruvate kinase (E.C.2.7.1.40)						
mouse	C57Bl/6J	M	6 vs. 23	0	protein	300
Pyrophosphatase (E.C.3.6.1.1)						
rat	Sprague-Dawley	F	12 vs. 24	+20	DNA	14
Succinic dehydrogenase (E.C.1.3.99.1)						
rat	Sprague-Dawley	F	12 vs. 24	-10	DNA	14
rat	Sprague-Dawley	M	12 vs. 24	0	DNA	14
rat	McCollum	F	13 vs. 26	-20	DNA	13
rat	McCollum	M	13 vs. 26	0	DNA	13
rat	wild	F	14 vs. 34	-20	DNA	13
Succinic dehydrogenase (E.C.1.3.99.1) in mitochondria						
rat	wild	F	14 vs. 34	0	protein	15,16
rat	McCollum	F	13 vs. 26	0	protein	15,16
rat	Sprague-Dawley	F	12 vs. 24	0	protein	14
Triosephosphate isomerase (E.C.5.3.1.1)						
mouse	C57Bl/6J	M & F	6 vs. 23	-10	protein	300

Table 12-10c. Age-Related Profiles of Enzyme Activities in the Heart

Enzyme Species	Strain	Sex	Compared in months	% change	Change given per unit	Reference
Cytochrome oxidase (E.C.1.9.3.1)						
rat	Wistar		13 vs. 26	-10	protein	89
Lactic dehydrogenase (E.C.1.1.1.27)						
rat	Wistar	F	12 vs. 24	0	protein	236
rat	Wistar	F	12 vs. 24	-10	protein	237
rat	not given		7 vs. 19	-40	weight	131
Succinic dehydrogenase (E.C.1.3.99.1)						
rat	Wistar	F	12 vs. 24	-10	weight	14
rat	McCollum	M & F	13 vs. 26	-10	DNA	13

Table 12-10d. Age-Related Profiles of Enzyme Activities in the Brain

	Strain	Sex	Compared in months	% change	Change given per unit	Reference
Acetylcholinesterase (E.C.3.1.1.7)						
rat	Wistar	F	10 vs. 29	0	DNA	116
mouse	C57Bl/6J	M & F	4 vs. 28	0	DNA	116

Adapted from Experimental Gerontology (78) with permission.

443

The effect of aging on the conversion of L-DOPA to monoamines in mice is similar to human Parkinson's disease in which the patients are generally in the middle age group (115). The lowered level of catecholamine may induce other age-related conditions both in humans and animals, for example, conditions such as Huntington's chorea (39), senile tremors (62), acyclic pituitary gonadotropin (270), and REM and State IV Sleep (74). Rats and mice are used as animal models for these and other similar conditions, for example, acyclic gonadotropin (268), spontaneous locomotor activity (92, 207), and impairment of thermoregulation (76, 203, 204).

CLINICAL AND PATHOLOGICAL CHANGES IN THE AGING PROCESSES

In recent years, considerable attention has been given to epidemiological studies in order to relate the cause of death with the aging process (Fig. 12-3). Life expectancy in humans may be increased significantly by elimination of various causes of death (Table 12-11). Fatal lesions commonly found during autopsy are listed in Table 12-12 (121, 269). Myocardial degeneration and atherosclerosis are frequent causes of death among the oldest age group patients (Table 12-13). The multiplicity of lesions found in geriatric patients makes it difficult to determine the exact cause of death. Some of the common lesions found at autopsy of these patients are osteoporosis, chronic osteoarthropathy, lesions in the gastrointestinal tract, and chronic renal diseases. Benign tumors of the prostate, chronic bronchitis, arteriosclerotic alterations in the heart, and varicose formation in the legs are the hallmarks of old age.

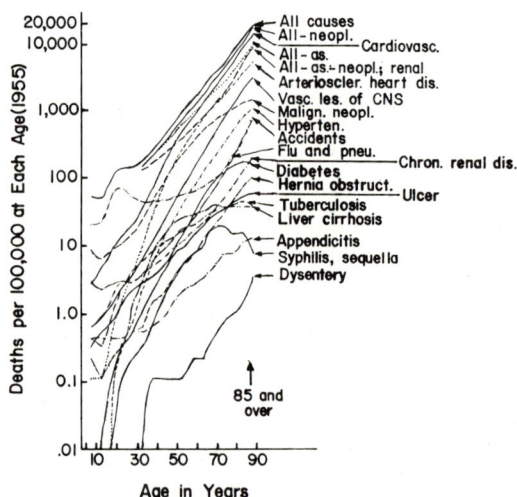

Fig. 12-3. Common causes of death by age in the United States. It is to be noted that life expectancy of adults would be increased by 1 to 3 years if malignant neoplasms could be cured—curve for all causes minus neoplasms (all-neopl.); by 7 years if atherosclerosis were abolished (all-as.); and by 10 to 11 years if both neoplasms and atherosclerosis could be cured (all-as.-neopl.). Reproduced from *Journal of Chronic Diseases* (135) with permission.

Table 12-11. Gain in Life Expectancy by Elimination of Various Causes of Death

Cause of death	Cause elimination	
	At birth	At age 65
Major cardiovascular and renal diseases	10.9	10.0
Heart disease	5.9	4.9
Vascular diseases affecting the central nervous system	1.3	1.2
Malignant neoplasms	2.3	1.2
Accidents excluding those caused by motor vehicles	0.6	0.1
Motor vehicle accidents	0.6	0.1
Influenza and pneumonia	0.5	0.2
Infectious diseases	0.3	0.1
Diabetes mellitus	0.2	0.2

Adapted from Some Demographic Aspects of Aging in the United States (272) with permission.

Table 12-12. Common Fatal Lesions in Old Men

1. Myocardial degeneration
2. Atherosclerosis
3. Cerebral hemorrhage
4. Carcinoma of stomach
5. Bronchopneumonia
6. Carcinoma of colon and rectum
7. Carcinoma of bronchus
8. Coronary thrombosis
9. Pyelonephritis
10. Carcinoma of the esophagus
11. Lobar pneumonia
12. Gastric ulcers
13. Cerebral thrombosis
14. Chronic bronchitis
15. Carcinoma of the biliary passage
16. Duodenal ulcer
17. Intestinal obstruction
18. Pulmonary tuberculosis
19. Carcinoma of the pancreas
20. Cirrhosis of liver
21. Cholecystitis
22. Carcinoma of the prostate
23. Pyonephrosis
24. Mitral stenosis
25. Carcinoma of bladder

Adapted from Developmental Physiology and Aging (269) with permission.

Table 12-13.   Common Fatal Lesions Found in Hospitalized Patients in Different Age Groups

Condition	General Hospital Age Group			
	65-69	70-74	75-79	over 80
	Percentage affected			
Cancer	29	27	27	24
Cardiovascular	25	25	32	36
Respiratory	14	12	13	10
Digestive	12	9	13	16
Nervous system	11	9	8	6
Renal tract	4	7	5	3
Others	5	1	2	5

	Geratic Unit Age Group	
	80-89	> 90
Atherosclerosis	21	30
Myocardial degeneration	19	10
Bronchopneumonia	17	25
Cancer	10	7
Cereberal thrombosis	9	–
Chronic bronchitis	7	–
Others	16	28

Adapted from Developmental Physiology and Aging (269) with permission.

Atherosclerosis

Atherosclerosis is a vascular disorder that may be caused by such factors as congenital defects of structure in the vascular system, and hypersensitivity or immunogenic diseases. Atherosclerotic lesions in humans may start appearing in the fourth decade of life. However, there may be a progressive change in the arterial wall from childhood, thus representing a model for the overall aging process. The atherosclerosis model proceeds as follows: fatty streaks in the arteries during the first decade of life, even with a month or year of life. The deposition of lipid also follows a definite pattern of aorta, coronary, and cerebral arteries. This is followed by fibrous (pearly) plaques with clinical consequences of cardiac infarction, stroke, gangrene, and aneurysm (174). The exact pattern and sequence in individuals varies. It has been reported that in humans the syndrome appears in "waves" as the young and old lesions can be found in the same artery in the same individual. Numerous studies have been conducted to correlate the morphological, ultrastructural and biochemical changes (Fig. 12-4, Table 12-14). Several theories have been developed to explain the mechanism involved in the genesis of atherosclerosis, including the lipid filtration theory (168, 256) and the thrombogenic theory (93, 181, 220). However, these theories do not explain the actual conditions. Recently an immunogenic theory of atherosclerosis has been described (Table 12-15).

In recent years, there has been an effort to develop an immunological model for atherosclerosis (211). The progressive process involves production of localized vascular wall lesion and the release of peptides. The peptides produce autoantibodies, thus forming complexes with the initial lesion. The antigenantibody reactions produce degradative changes resulting in the enlargement of lesions and eventual production of atheromatous plaque (210).

Animal Models of Atherosclerosis

Lesions similar to human atherosclerosis are found in almost all animals. Therefore, animals present unique systems for the study of spontaneous and induced atherosclerosis. However, the lesions in animal models are less severe, probably due to lesser complexity of the environment and absence of social pressures (205). The characteristics of atherosclerosis in various animals vary (157). The lesions can be seen in wild animals as well as animals in captivity; however, the frequency and extent of lesions are greater in domesticated animals (206). In nonhuman primates the lesions are similar to those found in humans (degenerative changes with abundant lipid infiltration). In birds (cockerels) and rabbits, the lesions consist of a primary lipid-induced lesion that may be associated with secondary fibrosis (98). Animals used for experimental atherosclerosis should include several species, at different ages, with varying environmental influences.

Some workers select animal models of atherosclerosis based on the similarity of lesions; others choose on the basis of phylogenic relation to humans. Important criteria to be kept in mind in the selection of a model are (a) if induced atherosclerosis is to be studied, the induction regimen should be physiologically reasonable; (b) the site of gross atherosclerosis is a distal rather than proximal portion of the aorta; (c) the process of atherosclerosis should involve not only the aorta but also renal, coronary, and cerebral arteries; (d) the lesions should be progressive from fatty streak to plaques with ulceration, calcification, hemorrhage, and thrombone formation; (e) in complicated atherosclerosis, a partial or total occlusion of the blood vessel should take place.

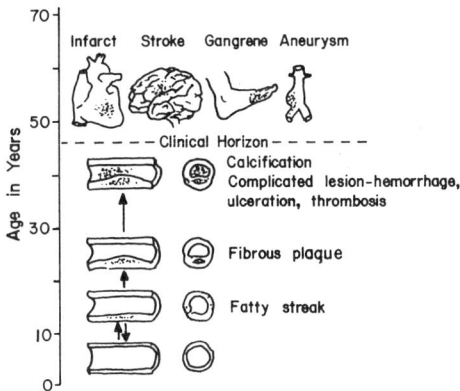

Fig. 12-4. Natural history of atherosclerosis shown in this diagrammatic concept of the pathogenesis of human atherosclerotic lesions and their clinical manifestations. Reproduced from Atherosclerosis and Its Origin (174) with permission.

Table 12-14. Distribution of Lipids in Various Lesions

	Lesions			
Lipid	Fatty streak	Fatty nodule	Fibrous plaques	Calcified fibrous plaques
Cholesterol ester	59.7	64.8	54.1	56.3
Free cholesterol	12.7	13.9	18.4	22.4
Triglycerides	10.0	8.7	11.1	6.5
Phospholipid	17.6	13.6	16.6	14.8

Adapted from <u>Journal Atherosclerosis Research</u> (247) with permission.

448

Table 12-15. Immunological Theory of Atherosclerosis

Primary changes
1. Localized degradation of elastic tissue (platelet elastase).
2. Formation of autoantibodies (to elastin, SGPs).
3. Reaction of circulating antibody with eroded arterial wall, fixation of complement etc.
4. Increased production of arterial degradation products, increased antibody formation and fixation.

Secondary changes
1. Lipid and calcium fixation, modification of metabolic pattern and macro molecular synthesis in arterial wall.

Symptoms
1. Antielastin (and anti GPS) antibodies in human sera.
2. Elastinolytic enzymes in blood platelets, leucocytes.
3. Presence of γ-globulins in arterial lesions.
4. Presence of elastin peptides in soluble extracts of aorta.
5. Calcium fixation on denuded microfibrils (SGP).
6. Production of arteriosclerotic lesions in rabbits by immunization with elastin and SGP.

Correlation with aging
1. Postive correlation between elastin degradation in aorta and dermis with age and pathological conditions.
2. Great frequency of lesions and steady increase with age.
3. Dependence on genetics, nutritional stress and other environmental factors.

Adapted from Gerontologia (210) with permission.

Rabbits. Rabbits show spontaneous atherosclerosis in the form of fatty streaking in young animals to full-fledged lesions in old animals (25, 198, 249). Rabbits were the first models used for the study of diet induced (milk, meat, and eggs) atherosclerosis (230). Later several workers confirmed that milk, meat, and eggs are atherogenic. It was found that cholesterol, when fed along with the diet, induces atherosclerosis (233), but the lesions were more severe with cholesterol alone (140, 143, 297). It has been shown that cholesterol mobilizes endogenous fat which is more saturated. Saturated fats alone induced lesions of atherosclerosis in rabbits (141, 142). Heating of oil in air (144) and ingested cholesterol-lactose produce a greater incidence of lesions in rabbits (284). The lesions have been studied in detail by Prior et al. (198).

Chickens. Natural lesions in domestic fowls vary with age and sex (57, 282). Chickens have been extensively used to study cholesterol induced lesions (118, 132). The gross lesions are seen in the aortic arch and in brachiocephalic arteries. The earlier lesions of "pure atheroma" are followed by more complicated lesions. However, ulceration and superimposed thrombosis of plaques are rare (193). Chickens are also sensitive to diethystilbesterol-induced atherosclerosis (117, 159).

Turkeys. The turkey represents one of the best spontaneous models not only for atherosclerosis but also for blood pressure studies. The condition is so severe that dissecting aneurysm of the aorta may result in death (38, 64, 176). The gross lesions most frequently occur in the posterior portion of the abdominal aorta (86). The lesions represent fibroelastic hyperplasia with moderate amount of lipid near the internal elastic lamine (97). A relationship between blood pressure, lesions, and death from aortic lesions has been reported (283).

Pigeons. These animals are highly susceptible to spontaneous atherosclerosis and many workers have used them as experimental models (44, 45, 46, 161, 199, 200). Microscopic lesions develop within 1 week in about 30 percent of pigeons under study and the same number show visual lesions within a year. Almost all pigeons have some lesions by the third year of their lives. Detailed studies show intra- and extracellular lipid accumulation, deposit of calcium, hemosiderin, hemorrhage in plaques, vascularization and medial elastic degeneration; thus mimicking the human condition. A considerable amount of natural morbidity and mortality in the pigeon is due to this condition.

Swine. A high percentage (35 percent) of swine, after the age of 3 years, show spontaneous plaques in the aorta (94, 245). The lesions in swine are age and sex dependent. Lesions in the coronary artery have also been reported, thus making swine an unusual human model (126). Several studies on induced atherosclerosis using diets high in butter (226) and cholesterol (180) have been reported. The aorta, coronary arteries, and carotid and cerebral arteries are involved.

Nonhuman Primates. These animals have been used because of their phylogenic relationship to humans. One study involving a 22-year-old zoo gorilla showed intimal fibrosis in the heart, nervous system, and aorta only (253). This study shows that the lesions are not similar to those in humans. Several authors have studied natural lesions (atheromatous protrusion at the opening of right anterior coronary artery) which vary with age and sex (88, 175). It is not easy to induce lesions in baboons by dietary manipulations (175). Chimpanzees are generally not very sensitive to atherosclerosis. In the rhesus monkey, atherosclerosis can be induced by hypertension and normal diet (173). The lesions in rhesus monkeys include frothy streaks and fibrous and pearly plaques. High doses of cholesterol (1.5 percent) and butter fat (15 percent) can induce atherosclerosis (52, 262); however, lower dosages are ineffective (123).

Dogs. The natural incidence of atherosclerosis in dogs is rather low (158). However, diet induced atherosclerosis can be induced with 50 to 60 days (252). Lesions are present in the thoracic aorta, and in the iliac, femoral, coronary, bronchial, renal and other arteries. The administration of ACTH prior to diet regimen produced severe lesions (222).

Rats and mice are not good models for studies on atherosclerosis.

Cholesterol Influx and Efflux

Animal experiments using rats show that there is a much greater influx of free cholesterol than of estercholesterol (58) and similar results were obtained in vitro with the rat aorta (109). The high concentration probably arises from increased endothelial permeability, an increase in reversible bond producing decreased reflux, or reduction in the rate of destruction of lipoproteins. However, lower values were obtained in atherosclerotic rabbits. Many workers have shown that atherosclerosis greatly influences the influx of free and esterified cholesterol in rabbits (19, 183, 238). It was proportional to the concentration of aortic cholesterol concentration. In atherosclerotic pigeons, the low concentration of lipoproteins, which is characteristic of human lesions, was not observed (162). It would appear that the pigeon is not a good model for duplicating the human condition of atherosclerosis.

EXPERIMENTAL ANIMALS FOR THE STUDY OF MECHANISMS OF AGING PROCESSES

Mammalian experiments involving changes in caloric restrictions (48, 223) and administration of hormones (20, 23, 82) indicate that the aging process can be modified. In humans, integral modification is also possible (229, 279). The aging process appears to involve information loss (from DNA) probably at the cellular level (49). An approach to modification is based on whether the information loss is predominantly in fixed or clonally dividing cells. This process may occur by the accumulation of undesirable material or by irreversible switching off of synthetic capacities with morphogenesis (28). Comfort (49) has suggested a broad approach to the study of the human aging process (Fig. 12-5). A number of hypotheses have come forth which may be tested by using experimental animal models.

Primary Error Hypothesis

The simplest hypothesis is the loss of information from DNA by mutation, macromolecular damage, or epigenetic masking. The mutation theory, which is based on radiation data, does not agree with radiosensitivity experiments with old mice (293). Nor does it correlate with the mathematical consideration (171) and with the failure of chemical mutagenic agents to cause radiation type aging in mice, unless they have crosslinking properties (3, 4). In bacteria, mitomicin-induced cross linkage is lethal (280). In mice and rats, the chromosomal abnormalities accumulate with age (54). In humans, it correlates with mental changes (125, 184). Using dogs as a model system, it has been shown that the coding for rRNA is lost at 10 years of age (129). Various studies indicate differences such as binding power and unwinding capacities between "old" and "young" DNA template activity (60, 275). However, the template activity seems to depend on RNA polymerase concentration rather than the age of the donor. Shirey and Sobel (241) found lack of transcriptional differences in dogs. Other theories indicate the role of immunity (31), lysosomal DNAases (5, 6), masking of histones (275), and age sensitivity of codons (177). Qualitative and quantitative changes in the tRNA of rat spleen have been reported. However, there are some important unanswered questions such as why some clones are more stable than others and why the rate of damage accumulation is fifty times greater in mice as compared to humans (229). The postmitotic can be modified by certain substances such as prednisolone (27, 28).

Non-DNA Error Theories

The popular hypotheses in non-DNA theories include (a) chromosomal (nuclear structure rather than molecular DNA); (b) transcription-translation (RNA, synthetases, and proteins); and (c) selective loss of ability to transcribe a particular codon (255). Loss of sex chromatin occurs with age in women (124) and chromatin loss has been associated with the organic brain syndrome (125). Most of the hypotheses are based on

AGING OF MAMMALS

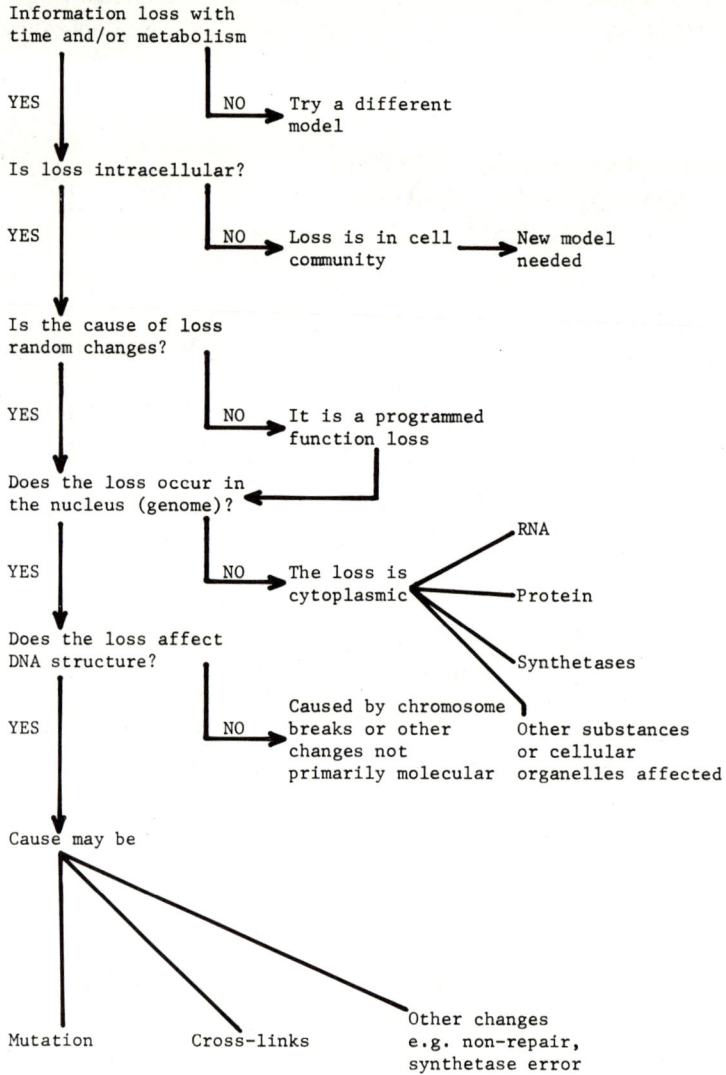

```
          Information loss with
          time and/or metabolism


     YES                   NO      Try a different
                                   model


          Is loss intracellular?


     YES                   NO      Loss is in cell ──→ New model
                                   community           needed


          Is the cause of loss
          random changes?


     YES                   NO      It is a programmed
                                   function loss


          Does the loss occur in
          the nucleus (genome)? ◄──


     YES                   NO      The loss is ───── RNA
                                   cytoplasmic
                                               ───── Protein

                                               ───── Synthetases

                         Caused by chromosome
     YES                 NO  breaks or other   Other substances
                             changes not       or cellular
                             primarily molecular  organelles affected


          Cause may be




     Mutation      Cross-links    Other changes
                                  e.g. non-repair,
                                  synthetase error
```

Fig. 12-5. Aging of Mammals. Reproduced from *Mechanism of Aging and Development* (49) with permission.

experiments with mice (54). Orgel (188, 189) has postulated that the primary target is RNA polymerases. Using rotifers as experimental models, it has been possible to prolong life span by inhibitors of RNA synthesis. The runaway syntheses of RNA, DNA, and protein as a possible mechanism to compensate for nonsense material production has been shown in drosophila (43) and in rat liver nuclei (231, 290). However, such a runaway synthesis has not been seen at the mitochondrial level in rats (178).

Table 12-16. Agents which Might Modify the Rate of Aging

Nature	Theoretical basis	Findings
Antioxidants	Scavenge free radicals, prevent attack on DNA or some other system	"Prolong life" (mouse) but may not alter specific age (105,107a,b).
Radioprotectants	Assume aging similar in nature to radiation damage	Antagonize radiation life-shortening (107a,b)
Protein synthesis inhibitors	Break 'vicious circle' if synthetase faults involved	Untested experimentally
Lysosome stabilizers	Prevent escape of enzymes (including lysosomal DNAse)	See prednisolone
Immunosuppressants	Limit aging effects due to autoimmune divergence	Azathioprine slightly prolongs mouse life (276)--splenectomy in old mice greatly prolongs it (2). Predict ALS may limit some age changes but increase clonal divergence
Anti-crosslinking agents	Aging reflects crosslinking in long-term molecules	BAPN may affect longevity (147) and tumour incidence (137). Penicillamine not yet tested
Hormonal agents	Modify chemical allometry, retard senescent programme	
Anabolics	Prevent decline in protein storage and muscle power, conserve Ca	Limited clinical effect in man
Somatotrophin	Maintains 'young' pattern of tissue chemistry	Fails to prolong rat life (68)
Prednisolone	Program-slowing, anti-autoimmune Lysosome stabilizer?	Doubles lifespan in short-lived mice (20). Increases amnion-cell survival (292).
Post-pituitary	Doubtful	Whole post-pituitary prolongs rat life (82). Active substance may be oxytocin (23)
Antimetabolic drugs	Stimulate calorie restriction	Untested
Enzyme inducers	Lifespan correlated with liver enzyme levels	Possible cause of lifespan prolongation by BHT and certain diets (223)

Adapted from Mechanisms of Aging and Development (49) with permission.

453

Protectant Substances

The concept is based on the fact that if radiation causes aging then radioprotectants should delay such a process in an intact animal. This hypothesis has been extensively tested with antioxidants using rats and mice as model systems (Table 12-16) (37, 50, 105, 107). Harman (105, 106) has proposed other ways to test the chemical-stochastic hypothesis. A vast amount of work has been done on the mechanism of free radical damage and its protection by antioxidants (12, 61, 133, 201, 260). Lipoperoxidation appears to be the most important factor (61) because it is known to occur with aging in rat brains (281). Oxidized feeds have been shown to produce several pathological conditions in farm animals, particularly chickens (7, 30, 167). However, experiments have not been performed using low level oxidants in farm animals. It has been reported that the α-tocopherol does not prolong the life of mice fed standard laboratory diet. However, it is not certain if tocopherol acts mainly as an antioxidant in this case (55, 95). At cellular level mitochondrial change has been studied in rats and mice (119, 169, 261).

Increase in life span has been reported in rats fed a restricted caloric diet with 2-mercaptoethylamine (2-MEA) (172, 223, 259), cysteine 2, 2'-diamino diethylsulfide (104), dibutyl hydroxytoluene (107), and nordihydro-guaiaretic acid (37), or ethoxyquin in mice (50). Kohn (136) reported that the life span of C57bl mice was prolonged only if the colony survival was suboptimal. Antioxidants (BHT, MEA) did not affect the life span under optimum conditions of feeding and housing. Increased oxygen tension decreased longevity of mice (84), and acute α-tocopherol deficiency in the diets of mice increased sensitivity (263). However, in a study reported by Sobel (248), increased oxygen tension did not decrease the life span in experimental mice.

IMMUNOLOGICAL BASIS OF AGING

The basic hypothesis put forward by Walford (277, 278) assumes (a) the origin of slow developing intolerance and the autoantibodies, (b) deleterious effect of autoantibodies; and (c) the relationship between the two and aging.

The appearance of autoantibodies is probably due to a programmed process by which derepression of the previously repressed clones or switching on of mutater gene results in immunogenic diversification which causes a prolonged low grade histocompatibility reaction and eventually autoimmune diseases (278). The autoantibody diversity is related to self-tolerance (127). The sequel to this concept of autoimmune diseases has been reported by several workers (91, 145, 210). Thus, aging is an increased intolerance of RES toward its own host. The concept is summarized in Table 12-17.

Many species of experimental animals have been used to develop these basic concepts. For example, using sheep as an experimental model (250, 251) and human or monkey isolated glomerular basement membranes, it was shown that sheep produce autoantibodies and eventually lethal glomerulonephritis. Rabbits have been used extensively to develop immunodeficiency by neonatal thymectomy combined with irradiation (258). The treatment causes high frequency of amyloidosis and autoimmune hemolytic anemia in the animals. The removal of lymphoid tissue and near lethal irradiation of adult rabbits resulted in autoimmune, Coombs positive, hemolytic anemia (134).

The work in the area of understanding the mechanisms of autoimmune diseases has advanced greatly due to special strains (NZB) of mice which develop autoimmune

Table 12-17. Immunoloogical Theory of Aging

I. Primary causes
    1. Somatic mutation; programmed modification of cell constituents.
    2. Accumulation of errors causing problems in copying mechanisms of DNA,
       RNA and proteins.
    3. Cross linking which may modify gene expression, antigenic specificity.

II. Erroneous cell--cell recognition mechanisms
    1. Modification in the cell--cell interaction due to modified cell coat
       and cell membrane synthesis (glycoproteins)
    2. Antigenic modulation, appearance of nonself motifs.
    3. Autoantibody formation: antibodies to modified cells, constituents
       and forbidden clones.

III. Diseases of aging
    1. Vascular diseases-Arteriosclerosis
    2. Diabetes
    3. Osteoarthrosis and other connective tissue diseases
    4. Neoplasia
    5. Amyloidosis
    6. Dystrophies (impaired function of teeth, muscle and hair)

Adapted from <u>Gerontologia</u> (210) with permission.

anemia spontaneously. The cause of anomaly was attributed to the abnormality in the structure of the thymus (34). Neonatal thymectomy in autoimmune susceptible mice accelerated not only the development of autoimmune hemolytic anemia but also produced anti-DNA and antinuclear antibodies (294). Several other strains of mice also produce these autoimmune diseases (264–267). Autoimmune phenomena were often accompanied by proliferation of plasma cells in lymphoid nodes and spleen (294). Immunodeficiency develops with aging at greater frequency in autoimmune susceptible strains of mice. In the NZB strain, the condition was fairly common in the second half of the first year (296). Conditions such as immunological deficiency, autoimmune hemolytic anemia, antinuclear antibodies, anti-DNA antibodies, hemolytic, hepatic, splenic and renal lesions seen in neonatal thymectomy are almost exactly duplicated in aging mice (218). The role of the thymus has been further emphasized by the "curing" of immunodeficiency syndromes with transplanted thymus or injection of lymphocytes (295). However, the production of immunodeficiency in aging mice by transplantation or otherwise has been partially achieved (266).

Role of Growth Hormone in Premature Aging

The involvement of immune system in aging is well established (33, 191, 277). The immunodeficiency may be due to an inadequate stimulation of lymphoid system by GH (72). Pituitary dwarf mice had a mean life span of 4.5 months as compared to the normal life span of 20 months. The dwarf mice had greying hair, cutaneous atrophy, bilateral cataracts, and low uptake of ^{3}H-thymidine (72). However, in another report (239), no such difference in life span or aging syndrome was observed despite the absence of GH and prolactin (240). Therefore, the role of dwarf mice as models for aging and the influence of GH on the aging process are not clear at the present time.

REFERENCES

1. Adelman, R. C., *J. Biol. Chem.,* **245**:1032 (1970).
2. Albright, J. F., T. Makinodan, and J. W. Deitchman, *Exp. Gerontol.,* **4**:267 (1967).
3. Alexander, P., *Sym. Soc. Exp. Biol.,* **21**:29 (1967).
4. Alexander, P., and D. I. Connel, *Radiat. Res.,* **12**:510 (1960).
5. Allison, A. C., *Proc. R. Soc. Med.,* **59**:867 (1966).
6. Allison, A. C., and G. M. Paton, *Nature,* **207**:1170 (1965).
7. Andrews, J. S., W. H. Griffith, J. F. Mead, and R. A. Stern, *J. Nutr.,* **70**:199 (1960).
8. Backerman, S., *Aging Life Processes,* Charles C Thomas, Springfield, Ill., 1969.
9. Bakerman, S., *Biochim. Biophys. Acta,* **90**:621 (1964).
10. Bal, H. S., and R. Getty, *Growth,* **34**:15 (1970).
11. Bal, H. S., R. Getty, *J. Gerontol.,* **28**:160 (1973).
12. Barber, A. R., F. Bernheim, *Adv. Gerontol. Res.,* **2**:355 (1967).
13. Barrows, C. H., J. A. Falzone, and N. W. Shock, *J. Gerontol.,* **15**:130 (1960).
14. Barrows, C. H., and L. M. Roeder, *J. Gerontol.,* **16**:321 (1961).
15. Barrows, C. H., L. M. Roeder, and J. A. Falzone, *J. Gerontol.,* **17**:144 (1962).
16. Barrows, C. H., L. M. Roeder, and D. A. Olewine, *J. Gerontol.,* **17**:148 (1962).
17. Beauchene, R. W., D. D. Fanestil, and C. H. Barrows, *J. Gerontol.,* **20**:306 (1965).
18. Beauchene, R. W., L. M. Roeder, and C. H. Barrows, *J. Gerontol.,* **22**:318 (1967).
19. Bell, F. P., H. B. Lofland, and N. A. Stokes, *Atherosclerosis,* **11**:235 (1970).
20. Bellamy, D., *Exp. Gerontol.,* **4**:327 (1968).
21. Bhatnagar, R. S., K. S. McCarty, *Fed. Proc.,* **24**:2345 (1965).
22. Bihari-Varga, M., T. Biro, and J. Levai, *Gerontologia,* **17**:148 (1971).
23. Bodanszky, M. and S. L. Engel, *Nature,* **210**:751 (1966).
24. Bouliere, F. and R. Molimard, *C. Rend. Soc. Biol.,* **151**:1345 (1957).
25. Bragdon, J. H., *Circulation,* **5**:641 (1952).
26. Brody, H., *Interdisciplinary Topics Gerontol.,* **7**:9 (1970).
27. Bullough, W. S., *Evolution of Differentiation,* Academic Press, London, 1967.
28. Bullough, W. S., *Nature,* **229**:608 (1971).
30. Bunyan, J., A. T. Diplock, E. E. Edwin, and J. Green, *Br. J. Nutr.,* **16**:519 (1962).
31. Burch, P. J. R., *An Inquiry Concerning Growth,* Oliver & Boyd, Edinburgh, 1968.
32. Burnet, F. M. L., *Clonal Selection Theory of Acquired Immunity,* Vanderbilt University Press, Nashville, 1959.
33. Burnet, F. M. L., *Immunological Surveillance,* Pergamon Press, Oxford, 1970.
34. Burnet, F. M., and M. Holmes, *J. Pathol. Bacteriol.,* **88**:229 (1964).
35. Bustad, L. K., *Lab. Anim. Dig.,* **3**:3 (1967).
36. Bustad, L. K., L. S. Rosenblatt, and J. L. Palotay, *Laboratory Animals in Gerontological Research,* National Academy of Sciences, Washington, D.C., 1968.
37. Buu-Hoi, N. P., and A. R. Ratsimamanga, *C.R. Soc. Biol.,* **153**:1180 (1959).
38. Carnaghan, R. B. A., *Vet. Rec.,* **67**:568 (1955).
39. Chandler, J. H., T. E. Reed, and R. M. De Jong, *Neurology,* **10**:148 (1960).
40. Church, R. B., and B. J. McCarthy, *J. Mol. Biol.,* **23**:459 (1967).
41. Chvapil, M., and Z. Hruza, *Gerontologia,* **3**:241 (1959).
42. Clark, A. M., and M. Rockstein, in M. Rockstein (ed.), *Physiology of Insecta,* Academic Press, New York, 1964.
43. Clarke, J. M., and J. Maynard Smith, *Nature,* **209**:627 (1966).
44. Clarkson, T. B., and H. B. Lofland, *Fed. Proc.,* **22**:385 (1963).
45. Clarkson, T. B., R. W. Prichard, H. B. Lofland, and H. O. Goodman, *Circul. Res.,* **11**:400 (1962).
46. Clarkson, T. B., R. W. Prichard, and A. F. Moreland, *Fed. Proc.,* **21**:220 (1962).
47. Clegg, J. S., and D. R. Evans, (a) *J. Exp. Biol.,* **38**:771 (1961); (b) *Science,* **134**:54 (1961).
48. Comfort, A., *Proc. Nutr. Soc.,* **19**:120 (1960).
49. Comfort, A., *Mechanism of Aging and Development,* **3**:1 (1974).
50. Comfort, A. I., I. Youhotsky-Gore, and K. Pathmanathan, *Nature,* **229**:254 (1971).
51. Corbin, K. B., and E. D. Gardner, *Anat. Rec.,* **68**:63 (1937).
52. Cox, G. E., C. B. Taylor, L. G. Cox, and M. A. Counts, *Arch. Pathol.,* **66**:32 (1958).

53. Culter, R. G., C. F. Blackman, Jr., B. Presoxitz, and A. Hilderbrandt, *Proc. 8th Int. Congr. Gerontol.,* Washington, D.C., 1969.
54. Curtis, H. J., *Symp. Soc., Exp. Biol.,* **21**:51 (1967).
55. Dam, H., *Carenzedivitamina E e rapporticon i lipidi alimentari,* Monog. Centr. Stud. Lip., Alliment, Paris, 1967.
56. Danielli, J. F., in F. Verzas (ed.), *Experimentelle Alternsforschung,* Birkauser Verlag, Basel, 1956.
57. Dauber, D. V., *Arch. Path.,* **38**:46 (1944).
58. Dayton, S., and S. Hashimoto, *Circ. Res.,* **19**:1041 (1966).
59. Denckla, W. D., *Fed. Am. Soc. Exp. Biol. 55th Ann. Meetg.,* Chicago, 1971.
60. Devi, A., P. Lindsay, P. L. Rainer, and N. K. Sankar, *Nature,* **212**:474 (1966).
61. Dormandy, T. L., *Lancet,* **2**:684 (1969).
62. Doshay, L. L., *Postgrad. Med.,* **30**:550 (1961).
63. Downie, H. G., J. F. Mustard, and H. C. Rowsell, *Ann. N.Y. Acad., Sci.,* **104**:539 (1963).
64. Durrel, W. B., B. S. Pomeroy, W. S. Carr, and A. C. Jerstad, *Proc. Book. Am. Vet. Med. Assoc.,* 1952, p. 280.
65. Edward, A. G., in Pacific Northwest Lab. (ed.), *Swine in BioMedical Research,* Richland, Washington, 1966.
66. Eisen, A. Z., *J. Invest. Dermatol.,* **52**:442 (1969).
67. Elens, A., and R. Wattiaux, *Exp. Gerontol.,* **4**:131 (1969).
68. Everitt, A., *J. Gerontol.,* **14**:415 (1959).
69. Everitt, A., *Gerontologia,* **17**:98 (1971).
70. Everitt, A. V., J. S. Giles, and A. Gal, *Gerontologia,* **15**:366 (1969).
71. Ewing, L. L., *Am. J. Physiol.,* **212**:1261 (1967).
72. Fabris, N., W. Pierpaoli, and E. Sorkin, *Nature,* **240**:557 (1970).
73. Fasting, M. F. W., and D. K. Blackmore, *Lab. Anim.,* **5**:179 (1971).
74. Feinberg, I., in A. Kales, (ed.), *Sleep Physiology and Pathology,* Lippincott, Philadelphia, 1969.
75. Festing, M. F. W., and D. K. Blackmore, *Lab. Anim.,* **5**:179 (1971).
76. Finch, C. E., *Cellular Activities During Aging in Mammals,* Ph.D thesis, Rockefeller University, New York, 1969.
77. Finch, C. E., in *Animal Models for BioMedical Research,* Vol. IV, National Academy of Sciences, Washington, D.C., 1971.
78. Finch, C. E., *Exp. Gerontol.,* **7**:53 (1972).
79. Finch, C. E., *Brain Res.,* **52**:261 (1973).
80. Finch, C. E., J. R. Foster and A. E. Mirsky, *J. Gen. Physiol.,* **54**:690 (1969).
81. Franklin, T. J., *Biochem. J.,* **82**:118 (1962).
82. Freidman, S. M. and C. L. Freedman, *Exp. Gerontol.,* **1**:37 (1964).
83. Fry, R. J. M., S. Lesher and H. I. Kohn, *Lab. Invest.,* **11**:289 1962.
84. Gerschman, R., *Proc. XXI Int. Congr. Cienc. Physiol.,* Buenos Aires, 1959.
85. Getty, R., in J. C. Roberts, and R. Straus (eds.), *Comparative Atherosclerosis,* Harper & Row, New York, 1965.
86. Gibson, E. A., and P. H. Gruchy, *Vet. Rec.,* **67**:650 (1955).
87. Giles, J. S., and A. V. Everitt, *Gerontologia,* **13**:65 (1967).
88. Gillman, J., and C. Gilbert, *Exp. Med. Surg.,* **15**:181 (1957).
89. Gold, P. H., M. V. Gee, and B. L. Strehler, *J. Gerontol.,* **23**:509 (1968).
90. Golub, L., B. Stein, M. Glimcher, and P. Goldhaber, *Proc. Soc. Biol. Med.,* **129**:465 (1968).
91. Good, R. A., and E. Yunis, *Fed. Proc.,* **39**:2040 (1974).
92. Goodrick, C. L., *J. Gerontol.,* **26**:58 (1971).
93. Gore, I., and F. J. Stare, *Circulation,* **25**:735 (1962).
94. Gottlieb, H., and J. J. Lalich, *Am. J. Pathol.,* **30**:851 (1954).
95. Green, J., *Ann. N.Y. Acad. Sci.,* **203**:29 (1972).
96. Gregerman, R. I., *Am. J. Physiol.,* **197**:63 (1959).
97. Gresham, G. A., and A. N. Howard, *J. Atheroscler., Res.,* **1**:75 (1961).
98. Gresham, G. A., and A. N. Howard, in *Le Role de la Paroi Arterielle dans l'Atherogenese,* Centre of Scientific Research, Paris, 1968.

99. Gross, J., *J. Exp. Med.,* **107**:265 (1958).
100. Hager, C. B., and F. T. Kenney, *J. Biol. Chem.,* **243**:3296 (1968).
101. Hall, D. A., *Exp. Gerontol.,* **3**:77 (1968).
102. Hall, D. A., *Nature,* **228**:1314 (1971).
103. Hall, D. A., *Age and Aging,* **1**:141 (1972).
104. Harman, D., *J. Gerontol.,* **12**:257 (1957).
105. Harman, D., *J. Gerontol.,* **16**:247 (1961).
106. Harman, D., in N. Shock (ed.), *Biological Aspects of Aging,* Columbia University Press, New York, 1962.
107. Harman, D., (a) *J. Gerontol.,* **23**:476 (1968), (b) *Gerontologist,* **6**:13 (1968).
108. Harris, E. D., Jr., and A. Sjoerdsma, *J. Clin. Endocrinol.,* **26**:358 (1966).
109. Hashimoto, S., and S. Dayton, *J. Atheroscler. Res.,* **6**:580 (1966).
110. Hayflick, L., in M. Rockstein, M. L. Sussman, and J. Chesky (eds.), *Theoretical Aspects of Aging,* Academic Press, New York, 1974.
111. Hayflick, L., *Fed. Proc.,* **34**:9 (1975).
112. Hayflick, L., and P. S. Moorhead, *Exp. Cell. Res.,* **25**:585 (1961).
113. Heikkinen, E., and E. Kulonen, *Experientia,* **20**:310 (1964).
114. Hershey, D., *Life Span and Factors Affecting It,* Charles C Thomas, Springfield, Ill., 1974. 1974.
115. Hoehn, H. M., and M. D. Yahr, *Neurology,* **17**:424 (1967).
116. Hollander, J., and C. H. Barrons, *J. Gerontol.,* **23**:174 (1968).
117. Horlick, L., and L. N. Katz, *J. Lab. Clin. Med.,* **33**:733 (1948).
118. Horlick, L., and L. N. Katz, (a) *J. Lab. Clin. Med.,* **34**:1427 (1949); (b) *Am. Heart J.,* **38**:336 (1949).
119. Horton, A. A., and L. Packer, *J. Gerontol.,* **25**:199 (1970).
120. Houck, J. C., and R. A. Jacob, *Proc. Soc. Exp. Biol. Med.,* **112**:446 (1963).
121. Howell, T. H., *A Student Guide to Geriatrics,* 2nd ed., Staples Press, London, 1970.
122. Howes, J. R., in *Laboratory Animals in Gerontological Research,* National Academy of Sciences, Washington, D.C., 1968.
123. Hueper, W. C., *Am. J. Pathol.,* **22**:1287 (1946).
124. Jacobs, P. A., M. Brunton, W. M. Court Brown, R. Doll, and J. Goldstein, *Nature,* **197**:1080 (1963).
125. Jarvik, L. F. and T. Kato, *Br. J. Psychiatry,* **115**:1193 (1969).
126. Jennings, M. A., H. W. Florey, W. E. Stanbens, and J. E. French, *J. Pathol. Bacteriol.,* **81**:49 (1961).
127. Jerne, N. K., in R. T. Smith, and M. Landy (eds.), *Immunological Surveillance,* Academic Press, New York, 1970.
128. Johnsen, S. G., *Acta Endocrinol.,* **31**:209 (1959).
129. Johnson, R., and B. L. Strehler, *Nature,* **240**:412 (1972).
130. Kang, A. H., K. A. Piez, and J. Gross, *Biochemistry,* **8**:3648 (1969).
131. Kanungo, M. S., and S. N. Singh, *Biochem. Biophys. Res. Commun.,* **21**:454 (1965).
132. Katz, L. N., and J. Stamller, *Experimental Atherosclerosis,* Charles C. Thomas, Springfield, Ill., 1953.
133. Kawashima, S., *Nagoya J. Med. Sci.,* **33**:303 (1970).
134. Kellum, M. J., D. E. R. Sutherland, E. Eckert, R. D. A. Peterson, and R. A. Good, *Intern. Arch. Allergy Appl. Immunol.,* **27**:6 (1965).
135. Kohn, R. R., *J. Chronic Dis.,* **16**:5 (1963).
136. Kohn, R. R., *J. Gerontol.,* **26**:378 (1971).
137. Kohn, R. R., and A. M. Leash, *Exp. Mol. Pathol.,* **7**:354 (1967).
138. Kohn, R. R., and E. Rollerson, *J. Gerontol.,* **15**:10 (1960).
139. Korenchevsky, V., *Physiological and Pathological Aging,* Hafner, New York, 1960.
140. Kritchevsky, D., A. W. Mayer, J. B. Logan, and R. F. J. McCandless, *Arch. Biochem. Biophys.,* **59**:526 (1955).
141. Kritchevsky, D., A. W. Moyer, W. C. Tesar, J. B. Logan, R. A. Brown, M. C. Davies, and H. R. Cox, *Am. J. Physiol.,* **178**:30 (1954).
142. Kritchevsky, D., A. W. Mayer, W. C. Tesar, R. F. J. McCandless, J. B. Brown, and M. E. Englert, *Am. J. Physiol.,* **185**:279 (1956).

143. Kritchevsky, D., J. L. Moynihan, J. Langan, S. A. Tepper, and M. Sachs, *J. Atheroscler. Res.,* **1**:211 (1961).

144. Kritchevsky, D., and S. A. Tepper, *J. Nutr.,* **77**:127 (1962).

145. Kunkel, H. G., and E. M. Tan, *Adv. Immunol.,* **4**:351 (1965).

146. Kurnick, N. B., and R. L. Kernan, *J. Gerontol.,* **17**:245 (1962).

147. Labella, F. S., *Canad. J. Physiol. Pharmacol.,* **46**:335 (1968).

148. Labella, F. S., S. Vivian, and D. P. Thornhill, *J. Gerontol.,* **21**:550 (1966).

149. Landowne, M., *J. Appl. Physiol.,* **12**:91 (1958).

150. Landowne, M., and J. Stanley, *Aging—Some Social and Biological Aspects,* Hara-Shafer, Baltimore, 1960, p. 159.

151. Lapina, Z. V., *Aknsherstvoi Ginekol.,* **33**:18 (1957).

152. Lasser, R. P., and A. M. Master, *Geriatrics,* **14**:345 (1959).

153. Lauson, H. D., J. B. Golden, and E. L. Sevringhaus, *Am. J. Physiol.,* **139**:396 (1939).

154. Lazarno, G. S., and H. M. Fullmer, *J. Invest. Dermatol.,* **52**:545 (1969).

155. Lescher, S., R. J. M. Ery, and H. J. Kohn, *Lab. Invest.,* **10**:291 (1961).

156. Levene, C. I., and J. Gross, *J. Exp. Medicine,* **110**:771 (1959).

157. Lindsay, S., and I. L. Chaikoff, in M. Sandler and G. H. Borune, *Atherosclerosis and Its Origin,* Academic Press, New York, 1963.

158. Lindsay, S., I. L. Chaikoff, and J. W. Gilmore, *Arch. Pathol.,* **53**:281 (1952).

159. Lindsay, S., C. W. Nicholos, and I. L. Chaikoff, *Arch. Pathol.,* **59**:173 (1955).

160. Loeb, L., V. Suntzeff, and E. L. Burns, *Am. J. Cancer,* **35**:159 (1939).

161. Lofland, H. B., and T. B. Clarkson, (a) *Proc. Soc. Exp. Biol. Med.,* **112**:108 (1963); (b) *Fed. Proc.,* **22**:385 (1963).

162. Lofland, H. B., and T. B. Clarkson, *Proc. Soc. Exp. Biol. Med.,* **133**:1 (1970).

163. Luginbuhl, H., in L. K. Bustad and R. O. McClellan, *Swine in Biomedical Research,* Pacific Northwest Laboratory, Richland, Wash., 1966.

164. Luginbuhl, H., and J. E. T. Jones, *Ann. N.Y. Acad. Sci.,* **127**:763 (1965).

165. Lumb, G. D., in L. K. Bustad and R. O. McClellan (eds.), *Swine in Biomedical Research,* Pacific Northwest Laboratory, Richland, Wash., 1966.

166. Maaske, C. A., N. H. Booth, and T. W. Nielsen, in L. K. Bustad, and R. O. McClellan (eds.), *Swine in Biomedical Research,* Pacific Northwest Laboratory, Richland, Wash., 1966.

167. Machlin, L. J., *J. Am. Oil Chem. Soc.,* **40**:368 (1963).

168. Marchand, F., *Verh. 21 Kongr. Inn. Med.,* **21**:23 (1904).

169. Marco, G. J., L. J. Machlin, and R. S. Gordon, *Arch. Biochem. Biophys.,* **94**:115 (1961).

170. Martin, G. R., J. Gross, H. A. Piez, and M. S. Lewis, *Biochim. Biophys. Acta,* **53**:599 (1961).

171. Maynard Smith, J., *Nature,* **184**:956 (1959).

172. McCay, C. M., M. F. Crowell, and L. A. Maynard, *J. Nutr.,* **10**:63, (1935).

173. McGill, H. C., M. H. Frank, and J. C. Geer, *Arch. Pathol.,* **71**:96 (1961).

174. McGill, H. C., Jr., J. C. Geer, and J. P. Strong, in M. Sandler, and G. H. Bourne (eds.), *Atherosclerosis and Its Origin,* Academic Press, New York, 1963.

175. McGill, H. C., J. P. Strong, R. L. Holman, and N. T. Werthessen, *Circ. Res.,* **8**:670 (1960).

176. McSherry, B. J., A. E. Ferguson, and J. Ballantyne, *J. Am. Vet. Med. Assoc.,* **124**:279 (1954).

177. Medvedev, Zh. A., *Exp. Gerontol.,* **7**:227 (1972).

178. Menzies, R. A., and P. H. Gold, *J. Biol. Chem.,* **246**:2425 (1971).

179. Mochow, R., and D. Olds, *J. Dairy Sci.,* **49**:642 (1966).

180. Moreland, A. F., T. B. Clarkson, and H. B. Lofland, *Arch. Pathol.,* **76**:203 (1963).

181. Morgan, A. D., *Pathogenesis of Cornary Occlusion,* Charles C. Thomas, Springfield, Ill., 1957.

182. Moyer, E. K., and B. F. Kalizewski, *Anat. Rec.,* **131**:681 (1958).

184. Nielsen, J., *Br. J. Psychiat.,* **114**:303 (1968).

185. Norris, A. H., T. Lundy, and N. W. Shock, *Ann. N.Y. Acad. Sci.,* **110**:623 (1963).

186. Norris, A. W., and N. Shock, in W. L. Johnson (ed.), *Science and Medicine of Exercise and Sports,* Harper, New York, 1960, p. 466.

187. O'Dell, B. L., D. F. Elsden, J. Thomas, S. M. Partridge, R. H. Smith, and R. Palmer, *Nature* (London) **209:**401 (1966).
188. Orgel, L. E., *PNAS* (USA) **49:**517 (1963).
189. Orgel, L. E., *PNAS* (USA) **67:**1476 (1970).
190. Page, R. C., and E. P. Benditt, *Lab. Invest.,* **15:**1643 (1966).
191. Pantelouris, E. M., *Exp. Gerontol.,* **7:**73 (1972).
192. Parks, R. D., P. P. Scheerer, and R. R. Greene, *Surg. Gynecol. Obstet.,* **106:**413 (1958).
193. Pick, R., L. N. Katz, D. Century, and P. J. Johnson, *Circ. Res.* **11:**811 (1962).
194. Piez, K. A., G. R. Martin, A. H. Kang, and P. Bornstein, *Biochemistry* **3:**3813 (1966).
195. Pincus, G., in E. T. Engle and G. Pincus (eds.), *Hormones and the Aging Process,* Academic Press, New York, 1956.
196. Pollard, M., and M. Kajime, *Am. J. Pathol.,* **61:**25 (1970).
197. Post, J., A. Klein, and J. Hoffman, *Arch. Pathol.,* **70:**314 (1960).
198. Prior, J. T., D. M. Kurtz, and D. D. Ziegler, *Arch. Pathol.,* **71:**672 (1961).
199. Prichard, R. W., T. B. Clarkson, H. B. Lofland, and H. O. Goodman, *Am. J. Pathol.,* **43:**651 (1963).
200. Pritchard, R. W., T. B. Clarkson, H. B. Lofland, H. O. Goodman, C. N. Herndon, and M. G. Netsky, *J.A.M.A.,* **179:**49 (1962).
201. Pryor, W. A., *Sci. Am.,* **223:**70 (1970).
202. Ramchandran, G. N., *Treatise on Collagen,* Academic Press, London, 1967.
203. Rapaport, A., *Gerontologia,* **13:**14 (1967).
204. Rapaport, A., *Gerontologia,* **15:**228 (1969).
205. Ratcliffe, H. L., *Ann. N.Y. Acad. Sci.,* **127:**715 (1965).
206. Ratcliffe, H. L., T. G. Verasimides, and G. A. Elliott, *Circulation,* **21:**730 (1960).
207. Richer, C. P., *Comp. Psychol. Monogr.,* **1:**11 (1922).
208. Ridge, M. D., and V. Wright, *Gerontologia,* **12:**174 (1966).
209. Robert, L., S. Derouette, and E. Moczar, *Gerontologia,* **17:**65 (1971).
210. Robert, L., and B. Mobert, *Gerontologia,* **19:**330 (1973).
211. Robert, L., B. Robert and A. M. Robert, *Exp. Gerontol.,* **5:**339 (1970).
212. Rockstein, M., in G. E. W. Wolstenholme and M. O'Connor, (eds.), *Life Span of Animals,* Little, Brown & Co., Boston, 1959.
213. Rockstein, M., in B. L. Strehler (ed.), *Biology of Aging,* Waverly Press, Baltimore, 1960.
214. Rockstein, M., in P. L. Krohn (ed.), *Topics in the Biology of Aging,* John Wiley & Sons, New York, 1966.
215. Rockstein, M. and P. L. Bhatnagar, *J. Insect. Physiol.,* **11:**481 (1965).
216. Rockstein, M., and K. F. Brandt, *Science,* **139:**1049 (1963).
217. Rockstein, M. and H. S. Liberman, *Gerontologia,* **3:**23 (1959).
218. Rodey, G. E., R. A. Good, and E. J. Yunis, *Clin. Exp. Immunol.,* **9:**305 (1971).
219. Rolle, G. K., and H. A. Charipper, *Anat. Rec.,* **105:**281 (1949).
220. Rokitansky, C., van, *A Manual of Pathological Anatomy,* Vol. 4, translated by G. E. Day, Sydenham Society, London, 1852.
221. Romanoff, L. P., M. N. Baxter, A. W. Thomas, and G. B. Ferrechio, *J. Clin. Endocrinol.,* **29:**819 (1969).
222. Rosenfield, S., J. Marmorston, H. Sobel, and A. E. White, *Proc. Soc. Exp. Med.,* **103:**83 (1960).
223. Ross, M. H., *J. Nutr.,* **75:**197 (1961).
224. Ross, M. H., *J. Nutr.,* **96:**563 (1969).
225. Ross, M. H. and J. O. Ely, *J. Franklin Inst.,* **258:**63 (1954).
226. Roswell, H. C., H. G. Downie, and J. F. Mustard, *Can. Med. Assoc. J.* **79:**647 (1958).
227. Roswell, H. C., H. G. Downie, and J. F. Mustard, *Can. Med. Assoc. J.,* **83:**1175 (1960).
228. Roswell, H. C., J. F. Mustard, M. A. Packham, and W. J. Dodds, in *Swine in Biomedical Research,* Pacific Northwest Laboratory, Richland, Wash., 1966.
229. Sacher, G. A., *Exp. Gerontol.,* **3:**265 (1968).
230. Saltykow, S., *Zentr. Allgem. Pathol. Anat.,* **19:**321 (1908).
231. Samis, H. V., V. J. Wulff, and J. A. Falzone, *Biochem. Biophys. Acta,* **191:**223 (1964).
232. Sant'Ambrogio, G., D. Frazier, and L. L. Boyarsky, *Am. J. Physiol.,* **200:**927 (1961).
233. Scebat, L., J. Renais, and J. Lengre, *Rev. Atheroscler.,* **3:**14 (1961).

234. Schaub, M. C., *Gerontologia,* **10:**137 (1965).
235. Schimke, R. T., *J. Biol. Chem.,* **239:**3808 (1964).
236. Schmuckler, M., and C. H. Barrows, Jr., *J. Gerontol.,* **21:**109 (1966).
237. Schmuckler, M., and C. H. Barrows, Jr., *J. Gerontol.,* **22:**13 (1967).
238. Schwenk, E., and D. F. Stevens, *Proc. Soc. Exp. Biol. Med.,* **103:**614 (1960).
239. Shire, J. G. M., *Nature,* **245:**215 (1973).
240. Shire, J. G. M., and E. A. Hambly, *Acta Pathol. Microbiol. Scand., A.,* **81:**225 (1973).
241. Shirey, T. L. and H. Sobel, *Exp. Gerontol.,* **7:**15 (1972).
242. Shock, N. W., In B. L. Strehler, J. D. Ebert, H. W. Glass, and N. W. Shock (eds.), *Biology of Aging,* American Institute of Biological Sciences, Washington, D.C., 1960.
243. Shuckla, S. P. and M. S. Kanungo, *Exp. Gerontol.,* **4:**57 (1969).
244. Singhal, R. L., *J. Gerontol.,* (a) **22:**77 (1967); (b) **22:**343 (1967).
245. Skold, B. H. and R. Getty, *J. Am. Vet. Med. Assoc.,* **139:**655 (1961).
246. Skold, B. H., and R. Getty, *Am. J. Vet. Res.,* **27:**257 (1966).
247. Smith, E. B., *J. Atheroscler. Res.,* **5:**224 (1965).
248. Sobel, H., *Aerospace Med.,* **41:**524 (1970).
249. Ssolowjew, A., *Zentr. Allgem. Anat. Path.,* **53:**145 (1932).
250. Steblay, R. W., *J. Exp. Med.,* **116:**253 (1962).
251. Steblay, R. W., *J. Immunol.,* **95:**517 (1965).
252. Steiner, A. and F. E. Kendall, *Arch. Pathol.,* **42:**433 (1946).
253. Steiner, P. E., T. B. Rasmussen, and L. E. Fischer, *Arch. Pathol.,* **59:**5 (1955).
254. Sto, G. G., *Changes With Age in the Canine Female Genitalia,* Ph.D. thesis, Iowa State University, Ames, Iowa, 1970.
255. Strehler, B. L., G. Hirsch, D. Gusseck, R. Johnson, and M. Bick., *J. Thero. Biol.,* **33:**429 (1971).
256. Stuart, M. and F. R. Magarey, *J. Pathol.,* **79:**319 (1960).
257. Sulkin, N. M., in *Laboratory Animals in Gerontological Research,* National Academy of Science, Washington, D.C., 1968.
258. Sutherland, D. E. R., O. K. Archer, R. D. A. Peterson, E. Eckert, and R. A. Good, *Lancet,* **1:**130 (1965).
259. Tannenbaum, A., *Ann. N.Y. Acad. Sci.,* **49:**6 (1947).
260. Tappel, A. L., *Geriatrics,* **23:**97 (1968).
261. Tappel, A. L. and H. Zalkin, *Arch. Biochem. Biophys.,* **30:**333 (1959).
262. Taylor, C. B., G. E. Cox, P. Manalo-Estrello, and J. Southworth, *Arch. Pathol.,* **74:**16 (1962).
263. Taylor, D. W., *J. Physiol.,* **140:**37 (1958).
264. Teague, P. O. and G. J. Friou, *Immunology,* **17:**665 (1969).
265. Teague, P. O., G. J. Griou and L. L. Myers, *J. Immunol.,* **101:**791 (1968).
266. Teague, P. O., G. J. Friou, E. J. Yunis, and R. A. Good, in M. Sigel and R. A. Good (eds.), *Tolerance Autoimmunity and Aging,* Charles C. Thomas, Springfield, Ill., 1972.
267. Teague, P. O., E. J. Yunis, G. Rodey, A. J. Fish, O. Stutman, and R. A. Good, *Lab. Invest.,* **22:**121 (1970).
268. Thung, P. J., L. M. Boot, and O. Muhlbock, *Acta Endocrinol.,* **23:**8 (1956).
269. Timiars, P. S., *Developmental Physiology and Aging,* MacMillan Co., New York, 1972.
270. Treloar, A. E., R. E. Boynton, B. G. Behn, and B. W. Brown, *Int. J. Fertil.,* **12:**77 (1967).
271. Troup, G. M. and R. L. Walford, *Transplantation,* **5:**43 (1967).
272. United States Public Health Service & United States Bureau of Census, *Life Tables,* 1973.
273. Veis, I. and J. Anesey, *J. Biol. Chem.,* **240:**3899 (1965).
274. Viidik, A., *Bio. Med. Eng.,* **2:**64 (1967).
275. Von Hahn, H. P., *Adv. Gerontol. Res.,* **3:**1 (1971).
276. Walford, R. L., *Gerontologist,* **4:**195 (1964).
277. Walford, R. L., *Immunological Theory of Aging,* Munksgaard, Copenhagen, 1969.
278. Walford, R. L., *Lancet,* **2:**1226 (1970).
279. Wallace, D. C., *J. Chronic Dis.,* **20:**476 (1967).
280. Waring, M. J., *Nature,* **219:**1320 (1968).
281. Weglicki, W. B., Z. Luna, and P. P. Nair, *Nature,* **221:**185 (1968).
282. Weiss, H. S., *J. Gerontol.,* **14:**19 (1959).

283. Weiss, H. S. and M. Sheahan, *Am. J. Vet. Res.,* **19:**209 (1958).
284. Wells, W. W., R. Quan-Ma, C. R. Cook, and S. C. Anderson, *J. Nutr.,* **76:**41 (1962).
285. Whitehead, R. G., and D. G. Coward, *Bibl. Nutr. Diet,* **13:**74 (1969).
286. Wilson, P. D., *Gerontologia,* **19:**79 (1973).
287. Woessner, F., Jr., in D. A. Hall (ed.), *Internation Reviews of Connective Tissue,* Academic Press, New York, 1965.
288. Woessner, J. F. and T. Brewer, *Biochem. J.,* **89:**75 (1963).
289. Wright, E. A. and J. M. Spink, *Gerontology,* **3:**277 (1959).
290. Wulf, V. J., H. Quastler, and F. G. Sherman, *PNAS* (USA) **48:**1373 (1962).
291. Yamauchi, S., *Jap. J. Vet. Res.,* **26:**107 (1964).
292. Yuan, G. C. and R. S. Chang, *J. Gerontol.* **24:**82 (1969).
293. Yuhas, J. M., *Exp. Gerontol.,* **6:**335 (1971).
294. Yunis, E. J. G., P. O. Fernandes, O. Stutman, and R. A. Good, in M. Sigel and R. A. Good (eds.), *Tolerance, Autoimmunity and Aging,* Charles C Thomas, Springfield, Ill., 1972.
295. Yunis, E. J., H. R. Hilgard, C. Martinez, and R. A. Good, *J. Exp. Med.,* **121:**607 (1965).
296. Yunis, E. J., R. Hong, C. Martinez, M. A. Grewe, E. Cornelius, and R. A. Good, *J. Exp. Med.,* **125:**947 (1967).
297. Yushchenko, N. A., *Bull. Exp. Biol. Med.,* (USSR) **47:**293 (1959).
298. Zorzoli, A., *J. Gerontol.,* **10:**156 (1955).
299. Zorzoli, A., *J. Gerontol.,* **17:**358 (1962).
300. Zorzoli, A., and J. B. Li, *J. Gerontol.,* **22:**151 (1967).

CHAPTER 13
ANIMAL MODELS IN TERATOLOGY

Teratology is the science of pathological embryology, which encompasses the study of abnormal development and congenital malformation of the embryo. The embryonic phase includes the period during which characteristic organs are being formed into definitive structures. It is followed by the fetal period, which is characterized by maturation of the organ systems. The defect or malformation in the development of the embryo may be produced by intrinsic factors, such as genetic, hormonal, uterine environment, metabolic, and nutritional status of the mother, or extrinsic factors, such as drugs, chemical agents, radiation, microbial agents (viruses, in particular) (225a) (Table 13-1). The congenital malformations at birth in humans are about 1 to 2 percent but follow-up studies indicate that the incidence may be much higher (Table 13-2a,b) (174). Therefore, it is important to understand the effects of certain physical, chemical, and biological agents on the developing embryos so that preventive measures can be taken to avoid the occurrence of abnormal development. Fetuses in vitro and in situ and embryos (chick embryos, in particular) have been used to test potential teratogens. However, experimental animals may be more useful as models for the study of not only teratogenic chemicals or drugs or other agents but also to study the interactions of the embryo, the mother, and the agents. This chapter describes uses of experimental animals for testing the possible teratogenic activity of certain physical, chemical, and biological agents.

SELECTION OF ANIMAL SPECIES FOR TERATOLOGICAL RESEARCH

Because of the unpredictable response shown by developing embryos, no single animal species is best suited for assessing the effects of any potential teratogens. However, teratogenic activity of certain chemical agents or drugs may be tested on certain animal species based on such paramenters as estrous cycle, gestation period, and exact fertilization or implantation data (71). Among the inappropriate animals for teratological testing are those that do not possess first line defense mechanisms inherent in higher mammals to developing tissues. The selective transport function of the placenta should also be present. Nonmammalian species are not very suitable experimental models for human applications, although chick embryos are used extensively for understanding the mechanisms of teratogen actions. One of the major differences is the relative static nutritional and excretory functions in the avian system as compared to the dynamic system of mammals. The test compound remains in the pool for a long time in avian embryos, whereas in mammalian embryos, the test compound has a short test life.

Rodents and Lagomorphs
These mammals possess a highly atypical yolk sac placenta during early embryogenesis and are, therefore, more sensitive to teratogens than other mammals, including

Table 13-1. Summary of Teratogenesis

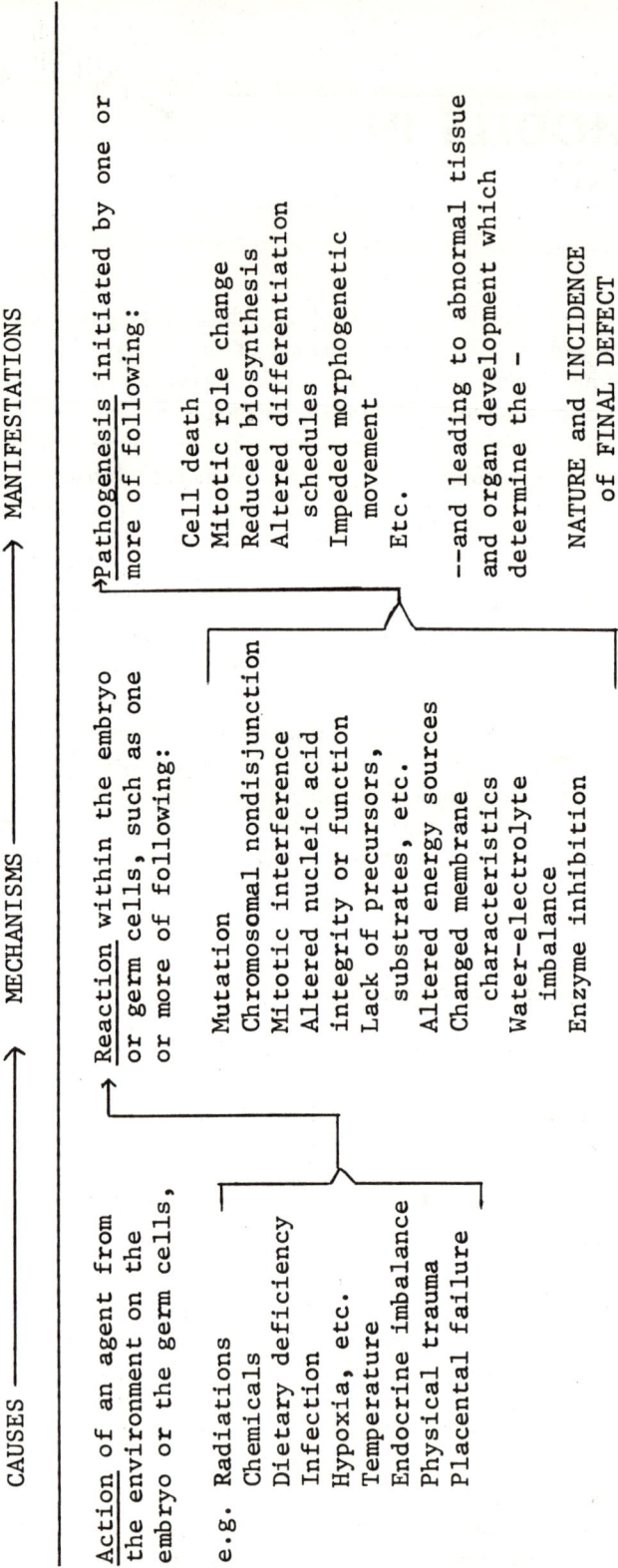

CAUSES ——→   MECHANISMS ——→   MANIFESTATIONS ——→

CAUSES	MECHANISMS	MANIFESTATIONS
Action of an agent from the environment on the embryo or the germ cells,	Reaction within the embryo or germ cells, such as one or more of following:	Pathogenesis initiated by one or more of following:
e.g. Radiations Chemicals Dietary deficiency Infection Hypoxia, etc. Temperature Endocrine imbalance Physical trauma Placental failure	Mutation Chromosomal nondisjunction Mitotic interference Altered nucleic acid integrity or function Lack of precursors, substrates, etc. Altered energy sources Changed membrane characteristics Water-electrolyte imbalance Enzyme inhibition	Cell death Mitotic role change Reduced biosynthesis Altered differentiation schedules Impeded morphogenetic movement Etc. --and leading to abnormal tissue and organ development which determine the - NATURE and INCIDENCE of FINAL DEFECT

Reproduced from Environment and Birth Defects (226) with permission.

Table 13-2a. Estimates of the Incidence of Congenital Defects in Human Embryos, Stillborn Children, and Liveborn Individuals.

```
Early Fetal Deaths
   Total incidence of abortions      20%
   embryonic defects                 25% total incidence 5%

Detected at Birth
   minor defects                      1%
   major defects                     0.5%
   lethal defects                    0.5% total incidence 2%

Detected in Follow-up Studies                          3%
                                     _____
                                     total incidence 10%
```

humans. However, the small size, short gestation period, large litter size, nonseasonal breeding, ready availability, and cost of rodents and lagomorphs make them the animals of choice for experimental teratology.

Mice

Highly developed inbred strains of mice are very useful for studies involving gene-chemical interactions (192). Special strains of mice have been developed which have a standard 4-day estrous cycle (141). Mice are very sensitive to some compounds, for example, 2,4,5-trichlorophenoxyacetate (2,4,5-T), cortisone, and so forth. The mouse is the animal of choice for testing mutagenicity because it is inexpensive (7, 27). However, some disadvantages in using mice for teratology research are (a) small size of the fetus; (b) small organ size for metabolic absorption and excretion studies; (c) erratic breeding performance and estrous cycle; (d) sensitivity to many chemicals and drugs. Nevertheless, mice are being used widely in teratological studies.

Table 13-2b. Follow-up Studies on the Incidence of Congenital Defects

Number of Children	Frequency at Birth	Observation Time	Follow-up (percent)	Frequency ("Final")
1. 672	1.5	4 years	75	10.1
2. 3179	2.4	3-6 years	91	16
3. 56760	1.73	5 years	95	2.31
4. 5531	3.5	6 months	89	6.0
		12 months	85	6.9
5. 63796	1.02	9 months	25	3.12

Adapted from <u>Congenital Birth Defects</u> (174) with permission.

Rats

Rats have certain unique characteristics which make them very useful species for teratological studies. These characteristics are (a) high fecundity; (b) low spontaneous malformation rate; (c) resistance to diseases; and (d) convenient size. Interspecies variation in rats is lower than other routinely used species (148). Rats are being used extensively in host-mediated tests for mutagenicity (116).

Hamsters

Hamsters have many of the same advantages as rats and mice. In addition, hamsters are larger in size and have more stable estrous cycles. Ovulation in hamsters can be easily determined (2, 144). The short organogenesis period has advantages as well as disadvantages, depending on the study.

Rabbits

Ovulation time and conception time in the rabbit can be precisely determined since the animals ovulate at the mating time. The relatively large size of certain breeds makes it possible to collect maternal and fetal samples for biochemical analysis. However, infectious diseases can create problems with reproductive performance of rabbits. The problem can be solved by using artificial breeding methods. A wide variety of studies have been reported on the genetic, and anatomical variations and spontaneous malformation in rabbits (146, 147, 172, 173). Teratogenic effect of thalidomide on rabbit embryosis is similar in many ways to that in humans; however, the lesions produced in rabbits are different (193).

Guinea Pigs

The guinea pig offers some interesting features for experimentation, particularly the long gestation period (68 days) and the production of histologically and functionally mature offspring. However, the small number of offspring (two to four litters) and the long period of the estrous cycle (16 days) are some of the disadvantages in using guinea pigs for teratological research. Information on genetics, embryology, and spontaneous malformations in guinea pigs are available in the literature (235).

Carnivores

This group of animals has not been used as extensively in teratology as the rodents. However, the carnivores have many similarities in the placenta to human beings. The three experimental species of carnivores, the dog, the cat, and the ferret, are seasonal breeders producing one to two litters a year. However, they are polytocons and their gestation period is not very long (about 60 days).

Dogs

Smaller breeds of dogs such as beagles are most often used in pharmaceutical and toxicological studies. Beagles breed well under laboratory conditions, and the incidence of spontaneous abnormalities is fairly low (26, 54, 74, 136, 147, 191). The use of dogs in teratology has been reported by many investigators (37, 43, 118, 171, 188). Studies have been made testing teratogenic activity of pharmaceuticals and drugs on dogs, for example, 6-diazo-5-oxonorleucine (56), thalidomide (37, 215), and oxytetracyclin (171). The teratogenic susceptibility of dogs cannot be fully established based on the limited available information.

Cats

Although cats have been used for teratological studies (84, 104a, 104b, 123, 206), lack of glucuronide forming enzyme in cats (which eliminates an important drug-conjugating system) makes it a rather difficult animal of choice. Also, cats are difficult to breed under laboratory conditions. However, the induced ovulation (24 hours after coitus), number of litters (2 to 3 a year), and the size and ease of handling are attractive features of cats as research subjects in teratology. Information on spontaneous malformation, anatomy, and embryology of cats is available (11, 185). Feline panleukopenia produces cerebellar hypoplasia and ataxia which should not be confused with abnormalities caused by the drugs.

Ferrets

The main advantages of using ferrets in teratology studies are small size, ease of breeding and maintenance, and susceptibility to most teratogens. However, they have only one estrous cycle. Not much information is available on pharmacological and toxicological studies with ferrets. A few studies reported the use of ferrets in teratological research (162, 198, 199).

Ungulates

The ungulates, particularly the pig and the sheep, are potentially useful animal species for the evaluation of human teratogens. Information on their breeding performance and management is available but pharmacological and toxicological studies are not extensively reported. Because of the large size of these animals, large quantities of test material are required.

Sheep

Several studies have reported on sheep for teratological research (41, 90, 91, 110, 202). Information on spontaneous malformations and intrauterine death of fetuses in sheep are also available (38, 47, 236). Usually the sheep has a low fecundity rate (one to two a year). However, certain breeds produce four or more lambs in a year (5, 6, 13, 34, 44, 70, 73, 99, 129, 150, 163, 180, 214, 237).

Pigs

Pigs are well suited for the evaluation of human teratogens because of their phylogenetic similarities to humans. Suitable management practices are necessary to reduce costs of maintainance of the litters, which may be as many as ten in number. Since the early work of Hale (73), the pig has been used for teratological studies by many workers (5, 6, 34, 44, 129, 163, 235). The organogenesis in the pig has been reported (150) and many studies have documented the incidence of malformations (70, 99, 180, 124).

Nonhuman Primates

Nonhuman primates offer the closest approximation to human teratological conditions because of polygenetic similarities and almost identical reproductive physiology in certain species. Several workers recommend the nonhuman primates as the animal of choice for teratological studies (189, 190, 201, 222, 223, 224, 231, 234). Among primates, *Papio* and *Macacca* species are most commonly used. Because of the high cost of purchase and maintenance and their dwindling supply, these animals can only be used for such critical studies as evaluations of pharmaceuticals and new drugs in pregnant women. Inasfar as the use of primates for predictive and drug safety evaluation studies is concerned, it has been suggested that a multiple animal testing procedure should be used. Compounds should be screened in a different animal at each stage of testing. However, a review of literature indicates that except for a few teratogens (sex hormones, thalidomide, radiation, etc.) the results in nonhuman primates are not comparable to those in humans (224). Nevertheless, phylogenetic relationship (225), metabolic similarities (190), placental structure and similarities in the vascular system (149, 157), and similarity in reactions with embryos make nonhuman primates the best test species available at this time. Specific examples of the use of nonhuman primates in teratological research have been reviewed (221, 222, 223, 224).

Chick Embryos

The use of chick embryos for testing teratogenic activity is not recommended because of its extreme sensitivity to teratogens. The chick embryo is not similar anatomically and physiologically to the mammalian embryo. Nevertheless, the chick embryo is an important research tool (232) and it has been widely used in many investigations (4, 102, 103). Gebhardt (61) tried to extrapolate the chicken data to the mammalian system. It appears that the teratogenic effects are to some extent similar. The chick embryo may therefore be reevaluated as an inexpensive test subject, particularly for LD50 determinations of highly toxic compounds (62, 208, 210).

COLLECTION AND INTERPRETATION OF DATA FROM EXPERIMENTAL ANIMAL STUDIES

A good experimental approach should be aimed at the collection of as much useful information as possible. In teratology, the effects of teratogens may be classified into the following four categories: (a) maternal effects, (b) embryonic effects, (c) effects on prenatal fetus, (d) effects on postnatal fetus.

Maternal Effects

The effect of the test drug on the body weight gain during pregnancy should be determined to indicate fetal development and the nutritional status of the pregnant female. Generally, the treated pregnant females do not gain weight to the same extent as the controls. This might be due to reaction to the test drug on the mother rather than fetal damage. Some authors have observed changes in liver weight (31a, 31b) and maternal anemia (219).

Embryonal Effects

Embryonic death is fairly common due to failure in implantation or implantation for a very short time with no recognizable implantation site. In the latter case it is difficult to interpret the results. The only way it can be recognized is to determine the number of corpus lutea of pregnancy in ovaries and compare these to implantation sites. This technique is useful with mice, rats, hamsters, and rabbits. In rhesus monkeys, the dead embryo is aborted followed by bleeding which cannot be differentiated from normal bleeding (229, 230). Observations are best made a day or a few days before parturition by hysterotomy or necropsy because most animals have a tendency to eat the dead fetus or fetus with malformation of the skin. Dead fetuses that show any degree of maceration should not be weighed or included in the examination of malformation because of edema, distortions, and artifacts. The fetuses should be examined by hand lens for varia-

Table 13-3. Examination of Traits at Autopsy for Teratological Findings

External examination
(before skeletonisation)

*Tail length
Abnormalities of the tail
Abnormal posture of the limbs
Polydactyly
Abnormal shape of head
Abnormal shape of jaw and mouth
regions
Macroglossia
Cleft palate
Cleft lip
Patent eyelids
Other abnormalities of the eye
(macrophthalmia, microphthalmia)
Exencephaly
Macroscopical spina bifida
Exomphalos
Ventral hernia
Generalised oedema
Hydramnios
Macroscopical degenerative changes
of the placenta

- -

Internal examination
(after skeletonisation)

Vertebral column
  Cervical region   Number of arches
    Ossification centres   *Absence
                            Fusion
                            Longitudinal splitting
                            Partial splitting
    Arches   Abnormal shape
                            Fusion
                            *Constriction
                            Diastasis

Thoracle region   Number of arches
  Ossification centres   *Reduction in size
                            Absence
                            Fusion
                            Longitudinal splitting
                            Partial splitting
  Arches   Abnormal shape
                            Fusion
                            *Constriction
                            Diastasis

Table 13-3. (Continued)

Lumbar region	Number of arches
Ossification centres	*Reduction in size
	Absence
	Fusion
	Longitudinal splitting
	Partial splitting
Arches	Abnormal shape
	Fusion
	*Constriction
	Diastasis
Intra-pelvic region	Number of arches
Ossification centres	*Number of ossification centres
	Fusion
	Splitting
Arches	Abnormal shape
	Fusion
	*Constriction
	Diastasis
Post-pelvic region	*Number of arches
Ossification centres	*Number of ossification centres
	Fusion
	Splitting
Arches	Abnormal shape
	Fusion
	*Constriction
	Diastasis
Abnormal shape of	Humerus
	Radius
	Ulna
	Femur
	Tibia
	Fibula
Number of ossification centres in paws	*Carpals
	*Metacarpals
	*1st phalanx
	*2nd phalanx
	*3rd phalanx
	*Tarsals
	*Metatarsals
	*1st phalanx
	*2nd phalanx
	*3rd phalanx
Ribs	*Number
	*Rudiment of 14th rib
	Abnormal costal angle
	Abnormal shape
	Branching
	Double ribs
	Fusion
	Partial absence
	Total absence

Table 13-3.   (Continued)

Sternum	Zig-zag formation
	*Tendency to zig-zag formation
	Total absence
Sternebrae	Fusion
	Splitting
	Other abnormalities of shape
	*Reduction in size
	Absence

* Features in which deviations from the normal have been interpreted, in the present study, as expressions of inhibition of embryonic growth.

tions in surface features, proportion, symmetry, missing parts, and intactness of the skin. The sex ratio should always be recorded (177).

Detection of Internal Abnormality

In small animals (mice, rabbits, rats, hamsters), it is best to divide the fetuses into two groups: one group for skeletal study using alizarin red S (175) and the other group for visceral examination by fixation in Bouins fluid. The direct examination of viscera is effective but time-consuming (135, 220). In large animals the skeletal examination is usually made radiologically, although alizarin procedure can also be used. The visceral examination is made postmortem and the weight of each organ is recorded. A summary of traits that should be examined is given in Table 13-3.

Postnatal Effects

For postnatal studies, weights should be measured weekly for 2 to 3 months. Deaths should be recorded and a detailed postmortem performed to determine malformations. Several workers have reported postnatal changes in behavior (58, 216) or structural defects (113, 114, 115, 145). Interpretation of experimental data is often difficult because of biological variability and human judgment of growth retardation, malformation, and functional deficiency. In the final analysis, the interpretation of data is based on statistical evaluations rather than absolute values.

TERATOGENS

Agents that may be potential teratogens and cause developmental defects in humans are radiation (therapeutic or nuclear); infections (rubella virus, cytomegalovirus, herpesvirus, toxoplasma, syphilis); maternal metabolic imbalance (endemic cretinism, diabetes, phenylketonuria, virilizing tumors); drugs and environmental chemicals; and known genetic transmission and chromosomal aberration. Some of these causes of malformations are briefly discussed in the following section.

Radiation

Congenital malformation of the mammalian embryo by radiation has been known for a long time. Early reports listed microcephaly, skull micromelia, club foot, and cleft palate in human embryos (69, 138). However, because of the possible adverse effects of radiation, therapeutic irradiation of the pelvis during pregnancy has been stopped. X-ray irradiation is not a major hazard to pregnant mammals. The main effects of irradiation include (a) intra- or extrauterine growth retardation; (b) embryonic, fetal, or neonatal death; and (c) congenital malformations.

Rats and mice are the most commonly used animals for studies to determine the effects of radiation. In these animals, irradiation almost always results in the death of the embryo or extensive malformations during the preimplantation period (15, 16, 168). The rate of aneuploidy (XO), which is about 1 percent, increases considerably in mice exposed to 100 rad (170) and the incidence of encephaly is also increased (164). However, irradiation on the first day of gestation does not produce any malformation (18, 19) but does result in mortality. Large numbers of malformations are produced during the organogenesis period [i.e., day .5 to day 12 in mice and day 9 to day 13 in rats (17), (Table 13-4)]. Low dosages at later stages of the organogenic period produce tissue hyperplasia and sometimes changes in the number of ribs (169). During the later stages of pregnancy in rats and mice (12 to 18 days), which is equivalent to the second and third trimesters in humans, defects in the nervous system are noticed. The x-ray irradiation does not produce gross malformations, but alterations in major organs and systems have been reported (3, 20, 23, 36). The severity is related to the magnitude of the dose. Major brain damage by irradiation includes irregular and bizarre neurons and discernible but subtle changes in adult brain cortex (81). The rat shows changes in the thyroid, diaphragm, and testes (14, 137).

It appears that rats and mice are good models for comparable studies in human pregnant females as similar findings have been observed by low dose x-ray irradiation between 14 days to full-term pregnancies in these animals (109, 131, 179).

The administration of low dosages of irradiation throughout pregnancy produces retardation of growth. Comparative data obtained by several workers are summarized in Table 13-5 (17). It seems that the results with animals are similar to those obtained with humans. Growth retardation may be a more sensitive criterion of radiation damage than malformations. The extent of retardation or its absence depends on the dose of radiation and the stage of gestation (119, 165, 166, 187). A dose of 60 rad is the lowest that may produce growth retardation in mice. Marked growth retardation occurs when neonatal and infant rats are irradiated to different levels (8, 152).

Examination of data on human children following atomic bomb irradation, as well as human neonates and infants, showed similar growth retardation patterns (29, 132, 133).

Effects of Therapeutic Radiations

The main effect of therapeutic or diagnostic levels of certain isotopes depends on (a) specific target organ, (b) ability to cross placenta barrier, (c) extent of distribution of radioisotopes, and (d) metabolism of radioisotopes.

Normal use of most of the radioisotopes for therapeutic purposes (Tables 13-6 and 13-7) (17) have not had deleterious effects except when given in ablative doses in pregnant females, which may cause complete destruction of fetal thyroid (200). Inorganic radioactive compounds, particularly phosphorus and strontium, may cause embryonic death in rats (16).

Infections

Cytomegalovirus, herpesvirus and rubella virus have been reported to produce congenital defects in human embryos (Table 13-8) (183). In addition, there are a number of

Table 13-4. Estimation of Acute Dose LD$_{50}$ and Minimal Malforming Doses of Irradiation for the Human Embryo Based on Compilation of Mouse, Rat, and Human Data

Age	Approximate Minimal Lethal Dose	Approximate LD$_{50}$	Minimum Dose for Non-rerecuperable Growth Retardation in Adult	Minimum Dose for Recognizable Gross Malformations	Minimum Dose for Induction of Genetic, Carcinogenic, and Minimal Cell Depletion Phenomena
	R	R	R	R	
			*	**	
Day 1	10	70-100			Unknown
Day 14	25	140	>25		Unknown
Day 18	50	150	>50	>25	Unknown
Day 28	>50	220	>50	25	Unknown
Day 50	>100	260	>50	50	Unknown
Late Fetus to term		300-400	>50	>50	Unknown

*Surviving embryos are not growth retarded even after high doses of irradiation.
**Malformation incidence is extremely low even after high doses of irradiation.
Adapted from <u>Pathology of Development</u> (17) with permission.

473

Table 13-5. The Effect of Low Exposures of Irradiation on the Embryo When Administered Throughout Pregnancy

Organism	Source	Approximate Exposure rate/min (R)	Exposure/Day	Exposure/pregnancy	Comments	Effects
Mouse	^{137}Cs	0.0086	12.4 R	171 R	Day 1–18	None except shortened breeding period in female
Mouse	^{137}Cs		8.4 R	170 R	During gestation and in some instances 20 days postpartum	None
Rat	^{60}Co	0.0017	2.5 R	—	Daily during pregnancy	None
Mouse	^{60}Co	0.0015	2.2 R	—	Continuous through 10 generations	None
Rat	^{60}Co	0.0014	2.0 R	—	Continuous through 11 generations	None
Mice	^{60}Co	0.007	10 R	180 R	Day 1–18	None
		0.014	20 R	360 R	Day 1–18	Growth retardation
					Day 6–13 only	None
Humans	Background		0.3 mR	0.1 R	Background radiation	Increased congenital malformation
Humans	Background		0.3 mR	0.1 R	Background radiation	Background radiation level not a factor in incidence of congenital malformations
Mouse	X-ray		2.5 R		Acute dose given daily	No influence
			5.0 R			No influence
			10 R			No influence
			20 R			Malformations, resorptions, growth retardations
Rat	X-ray		1.0 R	20 R	Acute dose given daily	Functional changes in behavior and motor activity (questionable)

Adapted from *Pathobiology of Development* (17) with permission.

Table 13-6 List of Isotopes and Their Uses

Isotope	Chemical Form	Use
Cesium 137	Encased in needles and/or applicator cells	Interstitial or intracavitary treatment of cancer
	Teletherapy source	Treatment of cancer
Chromium 51	Sodium chromate	Spleen imaging, placenta localization, and RBC labeling and survival studies
	Labeled human serum albumin	GI protein-loss studies, placenta localization
	Labeled RBCs	Placenta localization
Cobalt 58 or cobalt 60	Labeled cyanocobalamin	Intestinal absorption studies
Cobalt 60	Teletherapy source	Treatment of cancer
	Encased in needles and/or applicator cells	Interstitial or intracavitary treatment of cancer
Gold 198	Colloidal	Liver imaging, intracavitary treatment of malignant effusions, interstitial treatment of cancer
	Seeds	Interstitial treatment of cancer
Iodine 131	Iodide	Diagnosis of thyroid function, thyroid imaging, treatment of hyperthyroidism and/or cardiac dysfunction and thyroid carcinoma
	Iodinated human serum albumin	Blood volume determination, brain tumor localization, placenta localization, cardiac imaging for determination of pericardial effusions, cisternography
	Rose bengal	Liver function studies
	Iodopyracet, sodium iodohippurate, sodium diatrizoate, diatrizoate methylglucamine, sodium diprotrizoate, sodium acetrizoate, sodium iothalamate	Kidney function studies and kidney imaging
	Labeled fat and/or fatty acids	Fat absorption studies
	Chlorografin	Cardiac imaging for determination of pericardial effusions
	Macroaggregated iodinated human serum albumin	Lung imaging
	Colloidal microaggregated human serum albumin	Liver imaging

Table 13-6. (Continued)

Isotope	Chemical Form	Use
Iodine 125	Iodide	Diagnosis of thyroid function
	Iodinated human serum albumin	Blood volume determination
	Rose bengal	Liver function studies
	Iodopyracet, sodium iodohippurate, sodium diatrizoate, diatrizoate methylglucamine, sodium diprotrizoate, sodium acetrizoate, sodium iothalamate	Kidney function studies
Iron 59	Labeled fats and/or fatty acids	Fat absorption studies
	Chloride, citrate and/or sulfate	Iron turnover studies
Iridium 192	Seeds encased in nylon ribbon	Interstitial treatment of cancer
Krypton 85	Gas	Diagnosis of cardiac abnormalities
Mercury 197	Chlormerodrin 197Hg	Kidney and brain imaging
Mercury 203	Chlormerodrin 203Hg	Brain imaging
Phosphorus 32	Soluble phosphate	Treatment polycythemia vera, leukemia, and bone metastasis
	Colloidal chromic phosphate	Intracavitary treatment of malignant effusions, interstitial treatment of cancer
Potassium 42	Chloride	Potassium space studies
Selenium 75	Selenomethionine	Pancreas imaging
Strontium 85	Nitrate or chloride	Bone immaging in patients with known or suspected cancer
Strontium 90	Medical applicator	Treatment of superficial eye conditions
Technetium 99M	Pertechnetate	Brain imaging, thyroid imaging, placenta localization, blood pool imaging, salivary gland imaging
	Iron-ascorbate-diethylenetriamine pentaacetic acid complex	Kidney imaging
Xenon 133	Free gas or in solution	Diagnosis of cardiac abnormalities, blood flow studies, pulmonary function studies

Adapted from *Pathobiology of Development* (17) with permission.

Table 13-7. Utilization of Radioisotopes in a University Hospital for a 1-Year Period

Type of Study	Isotope	No. of Studies
Radioisotope scans		
Metastasis for thyroid cancer	^{131}I	16
Abdomen	^{198}Au	3
Chest	^{32}P	1
Brain	^{99m}Tc	645
Chest (thyroid metastasis)	^{131}I	51
Cisternography	^{131}I albumin	34
Liver	^{99m}Tc sulfur colloid	398
	^{131}I rose bengal	
Lung	^{131}I albumin macroaggrega-tes	152
Pancreas	^{75}Se	20
Parathyroid	^{75}Se	1
Spleen	^{51}Cr	
	^{99m}Tc sulfur colloid	13
Thyroid	^{131}I	358
Lymphangiogram	^{198}Au	7
CSF leak	^{131}I albumin	11
Shunt potency	^{131}I albumin	3
Blood studies		
Conversion ratio	^{131}I	81
Protein-bound iodine 131	^{131}I	77
Plasma volume	^{131}I albumin	31
RBC volume	^{51}Cr	7
RBC survival	^{51}Cr	6
T$_3$ uptake	^{131}I	488
T$_4$ uptake	^{125}I	17
^{131}I levels in blood, urine	^{131}I	15
Other studies		
Thyroid uptake	^{131}I	470
Placentogram	^{99m}Tc colloid	0
	^{131}I albumin	24
Renogram	^{131}I hippuric acid	60
Shilling test	^{57}Co	49
Therapy for thyroid cancer	^{131}I	18
Therapy for thyrotoxicosis	^{131}I	23
Therapy for polyeythemia	^{32}P	5
Therapy for visceral effusions	^{32}P	4
	^{198}Au	
Eye tumor localization	^{32}P	3
Bladder tumor therapy	^{198}Au	2

*Total studies--3,134; total patients--2,641; total patients pregnant (placenta location)--24; percentage of pregnant patients administered isotopes-- <1%; total deliveries--3,400; percentage of pregnant patients that had placental localization--0.7%.
Adapted from Pathobiology of Development (17) with permission.

477

Table 13-8. Viruses Teratogenic for Humans

	Cytomegaloviruses	Herpesvirus hominis	Rubella
Birth defects	Microcephaly, choriorentinitis, deafness Mental retardation, hepatosplenomegaly Epilepsy, hydrocephalus, cerebral palsy Death	Microcephaly, hepatosplenomegaly Microphthalmos, retinal dysplasia Mental retardation Death	Malformations of heart and great vessels Microcephaly, deafness, cataracts, mental retardation Newborn bleeding, hepatosplenomegaly, pneumonitis hepatitis, encephalitis Death
Detection Mother	No clinical symptoms Virus isolation from urine and cervix Seroconversion (neutralization test, complement fixation, hemagglutination inhibition)	Research needed Most asymptomatic and no herpetic lesions Vaginal-cervical infection; some ulcerative lesions; husband may also have infection Laboratory-papanicolaou smear often shows cells with inclusions; virus isolation from cervix	Exposure, rash, nodes Lab tests—antibody response (hemagglutination inhibition); virus isolation from nasopharynx
Child	Research needed Wide spectrum of clinical findings Only severely affected are usually recognized	Research needed Often difficult to recognize Skin lesions sometimes develop Lab tests—elevated IgM and	Congenital rubella syndrome Lab tests—elevated IgM and rubella-specific IgM; persisting rubella antibody after 6 months of age; virus isolation from nasopharynx

478

Table 13-8. (Continued)

	Cytomegaloviruses	Herpesvirus hominis	Rubella
Child	Lab tests—elevated IgM and CMV-specific IgM; virus isolation from nasopharynx and urine; large cells in urinary sediment and tissues	specific herpes type II IgM; isolation of virus from skin lesions or tissues	or cerebrospinal fluid
Prevention	Research needed Chemotherapy (?) Vaccines (?)	Research needed Chemotherapy (?) Vaccines (?) Delivery by cesarean section (?)	Rubella vaccine Gammaglobulin—for exposed susceptible women use hemagglutination inhibition test to detect inapparent infection Abortion

Adapted from Pathobiology of Development (183) with permission.

other viruses that produce teratogenic effects in domestic and experimental animals. The main characteristic of the infective viruses is their ability to cross the placental barrier.

Cytomegalovirus (CMV) Infection

CMV is widely distributed in nature. About one-third or two-thirds of women have antibodies against this virus (57, 181) and 5 percent carry it in the urine or cervix (182, 195). There is an increased number of infections in late pregnancy, but the majority of women are asymptomatic. Congential infection occurs in 0.5 to 1.5 percent of births (10), and the virus can be isolated from nasopharynx and urine. The IgM level is greatly increased to 22/mg percent. The affected children show permanent motor damage in the form of mental motor retardation and microcephaly.

Although no satisfactory animal model is available, mice have been used to study the teratogenic effects of CMV infection (79, 127).

Herpesvirus Hominis (HVH) Infection

HVH is generally acquired at birth through venereal infection and may take place late in gestation (181). The lesions, if present, may be confirmed by the presence of inclusion bodies in Papanicolaou smears (183). The disease may affect the skin, throat, or conjunctiva, the central nervous system, or it may occur as systemic disease manifested by anemia, hepatitis, jaundice, and other symptoms (139).

Rubella Virus Infection

The disease is primarily associated with infection during the first trimester; infection during the second trimester results in greater abnormalities. Various defects due to rubella infection have been frequently reported (30, 83).

Several animal models have been used for rubella, CMV, and herpesvirus infection. Many of these are unique and bear no resemblance to the human condition (Table 13-9) (182). For experimental viral teratology, the following guidelines were suggested by Sever (183).

1. Infection should be produced in adult, fetus, and newborn;
2. Virus must be passaged through animals before use;
3. Several different intervals during gestation should be studied;
4. The indirect effect on mother and infant should be considered;
5. An effect should be preventable by neutralization of the virus.

In humans, influenza A, Coxsackie B, mumps virus, reoviruses, and ECHO viruses are under investigation for their teratogenic effects. Initial studies indicate that the primary effects of these viruses are malformation of the CNC, congenital heart disease, and endocardial fibroelastosis (22, 72, 196).

Experimental animals have been used as models for many other viral teratogenicity studies. The attenuated blue tongue virus vaccine produced intrauterine death in sheep and blindness, ataxia, and dummies among the survivals (176). In swine, attenuated hog cholera virus caused cerebral hypoplasia, poor myelination, facial and renal edema, and ascites (46, 237). Two herpes-like viruses are teratogenic to horses and cattle (28, 42). Several laboratory animals (hamsters, mice, rats, and cats) have been used to test parvovirus, H-1, RV viruses. In most cases the viruses result in exencephaly, facial clefts, abdominal herniation, funny faces, hepatitis, and cerebral hypoplasia (53, 106, 107). Reovirus type I can produce hydrocephalus in hamsters, ferrets, rats, and mice (108).

Table 13-9. Experimental Intrauterine and Newborn Infections

Virus	Animal	Effects
Rubella	Monkey (*Macaca mulatta*) (*Macaca irus*)	Newborns and adults infected. Limited intrauterine infection. Reported lense, optic, cutaneous, chorion, and osseous lesions.
	Ferret	Limited intrauterine infection, persistent newborn infection, lense opacity in 10% of newborns following IC infection.
	Rat	Reported chronic intrauterine infection, purpura, lense opacities, malformations (rare), frequent neonatal deaths.
	Hamster	No transplacental infection.
Cytomegalovirus	Mouse	Rapidly fatal, generalized infection in newborns. No congenital infection but increased intrauterine deaths.
Herpes simplex	Rabbit	Intrauterine death and readsorption.
	Hamster	Direct inoculation in utero gives generalized fetal infection with rapid death.
Attenuated blue tongue virus	Sheep	Congenital anomalies of central nervous system (hydroanencephaly, subcortical cysts).
Attenuated hog cholera (Myxo virus like)	Swine	Congenital deformed noses, kidneys, edema, ascities. Inrauterine infection. Cerebral hypoplasia and poor myelination (tremor, ataxia).
Infectious bovine rhinotracheitis (Herpes like)	Cows	Congenital infection and fetal death.
Equine abortion virus (Herpes like)	Horse	Congenital infection and abortion.
H-1, RV viruses (DNA, heat and ether resistant. Intra nuclear inclution)	Hamster (mouse, cat, rat)	Intrauterine and new born infection and effects exencephalus, microcephalus, facial clefts, abdominal herniation, "Mongoloid faces", hepatitis and cerebellar hypoplasia, virus destroys mitotic cells. Not found in association

Table 13-9.   (Continued)

Virus	Animal	Effects
		with similar human malformations.
Feline ataxia virus (similar to H-1, RV viruses)	Cats	Congenital infection and cerebellar hypoplasia, ataxia
Mumps	Hamster	Newborn infection produces hydrocephalus due to aqueductal stenosis.

Adapted from Methods for Teratological Studies in Experimental Animals and Man (182) with permission.

Nutrition

In humans, nutritional deficiency seldom results in congenital defects unless the deficiency is very severe (Chapter 5). However, several workers have shown that protein and caloric deficiency during pregnancy may result in retarded growth and mental development and death (89, 153, 158, 217, 232, 239). The relationship between protein intake during early adolescence and reproductive performance and mental defects in later years was reported by Valadian and Reed (207), Dobbing (40), and Hertig et al. (80). In addition to proteins, vitamin A, B complex, ascorbic acid, and folic acid have been implicated as causes of certain defects (151).

Most of the work on the effects of nutrition on teratogenesis have been done using experimental animals because of the ease in manipulations of composition and amount of feed. However, unique animal conditions may not represent models for human disease (48, 65, 85, 92, 101). By and large, most of the nutrients produce deficiency symptoms rather than any malformations. However, malformations of eyes from vitamin A deficiency in pigs has been reported (73, 74). The role of dietary deficiency of riboflavin in skeletal defects in rats has also been reported (213). Another approach followed in teratogenic studies by several workers is to use vitamin antimetabolites including galactoflavin (riboflavin), deoxypridoxine (pyridoxine) methyl pteroglutamic acid and aminopterin (folic acid), methyl pantothenate and pantoyltaurine (panthothenic acid), and 6-amino nicotinamide (niacin) (233). Teratogenic changes by manganese in the form of ataxia has been reported in rats and guinea pigs. Other skeletal defects and defects of cranium have also been reported (86, 87, 88). Trace elements such as zinc, cobalt, copper, and molybdenum have also been studied using golden hamsters and rats (52) (Table 13-10). Teratogenic effects of other metals like lead, cadmium, arsenic, mercury, and indium have been reviewed by Fern (52).

Endocrines

Hormone levels in individuals vary with physiological (nutritional, psychological and physical stresses) and pathological conditions. In recent years, the use of hormones as therapeutic drugs and as contraceptives has considerably increased.

Table 13-10. Embryonal Effects of Molybdenum, Copper, and Nickel on Pregnant Golden Hamsters

	Amount injected (mg/kg*)	Number of pregnant mothers	Total number of embryos	Number of embryos resorbed	Total number of surviving embryos	Number of normal embryos	Number of abnormal embryos
Sodium molybdate	40	3	38	2	36	36	0
	80	2	28	1	27	27	0
	100	2	28	1	27	27	0
Manganese chloride	10	3	38	1	37	36	1
	20	7	90	17	73	71	2
	30	7	91	35	56	54	2
	35	3**	10	1	9	9	0
Copper sulphate	2	1	12	0	12	12	0
	5	14	176	64	112	105	6
	7.5	3	30	30	0	–	–
	10	2	13	13	0	–	–
Nickelous acetate	2	2	24	0	24	22	2
	5	2	22	1	21	20	1
	10	5	56	22	34	32	2
	20	5	55	10	45	44	1
	25	6	68	59	9	5	4
	30	3	33	33	0	–	–

*All animals injected intravenously on the eighth day of gestation. ** Two maternal animals died immediately after injection.
Reproduced from <u>Advances in Teratology</u> (52) with permission.

Administration of hormones as contraceptive medications or use of oral hormonal pregnancy test tablets may to some extent cause congenital malformation (59, 60). The effects may be due to nonspecific imbalances rather than brought about by the specific hormone under study. Pathological conditions in pregnant females affect hormonal imbalance thereby causing malformation. Various experimental animal studies and data on human subjects are available (204).

Large doses of female sex hormones have proved to be teratogenic in experimental animals; however, the conclusions from these animal studies may not be applicable to humans where the level of hormone intake is low (55, 100).

Genetic Transmission

The study of sex differentiation can ideally be carried in animals because the X chromosome appears to have the same genetic constitution in all species (142, 143). However, the sex-linked genes specify few products not quite linked with sex differentiation. In reality, the sex differentiation on a single locus on the X chromosome is a regulatory locus specifying for nuclear-cytosol and rogen-receptor protein. A female develops when this system is in the noninduced form (androgen induced), and a male develops when the system is in the induced state (143, 120). This may explain the differentiation into male at a very early age and sex differences are discernible though female differentiation has not yet taken place (95, 96a, 96b, 97). The exact time schedule for embryonic seminiform cords in the human embryo is 6 weeks (95), 39 to 40 days in the calf (98), and 13 days in the rat (94, 96b). In fact, the male hormones impose the differentialities of the male genital tract before the organogenesis for females can start. In the study of freemartins, it appears that the abnormality is produced by the influence of male hormones (98). The hormonal disturbances during embryogenesis can affect sexual differentiation; these errors may be of genetic or nongenetic type (68). The eventual phenotypic expression is under the control of three hormones. Two of these, müllerian regression factor and testosterone, are under the control of the embryonic testes (95). The müllerian factors suppress the müllerian ducts; thereby the uterus and fallopian tubes are not developed (93). Testosterone promotes virilization by directly affecting the development of the epididymis, vas deferens, and seminal vesicles. It acts as prohormone for dihyrotestosterone to induce the formation of the prostate and external genitalia (186, 224, 225). In contrast, the female development is passive and no action of fetal hormone is necessary.

Disturbance in this normal process of embryogenesis resulting in clinical abnormalities may be due to environmental influences, sex chromosome aberrations, birth defects and a single gene mutation (Table 13-11) (68). The disorders of phenotypic sex are due to defective androgen synthesis (12, 156, 238), defective androgen action (49, 66, 67, 224, 225), or defective müllerian duct regression (21, 35).

Drugs

In recent years, there has been great concern over the teratogenic properties of drugs. Certain drugs may be directly associated with teratogenicity, and others may be potentially teratogenic by interactions with some pharmaceutical compounds (Table 13-12) (226). Three types of drugs have been shown to have direct teratogenic effects (226, 227): steroid hormones with androgenic activity causing virilization (25, 209, 218); thalidomide with known facial and limb defects (117); and folic acid antagonists (45, 128, 134, 155, 184).

Table 13-11. Hereditary Disorders of Human Sexual Development Classified in Terms of the Site of Action of the Abnormal Gene During Embryogenesis

---

I. Errors of genetic sex
    No monogenic disorders have been identified to date

II. Errors of gonadal sex
    Familial true hermaphroditism
    Sex reversal syndrome
    Pure gonadal dysgenesis, familial XX type
    Pure gonadal dysgenesis, familial XY type

III. Errors of phenotypic sex
    A. Familial male pseudohermaphroditism
       Defective virilization
       a. Defective androgen synthesis
          21, 22-Desmolase deficiency
          3B-Hydroxysteroid dehydrogenase deficiency
          17-Hydroxylase deficiency
          17-Ketosteroid reductase deficiency
          17, 20-Demolase deficiency
       b. Defective androgen action
          1. Testicular feminization, complete and partial types
          2. Incomplete male pseudohermaphroditism, Type I
             Gilbert-Dreyfus syndrome
             Lubs syndrome
             Reifenstein syndrome
             Rosewater syndrome
          3. Incomplete male pseudohermaphroditism, Type II
             Familial pseudovaginal perineoscrotal hypospadias
       Defective müllerian duct regression
          Persistent müllerian duct syndrome
    B. Familial female pseudohermaphroditism
       21-Hydroxylase deficiency
       11-Hydroxylase deficiency
    C. Familial defects of wolffian-müllerian development
       Rokitansky-Küster-Hauser syndrome
       Cystic fibrosis

Adapted from Birth Defects (68) with permission.

The drug should be tested in more than one animal species to evaluate its teratogenic effects in humans. For example, Imipramine was initially thought to cause defects like thalidomide in humans based on observations in rats and rabbits (1, 125, 161). However, during administration of imipramine in the rhesus monkey, at a dosage several times larger than that given in humans, no embryotoxicity was reported (228). Later, the human report was withdrawn from the imipramine studies. It seems the rhesus monkey is a better animal model for evaluation of teratogenicity of drugs than any rodent. Anticonvulsant drugs have been implicated to cause birth defects in epileptic women (122, 194). The use of diphenylhydantoin at very high concentrations produced toxicity to embryo in the monkey at 100 days of gestation. The defects were minor in the

Table 13-12.   Some Drugs Associated with Human Teratogenicity

---

I. **Drugs suspected of some teratogenic potential**
   1. Anticonvulsants
      Diphenylhydantoin and/or phenobarbital
      Trimethadione
   2. Neurotropic-anorexigenic drugs
      Dextroamphetamine
      Phenmetrazine
   3. Oral hypoglycemics
      Tolbutamide and other sulfonylureas
   4. Alkylating agents generally

II. **Drugs possibly teratogenic under some conditions**
   1. Aspirin
   2. Some antibiotics
   3. Antituberculous agents
   4. Quinine
   5. Imipramine (?)
   6. Insulin
   7. Female sex hormones

Adapted from Environment and Birth Defects (226) with permission.

prenatal fetus and did not interfere with the postnatal life (228). Combination of diphenylhydantoin with phenobarbital did not increase the incidence of defects. Trimethadione, another anticonvulsant, has been implicated in human teratogenesis (64). This drug has also been shown to possess teratogenic property in monkeys and the extent of damage may be more pronounced (154).

Hypoglycemic agents, such as tolbutamide, have been shown to be teratogenic in rats and rabbits (204, 227). However, the effects of tolbutamide in the rhesus monkey is not yet confirmed. Folic acid antagonists, such as methotrexate, have been involved in human cases but monkeys appear to tolerate the drug. On the other hand retionic acid (vitamin A) has not been implicated in humans but it is highly effective in rodents and monkeys (224, 225). Another drug, hydroxyurea, (used in humans as an anticancer agent because of its low toxicity) has been shown to possess teratogenic acid. It may be related to inhibition of DNA synthesis activity (178). Several other drugs like cortisone, meclizine (antihistamine), tranquilizers, antiemetics, and sulfonamides are considered apparently safer drugs for therapeutic purposes and are being extensively used in humans. However, some of these drugs have been reported to produce teratogenic effect in rodents. Therefore, further studies should be carried out to reevaluate these drugs. A controversial drug, LSD, has been much in the news and its cytogenetic and teratogenic effects are also controversial. However, it appears that, except for drug abuse cases, it does not possess teratogenic properties. The animal reports on the effects of LSD in mice, rats, and hamsters are inconsistent, whereas monkeys are mostly refractible (39, 121, 227). The literature dealing with drugs as teratogens has been recently reviewed (225b).

Environmental Chemicals

In recent years, the role of increasing chemicals in the environment is being examined for their potential teratogenic effects (212). For example, a recent case involving the illicit use of whiskey resulted in a child with neuromuscular abnormalities and retarded growth (148). A high absenteeism rate and the birth of children with club feet among pregnant women working in a media preparation laboratory was related to selenium (160). Other examples are the teratogenic effect of methylmercury in humans expressed as multiple neurological symptoms resembling cerebral palsy. Intake of methylmercury is found in workers in hazardous areas (24) and in people using some food products (76, 124, 203). Organomercurials also produce teratogenic effects in mice (195) and hamsters (77).

Industrial Solvents

Many solvents have been shown to be teratogenic in laboratory animals, for example, acetamides and formamides in rats (205), dimethyl sulfoxide in hamsters (51), benzene in mice (211), and alkane sufonates in rats (78). Except for one report (111) with fat solvents, no other human cases are known.

Table 13-13. Tertogenic Effects of Various Agents on the Embryo of Humans and Animals

Agent	Chicken	Rat	Mouse	Rabbit	Monkey	Man
			Test Species			
Acetazolamide	t	t	p.n.t.	p.n.t.	p.n.t.	p.n.t.
Actinomycin D	t	t	t	p.t.	p.t.	p.t.
Aminopterin	t	t	p.t.	p.t.	p.t.	t
6-Aminocotinamide	t	t	t	t	p.t.	p.t.
Beta - aminopropionitril	t	t	p.t.	p.t.	p.t.	p.t.
Colchicine	t	p.t.	t	p.t.	p.t.	p.t.
Cortisone	t	n.t.	t	t	?	?
Cyclophosphamide	t	t	t	t	p.t.	t
5-Fluorouracil	t	p.t.	t	p.t.	p.n.t.	?
Hypoxia	t	t	t	p.t.	?	?
Insulin	t	t	t	?	?	?
Meclizine	?	t	t	?	?	n.t.
Methionine Sulfhoximine	t	t	t	?	?	?
Nitrogen mustard	t	t	t	p.t.	p.t.	p.t.
Propylene glycol	t	p.n.t.	p.n.t.	n.t.	p.n.t.	p.n.t.
Salicylates	t	t	t	n.t.	p.n.t.	p.n.t.
Sulfonamides	t	t	p.n.t.	p.n.t.	p.n.t.	p.n.t.
Tetracyclines	t	t	?	?	?	p.t.
Thalidomide	?	t	?	t	t	t
Triethylene melamine	t	p.t.	t	p.t.	p.t.	p.t.
Trypan blue	t	t	t	p.t.	p.n.t.	p.n.t.
X-Rays	t	t	t	t	t	t
Rubella virus	?	t	?	t	t	t

t = teratogenic; p.t. = probably teratogenic; p.n.t. = probably not teratogenic; n.t. = not teratogenic.
Adapted from Advances in Teratology (61) with permission.

Pesticides

The greatest concern has been about the defoliant 2, 4, 5-T and its alleged embryotoxicity (9, 33, 130). However, it appears that it is nonteratogenic to humans and only mildly effective in mice. Rabbits, sheep and rhesus monkeys showed no effects. Rat and hamster embryos were affected by doses that were very high and at maternal toxicity levels. Other pesticides give variable results in laboratory animals: captan caused malformations in rabbits (126); carboryl produced teratogenic effects in high dosage in guinea pigs but rats and hamsters showed no effect; thiram was teratogenic in hamsters (159). DDT does not show teratogenic activity in any of the laboratory animals tested (105). Captan PCNB, folpet 2, 4, 5-T, 4-D in its isocytl, butyl and isopropyl ester produced increased numbers of abnormal young in mice as compared to the normal population (31a 31b).

Cigarette Smoking

Cigarette smoking during pregnancy may have a teratogenic effect (50). The major malformation reported is congenital heart disease. However, other workers have shown increased chances of perinatal death and decreased body weight due to cigarette smoking (23, 167). Nicotine has been found teratogenic to chickens (112) and mice (140), but not in rats (63).

A partial list of teratogenic agents is given in Table 13-13 (61) comparing their effects on the embryos of certain animal species and humans.

REFERENCES

1. Aeppli, V. L., *Arzneim Forsch,* **19:**1617 (1969).
2. Alleva, J. J., M. V. Waleski, F. R. Alleva, and E. G. Umberger, *Endocrinology,* **82:**1227 (1968).
3. Altman, J., W. Anderson, and K. Wright, *Anat. Rec.,* **163:**453 (1968).
4. Ancel, P., *La Chimioteratogenese,* G. Dolin, Paris, 1950.
5. Barker, C. A., *Can. Vet. J.,* **11:**39 (1970).
6. Barker, C. A., *Can. Vet. J.,* **12:**125 (1971).
7. Bateman, A. J., and S. S. Epstein, in A. Hollanender (ed.), *Chemical Mutagens,* Vol. 2, Plenum Press, New York, 1971.
8. Billings, M. J. Yamazaki, L. Bennett, and B. Lamson, *Pediatrics,* **38:**1047 (1966).
9. Binns, W., C. Cento, B. C. Eliason, H. E. Heggestad, G. H. Hepting, and P. F. Sand, *Investigation of Spray Project Near Globe Arizona,* 1970.
10. Birnbaum, G., J. I. Lynch, A. M. Margileth, W. M. Lonergan, and J. L. Sever, *J. Pediat.,* **75:**789 (1969).
11. Bloom, F., in W. E. Ribelin, and J. R. McCoy (eds.), *Pathology of Laboratory Animals,* Charles C Thomas, Springfield, Ill., 1965.
12. Bongiovanni, A. M., in J. B. Stanbury, J. B. Wyngaarden, and D. S. Fredrickson, *Basis of Inherited Diseases,* 3rd ed., McGraw-Hill Co., New York, 1972.
13. Bradford, G. E., *J. Reprod. Fertil. Suppl.,* **15:**23 (1972).
14. Brent, R. L., *Clin. Obstet. Gynecol.,* **3:**928 (1960).
15. Brent, R. L., in, R. J. M. Fry, D. Grahn, M. L. Greim, and J. H. Rust (eds.), *Late Effects of Radiation,* Taylor & Francis Ltd., England, 1970.
16. Brent, R. L., *Radiat. Res.,* **45:**127 (1971).
17. Brent, R. L., in, E. V. Perrin, and M. J. Fingold (eds.), *Pathobiology of Development,* Williams & Wilkins Co., Baltimore, 1973.
18. Brent, R. L., and B. T. Bolden, *Radiat. Res.,* **30:**759 (1967).

19. Brent, R. L., and B. T. Bolden, *Radiat. Res., 36:*563 (1968).
20. Brizzee, K., L. Jacobs, and X. Kharetchko, *Radiat. Res., 14:*96 (1961).
21. Brook, C. G. D., H. Wagner, M. Zachmann, A. Prader, S. Armendares, S. Frank, P. Aleman, SS. Najjar, M. S. Slim, N. Genton, and C. Bozic, *Br. Med. J., 1:*771 (1973).
22. Brown, G. C. and R. S. Karunas, *Am. J. Epidemiol., 95:*207 (1972).
23. Butler, N. R., H. Goldstein, and E. M. Ross, *Br. Med. J., 2:*127 (1972).
24. Butt, E. M., and D. G. Simonsen, *Am. J. Clin. Pathol., 20:*716 (1950).
25. Cahen, R. L., *Adv. Parmacol., 4:*263 (1966).
26. Calkins, E., D. Kahn, and W. C. Diner, *Ann. N.Y. Acad. Sci., 64:*410 (1956).
27. Cattanach, B. M., in A. Hollaender (ed.), *Chemical Mutagens,* Vol. 2, Plenum Press, New York, 1971.
28. Chow, T. L., J. A. Melello, and N. V. Owen, *J. Am. Vet. Med. Assoc., 144:*1005 (1964).
29. Conard, R. and A. Hicking, *J.A.M.A., 192:*457 (1965).
30. Cooper, L. Z., P. R. Ziring, A. B. Ockerse, et al., *Am. J. Dis. Child. 118:*18 (1969).
31. Courtney, K. D., D. W. Galyor, M. D. Hogan, H. L. Falk, R. R. Bates, and I. Mitchel, *Teratology, 3:*199 (1970), *Science, 168:*864 (1970).
32. Cowen, D., and L. M. Geller, *J. Neuropathol. Exp. Neurol., 19:*488 (1960).
33. Cutting, R. T., T. H. Phuoc, J. M. Ballo, M. W. Beneson, and C. H. Evans, *Congenital Malformations, Hydatidiform Moles, and Stillbirths,* Rep. of Vietnam Govt. Printing Office, Washington, D.C., 1970.
34. Davey, J. R., and J. W. Stevenson, *J. Anim. Sci., 22:*9 (1963).
35. David, L., J. M. Saez, and R. Francois, *Acta Pediat. Scand., 61:*249 (1972).
36. Dekaban, A., *J. Nucl. Med., 9:*471 (1968).
37. Delatour, P., R. Dams, and M. Favre-Tissot, *Therapie, 20:*573 (1965).
38. Dennis, S. M., and H. W. Leipold, *Am. J. Vet. Res., 33:*339 (1972).
39. Dishotsky, N. I., W. D. Loughman, R. E. Mogar, and W. R. Lipscomb, *Science, 172:*431 (1971).
40. Dobbing, J., *Am. J. Dis. Child., 120:*411 (1970).
41. Dolnick, E. H., I. L. Lindahl, and C. E. Terrill, *J. Anim. Sci., 31:*944 (1970).
42. Domok, W. V., *J. Am. Vet. Med. Assoc., 96:*665 (1940).
43. Earl, F. L., E. Miller, and E. J. Von Loon, *Teratogenic Research in Beagle Dogs and Miniature Swine,* Proc. 5th Meeting Int. Comm. Lab. Anim., 1973.
44. Edmonds, L. D., L. A. Selby, and A. A. Case, *J. Am. Vet. Med. Assoc. 160:*1319 (1970).
45. Emerson, D. J., *Am. J. Obstet. Gynecol., 84:*356 (1962).
46. Emerson, J. L., and A. L. Delez, *J. Am. Vet. Med. Assoc., 147:*47 (1965).
47. Ercanbrack, S. K., and D. A. Price, *J. Hered., 62:*223 (1971).
48. Fave, A., *Therapie, 19:*43 (1964).
49. Federman, D. D., *Abnormal Sexual Development,* W. B. Saunder & Co., Philadelphia, 1967.
50. Fedrick, J., E. D. Alberman, and H. Goldstein, *Nature* (London) **231:**529 (1971).
51. Ferm, V. H., *J. Embryol. Exp. Morphol., 16:*49 (1966).
52. Ferm, V. H., *Adv. Teratology, 5:*51 (1972).
53. Ferm, V. H. and L. Kilham, *Proc. Soc. Exp. Biol. Med., 112:*623 (1963).
54. Fox, M. W., *Am. Vet. Med. Assoc., 143:*602 (1963).
55. Fraser, F. C., *Congenital Malformation,* Symposium of 2nd Int. Conf., Churchill, London, 1960.
56. Friedman, G. D., *J. Chronic Dis., 25:*11 (1972).
57. Fuccillo, D. A., F. L. Moder, R. G. Traub, S. Henson, and J. L. Sever, *Appl. Microbiol., 21:*104 (1971).
58. Furchtgott, E. and S. F. Walker, in Sikov, M. R. and D. D. Mahlum (eds.), *Radiation Biology of the Fetal and Juvenile Mammal,* U.S. Atomic energy Commission, Washington, D.C., 1969.
59. Gal, I., B. M. Kirman, and I. Stern, *Nature* (London) **216:**5110 (1967).
60. Gal, I., *Adv. Teratol., 5:*161 (1972).
61. Gebhardt, D. O. E., *Adv. Teratol., 5:*97 (1972).
62. Gebhardt, D. O. E., and M. J. Van Logten, *Toxic. Appl. Pharmacol., 13:*316 (1968).
63. Geller, L. M., *Science, 129:*212 (1959).

64. German, J. A. Kowal, and K. H. Ehlers, *Teratology,* **3**:349 (1970).
65. Giroud, A., *Fed. Proc.,* **27**:163 (1968).
66. Goldstein, J. L., and A. G. Motulsky, in R. H. Williams (ed.), *Textbook of Endocrinology,* 5th ed., W. B. Saunders, Philadelphia, 1974.
67. Goldstein, J. L., and J. G. Wilson, *J. Clinc. Invest.,* **51**:1647 (1972).
68. Goldstein, J. L., and J. G. Wilson, in A. G. Motulsky and W. Lenz (eds.), *Birth Defects,* Excerpta Medica, Amsterdam, 1974.
69. Goldstein, L. and D. P. Murphy, *Am. J. Obstet. Bynecol.,* **18**:189 (1929).
70. Gustafsson, B. and B. L. Gledhill, in G. M. Dalling, et al. (eds.), *International Encyclopedia of Veterinary Medicine,* 1966.
71. Hafez, E. S. E., *Reproduction and Breeding Techniques for Laboratory Animals,* Lea & Feiberger, Philadelphia, 1970.
72. Hakosalo, J., and L. Saxen, *Lancet,* **2**:1346 (1971).
73. Hale, F., *J. Hered.,* **24**:105 (1933).
74. Hale, F., *Am. J. Opthalmol.,* **18**:1087 (1935).
75. Hamlin, R. L., D. L. Smelzer, and R. C. Smith, *J. Am. Vet. Med. Assoc.,* **145**:331 (1964).
76. Harada, Y., *Minamata Disease,* Kumanoto University, Japan, 1968.
77. Harris, S. B., J. G. Wilson, and R. H. Printz, *Teratology,* **6**:139 (1972).
78. Helmswerth, B. N., *J. Reprod. Fertil.,* **17**:325 (1968).
79. Henson, D., R. D. Smith, and J. Gehrke, *Am. J. Pathol.,* **40**:871 (1966).
80. Hertzig, M. E., H. G. Birch, S. A. Richardson, and J. Trzard, *Pediatrics,* **49**:814 (1972).
81. Hicks, S. P., and C. J. D'Amato, in D. H. M. Wollam, *Adv. Teratol.,* Academic Press, New York, 1966.
82. Hilberbrant, J. R., J. L. Sever, A. M. Margileth, and D. A. Gallagan, *Am. J. Obstet. Gynecol.,* **98**:1125 (1967).
83. Ainman, A. R., in D. T. Janerich, R. G. Shalko, and I. H. Porter, (eds.), *Congenital Defects,* Academic Press, New York, 1974.
84. Hoover, E. A., and R. A. Grusemer, *Am. J. Pathol.,* **65**:173 (1971).
85. Hurley, L. S., *J. Nutr.,* **91**:27 (1967).
86. Hurley, L. S., *Fed. Proc.,* **27**:193 (1968).
87. Hurley, L. and G. J. Everson, *Proc. Soc. Exp. Biol. Med.,* **102**:360 (1959).
88. Hurley, L. S., E. Wooten, G. J. Everson, and C. W. Asling, *J. Nutr.,* **71**:15 (1960).
89. Jacobson, H. N., *1st Int. Conf. on Prematurity,* AMA, 1968.
90. James, L. F., *Am. J. Vet. Res.,* **33**:835 (1972).
91. James, L. F. and R. F. Keeler, *Teratology,* **1**:407 (1968).
92. Johnson, E. M. in J. G. Wilson, and J. Warkany (eds.), *Teratology: Principles and Techniques,* University of Chicago Press, 1965.
93. Josso, N., *Biol Neonate* (Basel), **20**:368 (1972).
94. Josta, A., in C. Cassano, M. Finkelstein, A. Klopper, and C. Conti (eds.), *Research on Steroids,* Vol. 3, North Holland Publishing Co., Amsterdam, 1968.
95. Jost, A., in H. W. Jones, and W. W. Scott (eds.), *Hermaphroditism, Genital Abnormalities and Related Endocrine Disorders,* Williams & Wilkins Co., Baltimore, 1971.
96a. Jost, A., *Johns Hopkins Med. J.,* **130**:38 (1972).
96b. Jost, A., *Arch. Anat. Microscop. Morph. Exp.,* **61**:415 (1972b).
97. Jost, A., in A. Motulsky, and W. Lenz (eds.), *Birth Defects,* Excerpta Medica, Amsterdam, 1974.
98. Jost, A., B. Vigier, J. Prepin, and J. P. Perchellet, *Recent Prog. Hormone Res.,* **29**:1 (1973).
99. Kalter, H., *Teratology of the Central Nervous System,* University of Chicago Press, Chicago, 1968.
100. Kalter, H., *3rd Int. Conference on Congenital Malformations,* Excerpta Medica Foundation, Amsterdam, 1969.
101. Kalter, H. and J. Warkany, *Physiol. Rev.,* **39**:69 (1959).
102. Karnofsky, D. A., in J. G. Wilson and J. Warkany (eds.), *Teratology: Principles and Techniques,* University of Chicago Press, Chicago, 1965.
103. Karnofsky, D. A., and C. R. Lacon, *Biochem. Pharmacol.,* **15**:1435 (1966).
104a. Khera, K. S., *Toxicol. Appl. Pharmacol.,* **24**:167 (1973).

104b. Khera, K. S., *Teratology*, **8:**293 (1973).

105. Khera, K. S., and D. J. Clegg, *Can. Med. Assoc. J.*, **100:**167 (1969).

106. Kilham, L., and V. H. Ferm, *Proc. Soc. Exp. Biol. Med.*, **117:**874 (1964).

107. Kilham, L., and G. Margolis, *Am. J. Pathol.*, **49:**457 (1966).

108. Kilham, L., and G. Margolis, *Lab. Invest.*, **21:**183 (1969).

109. Kinlen, L. J., and E. D. Acheson, *Br. J. Radiol.*, **41:**648 (1968).

110. Knelson, J. H., *Arch. Int. Med.*, **127:**421 (1971).

111. Kucera, J., *J. Pediat.*, **72:**857 (1968).

112. Landauer, W., *J. Exp. Zool.*, **143:**107 (1960).

113. Langman, J., and M. Shimada, *Teratology*, **3:**204 (1970).

114. Langman, J., and M. Shimada, *Anat. Rec.*, **169:**363 (1971).

115. Langman, J., M. Shimada, and W. Webster, *Teratology*, **5:**260 (1972).

116. Legator, M. S. and H. V. Malling, in A. Hollaender, *Chemical Mutagens,* Vol. 2, Plenum Press, New York, 1971.

117. Lenz, W., *Congenital Malformations,* Int. Medical Congress, New York, 1964.

118. Letavet, A. A. and E. B. Skaya, *Toxicology of Radioactive Substances,* Vol. 4, Pergamon Press, Elmsford, N.Y., 1970.

119. Levy, B., R. Rugh, L. Lunin, N. Chilton, and M. Moss, *J. Morphol.*, **93:**561 (1953).

120. Liao, S., T. Liang, and J. L. Tymoczko, *Nature (New Biol.)*, **241:**211 (1973).

121. Long, S. Y., *Teratology*, **6:**75 (1972).

122. Lowe, C. R., *Lancet*, **1:**9 (1973).

123. Malorny, G., *Ernaehrungswiss*, **9:**332 (1969).

124. Matsumoto, H. G., K. Goyo, and T. Takevchi, *J. Neuropathol. Exp. Neruol.*, **24:**563 (1965).

125. McBridge, W. G., *Teratology*, **5:**262 (1972).

126. McLauglin, J., E. F. Reynaldo, J. K. Lamar, and J. P. Marliac, *Toxicol. Appl. Pharmacol.*, **14:**59 (1969).

127. Medearis, D. N., *Am. J. Hyg.*, **80:**113 (1964).

128. Meltzer, H. J., *J.A.M.A.*, **161:**1253 (1956).

129. Menges, R. W., L. A. Selby, C. J. Marienfeld, W. A. Aue, and D. L. Greer, *Environ. Res.*, **3:**285 (1970).

130. Meselson, M. S., A. H. Westing, and J. D. Constable, *Herbicide Assessment Comm.*, American Assoc. for the Advancement of Science, 1970.

131. Meyer, M., E. Diamond, and T. Merz, *Johns Hopkins Med. J.*, **123:**123 (1968).

132. Miller, R. W., *Pediatrics*, **41:**257 (1968).

133. Miller, R. W., in R. J. M. Fry, D. Grahn, M. L. Griem, and J. H. Rust (eds.), *Late Effects of Radiation,* Taylor & Francis, Ltd., London, 1970.

134. Milunsky, A., J. W. Graef, and M. A. Gaynor, *J. Pediat.*, **72:**790 (1968).

135. Monie, I. W., K. Kho, and J. Morgan, *Proc. Teratol. Workshop,* 2nd ed., Berkely, 1965.

136. Mulvihill, J. J., and W. A. Priester, *Teratology*, **4:**236 (1971).

137. Murphree, R., and H. Page, *Radiat. Res.*, **12:**495 (1960).

138. Murphy, D. P., *Am. J. Obstet Gynecol.*, **18:**179 (1929).

139. Nahmias, J. J., C. A. Alford and S. B. Korones, *Adv. Pediat.*, **17:**185 (1970).

140. Nishimura, H. and K. Nakai, *Science*, **127:**877 (1958).

141. Nobunaga, T., *Lab. Anim. Sci.*, **23:**803 (1973).

142. Ohno, S., in A. Labhart, T. Mann, and T. Samuels, (eds.) *Chromosomes and Sex Linked Genes,* Vol. I, Springer-Verlag, Heidelberg.

143. Ohno, S., *Nature (London)*, **444:**259 (1973).

144. Orsini, M. W., *Proc. Anim. Care Panel*, **11:**193 (1961).

145. Palludan, B., *Teratology*, **3:**207 (1970).

146. Palmer, A. K., *Lab. Anim.*, **2;**195 (1968).

147. Palmer, A. K., in A. Bertelli and L. Donati, (eds.), *Teratology,* Excerpta Med. Found., Amsterdam, 1969, p. 55.

148. Palmisano, P. A., R. C. Sneed, and G. Cassady, *J. Pediat.*, **75:**869 (1969).

149. Panigel, M., in Tuchmann-Duplessis, H. (ed.), *Malformations Congenitales des Mammiferes,* Masson, Paris, 1971.

150. Patten, B. M., *Embryology of the pig,* 3rd ed., Blakiston, Philadelphia, 1948.

151. Peer, L. A., H. W. Gordon, and W. G. Bernhard, *Plastic Reconstructive Surg.,* **34**:358 (1964).
152. Petrosyan, S., and I. Pereslegin, *Med. Radiol.,* **5**:38 (1962).
153. Pilt, D. B., and P. E. Samson, *Austral. Ann. Med.,* **10**:268 (1961).
154. Poswillo, D. E., *Ann. R. Coll. Surg. Engl.,* **50**:367 (1972).
155. Powell, H. R. and H. Ekert, *Med. J. Aust.,* **2**:1076 (1971).
156. Prader, A. and R. E. Siebenmann, *Helv. Paediat. Acta,* **12**:569 (1957).
157. Ramsey, E. M., and J. W. S. Harris, *Carnegie Contrib. Embryol.,* **38**:59 (1966).
158. Revelle, R., *Prospects of the World Food Supply Symposium,* National Academy of Sciences, Washington, D.C., 1966.
159. Robens, J. F., *Toxicol. Appl. Pharmacol.,* **15**:152 (1969).
160. Robertson, D. S. F., *Lancet,* **1**:518 (1970).
161. Robson, J. M., and F. M. Sullivan, *Lancet,* **1**:638 (1963).
162. Rorke, L. B., A. Fabiyi, T. S. Elizan, and J. L. Sever, *Lancet,* **2**:153 (1968).
163. Rosenkrantz, J. G., F. P. Lynch, and W. W. Frost, *J. Pediat. Surg.,* **5**:232 (1970).
164. Rough, R., *J. Neuropath. Exp. Neurol.,* **18**:468 (1959).
165. Rugh, R., L. Duhamel, A. Chandler, and A. Varma, *Radiat. Res.,* **22**:519 (1964).
166. Rugh, R., L. Duhamd, and L. Skaredoff, *Proc. Soc. Exp. Biol. Med.,* **121**:714 (1966).
167. Rush, D. and E. H. Kass, *Am. J. Epidemiol.,* **96**:183 (1972).
168. Russel, L. B., in A. P. Hollaender, (ed.), *Radiation Biology Part II,* McGraw-Hill, New York, 1954, p. 861.
169. Russell, L. B., *Proc. Soc. Exp. Biol. Med.,* **95**:174 (1957).
170. Russel, L. B. and C. S. Montgomery, *Cint. J. Radiat. Biol.,* **10**:151 (1966).
171. Savini, E. C., M. A. Moulin, and M. F. Herrou, *Therapie,* **23**:1247 (1968).
172. Sawin, P. B., *Adv. Genet.,* **7**:183 (1955).
173. Sawin, P. B. and D. D. Crary, *Clin. Orthop. Relat. Res.,* **33**:71 (1964).
174. Saxen, L., and J. Rapola, *Congenital Birth Defects,* Holt, Reinhart & Winston, Inc., New York, 1969.
175. Schnell, V. and J. W. Newberne, *Teratology,* **3**:345 (1970).
176. Schultz, G. and P. D. Delay, *J. Am. Vet. Med. Assoc.,* **127**:224 (1955).
177. Scott, W. J., R. L. Butcher, C. W. Kindt, and J. G. Wilson, *Teratology,* **6**:239 (1972).
178. Scott, W. J., E. J. Ritter and J. G. Wilson, *Develop. Biol.,* **26**:306 (1971).
179. Segall, A., B. MacMohan, and M. Hannigan, *J. Chronic Dis.,* **17**:915 (1964).
180. Selby, L. A., H. C. Hopps, and L. D. Edmonds, *J. Am. Vet. Med. Assoc.,* **159**:1485 (1971).
181. Sever, L., *Public Health Service Publication 1692,* Dept. of Health, Education & Welfare, Washington, D.C., 1966.
182. Sever, J. L., in H. Nishimura and J. R. Miller (eds.), *Methods for Teratological Studies in Experimental Animals and Man,* Igaku Shoin, Ltd., Tokyo, 1969.
183. Sever, J. L., in E. V. Perrin and M. J. Finegold (eds.), *Pathobiology of Development,* Williams & Wilkins Co., Baltimore, 1973.
184. Shaw, E. B., and H. L. Steinbach, *Am. J. Dis. Child.,* **115**:477 (1968).
185. Sheppard, M., *Vet. Rec.,* **63**:685 (1951).
186. Siiteri, P., and J. D. Wilson, *J. Clin. Endocrinol.,* **38**:113 (1973).
187. Skreb, N., N. Bijelic, and G. Lukovic, *Experientia,* **19**:1 (1963).
188. Smalley, H. E., J. M. Curtis, and F. L. Earl, *Toxicol. Appl. Pharmac.,* **13**:392 (1968).
189. Smith, C. C., in H. Vagtborg (ed.), *Use of Nonhuman Primates in Drug Evaluation,* University of Texas Press, Austin, 1968.
190. Smith, C. C., *Ann. N.Y. Acad. Sci.,* **162**:604 (1969).
191. Smith, C. K. A., and L. P. Scammell, *Lab. Animal,* **2**:83 (1968).
192. Smithberg, M., *Adv. Teratol.,* **2**:258 (1967).
193. Somers, G. F., *Lancet,* **1**:912 (1962).
194. Speidel, B. D., and S. R. Meadow, *Lancet,* **2**:839 (1972).
195. Spyker, J. M., and M. Smithberg, *Teratology,* **5**:181 (1972).
196. St. Geme, J. W. Jr., G. R. Noren, and P. Adams, Jr., *N. Engl. Med.,* **275**:339 (1966).
197. Starr, J. G., and E. Gold, *J. Pediat.,* **73**:830 (1968).
198. Steffek, A. J., A. Fabiyi, and C. T. G. King, *Arch. Oral. Biol.,* **13**:1281 (1968).
199. Steffek, A. J., and A. C. Verrusio, *Teratology,* **5**:258 (1972).

200. Sternberg, J., *Am. J. Obstet. Gynecol.,* **108:**490 (1970).
201. Streeter, G. L., *Carnegie Contrib. Embryol.,* **30:**211 (1942).
202. Suttle, N. F., A. C. Field, and R. M. Barlow, *J. Comp. Pathol.,* **80:**151 (1970).
203. Synder, R. D., *N. Engl. J. Med.,* **284:**1014 (1971).
204. Takano, K., in H. Nishimura and J. R. Miller, *Methods for Teratological Studies in Experimental Animals and Man,* Igaku Shoin, Ltd., Tokyo, 1969.
205. Thiersch, J. B., *J. Reprod. Fertil.,* **4:**219 (1962).
206. Tuchmann-Duplessis, H., *Teratology,* **5:**271 (1972).
207. Valadian, I. and R. B. Reed, in D. T. Janerich, R. G. Skalko, and I. H. Porter (eds.), *Congenital Defects,* Academic Press, Inc., New York, 1974.
208. Verret, J. M., J. P. Marliac, and J. McLaughlin, *J. Assoc. Agr. Chem.,* **47:**1003, 1964.
209. Voorhess, M. L., *J. Pediat.,* **71:**128 (1967).
210. Waddington, C. H., *Ciba Symposium on Preimplantation of Pregnancy,* Churchill, London, 1965.
211. Wantanabe, G., S. Yoshida and K. Hirose, *Proc. 8th Cong. Anom. Res. Assoc.,* Tokyo, 1968.
212. Warkany, J., *Congenital Malformations,* Year Book Medical Publishers, Inc., Chicago, 1971.
213. Warkany, J. and E. Schraffenberger, *Proc. Soc. Exp. Biol. Med.,* **54:**92 (1943).
214. Warwick, E. J., A. B. Chapman, and B. Ross, *J. Hered.,* **34:**349 (1943).
215. Weidman, W. H., H. H. Young, and P. E. Zollman, *Mayo. Clin. Proc.,* **38:**518 (1963).
216. Werboff, J., and J. Havlena, *Exp. Neurol.,* **6:**263 (1962).
217. Wilcke, H. *Adv. Chem. Series 57,* American Chemical Soc., Washington, D.C. 1966.
218. Wilkins, L., *J.A.M.A.,* **172:**1028 (1960).
219. Wilson, J. G., *Proc. Soc. Exp. Biol. Med.,* **84:**66 (1953).
220. Wilson, J. G., in J. G. Wilson, and J. Warkany (eds.), *Teratology: Principles and Techniques,* University of Chicago Press, Chicago, 1965.
221. Wilson, J. G., in B. R. Fink (ed.), *Toxic Effect of Anesthetics,* Williams & Wilkins, Co., Baltimore, 1968.
222. Wilson, J. G., *Anat. Rec.,* **166:**398 (1970).
223. Wilson, J. G., *Fed. Proc.,* **30:**104 (1971).
224. Wilson, J. G., in E. Diczfalusy, and C. C. Standley (eds.), *Use of Nonhuman Primates in Res. on Human Reproduction,* WHO Res. & Training Center on Human Reproduction, Stockholm, 1972.
225. Wilson, J. G., *Acta Endocrinol.* (Kbh) Suppl., **166:**261 (1972).
226. Wilson, J. G., *Environment and Birth Defects,* Academic Press, New York, 1973.
227. Wilson, J. G., *Teratology,* **7:**3 (1973).
228. Wilson, J. G., in A. G. Motulsky, W. Lenz, and F. J. G. Ebling, (eds.), *Birth Defects,* Excerpta Medica, Amsterdam, 1968.
229. Wilson, J. G., and R. Fradkin, *Anat. Rec.,* **163:**286 (1969).
230. Wilson, J. G., R. Fradkin, and A. Hardman, *Teratology,* **3:**59 (1970).
231. Wilson, J. G., and J. A. Gavan, *Anat. Rec.,* **158:**99 (1967).
232. Winick, M., *Pediatrics,* **47:**967 (1971).
233. Wooley, D. W., in R. M. Hockster and J. H. Quastel (eds.), *Metabolic Inhibitors,* Academic Press, New York, 1963.
234. *WHO Techn. Rep. Ser. No. 364,* World Health Organization, 1967.
235. Wright, S., *Cold Spring Harbor Symp. Quant. Biol.,* **2:**137, (1934).
236. Young, G. A., *ARC Anim. Breeding Res. Organization,* Rep. 35–41, 1967.
237. Young, G. A., R. L. Kitchell, A. J. Leudke, and J. H. Sautter, *J. Am. Vet. Med. Assoc.,* **126:**165 (1955).
238. Zachmann, M., J. A. Vollmin, W. Hamilton, and A. Prader, *Clin. Endocrinol.,* **1:**369 (1972).
239. Zamenof, S., E. Van Marthens, and L. Granel, *Science,* **174:**954 (1971).

CHAPTER 14

ANIMAL MODELS FOR THE STUDY OF ENVIRONMENTAL HEALTH HAZARDS

The study of environmental hazards relates to the determination of physical, chemical, and biological agents that affect normal health conditions of human beings. The quality of air, water, soil, and food is affected by the activities of many forms of life, including humans. The quality of the environment has deteriorated ever since the process of human civilization began. The increased growth rate of human population and the development of modern technology and industrialization in this century have contributed toward the rapid changes in our environment, particularly in the pollution of natural air and water resources. The nature of pollutants varies considerably from physical agents such as radiation and noise to chemicals such as industrial waste products, fertilizers, pesticides, and biological agents such as microorganisms. Experimental animals have been used to test the long-range effects of many of these potential health hazards. A detailed discussion of all these factors (physical, chemical, and biological) is beyond the scope of this chapter. Therefore, the effects of certain pollutants (in the environment) on experimental animals are described to illustrate the usefulness of these animal models for the study of human health problems.

AIR POLLUTION

The main sources of air pollution are automobiles, power generators, industrial operations, heating of buildings, and refuse disposal (Table 14-1) (67). The mechanism of origin of pollutants either by combustion, or by emitting pollutants, (279) or by photochemical reactions (132) with ultraviolet light and nitrogen dioxide produces ozone:

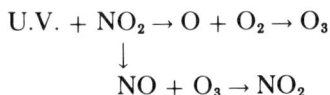

$$U.V. + NO_2 \rightarrow O + O_2 \rightarrow O_3$$
$$\downarrow$$
$$NO + O_3 \rightarrow NO_2$$

Molecular oxygen is a by-product of the reaction which mediates the continuation of the reaction. The presence of carbon monoxide accelerates the photochemical reactions (307).

Carbon Monoxide

It is a ubiquitous air pollutant and acts by binding tightly with hemoglobin (CO affinity is 210 times O_2 affinity). The human effects of increased CO concentration include severe headache, weakness, dim vision, nausea, and collapse. At low concentration, both the nervous system and cardiac functions are impaired (21, 27). The animal models most commonly used are rats, dogs, and monkeys. In rats, concentration of 50–100 ppm

Table 14-1. Main Sources of Air Pollution

Source	Carbon monoxide	Particulates	Hydrocarbons	Nitrogen oxides	Sulfur oxides	Total
Transportation	64.5	1.2	17.6	7.6	0.4	91.3
Fuel Combustion in Stationary Sources	1.9	9.2	0.7	6.7	22.9	41.4
Industrial Processes	10.7	7.6	3.5	0.2	7.2	29.2
Solid Waste Disposal	7.6	1.0	1.5	0.5	0.1	10.7
Miscellaneous	9.7	2.9	6.0	0.5	0.6	19.7
Total	94.4	21.9	29.3	15.5	31.2	192.3
Forest Fires	7.2	6.7	2.2	1.2	N	17.3
Total	101.6	28.6	31.5	16.7	31.2	209.6

Adapted from Advances in Environmental Science and Technology (66) with permission.

CO causes impairment in judgment of time intervals (29) and alternation in electrical responses of the visual nervous system (329). Dogs exposed to similar concentration (50–100 ppm) show changes in electrocardiograms and in brain cells (27, 218). Young healthy monkeys (*Macaca*) showed no physiological abnormalities or anatomical signs of the brain or heart pathology when exposed to 20–65 ppm of CO for 22 hours per day for 2 years (89).

Nitrogen Dioxide

Combustion sources emit NO and NO_2, but at present only NO_2 is a problem. This pungent, reddish brown gas is responsible for rusty chemical smog. The primary effects in humans (5 ppm) are increased airway resistance and destruction of terminal bronchi and alveoli (1).

The effect of NO_2 has been studied in several animal species. The pathological effects in rats depend on the concentration and length of time of exposure. Symptoms may range from peroxidation of lung lipids and structural changes in lung tissue after exposure for 4 hours at 1 ppm (287). At higher concentrations (5 ppm for 9 months), there was increased amount of water in lungs, decrease in lung volume, and alterations of pulmonary surfactants (17). Continuous exposure for 2 years at 2 ppm produced loss of cilia. Other symptoms such as thickening of bronchioles, and increased rate of cell division were temporary; recovery took place within 2 weeks of exposure (96, 278). Mice show increased susceptibility to bacterial pneumonia, distended alveoli, emphysema, and inflammation of bronchial mucosa and cilia (39, 92). In guinea pigs, 5 ppm NO_2 produces increased airway resistance, and the appearence of antibodies to lung tissue (22, 284). Squirrel monkeys show increased susceptibility to pneumonia when exposed to NO_2 (143). The structural changes in the proteins, particularly collagen, are seen in rabbits (48).

Ozone

Symptoms of headaches, chest constriction, throat irritation, eye irritation, tightness of facial skin, and dryness of nasal mucous membrane are attributed to ozone inhalation in humans (123, 170, 178). Most of the animal species, rats, guinea pigs, hamsters, and mice show distortion and thickening of the terminal airways and moderate emphysema (280). The susceptibility to pneumonia is increased in rats and hamsters (207). Rats show unique changes in lung lipids, particularly those with vitamin E deficiency. The arterial wall thickens and serotonin is released in rabbits exposed to 0.2–0.4 ppm of ozone (231, 232).

Other air pollutants such as SO_2, particulate matter and hydrocarbons are important but have not been thoroughly studied in animal species.

Animal Models for the Study of Air Pollution

Rodents, particularly rats, mice, and guinea pigs, have been used for experiments in evaluating air pollution. The heavy smog produces ultrastructural changes in alveolar tissues of rodents (35, 37). The change is most severe in older mice, who lose virtually all the mitochondria in the alveolar cells during exposure to heavy smog (43). Greater incidence of bronchial infections is seen in mice inhaling heavy smog (119). Rats had a greater incidence of chronic kidney disease when they inhaled smog (118). Certain strains of mice and guinea pigs secrete large amounts of adrenal hormone, indicating physiological stress (304). Smog inhalation appears to decrease the life span of male mice (36, 94, 304). Guinea pig colonies in Los Angeles have been used as sentinels for air pollution, particularly from smog, as the pulmonary resistance measurement showed a marked effect (284). However, similar tests conducted in Detroit did not produce similar results (284). Free living animals have been suggested as test systems for air pollution; for example, pigeons that dwell in the urban areas and are constantly exposed to air pollutants may be used as models. Similarly, house martins are sensitive models for testing air pollution (90). Rats, ground squirrels, and other wild rodents with limited ecological niches represent good models for microecological studies (66). Dairy cattle, because of their proximity to human population, may represent suitable models particularly when detailed records of their pedigree, feed, and performance (milk yield) are kept. A correlation between cattle and human conditions has been reported (59, 208).

PNEUMOCONIOSES AS MODELS OF ENVIRONMENTAL POLLUTION

Three major injurious agents that affect the lungs are silica, coal dust, and asbestos. Boiler scalers, sand blasters, addict sniffing, and flavored abrasive rich in silicon dioxide are common sources of human silicoses. The primary cytotoxic action of silicon dioxide results in eventual unique nodular and fibrotic response (328). Silicosis is an occupational disease associated with tunneling, open-pit mining, and packing fine silica. Generally, the lack of specific anatomical changes has for a long time prevented the recognition of silica (silicon dioxide) as an etiological factor in pneumoconiosis.

Lung of the Coal Worker

Coal has been shown to cause distinctive progressive pulmonary disability (124, 216). Two generalized categories exist. (a) Simple pneumoconiosis with little ventilatory distress. It may take as much as 12 years of exposure. It is estimated that about 1.5 to 3.0 percent of the miners are affected with this condition. (b) Progressive massive fibrosis (PMF) which affects about one-tenth of all cases of simple coal pneumoconiosis. It is generally recognized that agents other than dust are needed for the development of PMF. Death risks in coal miners double the general population and are higher than any occupational group in the United States. In addition to PMF, coal miners have a greater tendency to develop other conditions such as bronchitis or emphysema (326, 327).

Asbestos Pneumonconiosis

In contrast to other conditions, asbestos hazards occur primarily at the manufacturing site. Generally, a long latent period of 20 to 30 years exists before the diagnosis of pulmonary fibrosis is recognized. Of the several varieties of asbestos, crysotile ($Mg_3Si_2O_5(OH)_4$) is most important; however, in manufacturing, a blend of asbestos is used. The exact role or risk from asbestos in the development of malignant mesothelial tumors is low (197). However, asbestos has polluted the water supplies of certain areas and may pose a potential health hazard to human life. (128, 324).

Studies among workers in high asbestos risk areas are important because they can be used as models for the detection of asbestos pollution in the general population. This is particularly important since asbestos bodies in the lungs of urban residents have been reported (128).

Environmental Contamination with Heavy Metals

A number of heavy metals enter the environment from various sources and contaminate air, water, and food supplies. The mode of their entrance and their concentration in the environment vary with the metals (Table 14-2). Heavy metals produce varying effects in humans, including mutagenicity (Table 14-3). Threshold limit values for airborne metals and their salts for work places are listed in Table 14-4 (142, 181, 260). A summary of approximate body content of common trace metals is shown in Table 14-5. However, more frequently occurring metallic pollutants of the environment are mercury, lead and cadmium.

Environmental Pollution with Mercury

The critical nature of mercury as a pollutant was brought to light because a high incidence of death occurred in the Minamata Bay area of Japan from "Minamata disease." The disease is characterized by irreversible and fatal neurological disorders. In Sweden, a high incidence in the decline of a certain population of birds and their suffering was also related to mercury.

Mercury may enter the environment from industrial pollution (plastics, paper, electronics, hospitals, and power) and pesticides (252, 319). It is an occupational hazard for some workers, for example, dentists and velvet fabric workers. However, the biggest problem is the conversion of inorganic mercury into organic mercury compounds such as alkyl derivatives (particularly methyl mercury). This conversion is carried out by a wide variety of biological marine life (177), aquarium sludge (153), liver homogenates (309), and microorganisms via methylcobalamine (40, 320). A variety of microorganisms have been isolated from nature which possess this system (24, 48, 249). The cell-free extracts have excellent donors for methyl mercury formation in the form of methylcobalamine (318), methyl factor B (322), and methyl factor III (321).

The general symptoms of mercury poisoning in humans are fatigue, headache, and irritability, followed by tremors, loss of feeling in fingers and toes, blurred vision, and poor muscular contraction. At later stages, there is difficulty in speech and hearing and eventually mental disorders. Mercury is distributed throughout the body, but significant amounts accumulate in the brain. Mercury can pass placental barriers and cause muscular, neurological, and brain damage syndrome at or soon after birth (32). Methyl mercury has also been associated with mental retardation and cerebral palsy.

Animal Models

Several species of wild birds and fish represent good animal models for the study of environmental pollution with mercury. They are part of the human food chain and hence represent important models, both for testing as well as controlling mercury pollution.

Fish are natural indicators of mercury levels in the environment and have been used in the laboratory for test purposes (155, 308). Studies conducted with rodents using methyl mercury showed effects on reproduction, embryonic development, and retarded postnatal growth (116, 248, 272). The male rats respond to a dose-related reduction in incidence of fertile matings; however, mice do not demonstrate postimplantation losses or reduction in fertility (165, 167). Chromosomal breakage in the leukocytes of animals as well as humans has also been reported (269). The decrease in fertility is related to effects of mercury on spermatogenesis (151) or through their effect on the enzymes in acrosomes by affecting the SH $\rightleftharpoons$ SS exchange (30, 273, 334). Transplacental passage of mercury has also been shown using mice and rats (33, 120, 282). Cats have been used as

Table 14-2. Toxic Metal Levels in U.S. Cities and Nonurban Areas

Item Evaluated	Metal						
	Nickel	Beryllium	Cadmium	Tin	Antimony	Lead	Mercury
Cities (58)–µg/m³	0.001–0.118	0.001–0.002	0.002–0.370	0.01–0.03	0.042–0.085	0.1–2.3	****
Nonurban–µg/m³	0.0006–0.012	0.00013	0.004–0.026	0.0002–0.0018	0.001–0.002	0.0001–0.0004	0.003–0.009
Intake (µg/day) Food and water	600	12	100	7,300	<100	300	20
Air	2.36	0.04	7.4	0.6	1.7	46	0.108
Total Body (ug)	10,000	30	50,000	5,800	7,900	120,000	13,000
Soil (ppm)	40	6	0.06	10	4	10	0.03–1.3
Principal Source	Coal Petroleum	Coal?	Zinc	?	?	Gasoline Additives	Coal Petroleum

Adapted from Advances in Experimental Medicine and Biology (260) with permission.

Table 14-3.   Mutagenesis by Metal Salts

Metal	Salt	Point Mutations Microorganisms	Chromosome Plant	Aberrations Mammalian
Aluminum	Chloride		+	
Arsenic	Arsenate			+
Cadmium	Nitrate		+	
Calcium Deficiency				+
Ferrous	Chloride	+		
Lead	Acetate			+
Magnesium Deficiency				+
Manganous	Acetate	+		
Manganous	Chloride	+		
Mercury	Methyl		+	+
Mercury	Methoxyethyl		+	
Mercury	Phenyl hydroxide		+	
Potassium Excess		+		

Adapted from Advances in Experimental Medicine and Biology (260) with permission.

a model for methyl mercury (165, 213). It appears that neurological signs associated with cerebral and cerebellar lesions in offspring of cats mimic the human condition. The concentration of mercury was found in the following decreasing order: fetal blood, maternal blood, and fetal brain (165). The frequency of abortions and neuronal degeneration in cats was similar to that in humans (65, 74). Certain invertebrates have been used as the test system, for example, drosophila species and sea urchins; polyspermy is seen in sea urchins (254). Also, plant systems showing polyploidy have been suggested as models (248). Teratogenic effects of organic mercury compounds such as phenylmercuric acetate and methylmercuric dicyandiamide have been reported (248, 272). A number of heavy metals including mercury produce teratogenic effects in hamsters (Table 14-6) (101, 115).

The Rat as an Experimental Animal for Acute Methyl Mercury Intoxication. Methyl mercury has become an important environmental health hazard since high concentration of the compound has been observed in the environment. Its possible

Table 14-4. Threshold Limit Values for Airborne Metals and Their Salts for Work Places in the U.S.

Substance	ppm	mg/M^3
Antimony and compounds (as Sb)	–	0.5
Arsenic and compounds (as As)	–	0.5
Arsine	0.05	0.2
Barium (soluble compounds)	–	0.5
Beryllium	–	0.002
Cadmium (metal dust and soluble salts)	–	0.2
Cadmium oxide fume	–	0.1
Calcium arsenate	–	1
Calcium oxide	–	5
Chromic acid and chromates (as CrO$_3$)	–	0.1
Chromium-metal and insoluble salts	–	1
Cobalt (metal fume and dust)	–	0.1
Copper (fume)	–	0.1
Hafnium	–	0.5
Iron salts, soluble (as Fe)	–	1
Iron oxide fume	–	10
Lead	–	0.2
Lead arsenate	–	0.15
Lead tetraethyl (as Pb)-(s)	–	0.075
Lead tetramethyl (as Pb)-(s)	–	0.075
Lithium hydride	–	0.025
Magnesium (oxide fume)	–	15
C Manganese	–	5
Mercury (s)	–	0.1
Mercury (organic compounds) (s)	–	0.01
Molybdenum (soluble compounds)	–	5
Molybdenum (insoluble compounds)	–	15
Nickel carbonyl	0.001	0.007
Nickel, metal and soluble compounds	–	1
Osmium tetroxide	–	0.002
Rhodium (metal fume and dusts)	–	0.1
-Soluble salts	–	0.001
Selenium compounds (as Se)	–	0.2
Selenium hexafluoride	0.05	0.4
Silver, metal and soluble compounds	–	0.01
Tantalum	–	5
Tellurium	–	0.1
Tellurium hexafluoride	0.02	0.2
Tetramethyl lead (TML) (as Lead) (s)	–	0.075
Thallium (soluble compounds) (s)	–	0.1
Thallium (soluble compounds) (as Ti) (s)	–	0.1
Tin (inorganic compounds, except oxide)	–	2
Tin (organic compounds)	–	0.1
Titanium dioxide	–	15
Tungsten and compounds (as W)		
-soluble	–	1
-insoluble	–	5

Table 14-4. *(Continued)*

Substance	ppm	mg/M^3
Uranium (soluble compounds)	–	0.05
Uranium (insoluble compounds)	–	0.25
Uranium (soluble & insoluble compounds) (as U)	–	0.2
C Vanadium (V$_2$O$_2$ dust)	–	0.5
C Vanadium (V$_2$O$_2$ fume)	–	0.1
Yttrium	–	1
Zinc chloride (fume)	–	1
Zinc oxide (fume)	–	5
Zirconium compounds (as Zr)	–	5

*The Threshold Limit Values refer to airborne concentrations of substances and represent conditions under which it is believed that nearly all workers may be repeatedly exposed, day after day; (s) = salts.

Adapted from Advances in Experimental Medicine and Biology (260) with permission.

Table 14-5. Concentration of Abnormal Trace Metals in the Human Body

Metal	Daily intake in Diet (mg)	Approximate Body Content (mg)	(μg/g)	(μmol/gx10^3)	Comments
Cadmium	0.22	50	0.7	6.2	28% in kidney and liver
Lead	0.45	120	1.7	8.2	92% in bone. Brain is important target organ.
Mercury	0.02	13	0.2	1	69% in fat and muscle. Brain is important target organ.
Beryllium	0.013	0.03	4x10^{-4}	0.04	75% in bone. Toxic action usually involves lung.
Arsenic	1.0	~18	~0.3	~4	

Adapted from Advances in Experimental Medicine and Biology (260) with permission.

Table 14-6. Effects of the Metal Combinations Mercury + Cadmium and Mercury + Zinc on Hamster Embryogenesis

Compound (Dose=mg/kg)	No. solutions	No. animals treated	No. implant. sites	Resorption sites# No. (%)	Normal embryos# No. (%)	Abnormal embryos# No. (%)	Small embryos** No. (%)	Retarded embryos** No. (%)
Group I Hg+Cd								
Hg(2)+Cd(2)	1	7	92	26(28)	5 (6)	61(66)	40(61)	40(61)
Hg(2)+Cd(2)*	2(S+D)*	6	92	36(39)	5 (6)	51(55)	26(46)	20(36)
Cd(2)+Hg(2)*	2(S+D)*	6	81	56(69)	6 (7)	19(24)	17(68)	11(44)
Hg(4)+Cd(2)	1	6	70	38(54)	1 (2)	31(44)	17(53)	20(63)
Hg(4)+Cd(2)*	2(S+D)*	6	82	57(70)	0 (0)	25(30)	24(96)	22(88)
Cd(2)+Hg(4)*	2(S+D)*	6	84	80(95)	0 (0)	4 (5)	4(100)	4(100)
Group II controls								
Cd(2)	1	9	125	69(55)	6 (5)	50(40)	25(46)	11(20)
Hg(2)	1	14	172	20(12)	123(71)	29(17)	29(19)	25(16)
Hg(4)	1	19	216	113(52)	41(19)	69(29)	58(56)	62(60)
H_2O(0.5 ml/100g body wt)	1	23	302	17 (6)	278(92)	7 (2)	0 (0)	0 (0)
Zn(2)	1	6	81	5 (6)	76(94)	0 (0)	8(11)	0 (0)
Group III Hg+Zn								
Hg(2)+Zn(2)	1	6	85	3 (4)	80(94)	2 (2)	19(23)	2 (2)
Hg(2)+Zn(2)*	2(S+D)*	6	81	31(38)	43(53)	7 (9)	13(26)	4 (8)
Zn(2)+Hg(2)*	2(S+D)*	6	54	3 (6)	51(94)	0 (0)	5(10)	0 (0)
Hg(4)+Zn(2)	1	6	74	12(16)	60(81)	2 (3)	24(39)	0 (4)
Hg(4)+Zn(2)*	2(S+D)*	6	76	13(17)	21(28)	42(55)	32(51)	11(17)
Zn(2)+Hg(4)*	2(S+D)*	6	69	55(80)	7(10)	7(10)	14(100)	0 (0)

#Per cent resorptions, normals or abnormals = No. of resorptions, normals or abnormals/No. of implantation sites X 100.
*S=Simultaneous injection; D=delayed injection.
**Per cent small or retarded embryos = No. of small or retarded/total No. of embryos X 100.
Adapted from Environmental Research (115) with permission.

genetic and teratological effect as well as synergistic effect with other neurotoxins has been reported (148, 179, 219).

Methyl mercury hydroxide when given subcutaneously (10 mg/kg) for 7 days produced weight loss, decreased activity, ataxia, weakness, and hind leg crossing symptoms in rats (169). Several histological changes in the nervous systems, including advanced peripheral neuropathy, vaculization and facal neurophagia, decreased neuron density and pyknotic neurons, have been observed. Other symptoms, including tubular degeneration of kidneys, atrophy of malpighian corpuscles, and increased hemosiderin stores, have also been reported (169). These syndromes represent a close relationship to acute Minamata disease in humans.

Environmental Pollution with Lead

The contamination of the environment with lead amounts to approximately 400 million pounds, 98 percent of which comes from leaded gasolines and other sources, including coal and fuel oil combustion, lead compounds manufacture, primary and secondary lead smelting, brass manufacture, and ammunition (217). Among the food items, canned milk appears to be particularly high in lead concentration and its use for children should be avoided (266). Household items such as paint mixtures may contain some lead compounds. Several review articles have appeared recently indicating great concern over lead poisoning (41, 97, 135).

Lead poisoning can result in clinically definable encephalopathies, neuropathies, and behavioral change from cerebral dysfunction (Table 14-7). Brain tissues of children from birth to 7 years are most sensitive to lead poisoning. However, toxicity is over-looked until recognized encephalopathy occurs (125, 127). Well-defined his-topathological changes are found in victims who do not survive acute lead poisoning (236). However, other systems such as renal and hematopoietic systems have been studied in detail (117, 134). There is a high incidence of permanent brain damage which sometimes leads to periodic convulsions, irritability, hyperactivity, retardation of normal development, emotional instability, behavior disorders, low attention span, impaired motor development, and antisocial behavior. Children who survive acute lead poisoning show a high incidence of permanent brain damage (270). Functional brain damage associated with lead relative to mental development has been investigated (34, 60, 75, 204, 210). Most of these studies are based on clinical observations and no studies are available during the most rapid phases of brain growth.

An experimental animal model was developed by Pentschew and Garrow (237) using neonatal rats; the mother of the animals was given lead in diet. Since that study, several other authors have used neonatal rats (64, 201, 202, 288) and mice (253). The lead-poisoned newborn animals show retardation of growth during the fourth week of life. Severe signs of encephalopathy including extensive histological lesions of cere-bellum, brain edema, and paraplegia were found in those animals. There was an 85-fold increase in lead concentration in cerebellum and cerebral cortex. However, the RNA or protein concentration of the developing cerebellum was not affected. There was also an increase in the volume of the brain and flattening of convolutions. Experimental toxicity with lead in dogs has been studied by several workers (83, 239, 250, 277). In one study, dogs were given 100 ppm of lead in a diet containing low calcium and phosphorus (281). The animals had anorexia, cachexia, anemia, normoblastocytosis, and leukopenia within 6 weeks. Hypoproteinemia was accompanied by decreased levels of α_1- and α_2-globulins, albumin, and many serum enzymes. Marked enlargement of the liver, kidney,

Table 14-7. Range of Health Effects of Lead

	No Demonstrable Effects	Minimal Subclinical Effects Detectable	Compensation	Functional Injury (Short Intense Exposure)	Functional Injury (Chronic or Recurrent Intense Exposure)
Metabolic effects	Normal	Urinary ALA may increase	Increase in several metabolites in blood and urine	Further increase in metabolites	Increase only in case of recent exposure
Functional effects: Blood	None	None	Reduced red cell lifespan increased production	Reduced red cell lifespan with or without anemia (reversible)	Possible anemia (reversible)
Kidney Function	Normal	None	Sometimes minimal dysfunction	Fanconi syndrome (reversible)	Chronic nephropathy (permanent)
Central Nervous system	None	None	?	Minimal to severe brain damage (permanent)	Severe brain damage, particularly in children (permanent)
Peripheral nerves	None	None	?	Possible damage	Impaired conduction (may be chronic)
Symptoms	None	None	Sometimes mild, non-specific complaints	Anemia, colic, irritability, drowsiness; in severe cases, motor clumsiness, convulsions and coma	Mental deterioration, seizures, coma, foot or wrist drop
Residual effects	None	None	None known	Range from minimal learning disability to profound mental and behavioral deficiency, convulsive disorders, blindness	Mental deficiency (often profound), kidney insufficiency, gout (uncommon), foot drop (rare)

Adapted from Advances in Environmental Science and Technology (97) with permission.

and brain were also observed. Lead lines in the distal radic and thoracic spinous process were accompanied by delayed closure of thoracic-vertebral epiphyses. By far, the largest (97 percent) amount of the tissue lead was present in skeletal tissues.

Many of these findings in experimental dogs were also found in humans with lead poisoning. The animal model may be useful in the study of chronic toxicity due to lead present in certain environmental conditions.

Environmental Pollution with Cadmium

Cadmium is an important industrial trace metal used in electroplating and manufacture of pigments, plastics, alloys, and batteries (141). The cadmium enters the soil and air primarily through the mining and smelting of ores; its concentration in the vicinity of smelting factories is generally very high (49, 171, 203). Some other areas of emission include coal combustion and cement production.

The effect of cadmium on humans and animals has been reviewed by Fassett (100), Flick et al. (105), and Friberg et al. (109). The primary source of cadmium in humans appears to be food rather than air or water (109, 215, 258); an average intake of cadmium in foodstuffs is calculated at approximately 50 mg/day (85).

The acute and chronic effects may result from oral or inhalation routes. The acute oral effects include gastroenteritis, nausea, and vomiting (55, 98). Inhalation of cadmium produces toxic effects of pulmonary edema (100, 109). Chronic exposure results in emphysema (108), renal disturbances with specific low molecular weight proteins in urine (109, 240, 241, 247, 270), renal stones (161), fractures (225), and insomnia (5). A chronic condition called Iati-Iati which is characterized by musculoskeletal syndrome and classical renal effects has been reported in elderly multiparous Japanese women (6, 330). The role of cadmium as an etiological factor in hypertension is still controversial (109, 212, 238, 257).

Animal Models of Cadmium Poisoning

Studies with rats show acute necrotic effects of cadmium following parenteral injection or large oral doses (100, 109, 233). However, zinc, which otherwise resembles cadmium in toxicity, has protective action against acute atrophy of the testes (131). Anatomical studies with rat testes show damage to the vascular system of the testes (193). Rapid destruction of the fetal placenta in pregnant females has been reported (234). A teratogenic effect of cadmium that is counteracted by zinc in hamsters has been reported (102). However, the interpretation of these results and their application to humans has not been established. Weight measurements of the children of cadmium-exposed mothers has been reported (72). The carcinogenic properties of cadmium have been reported in rats generally at the sites of injection (133, 138, 139), and metastasis has been reported in other cases (161, 162). Interstitial cell neoplasm of rat testes has been reported (130, 186) and the increased incidence of neoplasm in rats but not in mice has also been reported (158, 259). Other studies (16, 267) did not find cadmium-related cancer in rats or dogs. The correlation of cadmium with cancer formation in the bladder and in the prostate has not been established (168, 191). Studies with rats have shown that very low doses provide protection against subsequent high or lethal doses of cadmium. (112, 286). Later studies have shown that the protective action is due to metallothionein in liver (265). The status of cadmium as an environmental health hazard has been recently reviewed (104).

Most of the metals do not exist alone and hence their interaction is important in toxicity studies. The action of some metals alone and in combination with other metals on embryogenesis is summarized in Table 14-6.

Guppies—A Possible Model for Pollution Studies

Guppies and other tropical fish, such as goldfish, paradise fish, and mollies, in small aquarium tanks have been suggested as models for testing pollutants in the waterways, streams, and ponds. Guppies and mollies are most sensitive to copper (214). Copper levels (10–100 ppm) were found to be lethal within 17 hours to all fish and the mortality was 100 percent at 1 ppm level. However, 10 ppb levels produced varying effects with mortality ranging from 0 to 60 percent. The procedure can be used as a bioassay and quick screening method for water soluble toxicants. Mollies can also be used as a model for "red tide" neurotoxin.

Environmental Pollution with Pesticides

Since the manufacture of DDT in 1940, the use of pesticides had steadily increased until 1966–1968. However, in recent years, because of greater concern over public health hazards, the development of safer pesticides has been sought.

Occupational hazards of pesticides in persons working in the pesticide manufacturing industries in the formation, packing, vector control, and crop control areas have been known for many years (160). However, indirect effects from the consumption of food and water contaminated with pesticide residues were not realized until many deaths in certain species of wildlife were investigated.

Animal studies have provided a life-death response relationship which is subjective depending on the test animal and solvent used (114, 256). However, ill effects have been reported at levels far below LD_{50} in animals (23). In addition to the pesticides, the vehicles and carriers for pesticides may be hazardous to human health. A few selected pesticides are discussed in this chapter to illustrate the role of animal models in understanding the biohazards from pesticides. Particular emphasis is placed on polychlorinated biphenyls, because of their relatively recent discovery as hazards and their similarities to DDT in many aspects.

Polychlorinated Biphenyls (PCB)

These chlorinated aromatic compounds have received great attention recently because of their wide distribution, persistence, tendency to accumulate in foodstuffs, and adverse effects. Despite their similarity in physical and chemical properties with DDT and DDE, the PCB behave differently in certain biological properties (335). As many as 210 different PCB can be distinguished by gas chromatographic analysis (275, 305). The physical and chemical properties of commercial PCB preparations have been reviewed (211). Because of the low volatility, nonflammability, high dielectric constant, and chemical stability of PCB, these compounds have been widely used in lubricating fluids, heat exchanger and dielectric fluids, and plasticizers for plastics and coatings. PCB has also been used as an ingredient in caulking compounds, adhesives, paints, printing inks, and carbonless duplicating paper. In Japan, the biggest single use is in the production of capacitors (150).

PCB contaminates the environment at the rate of approximately $1.5–2.5 \times 10^3$ tons/year in the atmosphere, 4.5×10^3 tons/year in water, and 1.8×10^4 tons/year in dumps and landfills (228, 276). Within the environment, transfer of PCB takes place by volatilization, aerial transport on particulates, fallout, and leaching from dumps (86, 91, 295, 296).

Human exposure to PCB is through air, water, and food. It is estimated that the average daily intake may be 1–3 μg/kg which is 100 times less than the lowest no-effect level in animal experiments (298) and at this time a tenfold accumulative safety is predicted (175). Human exposure to PCB from composite food samples is up to 0.38 ppm (174). An estimated mean daily intake is about 5 μg per adult as compared to about 40 μg/day for DDT (85). Of the various foods, fish show the greatest (70 percent) incidence of contamination (174, 310). However, a number of incidences of local contamination of other foods such as milk, poultry, eggs and cereal products has been reported (174). Besides food, certain occupational hazards to PCBs can be expected among workers in manufacturing plants for PCB. It appears that dermal exposure is more important than inhalation (69) in workers handling sealants, plastics, paints, vapors from air compressors, and paper coatings. No information is available on the extent of exposure except for carbonless papers in which individuals may pick up as much as 30 μg/100 sheets (194), and only one-third is washed off with soap and water. An unusual dust hazard (200 ppm) in certain houses has been reported (246).

Occurrence of PCB in human adipose tissue is approximately at the level of 1 ppm (246, 331); however, occasional samples up to 115–240 ppm (246) and 200–600 ppm in body fat (38) have been reported. Human milk samples in California contained 60 ppb whereas the levels in Sweden and Germany were 16 ppb and 100 ppb, respectively (4, 311). Infants taking 150 g/kg milk may have an ingestion level of 9 μg/kg/day.

The symptoms of PCB intoxication consist of acneiform eruptions; pigmentation of the skin and nails, and hypersecretion of the meibomian glands. The condition is often referred to as chloracne (69) involving the face, ears, abdomen, back, thighs, forearms, and buttocks. Systemic effects in severe cases include nausea, lassitude, anorexia, digestive disturbances, impotence, and hematuria (157, 262). Deaths have been reported from involvement of liver in conjunction with chlorinated napthalenes (262). Outbreaks of "Yusho" (rice oil disease) in Japan involving 325 cases occurred in 1968 (175). Detailed study of this episode has provided much information on the effects of PCB in humans. All ages and both sexes were equally affected; however, there was a lower risk for males and females over 60 years of age. The most common symptom in this episode was increased eye discharge (88.8 percent), followed by acne-like eruptions (87.6 percent) and pigmentation of nails (83 percent). Spasms in limbs were seen in about 8 percent of the cases. Certain fetal abnormalities, stillbirths, and retarded growth for males were also reported (111, 206).

Toxic effects have been observed from some of the trace contaminants such as chlorodibenzofurans (CDBF) (301). These contaminants are definitely chloracnegenic (261, 299) and are potent toxicants for animals (156).

Animal Models of PCB Toxicity. Experimental animals have been used to study the biological effects of the absorption, distribution, metabolism, and toxicity of PCBs.

Absorption studies with animals have not been carried out to any great extent. Earlier studies in guinea pigs showed that these compounds had fatal toxicity effects (205). Inhalation at 5.4 mg/cu m was toxic to guinea pigs (290). In rats, continuous accumulation of PCB was observed and a plateau was not reached even after exposure with PCB for 240 days (70). The major storage tissue was body fat with the concentration 10 to 100 times more than in any other tissues (126). Whole blood and plasma had lowest concentrations (Table 14-8). Despite the fact that skin lesions are very important in PCB poisoning, detailed information is not available. A greater persistence of PCB compounds in the skin of mice as compared to liver and kidney has been reported

Table 14-8. Distribution of PCB Among Various Tissues in Experimental Animals (ppm)

	Rat	California gull	Bengalese finch	Spot	Chicken
Whole body			61.	37.	
Liver	116.	4.3	25.	210.	6.
Fat	1000.	69.	574.		29.
Brain	40.	6.9	14.	8.4	
Muscle				7.4	0.44
Kidney	31.				
Heart	24.			12.	
Blood	2.				
Spleen	29.				
Skin					16.
Testes	19.				
Ovary		11.4			
Eggs		7.8			1.9
Gills				40.	

Adapted from <u>Environmental Research</u> (229) with permission.

(332). The distribution of PCBs in various edible tissues of poultry (263) and pheasants has been reported. In pheasants, a concentration of 300–400 ppm was observed in the brain of birds dying of PCB poisoning, whereas in cockerels under similar conditions the concentration was 70–700 ppm (300). The brain concentration was generally 85 percent of coexistent liver concentration and the concentration in fat was twenty to twenty-five times more than in the liver (245). Fish and aquatic invertebrates have been widely used for the determination of effects of PCB accumulation (Table 14-9) (226, 227, 313). One species of fish (spot) when exposed continuously to PCB for 56 days had accumulated a high concentration of PCB in 28 days (137). The fate of PCB in the body and its excretion has not been fully documented. Studies using rats as models indicate that the PCB peaks in gas chromatography either decrease or disappear after 4 days of feeding (70, 126, 172). Greater persistence of penta- and hexachlorobiphenyl isomers was found than of the tetrachloro isomer in mice (332). No metabolites have been definitely identified but phenolic degradation products have been observed in rats (333). In fish, the early peaks in GC were also decreased (274). The degree of retention, in general, related to the degree of chlorination. The overall clearance rate depends on the test species. In rats, the $T\frac{1}{2}$ for most organs was 6 to 9.5 days, whereas for fat it was 36 days (126). In cattle, the $T\frac{1}{2}$ of clearance for fat was 43 days (111). Most of the excretion (70 percent) in rats was in feces but about 5 percent was excreted in urine (333).

Toxicity studies of various aroclors (PCB) have been made in rats and rabbits. It appears that PCB has a low order of toxicity in these animals but rabbits are more sensitive to these compounds than rats. In ducks the oral LD_{50} was in excess of 2000 mg/kg (291). The subacute toxicity studies in rats show that 100 ppm in diet fed up to 15 months was not lethal. However, a dose of 1000 ppm in the diet was fatal to about 75 percent of male rats in 43 days (291). Minks appear to be the most sensitive mam-

Table 14-9. Effect of PCB on Aquatic Invertebrates

Species	Aroclor	Concentration (ppb)	Effect
Oyster	1254	10	Inhibited shell growth 59%
Pink shrimp	1254	5	Lethal in 20 days
Blue crab	1254	5	19/20 survived 20 days
Gammarus oceanicus	1254	10	Lethal threshold
Gammarus oceanicus	1254	1	Sublethal necrosis of bronchiae
G. pseudolimneus	1248	5	Highest safe concentration for
Daphnia magna	1248	2.2	survival, growth and
Daphnia magna	1254	3.3	reproduction
Tanytarsus dissimilis	1248	10	
Tanytarsus dissimilis	1254	10	
Anopheles quadrimaculatus larvae	1242	1	50% mortality in 48 hours

Adapted from Environmental Research (229) with permission.

511

mal; 30 ppm in diet was fatal to all animals in 2 to 4 months. In 1-day-old chicks, concentration of 250 ppm was fatal within 13 weeks (242). Other workers have reported deaths within 8 weeks at 300 ppm (106) and 50 percent death occurring within 4 weeks at 400 ppm (196).

The general symptoms of toxicity vary considerably with test species. Rats poisoned with aroclor 1254 showed ataxia, blanched retinas, ptosis of eyelids, and porphyrin-like nasal exudates. The symptoms varied with the type of aroclor (291). In chickens, symptoms of dyspnea, gasping, wadding, ataxia, anorexia accompanied by sudden death were frequently observed (173). Modified behavior in migrating robins (292), pheasants, chicks, and rats have been reported (73).

Deaths in rats, rabbits and guinea pigs following experimental poisoning with PCB show liver lesions, (fatty infiltration) atrophy, and necrosis (205, 300). Hyaline bodies occur in rats but not in guinea pigs and rabbits (31, 205). In avian species, hydropericardium, ascites, edema, liver damage, kidney damage, enteritis, and intestinal hemorrhage are common symptoms (106, 140, 196, 300). Embryotoxic effects have been reported in rabbits (297) and rats. Hatchability decreased considerably with 10–20 ppm of aroclor 1254 (263) and teratogenic effects were seen when 5 ppm were injected into the yolk sac (199). Using ring doves as the test species, it was reported that the embryonic mortality in the second generation fed 10 ppm was very high (235).

PCB has not been thoroughly tested for carcinogenicity. In mutagenic studies a high degree of chromosomal aberration was observed in second generation doves (235). However, dominant lethal assay in rats showed no mutagenic effects (163). Human lymphocyte cultures showed no effect (146). Invertebrates appear to be very sensitive and are affected by 1–10 ppb (86, 312). In a few studies, immunosuppressive activities of PCB have been reported in rats, rabbits, and guinea pigs (298, 299, 300). Because of the coexistence of PCB and DDT and its analogs, it was thought that some synergistic action may exist; however, no such action has been reported (140), except in horseflies and drosophila.

Most of the animal studies do not show the characteristic human syndrome represented by "chloracne." Hence animal models primarily provide basic and comparative biological activity of PCBs.

Organic Phosphate and Carbamate Pesticides

This group of insecticides and fungicides is quick acting and decomposes easily with a half-life of 180 days as compared to 3 to 10 years for DDT (200). The toxicity of these compounds varies; for example, parathion has an LD_{50} of 5 mg/kg and malathion 1375 mg/kg for rats (145).

The toxic symptoms in humans include headache, fatigue, mild indigestion; in serious cases pain in the chest and abdomen, incapacitating respiratory difficulty, fasciculation of muscles, and terminal convulsions and coma are seen (323). The principal action is the inhibition of acetylcholinesterase (315). Psychiatric sequelae has been reported (82, 121).

The most frequently used experimental species is the rat in which certain organophosphates produce narcotic and anesthetic effects (45, 145, 294). Behavioral studies in rats indicate strain-dependent changes (62, 271). Strain differences are genetically determined in sheep (182) and in humans (159). Teratogenic effect of these compounds has not been established but fetal and placental abnormalities have been reported in mice and rats (122). However human teratogenic effects have not been reported (166).

Bipyridyl Herbicides

Bipyridyls are highly effective phytological agents that do not persist in soil. Therefore, these compounds are preferred over 2,4D and 2,4,5,T pesticides. The toxicity studies in parakeets show an LD_{50} of 125 mg/kg. The symptoms of human toxicity include damage to skin, mucous membrane, lungs, corneal epithelium, nasal mucosa, and fingernails (56, 68).

Rats and mice have been used as models for the study of bipyridial toxicity. Lungs of rats and mice show proliferative fibrosis and alveolitis, by oral and intraperitoneal administration (63, 192, 251). However, the symptoms, except for lung irritation, are not produced in rats exposed to inhalations (113).

Carcinogenic Effects of Pesticides

This subject has been extensively reviewed (77, 87, 99). Tumorgenic activity of aldrin and dieldrin has been reported in mice (76), rats, and dogs (103); however, other workers have reported negative results (78, 303). High doses of DDT increase the incidence of tumors in susceptible mice (149). Multigeneration studies in mice show elevated incidence of leukocytosis, leukemia, and malignant tumors (285). High dosage of DDT administered in hamsters did not produce malignant growths (7). Some workers have assigned antitumorgenic activity to DDT in transplanted tumors in rats (180) and mammary tumors in mice and rats (230).

Environmental Pollution with Microbial Products

In the overall biosphere, biological hazards include certain varieties of fish, mushrooms, microorganisms including viruses, and plants. Most of these biological species produce toxins that have deleterious or toxic effects in humans. In recent years, there has been much concern about a variety of toxins produced by certain species of fungi and molds, for example, mycotoxins. Aflatoxins, a metabolite of certain species of fungi, are potent hepatocarcinogens. These compounds are widely distributed in nature and pose public health problems due to their contamination of environment. The effects of aflatoxin on various animal species have been studied to assess the potential dangers of these compounds to human health.

The first serious outbreak causing death in turkey pullets was reported by Blount (42). Among poultry, ducklings are most sensitive to aflatoxins, followed by turkeys, pheasant, quail, and chickens (129). Among various breeds of chicken, the New Hampshire strain is most sensitive. Pigs are sensitive to approximately 800 ppm dosage, but no toxicity is produced at 450 ppb (144). However, higher concentrations (690 ppb) cause growth retardation in growing pigs accompanied by liver degeneration (88). At later stages, the same feed showed substantial improvement in weight and regression of lesion. Histopathological changes include karyomegaly, cytoplasmic degeneration, proliferation of fibrous tissue, and bile duct epithelium. At higher dosage, formation of tubular structures and centrilobular plexuses have been reported in the liver (164).

Cattle, particularly younger animals, are highly sensitive to the effects of aflatoxin (185), showing signs of retardation of growth and tenesmus (10). Autopsy of the animals reveal liver fibrosis, ascites, and visceral edema. In cattle, the aflatoxin is cleared rapidly through milk (11).

Sheep and goats appear to be less sensitive as compared to other animals (3a, 3b, 79). Long-term experiments (5 years) showed lowered fertility, but there was no change in growth or liver biopsy samples throughout the study (183).

The effect of aflatoxins in dogs is often referred to as hepatitis, which is characterized by hepatic cell necrosis with fatty metamorphosis, bile duct proliferation, and cirrhosis (198, 268, 314). A single massive dose has much greater toxicity than multiple small doses over a long period of time (222).

Aflatoxin studies have also been carried out in nonhuman primates. Daily doses in the range of 0.1–0.5 mg/kg cause death in the animals within 30 to 40 days (71, 190). Pathological findings are influenced by the protein content of the feed. In recent studies, liver damage was not influenced by dietary protein (12, 80).

Rodents

The hepatocarcinogenic property of aflatoxin in rats has been observed by several workers (26, 57, 179, 317). Aflatoxin, in addition to liver damage, may cause carcinoma of the glandular stomach (53) and renal epithelial neoplasms (95). Long-term multigeneration feeding of aflatoxin showed no adverse affect on fertility (8). The acute toxicity studies indicate an LD_{50} of 5–7 mg (50, 317). Male rats are more susceptible than female (54) and there is a strain and feed difference also (9). Histopathological changes induced by aflatoxin follow a sequential order from precancerous to cancerous condition (224, 283, 293). Chronic changes in the nuclei and nucleoli are also found. Dietary proteins (187, 188, 221, 223) modify the histopathological changes in aflatoxin treated rats.

Guinea pigs are more sensitive (LD_{50} is 1.4 mg/kg) to acute toxicity than rats (25, 52). The acute liver lesions include centrilobular necrosis with biliary proliferation. In addition to liver damage, pathological changes in the kidneys have been reported (20, 189).

Teratogenic effects of aflatoxin have been reported in hamsters particularly when it is injected on the eighth day of pregnancy (93). Rabbits have not been widely used in aflatoxin studies; limited studies indicate that 65 ppm/kg can cause death in rabbits (209). In younger animals, 0.04 mg/day causes anorexia, growth retardation, weight loss, and death in 4 or 5 weeks (79). Skin application of aflatoxins may cause intraepidermal vesicles and bullae filled with leukocytic exudate and necrosis of the epidermis (154).

Fish, particularly trout, have been used as animal models for studies on aflatoxin associated hepatomas (Table 14-10) (81), because of their high sensitivity to aflatoxins (18, 19). Hyperplasia of bile duct epithelium is a characteristic finding (136, 325). Cellular changes include enlargement of nuclei, prominent nucleolus, and chromatin margination (51, 220, 302).

Houseflies appear to be very sensitive to the action of aflatoxicosis (13, 28). Other insects also show signs of toxicity with aflatoxin, for example, honey bees (107), mosquito larvae (264), and fruitflies (195).

Animal models not only provide insight into possible effects of aflatoxins in humans but also are useful in bioassay procedures. Among the more frequently used models are ducklings (58, 255, 316) (LD_{50} 18.2 μg for aflatoxin B_1); brine shrimp (44, 46), zebra fish and their larvae (2); and embryonated eggs (61, 152, 243) and mollusk eggs (289) with LD_{50} ranging from 0.025 μg to 0.048 μg depending on route of inoculation. Other bioassay systems include tissue cultures, plant albinism, and microorganism.

Antibacterial Drugs

Antibacterial drugs are widely used in humans, animals, and crops. Antibiotics are also being used to increase feed efficiency, rate of weight gain in animals, and egg production in poultry; they are also used in crop protection and food preservation. It is

Table 14-10. Liver Tumor Incidence in Rainbow Trout—Induction by Various Levels of Aflatoxin B$_1$ and Other Substances

Substance	Days fed	Total observation period (days)	Tumor frequency (%)
Aflatoxin B$_1$ (ppb)			
7.9	365	365	42
4.0	365	365	15
0.8	365	365	0
0.8	605	605	10
42	14	270	60
42	28	270	75
Dimethylnitrosamine (DMN) (ppm)			
300	365	365	42
19,200	485	485	100
2-Acetylaminofluorene (2-AAF) (ppm)			
150	365	365	7

Adapted from <u>Microbial Toxins</u> (81) with permission.

estimated that almost one-half of the antibiotics manufactured are primarily used as feed additives (244). The extensive usage of antibiotics increases their levels in the environment posing such problems as drug sensitization or intoxication, ecological imbalance due to residual chemicals, and development of antibiotic resistant organisms (14a, 14b). The exact hazards due to these factors have not been fully investigated. It is estimated that there are 17–20 million people in the United States who are sensitive to antibiotics and other chemotherapeutic agents (306). As little as 2 ppm of antibiotics have been shown to cause antibiotic sensitivity in gram negative organisms (15). In recent years there is growing concern over the indiscriminate use of antibiotics in feed and food products. In the final analysis, the cost-effect relationship may decide the future outcome of the use of these drugs as additives. This area has been reviewed by Huber (147).

REFERENCES

1. Abe, M., *Bull. Tokyo Med. Dental Univ.,* **14:**2 (1967).
2. Abedi, Z., and W. P. McKinley, *J. Assoc. Offic. Anal. Chemists,* **51:**902 (1968).
3. Abrams, L., *J. S. African Vet. Med. Assoc.,* **36:**5 (1965); *S. African Med. J.,* **39:**767 (1965).
4. Acker, L., and E. Schulte, *Natur Wissenschaften,* **57:**497 (1970).
5. Adams, R. G., and N. Crabtree, *Br. J. Ind. Med.,* **18:**216 (1966).
6. Adams, R. G., J. F. Harrison, and P. Scott, *Q. J. Med.,* **38:**425 (1969).
7. Agthe, C., H. Garcia, L. Shubik, L. Tomatis, and E. Wenyon, *Proc. Soc. Exp. Biol. Med.,* **134:**113 (1970).

8. Alfin-Slater, R. B., L. Aftergood, H. J. Hernandez, E. Stern, and D. Melnick, *J. Am. Oil Chemists Soc.*, **46**:493 (1969).
9. Alfin-Slater, R. B., L. Aftergood, P. Wells, and D. Melnick, *Fed. Proc.*, **31**:733 (1972).
10. Allcroft, R., in G. N. Wogan (ed.), *Mycotoxins in Foodstuffs*, MIT Press, Cambridge, 1965.
11. Allcroft, R., and B. A. Roberts, *Vet. Rec.*, **82**:116 (1968).
12. Alpert, E., A. Serck-Hanssen, and B. Rajagopolan, *Arch. Environ. Health*, **20**:723 (1970).
13. Amonkar, S. V., and K. K. Nair, *J. Invertebrate Pathol.*, **7**:513 (1965).
14a. Anderson, E. S., *Br. Med. J.*, **3**:333 (1968).
14b. Anderson, E. S., *Am. Rev. Microbiol.*, **22**:1506 (1968).
15. Anderson, E. S., and M. J. Lewis, *Nature*, **215**:89 (1967).
16. Anwar, R. A., *Arch. Environ. Health*, **3**:456 (1961).
17. Arner, E. C., and R. A. Rhoades, *Arch. Environ. Health*, **26**:156 (1973).
18. Ashley, L. M., J. E. Halver, W. K. Gardner, and G. N. Wogan, *Fed. Proc.*, **24**:627 (1965).
19. Ashley, L. M., J. E. Halver, and G. N. Wogan, *Fed. Proc.*, **24**:105 (1964).
20. Asplin, F. D., and R. B. A. Carnaghan, *Vet. Rec.*, **73**:1215 (1961).
21. Ayers, S. M., H. S. Mueller, J. J. Gregory, S. Gianelli, and J. L. Penny, *Arch. Environ. Health*, **18**:699 (1969).
22. Balchum, O. J., R. M. Buckley, R. Sherwin, and M. Gardner, *Arch. Environ. Health*, **10**:274 (1965).
23. Ball, W. L., K. Kay, and J. W. Sinclair, *Arch. Ind. Hyg. Occup. Med.*, **7**:292 (1953).
24. Barker, H. A., *Antonie van Leenwenhoek J. Microbiol. Sero 1*, **6**:20 (1940).
25. Barnes, J. M., *Trop. Sci.*, **9**:64 (1967).
26. Barnes, J. M., and W. H. Butler, *Nature*, **202**:1016 (1964).
27. Bear, R. R., *J. Air Poll. Control Assoc.*, **19**:722 (1969).
28. Beard, R. L., and G. S. Walton, *J. Invertebrate Pathol.*, **7**:522 (1965).
29. Beard, R. R., and G. A. Wertheim, *Am. J. Public Health*, **57**:2022 (1967).
30. Bedford, J. M., and L. Nicauder, *J. Anat.*, **108**:527 (1971).
31. Bennett, G. A., C. K. Drinker, and M. F. Warren, *J. Ind. Hyg. Toxicol.*, **20**:97 (1938).
32. Berglund, F., and M. Berlin, in M. W. Miller and G. C. Berg (eds.), *Chemical Fallout*, Charles C Thomas, Springfield, Ill. (1969).
33. Berlin, M., and S. Ullberg, *Arch. Env. Health*, **6**:589 (1963).
34. Beyers, R. D., and E. E. Lord, *Am. J. Dis. Child.*, **66**:471 (1943).
35. Bils, R. F., *Arch. Environ. Health*, **12**:689 (1966).
36. Bils, R. F., *Rev. Allergy*, **23**:471 (1969).
37. Bils, R. F., *Arch. Environ. Health*, **20**:468 (1970).
38. Biros, F. J., A. C. Walker, and A. Medbery, *Bull. Environ. Contam. Toxicol.*, **5**:317 (1970).
39. Blair, W. H., M. C. Henry, and R. Ehrlich, *Arch. Environ. Health*, **18**:196 (1969).
40. Blaylock, B. A., *Arch. Biochem. Biophysis*, **124**:314 (1968).
41. Blockker, P. C., *Atm. Env.* **6**:1 (1972).
42. Blount, W. P., *Turkeys*, **9**:52 (1961).
43. Brinkman, R., and H. B. Lamberts, *Nature*, **181**:1202 (1958).
44. Brown, R. F., *J. Am. Oil Chemists' Soc.*, **46**:119 (1969).
45. Brown, D. R., and S. D. Murphy, *Toxicol. Appl. Pharmacol.*, **18**:895 (1971).
46. Brown, R. F., J. D. Wildman, and R. M. Eppley, *J. Assoc. Offic. Anal. Chemists*, **51**:905 (1968).
47. Bryant, M. P., E. A. Wolin, M. J. Wolin, and R. S. Wolfe, *Arch. Mikrobiol.*, **59**:20 (1967).
48. Buell, G. C., Y. Toiwa, and P. K. Mueller, *J. Air Poll. Control Assoc.*, **19**:640 (1969).
49. Burkitt, A. P. Lester, and G. Nickless, *Nature*, **238**:327 (1972).
50. Butler, W. H., *Br. J. Cancer*, **18**:756 (1964).
51. Butler, W. H., *Am. J. Pathol.*, **49**:113 (1966).
52. Butler, W. H., and J. M. Barnes, *Br. J. Cancer*, **17**:699 (1963).
53. Butler, W. H., and J. M. Barnes, *Nature*, **209**:90 (1966).
54. Butler, W. H., and J. M. Barnes, *Food Cosmet. Toxicol.*, **6**:135 (1968).
55. California State Water Pollution Control Board, *Water Quality Criteria*, 2nd. Ed., Sacramento, Calif., 1957.
56. Cant, J. S., and D. R. H. Lewis, *Br. Med. J.*, **2**:224 (1968).

57. Carnaghan, R. B. A., *Br. J. Cancer,* **21**:811 (1967).
58. Carnaghan, R. B. A., R. D. Harley, and J. O'Kelly, *Nature,* **200**:1101 (1963).
59. Catcott, E. J., *Air Pollution,* World Health Organization, Geneva, 1961.
60. Chisolm, J. J. Jr., and H. E. Harrison, *Pediatrics,* **18**:943 (1956).
61. Choudhary, P. G., and S. L. Manjrekar, *Indian Vet. J.,* **44**:543 (1967).
62. Clark, G., *Aeroscope Med.,* **42**:735 (1971).
63. Clark, D. G., J. F. McElligot, and E. Weston-Hurst, *Br. J. Ind. Med.,* **23**:126 (1966).
64. Clasen, R. A., et al., *A n. J. Pathol,* **66**:1a (1972).
65. Clegg, D. J., *Proc. of Symposium Mercury in Man's Environment,* Royal Society of Canada, Ottawa, 1971.
66. Coffin, D. L., *Adv. Env. Sci. Technol.,* **2**:1 (1971).
67. Committee on Pollution, National Academy of Sciences, *Waste Management and Control,* Publication 1400, NAS-NRC, Washington, D.C., 1966.
68. Conning, D. M., K. Fletcher, and A. A. B. Swann, *Br. Med. Bull.,* **25**:245 (1969).
69. Crow, K. D., *Trans. St. Johns Hosp. Dermatol. Soc.,* **56**:79 (1970).
70. Curley, A., V. W. Burse, M. E. Guin, R. W. Jennings, and R. E. Lindner, *Env. Res.,* **4**:481 (1971).
71. Cuthbertson, W. F. J., A. C. Laursen, and D. A. H. Pratt, *Br. J. Nutr.,* **21**:893 (1967).
72. Cvetkova, R. P., *Gig. Tr. Prof. Zabol.,* **12**:31 (1970).
73. Dahlgren, R. B., and R. L. Linder, *J. Wild. Mgt.,* **35**:315 (1971).
74. Dales, L. G., *Am. J. Med.,* **53**:219 (1972).
75. David, O., J. Clark, and K. Voeller, *Lancet,* **2**:900 (1972).
76. Davis, K. J., and O. G. Fitzhugh, *Toxicol. Appl. Pharmacol.,* **4**:187 (1962).
77. Deichmann, W. B., *Ind. Med.,* **41**:15 (1972).
78. Deichmann, W. B., W. E. MacDonald, A. G. Beasley, and D. Cubit, *Ind. Med. Surg.,* **40**:10 (1971).
79. Delage, J. and P. Fehr, *Ind. Aliment. Prod. Animal,* **183**:44 (1967).
80. Deo, M. G., M. Dayal, and V. Ramalingaswami, *J. Pathol. Bacteriol.,* **101**:47 (1970).
81. Detroy, R. W., E. B. Lillehoj, and A. Ciegler, in A. Ciegler, S. Kadis and S. J. Ajl, (eds.) *Microbial Toxins,* Vol. 6, Academic Press, New York, 1970.
82. Dille, J. R., and P. W. Smith, *Aerospace Med.,* **35**:474 (1964).
83. Dodd, D. C., and E. L. J. Staples, *N. Z. Vet. J.,* **4**:1 (1956).
84. Duggan, R. E., *Ann. N.Y. Acad. Sci.,* **160**:173 (1969).
85. Duggan, R. E., and G. Q. Lipscomb, *Pesticide Monit. J.,* **2**:153 (1969).
86. Duke, T. W., J. I. Lowe, and A. J. Wilson, Jr., *Bull. Env. Contam. Toxicol.,* **5**:171 (1970).
87. Durham, W. F., and C. H. Williams, *Ann. Rev. Entomol.,* **17**:123 (1972).
88. Duthie, I. F., M. C. Lancaster, J. Taylor, E. B. Lomax, and H. M. Clarkson, *Vet. Rec.,* **82**:427 (1968).
89. Eckart, R. E., H. N. MacFarland, Y. C. E. Alarie, and W. M. Busey, *Arch. Env. Health,* **25**:381 (1972).
90. Editorial: "House Martins Return to London," *Sci. News,* **95**:569 (1968).
91. Edward, C. A., *Persistent Pesticides in the Environment,* Chemical Rubber Company, Cleveland, 1971.
92. Ehrich, R., and M. C. Henry, *Arch. Env. Health,* **17**:860 (1968).
93. Elis, J., and J. A. DiPaolo, *Arch. Pathol.,* **83**:53 (1967).
94. Emik, L. O., R. L. Plata, K. I. Campbell, and G. L. Clarke, *Arch. Env. Health,* **23**:335 (1971).
95. Epstein, S. M., B. Bartus, and E. Farber, *Cancer Res.,* **59**:1045 (1969).
96. Evans, M. J., R. J. Stephens, L. J. Cabral, and G. Freeman, *Arch. Env. Health,* **24**:180 (1972).
97. Ewing, B. B. and J. E. Pearson, *Adv. Env. Sci. Technol.,* **3**:1 (1974).
98. Fairhall, L. T., *Industrial Toxicology,* Williams & Wilkins, Baltimore, 1957.
99. Falk, H. L., S. J. Thompson, and P. Kotin, *Arch. Env. Health,* **10**:847 (1965).
100. Fassett, D. W., in D. H. K. Lee (ed.), *Metallic Contaminants and Human Health,* Academic Press, New York, 1972.
101. Ferm, V. H., *Adv. Teratol.,* **5**:51 (1972).
102. Ferm, V. H., and S. J. Carpenter, *Nature,* **216**:1123 (1967).

103. Fitzhugh, O. G., A. A. Nelson, and M. L. Quaife, *Food Cosmet Toxicol.,* **2:**551 (1964).
104. Fleischer, M., A. F. Sarofim, D. W. Fassett, P. Hammond, H. T. Shackette, I. C. T. Nisbet, and S. Epstein, *Env. Health Persptv.,* **7:**253 (1974).
105. Flick, D. F., H. F. Kraybill, and J. M. Dimitroff, *Env. Res.,* **4:**71 (1971).
106. Flick, D. F., R. G. O'Dell, and V. A. Childs, *Poultry Sci.,* **44:**1460 (1965).
107. Foot, H. L., *Am. Bee. J.,* **106:**126 (1966).
108. Friberg, L., *Acta Med. Scand.,* **138:**124 (1950).
109. Friberg, L., M. Piscator, and G. Nordberg, *Cadmium in the Environment,* Chemical Rubber Co., Cleveland, 1971.
110. Fries, G. F., *Env. Health Perspt.,* **1:**255 (1972).
111. Funatsu, I., F. Yamashita and T. Yoshikane, *Fukuoka, Acta Med.,* **62:**139 (1971).
112. Gabbiani, G., D. Baic, and C. Deziel, *Can. J. Physiol. Pharmacol.,* **45:**443 (1967).
113. Gage, J. C., *Br. J. Ind. Med.,* **25:**304 (1968).
114. Gaines, T. B., *Toxicol. Appl. Pharmacol.,* **14:**514 (1969).
115. Gale, T. F., *Env. Res.,* **6:**95 (1973).
116. Gale, T. F., and V. H. Ferm, *Life Sci.* Part 2, **10:**134 (1971).
117. Galle, P., and L. Morel-Maroger, *Nephzon,* **2:**273 (1965).
118. Gardner, M. B., C. G. Loosli, B. Hanes, W. Blackmore, and D. Teebken, *Arch. Env. Health,* **19:**637 (1969).
119. Gardner, M. B., C. G. Loosli, B. Hanes, W. Blackmore, and D. Teebken, *Arch. Env. Health,* **20:**310 (1970).
120. Garrett, N. E., R. J. B. Garret, and J. W. Archdeacon, *Toxic Appl. Pharmacol.,* **22:**649 (1972).
121. Gershon, S. and F. H. Shaw, *Lancet,* **I:**1271 (1961).
122. Gofmekler, V. A. and S. A. Tabakova, *Farmakol. Toksikol.,* **33:**737 (1970).
123. Goldsmith, J. R., and J. A. Nadel, *J. Air Poll. Control. Assoc.,* **19:**329 (1969).
124. Gough, J., in D. H. Collins, (ed.), *Modern Trends in Pathology,* Butterworth & Co., New York, 1959.
125. Goyer, R. A., and B. C. Rhyne, *Int. Rev. Pathol.,* **2:**2 (1973).
126. Grant, D. L., W. E. J. Phillips, and D. C. Vulleneuve, *Bull. Env. Contam. Toxicol.,* **6:**102 (1971).
127. Greenberg, M., et al., *Pediatrics,* **22:**756 (1958).
128. Gross, P., R. T. P. de Treville, and M. N. Haller, *Arch. Env. Health,* **19:**186 (1969).
129. Gumbmann, M. R., S. N. Williams, A. N. Booth, P. Vohra, R. A. Ernst, and M. Bethard, *Proc. Soc. Exp. Biol. Med.,* **124:**266 (1967).
130. Gunn, S. A., Gould, T. C., and W. A. D. Anderson, *J. Natl. Cancer Inst.,* **35:**329 (1965).
131. Gunn, S. A., T. C. Gould, and W. A. D. Anderson, *J. Reprod. Fertil.,* **15:**65 (1968).
132. Haagen-Smit, A. J., and L. G. Wayne, in A. C. Stern (ed.), *Air Pollution,* Vol. I, 2nd ed., Academic Press, New York, 1968.
133. Haddow, A., *Br. J. Cancer,* **18:**667 (1964).
134. Haeger-Aronsen, B., *J. Clin. Lab. Inv.* (Sup. 47), **12:**6 (1960).
135. Hall, S. K., *Env. Sci. Technol.,* **6:**31 (1972).
136. Halver, J. E., *Res. Repts. U.S. Fish and Wildlife Serv.,* **70:**78 (1967).
137. Hansen, D. J., P. R. Parrish, and J. I. Lowe, *Bull. Env. Contam. Toxicol.,* **6:**113 (1971).
138. Heath, J. C., *Nature,* **193:**592 (1962).
139. Heath, J. C., and M. R. Daniel, *Br. J. Cancer,* **18:**124 (1964).
140. Heath, R. G., J. W. Spann, and J. F. Kreitzer, *XV Congr. Int. Ornith.,* Den Haag, 1970.
141. Heindl, R. A., *U.S. Bur. Mines Bull.,* **650:**515 (1970).
142. Hemphill, D. D., *Trace Substances in Environmental Health,* University of Missouri, Columbia, Mo., 1972.
143. Henery, M. C., J. Findlay, J. Spangler, and R. Ehrich, *Arch. Env. Health,* **20:**566 (1970).
144. Hintz, H. F., H. Heitman, Jr., A. N. Booth, and W. E. Gagne, *Proc. Soc. Expt. Biol. Med.,* **126:**146 (1967).
145. Holmsted, B. O., *Pharmacol. Rev.,* **2:**567 (1959).
146. Hoppingarner, R., A. Samuel, and D. Krause, *Env. Health Perspec.* **1:**155 (1972).
147. Huber, W. G., *Adv. Env. Sci. Technol.,* **2:**289 (1971).
148. Hunter, D., R. R. Bomford, and D. S. Russel, *Q. J. Med.,* **9:**1913 (1940).

149. Innes, J. R. M., B. M. Ulland, M. G. Valerio, L. Petrucelli, L. Fishbein, E. R. Hart, et al., *J. Natl. Cancer Inst.,* **42:**1101 (1969).
150. Isono, N., *Critical News,* **1:**1 (1971).
151. Jackson, H., *Br. Med. Bull.,* **20:**107 (1964).
152. Jayaraman, A., E. J. Herbst, and M. Ikawa, *J. Am. Oil Chemists Soc.,* **45:**700 (1968).
153. Jensen, S. and A. Jernlöv, *Nordforsk,* **14:**3 (1968).
154. Joffe, A. Z. and H. Ungar, *J. Invest. Dermatol.,* **52:**504 (1969).
155. Johnels, A. J., M. Olsson, and T. Westermark, *Var Föda,* **7:**187 (1967).
156. Johnson, J. E., *Bio. Sci.,* **21:**899 (1971).
157. Jones, A. T., *J. Ind. Hyg. Toxiol.,* **23:**290 (1941).
158. Kanisawa, M., and H. A. Schroeder, *Cancer Res.,* **29:**892 (1969).
159. Kay, K. *Ind. Med. Surg.,* **35:**1068 (1966).
160. Kay, K. *Ann. Occup. Hyg.,* **10:**189 (1967).
161. Kazantzis, G., *Nature,* **198:**1213 (1963).
162. Kazantzis, G., and W. J. Hanbury, *Br. J. Cancer,* **20:**190 (1966).
163. Keplinger, M. L., O. E. Fancher, and J. C. Calandra, *P.C.B. Conference,* Quail Roost Conference Center, Rougemont, N.C., 1971.
164. Keyl, A. C., M. S. Masri, Booth, M. R. Gumbmann, and W. E. Gagne, *Proc. 1st U.S. Japan Conf. Toxic Microorganism,* Honolulu, Hawaii, U.S. Dept. of Interior, Washington, D.C., 1968, p. 72.
165. Khera, K. S., *Teratology,* **8:**293 (1973).
166. Khera, K. S., and D. J. Clegg, *Can. Med. Assoc. J.,* **100:**67 (1969).
167. Khera, K. S., and S. A. Tabacova, *Food Cosm. Toxic.,* **11:**245 (1973).
168. Kipling, M. D., and J. A. H. Waterhouse, *Lancet,* **1:**730 (1957).
169. Klein, R., S. D. Herman, P. E. Brubaker, G. W. Lucier, and M. R. Krigman, *Arch Pathol.,* **93:**408 (1972).
170. Kleinfeld, M., C. Giel, I. R. Tabershaw, *Arch. Ind. Health,* **15:**27 (1957).
171. Kobayashi, J., *Proc. 5th Ann. Conf. Trace Substances in Env. Health,* University of Missouri, Columbia, Mo., 1972.
172. Koeman, J. H., M. C. TenNoever de Brauw, and R. H. de Vos, *Nature,* **221:**1126 (1969).
173. Kohanawa, M., S. Shoya, and Y. Ogura, *Natl. Inst. Anim. Health Q.* **9:**213 (1969).
174. Kolbye, A. C., Jr., *Env. Health Perspec.,* **1:**85 (1972).
175. Kuratsune, M., T. Yoshimura, J. Matsuzaka, and A. Yamaguchi, *Env. Health Perspec.,* **1:**119 (1972).
176. Kurland, L. T., *World Neurol.,* **1:**370 (1960).
177. Kurland, L. T., S. N. Faro and H. Siedler, *World Neurol.,* **1:**370 (1966).
178. Lagerwerff, J. M., *Aerospace Med.,* **34:**479 (1963).
179. Lancaster, M. C., F. P. Jenkins, and J. McL. Philp, *Nature,* **192:**1095 (1961).
180. Laws, E. R., Jr., *Arch. Env. Health,* **23:**181 (1971).
181. Lee, D. H. K., *Metallic Contaminants and Human Health,* Academic Press, New York, 1972.
182. Lee, R. M., *Biochem. Pharmacol.,* **13:**1551 (1964).
183. Lewis, G., L. M. Markson, and R. Allcroft, *Vet. Rec.,* **80:**312 (1967).
184. Lichtenstein, E. P., *Env. Health Perspec.,* **1:**151 (1972).
185. Loosemore, R. M., and L. M. Markson, *Vet. Rec.,* **73:**813 (1961).
186. Lucis, O. J., R. Lucis, and K. Aterman, *Oncology,* **26:**53 (1972).
187. Madhavan, T. V., *J. Pathol. Bacteriol.,* **93:**443 (1967).
188. Madhavan, T. V. and C. Gopalan, *Arch. Pathol.,* **80:**123 (1965).
189. Madhavan, T. V., and K. S. Rao, *J. Pathol. Bacteriol.,* **93:**329 (1967).
190. Madhavan, T. V., P. G. Tulpule, and C. Gopalan, *Arch. Pathol.,* **79:**466 (1965).
191. Malcolm, D., *Ann. Occup. Hyg.,* **15:**33 (1972).
192. Manktelow, B. W., *Br. J. Exp. Pathol.,* **48:**366 (1967).
193. Mason, K. E., *Anat. Rec.,* **149:**135 (1964).
194. Masuda, Y., R. Kagawa, and M. Kurastsune, *Polychlorinated Biphenyl in Carbonless Carbon Paper,* Dept. of Public Health, Kyushu University, Japan, 1971.
195. Matsumura, F., and S. G. Knight, *J. Econ. Entomol.,* **60:**871 (1967).
196. McCune, E. L., J. E. Savage, and B. L. O'Dell, *Poultry Sci.,* **41:**295 (1962).

197. McDonald, A. D., O. A. Harper, A. El Attar, and J. C. McDonald, *Cancer,* **26:**914 (1970).
198. McKing, J. K., Ph.D. thesis, Cornell University, Ithaca, N.Y., 1965.
199. McLaughlin, J., Jr., G. P. Marliac, and M. J. Verret, *Toxicol. Appl. Pharmacol.,* **5:**760 (1963).
200. Menzies, C. M., *Ann. Rev. Entomol.,* **17:**199 (1972).
201. Michaelson, I. A., *Toxicol. Appl. Pharmacol.,* **26:**539 (1973).
202. Michaelson, I. A., and M. W. Sauerhoff, *Env. Health Perspec.,* **7:**201 (1974).
203. Miesch, A. T., and C. Huffman, Jr., *Area Environmental Pollution Study,* U.S. Environmental Protection Agency, Office of Air Programs, Publ. No. AP-91, 1970.
204. Millar, J. A., V. Battistini, R. L. C. Cumming, F. Carswell, and A. Goldberg, *Lancet,* **2:**695. (1970).
205. Miller, J. W., *U.S. Publ. Health Rep.,* **59:**1085 (1944).
206. Miller, R. W., *Teratology,* **4:**211 (1971).
207. Miller, S. and R. Ehrlich, *J. Infect. Dis.,* **103:**145 (1958).
208. Mills, C. A., *Air Pollution and Community Health,* Christopher, Boston, 1954.
209. Minne, J. A., L. F. Adelaar, M. Terblanche, and J. D. Smit, *J. S. African Vet. Med. Assoc.,* **35:**7 (1964).
210. Moncrieff, A. A., et al., *Arch. Dis. Child.,* **39:**1 (1964).
211. Monsanto, Co., *Aroclor Plasticizers,* Tec. Bull. PL-306, Monsanto Co., St. Louis, 1968.
212. Morgan, J. M., *Arch. Intern. Med.,* **123:**405 (1969).
213. Morikawa, N., *Kumamoto Med. J.,* **14:**87 (1961).
214. Morse, E. V., *J. Am. Vet. Med. Assoc.,* **162:**184 (1973).
215. Murthy, G. K., V. Rhea, and J. T. Peeler, *Env. Sci. Tech.,* **5:**436 (1971).
216. Naeye, R. L., *Am. J. Pathol.,* **59:**104a (1970).
217. National Academy of Science, *Committee on Biological Effects of Atmospheric Pollutants,* Washington, D.C., 1972.
218. National Air Pollution Control Administration, *Air Quality Criteria for Carbon Monoxide,* Publ. No. AP-62, 1970.
219. Nelson, N., T. Byerly, A. C. Kolbye, et al., *Env. Res.,* **4:**69 (1971).
220. Newberne, P. M., in G. N. Wogan (ed.), *Mycotoxins in Food Stuffs,* MIT Press, Cambridge, 1965.
221. Newberne, P. M., *Cancer Res.,* **28:**2327 (1968).
222. Newberne, P. M., D. H. Harrington, and G. N. Wogan, *Lab. Inves.,* **15:**962 (1966).
223. Newberne, P. M., A. E. Rogers, and G. N. Wogan, *J. Nutr.,* **94:**332 (1968).
224. Newberne, P. M., and G. N. Wogan, *Cancer Res.,* **28:**770 (1968).
225. Nicand, P., A. Lafitte, and A. Gros, *Arch. Mel. Prof. Med. Trav. Secur. Soc.,* **4:**192 (1942).
226. Nimmo, D. R., R. R. Blackman, and A. J. Wilson, *Marine Biol.,* **11:**191 (1971).
227. Nimmo, D. R., P. D. Wilson, R. R. Blackman, *Nature,* **231:**50 (1971).
228. Nisbet, I. C. T. and A. F. Sarofim, *Env. Health Perspec.,* **1:**21 (1972).
229. Norton, N., *Env. Res.,* **5:**253 (1972).
230. Okey, A. B., *Life Sci,* Part I, **11:**833 (1972).
231. Pan, A. Y. S., J. Beland, and Z. Jegier, *Arch. Env. Health,* **24:**229 (1972).
232. Pan, A. Y. S. and Z. Jegier, *Arch. Env. Health,* **24:**233 (1972).
233. Parizek, J., *J. Endocrinol.,* **15:**56 (1967).
234. Parizek, J., *J. Reprod. Fert.,* **7:**263 (1964).
235. Peakall, D. B., J. L. Lincer and S. E. Bloom, *Env. Health Perspec.,* **1:**1031 (1972).
236. Pentschew, A., *Acta Neuropath.,* **5:**133 (1965).
237. Pentschew, A. and F. Garro, *Acta Neuropath.* (Berlin), **6:**266 (1966).
238. Perry, H. M., *2nd Annu. Conf. Trace Substances in Environmental Health,* University of Missouri, Columbia, Mo., 1969.
239. Pettit, G. D., L. W. Holm, and W. E. Rushworth, *J. Am. Vet. Med. Assoc.,* **128:**295 (1956).
240. Piscator, M., *Arch. Env. Health,* **4:**607 (1962).
241. Piscator, M., *Arch. Env. Health,* **12:**335 (1966).
242. Platonow, N. S., and H. S. Funnell, *Vet. Rec.* **89:**109 (1971).
243. Platt, B. S., R. J. C. Stewart, and S. R. Gupta, *Proc. Nutr. Soc.* (Eng. Scot.), **21:**30 (1962).
244. Plumlee, M. P., *Proc. Ill. State Vet. Med. Assoc.,* Chicago, 1968.
245. Prestt, I., D. J. Jefferies, and N. W. Moore, *Env. Poll.,* **1:**37 (1970).

246. Price, H. A., *Env. Health Perspec.*, **1:**73 (1972).
247. Princi, F., *J. Ind. Hyg. Toxicol.*, **29:**315 (1947).
248. Ramel, C., *Hereditas*, **57:**445 (1967).
249. Reddy, C. A., M. P. Bryant, and M. J. Wolin, *Bacteriol Proc.*, **142:**164 (1969).
250. Robertson, B., *Vet. Rec.*, **86:**195 (1970).
251. Robertson, B., G. Einhorning, B. Ivemark, E. Malmquist, and J. Modee, *J. Pathol.*, **103:**239 (1971).
252. Rosen, C. G., H. Ackerfors, and R. Nilsson, *Svensk Kemtidskr*, **78:**8 (1966).
253. Rosenblun, W. I., and M. G. Johnson, *Arch. Pathol.*, **85:**640 (1968).
254. Runnstrom, J., and H. Manelli, *Exp. Cell Res.*, **35:**157 (1964).
255. Sargeant, K., R. Allcroft, and R. B. A. Carnaghan, *Vet. Rec.*, **73:**865 (1961).
256. Schafer, E. W., *Toxicol. Appl. Pharmacol.*, **21:**315 (1972).
257. Schroeder, H. A., *Chronic Dis.*, **18:**647 (1965).
258. Schroeder, H. A., and J. J. Balassa, *J. Chron. Dis.*, **14:**236 (1961).
259. Schroeder, H. A., J. J. Balassa, and W. H. Vinton, Jr., *J. Nutr.*, **86:**51 (1965).
260. Schubert, J., *Adv. Exp. Med. Bio.*, **40:**239 (1973).
261. Schulz, K. H., *Arbeitsmed. Sozialmed. Arbeitshyg.*, **3:**25 (1968).
262. Schwartz, L., *U.S. Publ. Health Bull. 229,* Washington, D.C., 1936.
263. Scott, M. L., D. V. Vadehra, and P. A. Mullenhoff, *Proc. 1971, Cornell Nutr. Conf. Feed Manuf.*, Cornell University, Ithaca, N.Y., 1971.
264. Shah, V. K., F. Matsumuya, and S. G. Knight, *J. Invertebrate Pathol.*, **10:**146 (1968).
265. Shaikh, Z. A., and O. J. Lucis, *Fed. Proc.*, **29:**298 (1970).
266. Shea, K. P., *Environment*, **15:**6 (1973).
267. Shubik, P., and J. L. Hartwell, *Public Health Service Publ. 149,* Dept. of HEW, Washington, D.C., 1969.
268. Sipell, W. L., J. E. Burnside, and M. B. Atwood, *90th Annu. Meeting Am. Vet. Med. Assoc.*, Toronto, 1957.
269. Skerfving, S., K. Hansson, and J. Lindsten, *Arch. Env. Health*, **21:**133 (1970).
270. Smith, J. P., J. C. Smith, and A. J. McCall, *J. Pathol. Bacteriol.*, **80:**287 (1960).
271. Smith, P. W., W. B. Stavinoha, and L. C. Ryan, *Aeroscope Med.*, **39:**754 (1968).
272. Spyker, J. M., and M. Smithberg, *Teratology*, **5:**181 (1971).
273. Srivastava, P. N., L. J. D. Zaneveld, and W. L. Williams, *BBRC*, **39:**575 (1970).
274. Stalling, D. L., *PCB Newsletter*, **3:**1 (1971).
275. Stalling, D. L. and J. N. Huckins, *J. Chromatogr.*, **54:**801 (1971).
276. Stanley, C. W., J. E. Barney, and M. R. Helton, *Env. Sci., Techn.*, **5:**430 (1971).
277. Staples, E. L. J., *N.Z. Vet. J.*, **3:**39 (1955).
278. Stephens, R. J., G. Freeman, and M. J. Evans, *Arch. Env. Health*, **24:**160 (1972).
279. Stern, A. C., *Air Pollution*, Vol. 3, Academic Press, New York, 1968.
280. Stokinger, H. E., W. D. Wagner, and O. J. Dobrogorski, *Arch. Ind. Health*, **16:**514 (1957).
281. Stowe, H. D., R. A. Goyer, and M. M. Krigman, *Arch. Pathol.*, **95:**106 (1973).
282. Suzuki, T., H. Matsumoto, T. Miyama, and H. Katsunuma, *Ind. Hlth. Japan*, **5:**149 (1967).
283. Svoboda, D., A. Racela, and J. Higginson, *Biochem. Pharmacol.*, **16:**651 (1967).
284. Swann, H. E., D. Brunol, and O. J. Balchum, *Arch. Env. Health*, **10:**24 (1965).
285. Tarjan, R., and T. Kemeny, *Food Cosmet Toxicol.*, **7:**215 (1969).
286. Terhaar, C. J., *Toxicol. Appl. Pharmacol.*, **7:**500 (1965).
287. Thomas, H. V., B. K. Milmore, G. A. Heidbreder, and B. A. Kogan, *Arch. Env. Health*, **15:**695 (1967).
288. Thomas, J. A., F. D. Dallenbach, and M. Thomas, *Virchows Arch. Path. Anat.*, **352:**61 (1971).
289. Townsley, P. M., and E. G. H. Lee, *J. Assoc. Offic. Anal. Chemists*, **50:**361 (1967).
290. Treon, J. F., F. P. Dleveland, and J. W. Cappel, *Am. Ind. Hyg. Assoc. Quart.*, **17:**204 (1956).
291. Tucker, R. K., and D. G. Crabtree, *Handbook of Toxicity of Pesticides to Wildlife*, U.S. Dept. of Interior, Bureau of Sport, Fisheries and Wildlife, Resources Publ. 84, Washington, D.C., 1970.
292. Ulfstrand, S., A. Sodergren, and J. Rabol, *Nature*, **231:**467 (1971).
293. Unuma, T., H. P. Morris, and H. Rusch, *Cancer Res.*, **27:**222 (1967).

294. Vandekar, M., *Nature,* **179:**155 (1957).
295. Veith, G. D., *Env. Health Perspec.,* **1:**51 (1972).
296. Veith, G. D., and G. F. Lee, *Water Res.,* **5:**1107 (1971).
297. Villeneuve, D. G., D. L. Grant, and W. E. J. Phillips, *Bull. Env. Contam. Toxicol.,* **6:**120 (1971).
298. Vos, J. G., *Env. Health Perspec.,* **1:**105 (1972).
299. Vos, J. G., and R. B. Beems, *Toxicol. Appl. Pharmacol.,* **19:**617 (1971).
300. Vos, J. G., and J. H. Koeman, *Toxicol. Appl. Pharmacol.,* **17:**656 (1970).
301. Vos, J. G., J. H. Koeman, and H. L. VanDer Maass, *Food Cosmet. Toxicol.* **8:**625 (1970).
302. Wales, J. H., *U.S. Dept. Interior Res. Report,* **70:**56 (1967).
303. Walker, A. I. T., D. E. Stevenson, J. Robinson, F. Thorpe, and M. Roberts, *Toxicol. Appl. Pharmacol.,* **15:**345 (1969).
304. Wayne, L. G., and L. A. Chamber, *Arch. Env. Health,* **16:**871 (1968).
305. Webb, R. G., and A. C. McCall, 162 National Meeting, American Chemical Society, 1971.
306. Welch, H. et. al. *Antibiot. Ann.,* 296 (1958).
307. Westberg, K., N. Cohen, and K. W. Wilson, *Science,* **17:**1013 (1971).
308. Westermark, T., and B. Sjostrand, *Int. J. Appl. Radiation Isotpes,* **9:**1 (1960).
309. Westöo, G., *Var Foda,* **19:**121 (1967).
310. Westöo, G., and K. Norén, *Var Foda,* **22:**93 (1970).
311. Westöo, G., K. Norén and M. Anderson, *Var Foda,* **22:**10 (1970).
312. Wildish, D. J., *Bull. Env. Cont. Toxicol.,* **5:**202 (1970).
313. Wildish, D. J., and V. Zitko, *Marine Biol.,* **9:**213 (1971).
314. Wilson, B. J., T. C. Campbell, A. W. Hayes, and R. T. Hanlin, *Appl. Microbiol.,* **16:**819 (1968).
315. Wilson, I. B., M. A. Hatch, and S. Ginsbury, *J. Biol. Chem.,* **235:**2312 (1960).
316. Wogan, G. N., *Mycotoxins in Food Stuffs,* MIT Press, Cambridge, 1965b.
317. Wogan, G. N., and P. M. Newberne, *Cancer Res.,* **27:**2370 (1967).
318. Wolin, M. J., E. A. Wolin, and R. S. Wolfe, *BBRC,* **12:**464 (1964).
319. Wood, J. M., *Adv. Env. Sci. Tech.,* **2:**39 (1971).
320. Wood, J. M., F. Scott Kenedy, and C. G. Rosen, *Nature,* **220:**173 (1968).
321. Wood, J. M. and R. S. Wolfe, *Biochemistry,* **5:**3598 (1966).
322. Wood, J. M., M. J. Wolin and R. S. Wolfe, *Biochemistry,* **5:**2381 (1966).
323. World Health Organization, *Pesticide Residues in Food,* WHO Tech. Report Ser. No. 391, WHO, Geneva, 1967.
324. Wright, G., *Am. Rev. Resp. Dis.,* **100:**467 (1969).
325. Wunder, W., and H. Otto, *Natur Wissenschaften,* **56:**352 (1969).
326. Wyatt, J. P., *Non-Silica Pneumoconiosis Fundamentals in Medical Progress,* Lea & Febiger, Philadelphia, 1953.
327. Wyatt, J. P., *Arch. Ind. Health,* **21:**445 (1960).
328. Wyatt, J. P., *Am. J. Pathol.,* **64:**197 (1971).
329. Xintaras, C., B. L. Johnson, C. E. Ulrich, R. E. Terrill, and M. F. Sobecki, *Toxicol. Appl. Pharmacol.,* **8:**77 (1966).
330. Tamagata, N., and I. Shigematsu, *Bull. Inst. Pub. Health,* **19:**1 (1970).
331. Yobs, A. R., *Env. Health Perspec.,* **1:**79 (1972).
332. Yoshimura, H. and M. Oshima, *Fukuoka Acta Med.,* **62:**5 (1971).
333. Yoshimura, H., H. Yamamoto, and J. Nagai, *Fukuoka Acta Med.,* **62:**12 (1971).
334. Zaneveld, L., P. N. Srivastava, and W. L. Williams, *Biol. Reprod.,* **2:**363 (1970).
335. Zitko, V., and P. M. K. Choi, *Fish Res. Board Can. Tech. Rep.,* **272:**1 (1971).

CHAPTER 15

UNIQUELY USEFUL ANIMAL SPECIES FOR BIOMEDICAL RESEARCH

B. M. Mitruka and M. J. Bonner

In recent years, many investigators have reported on a wide variety of mammals, amphibia, fish, and birds that may be quite useful in biological and clinical research studies. Some of these uniquely useful animal species are exact models of human diseases, for example, the nine-banded armadillo in leprosy research and the sand rat for the study of diabetes mellitus. As a result of these studies, biological and clinical data are available on many species of potentially useful experimental animals. New animal models of human diseases are being investigated at the present time in many laboratories, which will certainly contribute to the knowledge of human pathophysiological processes. This chapter briefly describes some of the animal species that seem to have definite usefulness in experimental medicine because of their anatomical, physiological, and/or biochemical similarities to humans. Although their usefulness as experimental subjects is emphasized in the following discussion, general characteristics and biological data of the animals are also given whenever possible. For further details on these and other animal species, the reader is referred to three comprehensive volumes on mammals (158), and reference works on birds (108), amphibia and reptiles (71), and biology data books (1, 2).

MAMMALS

Multimammate Mouse (*Praomys natalensis*; syn. *Rattus natalensis, Mastomys coucha*)

Adult male and female Mastomys may weigh 100 g and 80 g, respectively. Head and body length may reach 115–135 mm; tail length may reach 112–120 mm. The coloring is between that of the house mouse and the black rat. Anatomically, Mastomys resembles the rat, for example, in the absence of a gallbladder. Cannibalism occurs within the species.

Unique Features

Mastomys possess several unique features such as (a) a well-developed female ventral prostate; (b) a cordate anterior process of the baculum (or penis); (c) absence of preputial glands. The incisor teeth are often yellow-brown due to ferric ions and hemosiderin. There are eight to ten pairs of teats with nipples more or less evenly spaced along the mammary line, hence the common name, multimammate.

Physiological Data

Mastomys attain maximum weight in 1 year and have a 3-year life span. Sexual maturity in the female occurs at 57 to 61 days of age. Attainment of 40 g weight is a rough indication of sexual maturity. Gestation is 23 days, with an average litter size of eight. Postpartum estrus occurs 2.6 ± 0.52 days after parturition. Rectal temperatures taken in fifty males averaged 96.7°F (35.9°C), 98.4°F) (36.9°C) in twenty-six nonlactating females, and 99.6°F (37.5°C) in twenty-four lactating females (99).

Uses in Medical Research

Mastomys are especially useful in cancer research because of their many spontaneous tumors, for example, carcinomas of the glandular stomach, papillomas, lipomas, and sarcomas of the skin and muscle. Varying tumor patterns have been reported both from different laboratories and from different inbred lines (31).

Deer Mouse (*Peromyscus maniculatus*)

The genus *Peromyscus* includes over 50 species and more than 200 described subspecies (92).

Physiological Data

Organ weights of *Peromyscus maniculatus bairdii,* deer mouse of southeastern Canada and northeastern United States are given in Table 15-1 (33).

The gestation period of a nonlactating female is 22 to 23 days, and the average litter size is four. There is a postpartum estrus which is delayed if the mother is nursing. Laboratory life span is 5 to 6 years.

Uses in Medical Research

The deer mouse is used in the study of bacterial and viral diseases and in genetic studies.

New Zealand Black (*NZB*) Mouse

The NZB/Bl strain of mice was developed by Dr. Marianne Bielschowsky at the University of Otago, Dunedin, New Zealand by breeding from an outbred colony of

Table 15-1. Organ Weights of *Peromyscus Mainiculatus Bairdii,* Deer Mouse of Southeastern Canada and Northeastern United States

Organ	Weight gms	Organ	Weight gms
Body wt.	19.33 (20)[1]	Brown Fat	0.14 (64)
Heart	0.11 (27)	Pelt	2.58 (19)
Lungs	0.19 (16)	Brain	0.59 (12)
Liver	0.71 (28)	Carcass	10.55 (28)
Spleen	0.02 (50)	Uterus	0.14 (43)
Kidneys	0.27 (15)	Testes	0.38 (16)
GI Tract	0.60 (27)	Total Body	
Adrenals	0.01 (30)	Parts[2]	16.19 (24)
Pancreas	0.07 (43)		

*Sample size: 18 (14 M, 4F)

[1]Mean in grams (coefficient of variation).

[2]Lost and collected blood, GI fill and urine excluded.

Adapted from Laboratory Animal Science (33), with permission.

mixed coat color. Several inbred strains such as NZB, NZC, NZO, NZW, and NZY have been evolved in Dunedin from the original colony (58).

Physiological Data

In the course of collecting laboratory data on NZB mice, Bielschowsky et al. (16) found evidence of an autoimmune hemolytic anemia which serves as a model of autoimmune disease. Mice of the NZB/Bl strain spontaneously develop serological and pathological evidence of autoimmune disease at about 16 weeks of age.

Uses in Medical Research

NZB mice may serve as animal models for systemic lupus erythematosus, Sjogren's syndrome, and Waldenstrom's macroglobulinemia. They are thought to develop spontaneous endocarditis as they age (105). Generally NZB mice may be used in genetic studies and studies involving autoimmunity.

Brown Norway, Hanovarian, or Gunn Rat (*Rattus norvegicus*)

There are three well-known strains of the Hanovarian rat: (a) hooded with a colored head and shoulders, sometimes with colored dorsal markings the color varying from cream to black; (b) Wistar, albino laboratory rat strain from the Wistar Institute, Philadelphia, Pennsylvania; (c) Sprague-Dawley albino rat from the Sprague-Dawley Farms, Wisconsin. The Gunn rat is a mutant of the Wistar strain in which unconjugated hyperbilirubinemia appears as a recessive trait with incomplete penetrance. The jaundice is not hemolytic in origin but results from an inability to conjugate bilirubin.

Physiological Data

The litter size averages eleven. Males and females are usually mature at 50 to 60 days of age. Range of estrous cycle is 4.1 to 5.2 days. Newborn weight is 6.0 g.

Uses in Medical Research

The rat is used for studies on pneumonia, middle ear disease (labyrinthitis), and neoplasms. The most commonly occurring tumor is the fibroadenoma of the female mammary glands. The Gunn rat is a model for comparative research on bilirubin excretion, jaundice, kernicterus, bilirubin nephropathy, and perinatal toxicology.

Egyptian Sand or Desert Rat (*Psammomys obesus*)

Sand rats are unique in that they are not able to cope with the increased caloric load of synthetic chow and respond by marked insulin production. The sand rat (Fig. 15-1) has a unique capacity to concentrate its urine.

Physiological Data

Blood glucose, immunoreactive insulin, and liver glucose phosphotransferase levels for three weight categories of the sand rat are given in Table 15-2 (32). Litter size is three to five young. This genus is distinguished from *Meriones* by its nongrooved incisors and rounded, short, thick ears.

Uses in Medical Research

The sand rat is a good model for diabetes mellitus. The animal develops necrotizing renal papillitis very much like that seen in human diabetes, cataracts, pylonephritis, and cytomegalic inclusions in the salivary glands.

Fig. 15-1. Egyptian sand rat *Psammomys obsecus.* From *Mammals of the World* (160) with permission.

Mink (*Mustela vison*)

There are three important subspecies of mink: *M. vison vison,* the Eastern mink (Fig. 15-2); *M. vison melampeplus,* the Kenai mink; and *M. vison ingens,* the Alaskan mink. The "off color" or gun-metal color mink received the name Aleutian after the fox that has a similar pelt color. The coat color is inherited as an autosomal recessive trait (a).

Unique Features

All minks are susceptible to Aleutian disease (AD) which is caused by a filterable agent, probably a virus. It is of interest as a chronic viral infection. Aa and AA mink

Fig. 15-2. Eastern mink (*Mustela vison*). From *Mammals of the World* (160) with permission.

Table 15-2. Blood Glucose, Immunoreactive Insulin, Liver Glucose-phosphotransferase Levels for Three Weight Categories of the Sand Rat

Weight (gm)	94	+ 10	134	+ 10	220	+	7
Age (days)	35	- 52	71	- 75	124	-	160

Test							
Blood glucose (mg/dl)	84	+ 10	63	+ 18	84	+	12
Immunoreactive insulin (μU/ml)	32	+ 8	50	+ 9	108	+	29
Hexokinase[1]	0.2 +	0.03	0.5 +	0.1	0.1 +		0.03
Glucokinase[2]	1.7 +	0.2	1.6 +	0.7	1.8 +		0.4
Glucose-6-Phosphatase[2]	7.0 +	0.5	6.5 +	0.8	7.2 +		0.5

*Three subjects per category.

[1]Units/min/gm liver
[2]μMoles P/min/gm liver

Adapted from *Diabetologica* (32), with permission.

usually live to pelting while all aa seen have the Chediak-Higashi syndrome (C-HS). The Aleutian (aa) mink possesses abnormal bodies in the cytoplasm of neutrophils, eosinophils, monocytes, and lymphocytes. It possesses less vigor and greater disease susceptibility than heterozygous (Aa) or homozygous dominant (AA) mink (81). The C-HS animal is susceptible to staphylococcal and corynebacterial abscesses, pasteurellosis, and brucellosis. In humans, mink, and cattle with C-HS syndrome, there is partial albinism (giant melanin granules).

Physiological Data

Mustela vison breeds in February and March. Delayed implantation may occur, giving a gestation period of 39 to 78 days. Males and females reach sexual maturity at 1 year of age. Life span is 10 years. Minks are nocturnal.

Uses in Medical Research

The mink may be used as a biomedical model for urinary incontinence, "wet belly" of the male mink. They are prone to urinary tract infections with and without calculi as a complicating factor. They have also been used as models for hereditary deafness, Ehlers-Danlos syndrome—a primary collagen defect, transmissible encephalopathy which resembles scrapie of the sheep, Wernicke's disease, Chvostek's paralysis—a thiamine deficiency and malabsorption syndrome giving "grey diarrhea" (104).

Voles (*Clethrionomys* and *Microtus*)

The genus *Clethrionomys* includes some seventy named forms distributed throughout the temperate and near-Arctic regions of the northern hemisphere of both the old and new worlds. They are herbivorous rodents. The British forms include the bank vole, *C. glareolus britannicus,* found on the British mainland; *C. glareolus erica,* found on the island of Raasay, off the western seaboard of Great Britain; and *C. glareolus skomerensis* found on the Island of Skomer, off the western seaboard of Great Britain. The bank vole is the smallest, weighing 17 g with a head and body length of 90 mm and tail length of 45 mm. The others average 30–35 g in weight with a head and body length of 105–110 mm and tail length of 23–35 mm.

The genus *Microtus* includes forty-four species. Generally, the members of this genus are more docile than those of the *Clethrionomys.* The short-tailed field vole, *Microtus agrestis,* is easily kept in the laboratory. The Orkney vole, *M. orcadensis,* is larger and lives longer.

Physiological Data

Life span of the vole is approximately 75 weeks and varies with the strain. Voles breed well if given at least 15 hours illumination daily; given a 9-hour day they almost cease to breed. Litters are born every month with seasonal variation in numbers. The largest litters are produced in April and May, the smallest in December. Gestation period is 20 to 21 days with an average litter size of four, ranging from one to seven. Postpartum estrus is common, though not universal. Female sexual maturity occurs at 21 days of age; male sexual maturity occurs at 69 days of age. Weight at birth is 2–3.5 g.

Concentrations of volatile fatty acids in the coecal fluid and the Somogyi blood glucose level of the adult meadow vole, *M. pennsylvanicus,* is given in Table 15-3 (84). Comparative organ weights for three species of voles, *M. oeconomus,* tundra vole of northern and eastern Alaska; *M. pennsylvanicus tananaensis,* meadow vole of south central Alaska; and *C. rutilus dawsoni,* red-backed vole of Alaska and northern Canada, are presented in Table 15-4 (33). The prairie vole, *M. ochrogaster,* is useful for com-

Table 15-3. Concentrations of Volatile Fatty Acids in the Cecal Fluid and the Somogyi Blood Glucose Level of Adult Meadow Voles, *M. Pennsylvanicus*

Parameter	Mean ± Std. deviation	
No. of animals	28	
Age (days)	176	± 68
Weight (gm) [2]	41	± 7.5
Acetic acid (mM/l)	22	± 6.6
Propionic acid (mM/l)	3.3	± 0.8
Butyric acid (mM/l)	3.9	± 1.6
Blood glucose (mg/dl)	89	± 14

[1]Diet: cellulose, 40%; casein, 21%; cornstarch, 37.4%; salt mix, 0.4%; corn oil, 0.8%.
[2]Average body weight: females, 35 gm; males 46 gm.

Adapted from Laboratory Animal Care (84), with permission.

Table 15-4.   Comparative Organ Weights for Three Species of Voles

	M. oeconomous	M. pennsylvanicus tananaensis	C. Rutilus dawsoni
Sample size	18 (18 M) [1]	15 (5 M; 10 F)	21 (15 M; 6 F)
Body weight	33.82 (29) [1]	28.89 (34) [1]	24.95 (23) [1]
Heart	0.17 (18)	0.13 (15)	0.17 (24)
Lungs	0.29 (28)	0.20 (25)	0.20 (25)
Liver	1.26 (43)	1.27 (45)	1.17 (32)
Spleen	0.06 (67)	0.03 (33)	0.03 (67)
Kidneys	0.46 (28)	0.27 (22)	0.29 (24)
GI Tract	1.33 (20)	1.44 (29)	1.18 (28)
Adrenals	0.03 (64)	0.01 (40)	0.01 (89)
Pancreas	0.07 (43)	0.05 (40)	0.05 (40)
Brown Fat	1.56 (68)	0.25 (68)	0.50 (64)
Pelt	6.01 (29)	4.83 (35)	3.93 (24)
Brain	0.37 (19)	0.59 (07)	0.55 (11)
Carcass	17.25 (35)	14.49 (39)	13.41 (26)
Uterus	–	0.07 (71)	0.11 (36)
Testes	0.20 (35)	0.40 (55)	0.53 (34)
Total Body Parts [2]	29.06 (31)	23.75 (34)	21.91 (24)

[1] Mean in grams (coefficient of variation).

[2] Lost and collected blood, GI fill and urine excluded.

Adapted from Laboratory Animal Science (33), with permission.

parative developmental studies of rodents. The postnatal growth and external morphology of eight litters of M. ochrogaster pups is given in Table 15-5 (77).

Uses in Medical Research

The vole is used in the study of virus diseases: Rift valley fever, Louping ill, climatic bubo, the neurotropic strain of yellow fever, and foot-and-mouth disease. M. agrestis is susceptible to human and bovine strains of Mycobacterium and to Toxoplasma.

Coypu or Nutria (Myocastor coypu)

The coypu is a large Hystroicoid rodent of South American origin. It is aquatic with webbed hind feet. Blood samples are difficult to obtain except by cardiac puncture.

Physiological Data

Gestation period averages 128 days. Weight at birth is 150–250 g. Adult weight is over 8 kg. Average litter number is five. The length of the estrous cycle is 28 days. Female sexual maturity occurs at 4 months. Resorption of embryos occurs. There is a postpartum estrus. Captive feral nutrias may produce four to five litters every 2 years for 5 years. The mammae are 4 + 4 or 5 + 5. Normal body temperature is 99.5°F (37.5°C).

Uses in Medical Research

The animals are useful in physiological and metabolic research, heat stroke studies and tuberculosis research. They are suitable for taking electrocardiograms, blood

Table 15-5.   Postnatal Growth and External Morphology of Eight Litters of *M. Ochrogaster* Pups

	Mean wt. (gm)	S. D.	S. E.	Range
34 neonate pups	3.1	0.27	0.05	2.4 - 3.5
27 at 21 days	19.0	2.61	0.51	13.9 - 24.1

Measure	Birth	1	2	3	4	5	6	7	8	9	10
Pinna	Attached	Free in 90%	Free								
External auditory meatus	Closed						Open in 25%	Open in 75%			
Eyes	Closed							Open in 10%	20% 50%	75% 90%	
Upper incisors	Not erupted	Erupted									
Lower incisors	Erupted					Free in 90%	Free				
Front digits	3/4 fused										
Rear digits	3/4 fused		1&5 Free 2,3,4 Fused					2,3,4 Free free in 50%			
Anus	Not patent	Patent in 25%	in 75%								

Adapted from <u>Laboratory Animal Science</u> (77), with permission.

pressure, and respiratory measurements (cannulation of trachea or carotid artery). Coypus can be used in typing strains of human tuberculosis (65).

Lemming (*Lagurus lagurus Pall,* Steppe lemming; *Dicrostonyx groenlandicus, D. stevensoni Nelson,* collared lemming)

The lemming belongs to the subfamily Microtinae of the family Cricetidae as does the vole. The Steppe lemming is a small rodent that inhabits the steppe regions in the west of Siberia, north of Kazakkstan and Mongolia, and east of the European part of the U.S.S.R. It is a nonhibernating, colonial animal. The collared lemming, *Dicrostonyx groenlandicus* or *D. stevensoni Nelson,* inhabits the Arctic coastal areas.

Physiological Data

For the Steppe lemming, sexual maturity occurs between 40 and 50 days of age with a life span of 3 years. The gestation period is 20 to 22 days. It is not easily bred in captivity. The reproductive period lasts 1.5 years. There is a decrease in breeding rate after 10 to 12 months. The average litter size is between five and six, and ten to twelve litters can be produced in 1 year. Newborn weight is 1.5 g. At 20 days of age, males weigh 9.1 g and females weigh 8.5 g. The diploid chromosome number in testicular tissue is fifty-four.

The collared lemming (34) has a basal metabolic rate, measured in the range of 25°–30°C, 40 percent higher than the standard metabolic rate for a mammal of similar size. Litter size at birth averages 2.8 in captivity. The mean life span of 254 colony-reared lemmings dying from natural causes was 189 days. The growth rate is rapid during the first 2 months of life, with moderate increases thereafter to an adult weight of approximately 70 g. Organ weights for the collared lemming are given in Table 15-6 (33).

Uses in Medical Research

Lemmings are used in experiments with tularemia (36), listeriosis, and poliomyelitis (111). Suckling animals are highly susceptible to both types (II and IV) of the virus after intracerebral injections. Adults are susceptible to type II virus but resistant to type IV virus. The lemming is also of potential value in cancer research.

Chinchilla (*Chinchilla laniger*)

The chinchilla (Fig. 15-3) has a number of features that make it a good laboratory animal. It is small in size and easy to handle. It is trainable in a conditioned avoidance situation. It has odorless urine and feces. The following points are favorable for hearing research: (a) large, easily accessible bulla; (b) a cochlea having, as in humans, three turns; (c) absence of presbycusis in long-term study; (d) lack of susceptibility to middle ear infections.

Table 15-6.   Organ Weights for the Collared Lemming, *Dicrostonyx Groenlandicus*

Organ	Weight gms	Organ	Weight gms
Body wt.	61.46 (27)[1]	Brown Fat	1.19 (86)
Heart	0.28 (18)	Pelt	9.92 (31)
Lungs	0.46 (30)	Brain	0.88 (12)
Liver	2.67 (43)	Carcass	32.57 (32)
Spleen	0.05 (20)	Uterus	0.11 (45)
Kidneys	0.68 (12)	Testes	0.23 (35)
GI Tract	3.05 (24)	Total Body	
Adrenals	0.02 (33)	Parts[2]	52.07 (29)
Pancreas	0.15 (27)		

Sample size: 16 (7 M, 9 F).

[1]Mean in grams (coefficient of variation).

[2]Blood, GI contents and urine excluded.

Adapted from Laboratory Animal Science (33), with permission.

Fig. 15-3. Chinchilla (*Chinchilla laniger*). From *Mammals of the World* (160) with permission.

Physiological Data

Life span is 12 to 20 years, while sexual maturity occurs at 5 to 8 months of age. Gestation period is 112 days with a litter size ranging from one to six. Adult weight is 0.5–1.0 kg.

Uses in Medical Research

The chinchilla is used for studies in gerontology, otology, experimental bacteriology, genetics, immunization procedures, hematology, parasitology, physiology, psychology, and radiation exposure.

Chinese Hamster (*Cricetulus griseus*)

Adult Chinese hamsters are generally solitary animals. They weigh up to 55 g and measure 9.0 cm in length from muzzle to base of tail. The Chinese hamster exhibits classic symptoms of diabetes mellitus. The recessive factor or factors for the diabetes mellitus also require a background of an ill-defined number of homozygous modifiers before the disease can appear in either the juvenile or adult forms.

Physiological Data

Life span is 2 to 4 years. Males reach sexual maturity when the testes measure about 12 mm in length. High temperature is detrimental to male fertility. The youngest sexually mature female reported was 48 days at the time of first successful mating. In a captured female, the length of the estrous cycle is 4.5 days. Gestation is 20 to 21 days with an average litter size ranging from four to six.

Uses in Medical Research

The Chinese hamster is susceptible to a number of infectious diseases: pneumococcal pneumonia, leishmaniasis, diphtheria, tuberculosis, rabies, influenza, and viral encephalitis. It exhibits classic symptoms of diabetes mellitus. The Chinese hamster is used for cytogenetic studies (in vivo) particularly in the area of radiation and chemical mutagenesis. Since the normal Chinese hamster karyotype consists of twenty-two chromosomes arranged in ten pairs of autosomes and two sex chromosomes, a more exacting identification of normal and aberrant chromosomes in bone marrow cells can be made using G-banding patterns (80).

Squirrels (*Glaucomys volans*, flying squirrel; *Spermophilus* spp., *Citellus* spp., ground squirrel; *Sciurus niger*, fox squirrel)

The California ground squirrel, *Spermophilus beecheyi*, is large, powerful, and difficult to handle. In the wild, the adult weighs 500–800 g. The laboratory adult is often in excess of 800 g because of a larger amount of fat. The Belding ground squirrel, *S. beldingi*, is easier to handle (87). *G. volans* (Fig. 15-4) is found throughout the eastern United States and small isolated areas of Mexico and Guatemala. *S. niger* is found in the eastern United States.

Physiological Data

Nonhibernating ground squirrels have a body temperature of 38°C and whole blood pH of 7.43 (7.38–7.52). Hibernating ground squirrels have a body temperature of 10°C and whole blood pH of 7.10 (7.01–7.20) (143). *S. niger* can produce two litters a year. Mating takes place in January and again in June or July. *G. volans* is nocturnal. Sexual maturity occurs at 1 to 2 years of age. Litter size ranges from two to six after a gestation period of approximately 40 days.

Uses in Medical Research

The fox squirrel has a physiological porphyria extending into adult life, with elaboration of type I porphyrins (154). Osteomalacia has been reported in captive flying squirrels (132).

Woodchuck (*Marmota monax*)

The woodchuck of southcentral Pennsylvania, *Marmota monax Linnaeus*, is a large ground squirrel. Unlike marmots in other parts of the world, which live together in small colonies, the eastern woodchuck leads a solitary existence except during the breeding season.

Physiological Data

The life span of the eastern woodchuck is 5 to 10 years. Reproduction is governed by length of daylight. Most young are born between April 1 and April 21 in the wild in southcentral Pennsylvania. Gestation period is 31 to 32 days with one litter of two to four young. Polyovulation is not a common occurrence. Lactation lasts approximately 44 days (135). Weight at birth of male and female is 27.2 g and, at birth, the young

Fig. 15-4. Eastern "flying" squirrel (*Glaucomys volans*). From *Mammals of the World* (160) with permission.

measure 85 mm (3.35 in.) from tip of nose to base of tail. In the wild, hibernation begins in October. Weights of adult males average 6.8 kg, females, 5.4 kg, prior to hibernation. There is a 20 to 38 percent weight loss during hibernation averaging 0.2 g/day/kg body weight (136, 137). Laboratory woodchucks do not hibernate.

Uses in Medical Research

Snyder and Ratcliffe (138) have proposed the woodchuck, *Marmota monax,* as a model for cardiovascular, cerebrovascular, and neoplastic diseases. They reported on thirty woodchucks (twenty-three males and seven females) caged from the age of 5 months. The animals died at ages ranging from 5 to 77 months. Among the causes of death were aortic rupture (5), mean age 28.8 months; cerebrovascular disease with hemorrhage (4), mean age 32.2 months; renal glomerular disease (4), mean age 42.2 months; coronary heart disease (4), mean age 55.5 months; and malignant hepatoma (9), mean age 39.3 months. The arterial lesions were reported to correspond in their development pattern to arterial lesions found in a wide range of birds and mammals including humans.

Mongolian Gerbil (*Meriones unguiculatus*)

The Mongolian gerbil (Fig. 15-5) is a desert rodent and native to Mongolia and northeastern China. It is a member of the family *Cricetidae,* subfamily *Gerbillinae.* *Meriones* is one of 12 genera with over 300 forms. It is also called sand rat, antelope rat, desert rat, and jird. The black Mongolian gerbil was derived from the agouti *M. unguiculatus* and has the same physical and personality characteristics as the agouti or brown gerbil.

Physiological Data

Smith and Kaplan (134) reported the range of body weight for fourteen males as 48–68 g, and for fourteen females as 45–60 g. Other parameters measured included pulse, 491.4 $\pm$ 60.5/min and 468.6 $\pm$ 68.1/min; respiratory rate, 142.0 $\pm$ 32.3/min and 142.3 $\pm$ 16.4/min; rectal temperature, 38.2 $\pm$ 0.9°C and 38.2 $\pm$ 0.6°C for males and females, respectively.

Mean life span of males is 110 weeks; of females, 139 weeks. Maximum weight is reached at 15–18 months of age with approximately 70 percent of the maximum attained by 3 months of age. Mean age of females at first litter is 130 days; at last litter,

Fig. 15.5. Mongolian gerbil (*Meriones unguicultatus*). From *Mammals of the World* (160) with permission.

487 days. Average litter size at birth is five young. The mean number of litters per female is 7.1 with a maximum of 12. Nine percent of the litters were produced within 25 to 29 days of the previous litter; 47 percent within 39 days (6).

Uses in Medical Research

Meriones is useful for bioassay of testosterone and the major female sex hormones, progesterone and estradiol-17, by excising and weighing the abdominal sebaceous gland pad. The animal is used in studies of plague, leptospirosis, brucellosis, salmonellosis, tuberculosis, rabies, poliomyelitis, anthrax, ornithosis, schistosomiasis, helminthiasis, filariasis, bartonellosis, and behavioral research. *Meriones* is of interest also because it is a lipemic animal that does not respond with vascular lipoidosis. Therefore, it is useful for studying lipid metabolism and the inherent protective mechanisms that inhibit deposition of fat in the arterial wall (119).

Opossum (*Marmosa mitis,* pouchless opossum; *Didelphis marsupialis Virginiana,* American opossum; *Caluromys derbianus,* South American woolly opossum)

This marsupial provides unique experimental opportunities in comparative and developmental biology and medicine. The opossum represents the only opportunity among mammals for direct observation and chemical or physical manipulation of developing embryonic and fetal tissue in the absence of a placental barrier and under minimal maternal metabolic influence (68). *Marmosa* have a high incidence of bacterial infections and, in particular, septicemia. *D. virginiana* (Fig. 15-6), while in captivity, have developed bacterial endocarditis, glomerulonephritis, abscesses, lesions of the gastrointestinal tract, and other abdominal organs. Bacteriologic cultures from infected animals have yielded *Proteus* sp., *Salmonella* sp., *Escherichia coli* and α and β streptococcus (152).

Physiological Data

The opossum *D. virginiana* has a laboratory life span of 2 to 3 years. Its breeding season is 6 months in duration, from late December to early July in the northern hemisphere. Gestation is 13 days $\pm$ 6 hours with a litter size averaging seven and one-half. Pregnancy cannot be diagnosed from the vaginal smear in the marsupial. The average length of pouch skin flaps is the most dependable criterion for selecting opossum breeding stock from the wild. The karyotype of *D. virginiana* differs from other didelphids. *D. virginiana* has a chromosome diploid number of twenty-two and a fundamental number of thirty-two (68, 69, 45).

The opossum, *M. mitis,* is a pouchless, mouse-sized mammal, adapted by grasping hands and tail to an arboreal life. It is insectivorous and frugivorous by nature, prolific by marsupial standards, and easily handled (10, 78).

Uses in Medical Research

D. virginiana is utilized in studies of wound healing, early development of antibody response, sexual and neurological development, and the effects of oxygen upon the development of the eye. It is potentially useful for research in adrenal structure and function. Because of apparent resistance to rabies virus, it may be used to study the pathogenesis of this disease (153).

C. derbianus is used for cytogenetic, teratogenic, and mutagenic studies. It is especially useful in genetic research because it has a diploid number of only fourteen chromosomes (126).

M. mitis, like *Didelphis,* may serve as a model to investigate the role of bacte-

Fig. 15-6. American opossum (*Didelphis marsupialis Virginiana*). From *Mammals of the World* (159) with permission.

remias and septicemias in the production of bacterial endocarditis and other chronic pathological conditions associated with infections (152).

Nine-banded Armadillo (*Dasypus novemcinctus mexicanus*)

The nine-banded armadillo (Fig. 15-7) has several unique features that enhance its usefulness in medical research. These features, described by Storrs and Greer (144), include: (a) regular production of monozygous quadruplet young; (b) possession of a simplex uterus similar to that of a human; (c) period of implantation delay of the blastocyst from 17 to 21 weeks; (d) low body temperature, 30°–35°C; (e) ability to withstand low oxygen tension; (f) scute and band patterns on the carapace which may mutate readily; (g) an immune response which may be primitive; (h) life span estimated at 12 to 15 years; (i) fetuses possess an enormous adrenal similar to that of humans.

Physiological Data

D. novemcinctus has a fused müllerian uterus that becomes globoid during pregnancy. After a four- to 5-month period of dormancy in the uterus, the blastocyst implants, usually in November in this country, and the embryonic cell mass undergoes two proliferations and segregations that lead to the formation of four like-sexed, identical embryos. Gestation period is about 260 days. Adult weight ranges from 4–8 kg.

Uses in Medical Research

The nine-banded armadillo is particularly useful in leprosy studies. It is the first unaltered animal model found which is capable of developing the lepromatous form of the disease. It is also susceptible to several other human diseases such as Chagas disease, relapsing fever, African sleeping sickness, exanthematic and murine typhus, schistosomiasis, and *Nocardia brasiliensis* infection.

Boxer Dog (*Canis familiaris*)

The boxer is a medium-sized sturdy dog. Adult males are 22.5–25 in. and females are 21–23.5 in. at the withers. The build is square, with short back, strong limbs, and short, tight-fitting coat. The boxer is of special interest because of its apparent susceptibility to canine lymphoma.

Uses in Medical Research

The boxer breed of dog shares with humans a heritable susceptibility to emotional stress, to colitis, and to spondylosis deformans. It is susceptible to a disseminated lymphosarcoma with involvement initially limited to lymphoid tissues. The canine lymphoma is similar to Hodgkin's disease. Granulomatous colitis is seen only in dogs 2 months to 2 years of age of either sex. There are similarities in the pathogenesis to Whipple's disease and Crohn's disease.

Basenji (*Canis familiaris*)

The basenji or "barkless dog" is one of the oldest breeds. It is native to the African Congo where it was prized for intelligence, speed, hunting power, and silence. It was introduced to the United States in 1937. It is small, lightly built, weighing approximately 10.4 kg. The coat is silky, light brown to black with white markings.

Uses in Medical Research

A small number of basenji dogs have an enzyme deficiency, suggestive of hemolytic anemia, which causes their red blood cells to break down rapidly. It results in severe anemia with no cure available. It appears to be hereditary. The clinical signs appear at 4 to 7 months of age. Death occurs within 3 years. The basenji was offered as the first animal model for this type of human disease (37). Efforts to implicate mechanisms analogus to hereditary spherocytosis of humans, hereditary glycolytic enzyme deficiencies, or hereditary hemoglobinopathies were unsuccessful.

Ferret (*Mustela putorius afura*, polecat; *M. nigripes*, black-footed ferret)

Two species of ferrets exist. The black-footed ferret is a native of western North America. The polecat is a native of northern Africa, Europe, and Asia; it was introduced to North America in the late nineteenth century. Two domestic varieties exist in breeding colonies in the United States—the common or fitch ferret with a brown coat and mask-like facies, and the partially albino or English ferret with a yellowish white fur color.

Fig. 15-7. Nine-banded armadillo (*Dasypus novemcinctus mexicanus*). From *Mammals of the World* (159) with permission.

Unique Features

The ferret has a single vessel originating from the aorta which transverses its long neck to divide into right and left carotid arteries. This anatomical relationship provides an accessible vessel with which to study cerebral blood flow or the various effects or pharmacological agents on the nervous system.

Physiological Data

In northern temperate latitudes, the breeding season extends from March to August. Onset of breeding is readily induced by prolonging the daily period of sunlight with artificial light. Ovulation does not occur spontaneously. The ferret can reproduce twice during a breeding season, producing ten to twelve young. Gestation period is 40 to 43 days. Weight at birth is 10 g. Adult weight is about 0.7 kg. Normal rectal temperature is 101.6°F (38.5°C).

Uses in Medical Research

The ferret is susceptible to canine distemper, "hard pad" disease of dogs, influenza, tuberculosis (bovine, avian, and human strains), and streptococcal pneumonia. Physiological studies center on the habit of seasonal breeding. It is of interest to embryologists and teratologists because of spontaneous malformations consisting mainly (3 percent) of exencephaly and gastroschisis. The ferret is used to study the role of nutritional factors in the development and structure of teeth and supporting dental tissues and in the maintenance of dental and gingival health.

Aoudad or Barbary Sheep (*Ammotragus lervio*)

The aoudad is a herbivorous ruminant, hardy and prolific in captivity. It is the only wild sheep indigenous to Africa. It is of interest because of naturally occurring atherosclerosis.

Physiological Data

Life span is about 15 years. Weight of the adult is 50–115 kg. Females are monestrus in captivity. Gestation is 154 to 161 days with one or two young produced.

Uses in Medical Research

The aoudad is potentially useful in artherosclerosis research.

Guanaco or Wild Llama (*Lama guanicor,* var. *huanachus*)

The llama is a member of the family Camelidae. *Llama huanachus* (*Molina*), the guanaco is the only wild species in this genus. As does *llama glama* and *llama pacos,* it lives in southern and western South America at altitudes ranging from sea level to 5000 m, and in a semidesert habitat.

Physiological Data

Adult weight range is 48–96 kg; head and body length is 1.2–1.75 m. The guanaco has an erythrocyte life span of 235 days, the longest of all mammals. Breeding occurs in August and September. Gestation period is 10 to 11 months, with one young produced every other year. Life span is 20 years. The effect of altitude on oxygen capacity of acclimatized llamas is presented in Table 15-7 (1).

Uses in Medical Research

The llama is used in erythrocyte studies.

Table 15-7.   Effect of Altitude on O_2 Capacity in Acclimitized Llamas.

Altitude (M)	O_2 Capacity (Vol %)	
	Blood	Erythrocytes
Sea level	23.5	60.9
2800	17.1	60.6
5300	14.9	57.8

Adapted from Blood and Other Body Fluids (1), with permission.

Pygmy Goat or West African Dwarf Goat (*Fouta djallon*)

The African pygmy goat, (Fig. 15-8), also known as Fouta djallon, originated in West Africa, south of latitude 14° north. It averages 40–50 cm in height at the withers and 18–20 kg in weight. The pygmy goat is a herbivore, lives naturally in herds and is easily trained.

Physiological Data

Females are sexually mature at 3 months of age. It is preferable to wait until 6 months of age before breeding. The estrous cycle is 19 to 21 days. Gestation period is 5 months, and the adult female usually delivers twins. The pygmy goat is a seasonal breeder (60, 125).

Uses in Medical Research

The pygmy goat is primarily used for studies of the maternal and fetal circulation, but it should find use also in studies of physiology, gnotobiotics, and endocrine research. The fetus is of comparatively large size (18 kg) and is used for obtaining adequate samples of fetal blood from a variety of locations. Metcalf (60, 125) has found the pygmy goat to tolerate surgery, repeated venipuncture and arterial punctures, and blood loss well.

White-tailed Deer (*Odocoileus virginianus*)

Odocoileus virginianus ranges from southern Canada throughout most of the United States southward to northern South America.

Unique Features

The white-tailed deer has seven adult and two fetal hemoglobins. The number in a given animal varies from one to three. Two kinds of α subunits, six kinds of β subunits, and one γ subunit are related to the whole hemoglobin molecule (76).

Physiological Data

Adult weight is 48–145 kg; head and body length is 1.5–2.1 m. Mating usually occurs at 2 years of age in November and December. Gestation period is 196 to 210 days with one fawn produced in the first litter and usually two in each subsequent litter. The fawn weighs 1.5–2.5 kg at birth. Life span in the wild is 10 years and in captivity 20 years.

Uses in Medical Research

The white-tailed deer is useful for hematological research, particularly for the study of cell sickling (76).

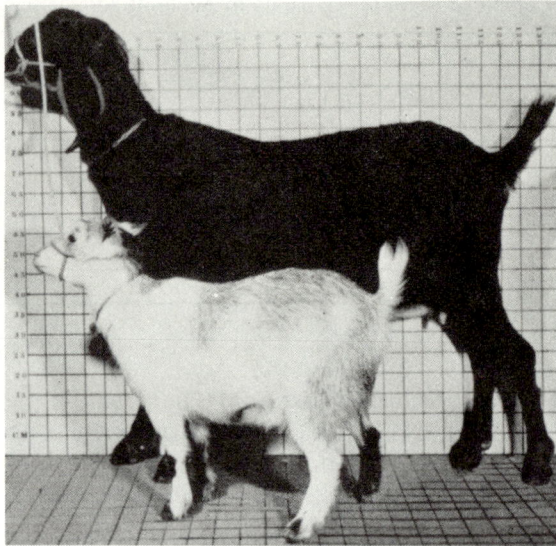

Fig. 15-8. West African pygmy goat. From *Laboratory Animal Care* (125) with permission.

Macaca (*Macaca arctoides*, stump-tailed macaque; *M. cyclopis,* Formosan rock or Taiwan macaque, *M. fascicularis* or *M. irus,* Crab-eating (cynomolgus) monkey; *M. mulatta,* rhesus monkey; *M. nigra,* black celebes ape (monkey); *M. nemestrina,* pig-tailed macaque)

These old world monkeys (Figs. 15-9 to 15-15) are found in Africa, Arabia, southern Asia, Indonesia, the Philippines, Formosa, and Japan.

The stump-tailed monkey is more docile than the rhesus and easier to train and handle. It is better suited for experiments involving intranasal or inhalation studies which require repeated dosing or extensive periods of restraint (103).

Physiological Data

Female macaques have a 27 to 52 day menstrual cycle. Gestation period is 5 to 9 months and one or two young are produced per season. Puberty occurs in 3 to 4 years and is indicated by a reddening of the buttocks. Life span is approximately 25 years.

Uses in Medical Research

The Rh blood factor was first demonstrated in rhesus monkeys. These monkeys are used in space flight tests. *M. fascicularis* was used in studies leading to the development of the vaccine against polio. In a closed breeding colony, *M. nigra* has a marked proclivity for a spontaneous diabetic-like state which is very similar to human diabetes (61). Generally, *Macaca* find uses in toxicological and nutritional studies.

Silvered-Leaf Monkey, Langur (*Presbytis cristatus*)

Langurs (Figs. 15-16 and 15-17) range from Ceylon and India to southern China, Indochina, and Indonesia. They have slender bodies, long tails, and long slender hands. They are essentially arboreal.

Physiological Data

Adult weight is 7–18 kg. Silvered-leaf monkeys are anemic in the wild state. The anemia is rapidly corrected once they are placed on a balanced diet. The anemia is of a normocytic normochromic type (161).

Uses in Medical Research

The silvered-leaf monkey is believed to have special potential as an animal model for human disease since minimal contact with humans has greatly reduced the possibility of prior exposure to human pathogens.

African Green Monkey (Cercopithecus aethiops)

The African green monkey (Fig. 15-18), also known as guenon or vervet, has several features that make it particularly useful for experimental studies. It has light temporal muscles and a thin cranium which makes the central nervous system easily accessible. The vagus and sympathetic outflow are fused in the neck of 75 percent of the animals. The abdominal viscera and nerves are readily accessible. It is the animal of choice for eye research because its eye is anatomically similar to that of humans. Also it has close similarity to humans in certain pharmacological responses. It accepts captivity well.

Physiological Data

Adult weight is 7 kg; head and body length is 325–700 mm. Tail length is 500–875 mm. Guenons are diurnal. Menstrual data for *C. aethiops* is given in Table 15-8 and newborn data is given in Table 15-9 (67).

Fig. 15-9. Stump-tailed macaque (*Macaca arctoides*). From *Taxonomic Atlas of Living Primates* (23) with permission.

Fig. 15-10. Rhesus monkey (*Macaca mulatta*). From *Laboratory Animal Care* (10) with permission.

Fig. 15-11. Crab-eating monkey (*Macaca fascicularis*). From *Taxonomic Atlas of Living Primates* (23) with permission.

Fig. 15-12. Bonnet monkey (*Macaca radiata*). From *Taxonomic Atlas of Living Primates* (23) with permission.

Table 15-8. Menstrual Data for *C. Aethiops*

	N	Range (Days)	Mean ± S.D. (Days)
Menses length	583	1 - 15	3.7 ± 2.3
Menstrual cycle length	392	16 - 50	30.9 ± 9.6
Amenorrheic cycles	95	51 - 240	84.2 ± 37.4
Post partum menses initiation[1]	37	24 - 108	56.9 ± 18.9

[1]Full term live births.

Adapted from Laboratory Animal Science (67) with permission.

Table 15-9. Newborn Data for *C. Aethiops*

	N	Mean gestation ± S.D. (Days)	Mean weight ± S.D. (gm)
Males[1]	24	164.1 + 6.6	328.2 + 57.8
Females[1]	14	161.6 + 5.5	308.1 + 51.7
Combination	38	163.2 + 6.3	321.3 + 56.9
Premature infants[2]	7	141.9 + 6.7	248.9 + 39.0

[1]All full term naturally delivered infants.

[2]Any live infant of less than 151 days gestation.

Adapted from <u>Laboratory Animal Science</u> (67), with permission.

Fig. 15-13. Celebes black ape (*Macaca nigra*). From *Taxonomic Atlas of Living Primates* (23) with permission.

Fig. 15-14. Pig-tailed macaque (*Macaca nemestrina*). From *Laboratory Animal Care* (10) with permission.

Fig. 15-15. Japanese macaque (*Macaca fuscata*). From *Taxonomic Atlas of Living Primates* (23) with permission.

Fig. 15-16. Silvered-leaf monkey, langur (*Presbytis cristatus*). From *Laboratory Animal Science* (161) with permission.

Fig. 15-17. Tree langur (*Presbytis entillus*). From *Taxonomic Atlas of Living Primates* (23) with permission.

Fig. 15-18. African green monkey (*Cercopithecus aethiops*). From *Taxonomic Atlas of Living Primates* (23) with permission.

Uses in Medical Research

The African green monkey is useful as a source for kidney cell cultures, and in studies concerning ophthalmology, protozoology, virology, toxicology, and cancer.

Chimpanzee (*Pan troglodytes*)

On a nutritionally adequate diet, the chimpanzee (Fig. 15-19) produces aortic atheroses histologically similar to those of humans. The developing atheromas in both chimpanzee and human show reduced 5′ nucleotidase and ATPase activity (20).

Physiological Data

The weight at birth of both male and female is 1.85 kg. Maximum life span is 55 to 70 years. The female has a 35-day menstrual cycle with ovulation occurring midway in the cycle. Body temperature is lower than that of the human with wide fluctuation within each day. Only 10 percent of rectal readings were over 100°F (38°C). Temperatures below 96°F (36°C) occur frequently during winter months (123).

Uses in Medical Research

The chimpanzee is used for research in artherosclerosis. Chimpanzees are also used in studies involving essential hypertension (146), ulcerative colitis (145), and cross-circulation studies. There is a 5 to 10 percent occurrence of O blood group in the chimpanzee population, whereas this blood group is not known to exist among baboons (21).

Fig. 15-19. Chimpanzee (*Pan troglodytes*). From *Taxonomic Atlas of Living Primates* (23) with permission.

Baboon (*Papio papio; P. cynocephlus; P. dogurea; P. ursinus; P. hamadryas*)

The baboon (Figs. 15-20 and 15-21) is confined to Africa and Arabia. Species vary in appearance, temperment and susceptibility to some human diseases. *P. papio,* the olive baboon from West Africa, *P. cynocephalus,* and *P. dogurea* from East Africa are popular for experimental research. Generally, baboons are easier to handle, tamer, and far less apprehensive and noisy than monkeys.

Physiological Data

Adult males may weigh 25 kg and reach 4 ft in height; the female is half the size of the male. Puberty is at 4 years of age; adulthood is reached at 7 years. Reproductive life of the female extends for 20 years with an estrous cycle of 30 days. One offspring may be produced yearly. Gestation period is 7 months. Life span is over 30 years.

Uses in Medical Research

Baboons are much used in studies of atherosclerosis. Their pattern of lipid metabolism is similar to that of humans and the blood vessel structure also resembles that of humans. Generally, they are useful in studies involving physiology, pharmacology, reproductive biology, endocrinology, and diseases of the eye. Their response to some disease agents is more like that of humans than of laboratory monkeys. *P. papio* is useful for trachoma investigations, and *P. hamadryas,* the sacred baboon, is good for work on human blood flukes. In reproductive biology, the baboon is especially useful for comparative studies since the pregnant baboon has a similar decrease in RBC, PCV, and Hgb as occurs in women during normal pregnancy with a return to normal blood values in the puerperium. It also has a leukocytosis and an increase in circulating neutrophils (13, 14, 15).

Fig. 15-20. Yellow baboon (*Papio cynocephlus*). From *Taxonomic Atlas of Living Primates* (23) with permission.

Fig. 15-21. Pig-tailed baboon (*Papio ursinus*). From *Taxonomic Atlas of Living Primates* (23) with permission.

New World Monkeys

There are six groups of new world monkeys: (a) marmosets and tamarins (Figs. 15-22 to 15-24); (b) squirrel monkeys (Fig. 15-25); (c) night or owl monkeys (Fig. 15-26); (d) howler monkeys (Fig. 15-27); (e) spider monkeys (Fig. 15-28); (f) white-faced monkeys (Fig. 15-29).

Fig. 15-22. Pygmy marmoset (*Callithrix pygmaea*). From *Taxonomic Atlas of Living Primates* (23) with permission.

Fig. 15-23. Cotton-topped tamarin (*Oedipomidas oedipus*). From *Primates Comparative Anatomy and Taxonomy III* (55) with permission.

Fig. 15-24. Tarmarin nigre (*Saquinus tamarin*). From *Taxonomic Atlas of Living Primates* (23) with permission.

Fig. 15-25. Squirrel monkey (*Saimiri sciurius*). From *Taxonomic Atlas of Living Primates* (23) with permission.

Fig. 15-26. Owl-faced night monkey (*Aotes trivirgatus*). From *Taxonomic Atlas of Living Primates* (23) with permission.

Fig. 15-27. Mantled howler (*Aloutta palliata aequatorialia*). From *Primates Comparative Anatomy and Taxonomy V* (56) with permission.

Fig. 15-28. Black-handed spider monkey (*Ateles geoffroyi pan*). From *Primates Comparative Anatomy and Taxonomy* (56) with permission.

Fig. 15-29. White-throated capuchin (*Cebus capucinus*). From *Taxonomic Atlas of Living Primates* (23) with permission.

Marmosets and tamarins belong to six genera of the family *Calithricidae* and extend from Panama to southern Brazil. Much experimental work has been accomplished with yellow fever as marmosets develop severe illness and high mortality. There is a single recognized species of squirrel monkey, *Saimiri sciurius*. Its range is Venezuela, the Guianas, Amazon areas of Brazil, Colombia, Pacific coast of western Panama, and adjacent areas in Costa Rica. It is extremely susceptible to yellow fever. There is a single species of night monkey, *Aotes trivirgatus*. Its range is from eastern Panama south to the Orinoco and Amazon basins. It is nocturnal and extremely susceptible to infection with yellow fever. Howlers belong in a single genus, *Alouatta*. Speciation in the group is not well understood. They range from Mexico south to Argentina and Bolivia. They are difficult to keep alive under laboratory conditions, and they have very high susceptibility to and morbidity and mortality from yellow fever. Spider monkeys belong to the single genus *Ateles*. *Ateles ater* or *A. panescus* of Brazil and *A. fusciceps* of Panama are black spider monkeys and are highly susceptible to yellow fever infection. *A. fusciceps* tolerates yellow fever infections well with light to moderate symptoms. White-faced monkeys, capuchin, belong to the genus *Cebus*. There are many species with a range from Honduras south to Brazil and Peru. They are susceptible to yellow fever and have asymptomatic infections (21).

Physiological Data

Marmosets comprise four genera. The pygmy marmoset, *Cebuella pygmaea*, is very timid. Adult weight averages 100–150 g. The diploid chromosome number is forty-four; the other three genera have forty-six chromosomes. Adult weight of genus *Leontideus*, lion-headed, averages 400–750 g. Eight species belong to genus *Callithrix*, short-tusked, and the adult weight range is 200–450 g. The genus *Saguinus*, long-tusked, has over twelve species with a weight range of 400–600 g. Two representatives of the species are the white-lipped marmoset, tamarin, and the cotton-topped pinché or marmoset. All marmoset species usually birth twins (27). Females are thought not to have a menstrual cycle. Gestation period is 130 to 150 days. Sexual maturity occurs at 12 to 15 months of age. Life span is about 16 years.

S. sciurius adult weight is 750–1100 g. Life span ranges 10 to 20 years. Adult weight of *A. trivirgatus* is 0.6–1 kg. Life span may be as long as 25 years. The weight of an adult howler monkey ranges from 7–9 kg. Gestation period is 140 days with a single young produced. The howler monkey is extremely difficult to maintain in captivity. Most likely a lack of defined nutritional requirements and stress are factors. The adult weight of spider monkeys is 6–8 kg. Gestation period is 139 days with a single young produced. The capuchin monkey has an adult weight range of 1.65–4 kg. Gestation period is 6 months with a single young produced. Life span is over 30 years.

Uses in Medical Research

Captive squirrel monkeys are susceptible to *Toxoplasma gondii*. They share with humans a limited ability to absorb cholesterol from the diet and the development of coronary atherosclerosis (91). Marmosets are useful in transplantation and tumor immunology. The marmoset can be kept and bred under laboratory conditions and is highly susceptible to several tumor viruses. *Callithrix jacchus*, pygmy marmoset, is used in studies of dental calculus, pigmentation, forms of malocclusion, and periodontal disease. It has a high incidence of dental abnormalities. The pygmy marmoset is also proposed as a subject for the study of aging and associated diseases. *Saguinus* (*Oedipomidas*) *oedipus*, the cotton-topped marmoset, is used in studies of low doses of radiation because the total body water phase is 68 percent of body weight (27). Spider monkeys are used as sentinels for the virus of yellow fever (91). The howler monkey is used in malarial studies. Human *Plasmodium vivax* malaria can be grown in the small *Aotes*, night or owl monkey. It will infect mosquitoes and be passed back to humans by mosquito bites. It can be successfully maintained in these monkeys by serial passage by the injection of infected blood (164).

Shrew (*Blarina breviacuda,* short-tailed shrew; *Tupaiaglis,* tree shrew; *Elephantalus myarus, E. intufi, E. rozeti, Petrodromus tetradactylus,* species of elephant shrews)

The short-tailed shrew and the elephant shrews belong to the order Insectivora. The tree shrew (Fig. 15-30) is a primate. The short-tailed shrew inhabits the island of Zanzibar. The tree shrew inhabits western China, Sikkim, Manipur, Assam, southeastern Asia, Sumatra, Java, Borneo, and the Philippines.

Unique Features

The tree shrew has several unique features. Female tree shrews have one, two, or three pairs of mammae. All other primates have two mammary glands. The penis is of the pendulous type, and a baculum is present in most forms. The testes are descended into the scrotum at birth.

Physiological Data

The adult weight of the short-tailed shrew is 15–30 g. Gestation period is 17 to 21 days with a litter size of three to nine. Birth occurs from early spring to early fall. Sexual maturity is at 1 year of age. Life span is 2.5 years.

Elephant shrews vary in weight from 40 g for *E. rozeti* to 230 g for *P. tetradactylus.* Gestation period is approximately 2 months with one or two young produced. Sexual maturity is attained at 5 to 6 weeks. Life span is 2 to 4 years.

Tree shrews bear no external resemblance to other primates but rather resemble squirrels in movement. Adult weight is 20–30 g. Females exhibit a physiological pattern similar to that of heat in nonprimates. Breeding may occur throughout the year. Gestation period is 46 to 50 days. Litter size is usually two.

Uses in Medical Research

Shrews are used in tissue culture and cytological work. The species is useful in ethological studies because of its pronounced stereotyped behavior and the ease with which it can be observed. Short-tailed shrews are used in studies involving hepatic microsomal enzymes, the cardiovascular system, and pesticide research. The tree shrew is susceptible to natural *Herpesvirus hominis* (*Herpes simplex*) infection (90).

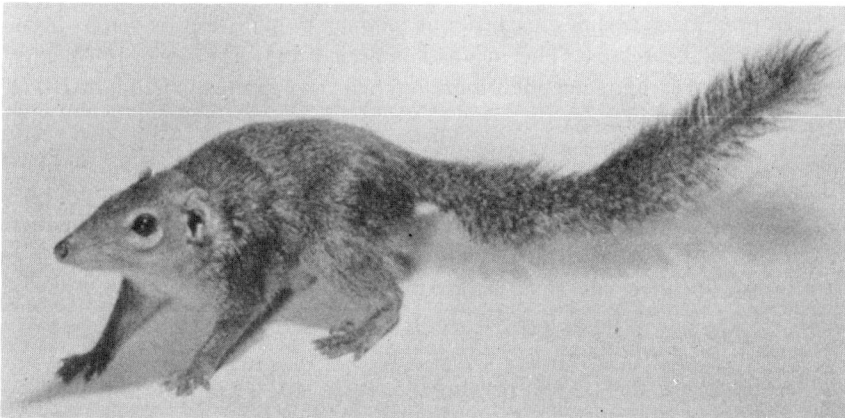

Fig. 15-30. Tree Shrew (*Tupia glis*). From *Mammals of the World* (159) with permission.

BIRDS

Canary (*Serinus canarius*)

Hardy breeds include Border, Yorshire and Norwich. Border is 5.5 in. (14 cm) in length. Yorshire and Norwich are larger (24 g) and more preferable for research. Roller canaries are kept primarily as pets (74).

Physiological Data

Canaries breed in the spring. Incubation period for eggs is about 13 days.

Uses in Medical Research

The canary is used in the study of malaria and other protozoal infections and for virological studies of pox. It is also used for toxicity tests on therapeutic substances.

Budgerigar or Shell Parakeet (*Melopsittacus undulatus*)

The budgeriger has several unique characteristics including a thick, fleshy, cylindrical tongue, an almost round spleen, no cecum and no gallbladder.

Physiological Data

Clutch size is five to six eggs every 2 days. Eighteen days are required for incubation. The chick at hatching weighs 2 g; adult weight is 45–50 g. Males and females mature at 3 to 4 months. Best breeding age begins at 11 months. Age limits are reported as 6 years for the male and 4 years for the female. Average rectal temperature of the hen is 41.2–41.3°C (106.2–106.4°F) (74). The shell parakeet has twenty-six macrosomes and an undeterminable number of microsomes (83).

Uses in Medical Research

The budgerigar is used in the study of neoplasms and nutritional research. Leav et al. (83) have reported a naturally occurring lipidosis in shell parakeets.

Turkey (*Meleagris gallopavo*)

The turkey is the only member of domestic poultry native to North America. The six recognized standard varieties of turkeys in the Unites States are bronze, Narragansett, White Holland, Bourbon red, black, and slate.

Physiological Data

Highest fertility occurs at springtime. Male sexual maturity occurs at 8 months of age. Semen production declines markedly in summer. Female fertility ranges from 1 day to over 7 weeks per season. The incubation period for turkey eggs is 28 days. The weights of young, yearling, and adult toms and hens is presented in Table 15-10 (157).

Uses in Medical Research

The turkey is used in studies on dissecting aneurysms, viral hepatitis, and muscular dystrophy. Experimental dissecting aneurysms have been produced in broad-breasted poults, male and female, fed high protein and fat rations. Field cases of dissecting aneurysms have been reported in the largest turkeys of well-managed flocks of broad-breasted bronze and Beltsville small white turkeys (112).

Pigeon (*Columbia livia*)

The pigeon has been domesticated since about 3000 B.C. There are many color variants and about 200 named strains comprising show pigeons, racing pigeons, and large food types.

Table 15-10.   Weight in Pounds of Young, Yearling, and Adult Tom and Hen Turkeys

	Tom			Hen		
Variety	Young	Yearling	Adult	Young	Yearling	Adult
Bronze	25	33	36	16	18	20
Narragansett	23	30	33	14	16	18
White Holland	23	30	33	14	16	18
Bourbon Red	23	30	33	14	16	18
Black	23	30	33	14	16	18
Slate	23	30	33	14	16	18

Adapted from Turkeys: Origin, History and Distribution (157), with permission.

Unique Features

Certain breeds of pigeons are of particular interest because they develop atherosclerosis spontaneously. Atherosclerosis develops spontaneously in nearly 100 percent of 3-year-old white carneaux pigeons. No sex differences have been noted in any of the measured aspects of aortic atherosclerosis in these birds.

Physiological Data

Two eggs are produced and incubated for 14 to 19 days. The white carneaux pigeon has a possible life span of 20 years. Serum osmotic pressure is 80–120 mm H_2O.

Uses in Medical Research

The white carneaux pigeon is of special interest in studies involving experimental aortic atherosclerosis.

White Pekin Duck (Anas platyrhynchos)

A strain of white Pekin duck was brought to the United States from China in 1873. It received the commercial designation "Long Island" duck, because it is propagated in northeastern areas of the United States, primarily in New York State.

Unique Features

The animal is unique in that myopathy occurs in approximately 90 percent of Pekin ducks, and 10 percent of these have contractures of the legs (122).

Physiological Data

Life span is over 5 years.

Uses in Medical Research

The white Pekin duck develops amyloidosis spontaneously; the frequency increases with age (122). There is apparent transfer of human hepatitis viruses to the Pekin duck (121). These ducks are used in experimental studies of malaria, vitamin A deficiency, cancer, and muscular dystrophy (115).

Quail (Coturnix coturnix japonica, Asiatic or Japanese quail; C. coturnix coturnix, European quail)

Coturnix (Fig. 15-31) is a migratory game bird. The oldest established lines in the United States are maintained by the Department of Poultry Science, Auburn University, Atlanta, Georgia, and the University of California at Davis.

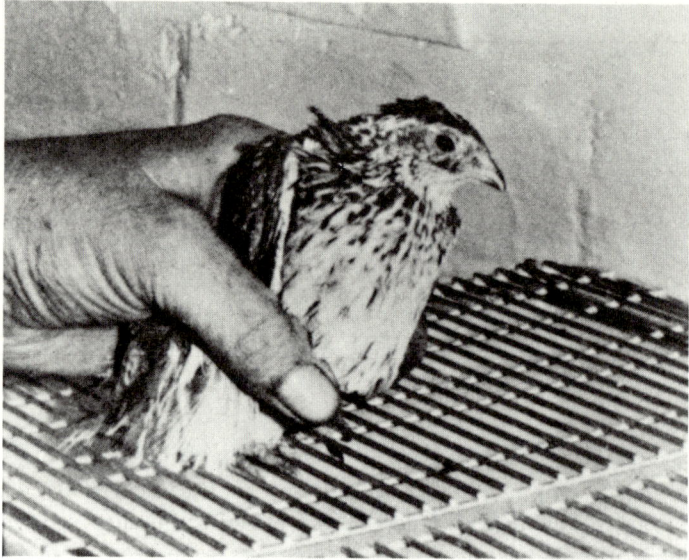

Fig. 15-31.   European quail (*Coturnix coturnix*). From *Handbook on the Care and Management of Laboratory Animals* (26) with permission.

Favorable qualities of the *Coturnix* quail include rapid growth, small adult size, early sexual maturity, prolific egg production, and adaptability to laboratory conditions. It is an inexpensive bird to maintain in the laboratory; twenty to thirty birds can be kept for the same cost as one chicken. The greatest limitation to laboratory use is the variation in fertility and hatching. Hens do not brood in captivity.

Physiological data
The *Coturnix* quail has a short life cycle, requiring 16 to 17 days for incubation. Males and females attain sexual maturity 6 to 7 weeks from hatching. Hatchability is reduced with birds over 8 months of age. Eggs average 9 g in weight. Adult male average weight is 100 to 140 g; adult female average weight is 110–160 g.

Uses in Medical Research
The *Coturnix* quail is particularly useful in the following research areas: embryological, physiological, nutritional, genetic, endocrinological, germfree, and cancer studies. Effects of light and photo periods on production and reproduction, learning, motivation, and social behavior can readily be studied using these birds. They are useful as a pilot animal for poultry husbandry, avian veterinary, and toxicology research. These quail are considered as intermediate in sensitivity to pesticides when compared with the bobwhite quail, ring-neck pheasant, and several other species of wild birds. *Cortunix* embryos have been infected with four strains of influenza, mumps, Newcastle disease, laryngotracheitis, fowl pox, infectious bronchitis, Rous sarcoma virus. Hens can be infected with avian encephalitis, Newcastle disease, fowl pox, and Rous sarcoma virus. Japanese quail are also naturally susceptible to visceral leukosis. Natural infections of lymphoid leukosis and fowl paralysis apparently are histologically comparable to those found in chickens (133).

AMPHIBIA, REPTILES, FISH, BATS

Frogs and Toads (*Rana temporaria,* common or grass frog, *R. pipens,* leopard frog; *R. esculenta,* water frog; *Bufo bufo,* common toad; *Xenopus laevis,* platanna or clawed toad)

Some species of frogs are permanently aquatic while other species become aquatic only during the breeding season. Some species burrow and are so adapted by means of an enlarged, sharp-edged tubercle at the base of the outer toe. Some species have tips of digits dilated into discs enabling them to climb trees. There are approximately 250 species of toads, genus *Bufo,* family *Bufonidae.* They are worldwide in distribution with the exceptions of Australasia and Madagascar. Distinctive features include toothless jaws and Bidder's organ which is thought to be a potential ovary with no known normal function.

Physiological Data

Some frogs grow to a large size. The bullfrog of North America grows to about 8 in. from snout to vent. In a study of fifteen male and fifteen female leopard frogs, Wass and Kaplan (162) reported body weight as 48–98 g and 69–101 g, respectively; pulse rate 50.0 ± 5.5/min; respiration rate 112.3 ± 9.0/min; cloacal temperature $21.2 \pm 0.5°C$.

Toads are stout bodied with short legs that limit them to hopping or walking rather than leaping as do frogs. Body size ranges from 0.75–9.0 in. Much of the dorsal skin and its warts contain poison-secreting glands; these are especially concentrated in two raised areas behind the eyes, the parotoid glands. Toads are mainly terrestrial but breeding takes place in standing or slowly moving water. The eggs are small and dark and are laid in two jelly tubes and number from 600 to 30,000, amount varying with the species. Tadpoles hatch in about 3 days and transform in 1 to 3 months. Toads mature in 2 to 3 years.

It is to be noted that temperature differences may cause profound differences in test responses of frogs and toads.

Uses in Medical Research

Much South American research utilizing the toad is performed on the process of spermination, ovulation, and secretions of the oviduct. Females of *Xenopus laevis* and males of a larger number of toads are used for tests of human pregnancy and assay of gonadotrophins. Changes in the skin color of the tree frog has been shown to follow the use of ACTH under standard conditions. The response of the individual melanophores in frog skin has been studied in vitro, and research has been carried out on the permeability of frog skin to various ions. Other uses of frogs and toads include work in pituitary, pancreas, water balance, and reproductive processes (38, 130).

Reptiles

The class *Reptilia,* cold-blooded or poikilothermic, has four orders: (a) *Chelonia,* of which tortoises are examples; (b) *Crocodila;* (c) *Rhynochocephalia,* containing only the lizard-like tuatara Sphenodon of New Zealand: (d) *Squamata,* lizards and snakes.

Generally, the reptiles are difficult to raise in captivity because of cyclical activities.

Uses in Medical Research

Reptiles have been used to study antibody formation, immunologic memory, structure of immunoglobulins, the organ cellular basis of the immune response, and the ontogeny of immunity. The alligator, *Mississippiensis,* has been used for studies involving metabolism of amino acids, testing of drugs that inhibit carbonic anhydrase, investigation of alkaline tide after feeding, and studies of insulin metabolism. There are many animal models for human diseases within this group as may be seen in Table 15-11 (71).

Fish (*Salmo trutta,* brown trout; *S. irideus,* rainbow trout; *Caraseius auratus,* goldfish)

There are more living species of true fish (over 30,000) than of any other class of vertebrate animals. They exhibit a great diversity of size, color, structure, and habitat.

Physiological Data

Fish are vocal particularly in the breeding season. They have vision, the extent of which probably varies with the species, and they utilize the chemical sense of smell. Many fish, such as the goldfish, can supplement oxygen obtained from the water by gills with air taken into the gullet. Life span of some species may be in excess of 50 years.

Uses in Medical Research:

Fish are mainly used for physiological research as subjects for the study of pollution and sometimes for testing the effects of various drugs (7, 120). Rainbow trout reared in hatcheries throughout the United States on processed commercial feed suffer from an epidemic of primary liver cancer, which does not seem to occur in brook or brown trout. Rainbow trout fed exclusively a diet of sheep liver did not have evidence of primary liver cancer (63).

Newts and Salamanders (*Ambystoma mexicanum,* Mexican axolotl; *Pleurodeles walthi,* Spanish salamander; *Triturus pyrrhogaster,* Japanese newt)

The following information on newts and salamanders is from Boterenbrood (18).

Only a few species of *Urodela* adapt themselves successfully to the laboratory. *A. tegrinum, P. walthi,* and *T. pyrrhogaster* can be adapted to a permanently aquatic mode of life. There are two established races of Axolotl, a black wild stock and a white stock which is caused by a single recessive Mendelian factor. In nature the Axolotl is found only in Mexico, in the Xochimilco and Chalco Lakes. The Spanish salamander is found in the southern half of the Iberian peninsula and in Morocco.

Physiological Data

Axolotl sexual maturity is reached in 1 year. The best breeding age is 2 to 5 years. It prefers cool temperature, 16°–22°C (61°–72°F), and thrives in open air ponds even when these are covered with ice in winter. In the laboratory, eggs can usually be obtained during 8 months of the year, November to June. One female lays hundreds of eggs over 2 or 3 days. The eggs have a diameter of 2.2 mm and are surrounded by a capsule consisting of one elastic and several jelly layers. In nature, spawning begins in February; the larvae grow rapidly and reach adult size in June. The axolotl has a high regenerative capacity and is consistently neotenous under natural conditions. Representatives of the species are *A. mexicanum, A. punctatum, A. tigrinum, A. opacum* and *A. jerrersonianum.*

P. walthii (Michah), the Spanish newt, is easily kept under laboratory conditions in which it is not allowed to come on land where it is difficult to feed. It is resistant to infection. Life span is 20 years. Its eggs have a soft jelly capsule and are easy to decapsulate. They can be obtained in large quantities, 200 to 800 eggs in 2 to 3 days, over a

Table 15-11. Reptiles: Animal Models for Human Disease

HUMAN DISEASE	ANIMAL MODELS
Aortic media necrosis	Komodo dragon
Arteriosclerosis	Iguana
Idopathic hepatitis	Red rattlesnake
Biliary cirrhosis	American alligator
Rickets	Tortoises; crocodiles
Osteomalacia	Spanish terrapin
Avitaminosis A	Lizards
Hypothyroidism	Tortoises
Osteoporosis	Spanish terrapin
Urate nephropathy	Iguana; king cobra
Gout	Iguana; gavial
Siamese twins	Terrapins; slow-worms
Hermaphroditism	Terrapins; tortoises; lizards
Double monsters	Terrapins
Sex reversal	Lizards
Albinism	Pythons
Melanism	Asp vipers
Dermal papilloma	Sand lizards; green turtle; musk turtle
Papilloma of gall bladder	Green turtle
Thyroid adenoma	Turtle
Pulmonary fibroadenoma	Horsfield's tortoise
Cardiac rhabdomyoma	Black terrapin
Bile duct adenoma	Black cobra
Chondroma	Monitor lizard
Rhabdomyoma	Pine snake
Osteoma	Green lizard; crocodile
Epithelioma	Puff-faced water snake; sand lizard; Gila monster; Ceylon terrapin
Carcinoma of thyroid	Ceylon terrapin
Carcinoma of stomach	Side-necked turtle
Adenocarcinoma of stomach	Bullsnake; giant tortoise
Carcinoma of pancreas	Pine snake; rattlesnake; water moccasin; black racer
Adenoma of bile duct	Water snake
Adenocarcinoma of kidney	Grass snake
Adenocarcinoma of colon	Bullsnake
Sarcoma of stomach	Water moccasin
Lymphosarcoma	Egyptian cobra; hognose snake
Malignant melanoma	Reticulated python; pine snake
Osteogenic sarcoma	Rufous-beaked snake

Adapted from CRC Handbook of Laboratory Animal Science (71), with permission.

long period of the year. Adult length is 20–25 cm. They have a rough yellow-grey skin with many dark spots. Sexual maturity is at 16 months of age. Breeding season varies in different stocks.

The Japanese newt, *T. pyrrhogaster* (Boie), adapts well to laboratory research. Adults are 12–17 cm in length and dark brown or black with a red-orange belly with black spots. Eggs are laid singly on water plants (*Vallisneria*). The female wraps a leaf around each egg for insulation. Decapsulation of eggs requires considerable skill because it is surrounded by a strong oval jelly capsule with a sticky outer layer. The Japanese newt hibernates for 6 weeks in the winter. Sexual maturity is at 2 years of age.

European newts, *Triturus* species, live in water only during the breeding season. The adult length varies from 6–10 cm with the species. Representatives of the species include *T. alpestris* (alpine newt), *T. vulgaris* (smooth newt), *T. helveticus* or *T. palmatus* (palmate newt), and *T. cristatus*. Eggs are laid singly within folded leaves of water plants. The eggs are 1.25–1.5 mm in diameter. The solid jelly capsule can easily be removed. In the laboratory, three to ten eggs are produced per day.

American newts, *Triturus* species, are amenable to permanent laboratory conditions. *T. viridescens* or *Diemiftylus irridescens,* the common spotted newt, is strictly terrestrial but transforms to an aquatic form after two to four seasons.

Uses in Medical Research
Newts and salamanders are widely used for experimental embryological research.

Seals and Sea Lions (*Pinnipadia: Otariidae,* Otariid seals, eared seals or sea lions, fur seals; *Odobenidae,* walrus; *Phocidae,* phocid or true seal)

Otariidae inhabit the coastline of western North America, South America, southern Africa, Australia, and New Zealand. An estimate of their population is 2 to 4 million. The walrus inhabits the waters of the Arctic Ocean. They are fewer than 100,000 in number. Members of the family *Phocidae* inhabit waters of most coastal areas and certain fresh water lakes. They are numerous in colder areas. The world population is estimated to be 10 to 22 million.

Physiological data
The *Otariidae* have slender, elongated bodies. The hind limbs can be turned forward as in the *Odobenidae* to support the body in land travel. The length of these pinnipeds is 150–350 cm and weight ranges to 1100 kg. Females have four mammae. The swimming mechanism is centered near the forepart of the body as it is in the *Odobenidae*. Males are polygamous. Gestation period is 250 to 365 days with delayed implantation occurring. Life span ranges 10 to 25 years.

The *Odobenidae* contains the single genus and species, *Odobenus rosmarus*. Bulls may be 3.7 m in length and weight 1260 kg. There is no external tail. Females have four mammae. Walruses are polygamous. Gestation period is 11 to 12 months. At birth, the calf weighs 45–68 kg and is about 1.25 m long. Cows are sexually mature at 4 to 5 years of age, bulls at 7 years of age. Life span is about 40 years.

Phocidae are not able to turn their hind limbs forward and must wriggle to travel over land. Their length is 125–650 cm and weight is 90 kg to 3.5 metric tons. The family has both the smallest (ringed seal, *Pusa*) and largest (elephant seal, *Mirounga*) pinnipeds. Females have two or four mammae. The swimming mechanism is centered near the hind part of the body. They have good sight and smell, but poor hearing. Gestation period is 270 to 350 days with delayed implantation occurring. Weight at birth ranges from 5–45 kg. Sexual maturity occurs between 2 to 8 years of age. Life span is 40 years.

Uses in Medical Research

Seals and sea lions are used in the study of intracellular metabolism, eye cataracts, shock therapy, and underwater communication. The sea lion is used as a model for study of disaccharide intolerance in human infants as lactose is toxic to the sea lion. They are also susceptible to cataracts. The diving reflex, an oxygen conserving mechanism, conserves oxygenated blood for brain and heart and is shared by humans. It is of importance as a study model because an improperly regulated oxygen conserving reflex in humans may be a cause of sudden death (62).

Bats (*Chiroptera*: *Eptesicus fuscus,* big grown bat; *Myotis sodalis, M. yumanensis,* brown; *M. myotis* common brown; *M. daubentonii,* Daubenton's; *Pipistrellus pipistrellus,* European brown; *Tadarida brasiliensis,* free tail; *Rhinolophus ferrumequinum,* greater horseshoe; *Myotis lucifugus,* little brown; *Plecotus auritus,* long eared; *Nyctalus noctula,* noctule, and *Antozous pallidus,* pallid.)

Bats may be loosely grouped into six categories according to diet: (a) insectivorous; (b) frugivorous; (c) nectarivorous; (d) vampire; (e) carnivorous; (f) fish-eating bats. Only rodents among the mammals exceed bats in the number of species. Bats are the only mammals that fly.

True vampires of the neotropics, *Desmodus* and *Diaemus,* feed exclusively on the blood of mammals and birds, respectively. *Desmodus* is hardy and may easily be kept in captivity on a diet of defibrinated blood. It carries forms of rabid myelitis.

Physiological Data

Most bats produce one young per year. Life span is about 20 years. Breeding generally takes place in the fall with hibernating forms. There is delayed implantation with ovulation occurring in the spring. In nonhibernating forms, breeding and ovulation occur without delayed implantation. *Glossophage soricina* is a nectarivorous and frugivorous member of the family *Phyllostomidae*. *G. soricina* menstruates and has a type of ovum implantation that bears many similarities to the process in humans (114). *Desmodus rotundus* has a gestation period of 90 to 120 days. Adults weigh 15 to 50 g; length of head and body is 75–90 mm. Rectal temperature of vampire bats is 37°–40°C.

Uses in Medical Research

At least 18 genera of bacteria have been identified from bats. Bacterial genera found in natural association with bats include: *Leptospira, Borrelia, Salmonella, Shigella, Escherichia, Enterobacter, Citrobacter, Staphylococcus, Enterococci,* and an unnamed gastric spirellum. These animals have been used to study induced infections with the plague bacillus, *Yersina pestis* (86).

Bats are used in reproductive research, physiological and behavioral research, and in the study of certain parasitic and fungal infections (e.g., Chagas' disease, histoplasmosis, dermatomycosis, and rabies).

GNOTOBIOTIC ANIMALS

As defined by The Institute of Laboratory Animal Resources of the National Academy of Sciences, Washington, D.C. (147), the gnotobiotic animal is one of an animal stock or strain derived by asceptic cesarean section (or sterile hatching of eggs), which is reared and continuously maintained with germfree techniques under isolator conditions and in which the composition of any associated fauna and flora, if present, is fully defined by accepted current methodology. The germfree animal (axenic animal) is a gnotobiote which is free from all demonstrable associated forms of life including bacteria, viruses, fungi, protozoa, and other saprophytic or parasitic forms. A defined flora animal is a gnotobiote maintained under isolator conditions in intentional association with one or more known types of microorganisms.

Biochemical Data

Rats and mice are most often used for gnotobiotic research. In germfree rats, the fecal fatty acids are typically unsaturated and of the long-chain, even-numbered type, versus saturated acids and cyclical and branched-chain fatty acids in controls (39, 40, 57). The amounts of trypsin and chymotrypsin are found to be consistently elevated in the bowel contents of feces of germfree animals (17, 85). Germfree chickens have undeveloped lymphatic tissue with fewer plasma cells and 25 to 50 percent lower globulin content in their blood serum in comparison to conventional controls (151).

Physiological Data

The guinea pig is particularly adapted to germfree studies because as a neonate it does not require a period of suckling but may be weaned directly onto semisolid natural diets.

In general, the weight of the heart, total blood volume, and cardiac output are reduced, whereas red blood cell count and hematocrit values are found elevated in germfree rats (46). Also there is a lower weight of the small intestine (118).

Uses in Medical Research

Gnotobiotic animals are used to study dietary effects, effects of closed environment, factors affecting microbial passage into the host, aging, tissue repair, enteric infections, and dental caries. Caries do not develop in germfree rats (102). Gnotobiotic lambs are used in experimental studies of rumen microbiology, digestion, and development. Gnotobiotic animals have potential value for studies of chronic and degenerative diseases where infectious agents may play a role. They may be of value in research on the etiology and pathogenesis of cancer and for cancer therapy. Animals used in gnotobiotics include rabbits, pigs, rats, mice, chickens, turkeys, quails, lambs, dogs, cats, and monkeys (47, 109).

For further details on gnotobiotic research, the reader is referred to the reviews by Gordon and Pesti (47) and Pleasants (109), and the bibliography compiled by Teah (150).

Certain uniquely useful animal models for the study of human diseases are listed in Table 15-12.

Table 15-12. Uniquely Useful Animal Models for Biomedical Research

ANIMAL MODEL	SPECIES	HUMAN EQUIVALENT	REFER- ENCE
Aleutian disease	Mink	Multiple myeloma	72, 80
Amyloidosis	Peking duck	Amyloidosis	121
Atherosclerosis	Squirrel monkey	Atherosclerosis	90
	Pigeon Aoudad		24
	Chimpanzee Baboon		20
Atrial septal defect	Chimpanzee	Interatrial septal defect	4
Autoimmune disease	New Zealand Black Mouse	Systemic lupus erythematosus	16
		Sjogren's syn- drome	57
		Waldenstrom's macroglobuli- nemia	104
Baldness, male pattern	Stumptail macaque	Baldness, male pattern	93
Cardiovascular disease	Woodchuck	Cardiovascular disease	137
Cataracts	Seals Sea lions Sand rat	Cataracts	61 31
Cerebellar hypoplasia	Ferret	Cerebellar hypoplasia	94
Cerebrovascular disease	Woodchuck	Cardiovascular disease	137
Chastek's para- lysis (a thiamine deficiency)	Mink	Thiamine deficiency disease	103
Chromosome errors, cause of	Frog	Cause of chromosome errors	162
Congenital eryth- rocytic porphyria (recessive)	Fox squirrel	Congenital erythrocytic porphyria	153

Table 15-12. (Continued)

ANIMAL MODEL	SPECIES	HUMAN EQUIVALENT	REFERENCE
Cystinuria	Blotched genet	Cystinuria	153
Diabetes	Black celebes ape	Diabetes	60 53
Diabetes mellitus	Chinese hamster Sand rat	Diabetes mellitus	116 31 25
Disaccharide (lactose) intolerance	Sea lion	Disaccharide intolerance	61
Diseases of eye	African green monkey Baboon Seal Sea lion	Diseases of eye	61
Dissecting aneurysms	Turkey	Aneurysm	111
Diving reflex	Seal Sea lion	Sudden death	61
Ehlens-Danlos syndrome (a primary collagen defect)	Mink	Ehlens-Danlos syndrome	103
Elliptical erythrocytes	Guanaco	Elliptocytosis	27
Encephalopathy, scrapie-like disease	Mink	Kuru, Crentzfeldt-Jakob syndrome	87 103
Essential hypertension	Chimpanzee	Essential hypertension	145
Familial anemia	Basenji	Hereditary spherocytosis	148 47
Fraternal twinning, regular	Marmosets	Twinning	139

Table 15-12. (Continued)

ANIMAL MODEL	SPECIES	HUMAN EQUIVALENT	REFER- ENCE
Fungal infections	Bats	Fungal infec- tions	
Gastrointestinal parasites and their treatment	Silver leaf monkey	Parasitic diseases	5
Gnotobiotic Research	Rabbit Pig Rat Mouse Turkey Chicken Quail Lamb Cat Dog Monkey	Pathogenesis of disease	46 108 149
Grand-mal seizures	Gerbil	Epilepsy	123
Granulomatosis colitis	Boxer dog	Ulcerative colitis	154
Heat stroke	Coypu	Heat stroke	64
Hepatitis	Marmosets	Viral hepatitis	88
Hepatitis, viral	Turkey Pekin duck	Viral hepatitis	147 40 120
Hepatoma	Rainbow trout	Hepatoma	62
Hereditary deafness	Mink	Deafness	128
Hereditary leukodystrophy	Mink	Familial meta- chromatic leukodystrophy	3
Hereditary leukomelanopathy	Mink	Chediak-Higashi syndrome	81 80
Hereditary muscular dystrophy	Pekin duck Turkey	Muscular dystrophy	121
Hereditary spherocytosis	Deer mouse	Sperocytosis	4

Table 15-12. (Continued)

ANIMAL MODEL	SPECIES	HUMAN EQUIVALENT	REFER-ENCE
Hermaphroditism	Mink	Hermaphroditism	96
Herpes	Marmosets Shrew	Herpes infection	41 89
Herpes-induced lymphoma	Marmosets	Lymphoma	74
Hydrocephalus	Mink	Hydrocephalus	47
Hyperostosis, polyostatic	Shell parakeet	Hyperostosis, polyostatic	130
Hyperbilirubinemia nonhemolytic	Gunn rat	Crigler-Najjar syndrome	28
Inflammation	Gnotobiotic animals	Inflammation	92
Influenza	Ferret	Influenza	109
Kuru, experi-mental	Chimpanzee	Kuru	43
Leprosy	Nine-banded armadillo	Leprosy	143
Lipemia	Mongolian gerbil	Lipemia	118
Lipidosis "foamy" macrophages	Shell parakeet	Lipid-storage disease	82
Listeriosis	Lemming	Listeriosis	110
Lymphoma	Gerbil	Lymphoma	51
Lymphosarcoma	Toad, Newt Boxer dog	Lymphosarcoma Hodgkin's disease	8 9 63 126
Malabsorption syndrome (grey diarrhea)	Mink	Malabsorption	103
Malaria	Penguin Canary	Malaria	48 73
Maternal and fetal circula-tion	Pygmy goat	Maternal and fetal circula-tion	59 124

Table 15-12. (Continued)

ANIMAL MODEL	SPECIES	HUMAN EQUIVALENT	REFER-ENCE
Molluscum con-tagiosum	Chimpanzee	Molluscum con-tagiosum	34
Multiple offspring identical	Nine-banded armadillo Trout	Heterokaryotic monozygous twins	11 141
Muscular dystrophy	Turkey Duckling	Muscular dystrophy	52 105 19 114
Neoplastic disease	Woodchuck Multi-mammate mouse	Neoplastic disease	137
Osteomalacia	Flying squirrel	Osteomalacia	131
Parasitic infections	Bat	Parasitic infections	85
Pigmentary liver disease	Howler monkey	Hepato-cellular melanosis	91
Plasmacytosis	Mink	Plasmacytosis	22 127
Poliomyelitis	Lemming	Poliomyelitis	110
Polyuria	Chinese hamster	Diabetes insipidus	65
Pulmonary adenomatosis	Chinchilla	Pulmonary adenomatosis	53
Rabies	Vampire bat	Rabies	58
Renal adeno-carcinoma	Frog	Renal adeno-carcinoma	140
Reproductive biology	Baboon Frog Toad Bats	Reproduction	13-15 37 129 113
Rheumatoid factor	Howler monkey	Rheumatoid factor	106

Table 15-12. (Continued)

ANIMAL MODEL	SPECIES	HUMAN EQUIVALENT	REFERENCE
Sarcoma	Citellus spp.	Sarcoma	42
Sex chromosome anomalies	Mink Marmosets	Sex chromosome anomalies	95 11
Sickling of erythrocytes in vitro	White-tailed deer	Sickle-cell anemia	75
Somatic segregation (antibody formation)	Peromyscus spp. Birds Fish	Somatic segregation (antibody formation)	99
Toxoplasmosis	Ground squirrel Squirrel monkey	Toxoplasmosis	155
Tularemia	Lemming	Tularemia	35
Ulcerative colitis	Chimpanzee	Ulcerative colitis	144
Urinary incontinence	Mink	Urinary incontinence	103
Urolithiasis	Mink	Urolithiasis	138
Viral induced connective tissue disease	Aleutian mink New Zealand mouse	Connective tissue disease	97

REFERENCES

1. Altman, P. L., and D. S. Dittmer, *Blood and Other Body Fluids,* Federation of American Societies for Experimental Biology, Bethesda, 1971.
2. Altman, P. L., and D. S. Dittmer (eds.), *Biology Data Book,* Second Edition Vols. I–III, 2nd ed., Federation of American Societies for Experimental Biology Bethesda, 1972.
3. Anderson, H. A., *Acta Neuropath.* (Berlin), **7:**297 (1967).
4. Anderson, R., R. R. Huestis, and A. G. Matulsky, *Blood,* **15:**491 (1960).
5. Arambula, P. V., J. B. Abass, and J. S. Walker, *Lab. Anim. Sci.,* **24:**299 (1974).
6. Arrington, L. R., T. C. Beaty, Jr., and K. C. Kelly, *Lab. Animal Sci.,* **23:**262 (1973).
7. Axelrod, M. R., and L. P. Schultz, *Handbook of Tropical Aquarium Fishes,* McGraw-Hill, New York, 1955.

8. Balls, M., *Cancer Res.*, **25**:3 (1965).
9. Balls, M., and L. N. Ruben, *Cancer Res.*, **27**:654 (1970).
10. Barnes, R. D., *Lab. Animal Care*, **18**:251 (1968).
11. Benirschke, K., J. M. Anderson, and L. E. Brownhill, *Science*, **138**:513 (1962).
12. Benirschke, K., M. M. Sullivan, and M. Marin-Badilla, *Obstet. Gynecol.*, **24**:819 (1964).
13. Berchelmann, M. L., T. E. Vice, and S. S. Kalter, *Lab. Animal Sci.*, **21**:564 (1971).
14. Berchelmann, M. L., T. E. Vice, and S. S. Kalter, *Lab. Anim. Sci.*, **21**:613 (1971).
15. Berchelmann, M. L., S. S. Kalter, and H. A. Britton, *Lab. Anim. Sci.*, **23**:48 (1973).
16. Bielschowsky, M. B., B. J. Helyer, and J. B. Howie, *Proc. Univ. Otago Med. School*, **37**:9 (1959).
17. Borgström, B., A. Dahlquist, B. E. Gustafsson, G. Lundh, and J. Malmquist, *Proc. Soc. Exp. Biol. Med.*, **102**:154 (1959).
18. Boterenbrood, E. C., in *University Federation for Animal Welfare Handbook on the Care and Management of Laboratory Animals*, Williams & Wilkins Co., Baltimore, 1967.
19. Bourne, G. H., and N. G. Ma, in *Muscular Dystrophy in Man and Animals*, Hafner, New York, 1963.
20. Bourne, G. H. and M. Sandler, in *The Chimpanzee*, Vol. 6, Karger, Basel, 1972.
21. Bourne, G. H. (ed.), *Non-human primates and medical research*, Academic Press, New York, 1973.
22. Chapman, I., and F. A. Jimenez, *N. Engl. J. Med.*, **269**:1171 (1963).
23. Chiarelli, A. B., *Taxonomic Atlas of Living Primates*, Academic Press, New York, 1972.
24. Clarkson, T. B., R. W. Prichard, M. G. Netsky, and H. B. Lofland, *Arch. Pathol.* (Chicago), **68**:143 (1959).
25. Coleman, D. E., and K. P. Hummel, *Diabetologica*, **3**:238 (1967).
26. Cooper, D. M., "The Japanese Quail," in UFAW (ed.), *Handbook on the Care and Management of Laboratory Animals*, 4th ed. Williams & Wilkins, Baltimore, 1972.
27. Cooper, R. W., *Lab. Animal Care*, **18**:267 (1968).
28. Cornelius, C. E., and J. J. Kaneka, *Science*, **137**:673 (1962).
29. Cornelius, C. E., and I. M. Arias, *Am. J. Pathol.*, **69**:369 (1972).
30. Datta, S. P., and H. Harris, *Ann. Eugen.* (London) **18**:107 (1953).
31. Davis, D. H. S., *S. Afr. J. Med. Sci.*, **28**:53 (1963).
32. DeFronzo, R., E. Miki, and J. Steinke, *Diabetologia*, **3**:140 (1967).
33. Dietrich, R. A., P. R. Morrison, and D. J. Preston, *Lab. Animal Sci.*, **23**:575 (1973).
34. Dietrich, R. A., *Lab. Anim. Sci.*, **25**:48 (1975).
35. Douglas, J. D., K. N. Tanner, J. R. Prine, D. C. Van Riper, and S. K. Derwelis, *J. Am. Vet. Med. Assoc.*, **151**:901 (1967).
36. Dunaeva, T. N., and N. G. Olsufiev, *Zool. Jb.* (Russ.) **3**:457 (1952).
37. Editorial, *J.A.V.M.A.*, **159**:1218 (1971).
38. Elkan, E., *Br. J. Herpet.*, **2**:177 (1960).
39. Evrard, E., P. P. Holt, H. Eyssen, H. Charlier and E. Sacquet, *Br. J. Exp. Pathol.*, **45**:409 (1965).
40. Evrard, E., E. Sacquet, P. Raibaud, A. Dickenson, H. Charlier, H. Eysson, and P. Holt, *Ernahrungsforschung*, **10**:257 (1965).
41. Fabricant, H., C. G. Richard, and P. P. Levine, *Avian Dis.*, **1**:256 (1957).
42. Felsburg, P. J., R. L. Heberling, M. Brack, and S. S. Kalter, *J. Med. Microbiol.*, **2**:50 (1973).
43. Finkelstein, E. A., and Z. I. Belagrudova, *Acta Unio. Intern. Contra Cancrum*, **20**:1587 (1964).
44. Gajdusek, D. C., C. J. Gibbs, Jr., D. M. Asher, and E. David, *Science*, **162**:693 (1968).
45. Gardner, A. L., Special Publ. No. 4, *Lubbock Mus. Texas Tech. Univ.*, (1973).
46. Gordon, H. A., B. S. Wostmann, and E. Bruckner-Kardoss, *Proc. Soc. Exp. Biol. Med.*, **114**:301 (1963).
47. Gordon, H. A. and L. Pesti, *Bact. Rev.*, **35**:390 (1971).
48. Gorham, J. R. (ed.), *Am. Fur Breeder*, **19**:20 (1947).
49. Griner, L. A., and B. W. Sheridan, *Am. J. Vet. Clin. Path.*, **1**:7 (1967).
50. Hackel, D. B., T. D. Kenney, and W. Wendt, *Lab. Invest.*, **2**:154 (1952).
51. Hackel, D. B., E. Mitkat, H. E. Lebovitz, K. Schmidt-Nielsen, E. S. Horton, and T. D. Kinney, *Diabetologia*, **3**:130 (1967).

52. Handler, A. H., S. I. Magalini, and D. Pav, *Cancer Res.*, **26**:844 (1966).
53. Harper, J. A., and J. E. Parker, *Poultry Sci.*, **43**:1326 (1964).
54. Helmboldt, C. F., E. L. Jungherr, and A. C. Caparo, *Am. J. Vet. Res.*, **19**:270 (1958).
55. Hill, W. C. O., *Primates Comparative Anatomy and Taxonomy III Pithecoidea, Platyr-rhini, Family Hapalidal,* Edinburgh at the University Press, 1957.
56. Hill, E. C. O., *Primates Comparative Anatomy and Taxonomy V Cebidal Part B,* Edinburgh at the University Press, 1962.
57. Holt, P. P., J. V. Joossens, E. Evhard, H. Eyssen, and P. DeSomer, in A. C. Frazer (ed.), *Biochemical Problems of Lipids,* Elsevier Publishing Co., Amsterdam, 1963.
58. Holmes, M. C., and F. M. Burnet, *Ann. Intern. Med.*, **59**:265 (1963).
59. Horst, R., and M. Langworthy, *J. Mammal.*, **53**:903 (1972).
60. Hoversland, A. B., Parer, J. T., and Metcalfe, J., *Fed. Proc.*, **24**:705 (1965).
61. Howard, C. F., Jr., *Diabetes*, **21**:1077 (1972).
62. Hubbard, R. C., and T. C. Poulter, *Lab. Animal Care*, **18**:288 (1968).
63. Hueper, W. C., and W. W. Payne, *J. Natl. Cancer Inst.*, **27**:1123 (1961).
64. Inoue, S., M. Singer, and J. Hutchinson, *Nature*, **205**:408 (1965).
65. Ippen, R., *Zbl. Bakt.*, **178**:195 (1960).
66. Jay, G. E., Jr., in W. J. Burdette (ed.), *Methodology in Mammalian Genetics,* Holden-Day, Inc., San Francisco, 1963.
67. Johnson, P. T., D. A. Valerio, and G. E. Thomson, *Lab. Anim. Sci.*, **23**:355 (1973).
68. Jurgelski, W., Jr., *Lab. Anim. Sci.*, **24**:376 (1974).
69. Jurgelski, W., Jr., and M. E. Porter, *Lab. Animal Sci.*, **24**:412 (1974).
70. Kaplan, H. M., in E. C. Melby Jr. and N. H. Altman (eds.), *Handbook of Laboratory Animal Science,* Vol. I, CRC Press, Cleveland, 1975.
71. Kaplan, H. M., in E. C. Melby Jr. and N. H. Altman (eds.), *CRC Handbook of Laboratory Animal Science,* Vol. I, CRC Press, Cleveland, 1975.
72. Katz, S., A. Gilardoni, N. Genovise, R. W. Wikinski, C. E. Cornelius, and M. R. Malinow, *J. Lab. Anim. Care*, **18**:626 (1968).
73. Kenyon, A. J., and C. F. Helmboldt, *Am. J. Vet. Res.*, **25**:1535 (1964).
74. Kegmer, I. F., in *University Federation for Animal Welfare. Handbook on the Care and Management of Laboratory Animals,* Williams & Wilkins Co., Baltimore, 1967.
75. King, N. W., and L. V. Melendez, *Lab. Invest.*, **26**:682 (1972).
76. Kitchen, H., F. W. Putman, and W. J. Taylor, *Blood*, **29**:867 (1967).
77. Kruckenberg, S. M., H. T. Gier, and S. M. Dennis, *Lab Anim. Sci.*, **23**:53 (1973).
78. Krupp, J. M., and R. Quellin, *Lab. Anim. Care*, **14**:189 (1964).
79. Lauris, V., and G. F. Cahill, Jr., *Diabetes*, **15**:475 (1966).
80. Lavappa, K. S., M. M. Fu, M. Singh, R. D. Beyer, and S. S. Epstein, *Lab. Anim. Sci.*, **23**:546 (1973).
81. Leader, R. W., G. A. Padgett, and J. R. Gorham, *Blood*, **22**:477 (1963).
82. Leader, R. W., G. A. Padgett, and J. R. Gorham, *NINDB Monograph*, No. 2, 1965.
83. Leav, I., A. C. Crocker, M. L. Petrak, and T. C. Jones, *Lab. Invest.*, **18**:433 (1968).
84. Lee, C. and D. J. Horvath, *Lab. Anim. Care*, **19**:88 (1969).
85. Loesche, W. J., *Proc. Soc. Exp. Biol. Med.*, **129**:380 (1968).
86. Macy, R. H., *Lab. Anim. Sci.*, **24**:530 (1974).
87. Marsh, R. E., and W. E. Howard, *Lab Anim. Sci.*, **21**:367 (1971).
88. Marsh, R. F., *Am. J. Pathol.*, **69**:209 (1972).
89. Mascoli, C. C., O. L. Ittensohn, V. M. Villarejos, J. A. Arguedas, P. J. Provost, and M. R. Hilleman, *Proc. Soc. Exp. Biol. Med.*, **142**:276 (1973).
90. McClure, H. M., M. E. Keeling, B. Olberding, R. D. Hunt, and L. V. Melindez, *Lab. Anim. Sci.*, **22**:517 (1972).
91. McKissick, G. E., H. L. Ratcliffe, and A. Koestner, *Pathol. Vet.*, **5**:538 (1968).
92. Miller, G. S., and R. Kellog, *U.S. Natl. Mus. Bull.* (1955).
93. Miyakawa, M., H. A. Gordon, and B. S. Wostmann, *Science*, **173**:171 (1973).
94. Montagna, W., *Am. J. Phys. Anthrop.*, **24**:71 (1966).
95. Myers, R., in J. R. Gorham (ed.), *Symposium of the International Academy Pathol. Puerto Rico,* Springer-Verlag, New York, 1967.
96. Nes, N., *Hereditas*, **50**:159 (1960).

97. Nes, N., *Hereditas,* **56:**259 (1966).
98. Norton, W. L., *Rheumatology,* **3:**1941 (1970).
99. Oettle, R. G., in *University Federation for Animal Welfare Handbook on the Care and Management of Laboratory Animals,* Williams and Wilkins Co., Baltimore, 1967.
100. Ohno, S., *In Vitro,* **2:**46 (1966).
101. Oregon Regional Primate Research Center Primate Newsletter **8**(1–10):(1970).
102. Orland, F. J., J. R. Blayney, R. W. Harrison, J. A. Reyniers, P. C. Thexler, M. Wagner, H. A. Gordon, and T. D. Luckey, *J. Dent. Res.,* **33:**147 (1954).
103. Oser, F., R. E. Lang, and E. E. Vogin, *Lab. Anim. Care.,* **20:**462 (1970).
104. Padgett, G. A., J. R. Gorham, and J. B. Henson, *Lab. Anim. Care,* **18:**258 (1968).
105. Pansky, B., and E. H. Freimer, *Arthr. Rheum.,* **17:**403 (1974).
106. Papperheimer, A. M., and M. Goettsch, *J. Exp. Med.,* **59:**35 (1934).
107. Persellin, R. H., and L. L. Wilson, *Bibl. Primat,* No. 7, Karger, Basel, 1968.
108. Petrak, M. L. (ed.), *Diseases of Cage and Avairy Birds,* Lea & Febiger, Philadelphia, 1969.
109. Pleasants, J. R., in E. C. Melby, Jr. and N. H. Altman (eds.) *CRC Handbook of Laboratory Animal Science,* Vol. I, CRC Press, Cleveland, 1974.
110. Potter, C. E., J. S. Oxford, S. L. Shore, C. McLaren, and C. Stuart-Harris, *Br. J. Exp. Pathol.,* **53:**153 (1972).
111. Povalishina, G. P. and L. L. Mironova, *Probl. Virol.,* (Russ.) **4:**402 (1960).
112. Pritchard, W. R., W. Henderson, and C. W. Beall, *Am. J. Vet. Res.,* **19:**696 (1958).
113. Prichard, R. W., T. B. Clarkson, H. O. Goodman, and H. B. Lofland, *Arch. Pathol.,* **77:**244 (1964).
114. Rasweiler, J. J., IV, and H. DeBonella, *Lab. Anim. Sci.,* **22:**658 (1972).
115. Ratcliffe, H. L., and J. A. Flick, *Am. J. Pathol.,* **39:**711 (1961).
116. Rauscher, F. J., J. A. Reyniers, and M. R. Saskateder, *Natl. Cancer Inst. Monograph,* **17:**211 (1962).
117. Renold, A. E., and W. E. Dulin, *Diabetologia,* **3:**63 (1967).
118. Reyniers, J. A., M. Wagner, T. D. Luckey, and H. A. Gordon, in J. A. Reyniers (ed.), *Lobund Reports No. 3,* University of Notre Dame Press, Notre Dame, Ind., 1960.
119. Rich, S. T., *Lab. Anim. Care,* **18:**235 (1968).
120. Riechenbach-Klinke, M., and E. Elkan, *The Principal Diseases of Lower Vertebrates,* Academic Press, New York, 1965.
121. Rigdon, R. H., *Am. J. Pathol.,* **39:**369 (1961).
122. Rigdon, R. H., *Ann. N.Y. Acad. Sci.,* **138:**28 (1966).
123. Riopelle, A. J., in *University Federation for Animal Welfare on the Care and Management of Laboratory Animals,* 3rd ed., Williams & Wilkins Co., Baltimore, 1967.
124. Robinson, D. G., *Sci. News,* **93:**16 (1968).
125. Rogers, L., F. Erickson, A. S. Hoverland, J. Metcalfe, and P. L. Clary, *Lab. Anim. Care,* **19:**181 (1969).
126. Rothstein, R., and D. Hunsaker, II, *Lab. Anim. Sci.,* **22:**227 (1972).
127. Ruben, L. N., and N. Balls, *Cancer Res.,* **27:**293 (1967).
128. Saison, R., L. Karstad, and T. J. Pridham, *Can. J. Comp. Med. Vet. Sci.,* **30:**151 (1966).
129. Saunders, L. Z., *Pathol. Vet.* (Basel), **2:**256 (1965).
130. Savage, R. M., *The Ecology and Life History of the Common Frog,* Putnam, London, 1961.
131. Schlumberger, H. G., *Am. J. Pathol.,* **35:**1 (1959).
132. Sheldon, W. G., W. C. Banks, and C. A. Gleiser, *Lab. Anim. Sci.,* **21:**229 (1971).
133. Shellenberger, T. E., *Lab. Anim. Care,* **18:**244 (1968).
134. Smith, S. M., and H. M. Kaplan, *Lab Animals,* **8:**213 (1974).
135. Snyder, R. L., and J. J. Christian, *Ecology,* **41:**785 (1960).
136. Snyder, R. L., D. H. Davis, and J. J. Christian, *J. Mamm.,* **42:**297 (1961).
137. Snyder, R. L., *Penna. Game News,* **40:**6 (1969).
138. Snyder, R. L., and H. L. Ratcliffe, *Acta Zool. Path. Antv.,* **48:**265 (1969).
139. Sompolinsky, D., *Cornell Vet.,* **40:**367 (1950).
140. Stellar, E., *J. Comp. Physiol. Psychol.,* **53:**1 (1960).
141. Stewart, H. L., K. C. Snell, and L. J. Dunham, *Am. Reg. Pathol. AFIP, fascie,* **40:**378 (1959).
142. Stockard, C., *Am. J. Anat.,* **28:**115 (1921).

143. Stormont, R. T., M. A. Foster, and C. Pfeiffer, *Proc. Soc. Exp. Biol.,* **42:**56 (1939).
144. Storrs, E. E., and W. E. Greer, *Lab. Anim. Sci.,* **23:**823 (1973).
145. Stout, C., and R. L. Snyder, *Gastroenterology,* **57:**256 (1969).
146. Stout, C., and W. B. Lemmon, *Exp. Mol. Pathol.,* **14:**151 (1971).
147. Subcommittee on Standards of Gnotobiotics, ILAR, in, *Gnotobiotes. Standards and Guidelines for the Breeding, Care and Management of Laboratory Animals,* Intl. Std. Book, No. 0-309-01858-7, Natl. Acad. Sci., Publishing Office, Washington, D. C. 1970.
148. Syoneyenbos, G. H., and H. I. Basch, *Avian Dis.,* **4:**477 (1960).
149. Tasker, J. B., G. A. Severin, S. Young, and E. L. Gillette, *J. Am. Vet. Med. Assoc.,* **154:**158 (1969).
150. Teah, B. A., in *Bibliography of Germfree Research,* 1885-1963 with Supplements, University of Notre Dame Press, Notre Dame Ind., 1964.
151. Thornbecke, G. J., H. A. Gordon, B. S. Wostmann, M. Wagner and J. A. Reyniers, *J. Infec. Dis.,* **101:**237 (1957).
152. Thrasher, J. D., M. Barenfus, S. T. Rich, and D. V. Shupe, *Lab. Anim. Sci.,* **21:**526 (1971).
153. Timmons, E. H., and P. A. Marques, *Lab. Anim. Care,* **19:**342 (1969).
154. Turner, W. J., *J. Biol. Chem.,* **118:**519 (1937).
155. Van Kruininger, H. J. L., *Gastroenterology,* **53:**114 (1967).
156. Van Pelt, R. W., and R. A. Dietrich, *J. Am. Vet. Med. Assoc.,* **161:**643 (1972).
157. Varton, O. A., in *Turkeys: Origin, History and Distribution,* North Dakota Agricultural College Extension Service, Fargo, N. D., 1939.
158. Walker, E. P., in *Mammals of the World,* Vols. I–III, 2nd ed., The Johns Hopkins Press, Baltimore, 1968.
159. Walker, E. P., in *Mammals of the World,* Vol. I, 2nd ed., The Johns Hopkins Press, Baltimore, 1968.
160. Walker, E. P., in *Mammals of the World,* Vol. II, 2nd ed., The Johns Hopkins Press, Baltimore, 1968.
161. Walker, J. S., F. C. Carigan, R. R. Sirimanne, and J. B. Abass, *Lab. Anim. Sci.,* **24:**290 (1974).
162. Wass, J. B., and H. M. Kaplan, *Lab. Anim. Sci.,* **24:**669 (1974).
163. Witschi, E., and R. Laguens, *Develop. Biol.,* **7:**605 (1963).
164. Young, M. D., J. A. Porter, and C. M. Johnson, *Science,* **153:**1006 (1966).

INDEX

}
}

R